AUTOMOTIVE BRAKING SYSTEMS

THIRD EDITION

es on the World Wide Web,

rvice of I(T)P®

AUTOMOTIVE BRAKING SYSTEMS

THIRD EDITION

Thomas W. Birch
Yuba College, Emeritus
Marysville, California

I(T)
Albany •
Melbourne •

NOTICE TO THE READER

Publisher does not warrant or guarantee any of the products described herein or perform any independent analysis in connection with any of the product information contained herein. Publisher does not assume, and expressly disclaims, any obligation to obtain and include information other than that provided to it by the manufacturer.

The reader is expressly warned to consider and adopt all safety precautions that might be indicated by the activities described herein and to avoid all potential hazards. By following the instructions contained herein, the reader willingly assumes all risks in connection with such instructions.

The publisher makes no representation or warranties of any kind, including but not limited to, the warranties of fitness for particular purpose or merchantability, nor are any such representations implied with respect to the material set forth herein, and the publisher takes no responsibility with respect to such material. The publisher shall not be liable for any special, consequential, or exemplary damages resulting, in whole or part, from the readers' use of, or reliance upon, this material.

Cover Design: Courtesy of Charles Cummings Art/Advertising, Inc.

Delmar Staff
Publisher: Alar Elken
Acquisitions Editor: Vernon R. Anthony
Developmental Editor: Denise Denisoff
Project Editors: Megeen Mulholland
Barbara Diaz
Production Coordinator: Karen Smith
Art and Design Coordinator: Cheri Plasse
Editorial Assistant: Betsy Hough

Printed in the United States of America

For more information, contact:

Delmar Publishers
3 Columbia Circle, Box 15015
Albany, New York 12212-5015

International Thomson Publishing Europe
Berkshire House
168-173 High Holborn
London, WC1V7AA
United Kingdom

Nels

Nelson Canada
1120 Birchmont Road
Scarborough, Ontario
M1K 5G4, Canada

International Thomson Publishing France
Tour Maine-Montparnasse
33 Avenue du Maine
75755 Paris Cedex 15, France

International Thomson Editores
Seneca 53
Colonia Polanco
11560 Mexico D. F. Mexico

International Thomson Publishing GmbH
Königswinterer Straße 418
53227 Bonn
Germany

International Thomson Publishing Asia
60 Albert Street
#15-01 Albert Complex
Singapore 189969

International Thomson Publishing Japan
Hirakawa-cho Kyowa Building, 3F
2-2-1 Hirakawa-cho, Chiyoda-ku,
Tokyo 102, Japan

ITE Spain/Paraninfo
Calle Magallanes, 25
28015-Madrid, España

Contents

Preface xi

Chapter 1 BRAKING SYSTEM FUNDAMENTALS 1

Objectives 1

1.1 Introduction 1

1.2 Brake System Components 2

1.3 Brake System Repair 8

1.4 Practice Diagnosis 9

Terms to Know 10

Review Questions 10

Chapter 2 PRINCIPLES OF BRAKING 13

Objectives 13

2.1 Introduction and Federal Requirements 13

2.2 Deceleration 14

2.3 Energy of Motion 17

2.4 Friction and Heat Energy 18

2.5 Heat Energy 18

2.6 Brake Lining Friction 19

2.7 Tire Friction 24

2.8 Brake Power 24

2.9 Weight Transfer 27

2.10 Skids 28

2.11 Stopping Sequence 30

2.12 Anti-Dive Suspensions 31

2.13 Practice Diagnosis 31

Terms to Know 32

Review Questions 32

Chapter 3 DRUM BRAKE THEORY 34

Objectives 34

3.1 Introduction 34

3.2 Shoe Energization and Servo Action 35

3.3 Brake Shoes 37

3.4 Lining Attachment 38

3.5 Backing Plate 39

3.6 Shoe Anchors 39

3.7 Brake Springs 40

3.8 Shoe Adjusters 42

3.9 Brake Drums 46

3.10 Practice Diagnosis 47

Terms to Know 47

Review Questions 48

Chapter 4 **DISC BRAKE THEORY** **50**

Objectives 50
4.1 Introduction. 50
4.2 Calipers. 51
4.3 Brake Pads 56
4.4 Caliper and Pad Mounting Hardware 57
4.5 Rotors 59
4.6 Rear-Wheel Disc Brakes. 61
4.7 Practice Diagnosis 62
Terms to Know 62
Review Questions. 62

Chapter 5 **PARKING BRAKE THEORY** **64**

Objectives 64
5.1 Introduction. 64
5.2 Cables and Equalizer 64
5.3 Lever and Warning Light 66
5.4 Integral Drum Parking Brake 67
5.5 Auxiliary Drum Parking Brake. 67
5.6 Disc Parking Brake. 68
5.7 Practice Diagnosis 69
Terms to Know 70
Review Questions. 70

Chapter 6 **HYDRAULIC SYSTEM THEORY**. **71**

Objectives 71
6.1 Introduction. 71
6.2 Hydraulic Principles 71
6.3 Master Cylinder Basics and Components. 75
6.4 Master Cylinder Basic Operation 75
6.5 Master Cylinder Construction. 76
6.6 Residual Pressure 78
6.7 Dual Master Cylinder Construction 79
6.8 Tandem Master Cylinder Operation 80
6.9 Diagonally Split Hydraulic System 82
6.10 Quick-Take-Up Master Cylinder 83
6.11 Miscellaneous Master Cylinder Designs 84
6.12 Wheel Cylinder Basics 85
6.13 Wheel Cylinder Construction. 86
6.14 Caliper Basics. 87
6.15 Caliper Pistons and Sealing Rings 88
6.16 Caliper Boot 90
6.17 Caliper Body and Bleeder Screw 91
6.18 Tubing and Hoses 92
6.19 Hydraulic Control Valves and Switches 95
6.20 Brake Fluid 101
6.21 Practice Diagnosis 106
Terms to Know 106
Review Questions. 106

Chapter 7 POWER BRAKE THEORY 109
Objectives 109
7.1 Introduction. 109
7.2 Vacuum Booster Theory 110
7.3 Vacuum Booster Construction 113
7.4 Vacuum Booster Operation. 115
7.5 Hydraulic Booster Theory 116
7.6 Hydraulic Booster Construction 117
7.7 Hydraulic Booster Operation 117
7.8 Electrohydraulic Booster Theory 118
7.9 Electrohydraulic Booster Construction. 120
7.10 Electrohydraulic Booster Operation 120
7.11 Practice Diagnosis 122
Terms to Know 122
Review Questions. 122

Chapter 8 ANTILOCK BRAKING SYSTEM THEORY 124
Objectives 124
8.1 Introduction. 124
8.2 Electronic Wheel Sensors 128
8.3 Electronic Controller 129
8.4 Electronic Modulator 132
8.5 Recirculation Pump 135
8.6 ABS Systems 135
8.7 ABS Brake Fluid. 142
8.8 Automatic Traction Control 142
8.9 Practice Diagnosis 143
Terms to Know 143
Review Questions. 143

Chapter 9 BRAKE SYSTEM SERVICE 145
Objectives 145
9.1 Introduction. 145
9.2 Brake Inspection 147
9.3 Troubleshooting Brake Problems 159
9.4 Brake Repair Recommendations 159
9.5 Special Notes on Surface Finish 163
9.6 Electrical Diagnosis and Repair 165
9.7 Practice Diagnosis 170
Terms to Know 171
Review Questions. 171

Chapter 10 DRUM BRAKE SERVICE 173
Objectives 173
10.1 Introduction. 173
10.2 Drum Removal 174
10.3 Drum Inspection 179

10.4 Drum Machining . . . 184
10.5 Shoe Assembly—Predisassembly Cleanup . . . 190
10.6 Brake Spring Removal and Replacement . . . 190
10.7 Brake Shoe Removal . . . 196
10.8 Component Cleaning and Inspection . . . 198
10.9 Component Lubrication . . . 199
10.10 Brake Shoe Preinstallation Checks . . . 201
10.11 Regrinding Brake Shoes . . . 204
10.12 Brake Shoe Installation . . . 204
10.13 Completion . . . 207
10.14 Practice Diagnosis . . . 208
Terms to Know . . . 208
Review Questions . . . 208

Chapter 11 DISC BRAKE SERVICE . . . 211
Objectives . . . 211
11.1 Introduction . . . 211
11.2 Removing a Caliper . . . 213
11.3 Rotor Inspection . . . 219
11.4 Rotor Refinishing . . . 223
11.5 Machining a Rotor Off-Car . . . 228
11.6 Machining a Rotor On-Car . . . 233
11.7 Resurfacing a Rotor . . . 236
11.8 Rotor Replacement . . . 236
11.9 Caliper Service . . . 238
11.10 Removing and Replacing Pads . . . 240
11.11 Caliper Installation . . . 245
11.12 Completion . . . 247
11.13 Practice Diagnosis . . . 247
Terms to Know . . . 248
Review Questions . . . 248

Chapter 12 HYDRAULIC SYSTEM SERVICE . . . 250
Objectives . . . 250
12.1 Introduction . . . 250
12.2 Working with Tubing . . . 251
12.3 Servicing Hydraulic Cylinders . . . 254
12.4 Master Cylinder Service . . . 256
12.5 Wheel Cylinder Service . . . 266
12.6 Caliper Service . . . 272
12.7 Bleeding Brakes . . . 284
12.8 Diagnosing Hydraulic System Problems . . . 296
12.9 Testing a Warning Light . . . 301
12.10 Brake-Light Switch Adjustment . . . 301
12.11 Practice Diagnosis . . . 304
Terms to Know . . . 304
Review Questions . . . 305

Chapter 13 POWER BOOSTER SERVICE 307
Objectives 307
13.1 Introduction. 307
13.2 Vacuum Supply Tests 308
13.3 Booster Replacement 310
13.4 Booster Service. 314
13.5 Practice Diagnosis 319
Terms to Know. 320
Review Questions. 320

Chapter 14 PARKING BRAKE SERVICE 321
Objectives 321
14.1 Introduction. 321
14.2 Cable Adjustment 322
14.3 Cable Replacement. 326
14.4 Parking Brake Warning-Light Service 328
14.5 Practice Diagnosis 329
Terms to Know. 329
Review Questions. 329

Chapter 15 ANTILOCK BRAKE SYSTEM SERVICE 331
Objectives 331
15.1 Introduction. 331
15.2 ABS Warning-Light Operation. 334
15.3 Warning-Light Sequence Test. 334
15.4 ABS Problem Codes and Self-Diagnosis 337
15.5 Electrical Tests 340
15.6 Hydraulic Pressure Checks. 348
15.7 Repair Operations. 349
15.8 Completion 352
15.9 Practice Diagnosis 352
Terms to Know. 352
Review Questions. 352

Chapter 16 WHEEL BEARINGS—THEORY AND SERVICE 353
Objectives 353
16.1 Introduction. 353
16.2 Bearings and Bearing Parts. 353
16.3 Bearing End Play—Preload 355
16.4 Seals 356
16.5 Nondrive-Axle Serviceable Bearings 356
16.6 Nondrive-Axle Nonserviceable Wheel Bearings 358
16.7 Solid-Axle Drive-Axle Bearings 358
16.8 Independent Suspension Drive-Axle Wheel Bearings 360
16.9 Nonserviceable Drive-Axle Wheel Bearings 362
16.10 Wheel Bearing Maintenance and Lubrication 362
16.11 Diagnosis Procedure for Wheel Bearing Problems 362
16.12 Repacking Serviceable Wheel Bearings. 364

16.13 Repairing Nonserviceable Wheel Bearings . . . 373
16.14 Repairing Solid-Axle Drive-Axle Bearings . . . 373
16.15 Repairing Serviceable FWD Front-Wheel Bearings . . . 379
16.16 Repairing Nonserviceable FWD Front-Wheel Bearings . . . 381
16.17 Practice Diagnosis . . . 382
Terms to Know . . . 383
Review Questions . . . 383

Chapter 17 TRAILER BRAKE SYSTEMS . . . 385
Objectives . . . 385
17.1 Introduction . . . 385
17.2 Electric Brakes . . . 385
17.3 Surge Brakes . . . 389
17.4 Slave Hydraulic Systems . . . 390
17.5 Trailer Brake Service . . . 391
Review Questions . . . 394

Chapter 18 BRAKE SYSTEMS FOR HIGH-PERFORMANCE VEHICLES . . . 395
Objectives . . . 395
18.1 Introduction . . . 395
18.2 Braking System Operating Limits . . . 396
18.3 Passenger-Car Brake Tuning . . . 396
18.4 Race-Car Brake Systems . . . 399
18.5 Conclusion . . . 411
Terms to Know . . . 412

GLOSSARY . . . 413

Appendix 1
ASE CERTIFICATION . . . 421

Appendix 2
ENGLISH-METRIC CONVERSION . . . 424

Appendix 3
BOLT TORQUE-TIGHTENING CHART . . . 426

Appendix 4
TORQUE-TIGHTENING CHART FOR LINE CONNECTIONS AND BLEEDER SCREWS . . . 427

Appendix 5
SHOE SIZE CHART . . . 428

INDEX . . . 429

Preface

Automotive Braking Systems is for technicians who will be repairing the braking systems not only of today's complex automobiles but of the even more complex cars of tomorrow.

The evolution of automotive braking systems has included the development of self-adjusting drum brakes, disc brakes, tandem hydraulic systems, diagonally split hydraulic systems, vacuum-operated boosters, hydraulically operated boosters, self-contained electrically powered hydraulic boosters, computer-controlled antilock braking systems, and self-adjusting parking brakes. The future is sure to bring further developments, such as computer-controlled radar systems that will automatically apply the brakes when a potential hazard is sensed.

As automotive systems become ever more complex and precise, the mechanic of yesterday needs to become the technician of today by learning repair methods that are more exacting and require far more knowledge and skill with specialized tools and equipment than did the methods of the past. Repair shops are changing also. New-car dealerships play much the same role they always have: they employ factory-trained technicians who have the specialized knowledge, tools, and equipment to repair the most complicated features of each new model. But many of the service stations and independent garages of the 1950s are gone, and in their place are shops that specialize in particular repair areas such as air conditioning or tune-ups or in particular makes of domestic or imported cars. The garage of yesterday, with a mechanic who could repair any car that was driven through the door, is gone. The evolution of the automobile has made that type of shop and mechanic extinct.

Automotive Braking Systems presents enough information about the systems of present-day cars to help in understanding, diagnosing, and repairing them. Some people will place some of this information in the "nice-to-know-but-not-really-necessary-for-fixing-cars" category. However, technicians should realize that a complete understanding of the why and how of a car's operation aids greatly in understanding why that car does not operate properly and how to go about repairing it. The highly desirable ability to solve problems is a direct result of a thorough understanding of basic operating principles. Many of our "new developments" are based on physical principles that have been understood for some time. Except for computer controls, there is not much that is really new in automotive systems.

This book provides a description of each of the various styles of braking systems and the components that make up these systems. All of the National Institute for Automotive Service Excellence (ASE) braking tasks are described or explained, and review questions in the style of ASE certification test questions are included. Additional ASE questions are included in the *Instructor's Manual* that accompanies this text, and the ASE task list for automotive brake systems, A5, is included as Appendix 1. The most common repair methods are described in a general manner.

Specialization is a definite advantage to technicians and repair shops. Rarely can any one person "know it all" about cars anymore; there is simply too much to know. It is also financially prohibitive for one shop to own all the special tools and equipment needed to make every possible car repair, if the equipment is not used on a regular basis. A general repair shop, such as a dealership, might pay rent of several thousand dollars a month for a desirable location and have a tool and equipment inventory worth over a hundred thousand dollars. That kind of shop has to do a large volume of work to stay in business. It must have technicians who can use their specialized skills and knowledge to perform complex tasks and ensure that customers' cars are operating safely and efficiently.

The most thorough and reliable sources of information about a particular car are its manufacturer's service manual and training centers. However, these cover only one make, and sometimes only one model, of car. Manuals are often followed up with specific service bulletins when a need for more information or a change in a new car part or systems makes them necessary. Broader coverage is available in technicians' service information (printed and on computer CDs) which is available from the major aftermarket brake component manufacturers. These manuals have varying amounts of information for particular cars; some of it is very thorough and specific to particular cars or systems, and some of it is very general. *Automotive Braking Systems* supplements these manuals by presenting the theoretical and practical knowledge needed for a complete understanding of the operation of these systems.

Technicians in different parts of the United States use different terms for the same item or task. Terms commonly used on the West Coast are slightly different from those used in the East. To the greatest extent possible, the terms used in this book are those used by automobile and aftermarket manufacturers. These terms are explained in Table 9-1 and also in the Glossary at the end of this text.

In describing repair procedures, the author has assumed that readers have a basic working knowledge of hand tools, fasteners, general automotive repair procedures, and the safety precautions that should be exercised during general service and repair operations. Space does not permit including them in a book of this nature, which concentrates on a few specialized systems. Wise technicians know that improper use of a wrench or other hand tool, or of a bolt or nut, often leads to a more difficult job, and that a violation of a commonsense safety rule can cause injury to the technician, the vehicle operator, or an innocent bystander.

Note: Answers to the end-of-chapter Review Questions can be found in the *Instructor's Manual* accompanying this text.

Acknowledgments

The author is sincerely grateful to the following individuals, companies, and businesses for their assistance in the preparation of this book.

Alston Engineering, Sacramento, CA
American Honda Motor Company, Gardena, CA
Ammco Tools, Incorporated, North Chicago, IL
ASE, The National Institute for Automotive Service Excellence, Herndon, VA
ATE/USA, Annapolis, MD
Bendix Brake, Allied Aftermarket Division, Jackson, TN
Kerry Birch, Yuba City, CA
Buick Motor Division, General Motors, Flint, MI
Carlson Quality Brake Division, International Brake Industries, Lima, OH
Chevrolet Division, General Motors, Warren, MI
Chrysler Corporation, Detroit, MI
CR Industries, Elgin, IL
Delco Moraine Division, General Motors, Dayton, OH
John M. Demko, Burmah-Castrol Inc., Edison, NJ
Dorman Products, Cincinnati, OH
Easco/K-D Tools, Lancaster, PA
EIS Brake Parts, Berlin, CT
Everco, Lincolnwood, IL
Federal-Mogul Corporation, Detroit, MI
Fluke Corporation
FMC Corporation, Automotive Service Equipment Division, Conway, AR
Ford Motor Company, Dearborn, MI
Forrest E. Folck, Motor Vehicle Forensic Service, San Diego, CA
Gramlich Tools, Ontario, Canada
Grigg Automotive Manufacturing, Huntington Park, CA
Hunter Engineering, Bridgeton, MO
JFZ Engineered Products, Chatsworth, CA
Kent-Moore Automotive Division, Warren, MI
Bruce D. Kirk, Smartville, CA
Stephen P. Klien, Yuba City, CA
Kwik-Way, Marion, OH
Lee Manufacturing, Cleveland, OH
Lisle Corporation, Clarinda, IA
Loctite Corporation, Cleveland, OH
Lucas Girling Limited, Englewood, NJ
Anthony Lux, Allied Signal/Bendix Brake, Rumford, RI
Mazda Corporation, Irvine, CA
Garland Moorehead, Colusa, CA
Neward Enterprises, Inc., Cucamonga, CA
Nilfisk of America, Malvern, PA
Nissan Motor Corporation, Gardena, CA
John Nissan, Williams, CA
Nuturn Corporation, Smithville, TN
Oldsmobile Division, General Motors, Lansing, MI
OTC Division, Sealed Power Corporation, Owatona, MN
Phoenix Systems, Tucson, AZ
Pontiac Division, General Motors, Pontiac, MI
Raybestos (Brake Systems Inc.), Franklin Park, IL
Raymond Savage, Robert H. Wager Co., Chatham, NJ
Rockwell International, Troy, MI
RSR Enterprises Ltd., Minneapolis, MN
Snap-On Tools, Kenosha, NJ
Stainless Steel Brakes Corporation, Clarence, NY
Storm-Vulcan, Dallas, TX
Stromberg-Hydramite Corporation, Carey, OH
Tapley Instrumentation Limited, Chatham, NJ
Tilton Engineering, Buellton, CA
McLane Tilton, Buellton, CA
The Timken Company, Canton, OH
Vacula Automotive Products, Williamsville, NY
Volkswagen of America, Troy, MI
Wagner Division, Cooper Industries Incorporated, Parsippany, NJ
Warner Electric Brake & Clutch Company, South Beloit, IL
Woody Woodward, Bendix Aftermarket Brake Division, Jackson, TN

Thomas W. Birch
September 1998

1 Braking System Fundamentals

Objectives

After completing this chapter, you should:

- ❑ Be able to identify the major components of the automotive braking system.
- ❑ Have an understanding of the general purpose of these components, how they relate to each other, and the role they play in stopping the car.

1.1 Introduction

The braking system provides the means to stop a car. Obviously, to be in full control of a machine, we have to be able to start it and stop it. There must be a stop for every start. To control an automobile, we need to be able to start it moving, make it turn, accelerate and decelerate and—of major importance—stop it. A car with a braking system that is not working properly is a candidate for the wrecking yard and may be a cause of injury to the driver and passengers as well as to others. Many drivers think about their braking systems only at the time of a panic stop—when it is too late to do anything but hope.

The braking system is considered by many people to be the most important system involved in the operation of a vehicle. There are more than a few federal and state requirements governing stopping ability, and many states make periodic inspections of vehicles' (especially busses') brake systems and stopping ability.

We do not normally use brakes at their maximum capability, but we want them to work flawlessly in emergencies, much like a parachute. Most people find hard braking unpleasant and avoid hard stops. Also, as we will see later, brakes waste energy. A vehicle is most efficient when the brakes are not used. People who normally do not stop hard, that is, most of us, seldom get a chance to really test their brakes. The braking systems of most cars are rarely tested to ensure that they are working at maximum efficiency. (I wonder how many pilots would jump out of an airplane using a parachute that had been assembled and packed five years ago and had deteriorated from being out in the weather.) Another problem related to brake operation is that most people are poorly trained in using their brakes in emergency situations. Many drivers are not sure how quickly a particular car can stop or how to make that car stop in the shortest distance.

A controversy occurs when it becomes time to repair braking systems. In the interest of saving money, many people will themselves replace only those items that are actually worn out—the brake shoes and sometimes a drum or rotor. This is the "do-it-yourself" or home-mechanic approach. At the other end of the spectrum is the technician who will replace or repair whatever items are necessary to achieve "new-car" braking performance. This second approach ensures that the "parachute" will work properly when needed.

The ideal braking system is one that will allow the driver to bring a vehicle to a stop in the shortest possible distance. To be able to do this, it should have enough power to lock up and skid all four tires while stopping on clean, dry pavement. The tire lockups should occur in a controllable fashion so that stops can be made without brake lockup. Also, stopping should occur with a moderate amount of pedal pressure so that even weaker drivers can achieve tire lockup. Brake lockup is not desirable because it results in a longer stopping distance, probable loss of vehicle control, and excessive tire wear. Brake lockup indicates that one or more brake assemblies are stronger than necessary. On the other hand, a brake that cannot achieve lockup is possibly not as strong as it should be. Another requirement for good brakes is a stable stop without pulling or darting to the side. Antilock braking systems (ABS) are used to prevent brake lockup so the driver can safely use stronger brakes.

It is the intent of this author to present enough information in this book so that you can rebuild brake systems to achieve this ideal braking performance. In the rapidly paced world of today, with liability lawsuits a common occurrence, it is best to always aim for "new-car" performance when making a repair. "Brake job" is an undefined term. No legislation specifies exactly what to do, and there are no commonly used test devices or instruments to assess exactly how good or bad a brake system is. Many customers do not know how brakes work, exactly what parts are involved in stopping a car, or how these parts can deteriorate and cause failure. Through this book, the reader should develop an understanding of the principles involved in stopping a car, how these principles can make stopping difficult, the interrelationship among braking system components, the things that can occur to reduce their performance, and the various methods you can use to check their operation and repair them as needed. Understanding these things makes diagnosing and correcting brake problems easy.

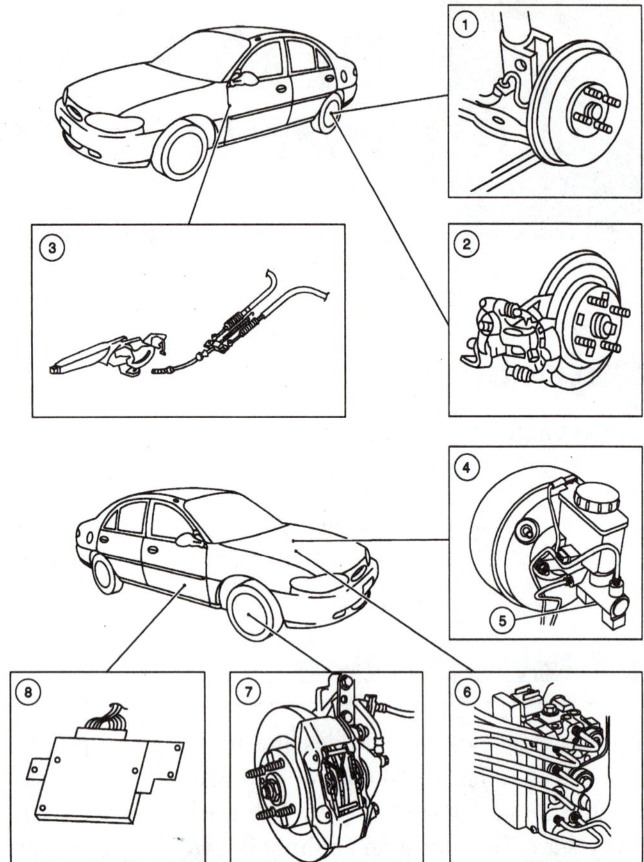

Figure 1-1 A typical automotive brake system consists of rear drum (1) or disc (2) brake units, a parking brake system (3), a power booster (4), a master cylinder (5), and front disc brake units (7). Cars with ABS will also have a hydraulic controller (6) and an electronic brake control module (8). (Courtesy of Ford Motor Company)

1.2 Brake System Components

Many of us are somewhat familiar with the parts of a braking system (Figure 1-1). To give us a common ground for discussion of the principles and to help us understand their interrelationship, we will briefly describe them here. Each part is related to the others, and proper operation of each part is necessary for correct operation of the whole system (Figure 1-2). The major parts of the **base** or **foundation brakes** plus ABS will be discussed in detail two more times: first in a chapter on the theory of their operation, and again in a chapter on service and repair.

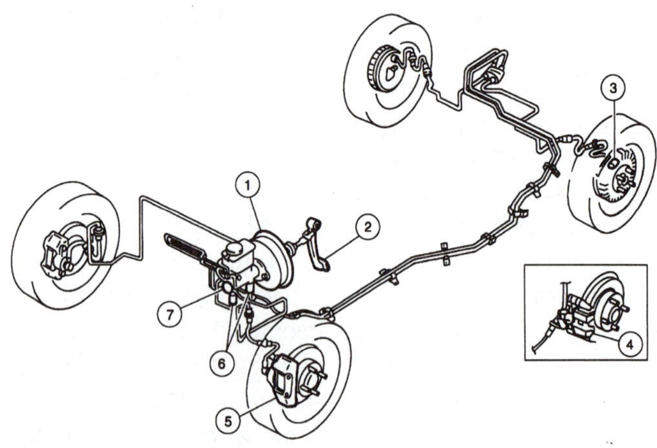

Figure 1-2 The basic brake system uses tubing to transmit hydraulic brake pressure from the master cylinder (7) to the front disc brake units (5) and rear drum (3) or disc (4) brake units. (Courtesy of Ford Motor Company)

1.2.1 Brake Shoes and Friction Materials

Brakes are heat machines. They provide stopping power by generating heat from the rubbing of a **friction material**, the **brake lining**, against a rotating **drum** or **rotor**. The car slows down as friction produced by this rubbing action converts the energy of the moving car into heat.

The brake lining is attached to the **brake shoes**, often called **pads** when disc brake linings are discussed (Figure 1-3). The lining must be able to rub against the drum or rotor without causing an excessive amount of wear to the drum or rotor surface. It also must be able to operate at very high temperatures (several hundred degrees) without failing.

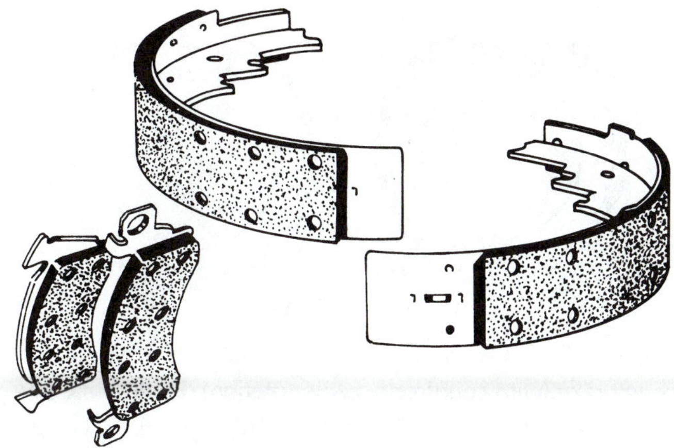

Figure 1-3 Brake lining is attached to the shoes in drum brakes (top) and to the shoes or pads in disc brakes (lower left). (Reprinted from Mitchell Anti-Lock Brake Systems, with permission of Mitchell Repair Information, LLC)

1.2.2 Drum Brakes

The drum brake is the traditional type of brake on older cars, and it was used on all four wheels of most cars prior to the 1970s. Drum brakes are currently used on the rear wheels of many cars (Figure 1-4).

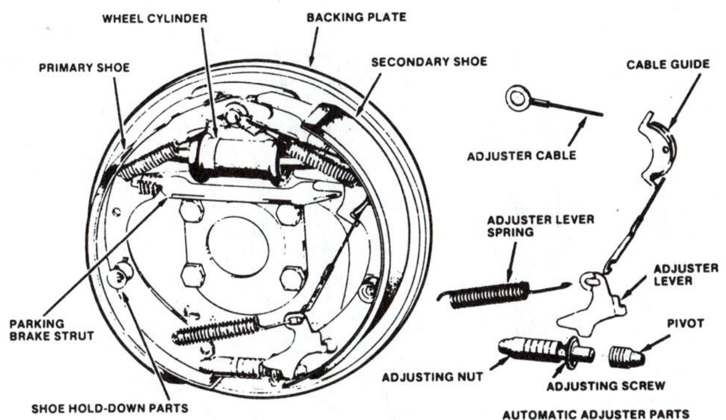

Figure 1-4 A drum brake assembly of the duo-servo type; this particular unit is a rear-wheel assembly with a parking brake and uses an automatic adjuster or self-adjuster. (Courtesy of Bendix Brakes, by AlliedSignal)

The pan-shaped drum is attached to the axle or hub flange, just inside the wheel, and it rotates directly with the wheel. The brake shoes are positioned just inside the drum and are mounted on the **backing plate**. The shoes are anchored to the backing plate so they can pivot into and out of contact with the drum but cannot rotate with it. The **anchors** can be arranged so an opening of the shoe is placed over a round anchor or so that the smooth end of the shoe butts up against a flat anchor block. Braking forces are transmitted from the shoes to the anchors, to the backing plate, and then to the suspension members.

To operate, the brake shoes are pushed outward so the lining is forced against the drum. This force can come from the hydraulic wheel cylinder or from the mechanical linkage of the parking brake. As the lining touches the drum, the rotation of the drum will tend to either pull the shoe along with it or push the shoe away, depending on the position of the shoe anchor relative to the direction of rotation. These actions are commonly referred to as **energizing** and **deenergizing** the shoes (Figure 1-5). The

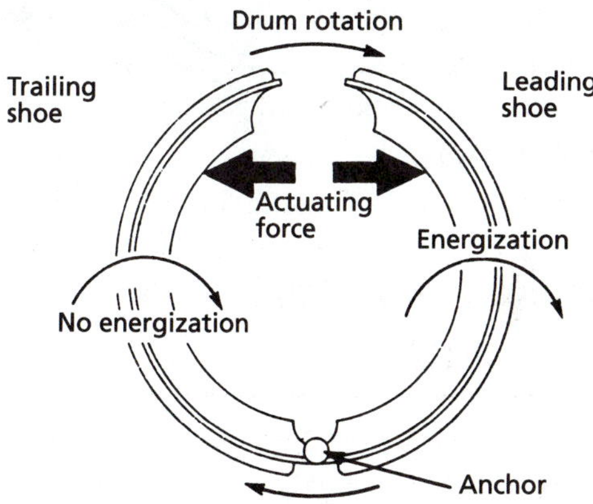

Figure 1-5 Depending on how the brake shoe is mounted, it can be energized or deenergized by the rotation of the drum. (Reprinted from Mitchell Anti-Lock Brake Systems, with permission of Mitchell Repair Information, LLC)

energized shoe is the **forward** or **leading shoe**, and the deenergized shoe is the **trailing shoe**.

If the two shoes are connected so that one shoe can transmit pressure to the other, and both shoes are using a single anchor, the leading shoe will apply pressure on the trailing shoe. This is called **servo action** (Figures 1-6 and 1-7). In **servo brakes**, the leading shoe is called the **primary shoe**, and the trailing shoe is called the **secondary shoe**. Most servo brakes are designed so that servo action can occur during both forward and reverse stops. This style of brake is called a **duo-servo brake**.

In **nonservo** designs, the brake shoes are mounted individually, so there is no interaction between them. There are several different styles of nonservo brakes, depending on how the shoes are positioned on the backing plate. Nonservo brakes are usually found in a leading-trailing shoe relationship. They can also be mounted in a two-leading shoe arrangement with both shoes in an energized position, or in a two-trailing shoe arrangement with

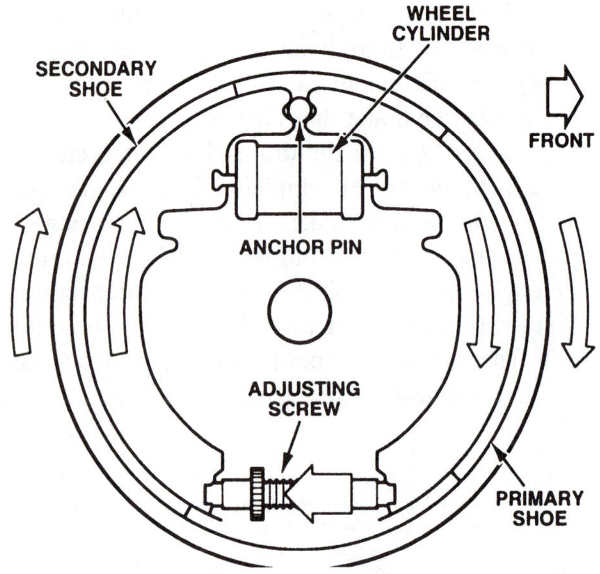

Figure 1-6 A servo brake. The primary shoe is energized by contact with the rotating drum and, by pushing through the adjuster screw, will apply pressure on the secondary shoe. (Courtesy of General Motors Corporation, Service Technology Group)

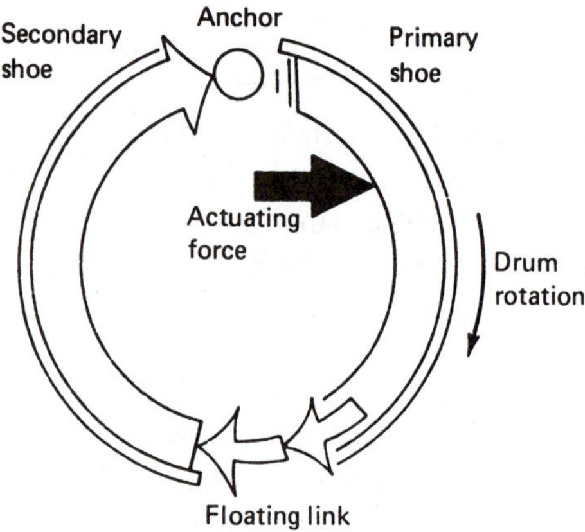

Figure 1-7 The actuating force from the wheel cylinder pushes the primary shoe against the drum. In turn, the primary shoe pushes the secondary shoe against the drum through servo action. (Reprinted from Mitchell Anti-Lock Brake Systems, with permission of Mitchell Repair Information, LLC)

both shoes in a deenergized position (Figure 1-8). Duo-servo and nonservo brakes will be discussed more completely in Chapter 3.

The reason for the different brake shoe arrangements is to create a brake that will operate more easily with less pedal pressure, to achieve a front-to-rear balance with a pair of brakes of different designs, or to create a more stable design. The duo-servo design generates the most stopping power, but it is also the most sensitive to pressure and frictional drag between the primary shoe and the drum. Severe pull can occur if there is a slight difference in power between the two primary shoes on one axle. A two-leading shoe brake is almost as powerful, but it is very sensitive to direction of rotation. There is much greater strength in the forward direction (Figure 1-9).

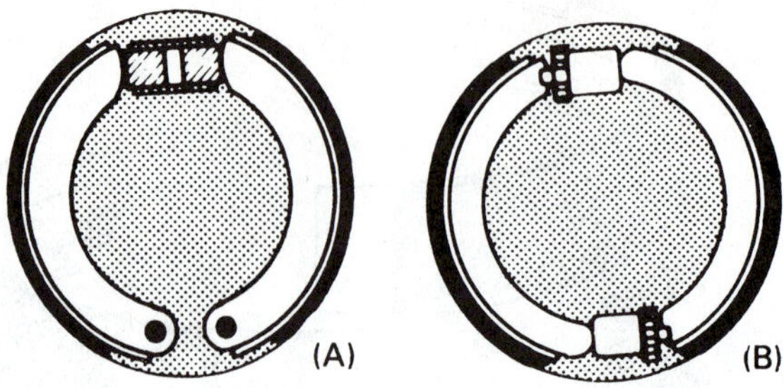

Figure 1-8 A leading-trailing shoe (A) and a two-leading shoe brake (with counterclockwise rotation) (B). Both are non-servo brake designs. (Courtesy of ITT Automotive)

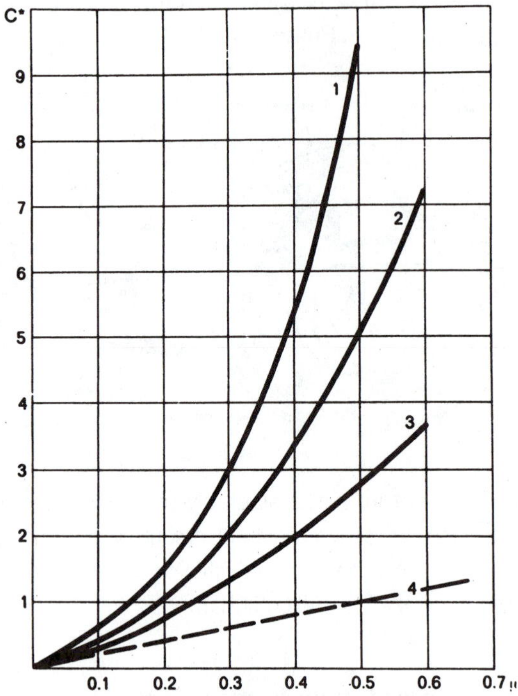

Figure 1-9 This graph compares the power of four brake designs. The vertical height represents the relative amount of stopping power, and the horizontal represents the amount of pedal pressure. (Courtesy of ITT Automotive)

1.2.3 Disc Brakes

A disc brake uses a flat, round disc, or rotor, attached to the wheel hub instead of a drum. The brake shoes, also called pads, are positioned on opposite sides of the rotor and are mounted in the brake **caliper**. The caliper contains the hydraulic piston(s) used to apply the shoes and to transmit the braking forces from the shoes to the suspension members. Disc brakes are found on the front wheels of most cars manufactured after 1970 and the rear wheels of many of the more expensive or higher-performance cars (Figure 1-10). More and more vehicles are being equipped with four-wheel disc brakes.

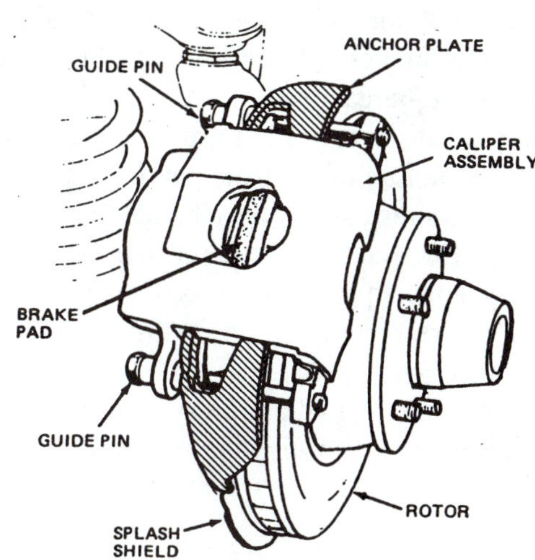

Figure 1-10 A sliding caliper disc brake assembly that is used on many front wheels. (Courtesy of Wagner Brake)

Most early disc brake designs used **fixed calipers**. This caliper is bolted solidly to the steering knuckle. Fixed calipers use a pair of pistons, with the inboard (of the rotor) piston applying the inboard shoe and the outboard piston applying the outboard shoe. The fixed calipers used on heavier vehicles have two pistons at each side to apply more pressure evenly over a longer shoe. Most recently designed calipers are of the **floating** or **sliding** style. A single piston applies the inboard shoe, and hydraulic reaction causes the caliper to move in an inward direction and apply the outboard shoe (Figure 1-11). Floating and sliding caliper designs use a slightly different type of attachment to the mounting bracket. Floating calipers use a caliper support, which allows the slight sideways motion required for brake operation and release, while holding the caliper in position and transmitting the braking forces from the caliper to the suspension members.

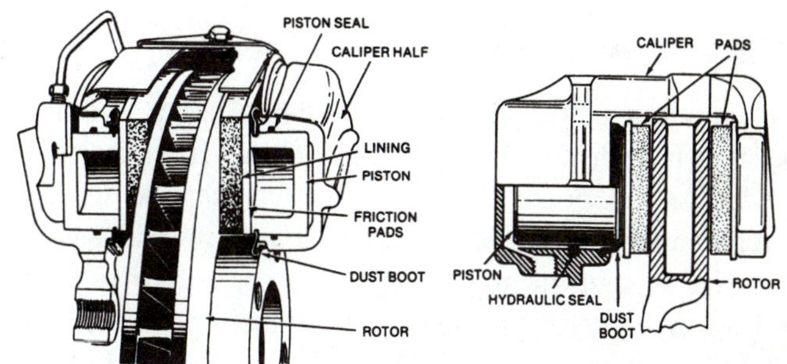

Figure 1-11 A fixed caliper (left) is mounted solidly on the steering knuckle and has pistons on each side of the rotor. A floating caliper (right) has a piston on the inboard side, and the caliper moves to apply the outboard pad. (Reprinted from Mitchell Anti-Lock Brake Systems, with permission of Mitchell Repair Information, LLC)

All disc brakes are nonenergized, nonservo brakes; lining pressure is directly proportional to brake pedal pressure. Because of this, disc brakes tend to be more directionally stable than drum brakes. Hard stops can be made with fewer brake-pull problems. A brake rotor will also stay cleaner—more free of water, dust, or dirt—than a drum brake. Centrifugal force will throw contaminants off a rotor, whereas a drum's inner friction surface tends to collect them. A disc brake will also have much cooler operations than a drum brake because of the increased area that is exposed to the air flowing past it. Still another advantage is that the clamping action of disc brake pads causes no distortion of the rotor, whereas the spreading action of drum brake shoes causes the drum to elongate into an elliptical or oval shape. This distortion of the drum will cause a lowering of the brake pedal and possible pinching of the ends of the shoes (Figure 1-12).

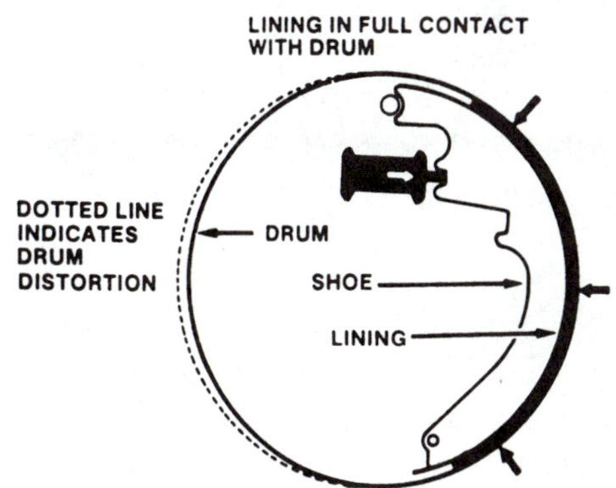

Figure 1-12 When a brake shoe is forced into the drum with sufficient force, the drum will distort to match the shape and curvature of the shoe. (Courtesy of Bendix Brakes, by AlliedSignal)

1.2.4 Brake Hydraulic Systems

All modern automotive brake systems use a hydraulic system to transmit the application forces from the brake pedal to the brake shoes. Hydraulics can be used to greatly increase force. A hydraulic system also has the benefit of transmitting an equal force to two or more places in the system at the same time, an important feature in brake systems. Both brake units on an axle must be applied with the same force at the same time during a stop. If not, a brake pull will be the likely result.

The brake's hydraulic system begins at the **master cylinder**. The master cylinder is basically a piston-type hydraulic pump operated by the brake pedal. Steel tubing and reinforced rubber hoses connect the master cylinder to the pistons in the disc brake caliper and the drum brake wheel cylinders. As the brake pedal is pushed, **brake fluid** is pumped through this tubing to the caliper or wheel cylinder piston. This fluid pushes on the pistons, which in turn push the brake shoes against the rotor or drum. Brake engineers vary the diameters of the master cylinder, caliper, or wheel cylinder pistons, and utilize a proportioning valve to provide a system that is balanced from front to rear.

In the past, a single master cylinder piston was connected to all four of the caliper or wheel cylinder pistons on the car. This provided a nice, simple system with a frightening possibility for failure. A fluid leak in any part of the hydraulic system could stop the operation of the whole system, and there would be a total loss of braking power if the fluid pressure escaped.

In the late 1960s, **tandem hydraulic systems** were introduced (Figure 1-13). The system was split into two parts: front and rear. A fluid leak in either part can take pressure only from that part of the system. Tandem master cylinders use two pistons, one in front of the other. One piston operates the front brakes, and the other operates the rear brakes. With a tandem system, one part of the system will still function if a leak develops in a wheel cylinder, caliper, or brake line. **Front-wheel-drive (FWD)** cars can use a **diagonal split system** (Figure 1-14). One piston of the master cylinder operates the right front and left rear brakes, while the other piston operates the left front and right rear brakes. A diagonal split system provides 50 percent of the braking power if a hydraulic failure occurs. A failure of the front part of a tandem split system will leave substantially less than 50 percent of the braking power. Diagonal split systems can be used only on cars having a negative scrub radius (the tire center line is inboard of the steering axis) on the front wheel, as on most FWD cars (Figure 1-15).

Automotive brake hydraulic systems will often include one or more of the following valves (Figure 1-16):

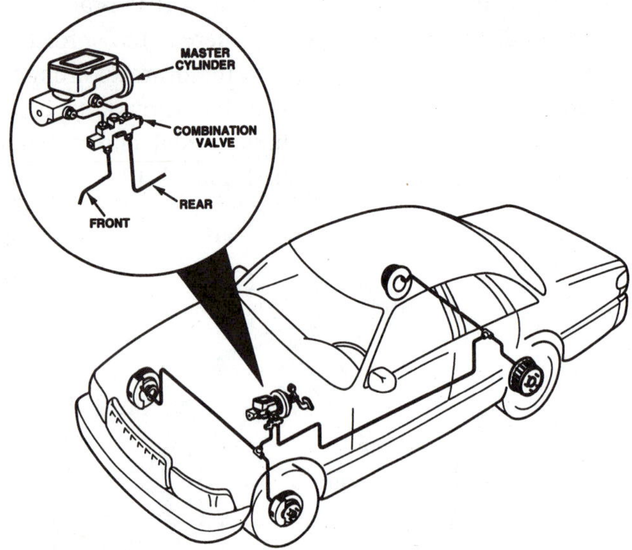

Figure 1-13 A tandem split hydraulic system. Note the position of the combination valve and that the two front brakes connect to one end of the master cylinder and the rear brakes connect to the other end. This system is used on most RWD vehicles. (Courtesy of General Motors Corporation, Service Technology Group)

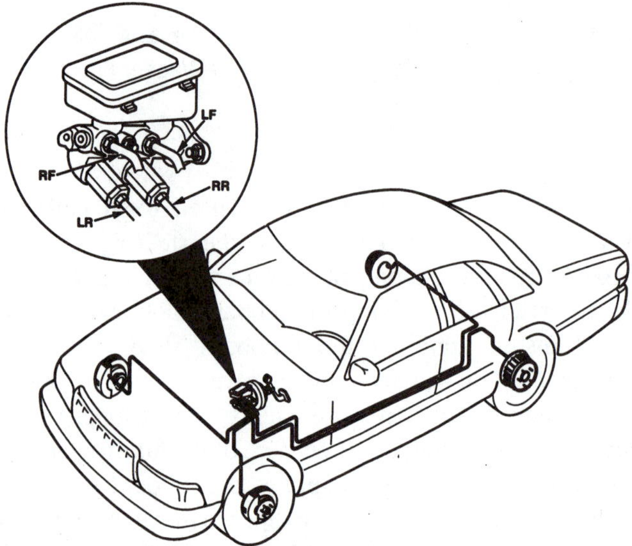

Figure 1-14 A diagonal split hydraulic system. Note that the brakes on the diagonal corners of the car are connected to one end of the master cylinder, and the other two corners are connected to the other end. This system is used on most FWD vehicles. (Courtesy of General Motors Corporation, Service Technology Group)

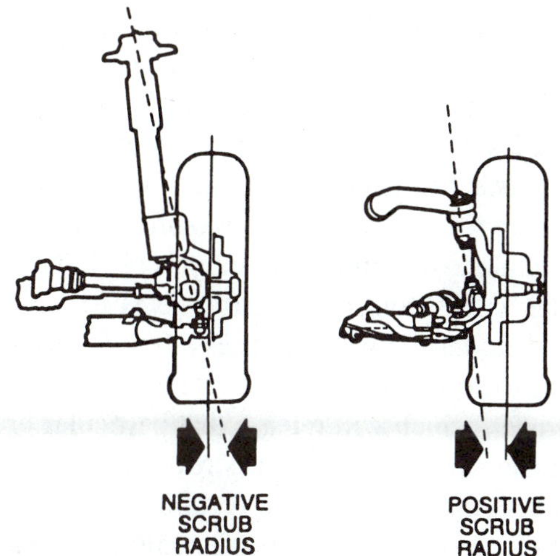

Figure 1-15 Scrub radius is the distance from the tire center to the steering axis at ground level. An FWD vehicle (left) has a negative scrub radius, while an RWD vehicle (right) has a positive scrub radius. (Courtesy of Chrysler Corporation)

■ A **pressure differential valve and switch**. This unit will turn on a **brake warning light** to inform the driver that a hydraulic failure has occurred in one part of a split system.

■ A **proportioning valve**. This unit reduces the hydraulic pressure to rear drum brakes to reduce the possibility of rear wheel lockup during hard braking.

■ A **metering valve**. This unit delays the application of the front disc brakes under light pedal operation so the front and rear brakes will apply at the same time.

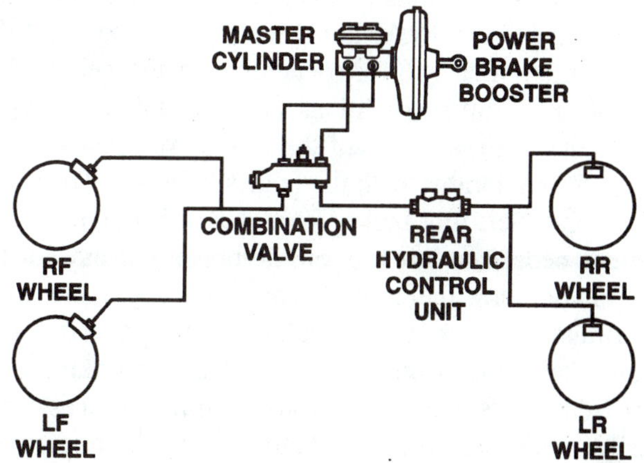

Figure 1-16 This tandem split brake system has a combination valve that includes the metering and proportioning valves and the brake failure warning light switch. This pickup also includes a rear hydraulic control unit for ABS. (Courtesy of Chrysler Corporation)

These valves can be used individually or in a group of two or three. They are often combined into a single assembly called a **combination valve.**

The entire hydraulic system will be described in more detail in Chapter 6.

1.2.5 Power Brakes

Power brakes use a booster to assist the driver in pushing on the master cylinder. Normally, this consists of a **vacuum booster** which fits into the linkage between the brake pedal and the master cylinder. The rest of the system differs very little from an ordinary, nonpower brake system. The booster can multiply the force of the driver's foot many times. It uses the pressure differential between atmospheric pressure and the intake manifold vacuum against a large diaphragm to obtain this boost (Figure 1-17).

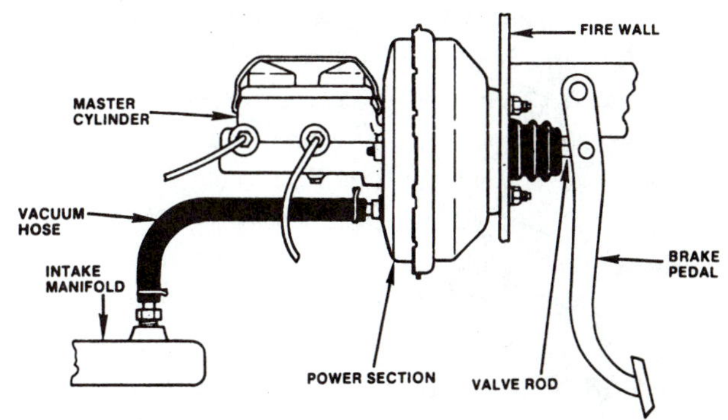

Figure 1-17 This power section is a vacuum power brake booster. The vacuum hose connects the booster to the intake manifold. (Courtesy of Bendix Brakes, by AlliedSignal)

Some cars use a **Hydro-boost** unit, a brake booster that uses power-steering-system hydraulic pressure for the increased force. Hydro-boost units are mounted in the same manner as vacuum boosters, but they are usually smaller and more powerful.

Some newer cars use an electric-motor-operated hydraulic booster. This unit operates much like a Hydro-boost unit, but it develops its own hydraulic pressure using a pump driven by the electric motor. This compact unit operates only when needed to provide an assist. With some ABS, the booster is integrated with the master cylinder and control valve assembly. All three styles of power boosters will be described in Chapter 7.

1.2.6 Parking Brakes

A mechanically operated **parking brake** is required on all motor vehicles with four or more wheels. This brake system normally uses the rear drums or rotors and shoes of the service brake system. It consists of the familiar pedal or lever in the passenger compartment, a means of holding the lever in the applied position, the cables that connect the pedal or lever to the linkage on the shoes, and the levers and struts that attach the cables to the shoes. This system is designed to lock two wheels of the vehicle—usually the rear ones—so that they will not rotate, holding the vehicle stationary while parked (Figure 1-18).

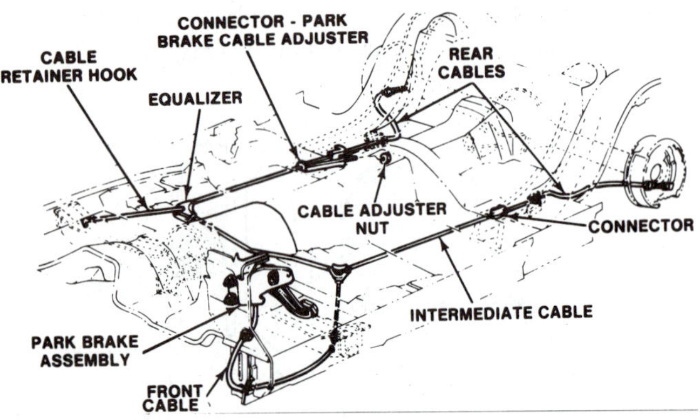

Figure 1-18 A parking brake system utilizes the brake shoes and drum or rotor of the rear brakes and uses cables and a pedal or lever to mechanically apply these brakes. (Courtesy of Chrysler Corporation)

The parking brake system is often called an **emergency brake**. But anyone who has tried to use the parking brake for a high-speed stop knows that it is not designed for true emergencies. Some drivers might be a little careless in maintaining their brake systems if they feel they have an effective system for use under emergency conditions. It is best to think of this system as a parking brake and to maintain the service brake so there will be no need for an emergency brake.

1.2.7 Antilock Braking Systems (ABS)

As we will discuss in the next chapter, it is quite difficult to make a very hard stop without locking up one or more of the brakes. Brake lockup causes tire skid, which in turn can cause loss of vehicle control. This can be especially severe when stopping on wet or icy roads or roads with loose material on the surface. An experienced driver can sense or feel brake lockup beginning to occur and can adjust the brake pedal pressure to apply the brakes just short of lockup. This will maintain vehicle control, but the reduction in pedal pressure reduces the power of all four brakes, not just the one that is beginning to lock up.

An **antilock braking system (ABS)** consists of electronic sensors for all four tires to monitor wheel speed, a control module to determine if lockup is beginning and to operate the valves, and a series of valves to control the brakes. If the control module senses a wheel decelerating or stopping too quickly, it will activate the valve assembly for that wheel, which will release that particular brake or pair of brakes. When the wheel is revolving again at the correct speed, the brake will be reapplied by the system. The electronic control module is a computer device that quickly determines the speed of each wheel; it then determines if the speed of any wheel is too slow and, if so, quickly releases and reapplies one or more of the wheel brakes. The brake can be cycled off and on many times per second, as necessary to keep the wheel rotating, so vehicle control can be maintained. All the driver needs to do in an emergency situation is push on the brake pedal and steer the car (Figure 1-19).

At this time, ABS is standard equipment on most luxury, high-performance cars, pickups, some vans and utility vehicles, and is an extra-cost option on most other cars. ABS is an effective and expensive addition to the brake system that is becoming quite common.

1.3 Brake System Repair

The brake lining, like the tires, wears out. Each time the brakes are used, a small amount of lining wears off during the rubbing action. Brake lining wear is to be expected. The rate of lining wear will vary depending on the weight of the car, its speed, traffic conditions, and, probably most importantly, the aggressiveness of the driver. We often see a car going through traffic with the brakes lights on. The highly aggressive "left-foot braker" with his or her right foot on the gas pedal and left foot on the brake will pay for this habit in both low fuel mileage and short brake lining life.

Brake lining wear is usually fairly slow and can be easily measured by pulling a brake drum or looking at the disc brake pads. Wear of the rotor or drum surfaces can be easily measured using a standard outside micrometer caliper or a brake drum micrometer. The size of these components can be compared to specifications indicated on the drum or rotor. At the same time the friction surfaces are wearing, the hydraulic system is also deteriorating. The rubber seals wear slightly as the pistons stroke in their bores and also become harder and less flexible as heat and time act on them. Also, most brake fluids absorb

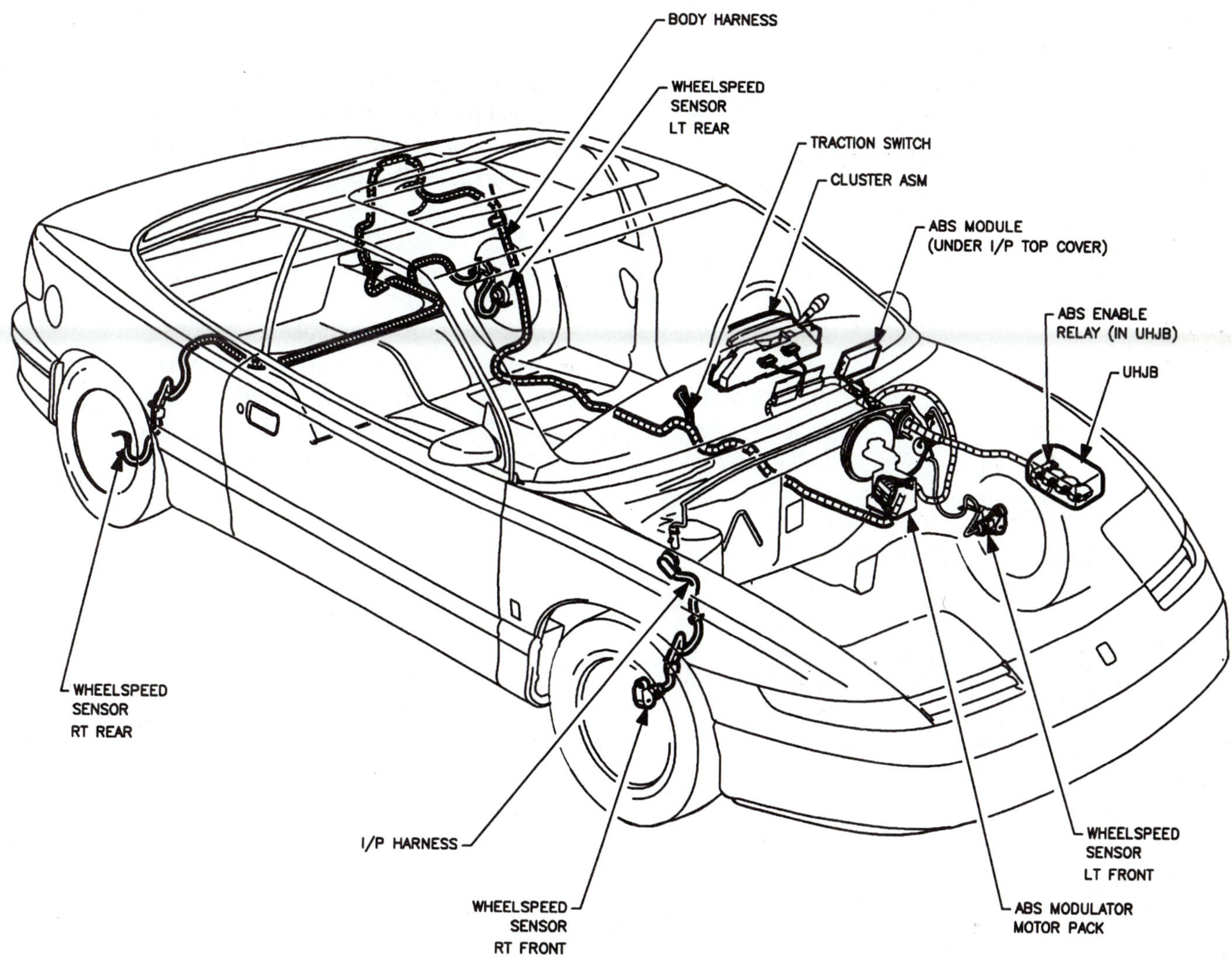

Figure 1-19 An ABS adds wheel-speed sensors and electronic controls to the basic brake system. (© Saturn Corporation, used with permission)

water, and this will cause rusting and etching of the cylinder bores as well as a lowering of the fluid's boiling point. A total loss of braking power will occur if braking heat causes the fluid to boil.

Worn-out brake lining is noisy and usually quite audible. The metallic grinding, grating noise is easily noticed by most drivers. At this point, the stopping power of the brake will be greatly reduced, and repair will probably be expensive because one or more of the drums or rotors will need to be replaced. A faulty wheel cylinder will either stick and fail to apply, fail to release, or leak. A leaky wheel cylinder will allow fluid to leak on the brake lining, which ruins the lining. Leaks can also cause a loss in the effectiveness of that portion of the brake system or, in some cases, a loss of all braking power.

Wise technicians will check the entire braking system when the brake lining is replaced. Both the friction surfaces and the hydraulic portions wear and a failure in the hydraulic system can have catastrophic results. The goal of the brake technician, just like that of the tune-up technician, engine overhaul technician, or any other technician, is to try to achieve "new-car" operation.

1.4 Practice Diagnosis

Imagine that you are working in a general repair shop as a technician specializing in brake repair.

CASE 1: The customer brought in a 1987 Ford Crown Victoria with a complaint that the left rear wheel locks up during a medium-to-hard stop. He says that the problem is probably caused by a faulty valve in the hydraulic system. Is his diagnosis correct? What should you do to confirm his diagnosis? What should you do to find the cause of the problem?

CASE 2: The customer has brought in her 1992 Saturn with a complaint that the emergency brake (parking) doesn't work. On checking it, you find the parking brake lever moves through its full travel with very little resistance. What is probably wrong? What should you do to find the cause of this problem?

CASE 3: The customer has brought in a recently purchased, two-year-old Camry. He complains that the rear tires chirp when he applies the brakes hard and he feels a slight shudder in the brake pedal. Could this be normal for an ABS-equipped vehicle? How can you determine if this car is equipped with ABS?

Terms to Know

anchors
antilock braking system (ABS)
backing plate
base brake
brake fluid
brake lining
brake shoes
brake warning light
caliper
combination valve
deenergizing
diagonal split system
drum
duo-servo brake
emergency brake
energizing
fixed calipers
floating caliper
forward shoe
foundation brake
friction material
front-wheel-drive (FWD)
Hydro-boost
leading shoe
master cylinder
metering valve
nonservo
pads
parking brake
pressure differential switch
pressure differential valve
primary shoe
proportioning valve
rotor
secondary shoe
servo action
servo brake
shoe
sliding caliper
tandem hydraulic system
trailing shoe
vacuum booster

Review Questions

The following quiz will allow you to check the facts you have just learned. Select the best answer that completes each statement.

1. The ideal brake is one that
   a. can lock up all four wheels.
   b. gives the driver a means of easily controlling brake lockup.
   c. will stop the car without pulling or darting to the side.
   d. all of the above.
2. Brake lining
   a. is called friction material.
   b. is attached to the brake shoes.
   c. rubs against the brake drum or rotor during a stop.
   d. all of the above.
3. Statement A: Most modern cars use drum brakes. Statement B: Most modern cars use disc brakes. Which statement is correct?
   a. A only
   b. B only
   c. a combination of A and B
   d. neither A nor B
4. In a duo-servo brake,
   a. the leading shoe increases the pressure on the trailing shoe.
   b. the leading shoe energizes the trailing shoe.
   c. the primary shoe applies the secondary shoe.
   d. neither the primary nor the secondary shoe is energized by drum rotation.

5. Which of the following brake designs will provide the most stopping power?
   a. two-leading shoe
   b. leading-trailing shoe
   c. two-trailing shoe
   d. duo-servo

6. In a disc brake, the piston used to apply the brakes is in the
   a. wheel cylinder.
   b. caliper.
   c. backing mount.
   d. either a or b.

7. Statement A: Floating calipers use a single piston, and sliding calipers use two pistons.
   Statement B: Fixed calipers use either two or four pistons. Which statement is correct?
   a. A only
   b. B only
   c. both A and B
   d. neither A nor B

8. Statement A: Disc brake designs can be classed as nonservo.
   Statement B: Disc brakes can be classed as duo-servo. Which statement is correct?
   a. A only
   b. B only
   c. both A and B
   d. neither A nor B

9. The major reason for using hydraulic brakes on modern cars is that hydraulic operation
   a. produces an increase in force between the brake pedal and the shoes.
   b. allows the driver to move the shoes in a slow, controlled manner.
   c. produces an equal amount of pressure at each wheel cylinder.
   d. keeps the brakes cooler.

10. Statement A: A tandem split brake system has two hydraulic circuits, one in front and the other in back.
    Statement B: A diagonal split brake system has one circuit with the right front and left rear brakes and one circuit with the other two brakes. Which statement is correct?
    a. A only
    b. B only
    c. both A and B
    d. neither A nor B

11. A pressure differential valve is used to
    a. delay the application of the front brakes.
    b. reduce pressure to the rear brakes.
    c. warn the driver of a hydraulic failure.
    d. ensure even application pressure at all the brake assemblies.

12. Front-wheel-drive cars use a diagonal split hydraulic system because
    A. the different front wheel geometry allows steering control while applying a single front brake.
    B. these cars would have very poor stopping ability with only the rear brakes.
    a. A only
    b. B only
    c. both A and B
    d. neither A nor B

13. Power brakes
    A. are essentially a standard brake system with the addition of a booster between the brake pedal and the master cylinder.
    B. will significantly reduce stopping distances when very high pedal pressures are used.
    a. A only
    b. B only
    c. both A and B
    d. neither A nor B

14. Power boosters provide a pedal assist through the use of
    a. a vacuum.
    b. hydraulic pressure.
    c. an electric motor.
    d. any of these.

15. Statement A: Cars use a hydraulically applied service brake system for normal stops.
    Statement B: Cars use a mechanically applied parking brake system to hold them stationary while parked. Which statement is correct?
    a. A only
    b. B only
    c. both A and b
    d. neither A nor B

16. A parking brake will usually use the rear brake shoes and drum or rotor and apply them through the use of a
    a. mechanical linkage.
    b. separate hydraulic system.
    c. series of vacuum lines and cylinders.
    d. none of the above.

17. An antilock braking system is designed to
    a. deliver more stopping power.
    b. operate with less pedal pressure.
    c. provide complete driver control over the car while stopping on wet or icy roads.
    d. all of the above.
18. An antilock braking system consists of
    a. a set of wheel speed sensors.
    b. pressure control valves.
    c. an electronic control module.
    d. all of the above.
19. Through time and use, as the car is driven, the
    a. brake lining will wear out.
    b. brake fluid will absorb moisture and deteriorate.
    c. hydraulic seals will harden and wear.
    d. all of the above.
20. When a brake job is done, the major criterion for the repair should be
    a. low cost.
    b. future driver and passenger comfort.
    c. "new-car" braking performance.
    d. both a and b.

2 Principles of Braking

Objectives

After completing this chapter, you should:

- ❑ Be familiar with the physical forces that act on a vehicle to produce a stop and the limitations that these forces place on stopping ability.
- ❑ Be familiar with the various methods used to measure braking power.
- ❑ Understand the purpose of friction material and the requirements placed on it.
- ❑ Understand the safety hazards that result from working around asbestos-based lining materials.
- ❑ Understand the effect of wheel lockup during a stop.

2.1 Introduction and Federal Requirements

Many aspects of slowing and stopping a car are controlled by simple physics. These are the natural laws, dealing with the deceleration of a body in motion, which cannot be violated. An understanding of these basic principles will aid greatly in thoroughly understanding brakes and the limitations placed on them by the laws of physics.

The federal (U.S.) laws pertaining to brakes apply to new car/vehicle braking systems. Federal Motor Vehicle Safety Standard (FMVSS) number 105 (developed by the Society of Automotive Engineers [SAE] and the National Highway Traffic Safety Administration [NHTSA]) requires that a passenger vehicle with a gross vehicle weight (GVW) of less than 8,000 pounds (3,630 kg) must be able to stop from a speed of 60 mph (100 km/h) in less than 216 feet (65.9 m) and stay within a 12-foot-wide (3.7 m) lane (Figure 2-1). With a loss of power assist, it must stop in less than 456 feet (139 m). This standard also requires that the parking brake must be able to hold the vehicle stationary for at least five minutes in both a forward and reverse direction while parked on a 30-percent grade. Federal standards also place requirements and a timetable for eliminating asbestos from OEM (original equipment manufacturer) brake lining. Other FMVSS standards that pertain to motor vehicle brakes are: 106, brake hoses; 108, lighting; 116, motor vehicle brake fluid; 121, air brake systems; 122, motorcycle brake systems; and 211, wheel nuts, disc and hub caps.

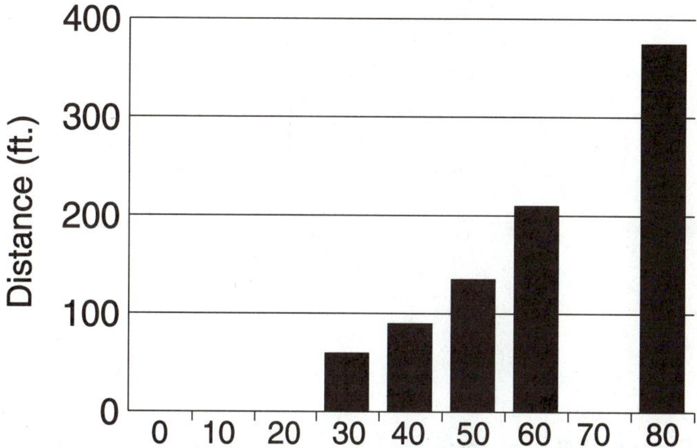

Figure 2-1 The stopping requirements of FMVSS No. 105 for a new vehicle with burnished brake lining. (Courtesy Wagner Brake)

Much of the discussion in this chapter refers to maximum stopping ability as encountered in panic stops. These are unusual conditions, but each brake system must be prepared for such emergency situations.

2.2 Deceleration

The brakes stop a car by decelerating it until its speed is zero. Deceleration is the opposite of acceleration, which is the action of increasing speed or velocity. The rate of deceleration or acceleration is commonly measured using **g** (an abbreviation for gravity) as the standard unit.

The force of gravity will cause an object to fall when dropped. While falling, the object will accelerate to a speed of 32.2 feet per second (fps) in one second; this speed is equal to 9.8 meters per second (mps), 21.95 miles per hour (mph), and 35.3 kilometers per hour (kph). The object will continue to accelerate at this rate for each second it falls until it either reaches the ground or aerodynamic drag prevents any further speed increase. For example, a falling steel ball will reach a speed of about 64.4 fps (19.6 mps) in 2 seconds and 322 fps (98 mps) in 10 seconds (Figure 2-2).

A reduction in speed or deceleration of 32.2 fps in 1 second (fps per s, fps/s, or fps^2) is referred to as a stopping rate of 1 **g**. The average passenger car can obtain a best stop of about 0.6 to 0.8 **g** (19.32 to 25.76 fps/s, 5.88 to 7/85 mps/s) with 0.7 **g** (22.54 fps/s, 6.87 mps/s) about the average. An Indianapolis-type or Formula 1 race car can stop at rates greater than 1.25 to 1.5 **g** (40.25 to 48.3 fps/s, 12.27 to 14.72 mps/s).

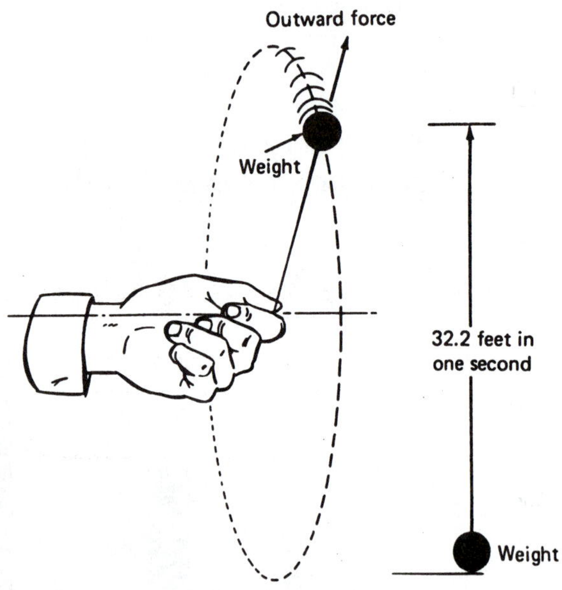

Figure 2-2 The force of gravity will cause an object to accelerate at a rate of 32.2 feet per second during each second that it falls (assuming no aerodynamic drag). If an object (left) is spun fast enough, it can generate enough centrifugal force to keep the string tight even in a vertical position. This force will be 1 **g** or greater.

The typical driver rarely encounters these stopping rates. The average stop made with a passenger car is less than 0.2 **g** (6.44 fps, 1.96 mps); higher stopping rates are considered unpleasant by most drivers. A 100-pound object will experience a forward pressure of 20 pounds at a 0.2-**g** stopping rate, 50 pounds at 0.5 **g**, 100 pounds at 1 **g**, and 150 pounds at 1.5 **g** (Figure 2-3). A 1-**g** stop creates a rather heavy load on the seat belt, and anything that is not fastened down will fly forward—relative to the car—during such a stop (Figure 2-4).

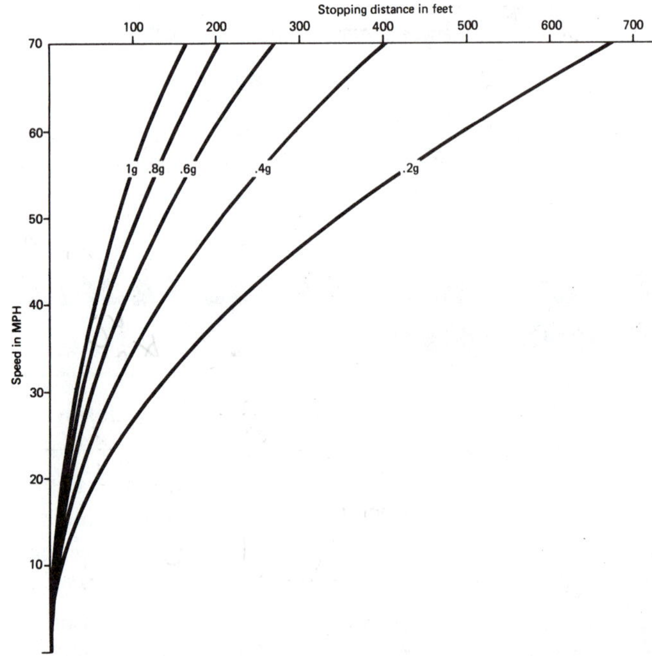

Figure 2-3 Relative stopping distances at five different deceleration rates. Note how the stopping distance increases as the speed increases.

If a stop is made at the rate of 0.5 **g** from 55 mph (88.5 kph, 80.6 fps, 24.6 mps), the speed will be reduced at the rate of 16.1 fps for each second of the stop. The speed of the vehicle at succeeding 1-second intervals will be 64.5, 48.4, 32.3, 16.2, and 0.1 fps (almost zero). This stop will take slightly more than 5 seconds and the car will travel about 161 feet (49.1 m) (Figure 2-5). A 0.2-**g** (6.44 fps, 1.96 mps) stop from the same speed would decelerate the

Figure 2-4 During a hard stop, inertia places a forward-acting force on everything in the car. The driver pushes on the steering wheel and floorboards to resist this forward motion. The seat belt also resists inertia forces.

car at 1-second intervals to speeds of 74.16, 67.72, 61.28, 54.18, 48.4, 41.96, 35.52, 29.08, 22.64, 16.2, 9.76, and 3.32 fps. This stop will take over 12 seconds, and the car will travel about 504 feet (153.72 m). It is easy to see that it takes less time and distance to stop a car at a higher stopping rate (Figure 2-6).

The term **braking efficiency** is often used when testing brakes. This refers to the brake's ability to stop a vehicle. The amount of braking force generated by the brakes is divided by the weight of the vehicle to determine braking efficiency. To achieve 100-percent efficiency, the braking force must equal the vehicle weight, which will produce a deceleration rate of 1 **g**. If the braking force were 60 percent of the weight of the vehicle, the braking efficiency would be 60 percent and the stopping rate would be 0.6 **g**. Like the deceleration rate, braking efficiency is affected by the condition of the road surface, the tires, and the brakes.

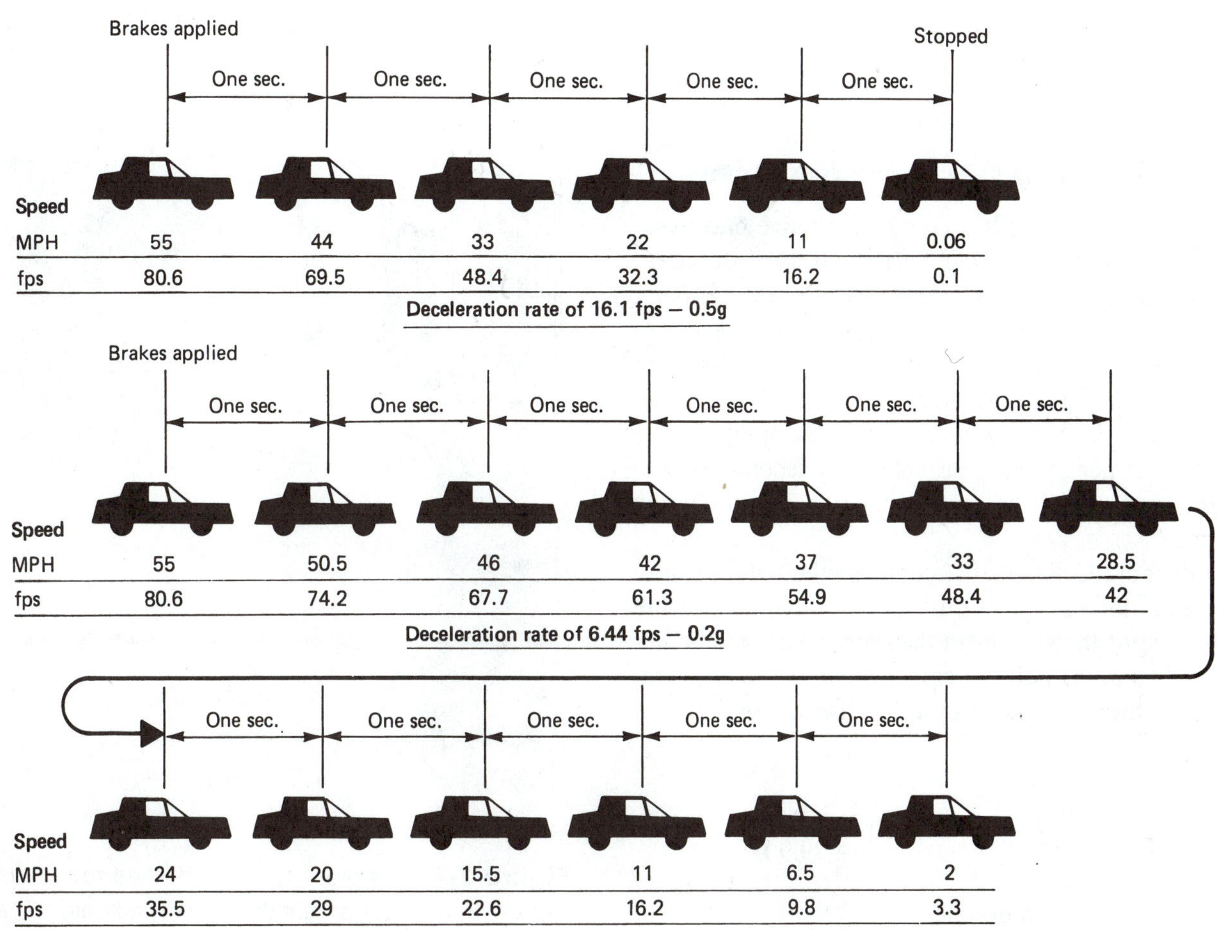

Figure 2-5 When a vehicle decelerates, the speed decreases at the rate of deceleration. The upper vehicle is decelerating at a rate of 0.5 **g**—16.1 feet per second per second. The lower vehicle is decelerating at a rate of 0.2 **g**—6.44 feet per second per second.

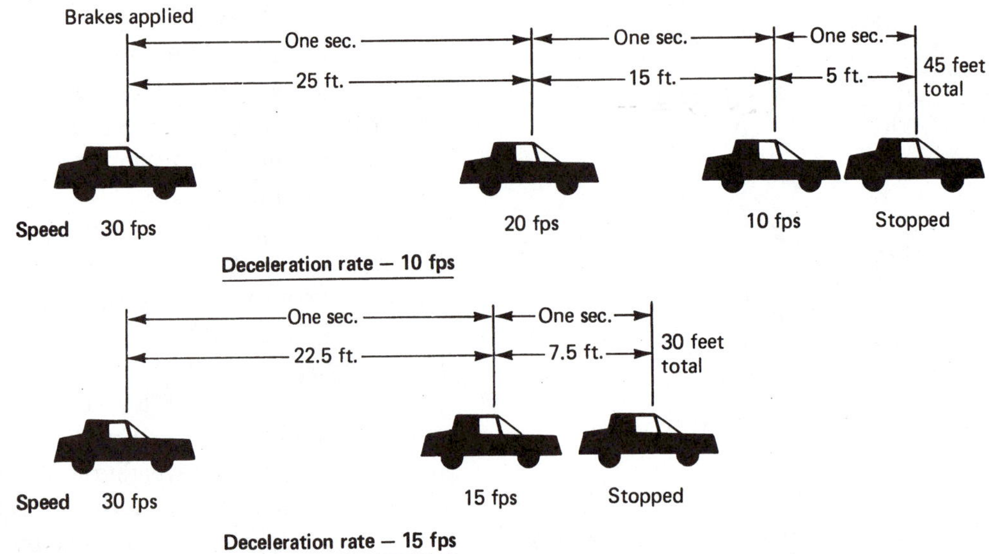

Figure 2-6 When a vehicle slows at a deceleration rate of 10 feet per second per second—about 0.3 **g**—it will stop from a speed of 30 feet per second in 3 seconds and will travel 45 feet. When stopping at a rate of 15 feet per second per second—almost 0.5 **g**—the lower vehicle will stop in 2 seconds and travel 30 feet.

2.2.1 Measuring Deceleration Rates

There are several methods used by automotive brake system repair technicians and engineers to measure deceleration rates in order to determine the quality of a braking system's overall operation. A device called a **decelerometer** can be mounted in the car. Some versions are attached to the windshield, others sit on the floor or seat. A decelerometer will give a readout of the stopping rate in **g** rate, feet per second^2, or meters per second per second; the readout will be on a meter or a liquid column (Figure 2-7).

A rather crude but fairly accurate method of determining deceleration rates is to time the stop, that is, measure the distance of the stop from the point where the brakes were applied to the point of complete stop. The deceleration rate is then calculated by using the appropriate formula:

$$\mathbf{DRg} = \frac{\mathbf{S}}{22\,\mathbf{T}} \qquad \mathbf{DRg} = \frac{\mathbf{S}^2}{29.9\,\mathbf{D}}$$

where **DRg** = deceleration rate (in **g**)
S = speed at the beginning of the stop (in mph)
T = time (in seconds)
D = distance traveled during stop (in feet)

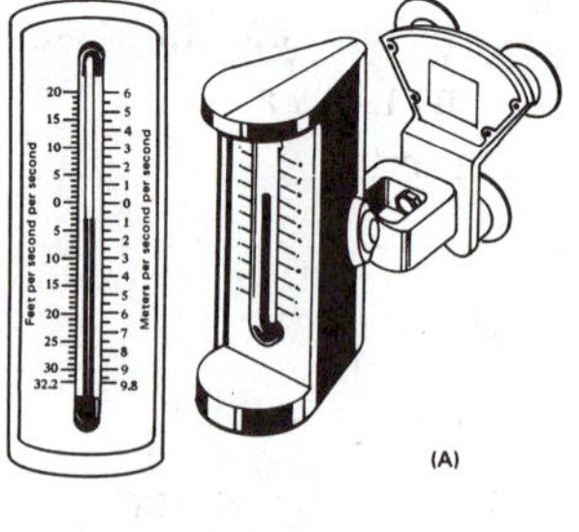

Figure 2-7 Decelerometers are attached to the windshield (A), set on the floor or seat (B), or clamped onto some part of the vehicle (C). As the vehicle makes a stop, the meter displays the deceleration rate in stopping distance, deceleration rate, or braking efficiency. (Courtesy of Ammco Tools, Inc.)

or

$$DRFPS^2 = \frac{1.476S^2}{T} \qquad DRFPS^2 = \frac{1.076S^2}{D}$$

where **DRFPS²** = deceleration rate (in feet per second per second)

A high number like 0.7 **g**, 22.5 fps, or greater means the brakes are working quite well, and a low number indicates that they are not as effective.

A brake technician seldom computes deceleration rates, because the numbers are not actually useful when repairing brake systems. Also, the time involved in meeting the requirements of a proper road surface with no traffic makes deceleration measurements difficult. Deceleration rates are measured from fairly high speeds and represent the maximum values a vehicle can achieve. This requires good tires, a stretch of clean, dry pavement with no traffic, and usually a passenger or observer to take the readings or measurements. These requirements, plus the time involved, make actual deceleration tests practical for the engineer developing a brake system for a particular car, but impractical for the repair technician. The brake technician usually measures stopping rates by the "seat of the pants" during a road test.

Most states require that a passenger car should be able to stop in a distance of 25 feet (7.6 m) from a speed of 20 mph (32 kph). This is a stopping rate of 0.53 **g** or 17.3 fps.

2.3 Energy of Motion

It takes energy to move a car. **Energy** is the ability or capacity to do work. **Work**, in this sense, is the process of moving something or changing the speed or direction of its travel. According to the laws of physics, energy cannot be created or destroyed. It can, however, be changed from one form to another. Energy is available in many forms. Heat, light, various fuels, and electricity are common examples. In an automobile, the potential energy of gasoline is converted to heat and then to mechanical motion within the engine. When the car is put to motion, this energy becomes **kinetic energy**: energy of motion.

The amount of energy in a moving object, such as a car, can be calculated by using the following formula:

$$\text{kinetic energy} = \frac{W \times S^2}{29.9}$$

where **W** = weight of car (in pounds)
S = speed (in miles per hour)

Unless you are an engineer, you will have little use for this formula other than to realize that a heavy car in motion will have more kinetic energy than a light car and, even more important, that the amount of energy will increase at a rate equal to the square of the speed. For example, a 1,000-pound (454.5-kg) car going 30 mph (48.27 kph) represents 30,100.3 foot-pounds (40,816 N-m) of kinetic energy. A 2,000-pound (909-kg) car going the same speed represents 60,200.6 foot-pounds (81,632 N-m) of energy (twice as much). At 60 mph (96.54 kph) the lighter car would represent 120,401.3 foot-pounds (163,264.16 N-m) of kinetic energy and the heavier car, 240,802.6 foot-pounds (326,528.3 N-m) of kinetic energy. Note that as the weight of the car is doubled, the amount of energy is also doubled, but as the speed is doubled, the amount of energy is increased about four times (Figure 2-8).

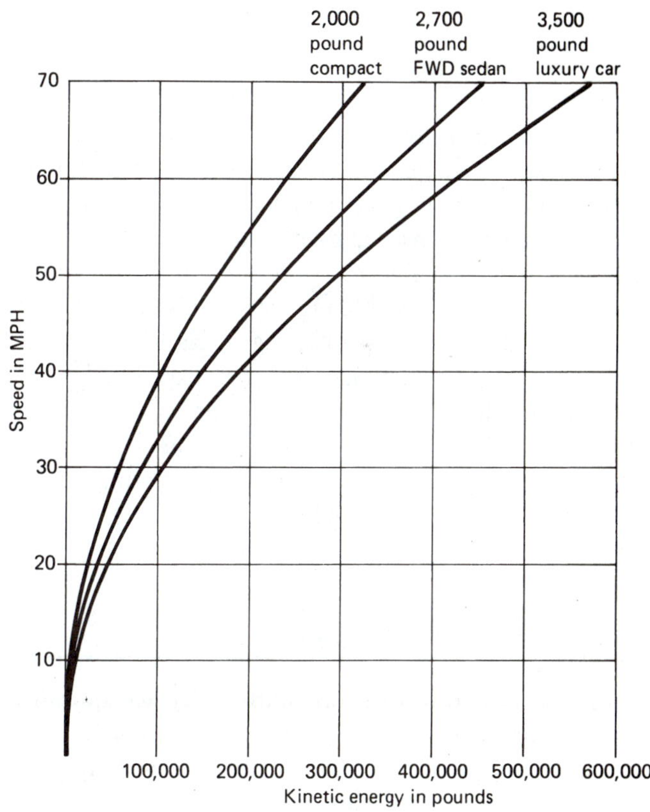

Figure 2-8 The amounts of kinetic energy of three different vehicles at various speeds. Note how the amount of energy increases with an increase in weight and speed.

Every moving object contains kinetic energy, and energy must be converted from some form and put into that object to start it moving. Because energy cannot be created or destroyed, the energy in an object must be removed in order to stop it. The kinetic energy needs to be converted to a different form of energy (usually heat), and as the speed increases, the amount of energy needed for this conversion greatly increases.

2.4 Friction and Heat Energy

The simplest way to stop a car is to convert the kinetic energy to heat. Heat is one of the more common, versatile forms of energy. This conversion occurs naturally when a car coasts to a stop. The rolling friction of the axle shafts and wheel bearings, the flexing friction of the sidewalls and treads of the tires, plus the molecular friction of the air drag over the body and undercarriage all create small amounts of heat. As the kinetic energy becomes heat energy, the car will decelerate and come to a stop. A car with very efficient bearings, tires, and aerodynamic design will coast for a long distance before using up its kinetic energy.

When we lack the time or distance needed to make coasting stops, we use the brakes. Brakes are essentially heat machines. They generate heat from friction by rubbing the lining against the rotating rotors or drums. Friction can generate large amounts of heat. Prove this to yourself by vigorously rubbing your hands together or by running, and then sliding on a rug or carpet. Be careful when you do this—anyone who has experienced "rug burn" knows why. Another good example of heat generated by friction is the smoke pouring off a tire that is locked up and skidding. This tire is generating enough heat to burn up the tread rubber.

2.5 Heat Energy

Heat is measured in two forms: **intensity** and **quantity**. Heat intensity is what we feel—how hot it is—and is measured in degrees. The two commonly used scales are **Fahrenheit (F)** and **Celsius (C)**. These two scales do the same thing, but they have different lengths and starting points. Easily remembered places on these scales are the melting point of ice, 32°F or 0°C, and the boiling point of water, 212°F or 100°C.

Heat quantity is the amount of heat in an object or area, or the amount of heat that an object can absorb or release. Heat quantity also has two measurement systems: **British thermal units (BTUs)** and **calories (c)**. One BTU is the amount of heat it takes to increase the temperature of 1 pound of water by 1°F. One calorie is the amount of heat it takes to increase the temperature of 1 gram of water by 1°C. A BTU is a much larger quantity: 1 BTU = 252 c.

Heat energy is important for the brake technician to understand. One BTU is equal to 778 foot-pounds of energy. Conversely, 1 foot-pound of energy is equal to 0.0013 BTUs. The energy of a 2,000-pound car going 30 mph is equal to 77.38 BTUs, and that of the same car going 60 mph is equal to 309.5 BTUs. If we were to remove that much heat energy from these two moving cars, they would stop.

Heat can be transferred from one object to any other object that is cooler. The object can be in solid, liquid, or vapor form, but it must be cooler. Heat will always flow from a warmer object to a cooler object. As the cooler object absorbs heat, its temperature increases, and it becomes hotter. As the hotter object loses its heat, it becomes cooler.

Heat generated at the brake lining and rotor or drum surfaces flows to the lining and into the rotor or drum and then to the cooler air passing over them (Figure 2-9). The temperature of the brake components increases during each stop. The amount of increase will be determined by the vehicle speed and weight, the rate of the stop, and the mass of the brake components, especially that of the drums and rotors. If the stop is gradual, from a slow speed, and short in duration, the heat will probably flow away as quickly as it is generated, and the components will not get much hotter than the **ambient** temperature (the temperature of the surrounding air). If the stop is rapid, from a high speed, if the car is heavy, or if there are several stops in a short period of time, the brakes will generate more heat than can be easily dissipated. Then the brake components will get very hot. Carefully feel the rotors or drums after a few hard stops; they can be very hot and could easily boil water. Under extreme conditions, temperatures can be high enough to cause brake lining fade or even high enough to boil the brake fluid in the wheel cylinders.

In the past, passenger-car brake temperatures stayed relatively low—about 250°F (120°C) for average usage, and 350°F (175°C) for heavy usage. In today's lighter cars the brake components have become substantially smaller, so there is less metal to absorb the heat and act as a heat sink during stops. An average brake temperature for a modern front-wheel-drive car is about 350°F (177°C) with heavy usage increasing it to as high as 500°F to 800°F (260°C to 425°C) (Figure 2-10). The brake components, especially the lining and fluid, of modern systems have to withstand considerably higher temperatures than those of the past.

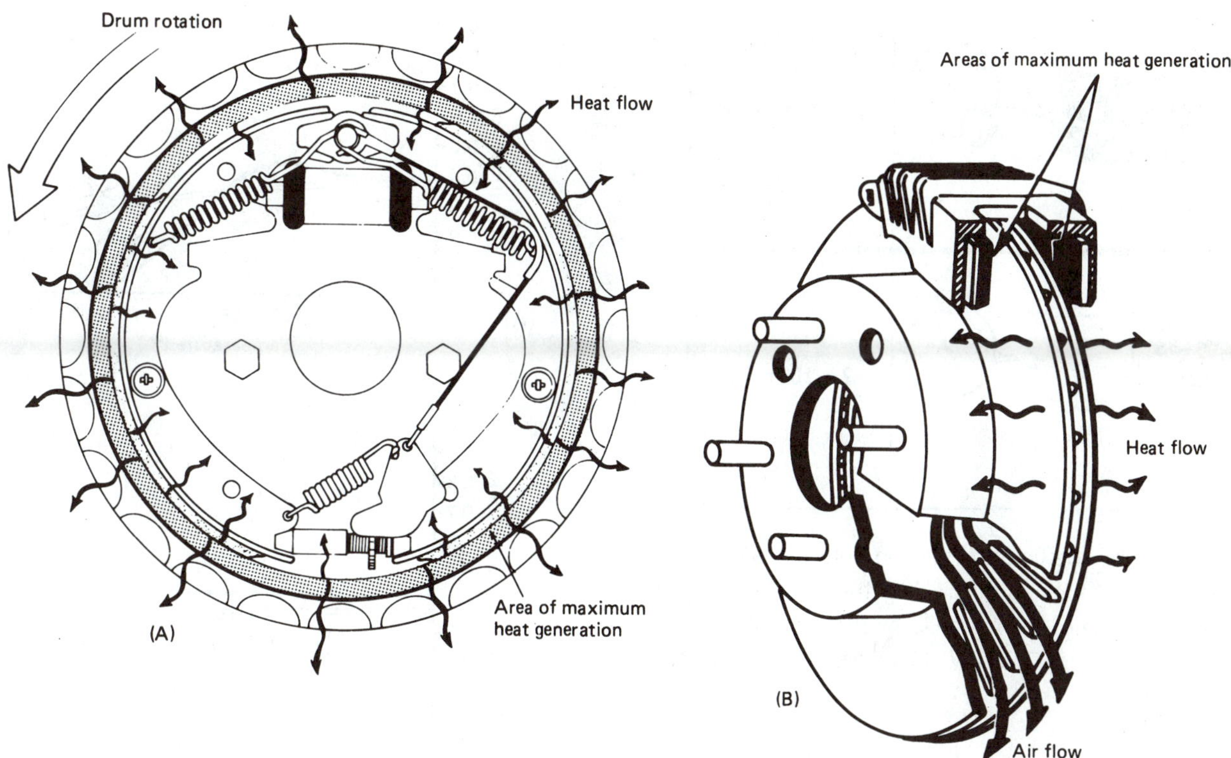

Figure 2-9 As the brakes are used, heat is generated where the lining rubs on the drum or rotor. This heat is either absorbed by the brake components or dissipated into the surrounding air.

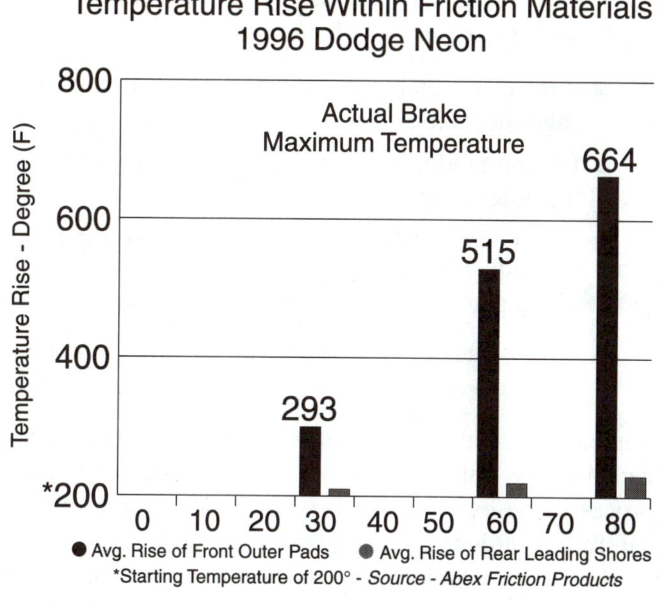

Figure 2-10 This graph shows the temperature increase of the front and rear brakes during stops at different speeds. (Courtesy of Wagner Division Brake)

2.6 Brake Lining Friction

Friction is the resistance as one surface slides over another. Friction produces heat. The brake lining is designed to produce heat from friction as it rubs against the rotor or drum and is commonly called **friction material**. The amount of heat produced is determined by the coefficient of friction between the lining and the rotor or drum, the amount of pressure pushing them together, and the relative speed of each.

The term **coefficient of friction** refers to the amount of resistance preventing one item from sliding across another. It is calculated by sliding one object across the surface of another, measuring the amount of force required, and then dividing that force by the weight of the moving object. For example, if it takes a 2-pound (0.9-kg) force to slide a 100-pound (45.45-kg) block of ice across a certain surface, the coefficient of friction will be 2 divided by 100, or a very low 0.02. If it takes 70 pounds (31.81 kg) of force to slide a 100-pound block of rubber across the same surface, the coefficient of friction will be 70 divided by 100, or a rather high 0.70 (Figure 2-11).

Friction material, normally called brake lining, has to provide the proper amount of drag or friction as it rubs

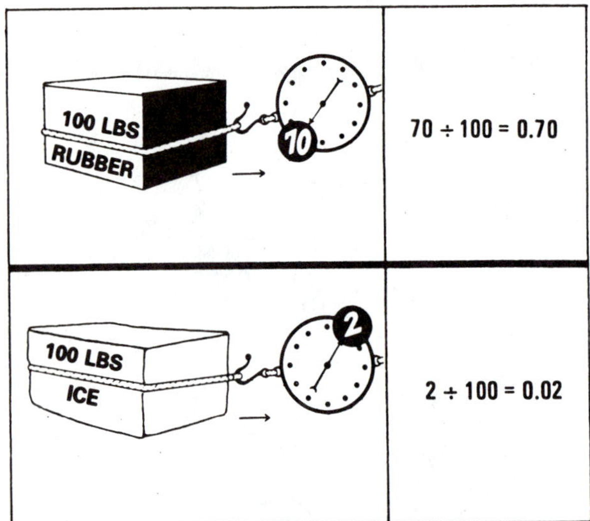

Figure 2-11 The coefficient of friction between two different materials is determined by dividing the amount of drag needed to move one of the materials by the amount of force pushing them together. (Courtesy of Ford Motor Company)

against the drum or rotor under various heat conditions from extremely cold—at the beginning of a stop in the winter—to extremely hot—at the end of a high-speed stop in the summer. At the same time, there must be a minimum amount of wear on the drum or rotor friction surfaces.

The coefficient of friction of most passenger-car brake linings is about 0.3. This represents a usable and controllable amount of friction. If the coefficient of friction is too low, the shoes will not produce enough friction and heat to stop effectively. The result of this will be a "hard" brake pedal with poor stopping power. If the coefficient of friction is too high, the brakes will be too "grabby" and will be very hard to control. Wheel lockup and skids will occur too easily.

The coefficient of friction of some lining blends will change as the lining heats up. An inferior lining will often fade—undergo a loss or lowering of the coefficient of friction—as it gets hotter. The stopping power will drop as the brake pedal gets harder. Some linings will experience an increase in the coefficient of friction and become more grabby as they get hotter. Quality lining materials will undergo a very minor change, if any, in their coefficient of friction as they heat and cool (Figure 2-12).

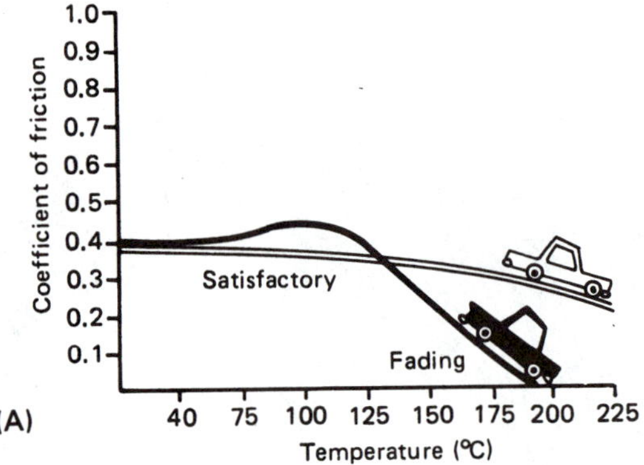

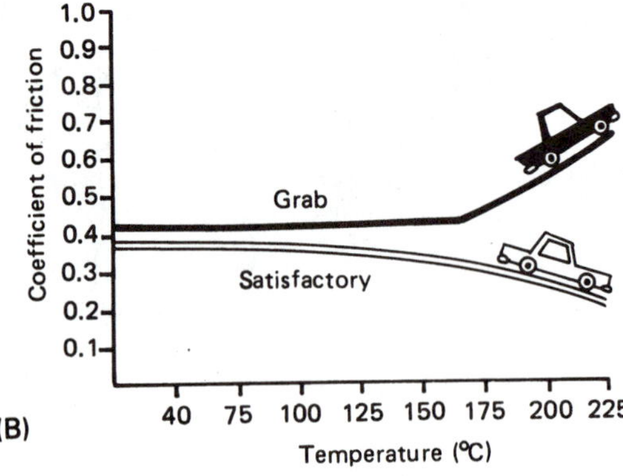

Figure 2-12 The dark brake lining curve (A) is showing severe fade as it gets hot. The dark curve in (B) shows a grabby condition as it gets hot. (Courtesy of Ford Motor Company)

2.6.1 Brake Lining

Brake lining is formed using different recipes and may contain the primary ingredient in several different forms. The major ingredient in the lining is the base frictional material. It may consist of a powder, short fibers, or longer fibers that are woven into cords or mats, and it provides the friction and the ability to resist heat. Other important ingredients combined in a lining are:

- binders—usually a resin which holds all the other ingredients together.
- friction modifiers—materials that change or adjust the coefficient of friction.
- fillers—materials that can improve the strength and the ability to transfer heat, or that provide quiet operation.
- curing agents—materials that provide the correct chemical reaction during manufacture and curing.

Some of the methods for producing brake lining material can involve an expensive manufacturing or curing process. The exact recipe for a lining type is usually a carefully guarded secret (Figure 2-13).

At one time, brake lining material was placed in one of the following two categories: **organic** and **inorganic**. Most so-called organic linings had **asbestos** as the base friction material, and most inorganic linings were **semimetallic**. These categories identified the primary friction material for the linings. This classification system seemed somewhat

Friction Material

Type	Ingredients	Advantage	Disadvantages
Organic	Early: wood & leather Recent: asbestos +	Quiet, cheap, low abrasiveness, good cold friction	Asbestos content, brake fade when hot
Metallic	Powdered metal	Fade-resistant	Poor cold friction, high pedal pressure, abrasive, noisy
Semimetallic	Combination	Fade-resistant, long wear life	Expensive, brittle, poor cold friction
Synthetic	Fiberglass	Good lining life, quiet, nonabrasive	Expensive, not good for very high temperatures
	Aramid	Very good lining life, quiet, nonabrasive	Poor cold performance

Figure 2-13 Some of the advantages and disadvantages of different lining types.

odd because both compounds—asbestos and metal—are inorganic. With the reduction in the use of asbestos (it has not been used on new domestic cars since 1993, but is still used in replacement linings), the major classifications for newer lining types are **nonasbestos organic (NAO)** and semimetallic. Most NAO linings use fiberglass or aramid fibers as a base. Asbestos is known to cause health hazards. Do not breathe the dust from any lining because the other materials can also cause problems (Figure 2-14).

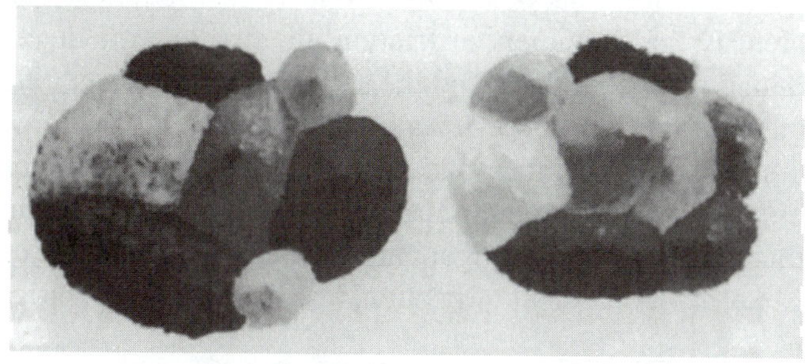

Friction King II semimetallic patented composition
- Coarse sponge iron
- Steel fibers
- Graphite and other friction modifiers
- Resin binders

Conventional (organic) composition
- Asbestos fibers
- Organic friction modifiers
- Resin binders

Figure 2-14 A brake lining is formed when the various ingredients are mixed and molded into shape. (Courtesy of Bendix Brakes, by AlliedSignal)

Semimetallic lining has become the standard **original equipment (OE)** lining for most new FWD cars. The brownish-black dust on many front wheels is evidence of semimetallic lining. The brown color comes from rust of the metal dust, and the black is from graphite, a friction modifier. Typical front rotor temperatures have risen from about 350°F (177°C) in the early 1970s to about 550°F (288°C) in the mid-1990s. Metallic-based linings have superior high-temperature wear characteristics. Most of the other lining types will wear quite rapidly at the higher temperatures (Figure 2-15). It is important that the replacement

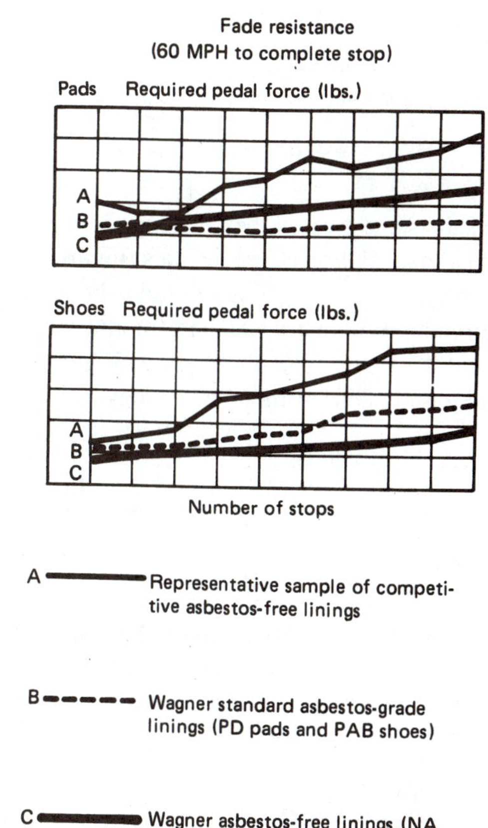

Figure 2-15 A comparison of the amount of pedal pressure required to stop a car for three different linings for a series of ten stops. Note the differing amounts of pedal pressure required to bring the car to a complete stop as the brakes heat up. (Courtesy of Wagner Brake)

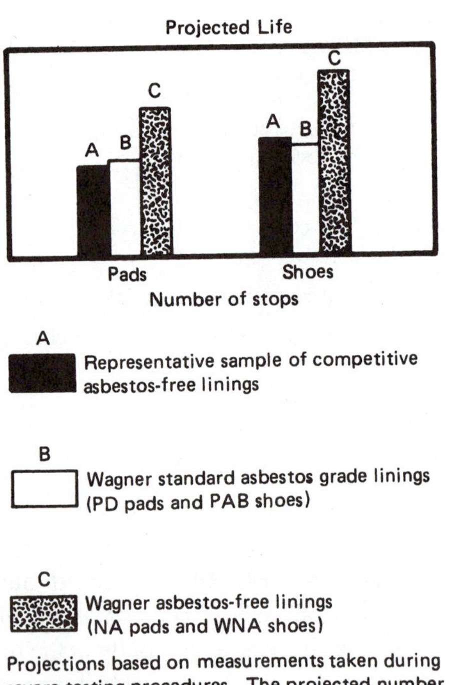

Figure 2-16 The projected life of three different linings calculated under severe operation conditions. (Courtesy of Wagner Brake)

lining be of the same type as the OE lining so it will have the same wear and friction characteristics (Figure 2-16).

Brake lining quality is difficult to judge because there is no government or industry standard for aftermarket lining. Well-known lining manufacturers with good reputations spend a good deal of time and money to ensure the quality of their product. Their reputation is usually a sufficient basis for selecting good lining. Most manufacturers produce several types of lining that differ in quality from an **economy** or **competitively priced** form that is equal to the OE lining to a **premium** or **heavy-duty** quality; these types are usually available in both an NAO and semimet (short for semimetallic) material. Some of the criteria used to judge the quality of a lining are

- high resistance to fade.
- a quick full recovery if fade should occur.
- long wear life.
- little or no rotor or drum wear.
- quiet operation.
- good friction characteristics when damp or wet (Figure 2-17)

The coefficient of friction of a lining can be easily determined by reading the **edge code** or **edge brand** printed on the edge of the lining (Figure 2-18). This code was established by the **Society of Automotive Engineers (SAE)**. It is composed of three groups of letters and numbers. The first group is a series of letters that identify the manufacturer of the lining. The second group is a series of numbers, letters, or both, that identify the lining compound or formula. The third group is two letters which identify the coefficient of friction; the first letter is determined from cold tests, and the second from hot tests. The meaning of these code letters is provided in Figure 2-19. It should be remembered that the friction code determines only the relative amount of drag that will be generated in that brake. Most passenger cars will have a code of EE or EF. A low coefficient of friction, less than 0.15, will prob-

Brake Lining Characteristics[a]

	Organic			*Metallic*			
Lining Material	*DM-5470*	*DM-5490*	*DM-8010*	*DM-8011*	*DM-8015*	*DM-8015A*	*DM-8032*
Duty	**All-purpose moderate**	**All-purpose moderate**	**All-purpose moderate**	**All-purpose moderate**	**All-purpose heavy-duty**	**All-purpose heavy-duty**	**All-purpose heavy-duty**
Noise characteristic[b]	**Superior**	**Superior**	**Excellent**	**Good**	**Good**	**Excellent**	**Good**
Life characteristic	**Superior**	**Superior**	**Excellent**	**Excellent**	**Superior**	**Superior**	**Superior**
Displacement characteristic	**Good**	**Excellent**	**Excellent**	**Excellent**	**Superior**	**Superior**	**Superior**
MVSS characteristic	**Good**	**Good**	**Excellent**	**Excellent**	**Superior**	**Superior**	**Superior**

[a]Based on comparable Delco Moraine product standards.
[b]Noise performance is affected by rotor design and internal rotor dampening. Ratings shown are based on "moderately dampened" systems.

Figure 2-17 This chart shows the operating characteristics for three organic and four metallic lining types. (Courtesy of General Motors Corporation, Service Technology Group)

ably produce a hard brake pedal with poor braking power in most cars. A high coefficient of friction, over 0.55, will probably produce a grabby brake. Other factors that determine lining quality such as fade resistance, wear on drums or rotors, wear life of the lining, moisture resistance, erratic behavior, quiet operation, and structural integrity are not included in this coding system.

Figure 2-18 The lettering on the edge of this lining is called an edge code. This code—DELCO 235FE—indicates that the lining was made by Delco Moraine and the friction class is FE. The 235 is the manufacturer's code identifying the lining type and composition.

Code Letter	*Coefficient of Friction*
C	Not over 0.15
D	Over 0.15 but not over 0.25
E	Over 0.25 but not over 0.35
F	Over 0.35 but not over 0.45
G	Over 0.45 but not over 0.55
H	Over 0.55
Z	Not classified

Figure 2-19 The last two letters of the edge code indicate the lining's coefficient of friction. The coding for the letters is shown here.

The terms "hard" and "soft" are often used to refer to the coefficient of friction and fade characteristics of a lining. A soft lining is one with a fairly high coefficient of friction. It tends to break in fairly quickly and easily, have quiet operating characteristics, and fade at relatively low temperatures. A hard lining has a lower coefficient of friction. It tends to take longer to break in, has a greater tendency to squeal, and has better high-temperature characteristics.

2.6.2 Concerns about Asbestos

It has been determined that asbestos can cause health problems. If the fibers are inhaled, the human body is not able to expel them, and these fibers can cause asbestosis or form the base for a cancerous growth.

Asbestos is a mineral which easily separates into long, flexible fibers. It makes a near-perfect friction material and heat insulator because these fibers are noncombustible, will not conduct electricity, and will not transmit heat. Most health problems are related to the amphibole type of asbestos which is used mostly for insulation. Brake and clutch friction material comes from the chrysotile asbestos family.

There are several forms of asbestos. The amphibole group contains silica as part of its chemical makeup; the outer layer of this fiber is totally resistant to human bodily fluids. It will not dissolve, so it can be expelled from the body. The outer layer of a chrysotile fiber is magnesia which is readily attacked by body fluids and expelled from the body. Fewer health problems come from friction materials than from insulation. This fact should ease the mind of the technician performing brake repairs, but some degree of concern for personal safety should still be exercised. The cases of health problems traced to brake repair involve a lag time of about fifteen to twenty years between exposure and the time that symptoms develop. In one case involving the child of a technician, the problem was traced to brake dust brought home on the technician's clothing. As brakes are used and heat up, the resins that bind the fibers together break down. The asbestos fibers which have worn off the shoe are released inside the drum. These fiber particles are very small—usually too small to be seen by the unaided eye. Breathing them along with other dust particles may cause injury. Many sources recommend that a face mask be worn while doing brake work and that dust from worn brakes be carefully collected and then carefully disposed of in the proper, approved manner.

The **Occupational Safety and Health Administration (OSHA)** has set standards concerning the acceptable level of asbestos fibers that can be present in a workplace or area where brakes are being repaired. OSHA has also determined safe methods that should be used to reduce exposure to these fibers. It has been determined that asbestos fibers will cling togather and mat easily with water and will not become airborne when wet. These fibers can also be captured using a vacuum cleaner equipped with a high-efficiency particulate air (HEPA) filter. This provides a relatively easy and safe method of cleanup. These procedures will be described in Chapter 10.

Traction n. gripping power to permit movement

2.7 Tire Friction

Tire-to-road friction—commonly called **traction**—allows the driver to start and stop a car and control where it goes. A rolling tire follows a path in the direction it is pointed. A car, of course, goes where the wheels are pointed. The amount of traction that a tire can produce is determined by the load on the tire and the coefficient of friction between the tire and the road. A large amount of traction will produce a car that can accelerate, brake, and turn corners quickly. Poor traction produces tire spinning on acceleration, lockup on braking, and skidding on cornering and braking maneuvers.

Interestingly, a tire will produce its highest traction when it is slipping slightly on the road surface, about 15 to 25 percent (Figure 2-20). This is a result of the elastic tread surface creeping into maximum contact with the road surface. This slippage occurs when an accelerating tire is turning slightly faster than the vehicle's speed or a decelerating tire is being braked to a speed slightly slower than that of the car. For example, maximum braking traction will occur for a car that is traveling 55 mph (88.5 kph) if the tires are slowed to a speed of about 45 mph (70 kph). At this slippage rate, about 20 percent, the tire usually produces a very light skid mark.

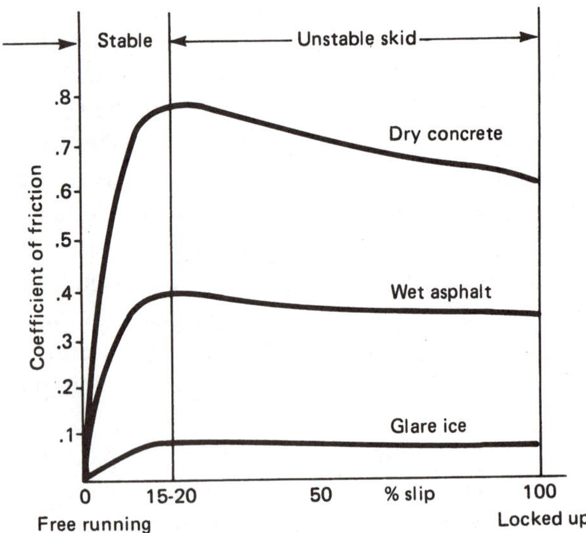

Figure 2-20 The road-to-tire coefficient of friction also varies with tire-to-road slip rate. The greatest amount of traction occurs at about 15- to 20-percent slip.

A tire that is slipping at a rate greater than 30 percent will begin to suffer a loss of traction as well as a loss of directional control. A skidding tire will skid sideways as easily as forward or backward.

Braking force cannot exceed the traction of the tires or lockup and a skid will result. A skidding tire will slow a car down because of the heat generated by the tire-to-road friction, but this is usually much less than the heat that can be generated by the shoes and the rotor or drum while the tire rotates. The highest deceleration rate that a car can attain is determined by the traction of the tires. The highest value of the tire-to-road coefficient of friction will be about 0.8 on dry pavement. Wet or icy pavement will greatly reduce this coefficient of friction and the amount of traction and thus greatly reduce stopping ability. The coefficient of friction between a tire and wet concrete pavement is about 0.5, and for ice it is less than 0.1 (Figure 2-21).

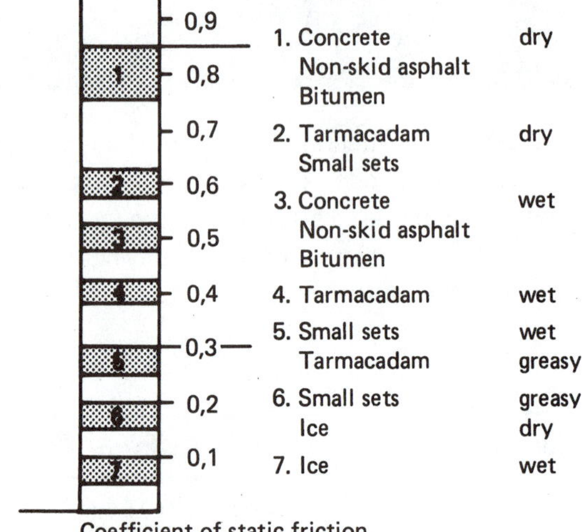

Figure 2-21 Tire traction depends on the tire-to-road coefficient of friction, which depends greatly on the type and condition of the road surface. (Courtesy of ITT Automotive)

2.8 Brake Power

The amount of power that a brake assembly can generate varies with the radius of the tire, the radius of the rotor or drum, the coefficient of friction of the lining, the application force at the lining, the weight on the tire, and the coefficient of friction between the tire and the road. Brake power is often called **brake torque** because it resists the turning of the rotor or drum (Figure 2-22). The tire-to-road factor enters into the amount of brake torque that can be generated because lockup will result if the amount of brake torque exceeds the amount of traction. The brake shoes develop a rotating retarding force (torque) between the tires and the suspension, measured in foot-pounds or newton-meters.

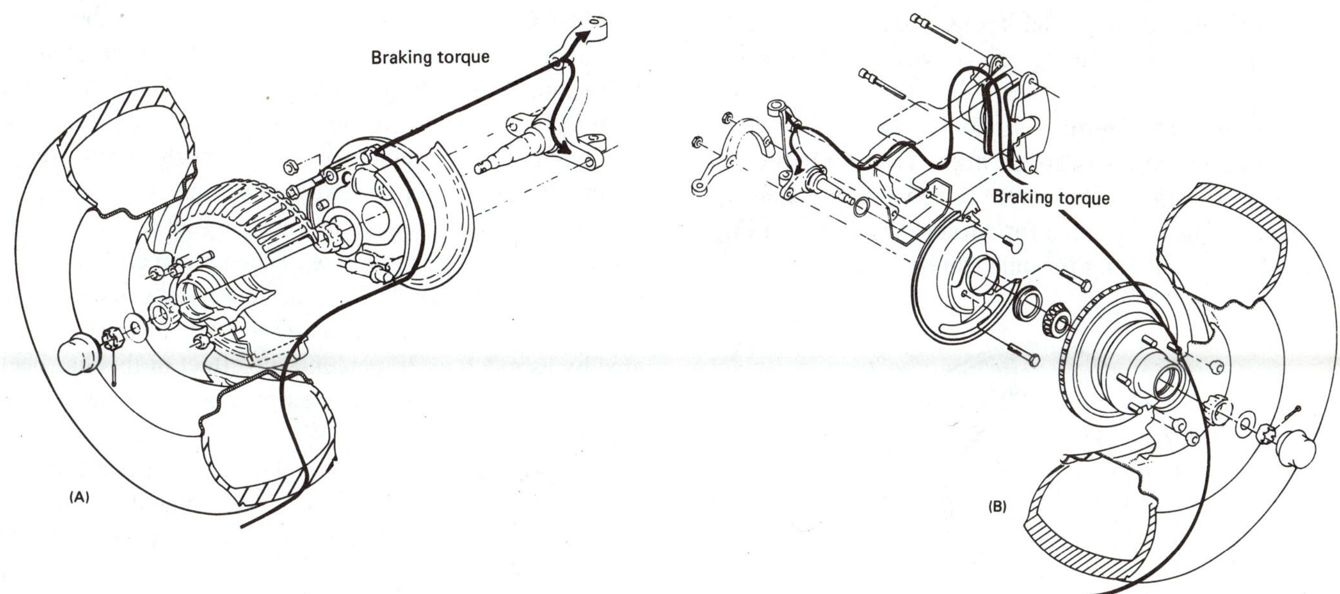

Figure 2-22 As the brakes are applied, a retarding force—between the tire's road contact and the car's suspension—is created. Since it is a rotating force, it is often called brake torque.

The actual amount of braking force for a given vehicle is calculated using a formula similar to that in Figure 2-23. The most accurate way to determine the amount of braking force is to work from the deceleration rate using the following formula:

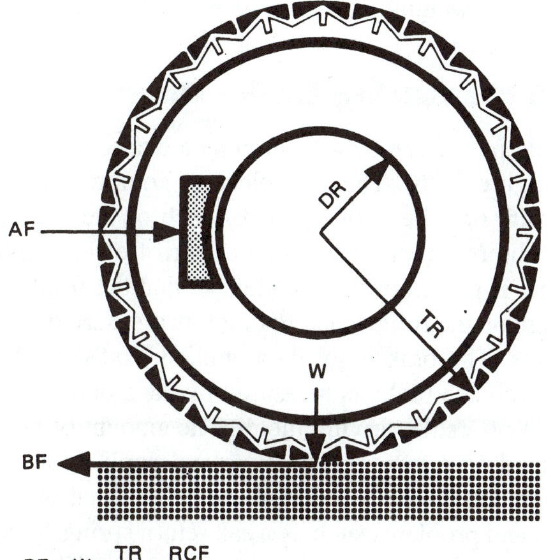

$$BF = W \times \frac{TR}{DR} \times \frac{RCF}{LCF}$$

BF = Brake force
W = Weight
TR = Tire radius
DR = Brake drum radius
RCF = Coefficient of friction, road
LCF = Coefficient of friction, lining

Figure 2-23 The amount of braking force that can be generated by a tire and brake assembly can be determined using this formula. (Courtesy of ITT Automotive)

$$\mathbf{F} = \mathbf{DRg} \times 2{,}000$$

where **F** = force, braking force per ton of vehicle weight (in pounds)
DRg = deceleration rate (in **g**)

As the brakes are used, the linings become hot. The amount of heat (the number of BTUs or calories) will be determined by the velocity change and the weight of the car. The amount of temperature increase in the lining and shoes will be determined by the amount of lining contact area, the amount of rotor or drum surface area, ambient temperature, and the amount of airflow present. The lining contact area—usually measured in square inches or square centimeters—is important because it influences the rate of heat absorption. If we were to reduce the lining contact area by one-half, we would double the rate of temperature rise in the lining that remains. Temperatures over a few hundred degrees can have a drastic effect on the lining's coefficient of friction and wear rate. Of concern to brake technicians, the lining contact area on newly replaced shoes should be as complete as possible to prevent the possibility of overheating. Good contact between the disc brake lining and the rotor is fairly easy to obtain because the surfaces are flat. Good drum-to-shoe contact is more difficult because both surfaces are curved, and the curvature of the shoe must match that of the drum. Poor contact will occur if the curves do not match. Hard stops with a newly replaced, improperly fitted lining can cause excessive temperatures in those areas of the lining that are making contact. Excessive temperatures can prematurely and permanently damage the lining. Methods for checking and correcting the lining contact will be described in Chapter 10.

An engineering term commonly used to describe potential brake power is **swept area.** This is the area of the rotors or drums that is swept or rubbed by the shoes. To calculate the swept area for a rotor (illustrated in Figure 2-24), the following formula can be used:

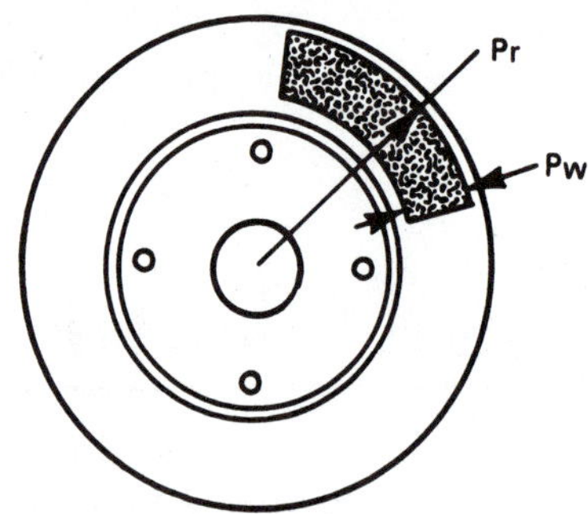

Figure 2-24 The amount of swept area of a disc brake can be determined by using these dimensions.

$$SA = Pr2 \times \pi - [(Pr - Pw)2 \times \pi]$$

where **SA** = swept area
Pr = pad radius to outer edge
Pw = pad width

To calculate the swept area of a drum (illustrated in Figure 2-25), use the following formula:

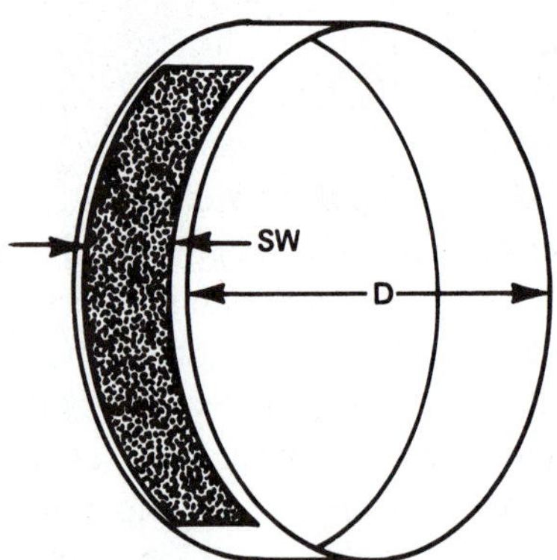

Figure 2-25 The amount of swept area of a drum brake assembly can be determined by using these dimensions.

$$SA = Dd \times \pi \times Sw$$

where **Dd** = drum diameter
Sw = shoe width

Most passenger cars use a swept area of about 200 square inches of lining per ton of vehicle weight, or about 10 pounds of weight per square inch (4.45 kg per 6.452 cm^2) of lining. Road racing cars that place a severe load on their brakes use about twice as much swept area—about 400 square inches per ton of vehicle weight. Recent changes in passenger cars to improve fuel economy have caused a reduction in the size of the brake components to help reduce the weight of the car. Reductions in the size of the rotors and drums have decreased the amount of swept area by as much as 30 percent; the effect of this has been an increase in the operating temperature of the shoes and rotors or drums.

Swept area of both drum and disc brakes is limited by the inside diameter of the wheel. The brake drum must fit inside the wheel, and there must be room between the rotor and the wheel for the caliper. One German car manufacturer developed a disc brake in which the caliper mounted through the rotor, thus allowing a larger diameter rotor, but the unit proved to be too expensive.

Clearly related to the swept area is the area of the lining surface. Some manufacturers provide this specification, which simply refers to the lining area that is in contact with the drums and rotors. At one time, most manufacturers produced cars with about 200 pounds (91 kg) of vehicle weight for each square inch (6.452 cm^2) of lining. Today some "down-sized" cars have over 350 pounds (159 kg) of weight per square inch of lining.

2.8.1 Measuring Brake Power

Braking power can be measured on a **brake dynamometer** (Figure 2-26). This is a piece of equipment that has two pairs of rollers, one pair for each of the tires on an axle. The rollers are driven by a pair of electric motors. The car is placed over the rollers so that the front or rear tires are on the rollers, and the motors are started to rotate the tires. The brakes are then applied, and the effective power of each brake is measured by the amount of resistance it offers the driving motor. The amount of power is indicated on a meter. The power and application rate of the right brake can be compared easily to that of the left brake, and problems such as weak return springs or sticky calipers or wheel cylinders will show up. The overall brake power is determined by applying the brakes completely. Some dynamometers are powerful enough to stress the brakes to the point of fade so that this aspect, as well as overall power, can be observed. After measuring the braking power on one axle, the car is moved to allow a similar measurement of braking power on the other axle.

Another style of dynamic brake tester is the **computerized plate tester**. This unit consists of four plates, one

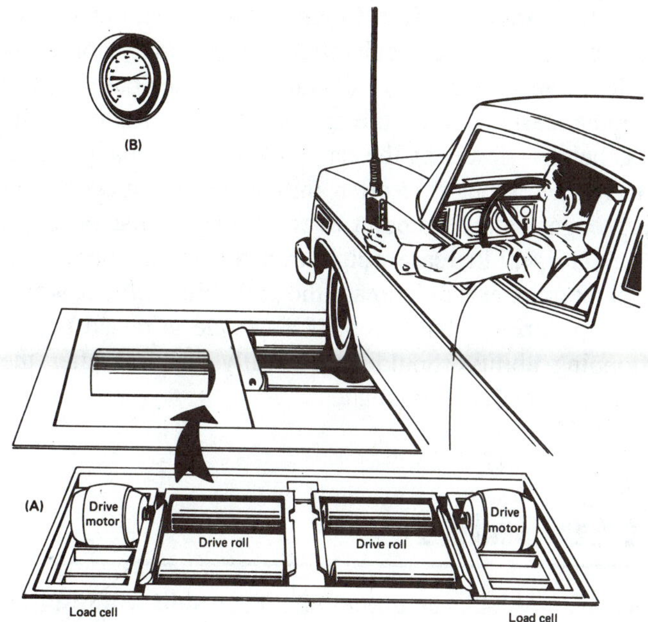

Figure 2-26 A brake dynamometer has two pairs of rollers that are driven by electric motors (A). The car is positioned with the front or rear tires on the rollers, and then the operator starts the drive motors, applies the brakes, and reads the amount of each wheel on the meter (B).

for each wheel of the car. The car is driven onto the plates at a low speed, 4 to 8 mph, and the brakes are applied while the wheels are on the plates. Braking force is measured, and the force generated by each wheel is displayed as a separate plot on a computer screen. Each plot shows the amount and the profile of the brake application, which allows the technician to diagnose many potential problems (Figure 2-27).

Brake dynamometers are too expensive to be used in most repair shops. Their use is generally limited to engineering and test labs and vehicle inspection or diagnostic lanes. The computerized plate tester is used in larger service shops.

2.9 Weight Transfer

Weight transfer is one of the most difficult forces to deal with in connection with brakes. It refers to the weight that moves from the rear tires to the front tires as the car decelerates (Figure 2-28). **Inertia** resists changes in motion. This physical law states that a body at rest tends to remain at rest, and a body in motion tends to remain in motion. Inertial forces act on the whole car, but to make things

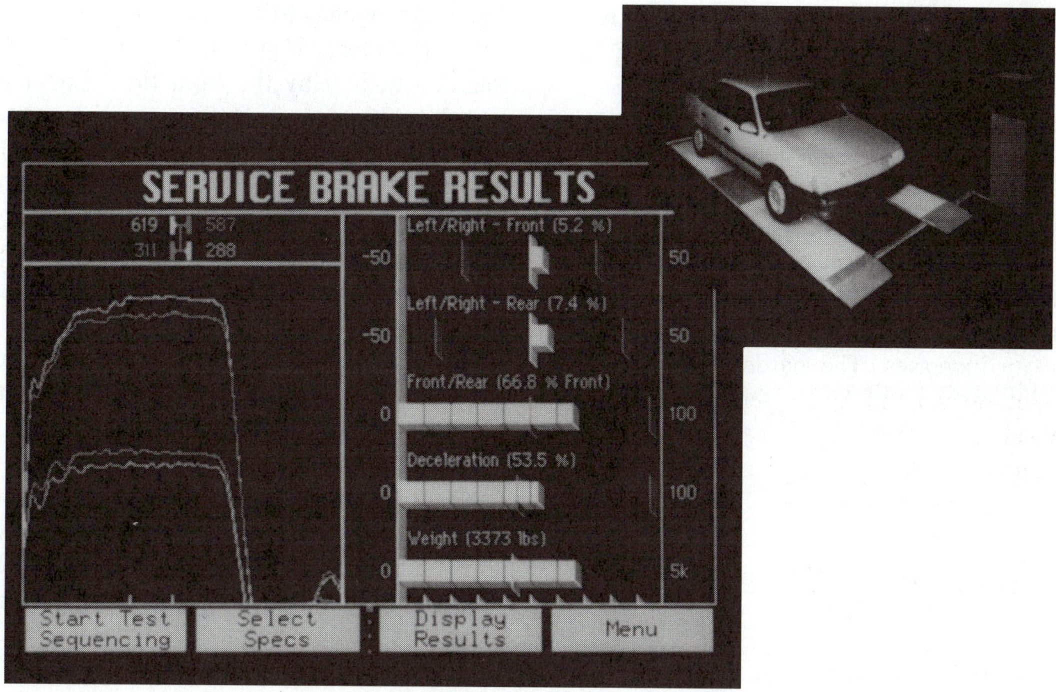

Figure 2-27 As the car is driven over the plates of the computerized plate brake tester (right), the brakes are applied. The computer screen then displays the amount of braking power of each wheel. Other tests display the deceleration rate, vehicle weight, parking-brake effectiveness, and any side slip at the front or rear tires. (Courtesy of Hunter Engineering Company)

easier to understand and calculate, we consider that inertia acts on the **center of gravity (CG)** of the car.

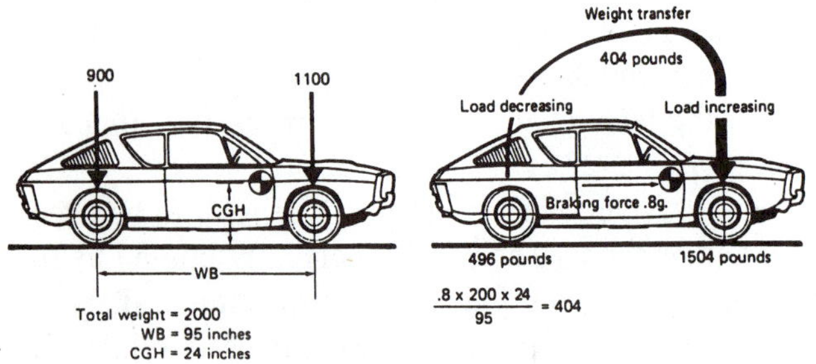

Figure 2-28 As a car stops, inertia will cause weight to transfer from the rear tires to the front tires.

A car's CG is its balance point—the point on which the car would balance if it were picked up. In a front-engined, rear-wheel-drive (RWD) car the CG is about 55 to 60 percent of the wheelbase distance in front of the rear axle, at the left-right (side-to-side) center of the car, and about as high as the upper one-third of the engine. In a modern FWD car, the CG is closer to the front axle than it is in a RWD car.

As the car is stopping, inertia creates a force equal to the deceleration force on the CG. This force acts against the traction force of the tires. The effect of the interaction of these two forces is to increase the load on the front tires, while the load on the rear tires is reduced by an equal amount. This is weight transfer.

Weight transfer produces two noticeable results, a lowering of the front end of the car along with a raising of the rear end (called **dive**) and a change in the relative amounts of traction. The transfer of weight will increase the available traction of the front tires, but it will also reduce the traction of the rear tires. As the loading and traction of the rear tires are reduced, the possibility of a rear-wheel lockup increases. The harder the stop, the greater the weight transfer and the greater the probability of a rear-wheel skid (Figure 2-29).

The formula used to calculate weight transfer is as follows:

$$\mathbf{WT} = \frac{\mathbf{DR} \times \mathbf{W} \times \mathbf{CGh}}{\mathbf{WB}}$$

where **WT** = lengthwise weight transfer (in pounds)
DR = deceleration rate (in **g** value)
W = car weight (in pounds)
CGh = height of the CG (in inches)
WB = wheelbase (in inches)

There are three ways to reduce the amount of weight transfer: reduce the deceleration rate (certainly not desirable in an emergency), lengthen the wheelbase (rather impractical), or reduce the height of the CG. Reducing the CG height—lowering the car—is a common change made on race cars but it is often impractical on street-driven passenger cars. The point to remember is that raising a car's CG will increase weight transfer, and an increase in weight transfer will increase the probability of rear-wheel lockup during a hard stop. If a vehicle is raised up, its stopping ability should be carefully checked after the modification has been made.

2.10 Skids

When a tire locks up during braking, a **skid** will result. A skidding tire will produce a loss of traction and stopping power and also a loss of directional control. A two-wheel, rear-end lockup is a fairly common occurrence. Front-to-rear balance on a brake system is a compromise between too little rear braking power on wet-pavement, low-rate stops, and too much power on dry pavement with harder stops.

If rear-wheel brake lockup does occur, the rear end of the car will slide out, either to the right or to the left, as road forces or road slope dictates (Figure 2-30). There is a natural tendency for the car to swing around and change ends when inertia is pushing forward on the CG and being retarded mainly by the front tires. Locking of the rear wheels reduces their ability to resist sideways motion. This effect is utilized in a "bootlegger's turn," a rapid U-turn made at a high speed which uses a quick application of the parking brake to cause a controlled rear-wheel lockup and skid as the steering wheel is flicked to make the turn. This maneuver requires quite a bit of skill and can be very dangerous.

A car is more stable if the front tires lock up, but the ability to steer will be lost. A skidding front tire has no turning power, so the car will travel in a direction dictated by inertia or the slope of the road.

If all four wheels lock up, the car will soon be in an extremely unstable skid. During the stop, it will probably spin sideways—completely at the mercy of inertia, the road surface, and any objects which might be encountered.

If a trailer is being towed and the brakes of the trailer lock up, the trailer will skid sideways—much like the rear end of a car with the rear brakes locked. If it is not stopped, the trailer will swing around far enough to run into the tow vehicle; this is called **jackknifing**.

Figure 2-29 As the stopping rate increases, the amount of weight transfer also increases. The loading for typical RWD and FWD cars is shown here.

When brake systems are designed, the engineers consider the following factors in balancing front and rear braking power: the type of rear drum brake used, the relative diameter of the front and rear brake rotors or drums, the width of the rear shoes and drums, the size of the wheel cylinder and caliper pistons, and the hydraulic pro-

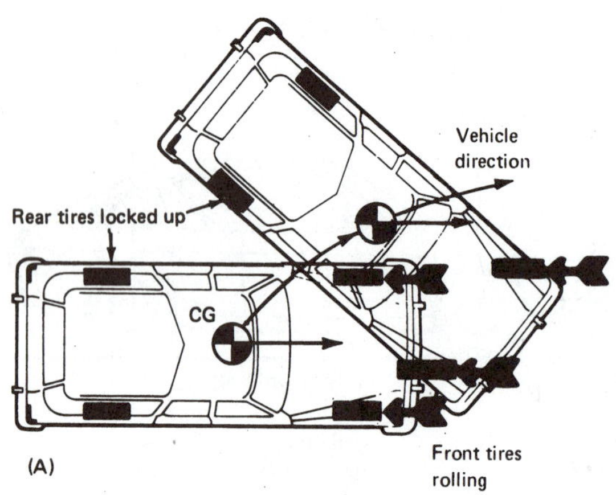

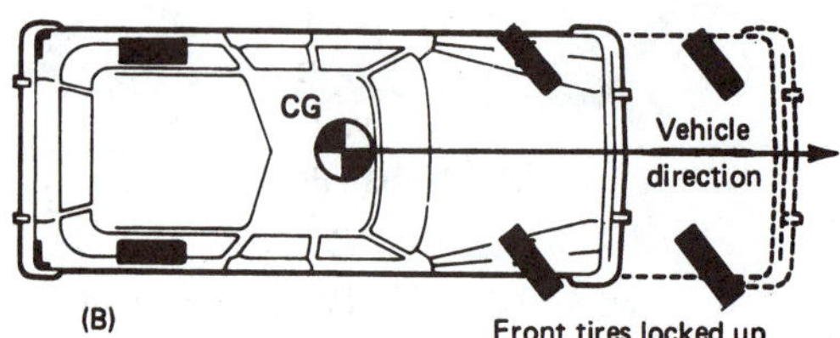

Figure 2-30 If the rear tires lock up (A) during a hard stop, the car will probably skid sideways because it is in an unstable condition—inertia is pushing forward on the CG and is being resisted primarily by the front tires. Front-tire lockup (B) will cause the front tires to lose turning power, and inertia will cause the car to travel in the original direction with no steering ability.

portioning valve. Normally, the correct balance produces a slight tendency to lock the rear tires on dry pavement while carrying a normal load.

Antilock braking systems are close to the ideal solution for wheel lockup. An ABS allows an increase in rear-wheel braking power to provide for fully loaded stops, and because of the electronic controls that automatically reduce or pulse the pressure in the wheel cylinders or calipers, wheel lockup and skids are prevented.

2.11 Stopping Sequence

Braking a car to a stop normally follows a sequence of events, which occur in a slow, relaxed manner during normal driving, and in the same order but with shorter time and distance intervals during panic stops.

A stop begins when the driver recognizes a danger or a reason to stop, makes a decision, and then takes action to move his or her foot from the throttle to the brake pedal. The time elapsed is referred to as **reaction time**. The average reaction time is slightly more than 1 second. The braking operation starts as the pedal is depressed. Pressure is built up in the hydraulic system, and the lining of the brake shoes is moved into contact with the rotor or drum surface (Figure 2-31). As braking force is generated at the brake assemblies, the car will begin undergoing weight transfer, and brake dive will occur as the car sets up for the actual stopping process. Then actual braking occurs and continues until the car comes to a stop or slows down sufficiently. The driver will often vary the pedal pressure and stopping rate to suit road and traffic conditions or to correct for a skid during a rapid stop.

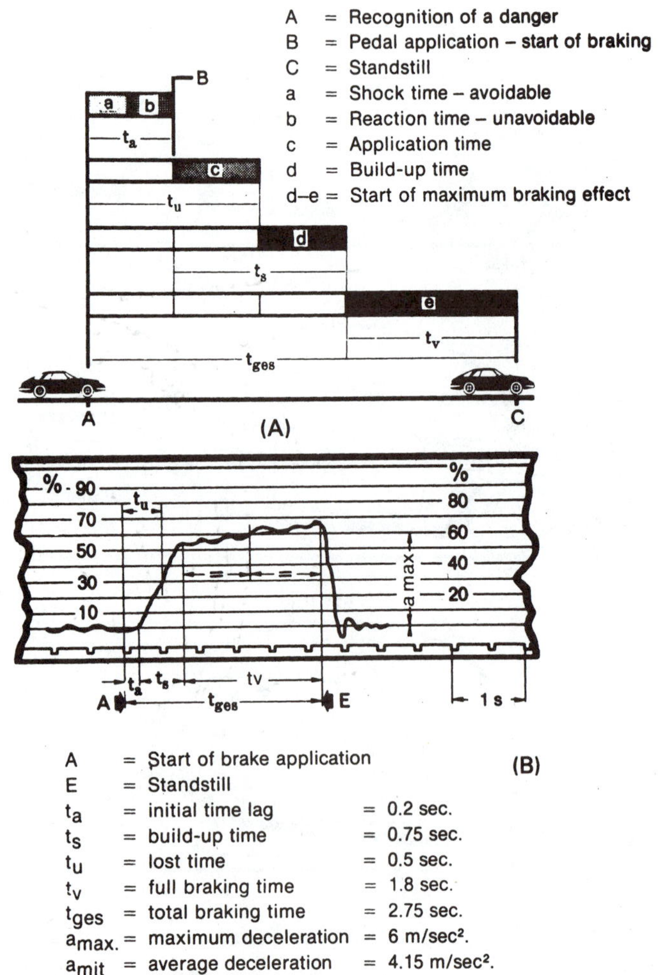

Figure 2-31 The normal sequence of events during a stop (A). The chart (B) shows the recorded time for these events during a fairly hard (60% to 65%) stop. (Courtesy of ITT Automotive)

2.12 Anti-Dive Suspensions

Many cars have a front suspension which is designed to resist the brake dive which results from weight transfer. Dive is not only annoying to many drivers as the front end drops, but can also cause severe load changes if it occurs so rapidly that it bottoms the suspension. An **anti-dive suspension** places the control arms of the front suspension in a position so that a lever arm of the suspension tends to lift the front of the car during braking. Braking force is thus used to counteract dive (Figure 2-32).

Production cars are designed to have about 50-percent anti-dive. A certain amount of dive is desirable as feedback to the driver, because the amount of dive tells the driver how effective the braking is. Too much anti-dive tends to bind suspension action during braking and also will cause excessive caster change during normal front-end action.

2.13 Practice Diagnosis

You are working in a tire shop that also specializes in brake repair and encounter the following problems.

CASE 1: The customer has brought in his 3/4-ton pickup that has an extended camper on it. He complains that when he goes downhill, the rig doesn't stop very well.

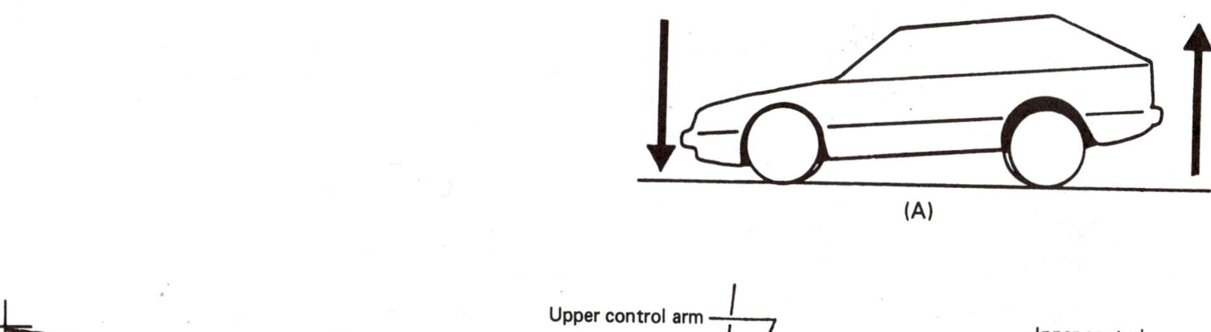

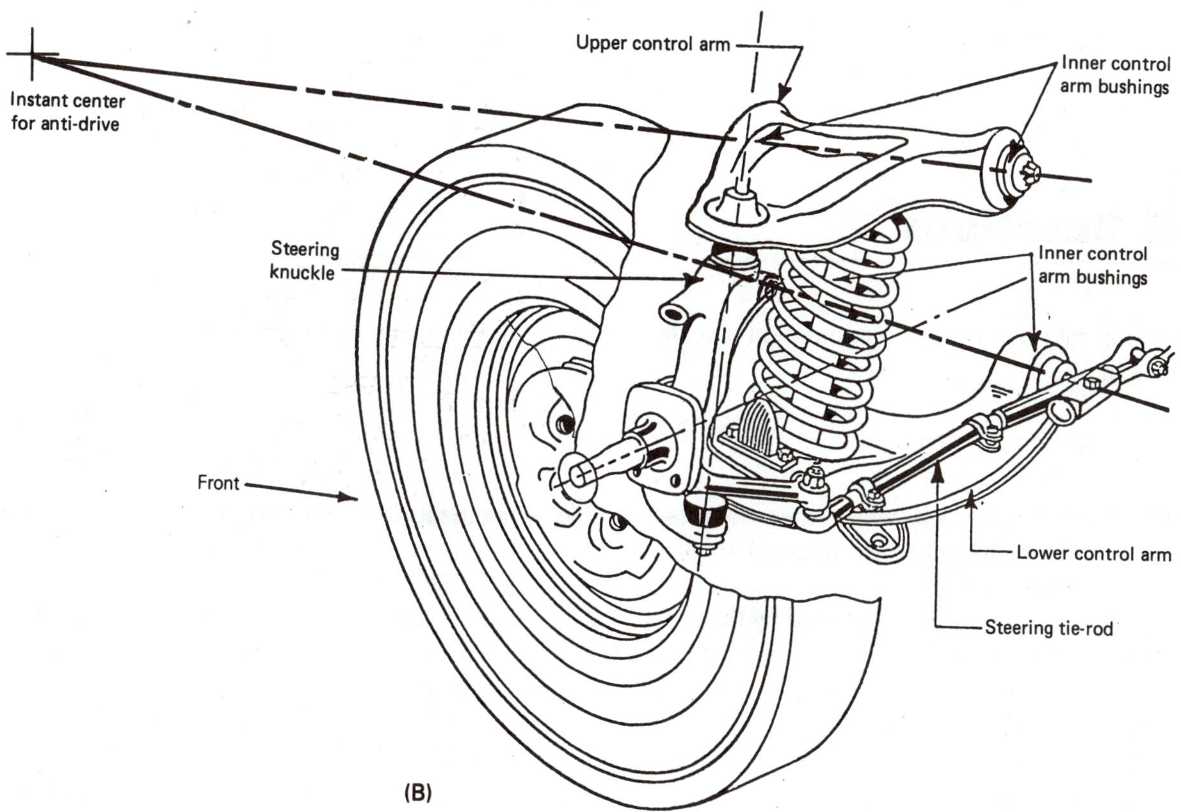

Figure 2-32 Brake dive causes the front of the car to lower (A) during braking. Anti-dive is achieved by placing the mounting points of the front control arms at an angle, creating a lever arm that creates a lifting force at the instant center (B).

The brakes get very hot (they smell bad) if he has to make several stops. Is this a typical problem? What should you do to help correct the problem? What should you tell this customer?

CASE 2: The customer has brought in her five-year-old compact pickup, V-6 with five-speed transmission, with a complaint that the rear tires lock up and skid during hard stops, especially on rainy days. On checking it, you notice the inside of the bed is in like-new condition. What should you do to help correct the problem? What should you tell the customer?

Terms to Know

ambient
anti-dive suspension
asbestos
brake dynamometer
brake torque
braking efficiency
British thermal unit (BTU)
calorie (c)
Celsius (C)
center of gravity
coefficient of friction
competitively priced
computerized plate tester
decelerometer
dive
economy
edge brand
edge code
energy
Fahrenheit (F)
friction
friction material
heavy-duty
inertia
inorganic
intensity
jackknifing
kinetic energy
nonasbestos
nonasbestos organic (NAO)
Occupational Safety and Health Administration (OSHA)
organic
original equipment (OE)
premium
quantity
reaction time
semimetallic
Society of Automotive Engineers (SAE)
skid
swept area
traction
work

Review Questions

1. Deceleration rate is normally measured on a scale calibrated in
   a. g values.
   b. feet per second per second.
   c. braking efficiency.
   d. any of the above.
2. Statement A: Stopping rates are affected by the strength of the brake assemblies.
   Statement B: Stopping rates are affected by the tire-to-road coefficient of friction. Which statement is correct?
   a. A only c. both A and B
   b. B only d. neither A nor B
3. Deceleration rates can be measured using
   a. a decelerometer.
   b. a stopwatch.
   c. specially sized rectangular blocks.
   d. any of the above.
4. The energy of a moving car is referred to as
   a. inertia.
   b. kinetic energy.
   c. horsepower.
   d. all of the above.
5. Statement A: As the weight of a vehicle increases, the amount of energy of motion will increase at the same rate.
   Statement B: As the speed of a vehicle increases, the amount of energy of motion will increase at the same rate. Which statement is correct?
   a. A only c. both A and B
   b. B only d. neither A nor B
6. Friction can be used to convert energy of motion into
   a. heat energy. c. calories.
   b. BTUs. d. any of the above.

7. Statement A: Heat always travels from a warmer object to a cooler object.
   Statement B: Heat intensity is measured using a Fahrenheit or Celsius scale. Which statement is correct?
   a. A only c. both A and B
   b. B only d. neither A nor B
8. Statement A: A stop from 55 mph will produce higher brake temperatures than a stop from 40 mph.
   Statement B: During a stop, the temperature of the brake lining is affected by the weight of the rotor. Which statement is correct?
   a. A only c. both A and B
   b. B only d. neither A nor B
9. If it takes 30 pounds of force to slide a 100-pound block of material, the coefficient of friction is
   a. 0.30. c. 0.7.
   b. 3.0. d. none of the above.
10. Statement A: If the lining coefficient of friction increases during a stop, this is called fade.
    Statement B: Fading lining will require that pedal pressure be decreased to prevent wheel lockup. Which statement is correct?
    a. A only c. both A and B
    b. B only d. neither A nor B
11. Statement A: A tire will have the greatest traction when it is slipping slightly relative to the road.
    Statement B: A skidding front wheel does not have the ability to steer the car. Which statement is correct?
    a. A only c. both A and B
    b. B only d. neither A nor B
12. Swept area refers to the size of the
    a. contact area between the tire and the road.
    b. rotor and drum area rubbed by the brake lining.
    c. bores in the master cylinder, calipers, and wheel cylinders.
    d. all of the above.
13. During a stop, weight transfer will
    a. reduce the weight on the rear tires.
    b. reduce the weight on the front tires.
    c. increase the traction of the rear tires.
    d. all of the above.
14. Statement A: Rear-wheel lockup can cause the rear end of the car to slide sideways.
    Statement B: Front-wheel lockup will cause the front end of the car to slide sideways. Which statement is correct?
    a. A only c. both A and B
    b. B only d. neither A nor B
15. Anti-dive is caused by the
    a. location of the center of gravity.
    b. location of the rear suspension arms.
    c. position of the front suspension arms.
    d. all of the above.

3 Drum Brake Theory

Objectives

After completing this chapter, you should:

- ❑ Be familiar with the different styles of drum brake units.
- ❑ Be familiar with the terms commonly used with drum brakes.
- ❑ Understand how brake shoe-to-drum pressure is increased or decreased because of self-energizing or deenergizing actions.
- ❑ Understand the operation of non-servo, duo-servo, and uni-servo brakes and how they differ.
- ❑ Understand the purpose of each of the components that is used in a drum brake assembly.

3.1 Introduction

Drum brakes used to be the type of brake used on most vehicles. The very earliest cars used **brake drums** attached to the rear wheels with a lined band wrapped around the drum. When the driver wanted to stop, the band was tightened onto the drum. This type of brake was called an *external contracting brake* because the lining was on the outside and contracted—was made smaller—to apply braking pressure (Figure 3-1). This brake design was adversely affected by road dust, dirt, and water, which could easily find their way between the lining and drum and severely reduce braking power or cause erratic brake operation along with severe wear.

A major advance in brake designs was the change from an external band to internal shoes. Being on the inside requires that the shoes expand to apply pressure, so this design is called an *internal expanding brake* (Figure 3-2). The shoes are mounted on the backing plate, also called a platform, and the edge of the backing plate is often fitted so it intermeshes with a groove in the brake drum. This fit is fairly effective at keeping water, dust, and dirt out of the brake assembly (Figure 3-3).

The first drum braking systems were applied through a mechanical linkage. Metal rods, or cables, and levers transmitted pressure from the brake pedal or lever to the shoes. Very early cars had brakes only on the rear wheels until a rather complicated linkage was developed to transmit motion to the steerable front wheels. All modern cars use hydraulics to easily transmit force equally to all four wheels. Hydraulic application will be described in Chapter 6.

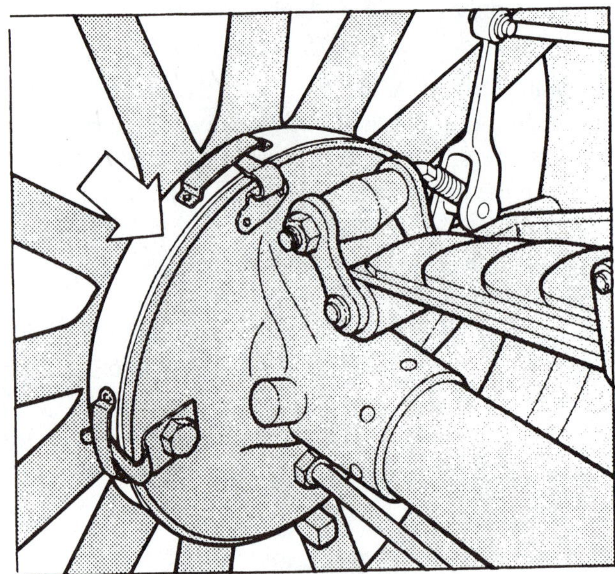

Figure 3-1 An early brake design using an external contracting brake. The brake band (arrow) wrapped around the drum and was pulled tighter to apply the brake. (Courtesy of Ford Motor Company)

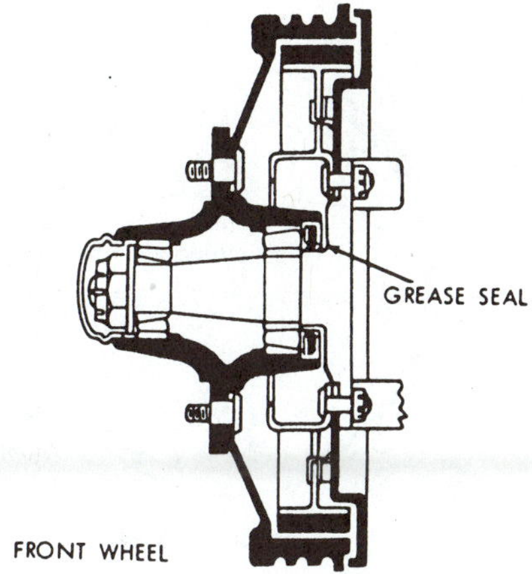

Figure 3-2 In most drum brake units, the backing plate and drum are arranged to form a labyrinth seal to help prevent the entrance of dirt and water. (Courtesy of John Bean Company)

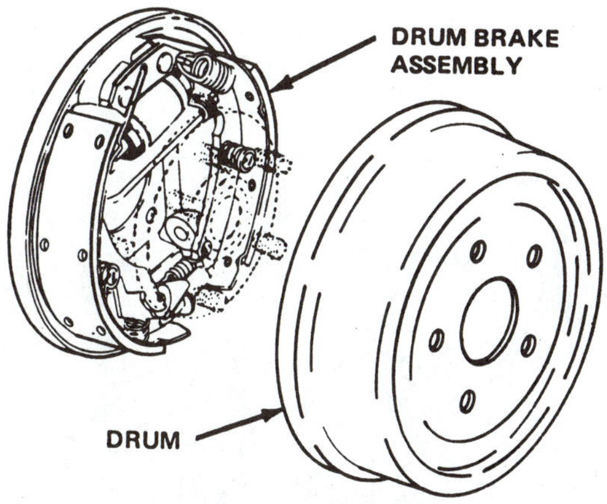

Figure 3-3 A drum brake assembly. The drum fits over the shoes and axle flange. (Courtesy of General Motors Corporation, Service Technology Group)

3.2 Shoe Energization and Servo Action

As mentioned earlier, brake shoes can be mounted onto the backing plate in different ways. In duo-servo designs, rotation of the drum can be used to help apply the pressure of the lining against the drum. During application, the frictional drag will tend to rotate the shoe around its anchor. Depending on the position of the anchor—at the leading or trailing end of the shoe relative to the direction of drum rotation—this rotational force will either increase or decrease the lining-to-drum pressure.

If the shoe is applied at the leading end and anchored at the trailing end—relative to drum rotation—the shoe will be pulled tighter into the drum. This occurs because the shoe attempts to rotate with the drum. It is called an **energized**, **forward**, or **leading shoe**. If the shoe is anchored at the leading end and applied at the trailing end, drum rotation will push the shoe away, reducing the application pressure. This is called a **deenergized**, **reverse**, or **trailing shoe** (Figure 3-4). A brake that uses a leading and a trailing shoe is often called a leading-trailing shoe or a nonservo brake (Figure 3-5).

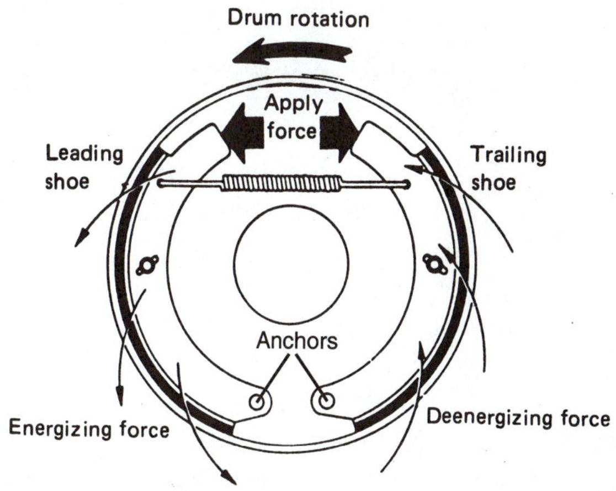

Figure 3-4 Because this drum is rotating in a counterclockwise direction, the leading shoe (left) is energized and pulled tighter against the drum. The trailing shoe (right) will be pushed away or deenergized.

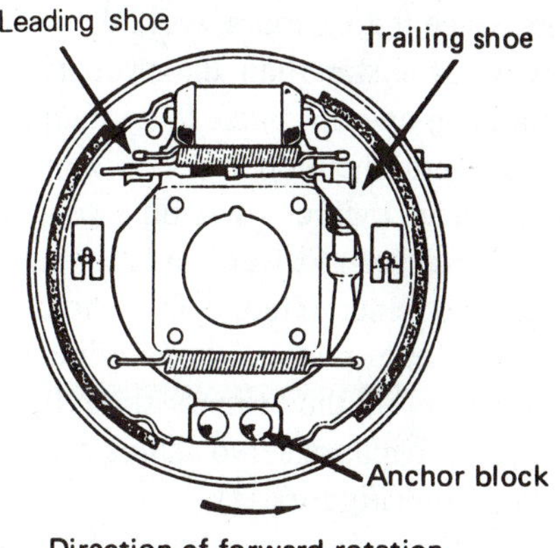

Figure 3-5 A leading-trailing shoe, nonservo brake. Note the anchor block is positioned between the two shoes. (Courtesy of LucasVarity Automotive)

Servo brakes use one anchor for both shoes; the shoes are arranged so that one can apply pressure on the other. With this arrangement, the leading shoe will apply pressure on the trailing shoe; this is called servo action. In servo brakes, the leading shoe is called the primary shoe and the trailing shoe is called the secondary shoe. The primary shoe is normally positioned toward the front of the car, and the secondary toward the rear. When the brakes are applied, the rotation of the drum energizes the primary shoe and it attempts to rotate with the drum. The motion of the primary shoe in turn applies the secondary shoe (Figure 3-6). About two-thirds of the stopping force from

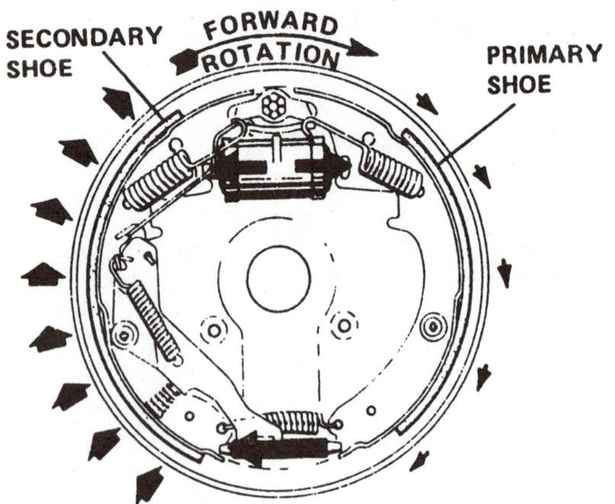

Figure 3-6 A duo-servo brake is designed so the energized primary shoe increases the application pressure on the secondary shoe. (Courtesy of General Motors Corporation, Service Technology Group)

this brake comes from the secondary shoe, and two-thirds of the force applying the secondary shoe comes from servo action. Since it does more work, the secondary shoe will usually wear faster than the primary shoe. Servo action produces a powerful brake that requires a relatively small amount of application force. When this brake is used with a wheel cylinder having two pistons, servo action can occur in both forward and rearward directions, and it is called a duo-servo brake. Occasionally this brake is used with a single-piston wheel cylinder which can apply pressure only on the primary shoe. It is then called a **uni-servo** brake because servo action will occur only in a forward direction (Figure 3-7).

The brake shoes on a nonservo brake are mounted independently, each with its own anchor and wheel cylinder piston; there is no interaction between them and, as mentioned earlier, usually one shoe is a leading shoe and the other a trailing shoe. This type of brake is often referred to as a **leading-trailing shoe** or *simplex* brake. Since it exerts a greater pressure against the drum, the leading shoe does more of the braking and will usually wear out faster than the trailing shoe. The shoes can also be arranged so that they are both leading shoes—called a

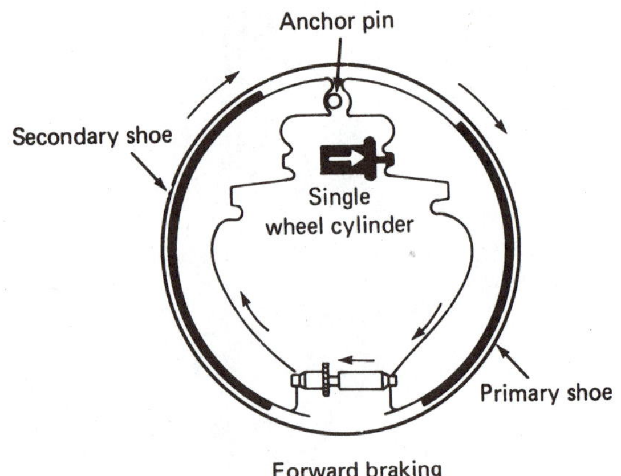

Figure 3-7 If there is only one wheel cylinder piston (uni-servo brake), servo action can only occur during forward-direction stopping.

two-leading shoe or *duplex* brake; or both trailing shoes—called a *two-trailing shoe* brake. Two-leading shoe and two-trailing shoe designs normally use two separate, single-piston wheel cylinders (Figure 3-8).

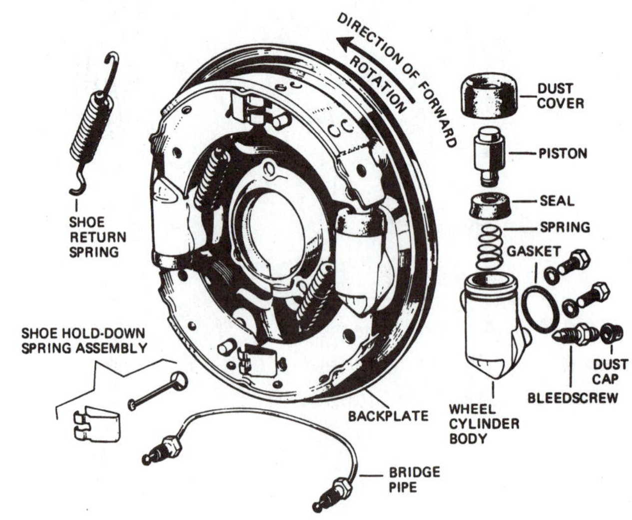

Figure 3-8 A two-leading shoe, nonservo brake. Note the two single-piston wheel cylinders and the relative positions of the shoes and anchors at the wheel cylinder bodies. (Courtesy of LucasVarity Automotive)

When duo-servo drum brakes are used on the rear wheels and disc brakes on the front wheels, there is a definite tendency for rear-wheel lockup during heavy braking. At the moment of braking, weight transfer is changing the traction balance, increasing the traction of the front tires and reducing the rear traction, while servo action is increasing the relative rear-brake strength. When the combination of rear drum brakes and front disc brakes is used with a tandem front-rear split of the hydraulic system, a fairly good front-to-rear balance of the two styles of brakes can be achieved by using a proportioning valve in the rear system. The proportioning valve is used to reduce the power of the rear brakes during a hard stop. On cars with a diagonal split hydraulic system, two proportioning valves are required, one between each of the master cylinder sections and a rear brake. Many FWD cars with diagonal split braking systems use nonservo rear brakes. The stopping force characteristics of a nonservo brake are quite similar to those of a disc brake, which reduces the tendency of rear-wheel lockup and reduces or eliminates the need for proportioning valves (Figure 3-9).

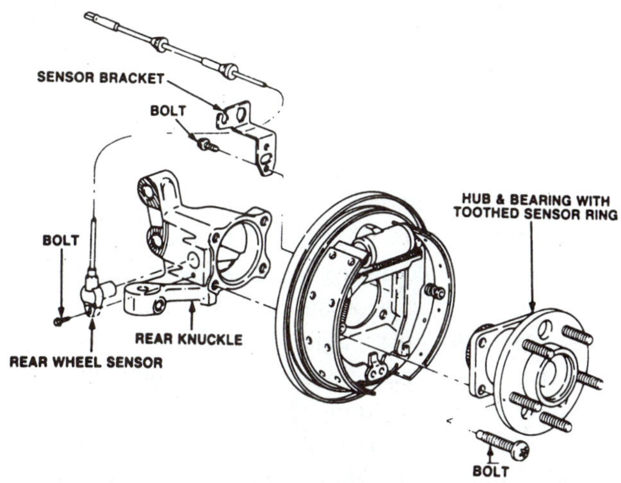

Figure 3-9 This late-model, full-sized passenger car uses leading-trailing brakes at the rear to help reduce the possibility of rear-wheel lockup. Note that it has a sensor and a toothed sensor ring for an ABS and also uses a unitized hub and wheel bearing assembly. (Courtesy of General Motors Corporation, Service Technology Group)

3.3 Brake Shoes

Most passenger-car brake shoes are fabricated from two pieces of stamped steel. The *shoe rim* is curved to match the curvature of the drum and is slightly narrower than the width of the drum's inner surface; the rim provides a surface for lining attachment. The *shoe web* is welded to the rim, reinforcing it and providing a place for the anchor, application force, hold-down and return springs, parking brake attachment, and adjusting mechanisms. The shoe rim usually has a series of nibs—bent areas—on the edges where the shoe rests against the backing plate. These nibs improve the bearing contact where the shoe slides on the backing plate during application and release (Figure 3-10).

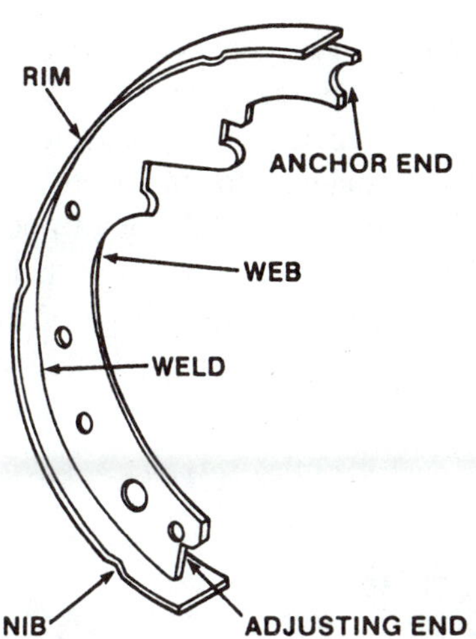

Figure 3-10 A steel brake shoe showing the various parts. (Courtesy of Bendix Brakes, by AlliedSignal)

Brake shoes are also made from aluminum castings. These shoes tend to be lighter in weight and conduct heat away from the lining better than steel shoes. They also tend to be weaker, especially when hot.

The application end of the shoe is often referred to as the **toe** while the end at the anchor is called the **heel**. The heel and toe of a nonservo brake shoe are easy to determine. The heel is at the anchor, and the toe is at the wheel cylinder. With duo-servo brakes, the heels and toes switch depending on the direction of drum rotation and the action of the shoe.

Brake shoes come in many different shapes and sizes (curvatures and widths) with webs of different shapes and different placement of the holes. The various types of shoes are identified by a shoe number assigned by the *Friction Materials Standards Institute (FMSI)* (Figure 3-11).

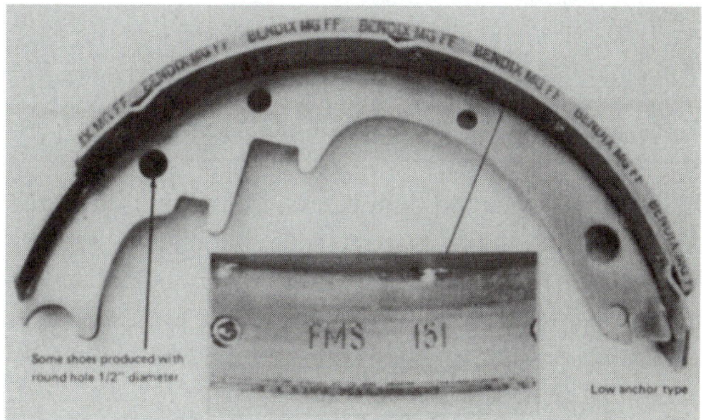

Figure 3-11 This shoe has an FMSI identification number of 151. It is a 10 x 1¾ inch shoe with the dimensions and shape as shown. Note that it has an edge code indicating Bendix lining MG with an FF friction code.

Normally, a shoe set can be ordered by using the make, model, and year for a particular car with a very good chance of getting an exact replacement. Occasionally it is necessary to specify the diameter or the width of the drum or both. Some manufacturers use drums of one diameter on their standard model and larger brakes on their sportier version, larger-size version, or station wagon. When replacing the lining, it is always a good practice to compare the new and old shoes to ensure a correct replacement (Figure 3-12).

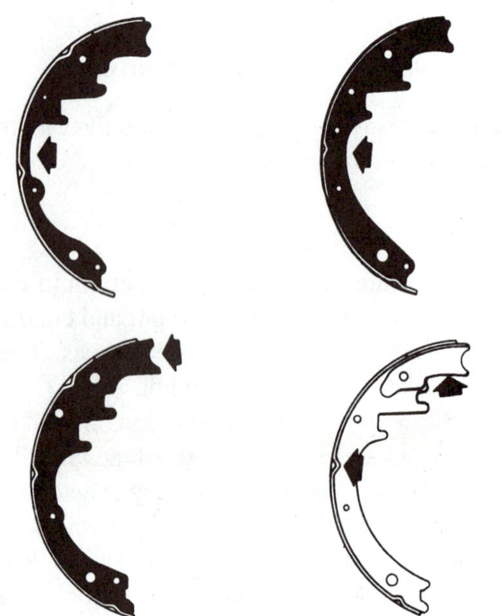

Figure 3-12 Brake shoes can have webs of different shapes and hole placement, and reinforcement or rims with differing nib positioning. Always compare replacement shoes with the old shoes to ensure proper replacement. (Courtesy of Wagner Brake)

3.4 Lining Attachment

On passenger cars and light trucks, the lining is attached to the shoe by one of two methods, bonding or riveting (Figure 3-13). Occasionally, the lining is bolted to the shoe on heavy trucks. Large truck shoes are expensive, and bolted linings, available in predrilled blocks, allow easy lining replacement in the field.

A **bonded lining** is secured to the shoe by a high-temperature adhesive. The lining is clamped onto the clean brake shoe rim with a layer of adhesive between the lining and the shoe. Then the shoe is placed in a high-temperature oven to set the adhesive. Some disc brake linings are *mold-bonded* or *integral-molded.* The pad backing is made with a series of holes into which the lining material is forced during the molding process. As the lining is molded, it is also bonded to the pad.

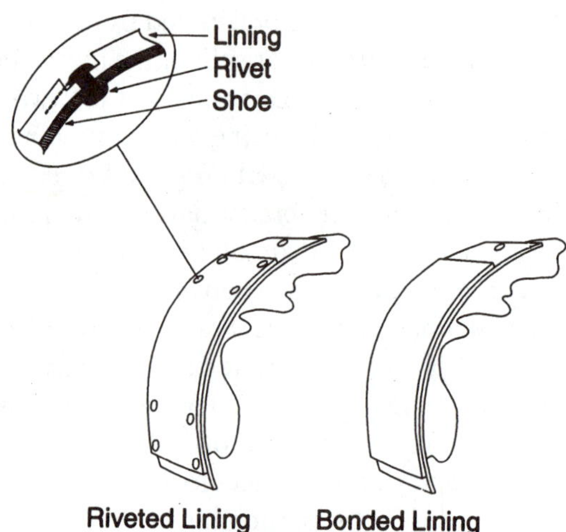

Figure 3-13 Passenger-car and light-truck brake lining is normally riveted or bonded onto the shoe rim. (Courtesy of John Bean Company)

A **riveted lining** is secured by a series of brass or aluminum rivets. They pass through holes which are drilled and countersunk in the lining. The rivets are upset or flattened inside the shoe rim to hold the lining tightly in place.

Bonded linings are often preferred because there is more usable lining on the shoe. When a riveted lining wears out, the rivets can contact the drum or rotor surface and cause severe scoring or grooving. When a bonded lining wears out, the shoe rim can also contact the rotor or drum and cause severe scoring, but bonding does allow the use of nearly all of the lining's thickness. In some linings—usually those of inferior grades—cracks can begin at the rivet holes, which can cause breakup and separation of the lining from the shoe. This is a rare occurrence and usually a result of high-heat conditions. Some technicians prefer riveted linings because they tend to be quieter. The bonding process laminates the lining to the shoe and creates a more rigid assembly which has a greater tendency to vibrate. In some circumstances, this more rigid shoe will produce vibrations which in turn can cause "squeal." A premium or heavy-duty lining is often riveted for cooler operation. Bonding adhesive tends to insulate the lining from the shoe, which results in poorer heat transmission than when rivets are used.

Normally, the lining is centered on the shoe rim, and it can be full length or shorter. Lining length is designed to obtain even wear characteristics in a pair of shoes, whether primary and secondary or leading and trailing. A secondary shoe will always have longer lining than a primary shoe. Sometimes a leading shoe will have longer lining than a trailing shoe. Occasionally, a lining shorter than full-length will be placed in a high or low position on the shoe to change the self-energization or servo characteristics of the shoe (Figure 3-14).

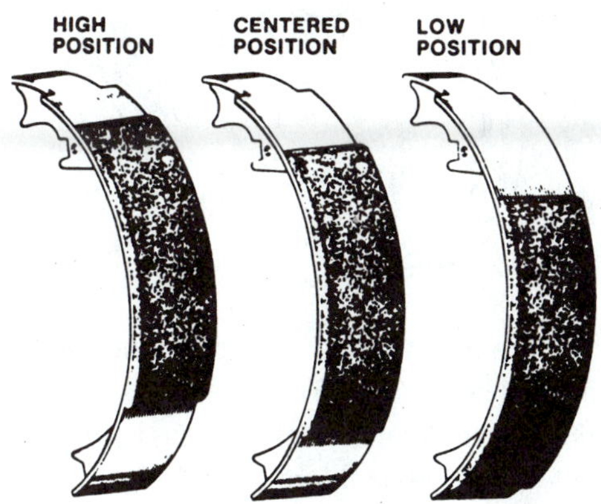

Figure 3-14 Brake lining can be attached to the shoe in different positions depending on the desired stopping characteristics. (Courtesy of Bendix Brakes, by AlliedSignal)

3.5 Backing Plate

The **backing plate** is the *foundation* on which the brake assembly is mounted. The backing plate is bolted securely to the rear axle or front steering knuckle. It locates the shoe anchor and is usually used to transmit the brake torque from the anchor to the axle or steering knuckle. An exception to this is the *direct torque* design used on late-model General Motors cars in which a flange on the axle contains mounting points for the wheel cylinder and anchor (Figure 3-15).

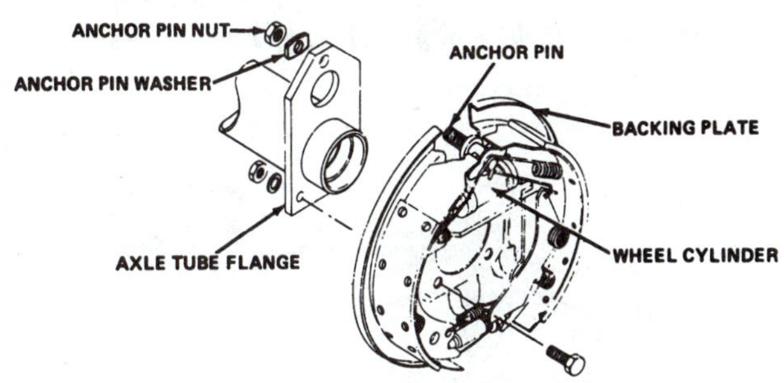

Figure 3-15 This direct-torque drum brake assembly attaches the anchor pin and wheel cylinder, along with the backing plate, to the axle tube flange. (Courtesy of General Motors Corporation, Service Technology Group)

Most backing plates have *ledges* or *platforms* on which the nibs of the brake shoes ride. These ledges support the shoes in a "square" position relative to the drum surface. In conjunction with the shoe anchor, they ensure proper positioning between the shoe and the drum. Also contained in the backing plate are the holes and bosses for attaching the wheel cylinder, shoe hold-down springs, and the parking brake cable (Figure 3-16).

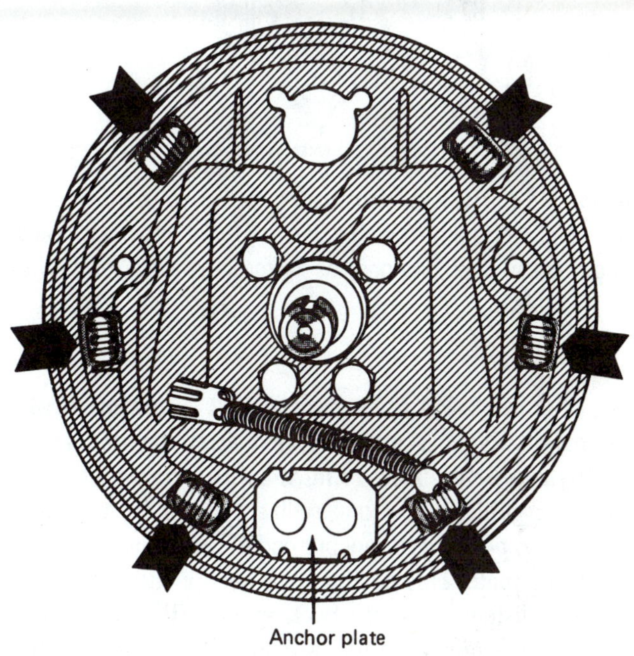

Figure 3-16 This shoe support/backing plate has six shoe platforms (arrows) and the anchor plate. (Courtesy of Chrysler Corporation)

3.6 Shoe Anchors

Traditionally, domestic cars (those from U.S. manufacturers) use a round shoe anchor pin, and the shoes have a semicircular opening which butts against the anchor. The primary purpose of the **anchor**, sometimes called a *support*, is to absorb braking torque from the shoes and transmit that force to the backing plate and the car's suspension (Figure 3-17).

Round anchor pins position the shoes vertically on the backing plate and prevent shoe rotation with the drum. The **anchor pin** is a steel pin welded or riveted solidly to the backing plate or threaded into the steering knuckle through the backing plate. These are called **fixed anchors**. At one time, some anchor pins were adjustable; they could be moved up or down to center the shoes in the

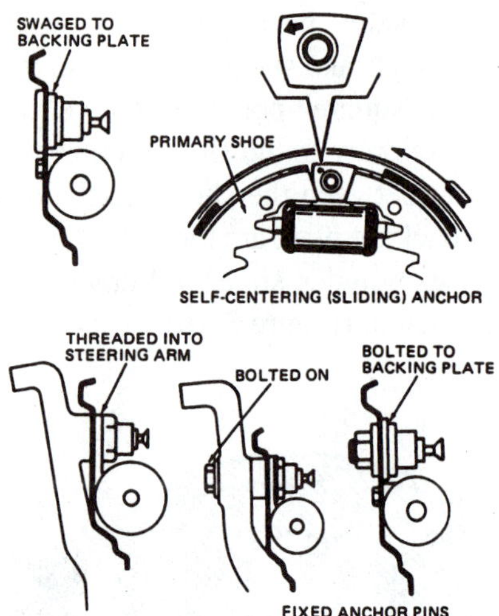

Figure 3-17 The anchor pin can be attached to the backing plate or steering knuckle by one of these methods. Most anchors are swaged, welded, or riveted in place. (Courtesy of Wagner Brake)

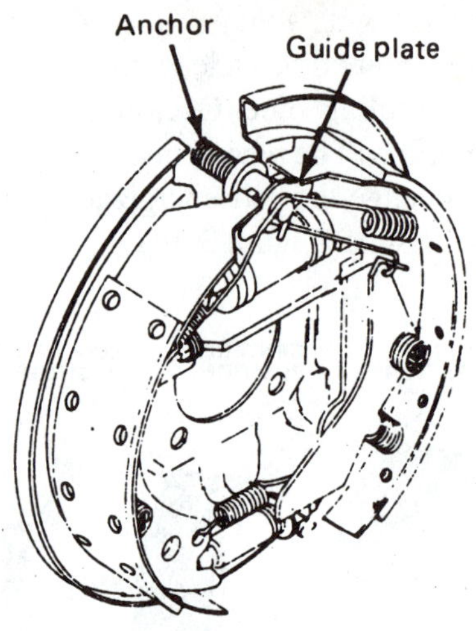

Figure 3-19 The end of the shoe is positioned vertically by the anchor pin and horizontally by the guide plate and backing plates. (Reprinted from Mitchell Anti-Lock Brake Systems, with permission of Mitchell Repair Information, LLC)

drum (Figure 3-18). This was called a **major brake adjustment**. No car manufactured after the mid-1950s uses **adjustable anchors** requiring this adjustment.

A shoe guide plate along with a step on the anchor is often used to position the anchor end of the shoe laterally on the backing plate. This positions the anchor end of the shoe the proper distance from the backing plate (Figure 3-19).

Many cars currently use a flattened, grooved anchor along with a shoe that is flat or slightly curved at the anchor or heel end. This design allows the shoe to slide up or down and center itself in the drum. The groove at the anchor, like the guide plate and steps on the anchor pin, serves to position the shoe laterally (Figure 3-20).

3.7 Brake Springs

Drum brake assemblies commonly use two sets of springs—one set returns the shoes to the released position

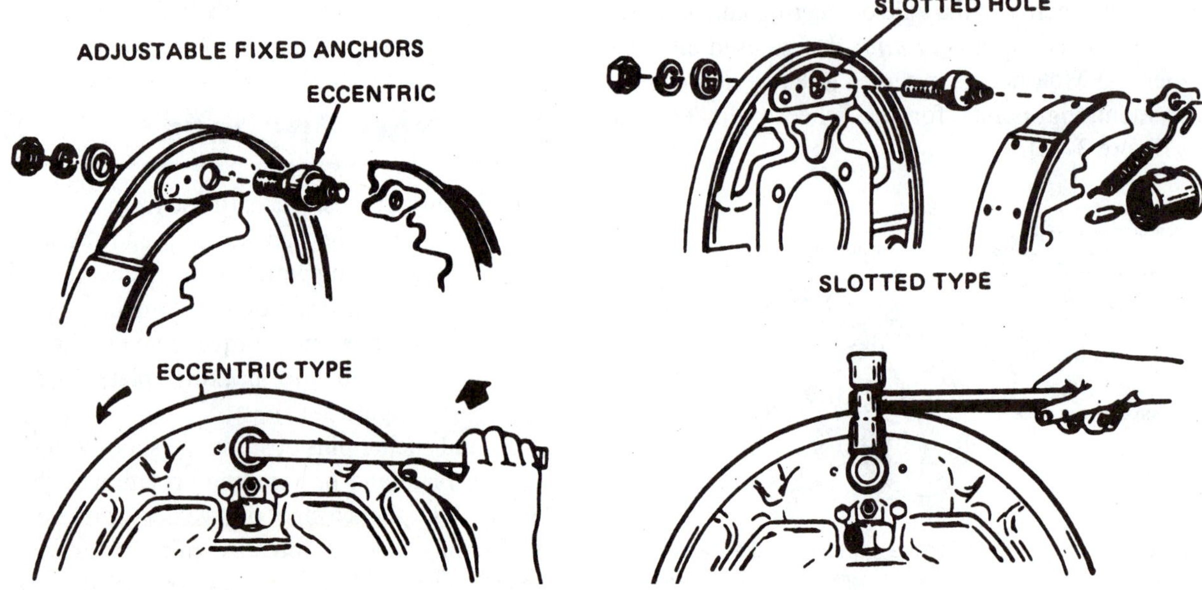

Figure 3-18 In the past, some brake designs used adjustable anchors so the shoe curvature could be aligned to the brake drum. Courtesy of Wagner Brake)

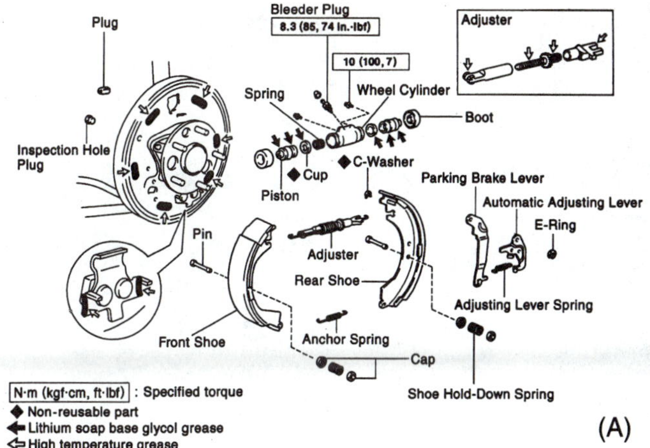

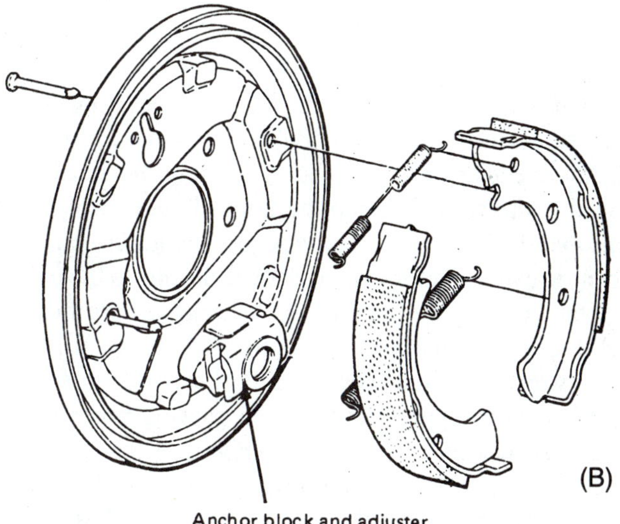

Figure 3-20 This anchor (A) allows the shoe to center itself with the drum by sliding up or down. Some sliding anchors include the adjuster mechanism (B). (B is reprinted from Mitchell Anti-Lock Brake Systems, with permission of Mitchell Repair Information, LLC)

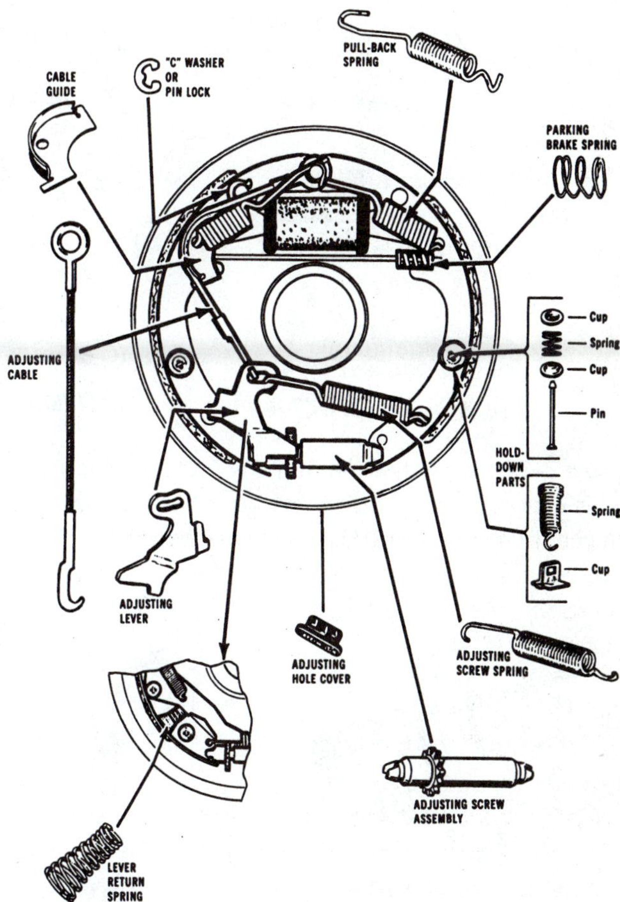

Figure 3-21 Some varieties of springs used in drum brakes. (Reprinted from Mitchell Anti-Lock Brake Systems, with permission of Mitchell Repair Information, LLC)

and the other set holds the shoes against the backing plate platforms. Additional springs are often used to operate the self-adjuster mechanism, to hold the ends of the shoes in a particular position, or to prevent looseness and rattles in the parking brake mechanism (Figure 3-21).

The shoe **return springs** have a very critical job, especially on servo brakes. During brake release, they pull the shoes back and push the wheel cylinder pistons inward as they return the shoes to the released position. A weak return spring can cause slow release or shoe drag during release. A weak return spring can also allow an earlier application on one wheel of an axle; this can cause a pull or even a grab or brake lockup. Remember that servo action multiplies the frictional drag between the primary shoe and the drum to apply the secondary shoe. A weak return spring will allow increased primary-shoe application pressure. Like other springs, return springs are designed for a particular installation. There are many shapes and sizes. Return springs are also called *pull-back springs* or *retracting springs* (Figure 3-22).

Several different shapes of shoe **hold-down springs** are used. These spring assemblies are used to provide a slight pressure between the shoe nibs and the backing plate ledges (Figure 3-23). They ensure that the shoes stay against the backing plate platforms so the lining is kept straight with the drum.

Checking and servicing of brake springs will be described in Chapter 10.

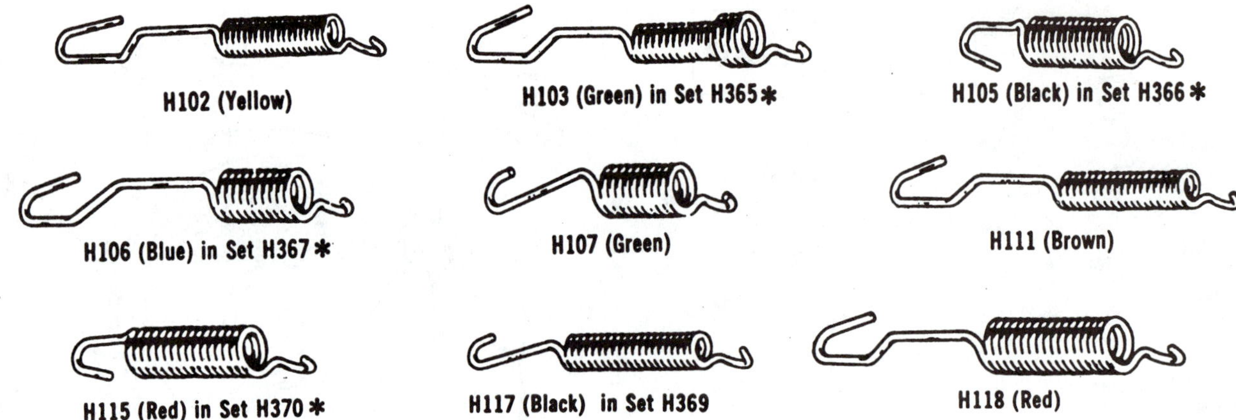

Figure 3-22 Brake shoe return springs come in many shapes and sizes. Be sure to compare any replacement so you are sure to use the correct spring. This supplier's part numbers and color coding are shown. (Reprinted from Mitchell Anti-Lock Brake Systems, with permission of Mitchell Repair Information, LLC)

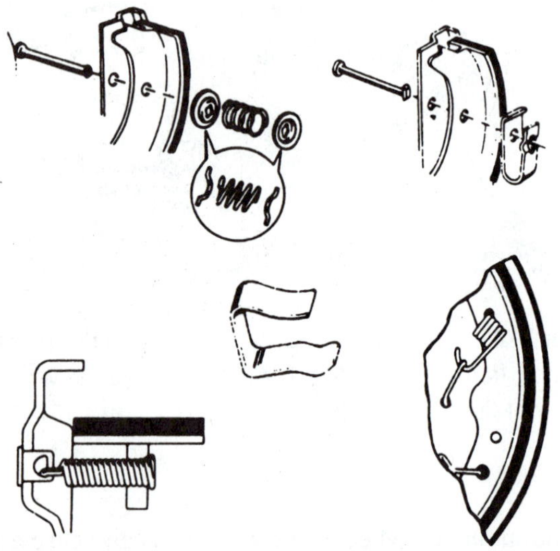

Figure 3-23 Five different styles of brake shoe hold-downs. The upper two are the most common. (Courtesy of Wagner Brake)

3.8 Shoe Adjusters

Brake shoes require periodic adjustment to keep the lining fairly close to the drum surface. The shoes need to be moved closer to the drum as the lining wears. Too much lining clearance requires more brake pedal movement as the shoes are applied. This will cause a "low brake pedal" or even a pedal which goes to the floor without applying the brakes.

At one time, brake shoe clearance was adjusted manually. The car was lifted off the ground, and a special tool, called a **brake spoon**, was used to adjust the shoe position. On servo brakes, the adjuster was a threaded adjuster assembly with a starwheel positioned between the lower ends of the two shoes. Turning the starwheel moved both shoes closer to the drum (Figure 3-24). For nonservo brakes, the most common adjuster was an eccentric cam positioned in back of each shoe in the backing plate; turning the cam moved the released position of each shoe closer to the drum (Figure 3-25). Other adjuster styles were a tapered wedge at the heels of the shoes, threaded shoe supports, a threaded parking brake strut, and threaded wheel cylinder caps (Figure 3-26).

Today, most cars use **self-adjusting** brakes. There are several different designs depending on the shoe design and manufacturer. Most duo-servo brakes use either *cable-style self-adjusters* or *lever-style self-adjusters* (Figure 3-27).

Figure 3-24 A typical starwheel-type adjuster commonly used to adjust the lining clearance on duo-servo brakes. (Courtesy of Wagner Brake)

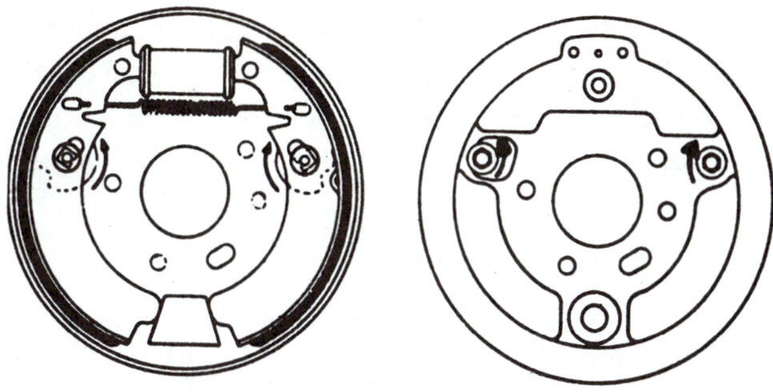

Figure 3-25 Nonservo brake designs without self-adjustment commonly use a pair of eccentric cams—which act as shoe stops—to adjust the lining clearance. (Courtesy of Wagner Brake)

General Motors cars traditionally use the lever style. The lever is attached to the secondary shoe by a bushing at the shoe hold-down spring. This lever is connected to the anchor pin by a wire link. During a stop in reverse, the secondary shoe forces the primary shoe against the anchor pin as a result of servo action. The secondary shoe and lever will rotate slightly with the drum, depending on the amount of clearance. More shoe clearance will cause more shoe rotation and movement during a stop. When this motion is far enough, it forces the adjuster lever downward far enough to turn the starwheel which in turn will reduce the shoe clearance (Figure 3-28).

Most other domestic cars with duo-servo brakes use a cable-type self-adjuster. The cable connects the adjuster lever, mounted on the secondary shoe, to the anchor and

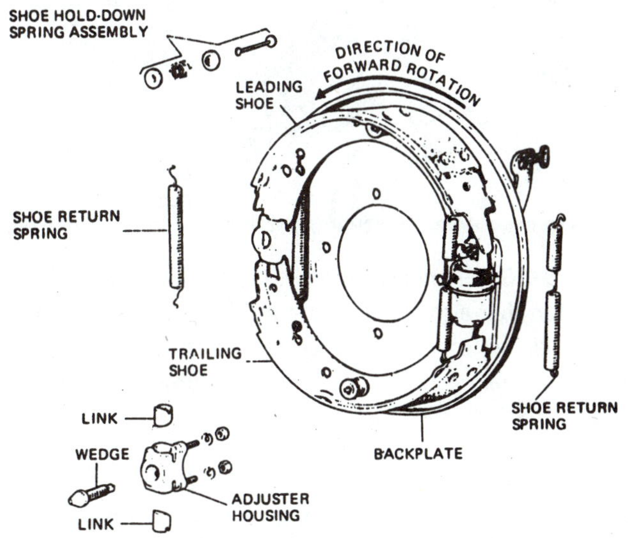

Figure 3-26 This nonservo brake uses a wedge-style adjuster. Threading the wedge inward causes the links to position the lining closer to the drum. (Courtesy of LucasVarity Automotive)

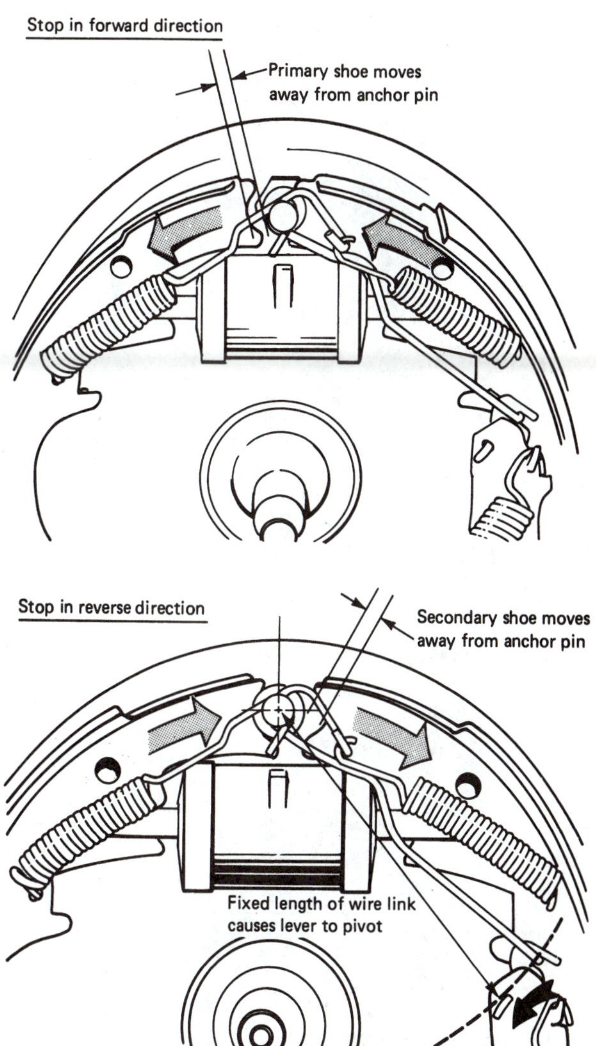

Figure 3-27 During a forward stop, servo action forces the secondary shoe against the anchor; during a stop in the reverse direction, servo action will move the secondary shoe away from the anchor pin. This movement is used to operate the self-adjuster mechanism. (Courtesy of General Motors Corporation, Service Technology Group)

passes over a guide which is also attached to the secondary shoe. During a stop in reverse, as the secondary shoe forces the primary shoe against the anchor pin, the rotation of the secondary shoe will cause the cable to lift the adjuster lever. As the brakes are released, the adjuster lever spring will pull the lever downward against the adjuster starwheel. If the travel is sufficient, the starwheel will be rotated, which in turn will reduce the shoe clearance (Figure 3-29). Some cable-type self-adjusters position the lever below the adjuster screw to provide a more positive adjustment. This type usually includes an override or overload spring at the cable-to-lever connection.

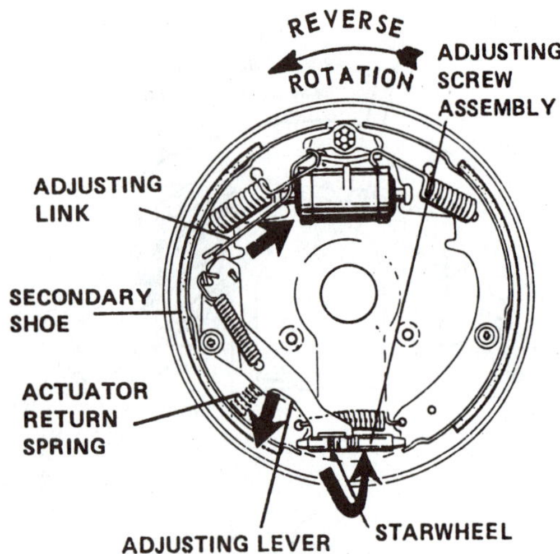

Figure 3-28 A lever-style self-adjuster. When the brakes are applied while backing up, the shoe movement after a certain point will cause the adjuster to operate. (Courtesy of General Motors Corporation, Service Technology Group)

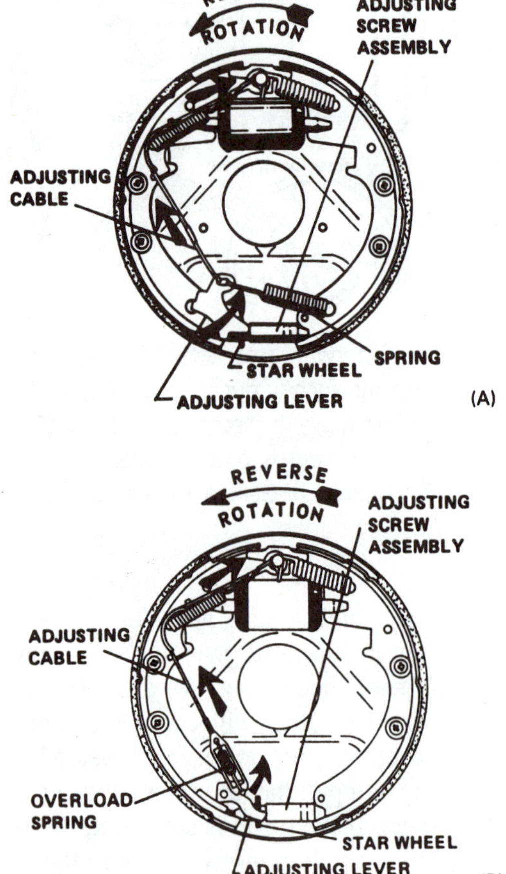

Figure 3-29 A cable type self-adjuster operates as the brakes are applied while backing up (A). Some cable type self-adjusters position the adjuster lever under the starwheel as shown in (B). (Courtesy of General Motors Corporation, Service Technology Group)

Several styles of self-adjusters are used with nonservo brakes. The most common leading-trailing brake self-adjuster is a *lever and starwheel* adjuster at the parking brake strut (Figure 3-30). Parking brake application pulls the lever away from trailing shoe, and if the movement is far enough, it will cause the starwheel to adjust, making the parking brake strut longer and reducing the lining clearance. The other common styles include a pair of *self-operating cams* (Figure 3-31), a *rotating ratchetlike adjuster* at the parking brake strut (Figure 3-32), and a

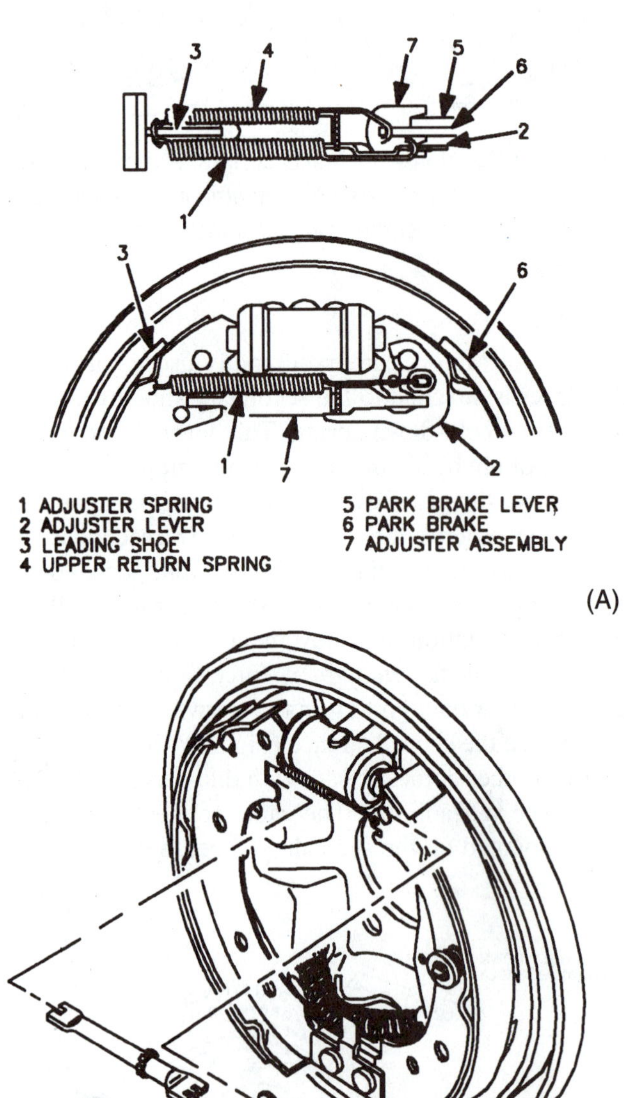

Figure 3-30 This nonservo brake design uses a self-adjuster mechanism that also serves as the parking brake strut. Applying the parking brake will cause the adjuster mechanism to lengthen if there is sufficient travel. (Courtesy of Saturn Corporation U.S.A., used with permission)

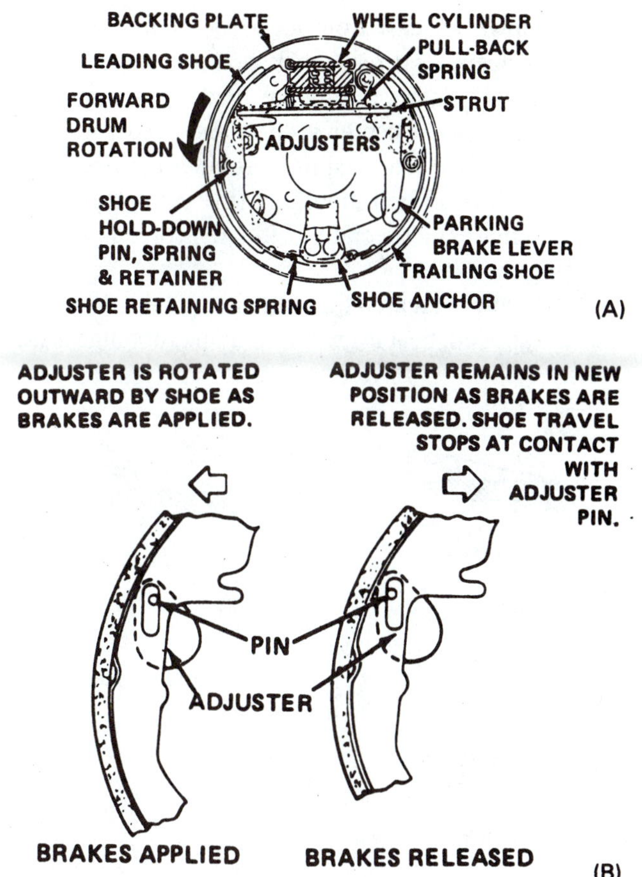

Figure 3-31 This nonservo brake uses an automatic cam adjuster (A). Movement of the shoe during brake application can reposition the adjuster (B). (Courtesy of General Motors Corporation, Service Technology Group)

telescoping rod and strut assembly (Figure 3-33). The first two of these styles make their adjustments whenever the shoe travels farther than the gap between a slot in the shoe and an adjuster pin. When this occurs, the adjustment ratchet or cam is moved or turned, moving the pin outward. The operation of the other self-adjuster style is based on the movement of the parking brake strut during application of the parking brake.

Each self-adjuster is designed to make an adjustment only when necessary to decrease excessive shoe clearance. An overadjustment that might cause shoe drag is very rare. Underadjustment, which causes a low brake pedal, is much more common, and this can be a fault of the adjuster mechanism or of the driver. The self-adjuster styles for duo-servo brakes operate when the brakes are applied while backing up. Some drivers of cars with automatic transmissions seldom apply the brakes after backing; they simply shift into drive and go forward. Because of this, the self-adjusters never get a chance to operate. A

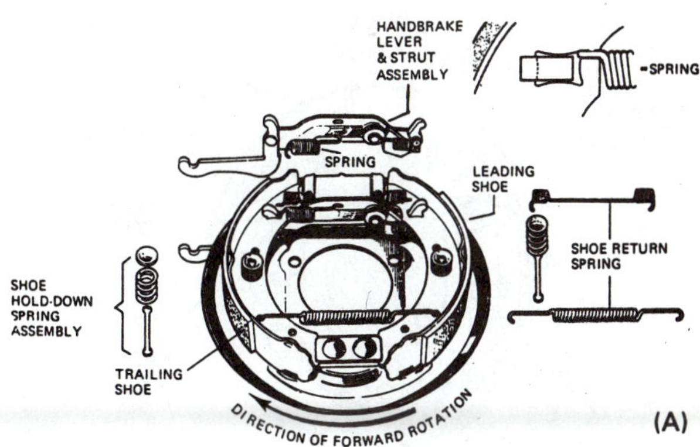

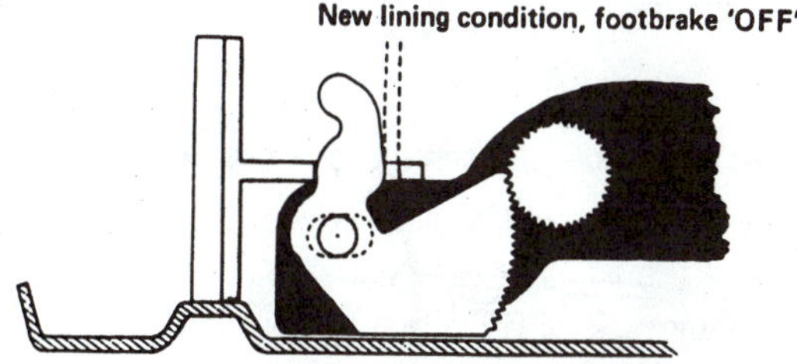

The quadrant lever is in contact with the knurled wheel, and the pull of the shoe return springs keeps the outer edge of the quadrant lever in contact with the rectangular slot in the shoe web. A clearance exists between the other edge of the quadrant lever and slot, which governs the gap between shoes and drum.

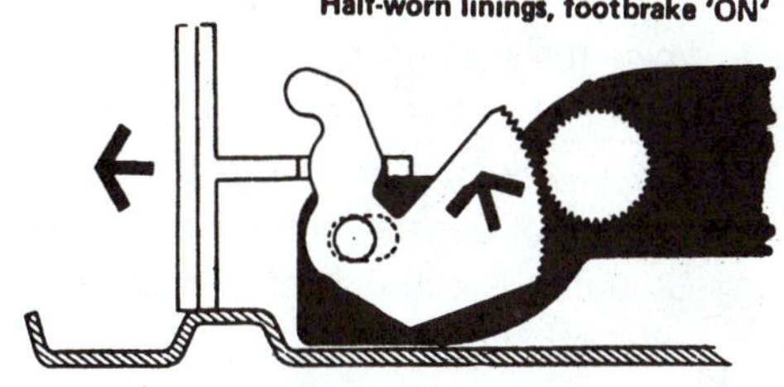

Hydraulic pressure from the master cylinder to the wheel cylinder pistons moves the shoes towards the drum. Once the clearance between the leading shoe and quadrant lever is taken up, further movement causes the teeth on the quadrant lever and knurled wheel to disengage and as the quadrant continues to move, it pivots sufficiently to take up the extra clearance caused by lining wear and so moves round the wheel. The knurled wheel is fixed and never moves.

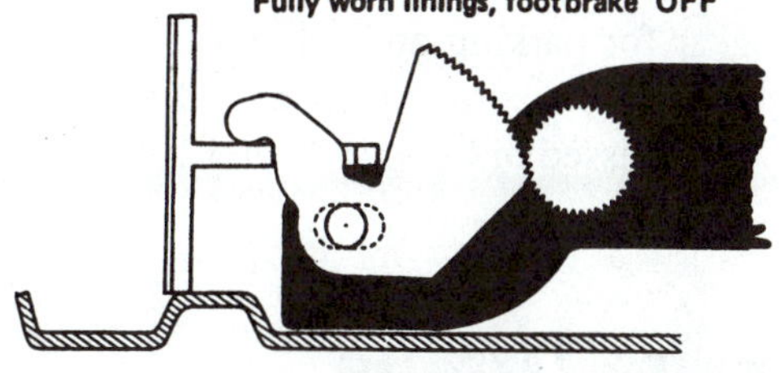

As the linings wear this movement is repeated until the tip of the quadrant lever abuts against the shoe web and the maximum adjustment condition is reached. From this point on, the quadrant lever does not pivot but slides in the slot in the strut and when the brakes are released, returns to the same position on the knurled wheel. As the linings continue to wear, the movement of footbrake pedal and handbrake lever increases, and this provides a warning that the lined shoes need replacing.

(B)

Figure 3-32 The nonservo, self-adjuster design is built into the parking brake lever and strut (A). As the lining wears, the quadrant lever will rotate and position the shoes closer to the drum (B). (Courtesy of LucasVarity Automotive)

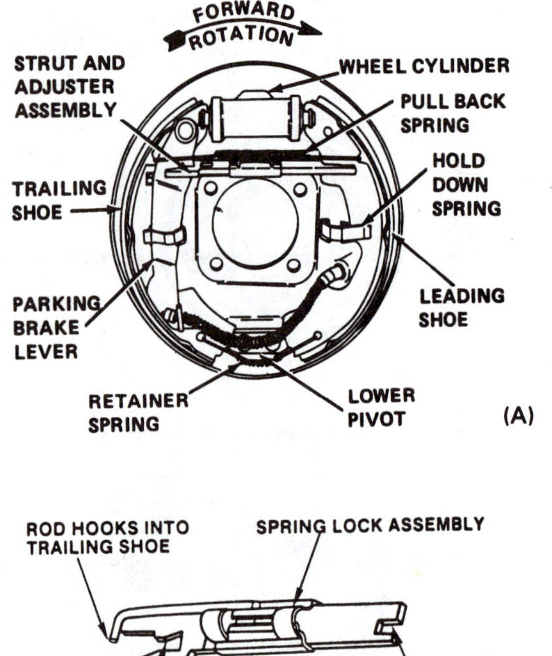

Figure 3-33 When the parking brake is applied on this design, the rod in the strut and adjuster assembly can pull the spring lock assembly to a longer position if the shoes need adjustment. (A is courtesy of General Motors Corporation, Service Technology Group; B is courtesy of Bendix Brakes, by AlliedSignal)

similar problem applies in the case of brakes that operate from the parking brake. Some drivers use an automatic transmission's park gear or a manual transmission's first or reverse gear for parking and seldom use the parking brake. Diagnosing and correcting poor self-adjuster operation will be discussed in Chapters 9 and 10.

3.9 Brake Drums

The **brake drums** have a rather simple job. They provide the rotating surface for the shoes to rub against. While doing this, they should present a hard, wear-resistant surface, have the physical strength to prevent excess deflection/distortion, and act as a heat sink.

The gray cast iron that most drums are made of is hard and quite wear-resistant, partially because of the high amount of carbon it contains. As far as wear resistance and strength are concerned, cast iron is an ideal drum material. However, cast iron does have some disadvantages. It is heavy and can crack or break fairly easily. Because of this, many drums are made of a composite with a stamped-steel center section and a cast-iron rim and friction surface. The cast iron provides the strength to resist distortion and bell mouthing (the size of the drum opening increases), especially if it is reinforced with ribs or fins (Figure 3-34).

A **heat sink** is an area that can absorb heat. When the brakes are applied, heat is generated at the rubbing surface of the shoes and drum. This heat will increase the temperature of these two surfaces or will be transmitted to cooler areas by conduction. Whatever heat is not conducted away will cause a rise in temperature. Cast iron can absorb heat at the rate of about 10°F per BTU per pound (5.5°C per 252 c per 0.45 kg); in other words, the temperature will increase about 10°F if 1 pound of cast iron absorbs 1 BTU of heat. A stop that generates 20 BTUs of heat will increase the temperature of a drum that weighs 5 pounds about 40°F. This same stop with a drum that weighs 1 pound will cause the drum's temperature to rise 200°F. A heavier drum is a much better heat sink; its temperature will not increase as rapidly as it absorbs heat. During a stop, some of this heat passes on to the air flowing past the drum, but heat transmission from metal to air is rather slow. Some of the features designed to improve brake drum cooling include finned drums to increase airflow;

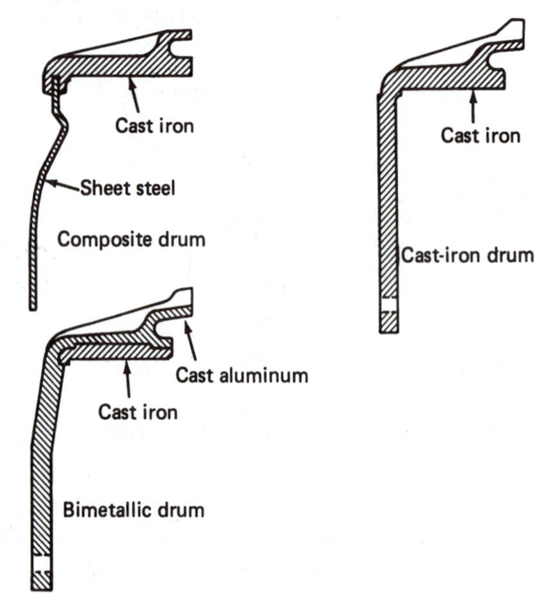

Figure 3-34 Most automotive brake drums are composites of stamped steel and cast iron. Some are all cast iron, and a few are cast aluminum with a cast-iron ring for the friction surface.

bimetallic—aluminum and cast-iron drums—to improve heat conductivity; and wheels with finned center sections to improve the airflow past the drum (Figure 3-35).

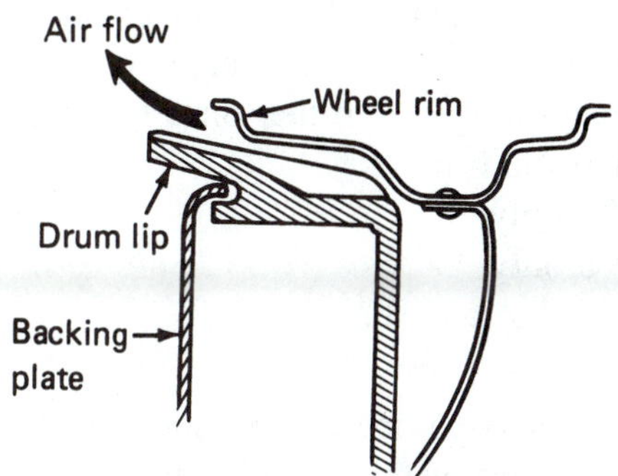

Figure 3-35 Some wheel and drum designs work together to cause airflow around the drum to help cool it.

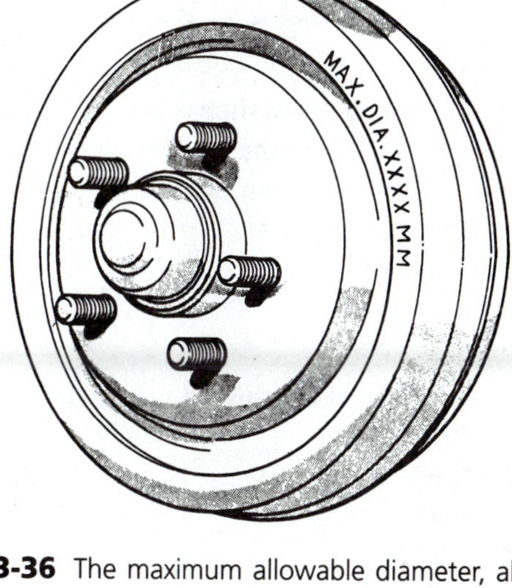

Figure 3-36 The maximum allowable diameter, also called the discard dimension, is indicated on all modern brake drums. (Courtesy of Chrysler Corporation)

As the brakes are used, the inner friction surface of the drum normally wears very slightly, but dirt or grit on the lining, or contact with a rivet or shoe rim, will cause very rapid drum wear. During lining replacement, this surface should be checked and, ideally, reconditioned to make sure it is true and has the right surface texture. All modern drums have the discard dimension indicated on one of the surfaces (Figure 3-36). Any drum with a diameter larger than its discard diameter should be thrown away. Using a drum that is too large can result in higher brake temperatures because of the heat-sink loss, or a spongy brake pedal because of the increased deflection that results from the reduced physical strength. Another potential drum problem is out-of-roundness. This can cause a pulsation of the brake pedal because the shoes will move in and out as they follow the fluctuating friction surface.

Brake drums are always checked during brake service for wear and cracks. These checks, along with their service procedures, will be described in Chapter 10.

3.10 Practice Diagnosis

You are working in a brake and front end shop as a service technician; these problems come to you:

CASE 1: A customer has brought in her 1993 Buick Roadmaster (large RWD car) with a complaint of a low brake pedal. On checking, you note that this car has a V-8 engine and automatic transmission. What is probably wrong? What should you do to confirm this? What should you tell this customer?

CASE 2: The customer has brought in his 1985 Ford Thunderbird complaining that the car swerves when he puts on the brakes. He does not blame the lining because he and his neighbor installed new shoes a couple of months ago. Your road test confirms the swerving problem. When you get the car back in the shop, you inspect the lining and find that the right rear has lining that is the full length of both shoes but the lining is shorter on both left shoes. What is wrong? What should you do to correct this problem? What should you tell the customer?

Terms to Know

adjustable anchors
anchor
anchor pin
backing plate
bonded lining
brake drums
brake spoon
deenergized
energized
fixed anchors
forward shoe
heat sink
heel
hold-down springs
leading shoe
leading-trailing shoe
major brake adjustment
return springs
reverse shoe
riveted lining
self-adjusting
toe
trailing shoe
two-leading shoe
uni-servo

Review Questions

1. If the anchor of a brake shoe is at the trailing end and the shoe is applied at the leading end, when the lining rubs against the drum, the shoe will be
   a. energized.
   b. deenergized.
   c. servoed.
   d. all of the above.
2. The brake shoe in Question 1 is called a
   a. leading shoe.
   b. trailing shoe.
   c. primary shoe.
   d. both a and c.
3. Servo brakes use
   a. an anchor for each brake shoe.
   b. two brake shoes with a single anchor.
   c. one brake shoe with two anchors.
   d. none of the above.
4. Statement A: A leading-trailing shoe brake is also called a simplex brake.
   Statement B: The trailing shoe of a leading-trailing shoe brake provides more stopping power than the leading shoe. Which statement is correct?
   a. A only
   b. B only
   c. both A and B
   d. neither A nor B
5. The major advantage of a duo-servo brake is that it
   a. operates well while backing up.
   b. offers a great deal of driver control.
   c. provides a large amount of stopping power for the amount of pedal pressure.
   d. all of these.
6. Statement A: The brake lining is attached to the shoe web.
   Statement B: The shoe rim is held against the backing plate by springs. Which statement is correct?
   a. A only
   b. B only
   c. both A and B
   d. neither A nor B
7. During the 1960s and 1970s, the most common ingredient in brake linings was
   a. asbestos.
   b. metal.
   c. fiberglass.
   d. Kevlar.
8. Statement A: The lining on a secondary shoe is usually longer than that on a primary shoe.
   Statement B: The lining length on a trailing shoe is usually longer than that on a leading shoe. Which statement is correct?
   a. A only
   b. B only
   c. both A and B
   d. neither A nor B
9. Brake service operations should be performed in a careful, planned manner because
   a. the dust released can cause health problems.
   b. vehicle accidents can result from an improper or incomplete repair.
   c. personal injury can result from improper use of tools and equipment.
   d. all of the above.
10. Statement A: Bonded lining is secured to the brake shoe by an adhesive which is cured at high temperatures.
    Statement B: Riveted lining is secured to the shoe by a group of hardened steel rivets. Which statement is correct?
    a. A only
    b. B only
    c. both A and B
    d. neither A nor B
11. Statement A: The backing plate is bolted securely to the rear axle.
    Statement B: The brake shoe anchor is usually welded or riveted solidly onto the backing plate. Which statement is correct?
    a. A only
    b. B only
    c. both A and B
    d. neither A nor B
12. The platforms on the backing plate are used to support the
    a. wheel cylinders.
    b. shoe rim.
    c. self-adjuster lever.
    d. hold-down springs.

13. Statement A: A weak return spring can cause a brake pull because that brake can apply early.
Statement B: A weak return spring can cause shoe drag and lining wear. Which statement is correct?
    a. A only
    b. B only
    c. both A and B
    d. neither A nor B
14. During brake release, brake shoe return springs are used to
    a. pull the brake shoes back against the anchor(s) or shoe stops.
    b. push the wheel cylinder pistons back in the bore.
    c. return brake fluid to the master cylinder reservoir.
    d. all of the above.
15. Statement A: The self-adjuster mechanism for most duo-servo brakes operates as the parking brake is applied.
Statement B: The self-adjuster mechanism for most nonservo brakes operates as the brakes are applied while backing up. Which statement is correct?
    a. A only
    b. B only
    c. both A and B
    d. neither A nor B
16. A brake drum
    a. provides a smooth surface for the lining to rub.
    b. provides a heat sink to absorb braking heat during a stop.
    c. should not be used if the inside diameter is larger than the number indicated on the drum.
    d. all of the above.

4 Disc Brake Theory

Objectives

After completing this chapter, you should:

- ☐ Be familiar with the different styles of disc brake units.
- ☐ Be familiar with the terms commonly used with disc brakes.
- ☐ Understand how a disc brake operates and its advantages as compared to drum brakes.
- ☐ Understand the purpose of each of the components that is used in a disc brake assembly.
- ☐ Be familiar with the different styles of lining wear indicators.

4.1 Introduction

The operation of disc brakes is simpler than that of drum brakes. The brake shoes or pads are squeezed against the disc or rotor during braking. For release, the pads merely relax their pressure. Though the terms are interchangeable, "pad" is used more often than "shoe," and "rotor" is used more often than "disc," when discussing disc brakes. Disc brakes have no self-energizing or servo action. Pad pressure increases in direct proportion to the brake application force.

Anyone familiar with the caliper brakes on a modern bicycle knows disc brakes in their simplest form (Figure 4-1). Two pads are tightened against the wheel rim by a simple, hinged, mechanically-operated caliper. Most important though, these caliper brakes are light in weight and very effective. They can usually lock up the wheel under very hard application.

Figure 4-1 The simple mechanical caliper used on a bicycle is an example of a disc brake. The pads will push directly on the wheel rim as the cable is pulled to apply the brake.

There are several advantages of disc brakes. The two pads press on each side of the rotor in opposition to each other. Because of this, there is no distortion such as elongation of a drum, which can change shoe-to-drum contact or cause a low brake pedal (Figure 4-2). A major portion of the rotor's friction surface is exposed directly to air. This surface will stay much cooler than a drum's friction surface. During rotation, centrifugal force will throw any contaminants off of a rotor's friction surface (Figure 4-3) but in the case of a drum, the contaminants are forced onto the friction surface. Also, the pads of a disc brake release to a position right next to the rotor's surface. This creates a wiping action which keeps dust, dirt, and water from entering between the lining and the rotor.

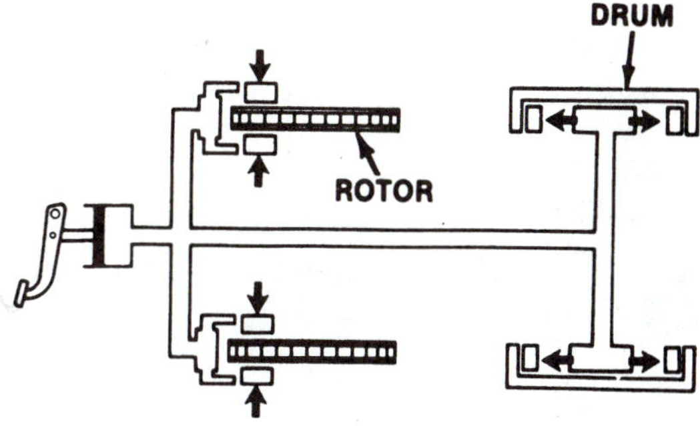

HYDRAULIC PRESSURE IS TRANSFORMED INTO MECHANICAL MOVEMENT BY CALIPERS OR WHEEL CYLINDERS

Figure 4-2 There is no distortion of a rotor during braking because the same pressure is applied on each side of a noncompressible surface by the two pads. A drum can expand slightly from the pressure of the shoes. (Courtesy of Brake Parts, Inc.)

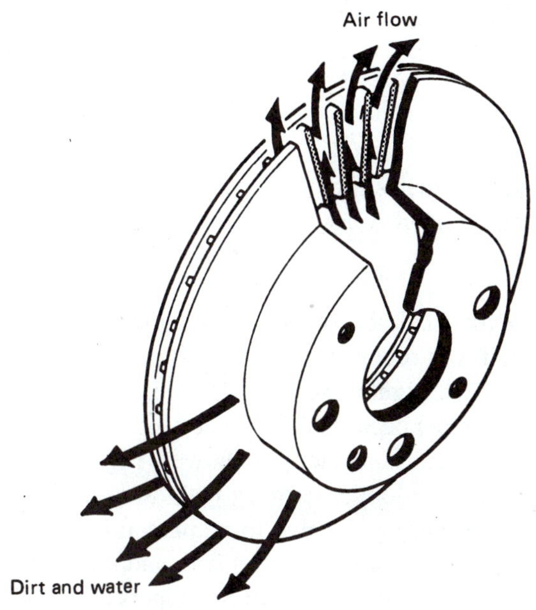

Figure 4-3 Friction surfaces of a rotor stay clean because centrifugal force throws off water and dirt. Centrifugal force also causes airflow through the internal ventilating fins to help cool the rotor. (Courtesy of American Honda Motor Company)

Next, and possibly most important, there is no self-energization or servo action so both brakes on an axle will usually generate the same braking power. With servo action, slight variations—common to high-mileage drum brakes—can change the frictional drag between the primary shoe and the drum, changing the amount of servo action, which in turn can cause a large difference in braking power. These variations can be a result of a weak return spring, sticky wheel cylinder, improperly fitting shoe, contaminated lining, faulty adjustment, and so on. Brake pull and uneven stops were fairly common when drum brakes were used on front wheels. Today, with front disc brakes, straight, even stops are easily achieved. An additional advantage of passenger-car disc brakes is that they are naturally self-adjusting.

Disc brakes do have a couple of disadvantages, however. Without servo action, a disc brake cannot develop the same braking power as a drum brake with the same hydraulic pressure. Much more force is necessary. Power boosters are required on most disc brake applications; they are an absolute necessity on mid-sized and larger cars. Also, it is difficult to include a mechanically operated parking brake in a caliper design. Several disc brake parking brake designs have been developed, but these tend to be expensive, complicated, heavy, weak, and prone to stick. The problems associated with the parking brake are probably the primary reasons why four-wheel disc brakes are not very common on lower-priced cars. These parking brake designs will be described in Chapter 5.

Another characteristic of disc brakes, and also of drum brakes to a lesser degree, is squeal. This is a highly annoying, high-frequency noise caused by vibration of a pad on the rotor or of a shoe on the drum. The large, flat rotor surface emits sound better than a drum's surface, making squeal a more common problem with disc brakes. The intensity of squeal is affected by the hardness of the lining, the rigidity of the shoe or caliper, the rigidity and speed of the rotor, and the amount of caliper pressure. Harder linings and thinner rotors increase the tendency to squeal, whereas firmer pad-to-caliper attachments and thicker pads, with dampening material on the backing, reduce vibrations and the tendency to squeal.

4.2 Calipers

The **caliper** is the casting that is mounted over the rotor. It contains the brake pads and the hydraulic piston(s) that apply the pads. It must be strong enough to transmit the high clamping forces needed and also to transfer the braking torque from the pads to the steering knuckle (Figure 4-4).

The pressure between the brake pads and each side of the rotor should be equal to prevent flexing and bind at the wheel bearings and flexing or distortion of the rotor or caliper. There are two major types of caliper design that are found on both front and rear brakes.

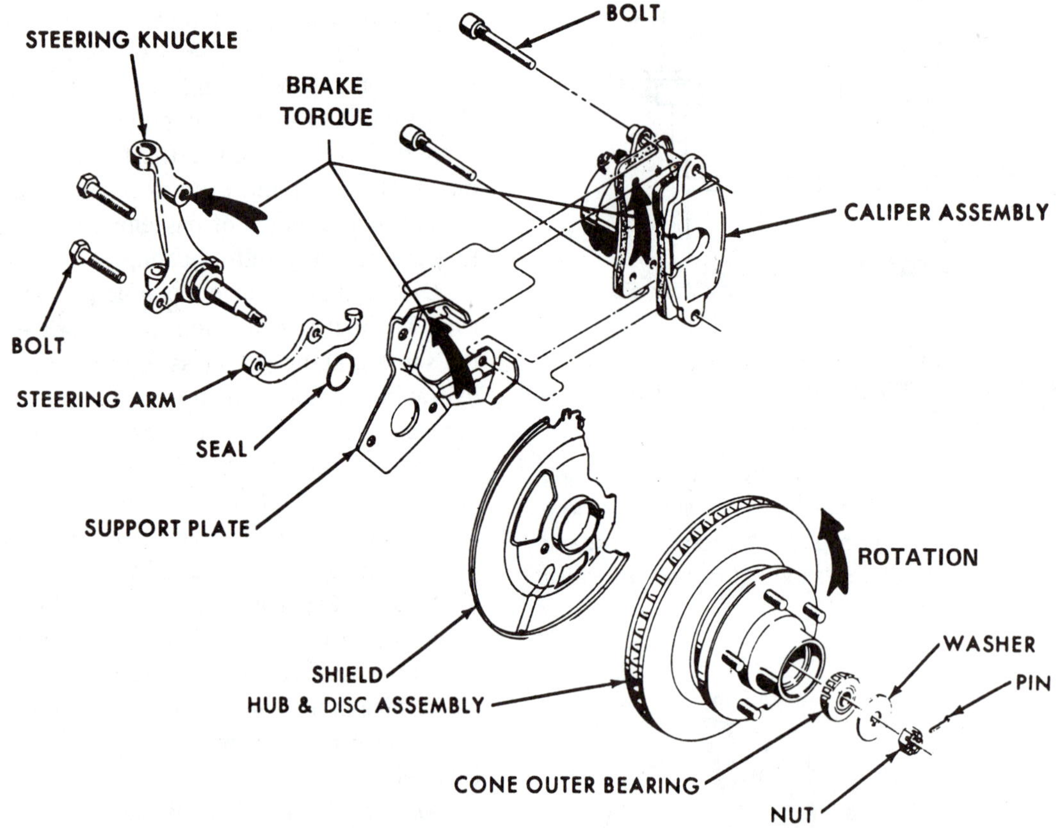

Figure 4-4 When the brakes are applied, the brake torque is transferred from the pads to the support plate and then onto the steering knuckle and front suspension. (Courtesy of General Motors Corporation, Service Technology Group)

4.2.1 Fixed Calipers

The first calipers used on passenger cars were of the **fixed-caliper** design. The caliper is fixed or fastened securely onto the steering knuckle. This caliper does not move relative to the steering knuckle. Fixed-caliper pistons are arranged in pairs with a piston on each side of the rotor. The inboard piston(s) applies the inboard pad, and the outboard piston(s) applies the outboard pad (Figure 4-5).

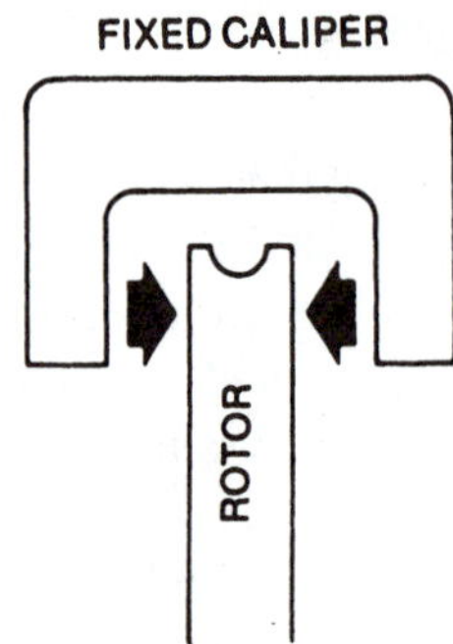

Figure 4-5 In a fixed-caliper brake, the caliper is stationary, and a piston on each side of the rotor puts pressure on the lining. (Reprinted from Mitchell Anti-Lock Brake Systems, with permission of Mitchell Repair Information, LLC)

The fixed calipers on heavier vehicles use four pistons (two pairs) in order to generate sufficient stopping force (Figure 4-6). Lighter-weight cars commonly use two pistons (one pair) because the lighter weight of the car does not require the same stopping power (Figure 4-7). Fixed calipers are no longer used on new domestic cars. They tend to be more expensive to manufacture than floating calipers because of their complexity and number of parts.

The brake pads of many fixed calipers can be slid in or out after removing a single retainer. The caliper does not usually have to be removed for pad replacement. The ends of the pads push against machined abutment surfaces in the caliper during braking. A slight clearance at the abutments allows pad movement during application and release.

4.2.2 Floating Calipers

Floating calipers are a simpler design. Normally only one piston is used. Some passenger cars and light trucks use a two-piston floating caliper. Floating calipers are used with a **caliper mount**, also called an **adapter** or *anchor* plate, which is bolted solidly to the steering knuckle. This mount transfers braking forces from the shoes in the caliper to the

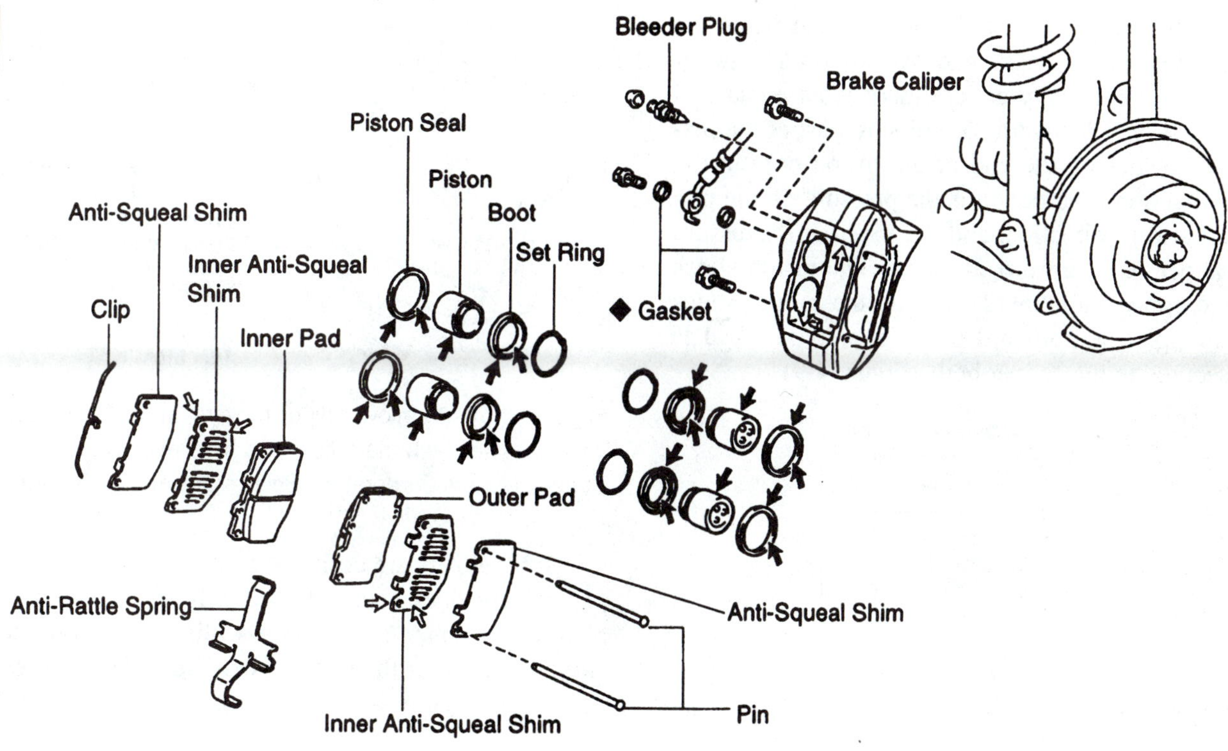

Figure 4-6 A four-piston, fixed-caliper brake is bolted securely to the steering knuckle and has two pairs of pistons.

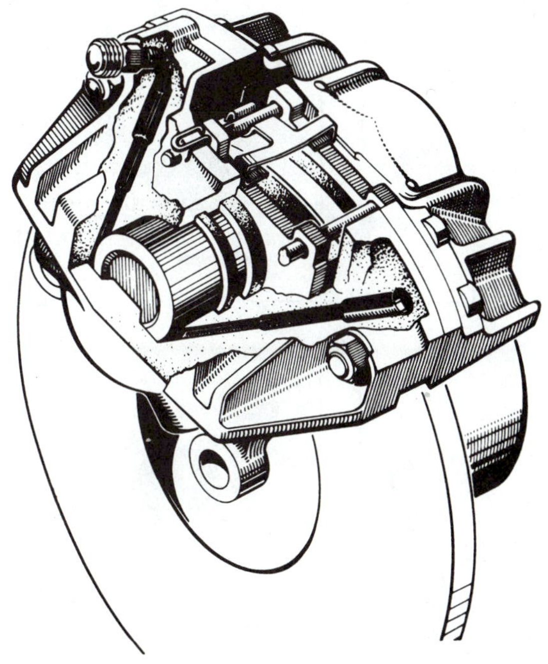

Figure 4-7 This fixed-caliper brake uses two pistons; note the internal passage to carry fluid to the outboard piston and air to the bleeder screw. (Courtesy of ITT Automotive)

steering knuckle. In some designs, the inboard shoe is fitted directly into the caliper mount. The caliper is fitted into the mount so that it can move sideways relative to the rotor and steering knuckle (Figure 4-8).

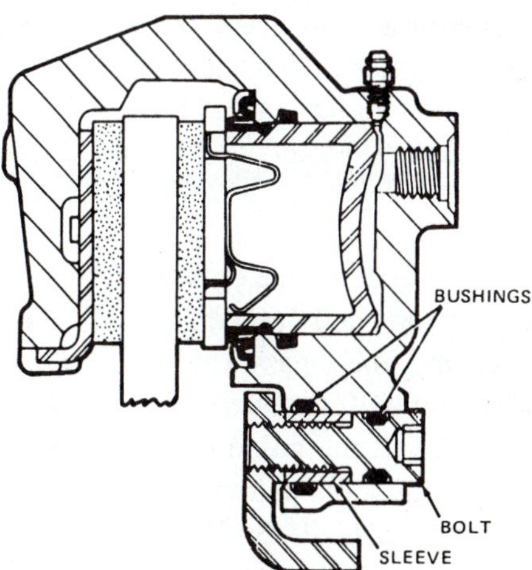

Figure 4-8 This floating caliper is designed to move sideways on its bushings, sleeve, and mounting pin. (Courtesy of General Motors Corporation, Service Technology Group)

The caliper piston applies the inboard pad while the caliper applies the outboard pad. An important law of physics states, "For every action, there is an equal and opposite reaction." As hydraulic pressure pushes the piston, it also pushes on the end of the piston bore in the opposite direction. The *action* is the pressure on the piston in an outward direction, and the *reaction* is pressure on the caliper in an inward direction. The piston slides outward to apply the inboard pad, and the caliper slides inward to apply the outboard pad. The outboard pad is usually secured to the caliper body (Figure 4-9).

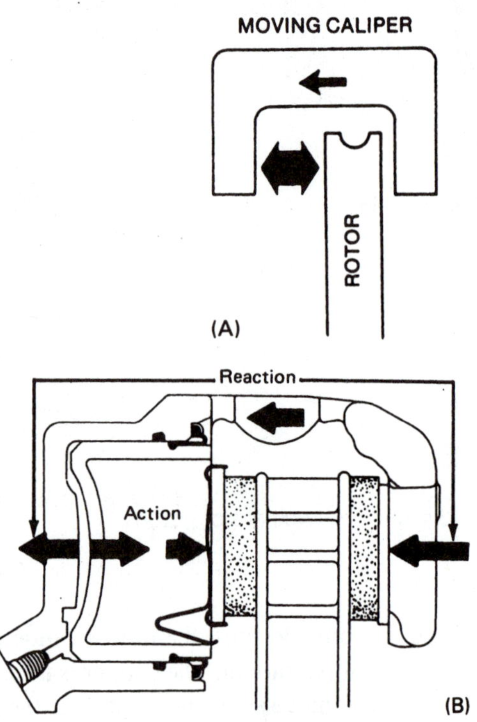

Figure 4-9 A floating caliper moves inward as the brakes are applied (A). The action of the piston puts pressure on the inboard pad while the caliper reaction puts pressure on the outboard pad. (A is reprinted from Mitchell Anti-Lock Brake Systems, with permission of Mitchell Repair Information, LLC; B is courtesy of General Motors Corporation, Service Technology Group)

Many calipers float on one or two **guide pins** or **bolts**. These pins are threaded into either the mount or the caliper body, and they pass through the other. Most designs use rubber or Teflon **sleeves**, also called *bushings* or *insulators*, around the guide pins for a better bearing surface. The guide pins allow the required sideways motion of the caliper body. Braking torque is transferred as the caliper body is forced against the abutment of the caliper mount. Normally, there is a slight clearance at the caliper-to-mount abutments to allow the necessary sideways motion, but not enough to cause a slap or knock on brake application (Figure 4-10).

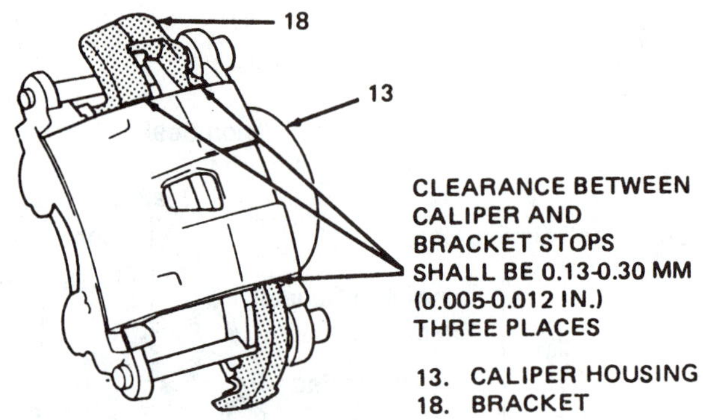

Figure 4-10 To allow caliper movement, a slight clearance must exist between the caliper and bracket stops. (Courtesy of General Motors Corporation, Service Technology Group)

Most calipers are designed so that a certain amount of pad release does occur. As we will explain more completely in Chapter 6, the piston-sealing O-ring produces a slight pullback of the piston when the hydraulic pressure drops off (Figure 4-11). Because of its relatively light

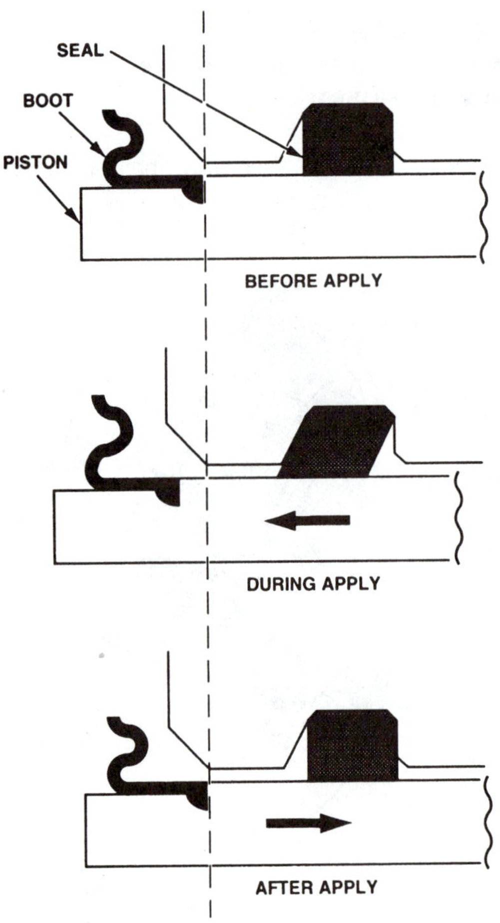

Figure 4-11 As the caliper piston moves to apply, the square seal ring becomes distorted. During release, the O-ring can then relax and pull the piston back a slight amount. (Courtesy of General Motors Corporation, Service Technology Group)

weight, the piston is fairly easy to release, much easier than the caliper. The release of the outboard pad is very slight. It occurs mainly from the relaxing of the caliper and the slight wobble or flexing of the rotor. This is aided by the sleeves and bushings around the guide pins. Proper floating of the caliper body is very important. Because of the piston size, a caliper can generate close to 10,000 pounds (4,545 kg) of clamping force. This much force will definitely press the pads against the rotor, even if the piston or caliper is sticky. Drag will probably occur if the piston or caliper is sticky, because there will not be nearly this amount of force to push the piston or caliper back to release the pads.

There are several variations of floating calipers. Some early domestic designs were called **sliding calipers**, also called a mechanical guide caliper. The body of these calipers is fitted between two V-shaped, machined grooves or ways in the caliper mount (Figure 4-12). These guides allow only a slight sideways motion, whereas the guide pins of a floating caliper also allow a slight twisting or flexing motion. Sliding calipers have more drag between the caliper body and the mount, and they tend to stick or drag unless the V-ways are properly lubricated.

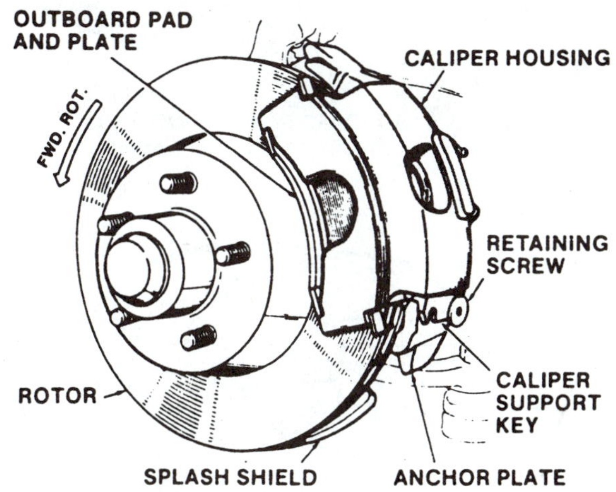

Figure 4-12 A sliding caliper. Note how the caliper fits in its mount. The braking action is like that of a floating caliper. (Courtesy of Bendix Brakes, by AlliedSignal)

Several important designs utilize an additional part called a **yoke**. One manufacturer calls this yoke a caliper and the piston body a cylinder. The yoke straddles the caliper or piston body and both the outboard and inboard pads. In one design, the caliper body is bolted to the steering knuckle, and one piston applies the inboard pad while a second piston pushes the yoke and the outboard pad inward (Figure 4-13). In a similar design, both the yoke and the piston body float on a retainer that is bolted onto the steering knuckle. Only one piston is used, as cylinder reaction pushes inward on the yoke to apply the outboard pad (Figure 4-14).

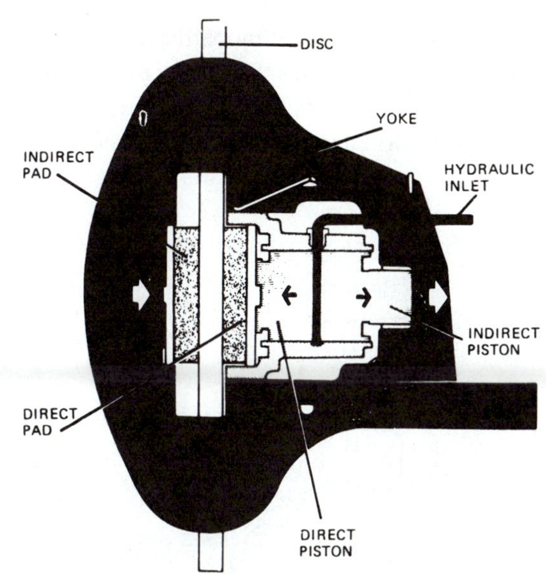

Figure 4-13 A caliper that uses a yoke straddling the rotor to transmit the application force from the indirect piston to the indirect (outboard) pad. (Courtesy of LucasVarity Automotive)

Another design uses a caliper plate that attaches to the mounting bracket on a pivot pin so that it can swing on the pivot during application and release. As in a floating

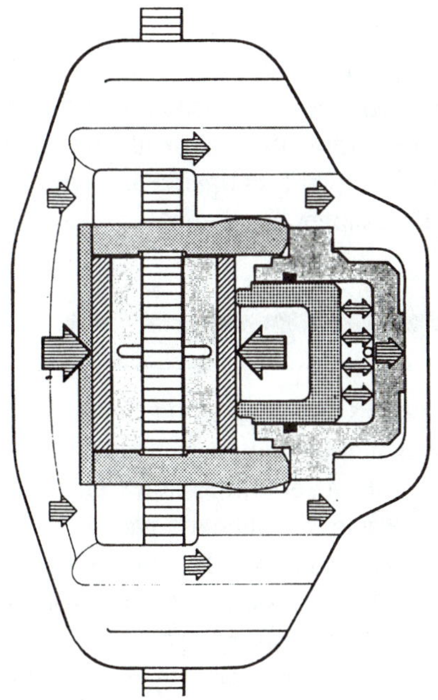

Figure 4-14 This caliper uses a yoke to transmit the force from caliper reaction to the outboard pad; the caliper body must float. (Courtesy of ITT Automotive)

caliper, the piston applies the inboard pad, and reaction force from the cylinder moves the caliper to apply the outboard pad. This caliper does not stay parallel to the rotor, and the lining is normally tapered or wedge-shaped (Figure 4-15).

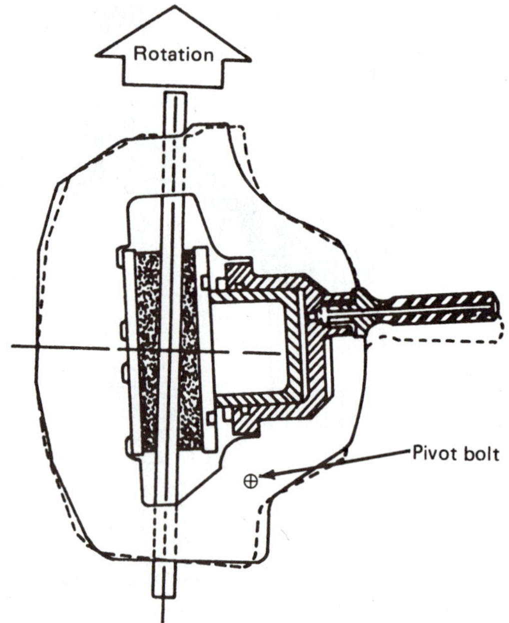

Figure 4-15 A swing or pivoting caliper. Note that the caliper pivots to apply the outboard pad. Also note the normal taper of the lining. (Courtesy of Wagner Brake)

The pads are fitted into most floating calipers from the open, inner portion of the caliper. Typically, the caliper must be removed from the mounting bracket to allow pad replacement. In most cases, the inboard pad floats in the caliper or mounting bracket while the outboard pad is secured tightly to the caliper. Pad replacement will be described in Chapter 11.

4.3 Brake Pads

Most disc brakes use a pad with a flat metal backing. Most pad backings include mounting ears, clips, or projections; some have only a mounting hole. The pads of a fixed-caliper brake and the inboard pad of a floating-caliper brake are normally designed to drop in place between two abutments with just enough clearance for application and release movement. Braking pressure is transferred to the abutments in the caliper or caliper mounting bracket. As mentioned earlier, most floating-caliper outboard pads are secured solidly to the caliper. Any motion between the outboard pad and the caliper might cause vibration and brake squeal (Figure 4-16).

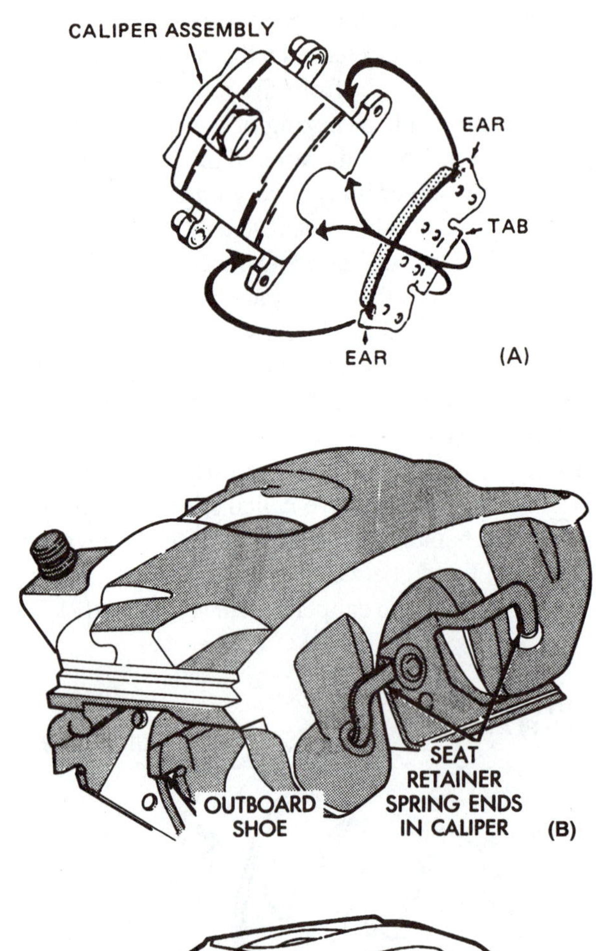

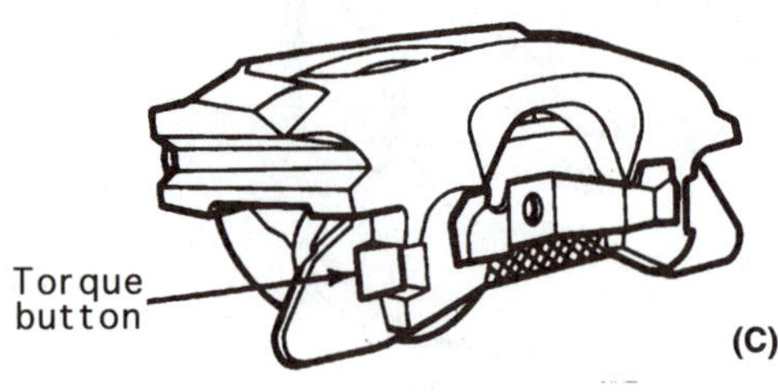

Figure 4-16 The outboard pads used with floating calipers are normally attached to the caliper by locking tabs and ears (A) or spring pressure (B). Torque is transmitted from the pad to the caliper by the mounting tabs and ears (A) or torque buttons (C). (A is courtesy of Wagner Brake; B is courtesy of Chrysler Corporation; C is courtesy of Ford Motor Company)

On most fixed-caliper designs, the inboard and outboard pads are identical and interchangeable. Most floating-caliper designs have an inboard pad that is noticeably different from the outboard pad. A few vehicles will have pads that fit in only one particular location, such as inboard left side or inboard right side (Figure 4-17). Some use *torque buttons* to transfer braking torque between the pad and the caliper and also to keep the pad properly positioned in the caliper.

Disc brake lining is essentially the same as that used on drum brakes. With many newer front-wheel-drive cars,

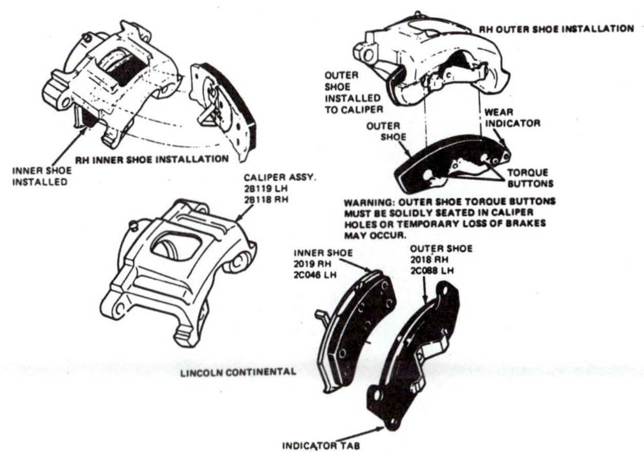

Figure 4-17 This brake pad set has four different shoes, each of which must be fitted in the correct location. Note the warning concerning proper installation. (Courtesy of Ford Motor Company)

semimetallic linings are fairly common because of their ability to withstand higher operating temperatures. The lining is riveted, bonded, or mold-bonded to the pad backing. In all cases the lining surface is flat and, except for swing-caliper designs, the lining surface is parallel to the pad backing. Pads used with swing calipers are tapered—thicker at one end than the other. In a few cases, the lining will be offset sideways on the shoe—closer to one end than the other—and this feature produces more even lining wear. The trailing end of a shoe is always hotter than the leading end, and therefore wears more rapidly. Excessive taper wear can lead to excessive caliper flex and a low brake pedal (Figure 4-18).

Many pads include **pad wear indicators**. The most common is a tab secured to the trailing edge of the outboard pad, which is called an *audible sensor*. When the lining is worn to the point where replacement is necessary, the wear indicator tab will rub against the outer edge of the rotor and produce an easily noticed, high-pitched squeal (Figure 4-19). Some cars use *visual sensors*. A warning light on the dash lights up when the pads are worn. In this system, each pad includes an electrical sensor that is connected to the warning light by an electrical wire. A few cars use a *tactile sensor*. Here a projection on the rotor strikes the pad backing place when the lining is worn out. This creates a pulsating action of the brake pedal during braking.

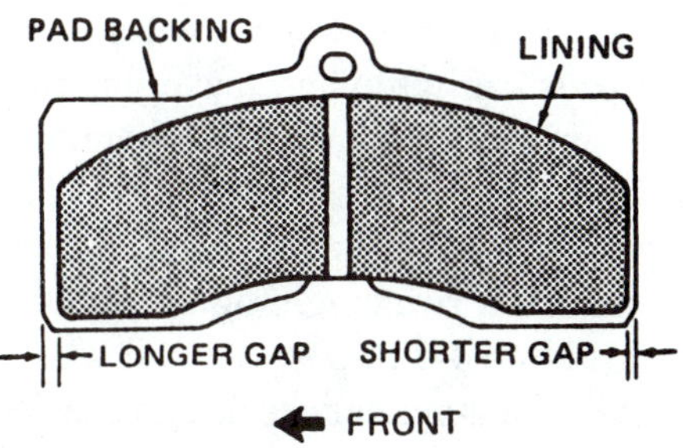

Figure 4-18 A few brake pads have offset lining; this places slightly more lining at the trailing edge to produce more even wear. (Courtesy of Wagner Brake)

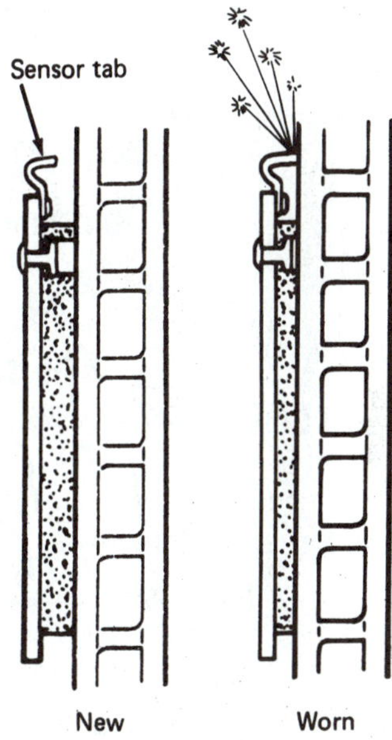

Figure 4-19 This pad wear sensor, also called an audible wear indicator, is attached to the trailing edge of the outboard pad. It will produce a high-pitched squeal if the lining wears to the point where the sensor tab contacts the rotor. (Courtesy of Wagner Brake)

4.4 Caliper and Pad Mounting Hardware

Every caliper design uses a few supplementary parts, commonly referred to as **mounting hardware**. Hardware, also called small parts, comes in various forms and shapes to suit each of the different calipers. Most hardware can be replaced in one of four categories: *anti-squeal, pad retention, pad antirattle*, and *pad* or *caliper positioning* (Figure 4-20).

As mentioned previously, the pads on most fixed calipers drop into place. They are usually retained by a steel pin that prevents them from working out or bouncing out. This pin passes through a hole in the pads and caliper body and is retained by a spring clip or cotter pin (Figure 4-21).

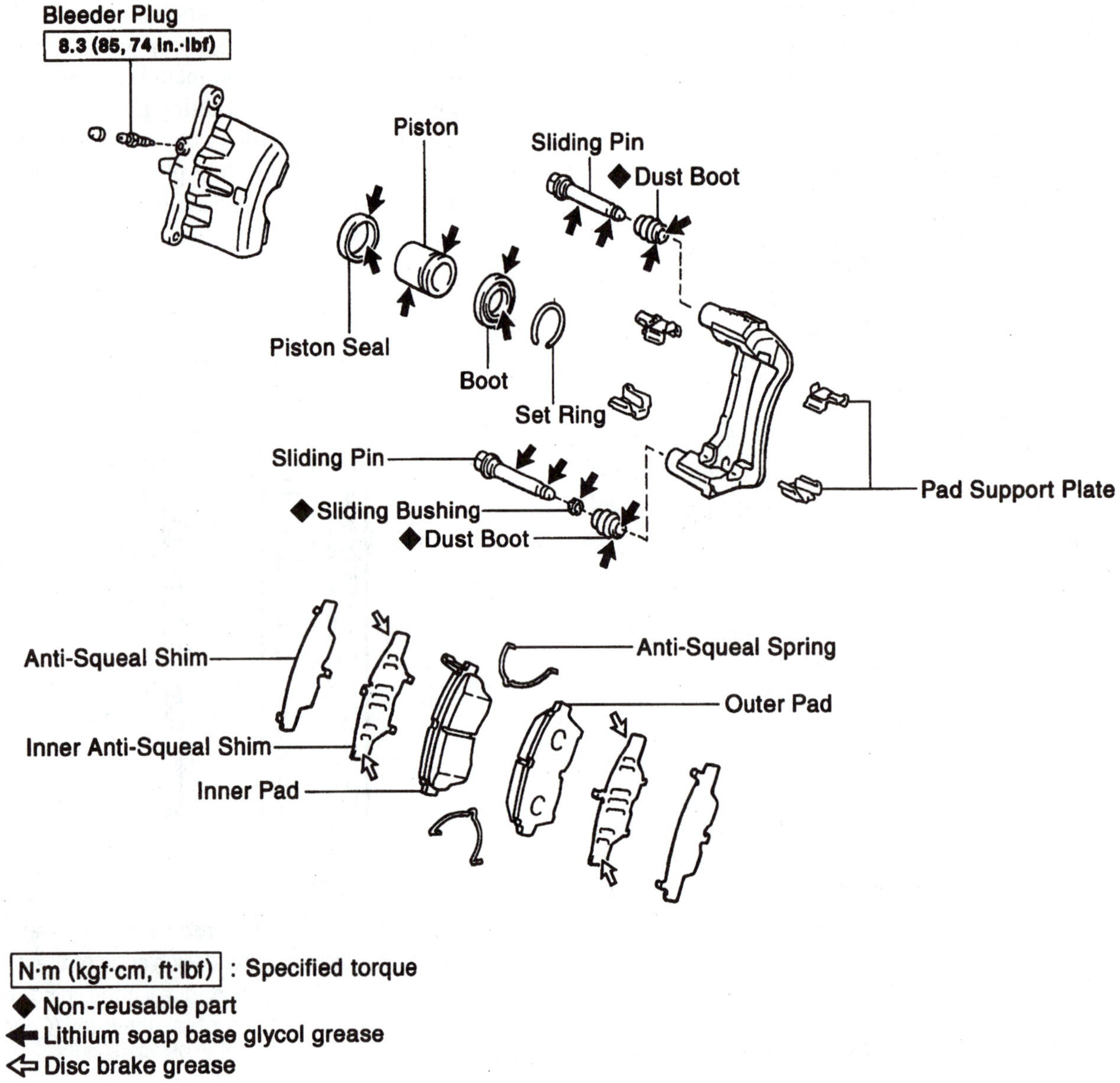

Bleeder Plug
8.3 (85, 74 in.·lbf)
Piston
Sliding Pin
◆Dust Boot
Piston Seal
Boot
Set Ring
Sliding Pin
◆Sliding Bushing
◆Dust Boot
Pad Support Plate
Anti-Squeal Shim
Anti-Squeal Spring
Outer Pad
Inner Anti-Squeal Shim
Inner Pad

N·m (kgf·cm, ft·lbf) : Specified torque
◆ Non-reusable part
← Lithium soap base glycol grease
⇦ Disc brake grease

Figure 4-20 This caliper uses several types of hardware: the anti-squeal shims and springs and the pad support plates ensure proper pad operation, and the sliding bushings and boots ensure proper caliper operation.

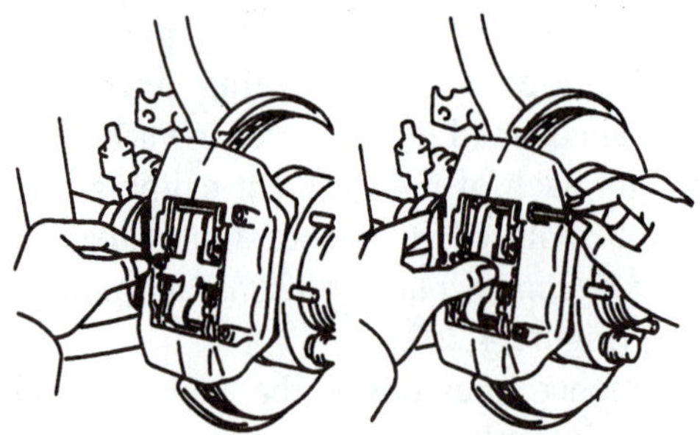

Figure 4-21 The pads on most fixed-caliper brakes are held in the caliper by retaining pins. Removal of the pins will allow the pads to slide out.

In many cases, the pads are loose while the brakes are released, free to move between the rotor and caliper. These loose pads can rattle and make jingling noises as the vehicle is driven. In some cases, a spring is used to apply a slight pressure to keep them quiet. The **antirattle spring** or **clip** is sometimes also used to move the pad away from the rotor, reducing drag or vibration that can cause squeal (Figure 4-22).

Almost every floating-caliper design uses bushings or insulators along with the guide pins or bolts. These help locate the caliper and pads, return the caliper to a released position, cushion the caliper movement, and prevent

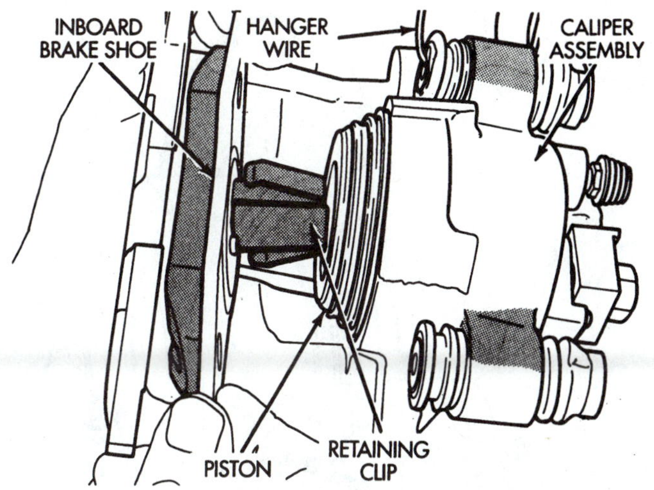

Figure 4-22 This inboard pad is connected to the piston by the retaining clip; this helps pull the lining away from the rotor and reduces rattles while released. (Courtesy of Chrysler Corporation)

metal-to-metal contact between the caliper and the mounting pins. Improper or worn caliper mounting hardware can cause excessive pad drag and wear or brake squeal (Figure 4-23).

In all cases, caliper and pad hardware should be serviced or replaced whenever the brake pads are replaced. The high heat produced by the brakes, plus contamination from dirt, sand, and salt from the road, as well as water and ice, have a drastic effect on these parts. Cleaning, replacement, and correct lubrication of them are mandatory to achieve a properly operating disc brake. These procedures will be described in Chapter 11.

4.5 Rotors

A rotor or disc, like a drum, provides the friction surface for the lining to rub against. Also, like a drum, it is made from gray cast iron for all of the same reasons. In some cases the rotor and the front hub are cast as one piece. Typically, they are two pieces so either the rotor or the hub can be replaced separately. Some rotors are a composite with cast iron braking surfaces and a much thinner, stamped steel center section (Figure 4-24). A composite rotor is lighter than a similar cast-iron rotor.

Two styles of rotors are used on passenger cars: solid and vented. **Solid rotors** are often used on smaller cars and are smaller in width, lighter, and less expensive

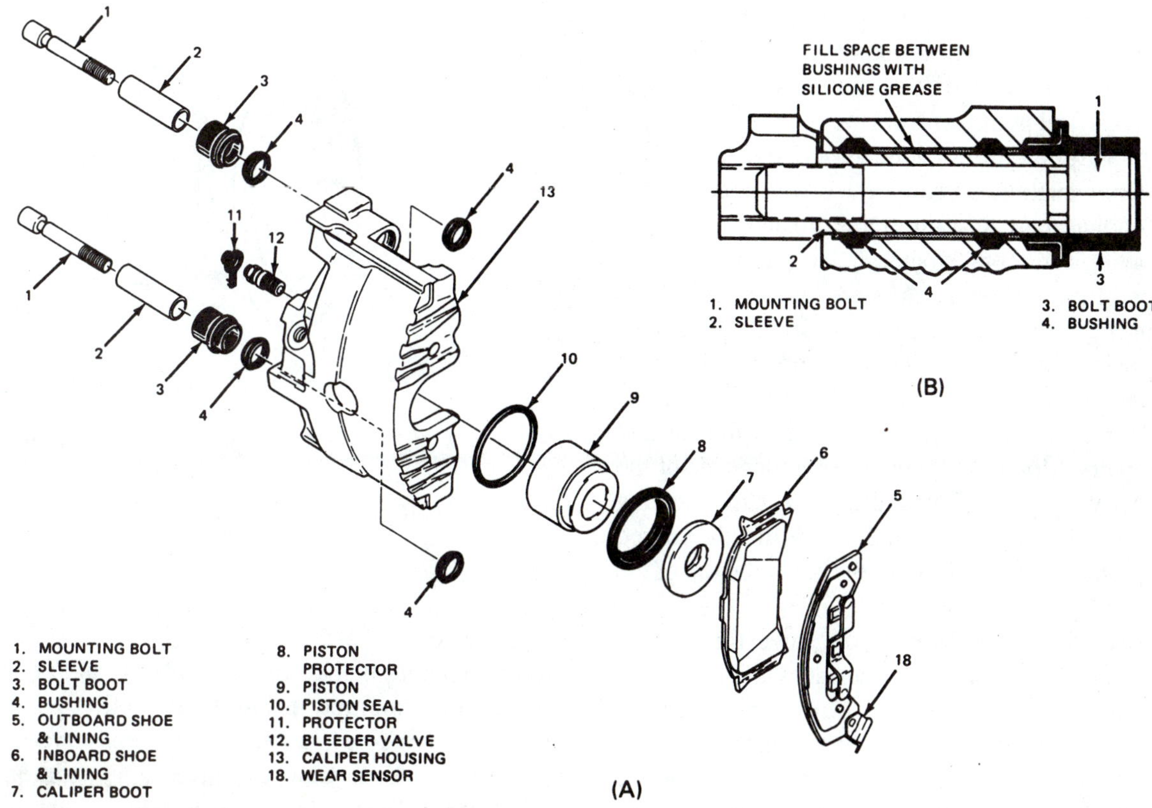

Figure 4-23 This caliper slides on sleeves (2); silicone grease at the sleeve and bushings ensures good caliper action for a long time. (Courtesy of General Motors Corporation, Service Technology Group)

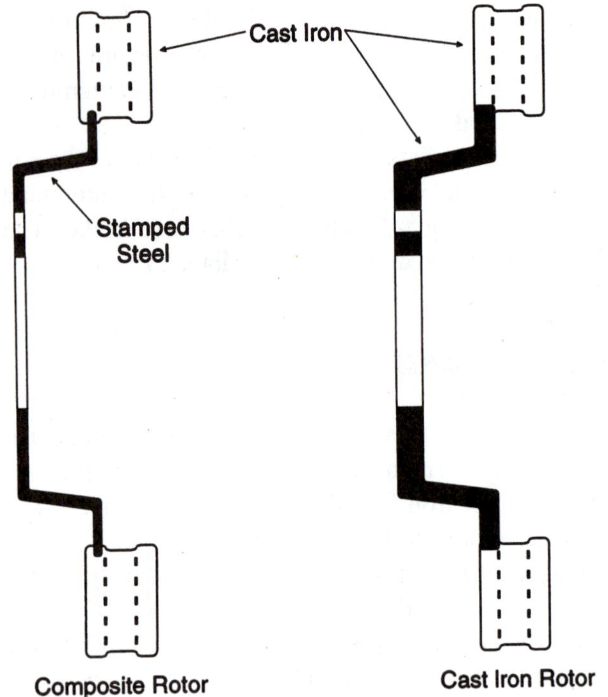

Figure 4-24 Many newer rotors are composites; the cast-iron braking surfaces are combined with a stamped-steel center section. Rotors are also all cast iron.

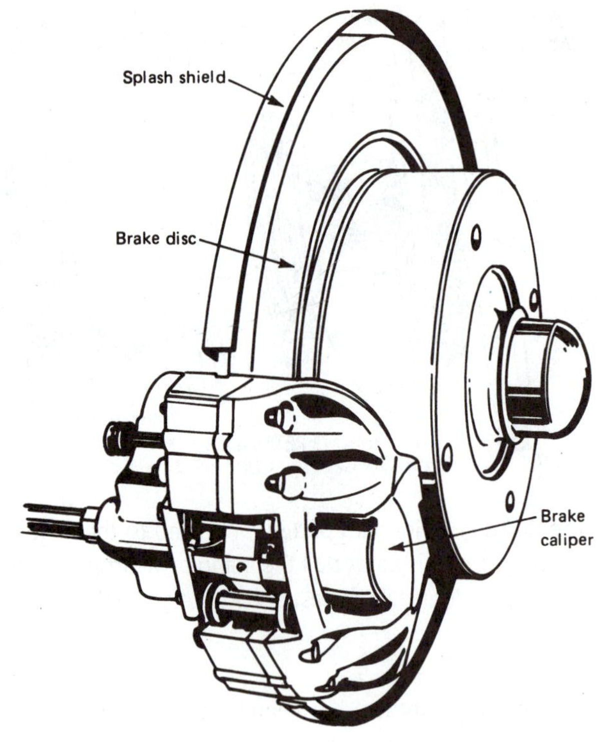

Figure 4-25 Many smaller cars use a solid, non-vented rotor. (Courtesy of Volkswagen of America)

(Figure 4-25). Heavier cars, which produce more braking heat, use **vented rotors**. These rotors are cast with cooling fins between the two friction surfaces. As the rotor turns, these fins will pump air from the inner eye (the center of the inboard side) through the rotor to the outer edge. Because of this airflow, a vented rotor will operate substantially cooler than a solid rotor. Vented rotors are thicker, heavier, and more expensive than solid ones.

The fins in most vented rotors are straight and point straight toward the center of the rotor. Some rotors have angled or curved fins; these will pump more air and therefore run cooler. Angled or curved fins are directional. They must be mounted on the proper side of the car and are used in pairs. Mounting them on the wrong side of the car will greatly reduce airflow and can cause overheating of the caliper and lining. A rule of thumb is that the curved fin should point toward the front of the car at the top of the rotor (Figure 4-26).

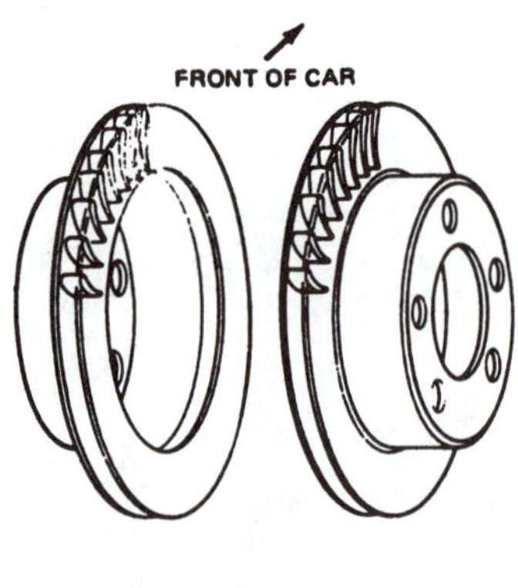

Figure 4-26 These rear rotors use curved internal fins; they must be mounted on the proper side of the car to achieve correct air flow. (Courtesy of Bendix Brakes, by AlliedSignal)

Like brake drums, modern rotors have the discard dimension indicated on one of the surfaces (Figure 4-27). Any rotor with a thickness less than the discard dimension should be thrown away. Using a rotor that is too thin will result in higher operating temperatures because of the heat-sink loss. This can cause higher lining temperatures which in turn will cause faster lining wear, possible fade, and possible brake fluid boiling. Another problem with too-thin rotors is that the pad, caliper, and/or piston will move to an improper position. Two other potential rotor problems are rotor runout and thickness variation, also called parallelism. **Lateral runout** is a condition in which the rotor surface(s) wobbles sideways. It can cause pad "knock back" and a low brake pedal, or pedal pulsation and possible grab (Figure 4-28). Thickness variation is a

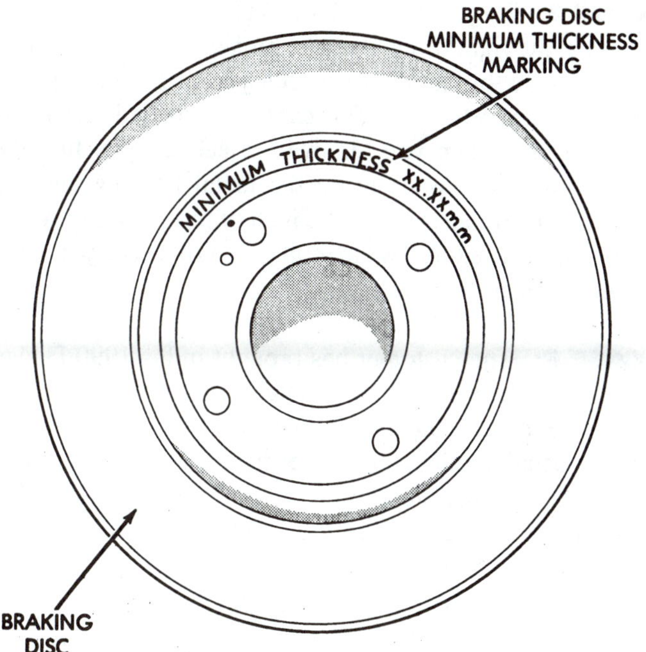

Figure 4-27 All modern rotors are marked with the minimum thickness dimension. (Courtesy of Chrysler Corporation)

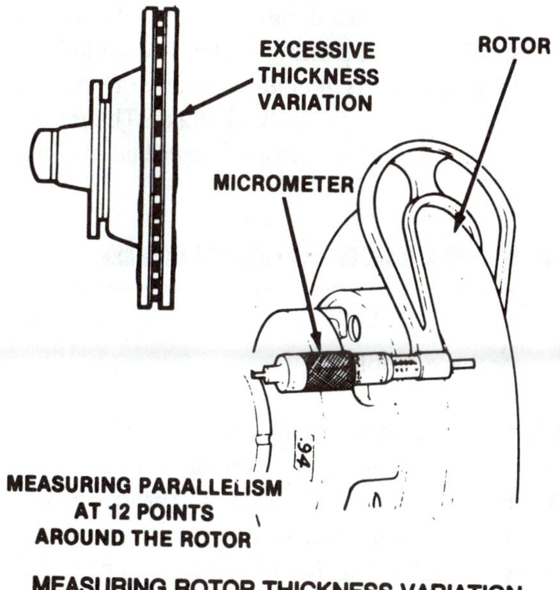

Figure 4-29 A rotor with excessive thickness variation will have different measurements at diferent points on its surface. (Courtesy of Brake Parts, Inc.)

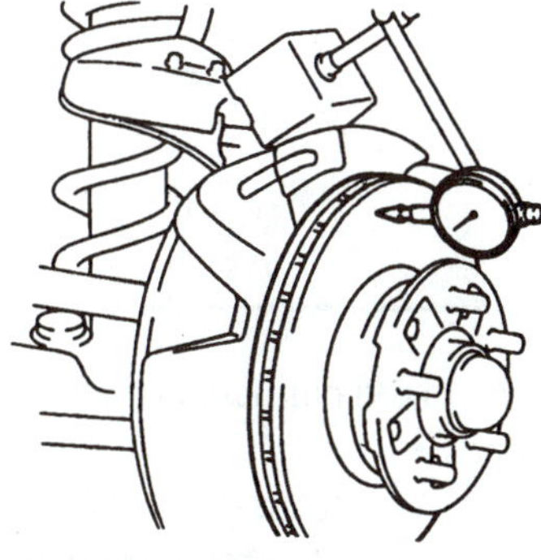

Figure 4-28 A rotor with runout will wobble back and forth laterally. The amount of runout is measured using a dial indicator.

condition in which some parts of the rotor wear faster than others, creating thick and thin areas, which causes pedal pulsations. Thickness variation is probably caused by overtightening or uneven tightening of the wheel lug nuts (Figure 4-29).

4.6 Rear-Wheel Disc Brakes

As mentioned earlier, it is difficult and expensive to incorporate a parking brake into a disc brake assembly. Since the need for equal stopping power is not as great on the rear brakes, most manufacturers choose to install drum brakes at the rear as a cost-saving compromise. A car with rear or four-wheel disc brakes is more expensive to build. Where cost is not a factor, four-wheel disc brakes have become standard or optional equipment. The Corvette has used them since 1965. Four-wheel disc brakes have become standard equipment on many domestic luxury cars and performance-oriented cars since the late 1970s.

Rear-wheel disc brakes are of both fixed- and floating-caliper design (Figure 4-30). They will normally use the same style of caliper that is used at the front. Some rear-wheel calipers are identical to the front calipers except for the piston diameter. Rear caliper pistons are usually

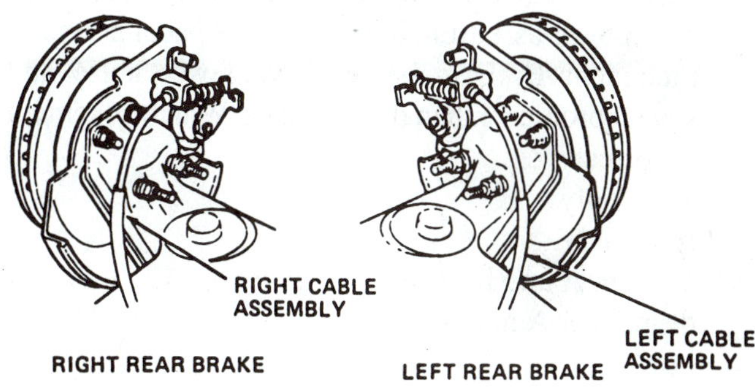

Figure 4-30 These rear-wheel disc brake assemblies use calipers with a mechanical parking brake mechanism. (Courtesy of General Motors Corporation, Service Technology Group)

smaller in diameter. Fixed-caliper units use a small drum brake assembly for a parking brake. Most modern rear-wheel floating calipers include a means of applying the pads mechanically for the parking brake. The mechanical operation of these calipers will be described in Chapter 5.

4.7 Practice Diagnosis

You are working in a brake and front-end shop and encounter these problems:

CASE 1: The customer has brought in her 1994 Honda Accord with a complaint of a very harsh noise at the right front when she applies the brakes. The car has 76,000 miles on it, and your road test confirms the noise. What is probably wrong? What should you do to confirm this?

CASE 2: The ten-year-old Mazda has a loud, unpleasant noise at the right front when the brakes are applied. When you remove the right front wheel, you find the outside of the rotor severely worn. This car has a single piston, floating caliper. When you remove the caliper, you find the outboard pad is wearing on the metal backing and the inside pad still has about 3/32 inch of usable lining. What probably caused this wear problem? What will you need to do to repair this car?

CASE 3: The customer has brought in his 1990 Dodge Dakota pickup in with a complaint of jerky brake pedal when he makes a hard stop. Your road test confirms a pulsating pedal problem that causes the pedal to move up and down during the stop. What is probably wrong? What should you do to confirm this? What will you probably have to do to fix this problem?

Terms to Know

adapter
antirattle clip
antirattle spring
caliper mount
fixed-caliper
floating caliper
guide bolts
guide pins
lateral runout
mounting hardware
pad retention
pad wear indicators
sleeves
sliding calipers
solid rotor
vented rotor
yoke

Review Questions

1. An advantage with disc brakes over drum brakes is that disc brakes
   a. cool better.
   b. have friction surfaces that stay cleaner.
   c. provide better stopping control.
   d. all of the above.
2. Statement A: A disc brake will have the same stopping power as a drum brake with less pedal pressure. Statement B: Disc brakes have fewer noise and squeal problems than drum brakes. Which statement is correct?
   a. A only
   b. B only
   c. both A and B
   d. neither A nor B
3. Statement A: Fixed calipers are bolted solidly to the steering knuckle.
   Statement B: Most fixed-caliper designs use one or two pistons. Which statement is correct?
   a. A only
   b. B only
   c. both A and B
   d. neither A nor B
4. In a floating-caliper design, the
   a. caliper has to move sideways during brake application and release.
   b. piston applies the inboard pad.
   c. outboard pad is applied by caliper reaction to hydraulic pressure.
   d. all of the above.
5. A floating caliper and a sliding caliper operate in the same manner except for the way the
   a. outboard pad is applied.
   b. caliper is located in the mounting bracket.
   c. piston and the inboard pad are connected.
   d. outboard pad is secured in the caliper.

6. Lining clearance of most disc brake units is adjusted by a
   a. starwheel self-adjuster.
   b. special cam.
   c. rubber O-ring.
   d. none of the above.
7. Statement A: Springs are used to release the brake shoes on floating-caliper designs.
   Statement B: The disc brake piston is moved to a released position by the rolling action of a rubber O-ring. Which statement is correct?
   a. A only
   b. B only
   c. both A and B
   d. neither A nor B
8. Statement A: Lining wear indicators produce a squealing noise when they rub on the rotor.
   Statement B: Some pad wear indicators turn on a warning light when the lining is worn. Which statement is correct?
   a. A only
   b. B only
   c. both A and B
   d. neither A nor B
9. Small parts are used with disc brake calipers and pads to
   a. keep the pads from rattling.
   b. retain the pads in the caliper.
   c. help position the caliper.
   d. all of the above.
10. Statement A: On most floating-caliper designs, the inboard pad is secured tightly to the piston.
    Statement B: In most floating-caliper designs, the outboard pad floats when the caliper is released. Which statement is correct?
    a. A only
    b. B only
    c. both A and B
    d. neither A nor B
11. On most floating-caliper brakes, the caliper mounting bracket is bolted securely to the
    a. rear axle.
    b. steering knuckle.
    c. lower control arm.
    d. caliper.
12. Statement A: A vented rotor runs cooler than a solid rotor.
    Statement B: Solid rotors are easier to replace but cost more than vented rotors. Which statement is correct?
    a. A only
    b. B only
    c. both A and B
    d. neither A nor B
13. Potential rotor problems include
    a. lateral runout.
    b. a variation in rotor thickness.
    c. rotors that are worn too thin.
    d. all of the above.
14. Statement A: Cars with four-wheel disc brakes use a special caliper that applies the pads hydraulically for parking brake usage.
    Statement B: Some cars with four-wheel disc brakes use a brake drum and shoes for a parking brake. Which statement is correct?
    a. A only
    b. B only
    c. both A and B
    d. neither A nor B
15. Excessive lateral runout of the rotor surface can cause
    a. pad "knock back."
    b. a low brake pedal.
    c. brake grab.
    d. all of the above.

Parking Brake Theory

Objectives

After completing this chapter, you should:

- ❑ Understand the purpose for which parking brakes are designed.
- ❑ Understand how parking brakes operate.
- ❑ Be familiar with the terms commonly used with parking brakes.
- ❑ Be familiar with the different styles of parking brake control units and how they produce parking brake operation.
- ❑ Be familiar with the different styles of parking brakes used with disc brake assemblies.

5.1 Introduction

The **parking brake** is a mechanically operated brake that is designed to hold the vehicle stationary when parked. The parking brake must be able to hold the vehicle while parked on any grade. In cases where there is poor traction, the parking brake should be able to lock the braked wheels of the parked vehicle. In passenger cars, the parking brake is normally applied by the muscular efforts of the driver using a hand or foot lever. The lever must include a latch or ratchet to hold it in the applied position. Parking brakes may share brake shoes and rotors or drums with the service brakes, but they must use a separate method of application. Also, the parking brake linkage must not interfere with the operation of the service brake.

A parking brake is often called an emergency brake, but this is a misnomer. There is no legislative or manufacturing design requirement concerning the ability of a parking brake to *stop* a car—only a requirement to hold the car after it has been stopped. As mentioned earlier, anyone who has tried to stop a car using only the parking brake realizes how inadequate it is for this purpose.

Potential problems can occur from drag of the parking brake if it is not released completely while driving. This causes premature wear and glazing of the lining, but, even worse, the heat can cause fluid boiling. This, in turn, can cause pedal fade, and *in a diagonal split system, the brake pedal can go to the floor with a total brake loss.*

5.2 Cables and Equalizer

All passenger-car parking brakes operate by a pulling force on a metal cable or rod. This pulling force originates at a hand-operated lever or a foot-operated pedal. The front cable, often called a **control cable**, is attached through an equalizer to a second cable or pair of cables, sometimes called *application cables*. The cable ends are attached to a lever at each brake assembly (Figure 5-1).

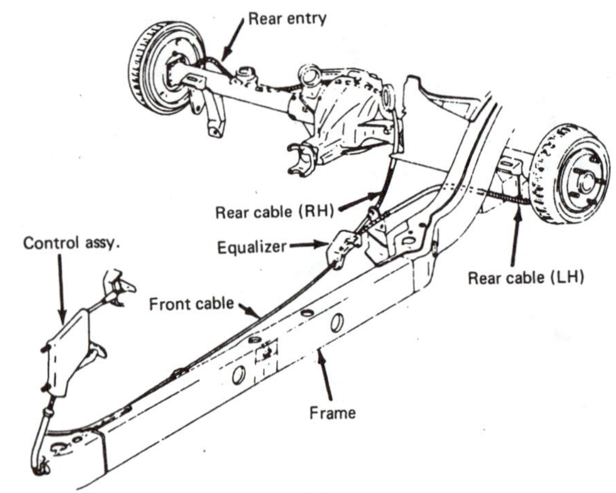

Figure 5-1 A typical RWD parking brake system consists of the control assembly and the cables for transmitting the application force to the rear brake shoes. (Courtesy of General Motors Corporation, Service Technology Group)

The **equalizer** allows equal force and motion to be exerted at each brake. It splits, or equalizes, the operating force so each wheel will have equal brake application.

The cable runs through a flexible *housing* or *conduit* at the axle end. It is usually exposed between the equalizer and the housing end. Metal guides are often used to route the cable around obstacles such as the drive shaft or an exhaust pipe. A cable functions by pulling, and when in operation, it is tight and tries to pull in a straight line. The cable housing permits the cable to operate with enough slack to allow vertical axle movement. In some cases, a single cable is used at the back. This cable starts at one brake, runs through the equalizer, and ends at the other brake. Many cars use a separate cable for each brake with the two cables connected at the equalizer (Figure 5-2).

The equalizer assembly is normally mounted under the floor of the passenger compartment, near the center of the car, under or slightly to the rear of the driver's seat (Figure 5-3). When the parking brake lever is mounted between the front seats, a metal rod is often used to connect it to the equalizer. Occasionally the equalizer is built into the lever assembly. When the lever is mounted under the instrument panel, a control cable and housing are normally used to attach it to the equalizer (Figure 5-4).

In some cars, the equalizer is at the rear axle. The front inner cable runs back to the axle and is connected to the cable from one of the brakes, and the cable housing is connected to the cable from the other brake. The action of pulling the cable produces an equal and opposite reaction in the cable housing (Figure 5-5).

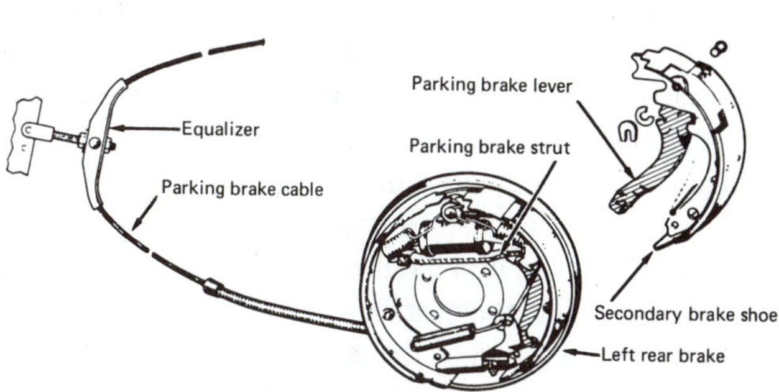

Figure 5-2 Most parking brake cables include an equalizer mechanism that ensures that the same application force is going to each of the rear brakes. (Courtesy of Bendix Brakes, by AlliedSignal)

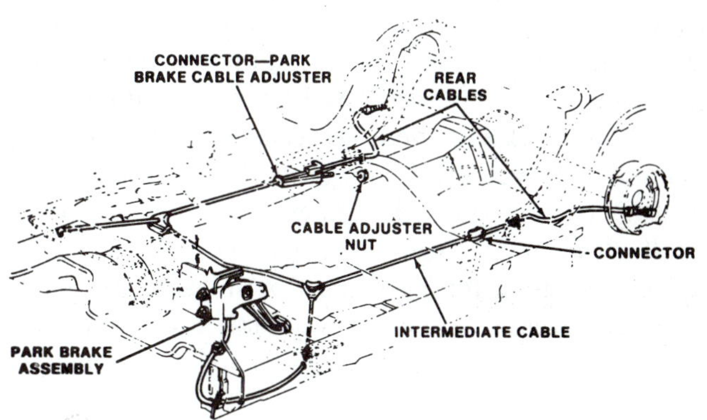

Figure 5-4 This parking brake system uses a foot-operated pedal to pull the front cable that, in turn, pulls the intermediate and then the rear cables. (Courtesy of Chrysler Corporation)

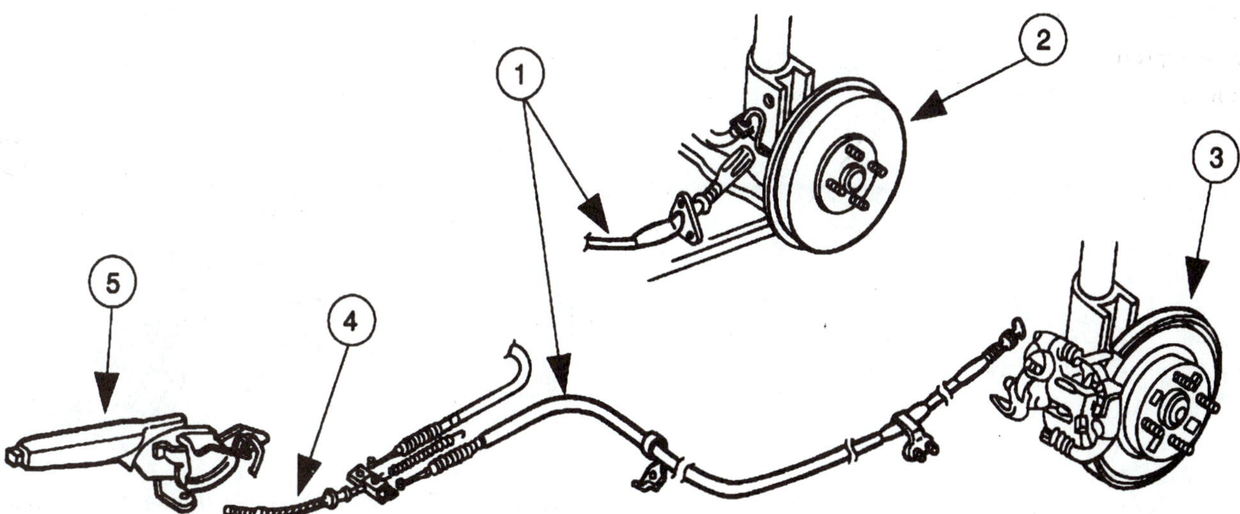

Figure 5-3 This parking brake system uses a hand lever or control (5). It pulls on the front cable (4), equalizer, and rear cables (1) to apply either the rear drum (2) or disc (3) brakes. (Courtesy of Ford Motor Company)

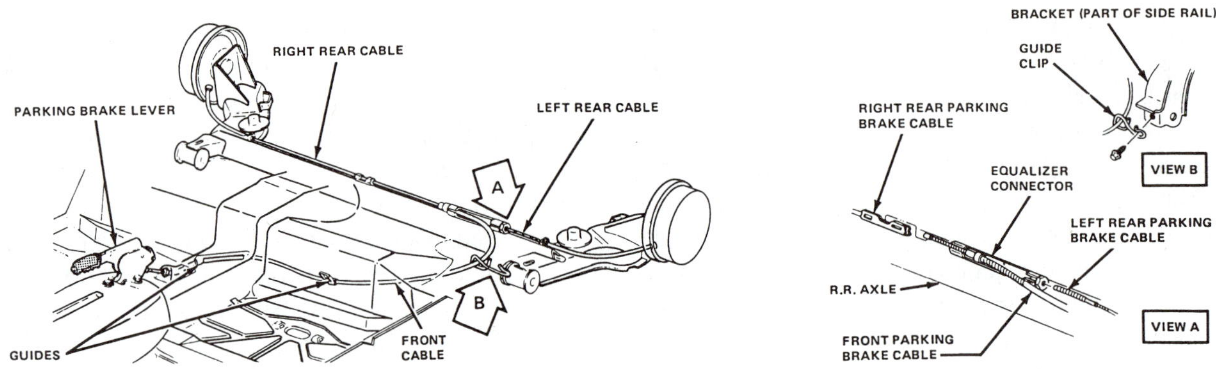

Figure 5-5 This parking brake uses an equalizer at the rear of the car. The inner cable of the front parking brake cable connects to the right rear brake cable, and the housing connects to the left rear brake cable. (Courtesy of General Motors Corporation, Service Technology Group)

5.3 Lever and Warning Light

Every driver should be familiar with the **parking brake lever**. This lever is designed to pull on the parking brake cable when the lever is depressed. As its name implies, the lever provides a leverage or a mechanical advantage that multiplies the driver's force substantially. The lever can be hand- or foot-operated (Figure 5-6). A latch is included in all levers so that they will be automatically held in the applied position; the latch must be released before the lever will release. In some luxury cars, this latch is automatically released when the car is started and shifted into drive or reverse.

Some modern parking brake systems include an **automatic adjuster** at the parking brake lever to remove any slack from the cables. When the lever is in the released position, a pawl is lifted from the ratchet mechanism, and this allows a clock spring to wind up, creating a 19-pound pull on the cables. Pulling the parking brake lever causes the pawl to engage the ratchet so normal action can occur (Figure 5-7).

Most cars include a **warning light** to remind the driver that the parking brake is applied. This light helps prevent the embarrassment of driving a car with the parking brake partially applied, which causes added wear, possible heat damage to the shoes and drums, and possible brake failure. The light can be a separate unit, but it usually shares the same bulb as the brake failure warning light. It is activated by a switch at the parking brake lever. The switch is opened or turned off when the lever is released and closed or turned on when the lever is applied (Figure 5-8). Most switches have a simple and easy adjustment for resetting if necessary (Figure 5-9).

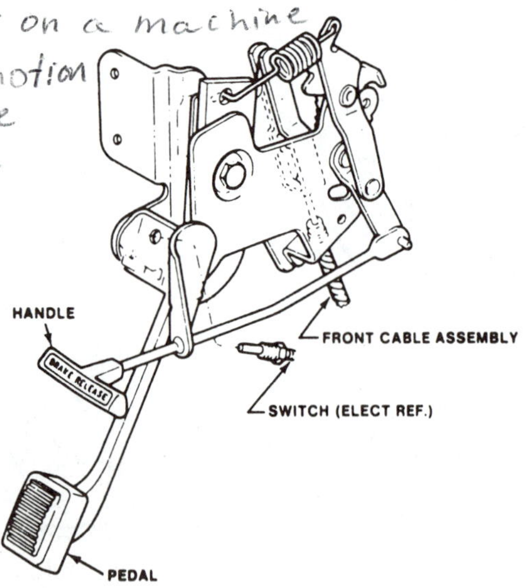

Figure 5-6 Pulling on the handle of this pedal-operated control assembly releases the latch and allows the pedal and parking brakes to release. (Courtesy of Chrysler Corporation)

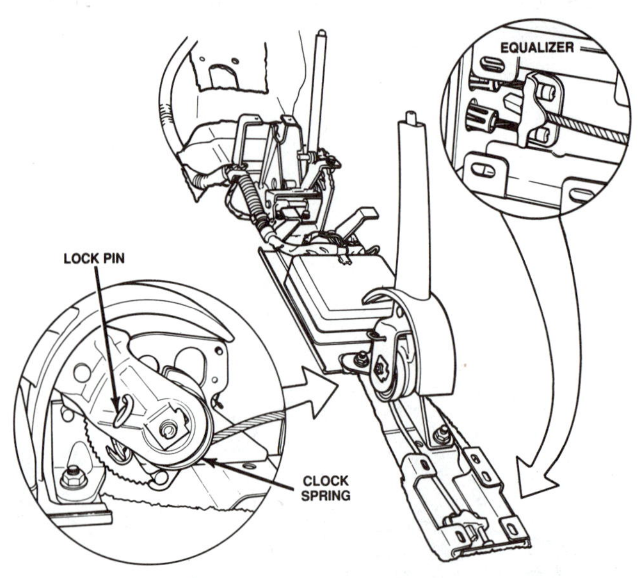

Figure 5-7 This control assembly includes a clock spring and ratchet that automatically take up any cable slack when the parking brake is released. The lock pin is installed to keep the clock spring from unwinding if a cable or control is disassembled. (Courtesy of Chrysler Corporation)

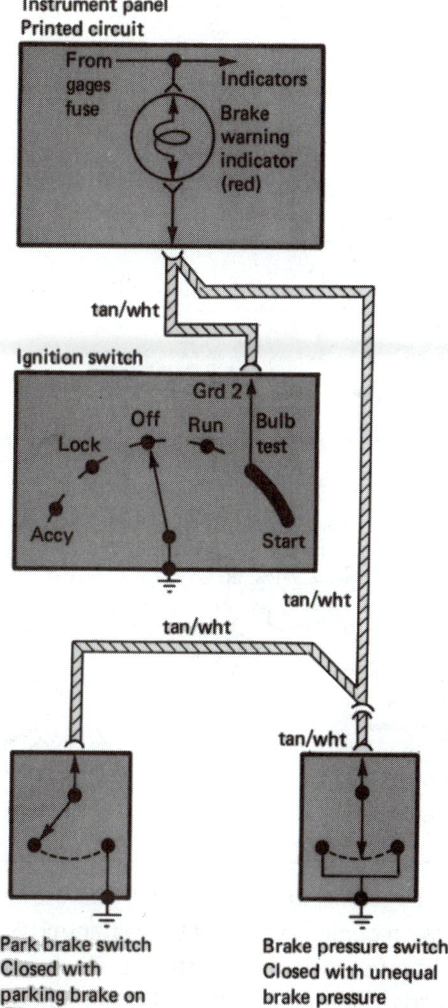

Figure 5-8 in most cars, the parking brake switch is open when the brake is released and is in a parallel circuit with the brake warning light switch. (Courtesy of General Motors Corporation, Service Technology Group)

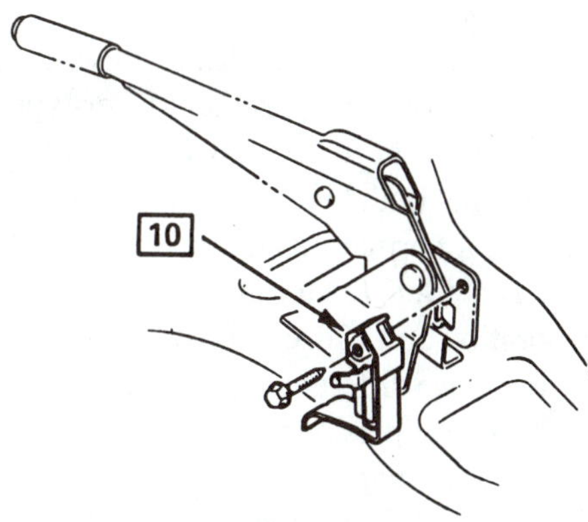

Figure 5-9 The parking brake switch (10) is usually mounted next to the lever or pedal so that movement of the lever will close the switch and turn on the warning light. (Courtesy of General Motors Corporation, Service Technology Group)

5.4 Integral Drum Parking Brake

Most of us are familiar with the common **integral parking brake** mechanism that is added to rear-wheel drum brakes. It consists of a lever that pivots on the web of the rear shoe and a strut that transmits force from the lever to the front shoe (Figure 5-10). Application of the parking brake produces a forward pull from the parking brake cable to the lever. This causes a rearward force at the lever-to-secondary-shoe pivot and a forward force at the strut-to-primary-shoe connection.

When the brakes are released, the shoe return springs return the shoes to the anchors or stops, and the spring at the end of the parking brake cable returns the parking brake lever against the rim of the secondary shoe. At this point, there should be clearance at the ends of the parking brake strut. The coil spring normally positioned between the strut and the primary shoe prevents rattles (Figure 5-11). In many cases, a spring washer is also included at the parking brake lever pivot to eliminate rattles.

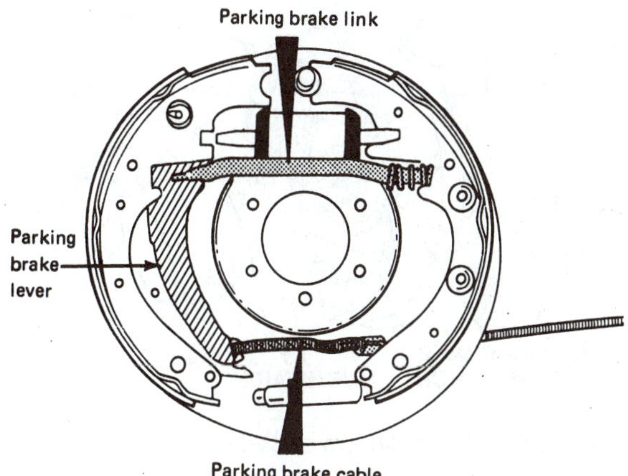

Figure 5-10 When the parking brake is applied, the cable pulls the parking brake lever forward, which pushes the parking brake link/strut to the right and the secondary shoe to the left. Note the spring at the right end of the link to keep the link from rattling while released. (Courtesy of Ford Motor Company)

5.5 Auxiliary Drum Parking Brake

Some cars use separate drum brake assemblies whose only function is as a parking brake (Figure 5-12). This arrangement is found on some vehicles with four-wheel disc brakes. On these cars, the drum for the parking brake is cast as the inner part of the rotor. The drum is usually

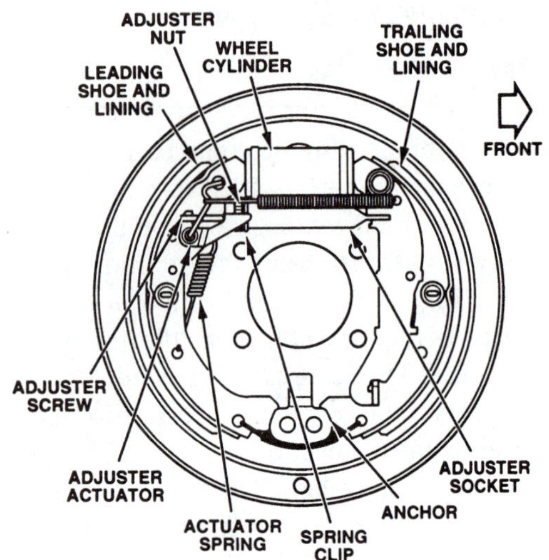

Figure 5-11 When this leading-trailing brake is applied, shoe travel can cause the adjuster actuator to turn the adjuster screw and adjust the brake clearance. The shoes will be right against this adjuster screw and socket while they are released. (Courtesy of General Motors Company, Service Technology Group)

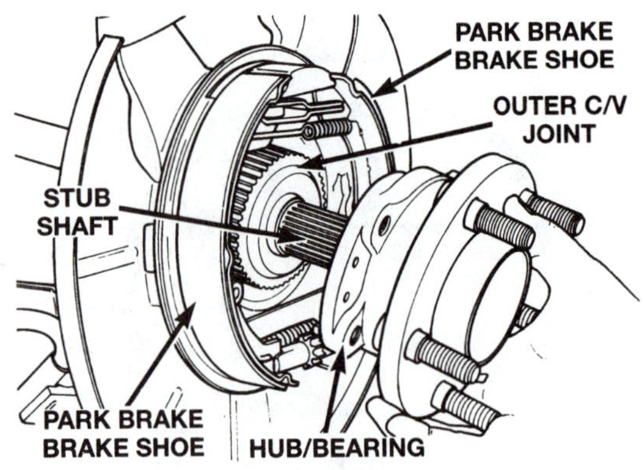

Figure 5-12 This small drum brake assembly is used only for a parking brake on this four-wheel disc brake vehicle. The drum is in the center of the rear rotors. (Courtesy of Chrysler Corporation)

rather small, only 6½ inches (165 mm) on the Corvette, but its only purpose is to hold the wheel stationary. In the past, some cars used a single auxiliary drum assembly mounted at the rear of the transmission. Auxiliary parking brakes are still used on many trucks (Figure 5-13).

Most auxiliary drum brakes are simply small versions of a duo-servo drum brake without the wheel cylinder. The wheel cylinder is replaced by an actuating lever or mechanical cam. Shoe adjustment is accomplished by a threaded starwheel, as in the mechanically adjusted duo-servo brakes of the past. Because the shoes are normally applied while the drum is stopped, there is very little rubbing of the lining. Since there is negligible lining wear, there is little need for a periodic shoe adjustment.

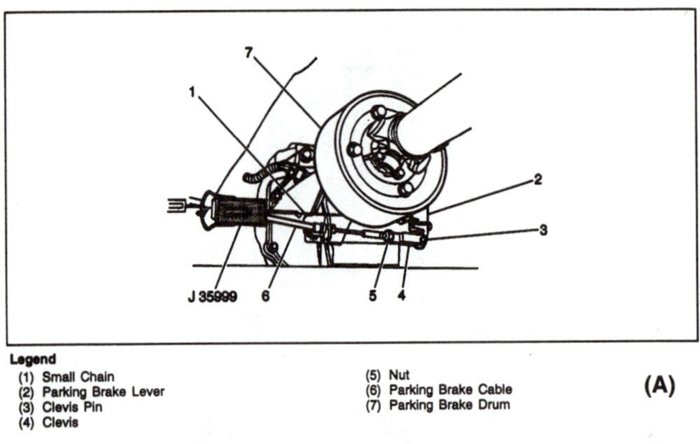

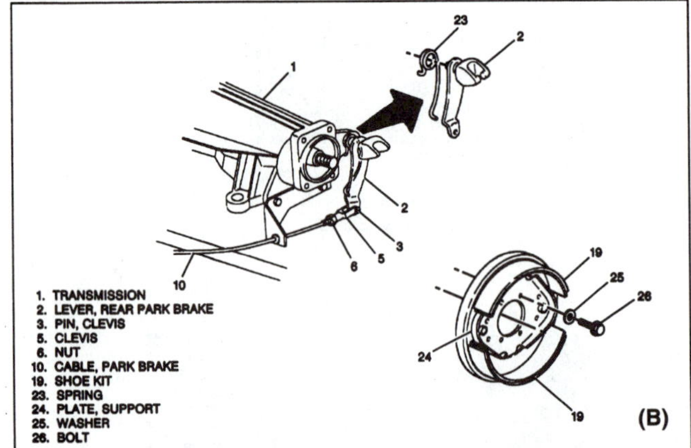

Figure 5-13 Some trucks and older cars use an auxiliary parking brake mounted at the rear of the transmission (A); tool J35999 is being used for an adjustment. (B) shows the cable and inner parts. (Courtesy of General Motors Corporation, Service Technology Group)

5.6 Disc Parking Brake

Compared to the integral drum parking brake, disc parking brake systems are rather complicated. There are essentially three different styles of parking brakes that operate through the brake caliper. Domestic cars use one of two different rotary action or lead screw mechanisms, and some import cars use a cam-and-lever-type arrangement (Figure 5-14). Caliper-type parking brakes must be self-adjusting so that both hydraulic and mechanical operation can occur in a normal manner and not be affected by lining wear. The Delco Moraine Division of General Motors Corporation and Kelsey-Hayes, two major domestic manufacturers, use different approaches in their designs.

Delco Moraine calipers operate and self-adjust hydraulically in the same manner as front, no-parking-brake calipers (Figure 5-15). This operation will be described in the next chapter. When the parking brake lever is applied, the cable movement rotates a lever that is attached to a high lead screw, which in turn will pass through the inboard side of the caliper cylinder. A high lead screw is threaded with a

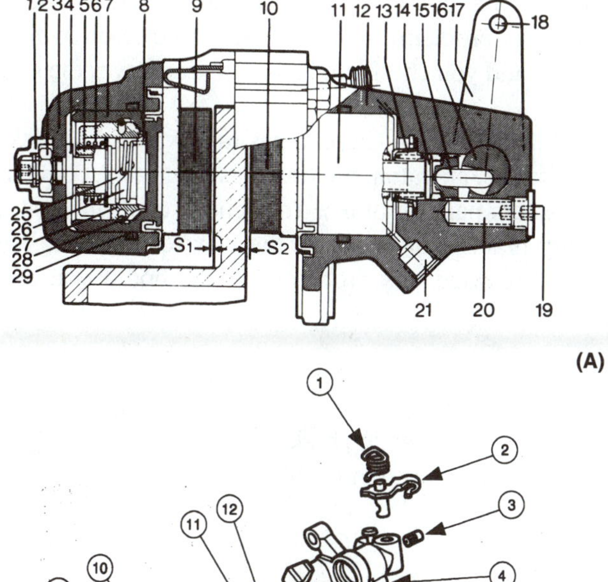

Figure 5-14 A rear fixed caliper (A) and floating caliper (B) showing the mechanical parking brake mechanism. (A is courtesy of ITT Automotive; B is courtesy of Ford Motor Company)

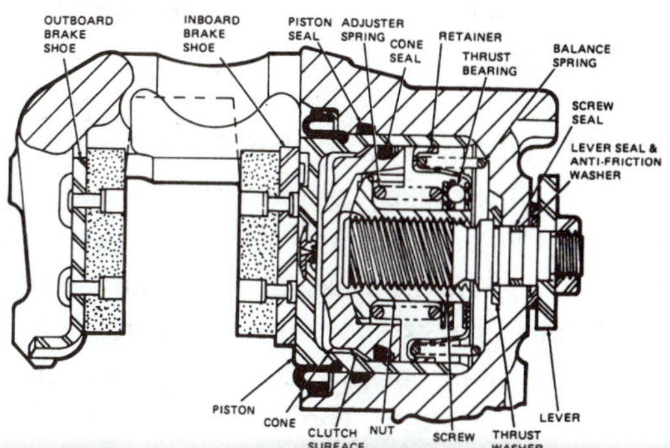

Figure 5-15 This cutaway view of a Delco Moraine rear caliper shows the relationship of the internal parts for both hydraulic service brake and mechanical parking brake operation. (Courtesy of General Motors Corporation, Service Technology Group)

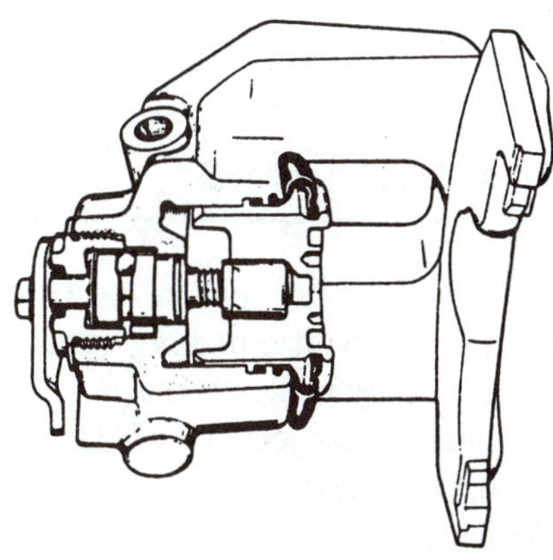

Figure 5-16 This cutaway view of a Kelsey-Hayes rear caliper shows the relationship of the internal parts for both hydraulic service brake and mechanical parking brake operation. (Courtesy of Bendix Brakes, by AlliedSignal)

very high pitch. It threads into an adjusting nut in the caliper piston assembly. Rotation of the high lead screw will cause a sideways movement of the caliper piston and application of the brake pads.

Mechanical self-adjustment occurs during hydraulic application of the piston whenever the piston travels farther than the design allows. When adjustment is necessary, the piston will pull on the high lead screw and adjusting nut. If it pulls hard enough, a gap will develop between the adjusting nut and the cone, which will allow the nut to rotate on the screw threads, changing the position of the nut and thereby making an adjustment. During mechanical application, the high pressure between the nut, the cone, and the piston prevents nut rotation.

Kelsey-Hayes parking brake calipers also use normal operation for hydraulic application and self-adjustment (Figure 5-16). However, the mechanical operation is different from that of the Delco Moraine design. Movement of the parking brake lever at the back of the caliper causes a rotation of the operating shaft. The operating shaft is like a cam in that it has three ramp-shaped detent pockets at its inner face; there are also three pockets at the outer face of the thrust screw. A ball is placed in each of the three pockets between the thrust screw and the operating shaft. The thrust screw is held from rotating by an antirotation pin. When the operating shaft is rotated by the operating lever, the detents in the face of the operating shaft and the thrust screw, plus the three balls, generate an inward motion of the thrust screw. The thrust screw will then push inward on the piston and apply the brake shoes.

Mechanical self-adjustment of this caliper also occurs during hydraulic operation. Whenever the piston travels farther than the preset amount, the pulling force between the piston and the thrust screw will cause the adjuster assembly in the piston to rotate, thereby making the adjustment.

5.7 Practice Diagnosis

You are working in a tire shop that also does brake and front-end repair, and encounter these problems:

CASE 1: While doing a brake inspection on a four-year-old pickup, you find the parking brake pedal goes almost to the floor before the brakes are applied. The rest of the brakes are normal. What is probably wrong? What will you need to do to correct it?

CASE 2: The customer brought in her Nissan Stanza with a complaint that the brakes completely failed; she was able to stop without an accident and noticed a burned smell at the time. After a while the brakes seemed okay so she drove the car in. Your checks show the brake pedal and front brakes to be normal, but the rear brake shoes are badly glazed with some cracking. The shoes and drums also show a bluish-gold color and heat damage. What probably caused this problem? What will you need to do to fix it? What should you check out before the car leaves the shop?

CASE 3: The 1994 Camaro has four-wheel disc brakes, a V-8 engine, an automatic transmission, and a complaint of a useless parking brake. Your check confirms that the parking brake pedal goes all the way to the floor. What probably caused this problem? What should you do to correct it?

Terms to Know

automatic adjuster
control cable
equalizer
integral parking brake
parking brake
parking brake lever
warning light

Review Questions

1. Statement A: Every car sold in the United States must be equipped with a braking system that will hold a car stationary while parked.
   Statement B: Every car sold in the United States must be equipped with a mechanically applied braking system that will stop a car within the distances established by the vehicle code. Which statement is correct?
   a. A only c. both A and B
   b. B only d. neither A nor B
2. A parking brake is normally applied by the driver's hand pressure or foot pressure on a pedal or lever, and the amount of effort is multiplied through
   a. a power booster.
   b. one or more levers.
   c. a series of cables.
   d. electrical magnetism.
3. As the parking brake is applied, the equal application of force at the two brake assemblies is ensured by the
   a. equalizer.
   b. cable guides.
   c. lever arrangements.
   d. all of the above.
4. Statement A: On most modern cars, the parking brake warning-light circuit consists of a separate circuit containing a fuse, a warning light, and a switch.
   Statement B: The parking brake warning-light switch is usually mounted close to the pedal or lever. Which statement is correct?
   a. A only c. both A and B
   b. B only d. neither A nor B
5. In a drum parking brake assembly, the spring at the end of the strut is used to
   a. help apply the brakes.
   b. ensure a complete release.
   c. prevent rattles.
   d. none of the above.
6. When the parking brake is released, the parking brake mechanism is returned to a released position by a spring
   a. around the cable and between the backing plate and the parking brake lever.
   b. between the parking brake lever and the secondary shoe.
   c. at the end of the parking brake strut.
   d. all of the above.
7. Statement A: Some cars with four-wheel disc brakes use small drums and shoes that function only as parking brakes.
   Statement B: Some trucks and older cars use a drum-type parking brake mount on the output shaft of the transmission. Which statement is correct?
   a. A only c. both A and B
   b. B only d. neither A nor B
8. On most domestic cars with four-wheel disc brakes, the parking brake
   a. is applied by a mechanical series of cables and levers.
   b. in an integral part of the rear calipers.
   c. has a self-adjusting mechanism built into the pistons.
   d. all of the above.

6 Hydraulic System Theory

Objectives

After completing this chapter, you should:

- ❑ Understand the purpose of the brake hydraulic system and how it operates.
- ❑ Be familiar with the components used in a brake hydraulic system.
- ❑ Be familiar with the terms commonly used with brake hydraulics.
- ❑ Understand how a hydraulic system can be used to transmit force from a brake pedal to the brake lining.
- ❑ Understand the reason for split hydraulic systems and how the two styles of split systems differ.
- ❑ Be familiar with the different styles of master cylinders used on automobiles.
- ❑ Understand how a master cylinder, wheel cylinder, and brake caliper operate.
- ❑ Be familiar with the purpose and operation of the various valves and switches that are used in a hydraulic brake system.
- ❑ Be familiar with the different types of brake fluid and their different characteristics.

6.1 Introduction

The **hydraulic brake system** is used to apply the brakes. It is designed to do three basic things: transmit motion from the driver's foot to the brake shoes, transmit force along with motion, and multiply that force by varying amounts to the different wheel assemblies (Figure 6-1). These operations could also be performed by a mechanical system of rods, cables, and levers, but hydraulic operation has the definite advantage in that an exactly equal force is applied to both brake assemblies on an axle (Figure 6-2). This equality of hydraulic pressure helps ensure even, straight stops.

The stopping power of the right and left sides of the vehicle is equalized by the hydraulics and the similarity of the brake units. The stopping power of the front and rear of the vehicle is balanced by the relative size of the caliper and wheel cylinder pistons, the type and size of the brake assemblies, and, sometimes, the hydraulic valves.

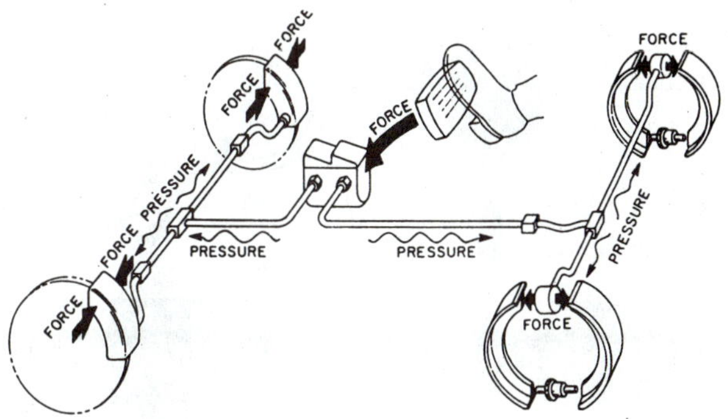

Figure 6-1 In hydraulic brakes, fluid pressure is used to transmit the force from the driver's foot to the brake shoes. (Reprinted from Mitchell Anti-Lock Brake Systems, with permission of Mitchell Repair Information, LLC)

6.2 Hydraulic Principles

Hydraulics, often called *fluid power*, is a method of transmitting motion or force. Hydraulics is based on the fact that liquids can flow easily through complicated

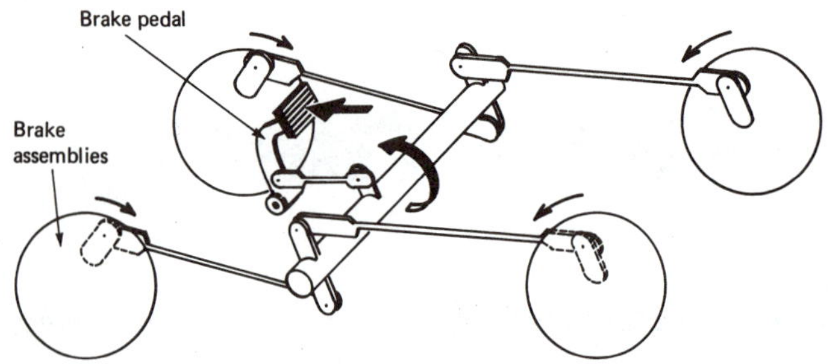

Figure 6-2 Early automobiles used mechanical brakes. A series of metal rods or cables, levers, and a cross shaft connected each of the wheel assemblies to the brake pedal. If the rods were of the wrong length, a brake would apply too early or too late.

paths, yet cannot be compressed (squeezed into a smaller volume) (Figure 6-3). Another important feature is that when liquids transmit pressure, that pressure will be transmitted equally in all directions. This is a simplified version of **Pascal's law** (Figure 6-4).

If we were to fill a strong container with liquid, we would find it impossible to add more liquid, even by force. The only way for more liquid to enter would be for the container to rupture and leak. Once the container is full, any added force on the fluid becomes fluid pressure. Pressure is defined as the amount of force pushing on a certain area. In the United States, pressure is measured using **pounds per square inch (psi)**; 10 psi means a force of 10 pounds acting on an area of 1 square inch. A smaller area would have a smaller force on it, and so on. Traditionally, in Europe and other parts of the world using the metric system, pressure was measured using bar or kilograms per square centimeter (kg/cm^2). Today pressure is also measured in kilopascals (kPa); 1 psi is equal to 6.895 kPa or 0.07 kg/cm^2.

Pressure can enter a hydraulic system in several ways. It is easier to describe and understand if we use a piston as the pressure input and one or more pistons for the output. This is similar to a brake hydraulic system. The amount of pressure in a system is a product of three factors: the ability of the system to contain the pressure, the size or area of the input piston, and the amount of force on the piston. The strength of the system is important, because if the pressure gets too high, the system will rupture and release the pressure. Imagine a brake system with the brake drums removed. We could not develop much fluid pressure because the wheel cylinders could move too far and pop out of their bores. When force is exerted on the piston of a closed system, that force becomes fluid pressure. The amount of pressure will be equal to the force divided by the area of the piston. A 200-pound (90.9 kg) force on a piston that is 1 square inch (6.45 m^2) in area will generate a force of 200 psi (13.8 bar or 14 kg/cm^2). This pressure can be converted to 1,379 kPa by multiplying 200 by 6.895 (Figure 6-5). The amount of pressure is determined by dividing the force by the area of the piston. The same 200-pound force acting on an area of 0.5 square inch (3.2 cm^2) would generate a pressure of 200 + 0.5 or 400 psi (27.6 bar or 28 kg/cm^2; 2,758 kPa).

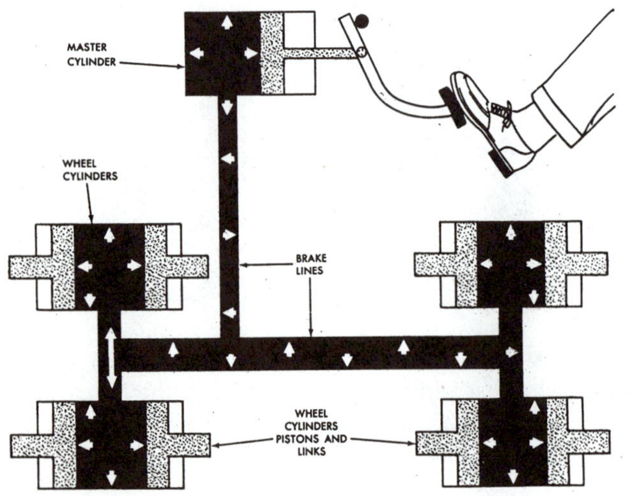

Figure 6-3 Fluid is forced out of the master cylinder as the driver applies the brake. The fluid will transmit the pressure equally throughout the system. (Courtesy of General Motors Corporation, Service Technology Group)

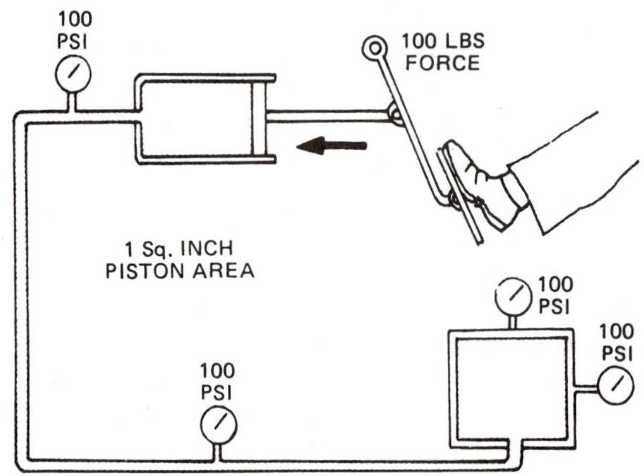

Figure 6-4 Fluid transmits pressure equally throughout a hydraulic circuit and acts with an equal force on each surface of the same size. This is an example of Pascal's law. (Courtesy of Wagner Brake)

400. PSI
.5 | 200.0

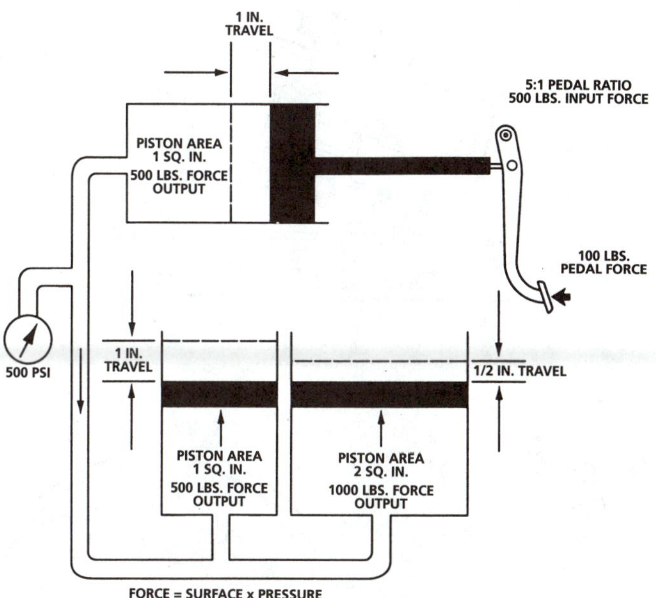

Figure 6-5 Dividing the application force by the area of the master cylinder piston tells us the fluid pressure. Multiplying fluid pressure by the output piston area tells us the output force. In this example, 500 pounds of force at the master cylinder piston will produce 500 psi (500 ÷ 1 = 500). The smaller output piston will exert 500 pounds of force (500 × 1 = 500), and the larger piston will exert 1,000 pounds of force (500 × 2 = 1,000). (Courtesy of General Motors Corporation, Service Technology Group)

When discussing hydraulic pistons and computing fluid pressures and forces, it is important to use the area rather than the diameter. The area of a piston or a circle can be easily determined using either of the following formulas:

$$\textbf{Area} = \pi \mathbf{r}^2 \quad \text{or} \quad \textbf{Area} = 0.785\mathbf{d}^2$$

where π (*pi*) = 3.1416
r = one-half diameter
d = diameter

The pressure in a hydraulic system becomes a force to produce work and make things move. The amount of force can be determined by multiplying the area of the output piston by the pressure. A pressure of 200 psi pushing on a piston with an area of 1 square inch will produce a force of 200 pounds. The same pressure on a 4-square-inch piston (26.17 cm^2) will produce a force of 800 pounds (363.6 kg) (200 × 4 = 800). The application force is multiplied whenever the output piston is larger than the input piston. Force is divided or made smaller if the input piston is larger (Figure 6-6).

The output piston motion is also related to the input piston. It should be remembered that a hydraulic system can only transmit the energy, force, and motion that are put into it; it cannot create energy. What goes in at one place is all that will come out at another. It is possible, however, to change force to motion and vice versa. Most brake systems multiply and increase force with a loss in motion, but some systems increase motion. As the input piston moves, it will displace or push fluid through the tubing to the system. The amount of fluid displaced will be equal to the piston area times the length of the piston stroke. A piston with a diameter of 1 inch (2.54 cm) has an area of 0.785 square inches (5.067 cm^2). If the piston strokes 2 inches (5.08 cm), 1.57 cubic inches (25.7 cm^3) of fluid will be displaced (0.875 in.^{2} × 2) (5.067 cm^2 × 5.08). This much fluid can move a 1-inch-diameter piston 2 inches. A piston with a diameter of 3 inches (7.62 cm) has an area of 7.07 square inches (45.6 cm^2). It will travel a distance of 0.22 inch—1.57 cubic inches ÷ 7.07 square inches = 0.22 inch (0.558 cm—25.7 cm^3 + 45.6 cm^2 = 0.558 cm) (Figure 6-7).

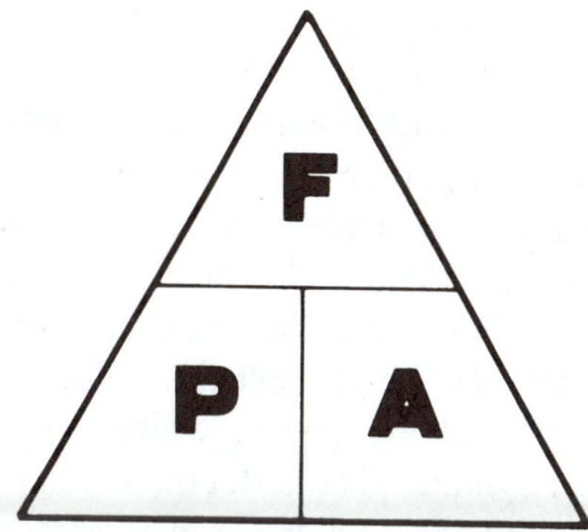

Figure 6-6 A memory triangle for working with hydraulic forces and pressures. Cover up the unknown quantity—F for force, P for pressure, or A for area—and the remainder of the triangle will indicate the solution. For example, P = F ÷ A.

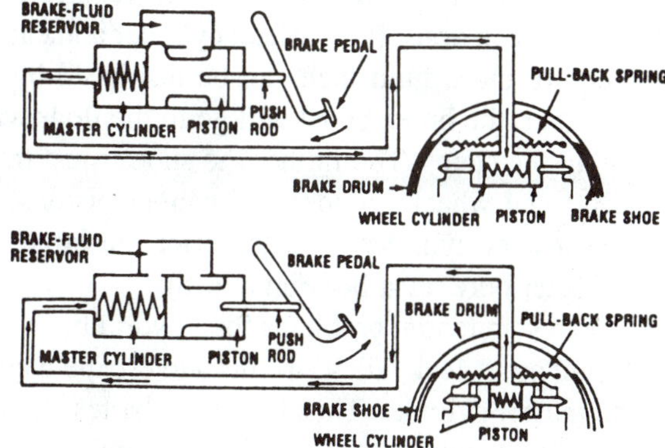

Figure 6-7 When the brakes are applied (top), fluid is displaced out of the master cylinder and into the wheel cylinders and calipers. When the brakes are released (bottom), the brake shoe pull-back/return springs force the fluid back to the master cylinder. (Courtesy of Chrysler Corporation)

A typical hydraulic brake system on a domestic passenger car will have a master cylinder piston with a diameter of about 1 inch (2.54 cm), two disc brake caliper pistons with diameters of about 2.5 inches (6.35 cm) each, and two rear-wheel cylinders with diameters of about 7/8 or 0.875 inches (2.22 cm) each. The actual diameters will depend on the weight of the car, whether power brakes are used or not, and the weight balance of the car. Heavier cars need more braking power, which often means a smaller master cylinder piston. But the smaller amount of fluid displacement will give only a small amount of brake shoe movement. Smaller cars can use a larger diameter piston in the master cylinder, which will displace more fluid and give more wheel cylinder and caliper piston movement. A typical car with a master cylinder diameter of 1 inch (2.54 cm)(0.78 sq in.)(5 cm^2) and a 6-to-1 brake pedal ratio will develop a fluid pressure of 382 psi (2,634 kPa) under a pedal pressure of 50 pounds (22.7 kg). If the master cylinder bore is increased to 1⅛ inch (2.8 cm), the area will increase to 0.99 square inches (6.4 cm^2), and this area increase will drop the fluid pressure to 303 psi (2,089 kPa) at the same pedal pressure.

Cars with power boosters will often use larger-diameter master cylinders. Larger wheel cylinder and caliper pistons can also be used to increase braking power, whereas smaller pistons will increase piston and brake shoe travel. When it is necessary to increase the rear-wheel braking power relative to the front, larger-diameter wheel cylinders will be used. Calipers with larger-diameter pistons will be used when it is necessary to increase the relative power of the front brakes. The relative power between the front and rear brakes is called balance.

If the size of both the front and the rear pistons is increased to gain more application force, the brake engineer must remember that the increase will require more fluid to move the pistons. If the wheel cylinder and caliper pistons require more fluid than can be displaced by the master cylinder, the brake pedal will go to the floor without applying the shoes. The farther the shoes must travel or the larger the wheel cylinder and caliper pistons, the farther the master cylinder must travel or the larger the master cylinder bore must be (Figure 6-8).

The greatest potential problems for a hydraulic system are leaks and air. A leak will allow fluid to escape when it leaves the master cylinder. Thus there will be less fluid to apply the wheel cylinder and caliper pistons, and the pedal will sink to the floor. Air is compressible; its volume will decrease under pressure. This volume reduction takes place at the expense of fluid displacement and will be experienced by the driver as a "spongy" brake pedal. The expansion and contraction of air gives a springiness to the brake pedal (Figure 6-9).

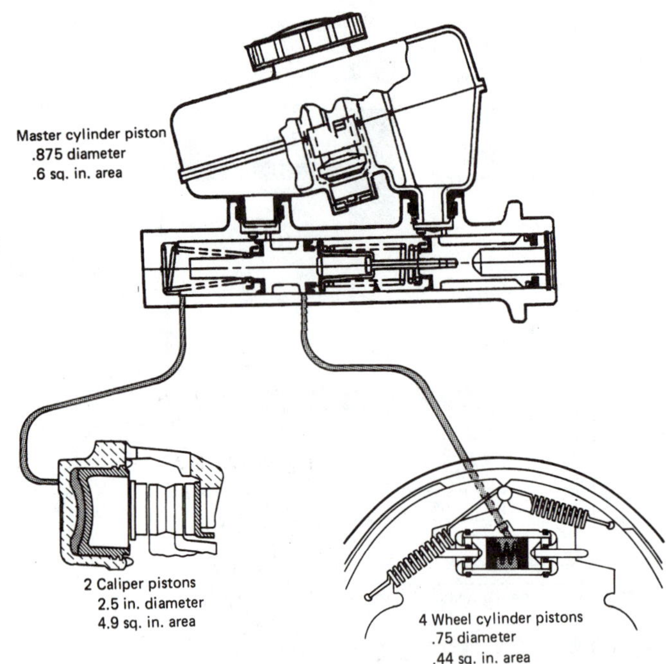

Figure 6-8 A typical manual brake master cylinder on an RWD, full-sized car has a bore diameter of 0.875 inches. In one inch of travel, the pistons will displace 2 × 0.6 cubic inches of fluid. The two caliper pistons and four wheel cylinder pistons have a combined area of 10.68 square inches. If they all moved an equal distance, 1.2 cubic inches of fluid would move each of them 0.10 inches.

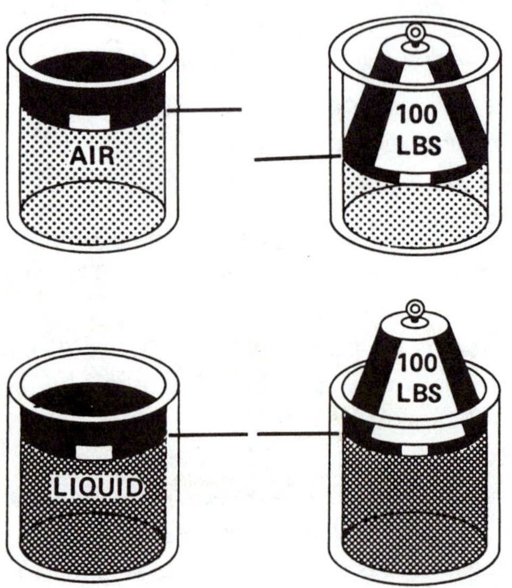

Figure 6-9 Air is compressible; its volume will become smaller under pressure, whereas the volume of a fluid will remain the same because it is non-compressible. (Courtesy of General Motors Corporation, Service Technology Group)

6.3 Master Cylinder Basics and Components

The **master cylinder** is the input piston for the car's brake system. It is connected to the brake pedal so that movement of the brake pedal is transmitted to the master cylinder piston by a pushrod. The **brake pedal** is a simple lever. The ratio is usually about 6 to 1 or 7 to 1. This means that the application force from the foot will be multiplied six or seven times, but the motion of the foot will be reduced by the same amount. With a 6 to 1 ratio, 50 pounds (22.7 kg) of pedal pressure will produce 300 pounds (136.4 kg) of pressure on the master cylinder piston, but it will take 6 inches (15.24 cm) of pedal motion to produce 1 inch (2.54 cm) of motion at the piston.

The first master cylinders used a single piston and cylinder bore with one outlet at the end of the bore (Figure 6-10). From this outlet, tubing branched out to each of the wheel cylinders. The master cylinder body has two major areas, the piston and cylinder bore and the reservoir. These two areas are connected by two passages, a small **compensating port** and a larger **bypass port**. The bypass port is also called an *intake* or *replenishing port*. The cylinder bore is a smooth, precision-sized bore in which the piston and seals slide. The piston moves inward when the brake pedal is pushed. A pushrod connects the brake pedal to the piston. When there is no pedal pressure, the piston return spring pushes the piston against a retaining ring or clip at the end of the cylinder bore. The piston uses two seals called cups: a **primary cup** at the inner face of the piston and a **secondary cup** at the outer end. The primary cup pumps fluid, and the secondary cup keeps fluid from leaking out of the end of the cylinder. The two passages between the cylinder bore and the reservoir are located close to the primary cup. The compensating port is just in front of it, and the bypass port is behind it (Figure 6-11).

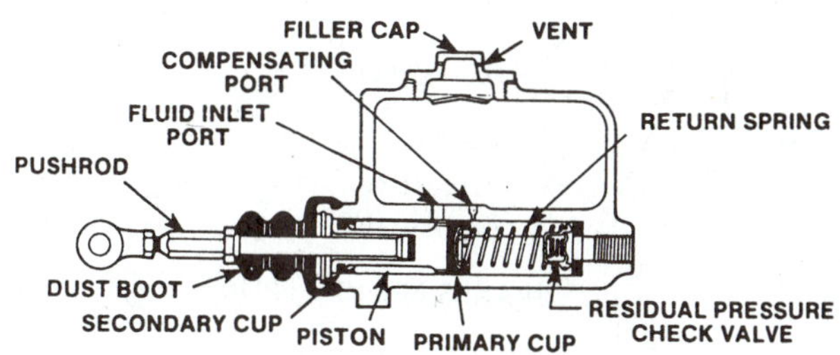

Figure 6-10 A cutaway view of a single master cylinder in the released position. Note the position of the primary cup relative to the compensating port. (Courtesy of Bendix Brakes, by AlliedSignal)

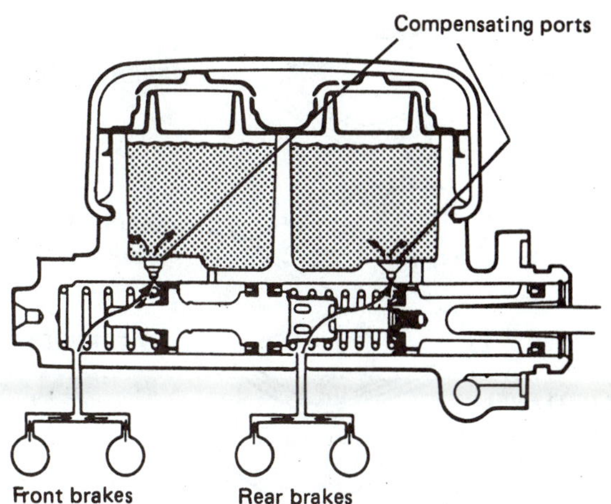

Figure 6-11 If the fluid in the calipers and wheel cylinders heats up and expands, the increased volume will flow through the compensating ports to the reservoir. (Courtesy of General Motors Corporation, Service Technology Group)

6.4 Master Cylinder Basic Operation

In a normal brake system, the whole system is interconnected and filled with fluid up to the upper part of the master cylinder reservoir. The reservoir is vented to the atmosphere so fluid can expand and contract and still stay at atmospheric pressure. When the fluid in the wheel cylinders and calipers gets hot it can expand and move through the open compensating port to the reservoir. When it cools and contracts, it can move the other way, keeping the system filled with fluid. Modern master cylinders use a rubber diaphragm over the reservoirs to separate the fluid from air and to help reduce fluid contamination (Figure 6-12).

When sufficient force is placed on the brake pedal to overcome the piston return spring, the master cylinder piston will move inward. Fluid from the bore will be displaced to the reservoir until the lip of the primary cup moves past and closes the compensating port. From this point further force and movement of the brake pedal will displace fluid to the system and move the brake shoes into contact with the rotors or drums (Figure 6-13). Lining contact will stop further movement of the shoes and wheel cylinder or caliper pistons. From this point, system pressure will increase. Any increase in pedal force will cause a corresponding increase in wheel cylinder and lining pressure.

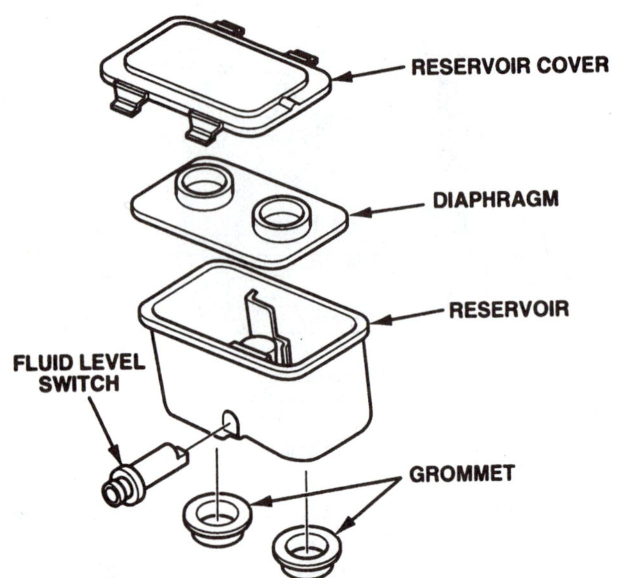

Figure 6-12 The rubber diaphragm under the reservoir cover helps keep the fluid clean and free from water while allowing atmospheric pressure to act on the fluid. (Courtesy of General Motors Corporation, Service Technology Group)

When the brake pedal is released, the piston return spring will move the piston very rapidly back to its stop at the retaining ring—much faster than the fluid will return from the wheel cylinder or calipers. During this motion, the primary cup will collapse or relax its wall pressure slightly and move through the fluid. There will be a flow of fluid past the edges of the primary cup (Figure 6-14). Some pistons have a series of holes in the primary cup face to improve this flow. The flow past the primary cup allows more fluid to pump into the system if the brake pedal is pumped or cycled rapidly. It also prevents low fluid pressure or a vacuum during brake release. A vacuum might cause air to enter the system past the secondary cup of the master cylinder piston or past one of the wheel cylinder cups.

6.5 Master Cylinder Construction

In the past, master cylinder bodies were made of cast iron. This strong material has a relatively low cost, can be cast into complex shapes, and is easily machined. Today, because of the need to reduce vehicle weight to improve fuel mileage, most master cylinders are made from cast aluminum and plastics. These materials can also be cast into complex shapes and are easily machined.

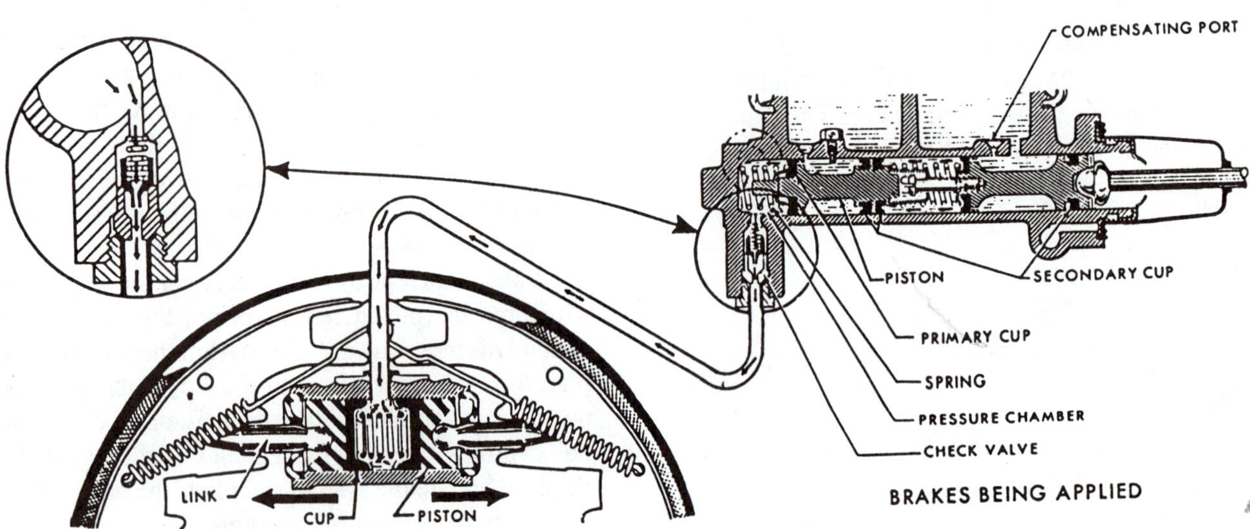

Figure 6-13 As soon as the primary cup moves past the compensating port during brake application, fluid will be forced out of the master cylinder. If a residual valve is used, fluid will flow through the valve and out to the wheel cylinder and caliper pistons. (Courtesy General Motors Corporation, Service Technology Group)

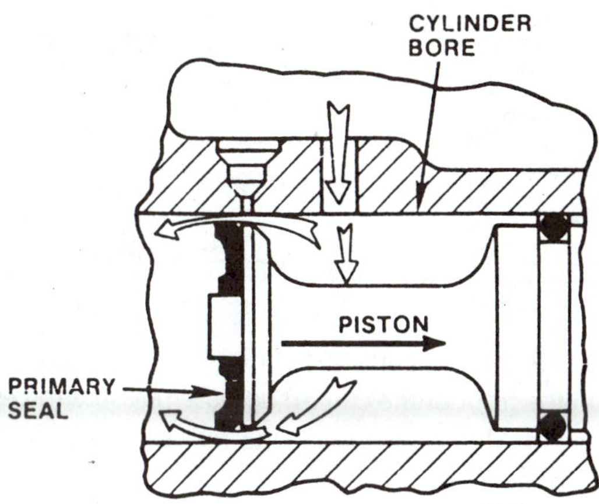

Figure 6-14 During master cylinder release, fluid will move from the reservoir through the bypass port and past the primary seal. (Courtesy of Bendix Brakes, by AlliedSignal)

Figure 6-15 A cast-iron master or wheel cylinder bore is given a very exacting finish by being "bearingized." The final manufacturing step is to force a hardened steel ball through the bore to smooth out any imperfections. (Courtesy of Brake Parts, Inc.)

The cylinder bore of a master cylinder has the following critical requirements. It must be strong enough to contain the pressure; it must be of an exact size, no larger than 0.006 inches (0.15 mm) greater than the piston diameter; it must be round and straight; and it must be smooth. In production, cylinder bores are given their final size by **roller burnishing**. A hardened-steel, precision-sized roller is forced through the bore that has been machined to a slightly undersized diameter. As the burnishing tool passes through the bore, the machine scratches are smoothed out as the bore is expanded to the correct size. This action also produces a harder bore because of the work-hardening of the metal (Figure 6-15). Aluminum bores are **anodized** to harden the bore after machining. Anodizing is an electro-chemical process that helps the normally soft aluminum resist corrosion, galling, and wear. The hard anodized layer is extremely thin, however, and when this layer wears through, the soft aluminum underneath will be exposed to rapid corrosion and wear.

The **reservoir** can be cast into the master cylinder body or attached to it by other means. Many modern units use a plastic or nylon reservoir that is clamped onto the body or connected to it by a pair of rubber grommets (Figure 6-16). In congested engine compartments, the reservoir can be mounted in a remote location and connected to the cylinder body by rubber or metal tubing (Figure 6-17).

Master cylinder pistons are cast from aluminum. They are often anodized to resist corrosion. They usually have one flat face for the primary cup and a groove for the ring-

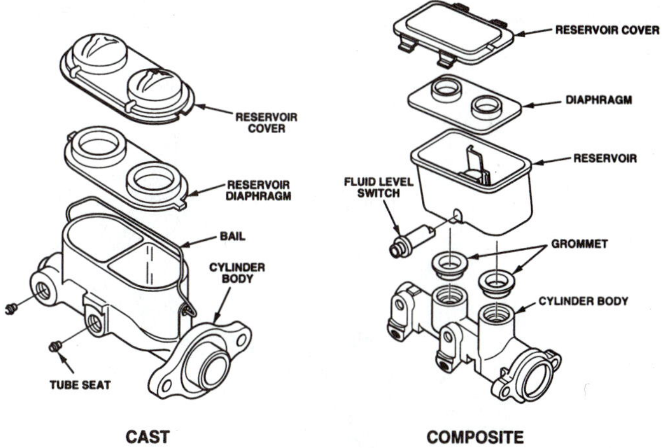

Figure 6-16 Most modern master cylinders (above right) use a cast-aluminum body with a plastic or nylon reservoir. (Courtesy of General Motors Corporation, Service Technology Group)

shaped secondary cup. The secondary end of the piston has a provision for the pushrod (Figure 6-18).

At one time, **cups** were made from natural rubber; now synthetic rubber is used. A rubber cup is a somewhat amazing sealing device. One expert has determined that the average American driver uses the brakes about 75,000 times a year. Each application requires that the master cylinder cups slide down the bore, and while this is occurring, no fluid or fluid pressure should escape past the cup. Normally, some wear of the cup occurs. This causes a blackish coloration of the fluid and residue at the bottom of the master cylinder reservoir. A cup is especially good at sealing pressure because of its shape. Pressure pushes the lips of the cups tightly against the bore, which improves the seal. The greater the pressure, the tighter the

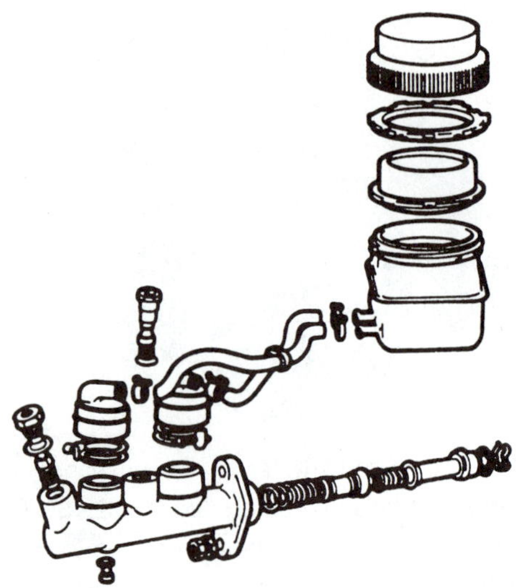

Figure 6-17 The reservoir of this master cylinder is remotely mounted and connected to the cylinder body by a pair of hoses. (Courtesy of Bendix Brakes, by AlliedSignal)

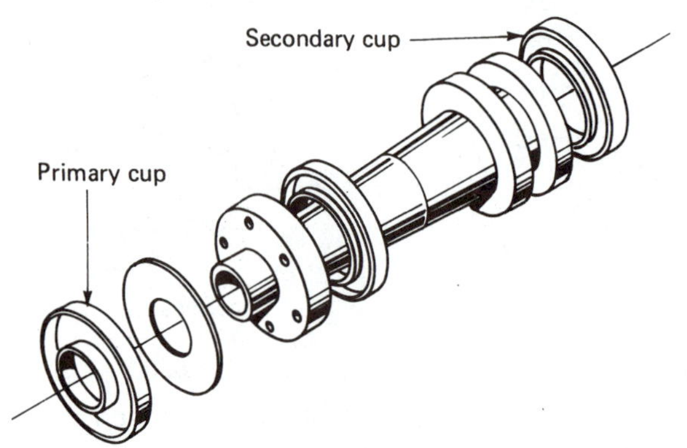

Figure 6-18 Some master cylinder piston faces are drilled so fluid can bypass the primary cup during release. Ring-type secondary cups are normally used.

seal. The primary cup has to slide down the bore as the system's pressure increases. There is very little pressure on the cup lips when the brakes are released. A cup is a single-direction seal, somewhat like a check valve. Pressure from the front will seal the lips tightly to the bore, but pressure from the back will cause flow past the cup. This can be a desirable part of the design as with a primary cup during release. It can be a problem if air enters past a cup at a wheel cylinder (Figure 6-19).

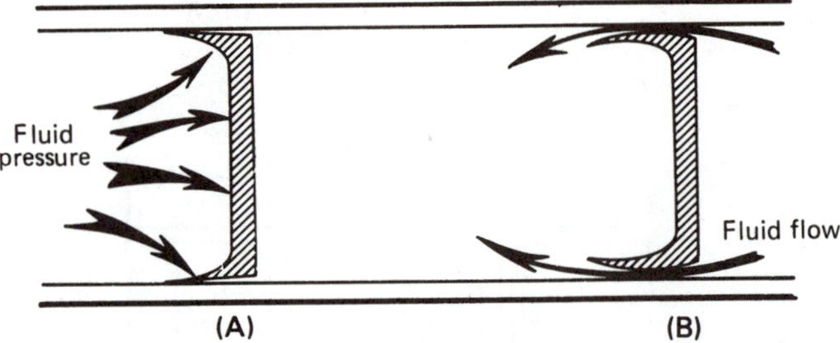

Figure 6-19 Rubber cups seal tighter when fluid pressure is exerted on the front of the cup (A) because pressure forces the lips against the bore. Pressure at the rear of the cup (B) will move the lips away from the bore.

6.6 Residual Pressure

At one time, each master cylinder contained one or two **residual pressure check valves.** A single master cylinder uses one valve placed at the outlet end of the cylinder bore (Figure 6-20). A **tandem master cylinder** has a valve under one or both outlet tube seats (Figure 6-21).

These valves are designed to allow a free flow of fluid out of the master cylinder but will stop some of the fluid from returning. They will shut off the return flow when the system pressure drops down to about 5 to 25 psi (34.5 to 172 kPa). This pressure stays in the tubing and wheel cylinders of the system while the brakes are released.

At one time, residual pressure was thought necessary to keep air from entering the system past the wheel cylinder cups. This pressure also kept the wheel cylinder cups against the pistons, the piston next to the pushrods, and the pushrods next to the brake shoes, so that there would be no lag or lost motion during brake application. Today the **cup expanders** and the springs in the wheel cylinders take care of this duty and residual pressure valves are no

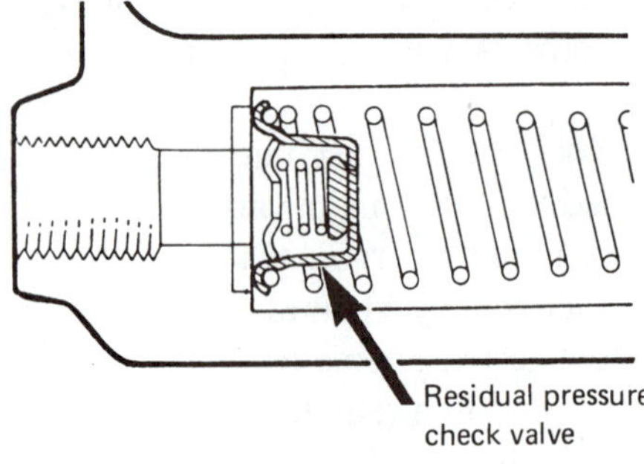

Figure 6-20 Most single master cylinders use a residual check valve at the outlet end of the cylinder bore. (Reprinted from Mitchell Anti-Lock Brake Systems, with permission of Mitchell Repair Information, LLC)

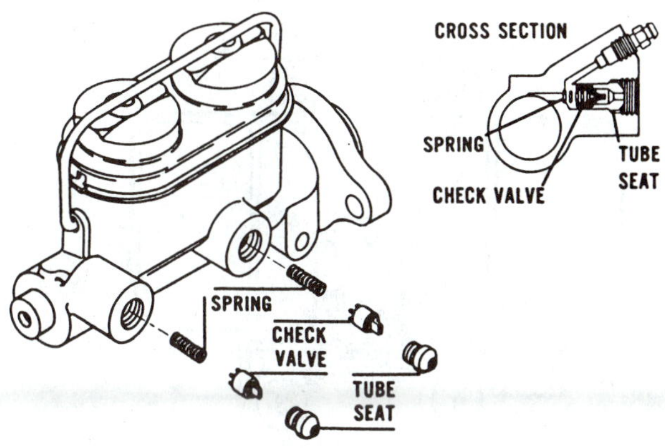

Figure 6-21 Residual check valves for a tandem master cylinder are mounted under the outlet port tube seats. Fluid flows through the center of the check valve on the way out and around the check valve on the way back in. (Courtesy of General Motors Corporation, Service Technology Group)

longer commonly used. Residual pressure is never used with disc brakes. It would provide enough pressure for partial application and cause lining drag.

6.7 Dual Master Cylinder Construction

One major drawback with a brake hydraulic system is that a sudden failure can result from a major fluid leak. If a tube, hose, wheel cylinder, or other component should rupture, there will be no braking power. One solution for this problem is to split the hydraulic system into two parts. This became a requirement for all cars sold in the United States in 1967. The hydraulic system and master cylinder were divided so that the front brakes operate from one piston of the master cylinder and the rear brakes operate from the other (Figure 6-22).

A dual **tandem master cylinder** has two pistons in tandem—one in front of the other—that operate in the same cylinder bore. The **primary piston** operates mechanically, directly from the pushrod, and the **secondary piston** operates hydraulically, from the hydraulic pressure in the primary section. Normally, there will be the same hydraulic pressure on each side, front and rear, of the secondary piston. The primary piston has two rubber cups, a primary and a secondary, just like the piston in a single master cylinder. The secondary piston usually has three cups, a primary cup and two secondary cups (Figure 6-23). Some designs use a secondary sealing ring in place of one of the cups. The primary cup of the secondary piston, like the other primary cups, is a pumping cup. The secondary cups on the secondary piston are slightly different. One prevents a fluid leak from the secondary system into the primary, and the other prevents a high-pressure fluid leak from the primary system into the secondary reservoir. The lip of the latter cup faces the primary piston, and all of the other cups face the other way (Figure 6-24). Some new master cylinders use slightly different secondary cup arrangements on the secondary piston.

A tandem master cylinder bore will have a compensating port and a bypass port for each piston. The reservoir is

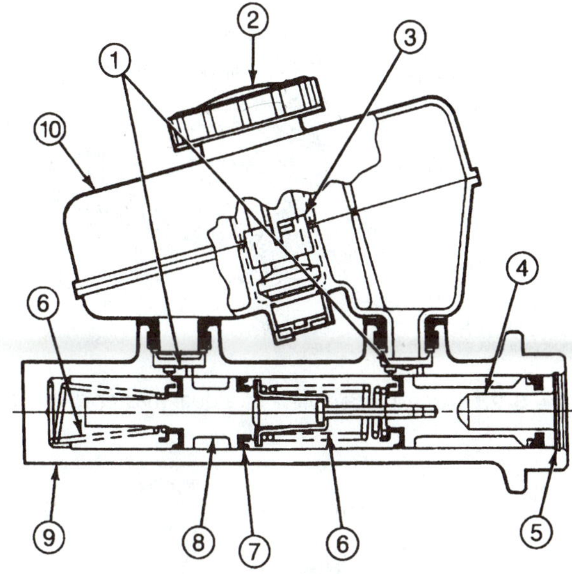

Item	Part Number	Description
1	—	Compensating Ports (Part of 2140)
2	2162	Brake Master Cylinder Filler Cap
3	—	Float Magnet Assembly (Part of 2K478)
4	—	Primary Piston (Part of 2140)
5	—	Bore End Seal (Part of 2140)
6	—	Spring (Part of 2140)
7	—	Seal (Part of 2140)
8	—	Secondary Piston (Part of 2140)
9	2140	Brake Master Cylinder
10	2K478	Brake Master Cylinder Reservoir

Figure 6-22 A cutaway view of a dual master cylinder showing the relative position of the internal parts. (Courtesy of Ford Motor Company)

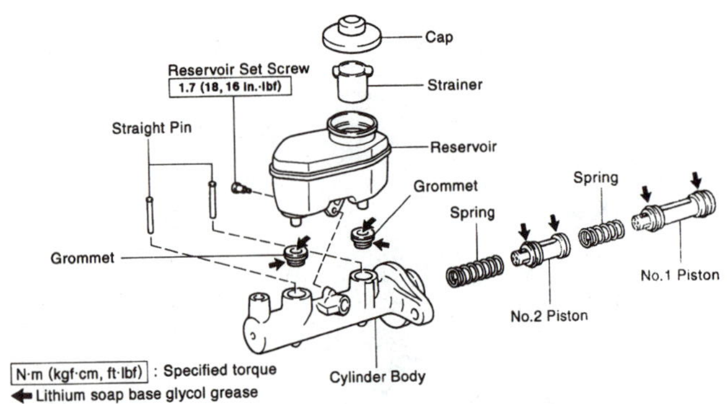

Figure 6-23 An exploded view of a dual master cylinder.

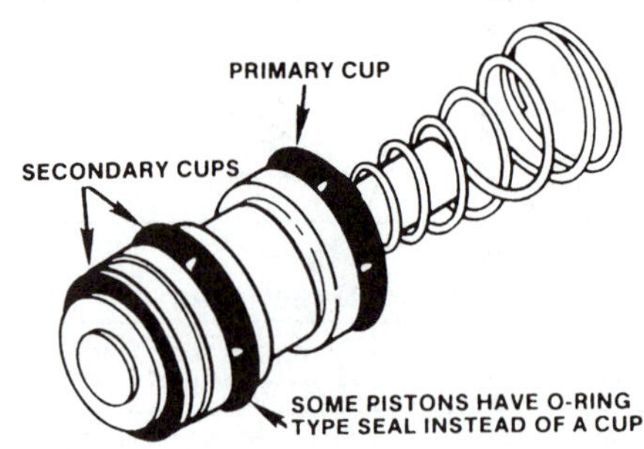

Figure 6-24 The secondary piston assembly from a dual master cylinder. Note the lip of the rear cup (left) faces to the left and the other two face to the right, toward the fluid to be sealed. (Courtesy of Bendix Brakes, byAlliedSignal)

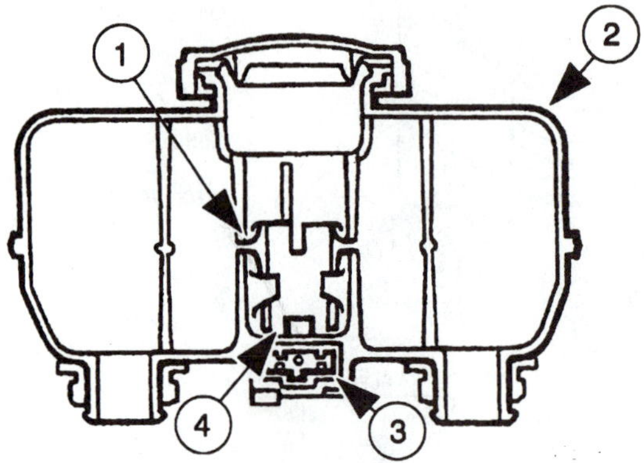

Figure 6-25 This reservoir is divided into two fluid cavities on each side of the well for the float (1), magnet (4), and fluid level switch (3). (Courtesy of Ford Motor Company)

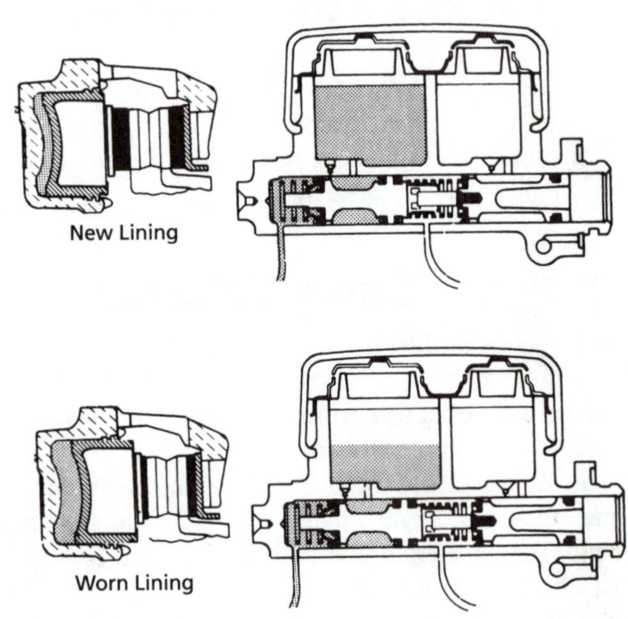

Figure 6-26 As disc brake lining wears, the caliper and piston move closer to the rotor, and the piston will move outward in the bore. The master cylinder fluid level will lower as this occurs.

divided into two parts so that a failure in one of the systems will not drain the fluid out of the other (Figure 6-25). In a master cylinder with four wheel drum brakes, both reservoir sections will be the same size. A master cylinder for a car with a disc-front and drum-rear combination will have a larger section for the disc brakes. Disc brake pistons do not return but gradually creep outward as the lining wears. The reservoir fluid level will drop in relation to the lining wear (Figure 6-26). This master cylinder will also have two fluid outlets. They will usually exit through the side of the cylinder body. These ports are normally drilled so that they intersect the cylinder bore at the top and at the end of each chamber. This feature helps when bleeding the air out of the cylinder and lines. As mentioned earlier, some master cylinders will incorporate a residual check valve under one or both of these line fittings.

6.8 Tandem Master Cylinder Operation

The operation of the primary section of a tandem master cylinder is the same as that of a single master cylinder. The secondary piston operates in a similar manner except that there is no pushrod. Any pressure buildup in the primary section will move the secondary piston so that it will build up the same pressure in the secondary section. The secondary piston floats according to the pressure in the primary and secondary sections (Figure 6-27).

When released, the primary piston returns to the retaining ring at the end of the bore. It is pushed there by both primary and secondary piston return springs. The released position of the secondary piston is determined by the length and strength of the primary and secondary

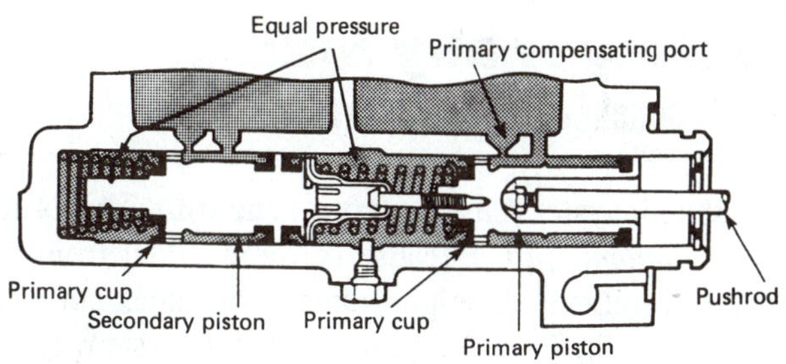

Figure 6-27 Normally, the secondary piston is operated by the hydraulic pressure in front of the primary piston. The pressure in front of the two pistons will be equal. (Courtesy of Bendix Brakes, by AlliedSignal)

return springs. The values of these two springs are adjusted during manufacture to ensure correct secondary piston positioning. The primary cup should be returned just past its compensating port. If the port is not uncovered during release, brake drag will occur because the fluid will not be able to flow back from the wheel cylinders or calipers. If the piston returns too far past the port, a low brake pedal can result from loss of the pump stroke. Pumping will not begin until the primary cup moves past the compensating port. In the past, a stop screw was threaded into the cylinder bore to ensure proper secondary piston positioning. The secondary piston stopped against this screw during piston return. This stop screw was threaded into the bore from the side, the bottom, or the top or reservoir side (Figure 6-28).

Some newer master cylinders used with ABS use two central valves instead of compensating ports. These valves, at the centers of the primary and secondary pistons, are open while the brakes are released to allow flow between the reservoirs and the area in front of the pistons, like a compensating port. When the brakes are applied, these valves are forced shut, blocking any flow to the reservoirs, so the fluid in front of the pistons can be forced to the calipers and wheel cylinders (Figure 6-29). During ABS actuation in some systems, the master cylinder pistons will shift back and forth rapidly with the lip of the primary cups sliding past the compensating ports and pressure in front of the cups; this can cause wear or etching of the lips of the cups and possible failure.

In case of hydraulic failure in one of the systems, there are two features built into this unit to ensure operation of the other half. The primary piston has an extension screw threaded into the forward, primary cup end. In case of a pressure loss in the primary section, this extension will push on and apply the secondary piston mechanically. There will be some lost motion until the extension meets the secondary piston. The driver will notice a lower brake

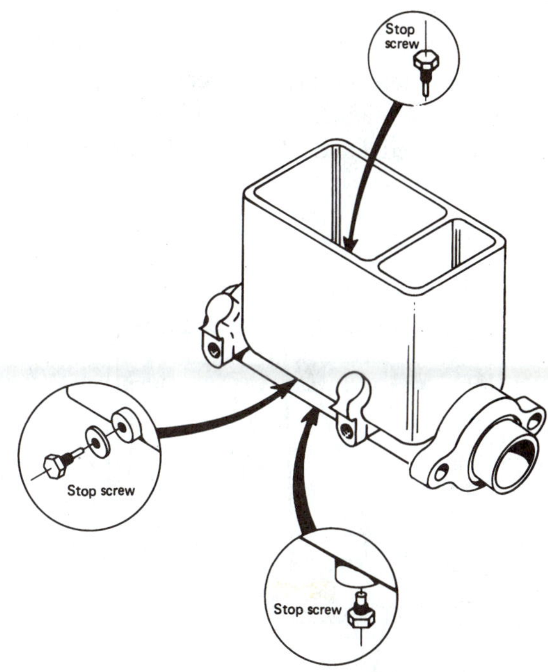

Figure 6-28 A dual master cylinder can use a stop screw or pin to locate the released position of the secondary piston. A stop screw/pin can enter the cylinder bore from the top, bottom, or side.

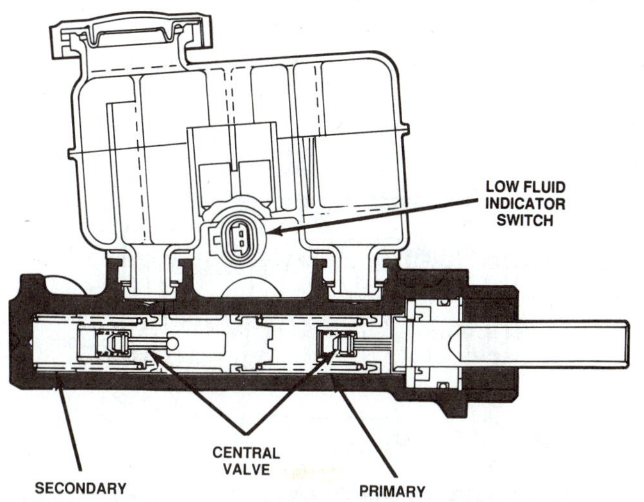

Figure 6-29 This ABS master cylinder uses a central valve in place of the compensating ports; the central valves are opened when the brakes are released. (Courtesy of Chrysler Coproration)

pedal and a loss in braking power. The brake failure warning light should also come on to indicate a hydraulic failure (Figure 6-30). There is also an extension on the forward end of the secondary piston. If a pressure loss occurs in the secondary system, the secondary piston will move easily until this projection strikes the end of the bore. From this point, the primary section can begin to operate in a normal manner. There will also be an obviously lower brake pedal and a loss in braking power (Figure 6-31).

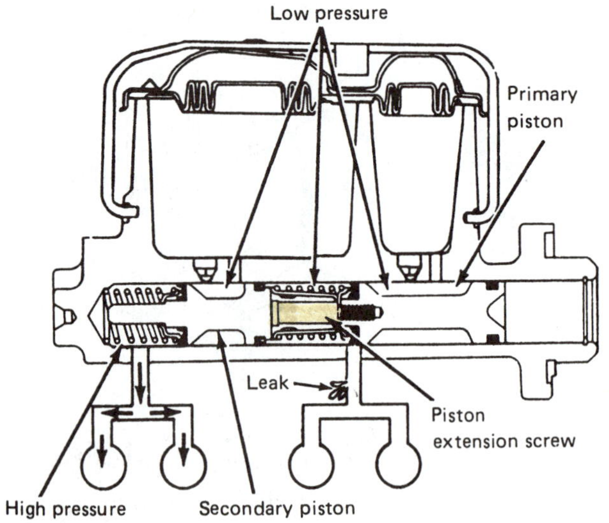

Figure 6-30 If a hydraulic failure occurs in the primary hydraulic section, the secondary piston will be operated mechanically by the piston extension screw. (Courtesy of General Motors Corporation, Service Technology Group)

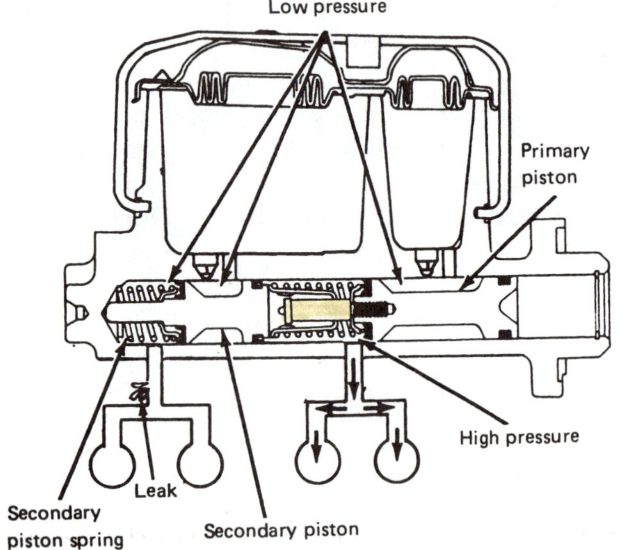

Figure 6-31 If a hydraulic failure occurs in the secondary hydraulic section, the secondary piston will move until its extension stops at the end of the bore. The primary section will then operate in the normal manner. (Courtesy of General Motors Corporation, Service Technology Group)

6.9 Diagonally Split Hydraulic System

The first split systems used a **front-rear split**. This is also called a *tandem split*. Because of the unequal front and rear axle weights and weight transfer, this does not mean a fifty-fifty split. In actual practice, a rear-system failure will leave 60 to 70 percent of the braking power, while a front-system failure will leave only 30 to 40 percent. The only way to produce a true fifty-fifty split was to divide the system in a right-side–left-side manner or diagonally. Trying to stop a car with only right-side or left-side brakes would produce a car that would be extremely hard to control. On a rear-wheel-drive car, this would occur with a diagonal split; the car would be very difficult to control while stopping with one front brake and the rear brake at the opposite corner.

Front-wheel-drive cars have a different wheel geometry. A negative scrub radius is used, while rear-wheel-drive cars use a positive scrub radius. The *scrub radius* is the distance between the center of the tire and the steering axis at the road surface. The steering axis is the pivot point for the front wheels when they steer or turn a corner. With a *positive scrub radius*, the tire is outboard of the steering axis, and with a negative scrub radius, the tire is inboard. With a positive scrub radius, a tire under braking will try to turn or steer outward. During stops, this tendency is balanced out between the two front tires. With a *negative scrub radius*, a braked tire will try to turn or steer inward, but this will be offset by the braking drag on the tire which tries to pull the car in the opposite direction. With a negative scrub radius, a controlled stop can be made with only one front brake. A **diagonal split**, sometimes called a *crisscross system*—right front with left rear and left front with right rear—will provide 50-percent braking if either system fails (Figure 6-32). Other versions of split hydraulic systems are used in some of the more expensive vehicles (Figure 6-33).

The master cylinder used with diagonal split systems is usually a seemingly ordinary appearing tandem master cylinder. Some of them have four outlet ports with a front and rear outlet for both the primary and secondary sections. In some of these master cylinders, a proportioning valve is installed in each of the two rear wheel outlets. These valves will be discussed later in this chapter. Also, in some designs, the pressure differential valve for the brake failure warning-light switch will be included in the master cylinder body.

A potentially serious problem with diagonal split systems is that heat from parking brake drag can cause total

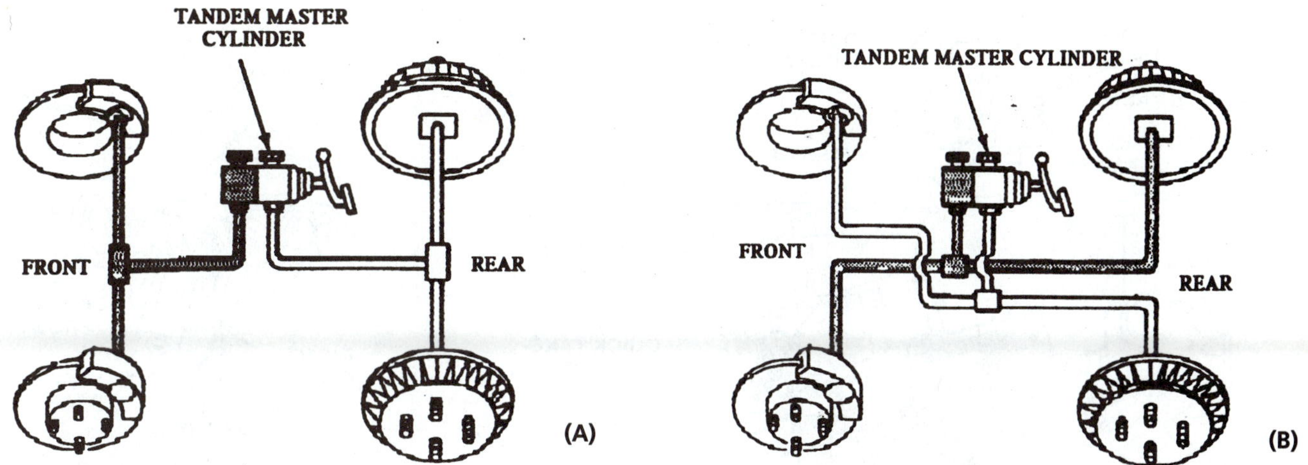

Figure 6-32 A tandem split system pairs the front and rear brakes at the primary and secondary ends of the master cylinder (A). A diagonal system (B) pairs up the right front with the left rear and the left front with the right rear. (Courtesy of Hunter Engineering Company)

Braking force apportioning in accordance with DIN 74000.

Version	Type of apportioning ← Direction of travel	Remarks
1		**Front-axle/rear-axle split.** One axle is braked in each circuit.
2		**Diagonal split.** One front wheel and the diagonally opposite rear wheel are braked in each circuit.
3		**Front-axle and rear-axle/ front-axle split.** One circuit brakes the front and rear axles, and one circuit brakes only the front axle.
4		**Front axle and rear-wheel/ front-axle and rear-wheel split** Each circuit brakes the front axle and one rear wheel.
5		**Front-axle and rear-axle/ front-axle and rear-axle split.** Each circuit brakes the front axle and the rear axle.

Figure 6-33 Five different styles of split or dual braking systems are shown here. Most vehicles use Version 1 (tandem split) for RWD vehicles and Version 2 for FWD vehicles. Versions 3, 4, and 5 provide better stopping control should failure occur, but they are complex and expensive. (Courtesy of Robert Bosch Corporation)

brake loss. If the car is driven with partially applied parking brakes, the heat generated can cause the fluid to boil in the rear wheel cylinders. *Since the boiling fluid is in both brake circuits, it can cause total brake failure.* After the brakes cool down, the brakes usually return to normal operation.

6.10 Quick-Take-Up Master Cylinder

A **quick-take-up master cylinder**, also called a *dual-diameter bore*, *step bore*, or *fast-fill* master cylinder, is used with newer, low-drag disc brake calipers (Figure 6-34). These calipers have a redesigned piston-sealing, square-cut O-ring and groove, which has the ability to pull the piston back a greater distance. This provides a slight clearance between the front pads and the rotor, reducing brake drag and improving fuel mileage, but it can also cause a very low brake pedal. More pad clearance requires more movement of the caliper piston during application. More movement of these two rather large pistons requires more fluid displacement from the master cylinder. A large-bore master cylinder is needed, but this will usually mean lower hydraulic pressures. Remember that hydraulic pressure equals application force divided by piston area.

A quick-take-up master cylinder has a step bore in the primary section and uses a primary piston which has a small, normal-diameter primary cup with a larger, oversized secondary cup. This secondary cup is a low-pressure pumping cup. It pumps the extra fluid needed to take up the clearance at the calipers. The stepped-down area between the primary and secondary cups is now called the low-pressure section.

One manufacturer uses a master cylinder that has a 7/8 or 0.875-inch (22.2 mm) bore which steps to a 1½ or 1.25-inch (31.75 mm) bore. A bulge at the outside of the cylinder body in this section usually identifies this mas-

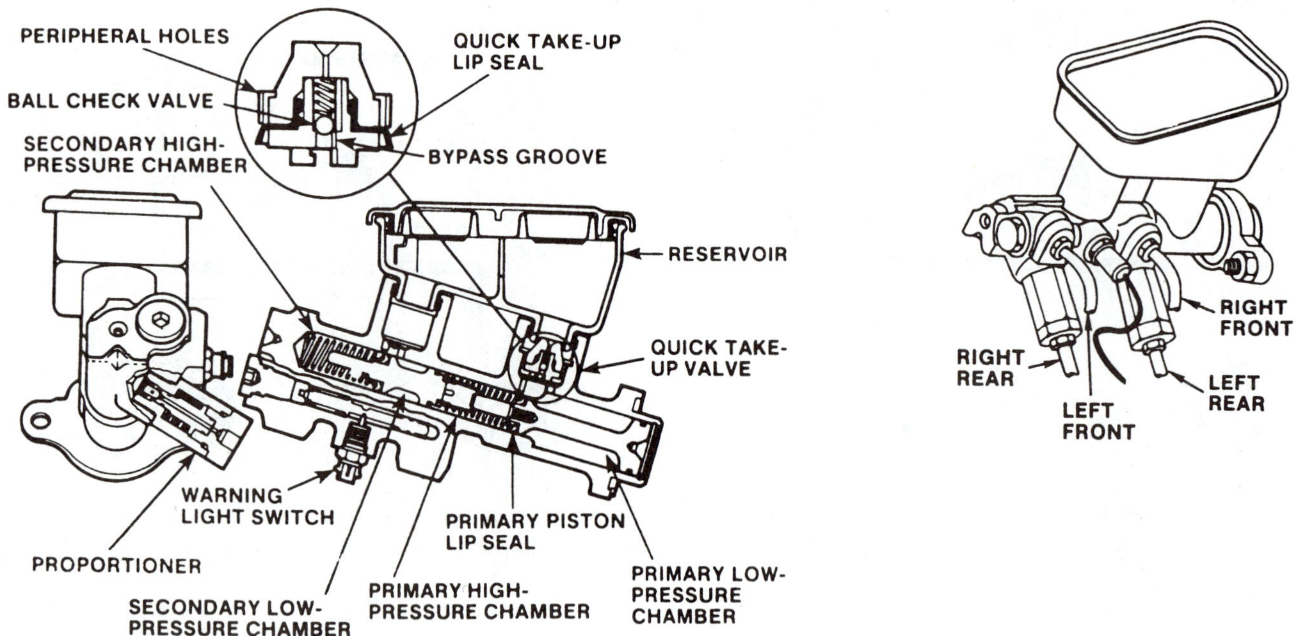

Figure 6-34 A quick-take-up master cylinder uses a larger diameter primary low-pressure chamber and a quick-take-up valve. Also note that this particular master cylinder has four outlet ports, proportioning valves at the rear ports, and a hydraulic failure warning light switch. (Courtesy of Bendix Brakes, by AlliedSignal)

ter cylinder type. The larger bore has twice the area, so it will displace twice the amount of fluid as the smaller bore.

During brake application, fluid is pumped out of the low-pressure chamber past the lips of the primary cup. This fluid goes into the high-pressure chamber of the primary system to move the caliper pistons and also to move the secondary piston in the master cylinder. As soon as the brake lining clearance is taken up, the hydraulic pressure will begin to increase. This could produce enough back-pressure on the larger-displacement piston to cause a hard brake pedal, but before this point is reached, the quick-take-up valve between the low-pressure section and the reservoir will open. This valve opens at a preset pressure to allow the extra fluid, which is no longer needed, to flow into the reservoir with little effort. From this point, the primary cups of the pistons will produce braking pressure in a normal manner. Fluid flows during release are also very similar to those in a conventional master cylinder (Figure 6-35).

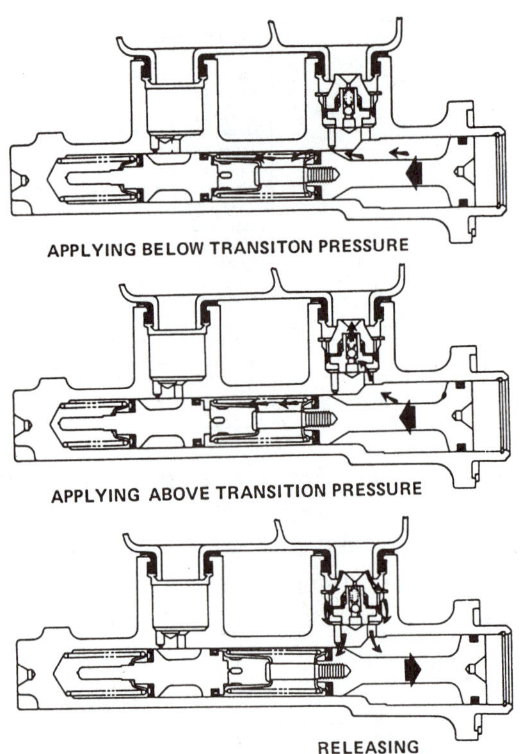

Figure 6-35 During brake application, fluid flows from the primary low-pressure chamber around the primary cup to move the lining into contact with the rotor. As pressure is generated (above transition), fluid goes from this section through the quick-take-up valve back to the reservoir. (Courtesy of General Motors Corporation, Service Technology Group)

6.11 Miscellaneous Master Cylinder Designs

There are additional master cylinder designs that have very limited usage.

The **dual master cylinder** is essentially two complete master cylinders, side by side in the same body casting.

One of them is used for the brakes, and the other operates the clutch slave cylinder. The two bores and two reservoirs are completely separate from each other so that possible failure of one will not affect the operation of the other. This design was used on some domestic pickups.

The *step bore master cylinder* is a tandem master cylinder with a bore that has two different diameters. The secondary piston and bore section are smaller. This design provides two different hydraulic pressures in the primary and secondary sections, which produces automatic brake proportioning (Figure 6-36).

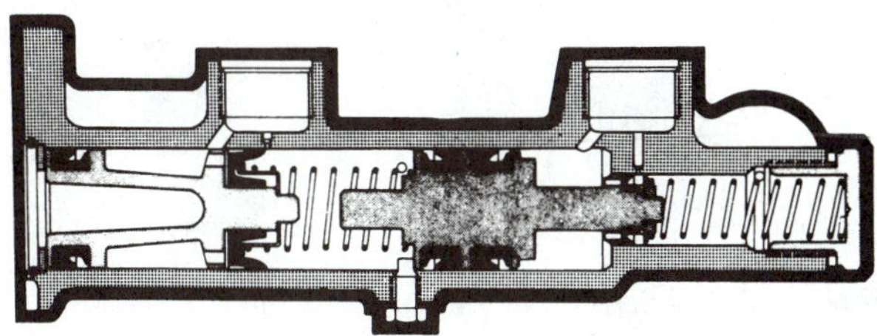

Figure 6-36 A step bore tandem master cylinder uses a smaller diameter primary cup on the secondary piston. This will produce higher pressure in the secondary hydraulic circuit. (Courtesy of ITT Automotive)

Lucas-Girling master cylinders are used on many British-made cars. Older, single-piston designs have a single cup on the piston. The compensating port to the reservoir is at the end of the master cylinder bore, and the flow through this port is controlled by a valve. When the brakes are released, this valve is open to allow compensation. It is closed as soon as the piston is moved for brake application (Figure 6-37).

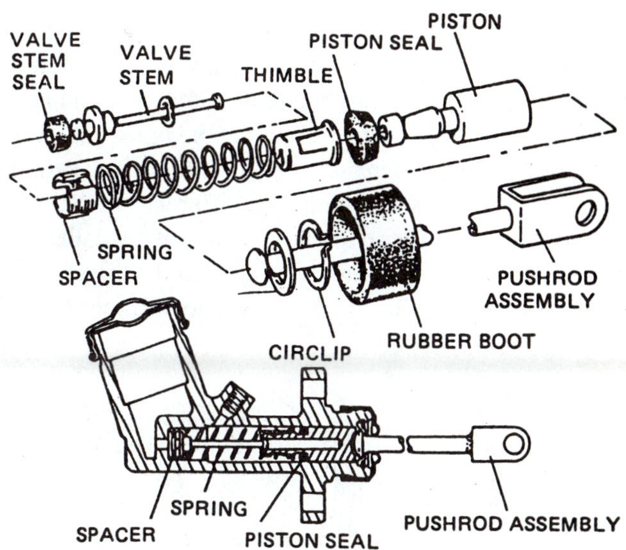

Figure 6-37 This single master cylinder design was used on many British-made cars. Note the outlet port is above the middle of the cylinder bore. (Courtesy of Wagner Brake)

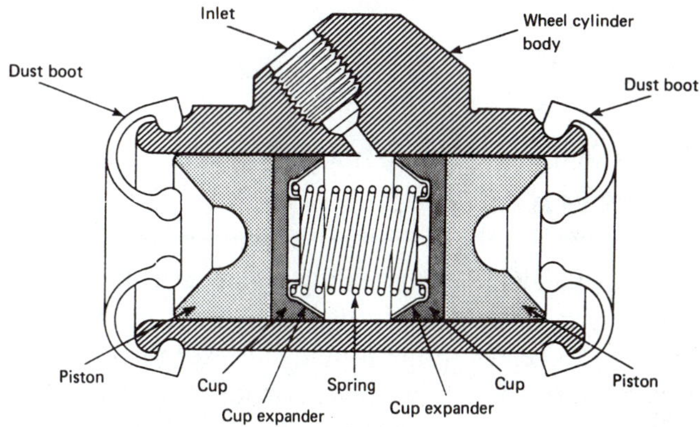

Figure 6-38 A cutaway of a typical wheel cylinder showing the internal parts. (Courtesy of Bendix Brakes, by AlliedSignal)

6.12 Wheel Cylinder Basics

Wheel cylinders are *output pistons* for the hydraulic system. Pressure generated in the master cylinder pushes against the cups in the wheel cylinders to develop the force to apply the brake shoes (Figure 6-38). The amount of force can be determined by multiplying the amount of hydraulic pressure by the area of the wheel cylinder cups or pistons. The diameter of many wheel cylinders is indicated by a number on the cylinder body, inner face of the cup, or piston.

Most wheel cylinders use a straight bore with a cup and piston at each end. Each piston and cup apply the same force to each brake shoe. Two-leading-shoe, two-trailing-shoe, and some uni-servo designs use a single wheel cylinder. One piston and one cup are used in a cylinder bore that is closed at one end (Figure 6-39). One

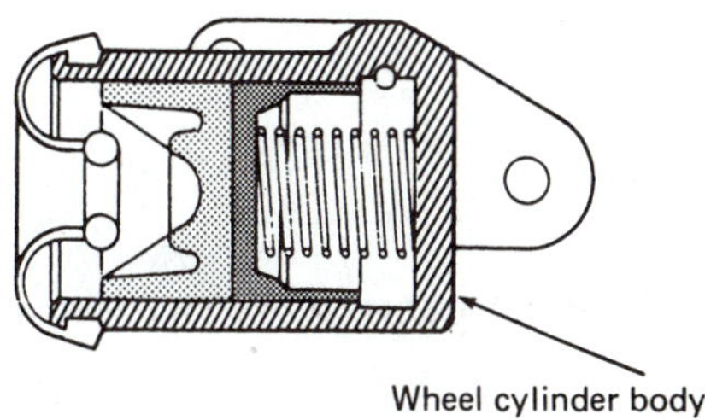

Figure 6-39 This single-piston wheel cylinder can apply pressure to only one brake shoe. It is used in a two-leading shoe or a uni-servo brake. (Courtesy of Bendix Brakes, by AlliedSignal)

nonservo design uses a sliding single wheel cylinder. It is mounted to the backing plate in such a way as to allow the cylinder bore to slide from hydraulic reaction and apply one shoe while the piston is applying the other one. A few brake designs use a *step bore wheel cylinder*. The cylinder bore has two diameters with two different-sized pistons and cups. The smaller cup and piston in the smaller bore will exert less pressure on one shoe than the larger piston at the other end applies on the other shoe (Figure 6-40).

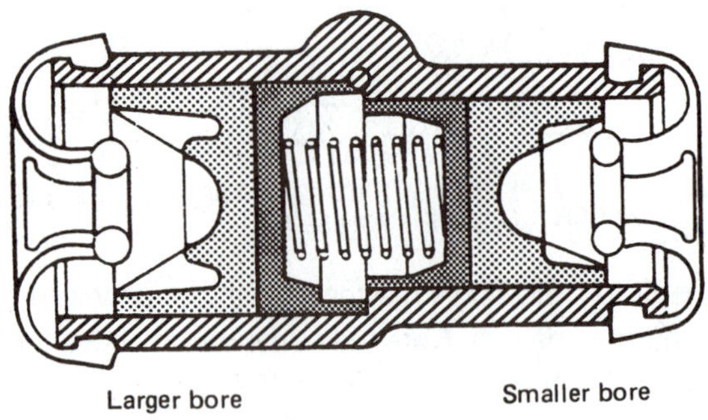

Figure 6-40 This step bore wheel cylinder will apply greater force on the shoe to the left than on the shoe to the right. (Courtesy of Bendix Brakes, by AlliedSignal)

6.13 Wheel Cylinder Construction

A common double-piston, straight-bore wheel cylinder consists of the cylinder body, two pistons, two piston cups, two boots, a center spring (often with expanders), and a bleeder screw. The cylinder body is made from cast iron or aluminum. As in a master cylinder, the bore must be straight, smooth, and of a precise size. Aluminum bores are anodized to resist corrosion and wear (Figure 6-41).

The cylinder body is normally manufactured with a mounting projection that extends through the backing plate. During brake application, there is a lot of sideways or shear force between the wheel cylinder and the backing plate. This projection ensures that the wheel cylinder will not move relative to the backing plate. Most wheel cylinders are secured to the backing plate by a pair of machine screws or bolts. A recent Delco Moraine design uses a special clip (Figure 6-42).

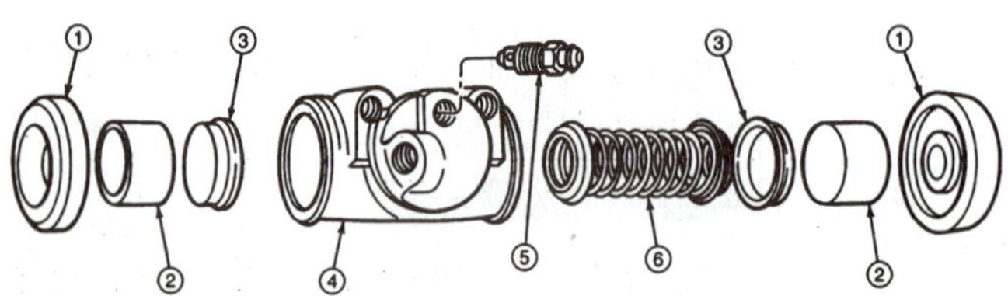

Figure 6-41 A disassembled wheel cylinder showing the two boots (1), pistons (2), cups (3), body (4), bleeder screw (5), and spring (6). (Courtesy of Ford Motor Company)

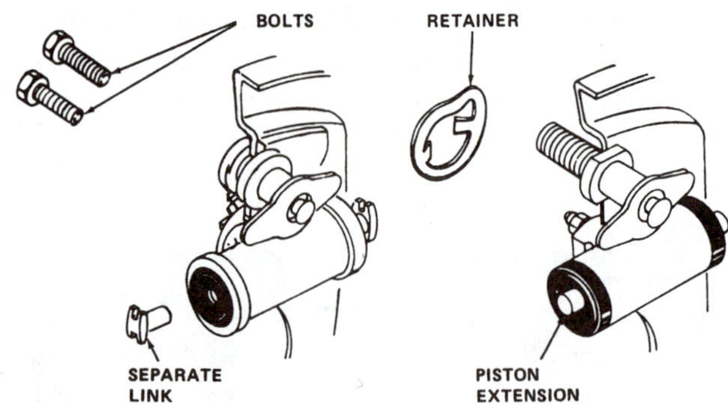

Figure 6-42 A wheel cylinder can be attached to the backing plate by a pair of bolts or a retainer. Note the left unit uses an internal boot and a separate link to contact the brake shoe; the other unit uses a piston extension and external boot. (Courtesy General Motors Corporation, Service Technology Group)

Wheel cylinder pistons are made from anodized cast aluminum, sintered iron, or plastic and are shaped to accept the end of the brake shoe link or pushrod, or a projection of the brake shoe. The inner side of the piston is flat and smooth for the cup to rest on and push against. Some wheel cylinders use a piston that is grooved to accept a ring-shaped cup similar to a master cylinder secondary cup (Figure 6-43).

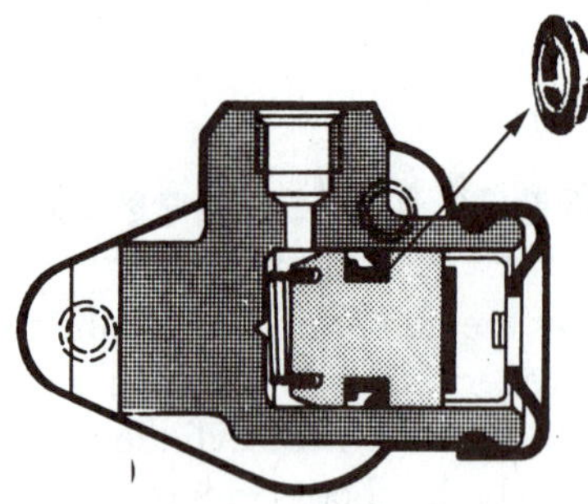

Figure 6-43 Some wheel cylinders use a ring-type seal. (Courtesy ITT Automotive)

A *wheel cylinder cup* operates just like a master cylinder cup. Internal pressure forces the lips of the cup against the cylinder, ensuring a pressure-tight seal. A wheel cylinder cup has a definite wear advantage over a master cylinder cup in that pressure moves the cup and piston away from the lip of the cup and not toward it. Also, when hydraulic pressure gets high enough to really pin the lips of the cup against the bore, the cup does not need to travel much farther. Most cups have tapered lips; they are much thicker at the inner section. When the hydraulic pressure gets very high, the thicker section can stretch to allow more cup movement without sliding the lip of the cup. Seals of the synthetic rubber cup or lip type can be easily damaged if the lip of the cup is cut or torn by faulty installation or rough bore surfaces. During hard stops, the piston-to-bore clearance becomes very critical. Excessive clearance provides enough room for hydraulic pressure to force the cup material between the piston and the bore. If this happens, brake drag can occur because the piston will not return, or else the edges of the cup will wear or be nibbled away. This condition is called **heel drag** (Figure 6-44).

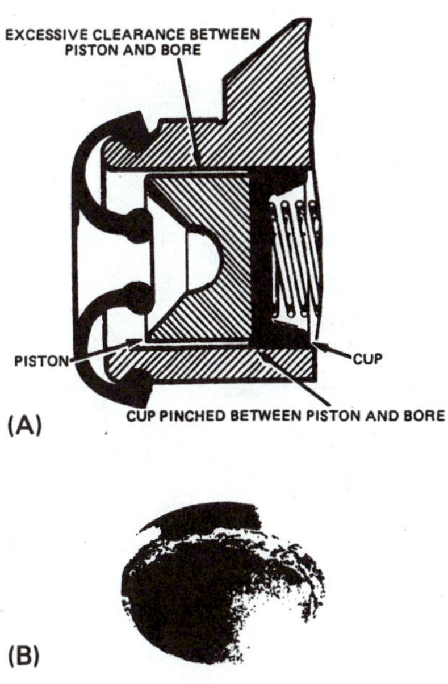

Figure 6-44 If there is excessive piston-to-bore clearance, the cup can be forced into this gap by very high fluid pressure (A). This will cause heel drag and nibbling of the cup edges (B). (A is Courtesy of Wagner Brake; B is courtesy of ITT Automotive)

The spring has a simple job. It pushes outward on the cups so everything between them and the brake shoes stays in contact. Many springs include a pair of **expander washers** to maintain a slight pressure between the lips of the cups and the cylinder bores. This helps prevent the entrance of air into the system when there is no hydraulic pressure on the cups.

Each end of the cylinder bore is enclosed in a rubber **dust boot**. This **boot** prevents contaminants from entering the bore and becoming wedged between the piston and bore or causing corrosion of the bore or piston. Either condition can cause the piston to stick. Most boots are *external*. They fit over a raised lip at the outer end of the cylinder body and are held in place by the elastic nature of the rubber material. Some boots are *internal*. They fit into a recess at the end of the bore and are held tightly in place by an internal metal ring. The center of either style of boot fits snugly over the brake shoe link, or over a projection of the piston, to make a seal.

The **bleeder screw** is fitted into a passage extending from the top center of the cylinder bore. It is actually a valve; the hardened, tapered end of the screw fits tightly against a tapered seat to close off this passage. When the screw is loosened, air and fluid can flow past the seat and through a passage in the screw, allowing air to be bled from the cylinder. The bleeder screw is normally closed tightly to keep fluid from escaping. Many include a rubber or plastic cup to keep dirt, water, or salt from entering the outer opening and plugging it up (Figure 6-45).

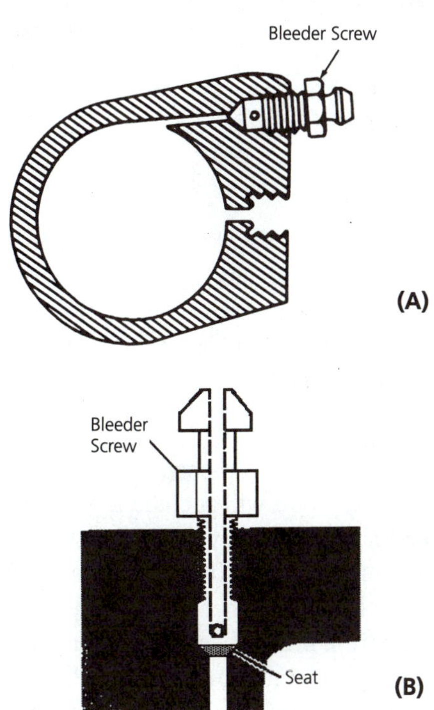

Figure 6-45 The bleeder screw is positioned so that the passage from the valve seat connects to the very top of the cylinder bore (A). The hardened end of the screw seals against the tapered seat in the cylinder (B).

6.14 Caliper Basics

As previously mentioned, two major styles of calipers are used on passenger cars: *fixed* and *floating*. The primary hydraulic differences include the number of pistons used, the diameter of the pistons, the type of piston seal used, and the fact that many fixed calipers need to be split for service. The caliper pistons, like wheel cylinder pistons, are output pistons; hydraulic pressure pushing on them applies the brake pads. The amount of force that they exert on the pads is determined by multiplying the area of the piston(s) by the hydraulic pressure. With floating calipers, this force is doubled because of the cylinder reaction (Figure 6-46).

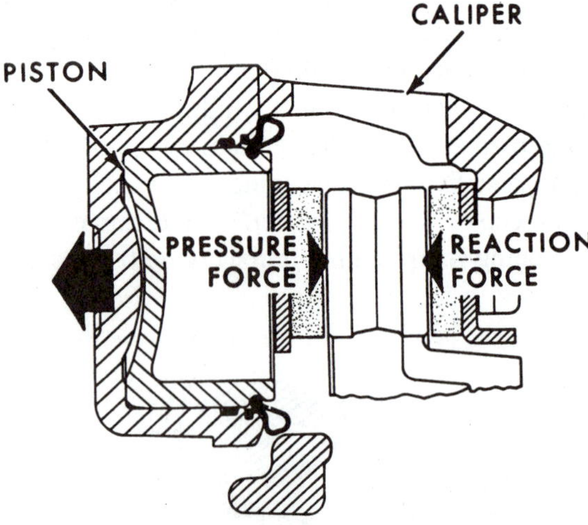

Figure 6-46 In a floating caliper, hydraulic pressure causes the piston to push on the inboard pad. The same hydraulic pressure at the end of the cylinder bore will develop a reaction force to move the caliper and apply the outboard pad. (Courtesy of Ford Motor Company)

Most calipers use a cup-shaped bore in a cast-iron caliper body. Each bore will include a piston with a seal and boot, and a bleeder screw. A few pistons include an insulator to block heat flow from the lining to the piston. Disc brakes experience higher brake fluid operating temperatures than drum brakes because the fluid in the caliper is fairly close to the friction material, and the caliper body and piston conduct quite a bit of heat to the fluid.

6.15 Caliper Pistons and Sealing Rings

Most calipers use a *square-cut* (also called a *lathe cut*), *O-ring seal* that has a square cross section for a sealing ring. This O-ring fits into a groove in the cylinder bore. The piston that fits through this sealing ring has smooth, straight sides. The O-ring seal has a static (stationary) contact with the cylinder groove and both a static and a dynamic (moving) contact with the piston.

In the released position, the elastic nature of the synthetic-rubber O-ring will cause it to assume its normal square shape in the groove with the piston held snugly inside it. During brake application, hydraulic pressure will move the piston outward in the bore as the fluid pushes on the face of the piston and on the inside of the O-ring seal. The O-ring seal will deflect or twist enough so that the piston and caliper can move to apply the brake pads. Release of the hydraulic pressure will allow the O-ring to relax and pull the piston back to the released position (Figure 6-47). The O-ring seal serves two major purposes. It seals the hydraulic pressure and returns the piston. Some newer calipers have a redesigned O-ring seal and groove to obtain more piston pull-back, which provides slight pad clearance instead of drag, and therefore allows better fuel mileage. As the brake lining wears, the piston will slip outward through the O-ring seal whenever necessary to compensate for the amount of wear. All disc brake calipers are self-adjusting.

The piston used with O-ring seals is precision sized, with straight, smooth sides. Traditionally, disc brake pistons are made from *steel* that has been stamped into the cup shape. After stamping, steel pistons are machined to size and plated to protect them from corrosion. A few cars use cast *aluminum* pistons as weight-saving measures. Many newer pistons are made from *phenolic* resin. These pistons are much thicker and have a brownish-gray color. Phenolic pistons have the advantage of being lighter in weight, corrosion-free, and very good heat insulators. The last advantage will reduce the fluid temperature in the caliper. Phenolic pistons earned a reputation of sticking or rocking in the bore when they were first introduced. After a slight revision of the bore clearance dimension, these pistons now have a very favorable reputation. It is not recommended that steel or aluminum pistons be used to replace damaged phenolic pistons because of the possibility of fluid overheating and boiling as a result of the increased heat transfer (Figure 6-48).

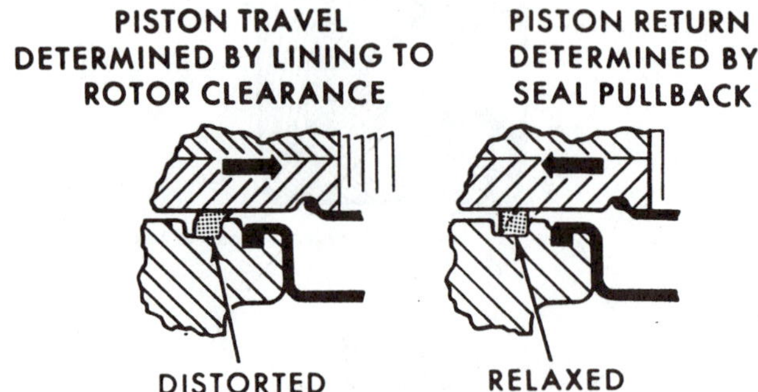

Figure 6-47 As the piston moves to apply the brake pads, the O-ring seal is pulled into a distorted position. It will relax and pull the piston back slightly when the hydraulic pressure is released. (Courtesy of Ford Motor Company)

Figure 6-48 The thick-walled, dark piston on the left is made of a phenolic material; the shiny, thin-walled piston on the right is a common chrome-plated steel piston.

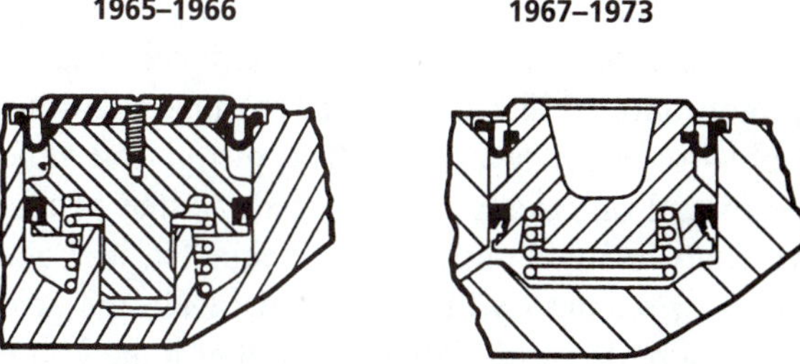

Figure 6-50 The calipers used on 1965 and 1966 Corvettes used pintle-type pistons that had a guide in the caliper. From 1967 to 1973, Corvettes used the piston style shown on the right. Note that the early pistons used a heat insulator next to the brake pad. (Courtesy of General Motors Corporation, Service Technology Group)

Some fixed-caliper designs use a cup or lip seal similar to those of wheel cylinders. A larger, ring-shaped cup is used. Like the cup in a wheel cylinder, it has a **dynamic** (or moving) **seal** with the bore and a **static** or **stationary seal** with the piston. This is often called a **stroking seal**. A caliper which uses stroking seals must have straight, smooth bores just like a master cylinder or wheel cylinder. A light spring is commonly used with these pistons because the seal does not have much wall drag, and vehicle vibrations can cause the piston to overrelease. This can cause too much pad clearance and a low brake pedal on the next application. No residual hydraulic pressure is used with disc brake calipers (Figure 6-49).

Some of the early pistons used with cup seals were of the *pintle* type (Figure 6-50). The piston had a round pintle—a projection on its inner side that fitted into a smaller bore in the cylinder. The pintle was used to hold the piston straight in the bore. Most calipers use a cup-shaped piston similar to the pistons used with O-ring seals, but with a groove for the piston cup. Most of these pistons are made from cast aluminum.

Calipers that use cup seals have a definite disadvantage in that the cylinder bore must be smooth and free from corrosion. Any imperfections can damage the sealing lip of a cup. A rusty bore can easily ruin an expensive caliper body. Several companies are remanufacturing fixed-caliper bodies with stainless-steel sleeves in the bore. These calipers are free from future corrosion and are much less expensive than new calipers (Figure 6-51). The condition of the cylinder bore is not as critical when O-ring seals are used. A positive seal will occur as long as the piston is clean and smooth and can slide in the bore.

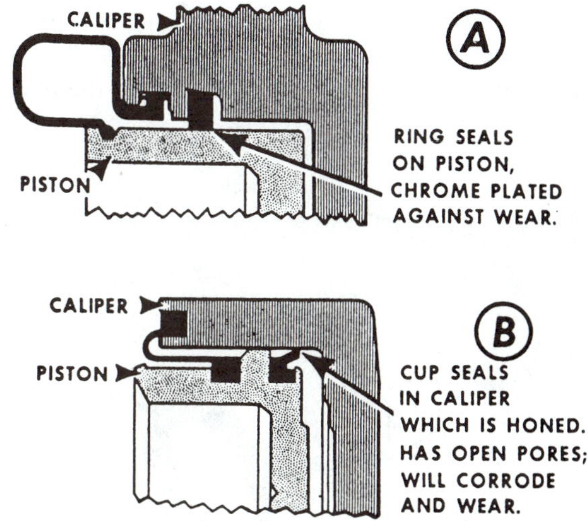

Figure 6-49 Modern calipers use a stationary, square-cut O-ring seal in the bore (A) that seals against the piston. Some older calipers use a stroking seal mounted on the piston (B) that seals against the bore. (Reprinted from Mitchell Anti-Lock Brake Systems, with permission of Mitchell Repair Information, LLC)

Figure 6-51 This caliper was saved and improved by machining the bore and installing a stainless-steel sleeve. (Courtesy of Stainless Steel Brakes Corp.)

Another possible problem with calipers using cup seals is entrance of air. While the brake is released, excessive rotor runout will cause the pad to move sideways—in and out—on each revolution. The piston and seal will move in and out along with the pad. As the piston seal moves outward without pressure behind it, a slight low pressure or vacuum will be created inside the seal lip. This pressure drop can pull air inward past the seal lip. If enough air enters, the next brake application will produce a low, spongy pedal as this trapped air is compressed (Figure 6-52).

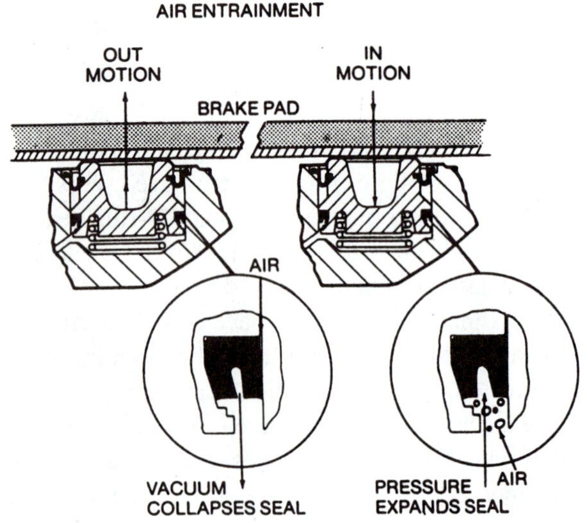

Figure 6-52 Excessive rotor runout can cause an in-and-out piston motion that can pump air past a stroking seal. (Courtesy of Stainless Steel Brakes Corp.)

6.16 Caliper Boot

The piston boot in a caliper serves the same purpose as a wheel cylinder boot, but it is somewhat more critical because it is directly exposed to road elements. Also, its proximity to pads results in very high operating temperatures. A faulty boot can allow dirt, water, or road salt to enter the cylinder bore, which will cause corrosion (Figure 6-53). The buildup of contaminants or the corrosion that they produce can cause the piston to stick, resulting in pad drag and excess lining wear. On a stroking-seal-type caliper, the corrosion will usually cause leakage to occur when pad wear allows the piston to move outward to the corroded area.

Boots are made from synthetic rubber and need to be securely attached to both the piston and the caliper body. Some designs experience problems during the heating and cooling cycles of the caliper. At this time some calipers "breathe." The expansion and contraction of the

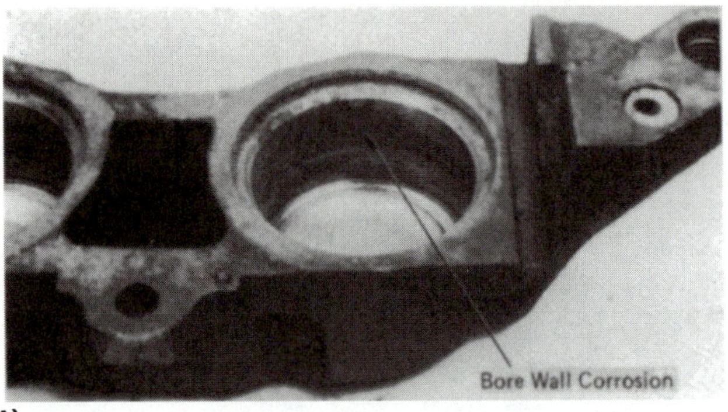

(A)

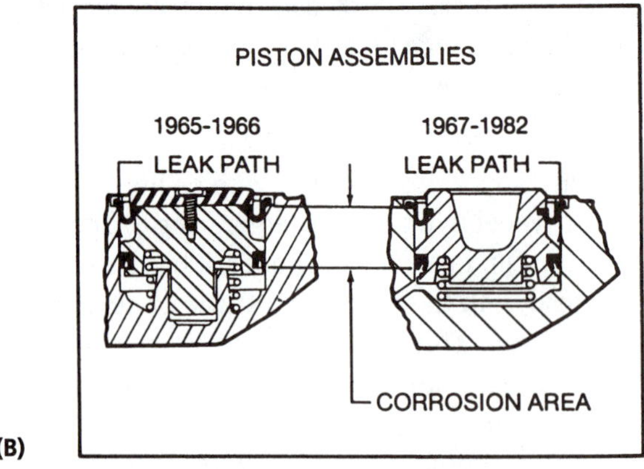

(B)

Figure 6-53 A caliper bore (A) can be ruined by moisture in the brake fluid or by moisture entering past the outer boot (B). (Courtesy of Stainless Steel Brakes Corp.)

air inside the boot can cause air to be expelled during the heat cycle and new air to be drawn back in during the cooling phase. This new air can bring in moisture, which in turn can cause corrosion buildup and damage to the cylinder bore (Figure 6-54).

In most cases, the boot is secured around the outer portion of the piston by the elastic nature of the rubber boot material. There are several different methods of

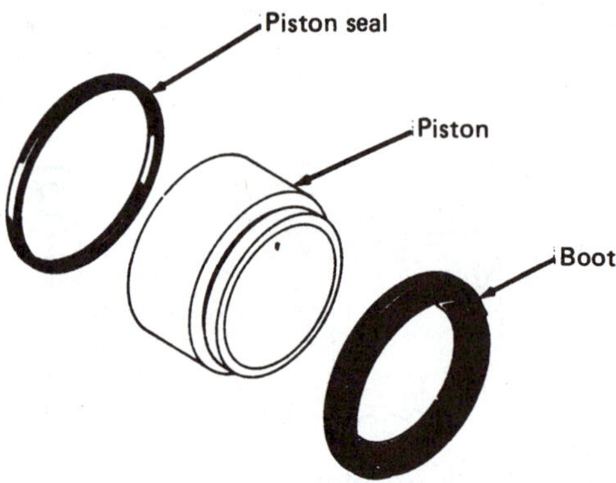

Figure 6-54 The O-ring seal prevents fluid pressure from leaking past the piston, and the boot keeps contaminants from entering the cylinder bore. (Courtesy of General Motors Corporation, Service Technology Group)

securing the boot to the caliper. One manufacturer, Kelsey-Hayes, places the end of the boot into a groove at the outer end of the cylinder bore. It is locked in place when the piston is installed. This boot style requires some skill to install and forms a rather effective seal as long as the piston and cylinder groove are clean (Figure 6-55). Another manufacturer, Delco Moraine, locks the boot into a groove outside the cylinder bore with a metal ring that is positioned in the boot. This style is fairly easy to install if the correct tool is used. Installing this boot without the correct tool will probably result in a poor seal and future failure (Figure 6-56). Several import caliper designs use a boot that fits into a groove or over a projection and is secured in place with a metal retaining ring. Servicing of these different boots will be described in Chapter 11. In each case, it is very important to obtain a clean, weather-tight seal to help prevent any future internal contamination (Figure 6-57).

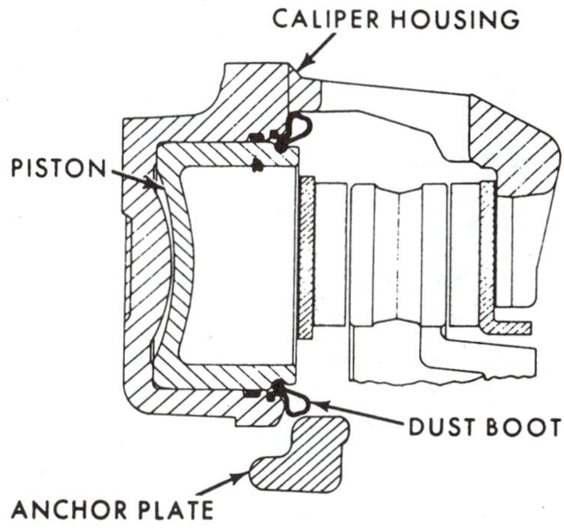

Figure 6-55 This dust boot is locked into the cylinder bore when the piston is installed. (Courtesy of Ford Motor Company)

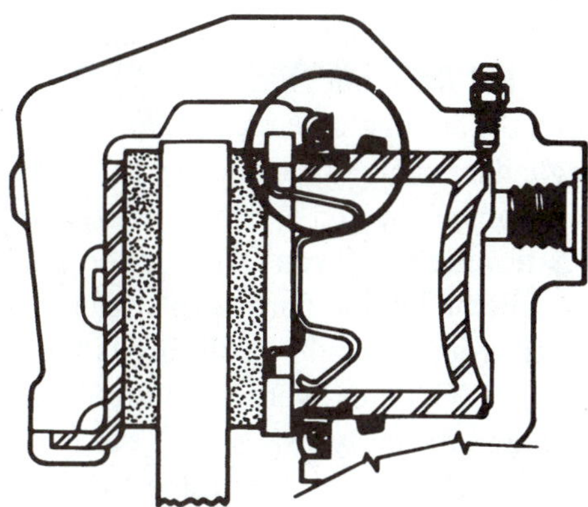

Figure 6-56 This boot is sealed to the cylinder bore by a soft metal ring that is part of the boot. (Courtesy of General Motors Corporation, Service Technology Group)

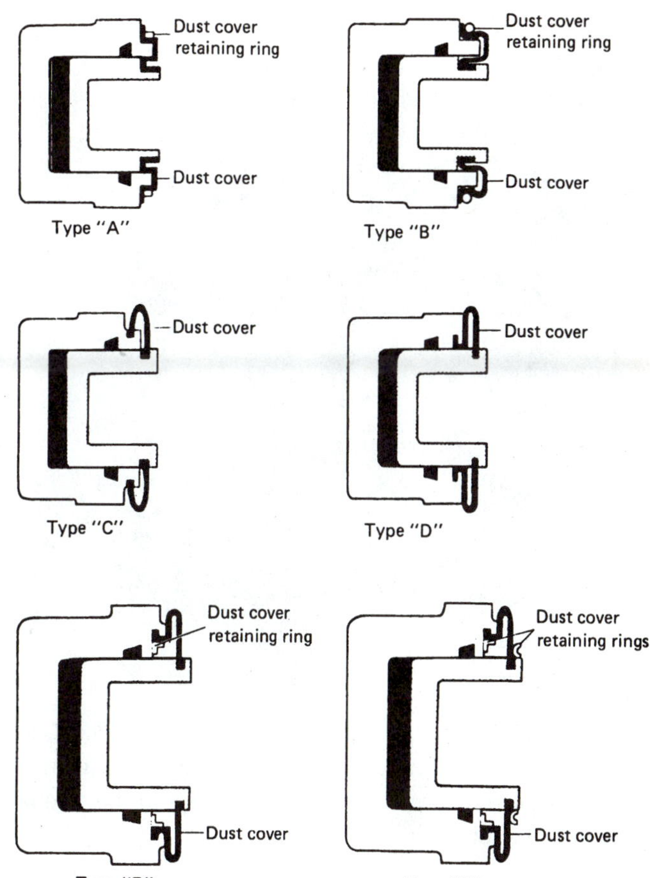

Figure 6-57 There are at least six different methods of sealing the piston to the caliper. (Courtesy of LucasVarity Automotive)

6.17 Caliper Body and Bleeder Screw

As mentioned earlier, most caliper bodies are made from gray cast iron. A caliper must be strong enough to resist distortion during high brake pressures. Caliper bodies are manufactured in pairs and are almost interchangeable side-for-side. The major difference between the right and left calipers, in most cases, is the location of the bleeder screw (Figure 6-58). As in a wheel cylinder, the passage to the bleeder screw must be drilled into the uppermost location of the cylinder bore. Switching the right and left calipers will position the bleeder screw and passage incorrectly and make it impossible to bleed all of the air out of the caliper.

The caliper is usually machined for the cylinder bore; the groove for the boot and the O-ring are machined in the bore; and the mounting pads for the outboard shoe and the abutments where the caliper body meets the caliper mount are machined on the outside. On fixed calipers, the two caliper body halves are machined where they join at the

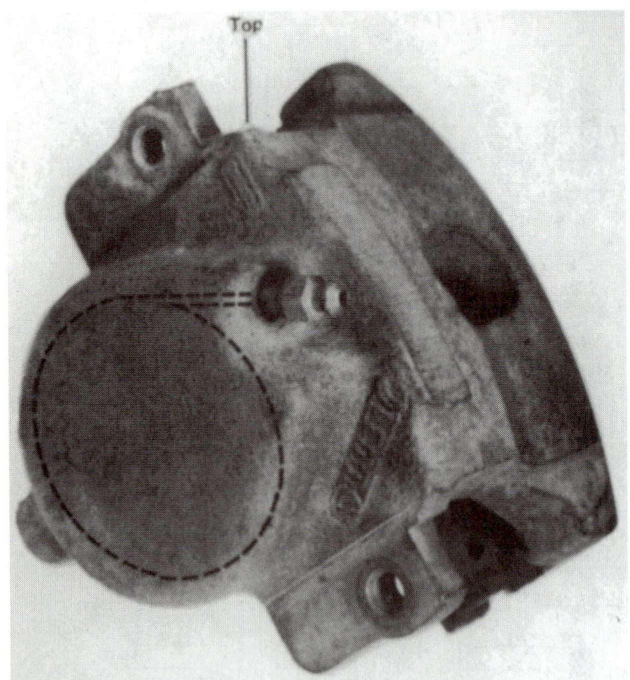

Figure 6-58 In many cars, the right and left calipers are identical with the exception of the bleeder screw location. A caliper is always mounted so that the passage to the bleeder screw is at the top.

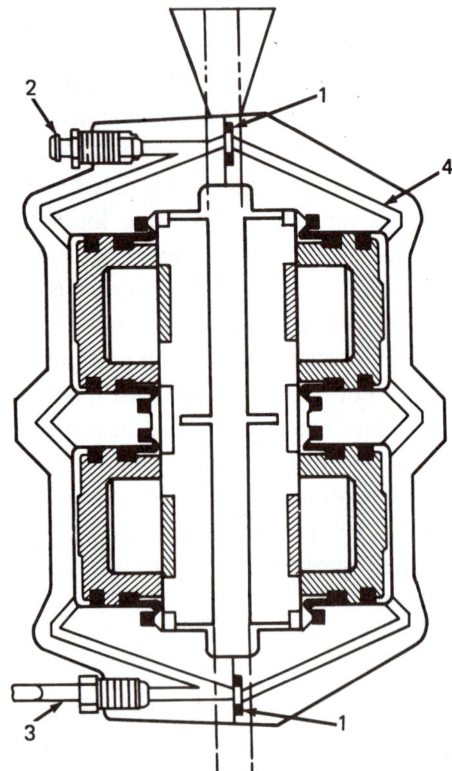

Figure 6-59 This fixed caliper uses internal passages to transfer fluid pressure from the inlet (3) to each of the bores, and the air up to the bleeder screw (2). Note the O-rings (1) which are used to seal the passages where the two caliper halves join. (Courtesy of General Motors Corporation, Service Technology Group)

center of the caliper. Two or more **bridge bolts** secure the two halves together. The mating surfaces of the two parts usually include a drilling to each of the caliper bore(s). These drillings are sealed by O-rings where they meet at the center. They are used to bring fluid to the outboard cylinder(s) and to bring air to the bleeder screw (Figure 6-59). Some caliper designs use two bleeder screws—one for each caliper half. Some caliper designs use an external tube to deliver fluid to the outboard cylinder(s).

6.18 Tubing and Hoses

Steel tubing is used for the lines that transfer fluid from the master cylinder to the wheel cylinders or calipers. Flexible hoses are used in those portions next to the front and rear suspensions where movement is required (Figure 6-60). Tubing is similar to pipe except it is sized by the outside diameter (OD), and does not use threads to join at the ends. The common sizes for automotive brake tubing are 3/16-inch (4.6 mm) for disc brakes and 1/4-inch (6.4 mm) for drum brakes. Occasionally 5/16-inch (7.9 mm) tubing is used. It is usually made by wrapping a steel sheet into a tubular shape and welding or brazing the seam closed (Figure 6-61).

Tubing must be flared to make leak-free high-pressure connections. Two different styles of flares are used. Traditionally, a standard SAE 45° **double flare** was made at the end of the tube, and this was locked onto the tube seat with an **inverted flare** nut (Figure 6-62). A double flare folds the end of the tube inward, toward the center of the tube. A single flare cannot be made in steel tubing because the tube will split as the end is stretched to make the flare. The **International Standards Organization (ISO) flare**, sometimes called a bubble flare, also stretches the tube a short distance from the end so splitting is not a problem (Figure 6-63). ISO flares were introduced on domestic cars beginning in the early 1980s.

Both the inverted flare nut and the ISO flare nut use male (or external) threads. Plain flare nuts are commonly used in household gas lines and use female (or internal) threads (Figure 6-64). The standard-sized nut used with

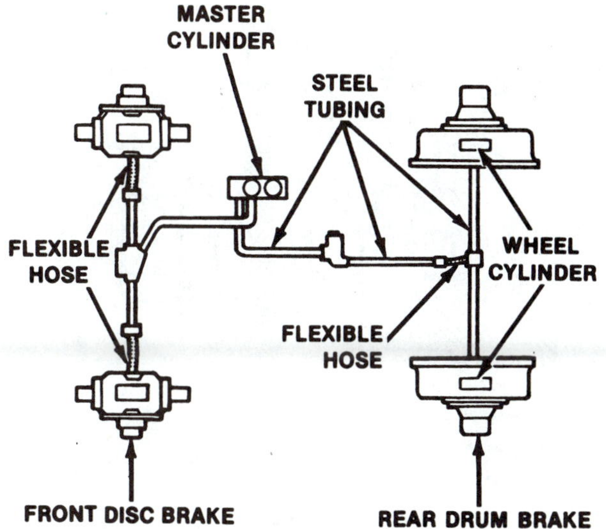

Figure 6-60 Most vehicles use a flexible hose at each front caliper and one (tandem split) or two (diagonal split) hoses at the rear axle or wheel assemblies. Steel tubing is used for the remaining lines. (Courtesy of Brake Parts, Inc.)

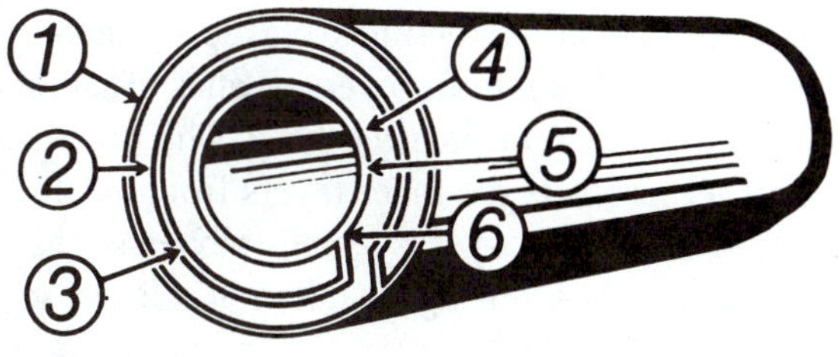

1. Tinned copper-steel alloy protects outer surface.
2. Long-wearing and vibration-resisting soft steel outer wall.
3. Fused copper-steel alloy center unites two steel walls.
4. Long-wearing and vibration-resisting soft steel inner wall.
5. Copper-steel alloy lining protects inner surface.
6. Bundyflex's beveled edges and single close-tolerance strip results in no inside bead at joint. The tubing is uniformly smooth, inside and out.

Figure 6-61 Quality brake tubing is made by wrapping a strip of steel into a two-layer tube, fusing it together, and applying corrosion protection. (Courtesy of Wagner Brake)

3/16-inch tubing uses a 3/8-inch-by-24 thread. With 1/4-inch tubing, a 7/16-inch-by-24 thread is used. ISO flare nuts use metric threads. Manufacturers commonly use tubing with oversized nuts for certain locations to ensure that the components are assembled correctly on the assembly lines. Occasionally this creates a problem when replacing a tube or a master cylinder. Step-up or step-down fittings are available to allow connection of standard-sized tube nuts to nonstandard-sized tube seats, or vice versa (Figure 6-65).

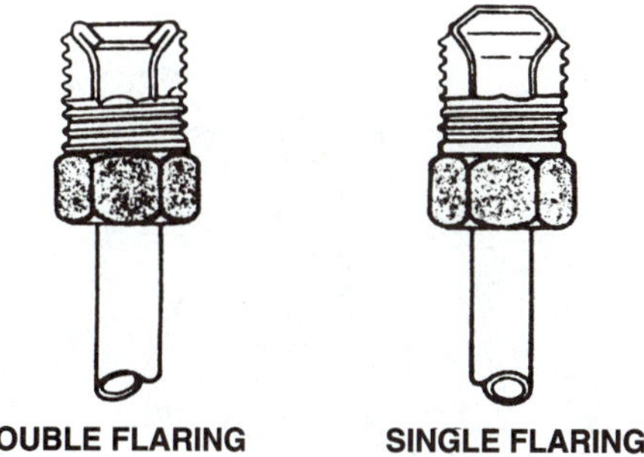

Figure 6-62 Brake tubing uses either a double flare or a single flare. The tube nuts look alike, but the end of the tube has a different shape. (Courtesy of Bendix Brakes, by AlliedSignal)

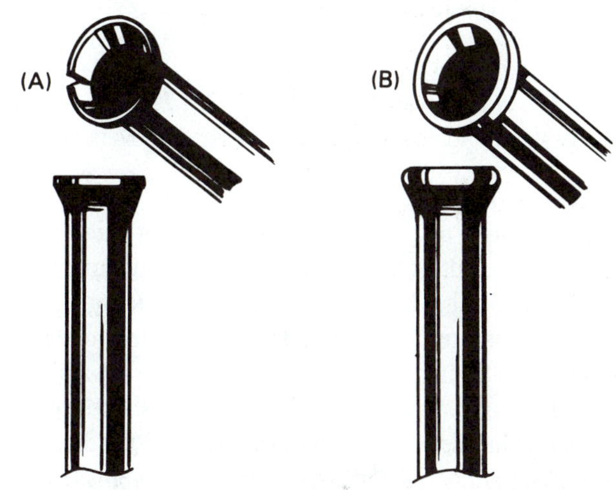

Figure 6-63 A single flare on steel tubing will usually split (A). A double flare (B) will not. (Courtesy of General Motors Corporation, Service Technology Group)

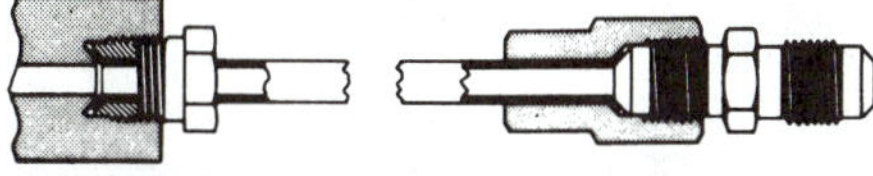

Figure 6-64 A standard flare nut (right) has female threads; an inverted flare nut (left) has male threads. (Courtesy of ITT Automotive)

Tubing failure and/or damage to the tube nut is usually caused by collapse of the tubing from impact, rusting through of the tubing, twisting of the tubing because of improper wrench technique, or kinking of the tubing from

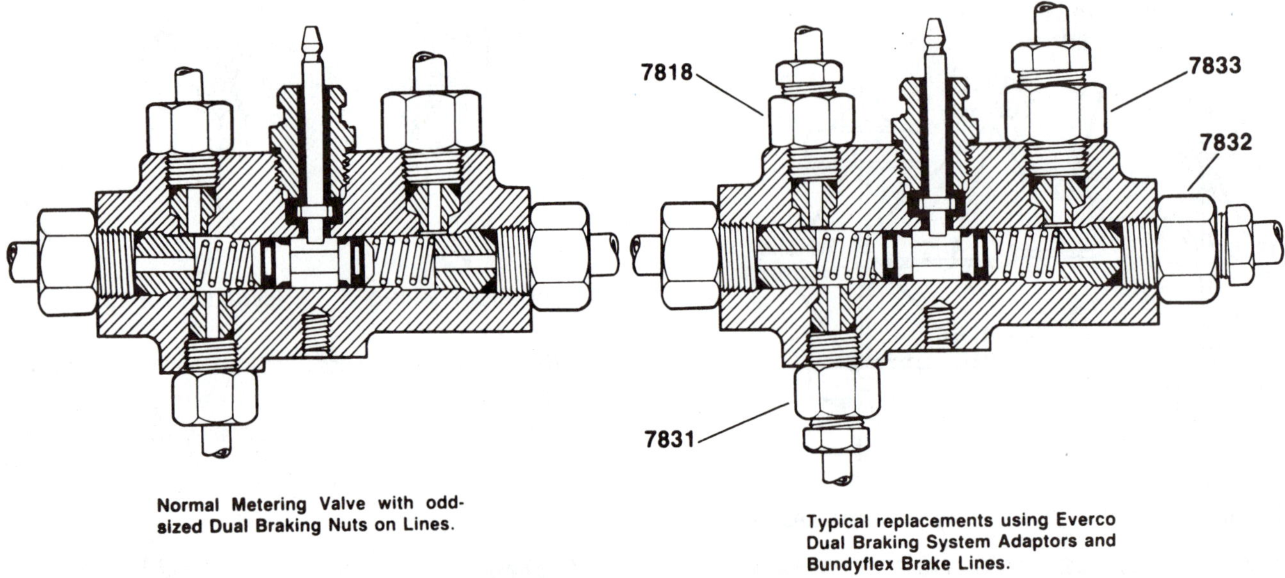

Figure 6-65 Fitting adapters (7818, 7833, 7832, 7831) allow the connection of an inverted-flare nut to a port of a different size. Many master cylinders and valve assemblies use oversize tube nuts to prevent improper assembly (Courtesy of Wagner Brake)

trying to bend it too tight. Note that most of these problems are the result of poor work habits. Replacement tubing is available in different diameters and lengths. It is usually sold as a unit, assembled with flares and tube nuts. Replacement tubing is often plated to resist rust and corrosion. It is often available with a spring-shaped guard around it. Guarded tubing is used where road hazards, abrasion, or salt corrosion are particular problems (Figure 6-66).

Copper tubing must never be used for brake lines. It might burst because its working pressure is less than that of steel tubing. Also, copper tends to work-harden from vibration, and this can cause cracks to form in the tubing.

Brake hoses are generally composed of rubber hose that is reinforced with woven fabric. The fabric strengthens the rubber to prevent expansion under pressure. The inner layer of rubber (sometimes plastic) contains and transfers the fluid. The middle layers or plies of fabric and rubber, usually two, reinforce the hose. The outer layer, called a jacket, protects the inner layers from weathering and abrasion. The outer layer is usually ribbed so that any twists are easily seen during installation. Hoses should not be twisted or kinked too tightly or internal damage will result (Figure 6-67).

Brake hoses are available in different lengths and with one of three different forms of end fittings: male threads,

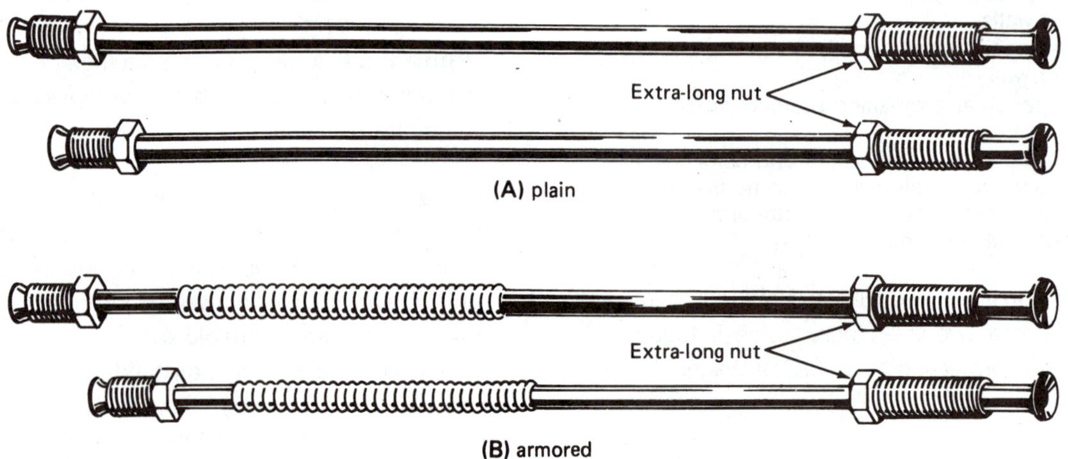

Figure 6-66 Replacement tubing is available in a variety of lengths for the more popular sizes and in a plain or an armored form. (Courtesy of Wanger Brake)

Figure 6-67 Flexible rubber hoses are usually of a two-ply construction. The fluid-carrying inner liner is reinforced by two layers of woven fabric. Note how the end of the hose is crimped tightly into the end fitting. (Courtesy of Wagner Brake)

female threads, or banjo fittings (Figure 6-68). The male thread can be a straight SAE or metric thread that is sealed by a copper gasket, or one that is sealed by a tapered seat at the end of the thread. The more common female threads include a seat for either a SAE or an ISO flare. A **banjo** fitting uses a ringlike nipple. A special hollow bolt is used to transfer fluid from the fitting to the unit where the hose is connected. A pair of copper gaskets, one on each side, is required to seal a banjo-type connection.

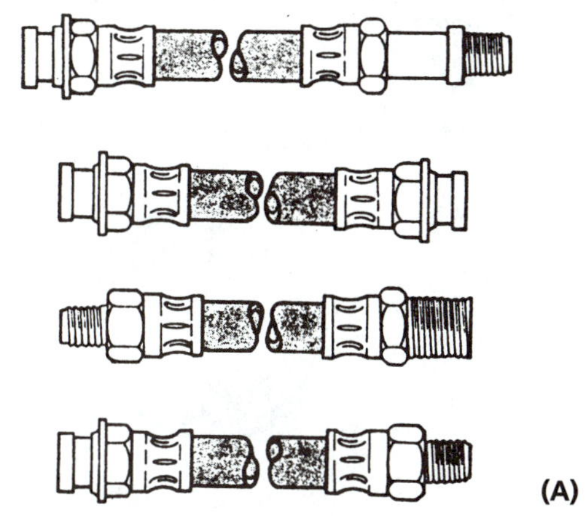

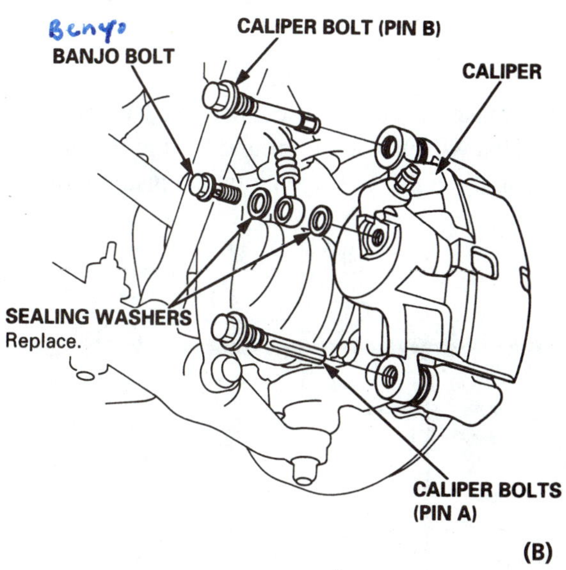

Figure 6-68 Typical brake hose ends (A); note the different thread types. Some hose ends are the banjo type that requires two sealing washers and a special drilled banjo bolt. (A is courtesy of Bendix Brakes, by AlliedSignal; B is courtesy of American Honda Motor Company)

The most common causes for hose failure are weather and abrasion that breaks through the jacket and weakens the hose. This can cause a burst hose. Other problems include a collapse of the inner liner or a torn flap in the inner liner. The first of these problems will cause constant fluid restriction, which usually results in brake drag. In the second case, the torn flap acts like a one-way valve, which usually causes brake drag also. Occasionally, a leak can form in the inner liner, causing bulges or bubbles in the jacket. Normal repair of a faulty hose involves replacing it with the same type and length of hose.

6.19 Hydraulic Control Valves and Switches

All modern brake systems include two or more valves and switches to help control the system pressure to the wheel cylinders or calipers; to warn the driver of a system failure, pad wear, or low brake fluid; and in some cases, to automatically turn on the stoplights. All tandem or diagonal split systems include a pressure differential valve and switch, which turns on the brake failure warning light to alert the driver if a hydraulic failure occurs in one of the systems (Figure 6-69). In all cars, stoplights are activated when the brakes are applied to warn following drivers that the brakes are being applied. Most cars have a switch on the parking brake, as described in Chapter 5, to remind the driver that the parking brake is on.

Many cars using a combination of disc and drum brakes include a **proportioning valve** to reduce the possibility of rear-wheel lockup under hard stops, and/or a **metering valve** to ensure both the disc and drum brakes will apply at the same time. These two valves are often combined into a single unit with the differential pressure valve. This assembly is called a **combination valve**.

Many cars also include a warning light to inform the driver that the brake fluid is low or that the front pads are excessively worn. The light(s) are controlled by a switch

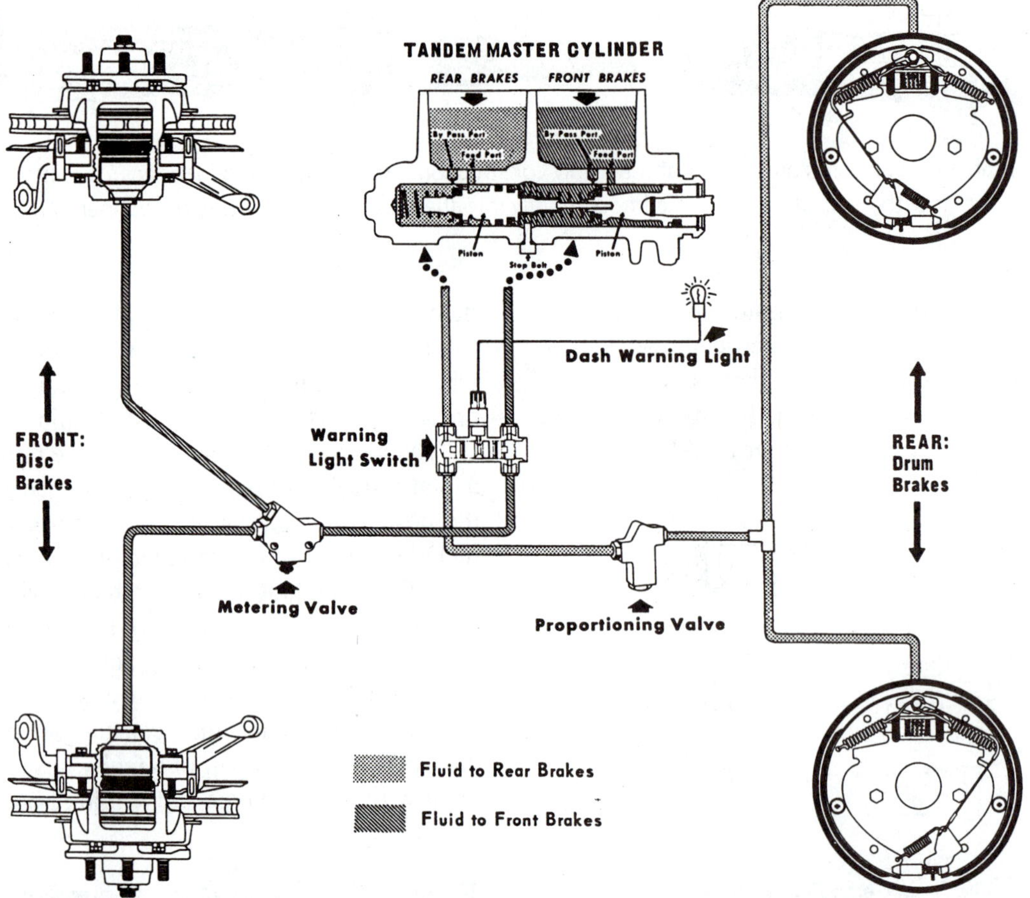

Figure 6-69 This tandem system uses three separate valves—a metering valve in the front brake line, a proportioning valve in the rear brake line, and a pressure differential valve to operate the warning light switch. (Reprinted from Mitchell Anti-Lock Brake System, with permission of Mitchell Repair Information, LLC)

at the master cylinder reservoir or by contacts in the brake pads as described in Chapter 4.

6.19.1 Pressure Differential Valve-Switch

This unit, also called a *warning light switch*, a *dash lamp switch*, or a *system effectiveness switch*, is a combination of a valve and a switch; the valve operates the switch (Figure 6-70). The valve is connected hydraulically to the primary and secondary hydraulic sections. The valve piston floats between these two pressures or two springs of equal pressure. When the brakes are released, the valve is held out or moved to the center position by the two springs. When the brakes are applied, the valve is still centered by the equal hydraulic pressure in the two systems.

If there is a pressure loss in the primary or secondary system the pressure remaining in the other section will move the piston off-center. The movement of the piston will then make an electrical connection with the switch contact. This will provide a ground for the warning light and turn on the light. In some units, the piston movement

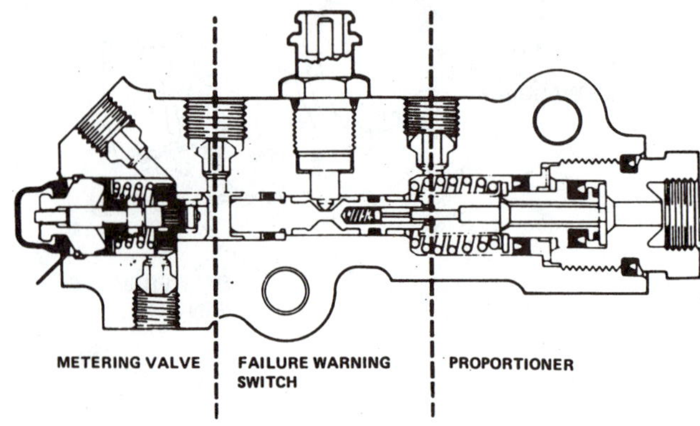

Figure 6-70 This combination valve contains the three valve sections. (Courtesy of General Motors Corporation, Service Technology Group)

operates a switch which also provides a ground for the warning light. When the brakes are released in most systems, the springs will recenter the piston and turn off light (Figure 6-71).

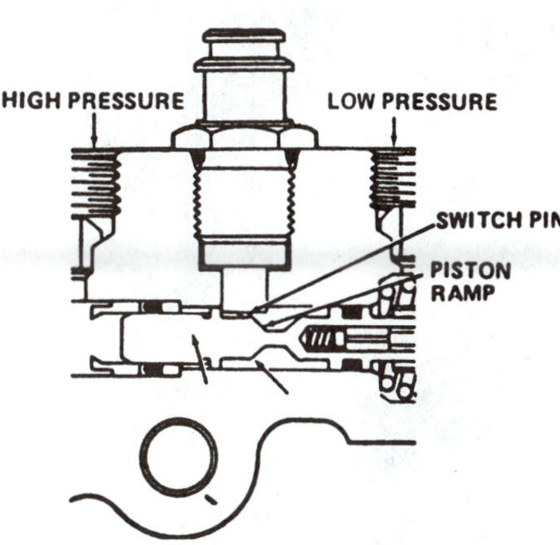

Figure 6-71 If there is a pressure loss—failure in one of the sections—the valve piston will move off-center and turn on the switch to illuminate brake warning light. (Courtesy of General Motors Corporation, Service Technology Group)

Some early-design valves, used mostly on Ford products in the late 1960s, had no centering springs. This type of valve would not automatically recenter (Figure 6-72). If a pressure loss occurred, the light would stay on until the valve was recentered and rearmed. To rearm the valve, a bleeder screw at the front or rear—whichever side that did not fail—is opened while the brake pedal is slowly applied. At the instant that the light goes out indicating a centered piston, the bleeder should be closed. Too much speed or pressure at the brake pedal will cause the valve to go past center. If this happens, the procedure will have to be repeated while opening a bleeder screw at the other end of the car.

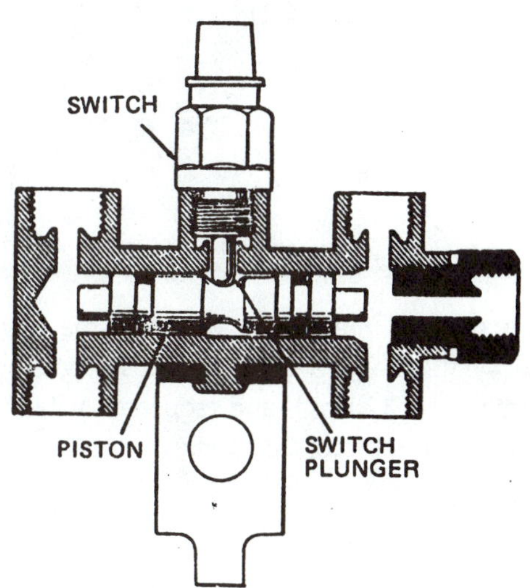

Figure 6-72 This pressure differential valve has no centering springs so it will remain in a noncentered position. (Courtesy of of Wagner Brake)

The *failure warning light* uses a rather simple circuit (Figure 6-73). An electrical wire is connected to the ignition terminal of the ignition switch, then to the fuse block, and then to one side of the warning light. The second side of the warning light is connected to the differential switch and also to the lamp test or *proof* terminal of the ignition switch. The same grounded side is usually also connected to the switch at the parking brake. The warning light will turn on when the ignition switch is on and the differential switch is grounded by an off-center piston, when the parking brake is applied (on most systems), and when the ignition switch is turned to the "start" position and the lamp test circuit is closed. This last circuit is used each time the car is started to inform the driver that the light works.

In a few cases, the differential valve is built into the body of the master cylinder with a passage from each end of the valve to the primary and secondary systems. In most cases, the outlets of the master cylinder will be connected to each end of the valve by steel brake tubing. Brake fluid will pass the ends of the valve on its way to the wheel cylinders or calipers.

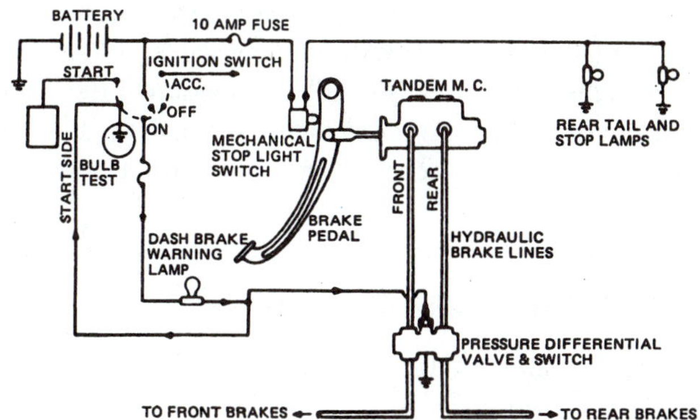

Figure 6-73 A basic brake warning light and a simple stoplight circuit. (Courtesy of Wagner Brake)

6.19.2 Metering Valve

The metering valve, also called the *hold-off* or *disc-balancing valve*, is used in some disc-drum combination systems. It is connected between the pressure differential valve and the front calipers.

Disc brakes can be applied at lower pressures than drum brakes because there are no brake shoe return springs. It takes a pressure of about 100 to 150 psi (690 to 1,034 kPa) in drum brake wheel cylinders to move the brake shoes into contact with the drum. The metering valve is designed to be closed at lower pressures and wide open at about 75 to 135 psi (517 to 931 kPa) so that disc brake application will be held off until drum brake application. A metering valve is also open at very low pressures, below 5 to 15 psi (34 to 103 kPa), so that compensation can take place for fluid expansion or contraction in the calipers (Figure 6-74).

Metering valves have an external stem so that they can be held open during operations using a pressure bleeder. Brake bleeder pressure is usually high enough to close off the metering valve. The valve stem can be of either the *push type* or the *pull type*. A pull-type stem is pulled outward to open the valve, and a push-type is pushed inward. If you touch this stem while the brakes are being applied, you can usually feel the valve operate (Figure 6-75).

Metering valves are not used on all modern systems. In mild climates, the worst thing that will occur without one is early disc pad wear on cars with low pressure or easy stops, or when drivers rest their left foot on the brake pedal while driving. In severe climates with icy roads, the metering valve holds back front brake application until the rear brakes can apply. This reduces the possibility of front brake lockup under light, gentle pedal application.

The metering valve can be a separate valve connected to the outlet of the pressure differential valve, or it can be built into the same assembly with the pressure differential valve and/or the proportioning valve. When used, a line to each front caliper is connected to the metering valve outlet.

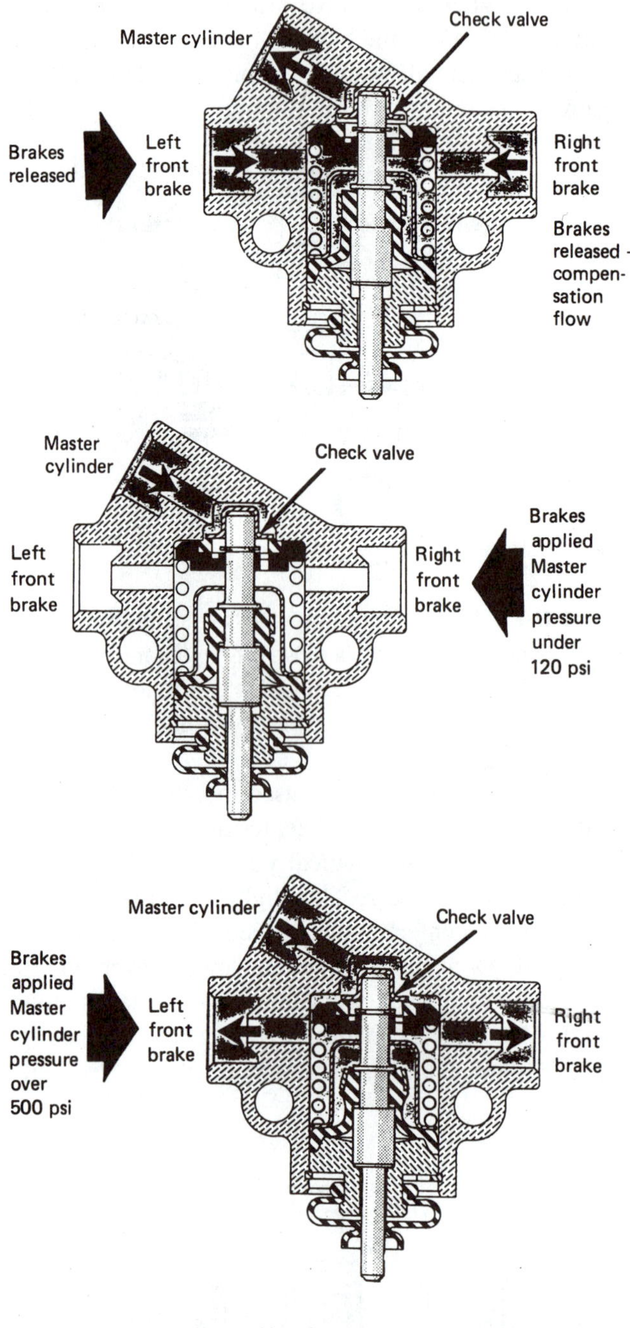

Figure 6-74 A metering valve. Note the valve is partially open while the brakes are released, closed when there is a low pressure (under 120 psi), and reopened at higher pressures. (Courtesy of Ford Motor Company)

6.19.3 Proportioning Valve

The proportioning valve, also called a *pressure-reducing valve*, a *pressure-ratio valve*, a *pressure-regulating valve*, a *pressure control valve*, or an *apportioning valve*, is used to reduce the hydraulic pressure going to the rear wheel cylinders. During a hard stop, while weight transfer is moving weight off the rear end, the rear wheels will tend to lock up and skid. This is especially true if the car has duo-servo brakes at the rear in combination with front disc brakes (Figure 6-76).

The proportioning valve is located between the pressure differential valve and the rear wheel cylinders. At a preset pressure, usually somewhere between 400 and 600 psi (2,758 and 4,137 kPa), the valve will begin to reduce the rate of pressure increase at the rear. This is called the **changeover** or **split point**. This is called the **slope**. After the changeover point, the rear pressure will increase at a rate slower than that of the front brakes. A slope of 0.50 will allow the rear brake pressure to increase at a rate of 50 percent of that of the front. Most passenger-car valves will let the rear pressure increase at somewhere between

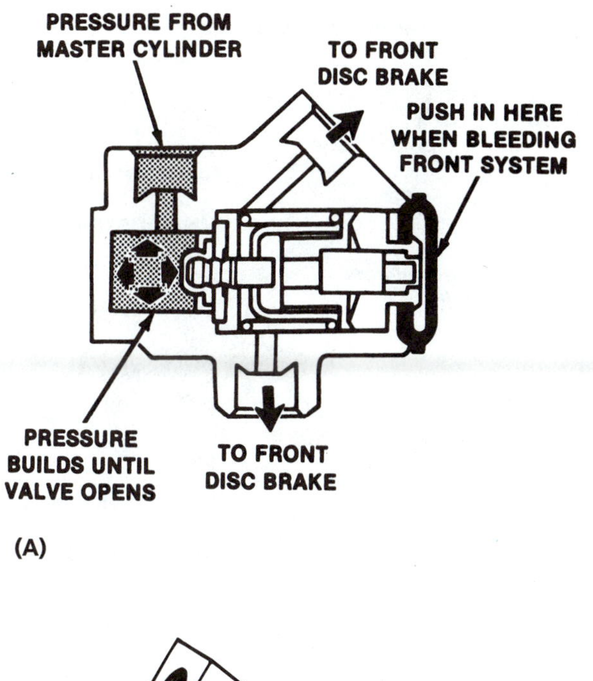

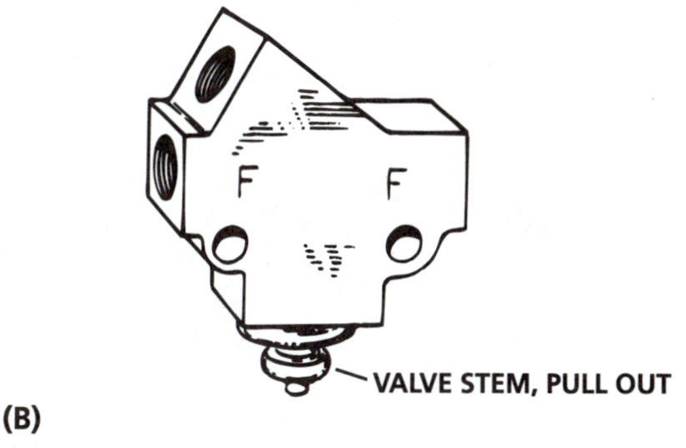

Figure 6-75 Push-type (A) and pull-type (B) metering valves. The name refers to the direction the valve stem must be moved to open it when using pressure brake bleeders. (A is courtesy of Brake Parts, Inc.; B is courtesy of Wagner Brake)

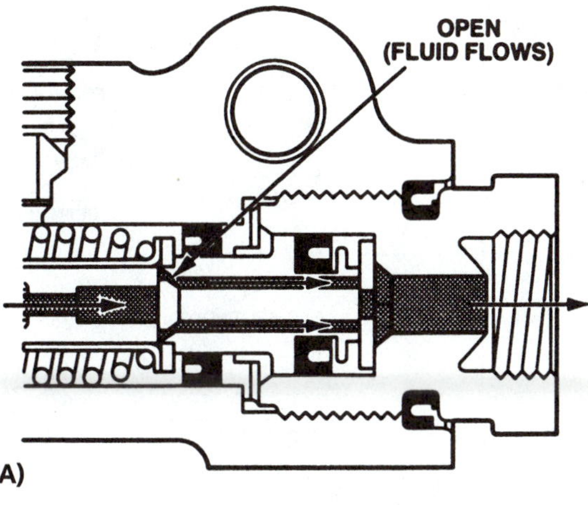

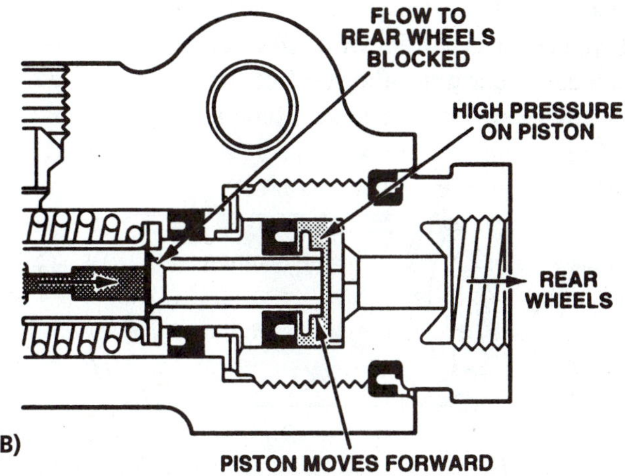

Figure 6-76 At lower pressures, the proportioning valve is open and does not change fluid pressures (A). Above the split point, the valve will reduce the pressure to the rear brakes (B). (Courtesy of General Motors Corporation, Service Technology Group)

25 and 60 percent of the front pressure increase. As mentioned earlier, the balance between the front and rear brakes during a hard stop is a compromise between too much rear brake and a skid, or too little rear brake and a loss in braking power. Engineers try to design the changeover point, and the ratio of the pressure increase from that point, according to the size, weight, and type of car (Figure 6-77). The car should be balanced just short of a rear-wheel lockup under heaviest-load conditions on dry pavement. Some proportioning valves are marked with the split point and slope (Figure 6-78).

Many cars with diagonal split systems will have a double proportioning valve arrangement. Some cars use two individual proportioning valves mounted directly on the master cylinder (Figure 6-79). Other cars use a pair of proportioning valves or a dual proportioning valve mounted remotely. The dual valve contains two separate valves mounted in the same housing (Figure 6-80). Two lines come from the master cylinder, and a line goes to each rear wheel cylinder.

Some cars, pickups, and light trucks are equipped with a **height-sensing proportioning valve**. A vehicle needs more rear braking power as weight is added to the back end. A height-sensing valve has an external lever that is attached by a link to the rear axle (Figure 6-81). Heavier loads, which lower the rear of the vehicle, will cause the valve to move the changeover point to a higher setting. Most valves are somewhat adjustable for fine-tuning of the brake balance.

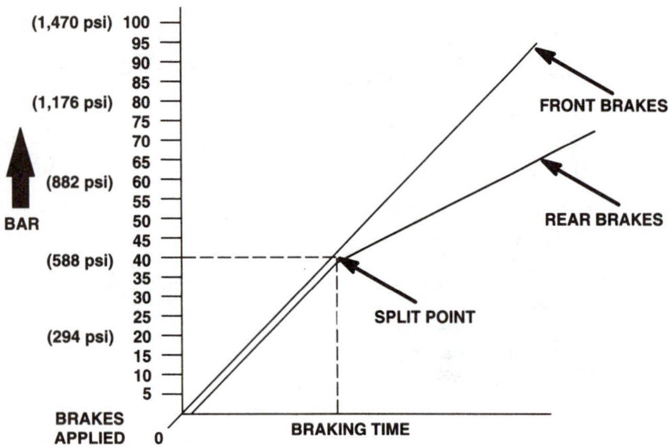

Figure 6-77 A typical brake pressure curve shows a split point at 40 BAR and a slope of 50 percent; the rear brake pressure is increasing at 50 percent of the front pressure rate of increase. (Courtesy of Chrysler Corporation)

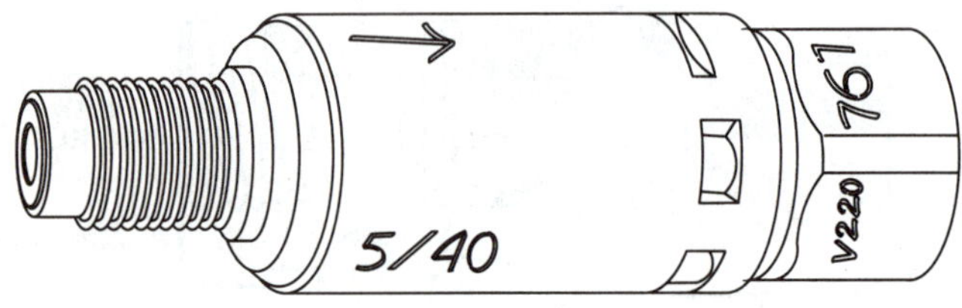

Figure 6-78 This proportioning valve is marked with the split point and slope calibration. (Courtesy of Chrysler Corporation)

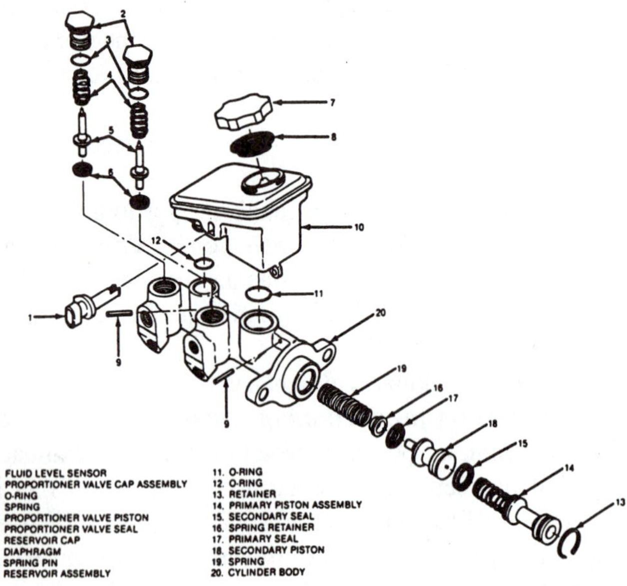

Figure 6-79 This diagonal split system master cylinder has a proportioning valve at each rear outlet port. (Courtesy of General Motors Corporation, Service Technology Group)

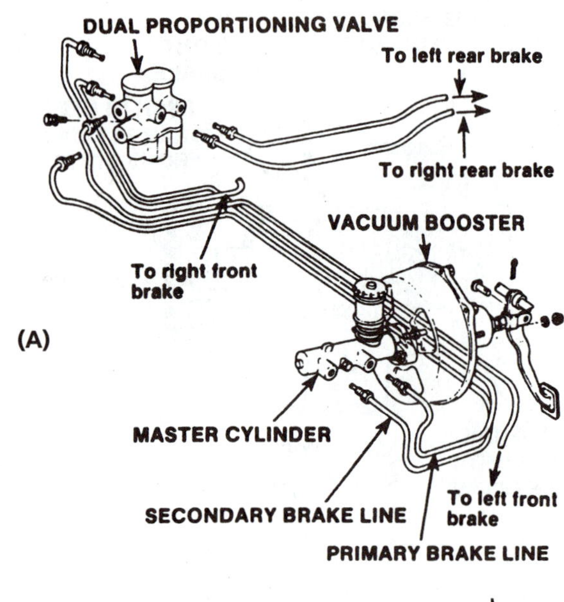

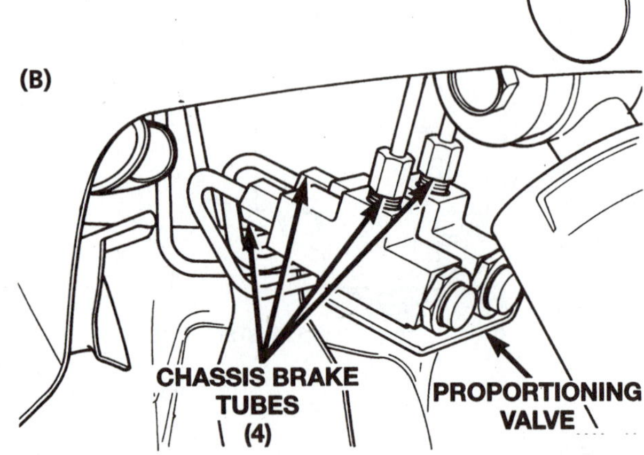

Figure 6-80 Diagonal split systems that use proportioning valves can use a dual proportioning valve (A) or a pair of proportioning valves (B). (A is courtesy of Bendix Brakes, by AlliedSignal; B is courtesy of Chrysler Corporation)

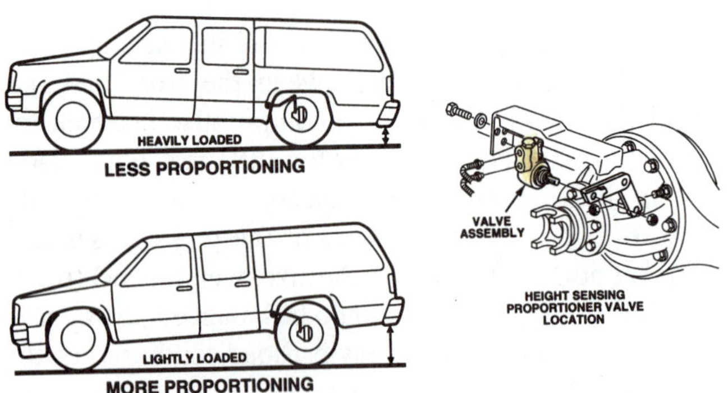

Figure 6-81 Many light trucks, vans, and wagons use a height-sensing proportioning valve connected between the body and rear axle. This valve allows more rear braking while heavily loaded, and less while lightly loaded, to reduce rear wheel lockup. (Courtesy of General Motors Corporation, Service Technology Group)

6.19.4 Stoplight Switches

The **stoplight switch**, also called the *brake light switch*, turns on the stoplights at the rear of the car when the brakes are applied. Modern cars use a mechanical switch that is activated by brake pedal movement. Older cars use a hydraulic pressure switch that is activated by system pressure. Some switches contain two circuits. The second circuit shuts off the cruise control when the brakes are applied.

Several styles of mechanical switches are used (Figure 6-82). They are all normally closed (completed circuit) switches that are opened (switched off) when the brake is released. Most of them are mounted on a bracket next to the pedal so that release of the pedal will open the switch and turn the brake lights off. Switch brackets and switches are often adjustable so that the point of switch operation can be changed. When adjusting a switch, it is important to make sure that the brake pedal is not held in a partially applied position by the switch.

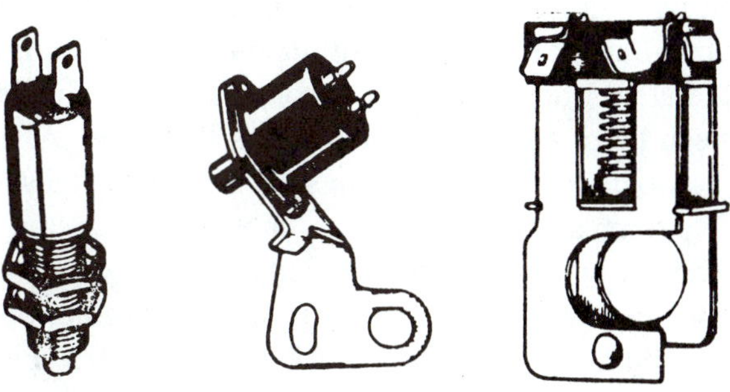

Figure 6-82 Three different mechanical stoplight switches. They are mounted so that brake pedal movement will close the switch and turn on the stoplights. (Courtesy of Wagner Brake)

Several styles of hydraulic switches are also available; the major difference between them is the size and type of electrical connections. This switch is usually threaded into a tee connection in the brake tubing. Sometimes it threads directly into the master cylinder body. It is usually a normally open (no circuit) switch that is closed (turned on) by system pressure. As the brakes are applied, hydraulic pressure closes the switch and turns the stoplights on.

Stoplights use a fairly simple electrical circuit. It begins at a fuse or circuit breaker, and a wire carries battery positive (B+) power to one terminal of the switch. The second wire of the stoplight switch, goes to the turn indicator switch which splits the circuit to the left and right stoplights. In cars with three rear lights—tail, turn, and stop—on each side, the wire from the stoplight switch connects directly to the two or three stoplights (Figure 6-83).

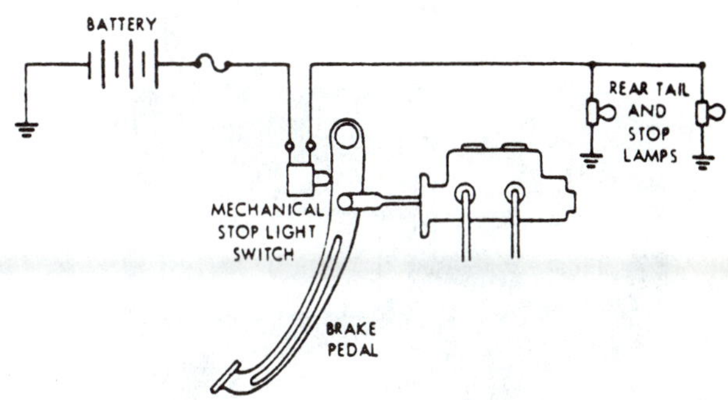

Figure 6-83 This simple stoplight circuit is for a system with separate turn indicators. (Reprinted from Mitchell Anti-Lock Brake Systems, with permission of Mitchell Repair Information, LLC)

6.19.5 Fluid Level Switches

Low-brake-fluid warning lights are used on many new cars. This light is controlled by a switch in the master cylinder reservoir. The switch is operated by a float. When the brake fluid level drops, the float lowers and closes the switch contacts. When the contacts close, the light is turned on (Figure 6-84).

The low-brake-fluid warning light also has a fairly simple electrical circuit. A wire connects B+ voltage from the ignition switch through the light to one terminal of the fluid level switch at the master cylinder. Another wire connects the other switch terminal to ground. If the switch closes because of low brake fluid, current will flow through the completed path to ground and the light will turn on.

An odd thing related to this switch can occur on some vehicles. Electrical radiation from a spark plug wire can turn on the low-brake-fluid warning light. Spark plug wires should be kept at least 2 inches away from the switch area to keep this from happening.

6.20 Brake Fluid

No hydraulic system can operate without fluid, and the automotive hydraulic brake system has some very serious requirements. Brake fluid quality is regulated in the United States by federal standards established by the Department of Transportation (DOT) and the National Highway Traffic Safety Administration (NHTSA) and also by various departments in many states. These standards were established by the Society of Automotive

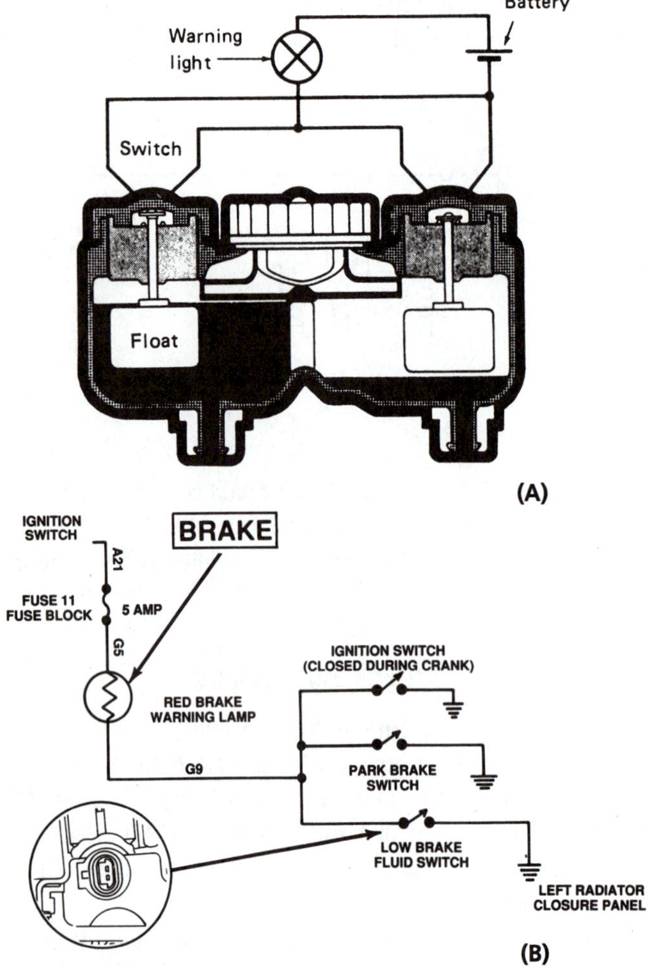

Figure 6-84 Low-brake-fluid warning circuits use a switch(es) in the master cylinder reservoir (A). The switch can be a parallel circuit to the parking brake switch (B). (A is courtesy of ITT Automotive; B is courtesy of Chrysler Corporation)

Engineers (SAE). Brake fluid must possess the following characteristics:

- **Viscosity:** It must be free-flowing at all temperatures.
- **High boiling point:** It must remain in the liquid state at the highest temperatures that are normally encountered.
- **Noncorrosive:** It must not attack plastic, rubber, or metal parts.
- **Water tolerance:** It must be hygroscopic—able to absorb and retain moisture that collects in the system.
- **Lubricating ability:** It must lubricate the pistons and cups to ensure free movement and reduce wear and internal friction.
- **Low freezing point:** It must stay fluid and flow at the lowest operating temperatures.
- **Compatibility:** It must be compatible with other brands of brake fluid as well as all components in the system.

In addition, there are specified minimum boiling points for the three classes of brake fluid as shown in Figure 6-85.

DOT 3 and DOT 4 fluids must be amber to clear in color and are generally made from a *polyglycol* base. DOT 5 fluid must be purple in color and uses a *silicone* base. DOT 3 and DOT 4 fluids can be mixed with each other, but neither one should be mixed with DOT 5 fluid. Silicone fluid is lighter and will float on glycol fluid. It is said that a water film will tend to form at the layer between them and cause corrosion at that level. Also, if these two fluids are mixed and then agitated, they will make a foam and encapsulate air bubbles.

The dry boiling points specified are for new fluid with no absorbed water (Figure 6-86). The wet boiling points are for fluid that has absorbed 2 percent water (Figure 6-87). The boiling points are often printed along with the term **Equilibrium Reflux Boiling Point (ERBP)**, which refers to the method of measuring the boiling point.

There are a few more types of brake fluids. A few import cars—Rolls Royce and Citroen—have seal materials that are designed to be used with a special hydrocarbon brake fluid called *hydraulic system mineral oil (HSMO)*. Many older British-make cars required the use of a special brake fluid. One of the original brake fluids used a castor-oil base. It is important to follow the fluid requirements of the manufacturer to ensure long life of the seal materials. Older glycol fluids had a boiling point much lower than that of today's fluids; at one time the

Boiling Point	*DOT 3*	*DOT 4*	*DOT 5*
Dry	401° F (205° C)	446° F (230° C)	500° F (260° C)
Wet	284° F (140° C)	311° F (155° C)	356° F (180° C)

Figure 6-85 The dry and wet boiling point requirements of DOT 3, DOT 4, and DOT 5 brake fluid.

Figure 6-86 Brake fluid containers are marked with the DOT rating and the wet and dry boiling points.

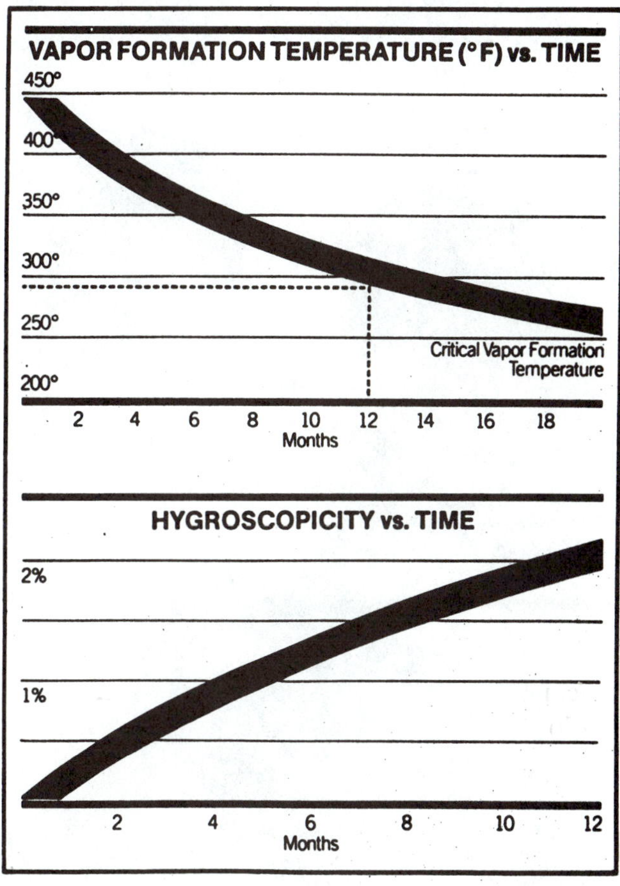

Figure 6-87 After about twelve months in humid areas, the average brake system will absorb about 2 percent water. Note that as water content increases, the fluid's boiling point lowers. (Courtesy of Stainless Steel Brakes Corp.)

brake fluid standards were SAE J1703 or 70R3—boiling point 374°F (190°C)—and SAE 70R1—boiling point 302°F (150°C). These older fluids are no longer used.

It is important to not mix industrial hydraulic oils with automotive brake fluid. They are petroleum based and will cause excessive expansion and swelling of the seal materials.

Glycol fluids have two serious drawbacks. They are **hygroscopic**, which means they will absorb water. Glycol fluids will also attack paint if fluid is spilled on the car's finish, and the paint will be discolored or even lifted. Spilled fluid should be immediately flushed with a lot of water.

Water enters most brake systems by permeating the rubber hoses and diaphragm at the reservoir. Water molecules (vapor) can pass through the pores of the rubber. The rate of water absorption will vary with the humidity of the environment. It has been found that the average automobile reaches 2-percent water in the brake fluid in about eighteen months. In humid areas, brake fluid will gain water at a rate of 2 percent or more per year. Water absorption by the fluid presents two serious problems: boiling point loss and corrosion. Two-percent water absorption has a serious effect on the boiling point, lowering it to about three-fourths of the dry boiling point (Figure 6-88). Cars operating under severe braking

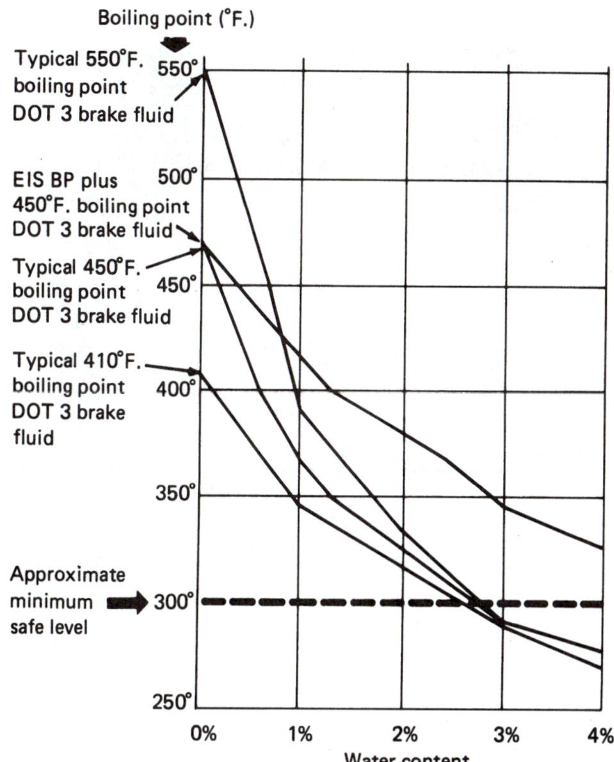

Figure 6-88 This chart shows the dry and wet boiling points for several brake fluids. (Reprinted from Mitchell Anti-Lock Brake Systems, with permission of Mitchell Repair Information, LLC)

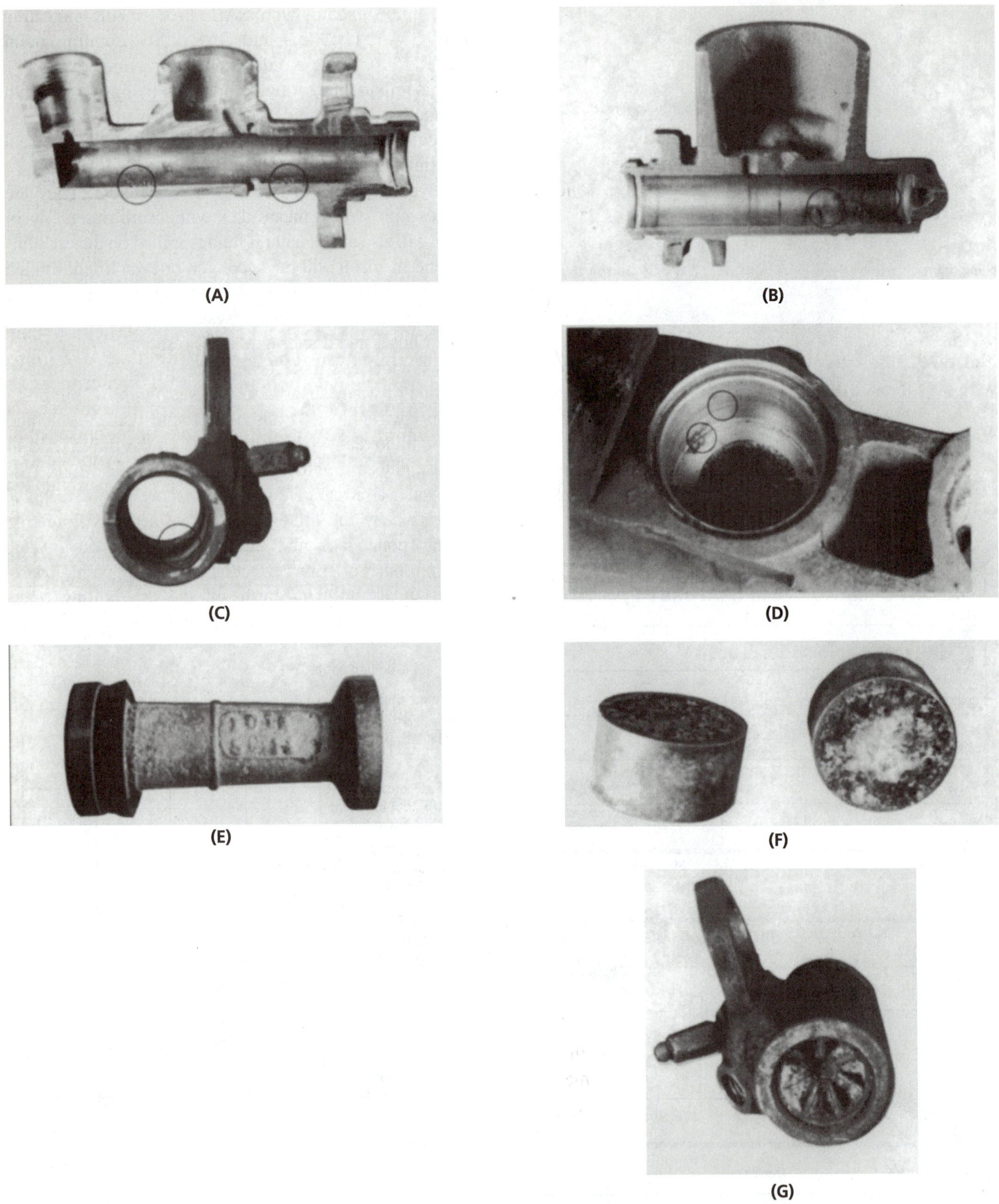

Figure 6-89 Contaminated fluid has caused corrosion and failure of these components. The aluminum master cylinder bore (A) has deep pits; the cast-iron master cylinder (B) and wheel cylinder (C) bore have mounds of rust; the caliper bores (D) have severe rust pits; and the aluminum pistons (E, F, G) show severe corrosion.

conditions, with fluid that is several years old, can experience fluid vaporization—which will cause a spongy pedal or even a total loss of braking power as the fluid heats up. During an overhaul of a system that is a few years old, the corrosive effects of brake fluid are seen as rust in the lower portions of cast-iron cylinders and corrosion in the lower portions of aluminum cylinder bores and on aluminum pistons. When the rust or corrosion is removed, the severe pitting that results still remains. It is this pitting that usually ruins these parts (Figure 6-89).

Silicone fluids do not share these problems. They are nonhygroscopic, so they will not absorb water; and they have no effect on painted surfaces. Silicone fluid, though, will not accept paint on top of it. It must be carefully cleaned from any surface to be painted. Since no water is absorbed, there will be very little or no corrosion in the system. Silicone fluid does have a few drawbacks. It is expensive, costing at least three to four times as much as glycol fluids; in some areas, it costs as much as ten times as much. Silicone fluid is also compressible when it gets hot. This will have the effect of causing a spongy brake pedal under severe operating conditions. Silicone fluid also has a greater tendency to foam when it is forced through small openings or experiences rapid vibrations. Because of this tendency to foam, silicone fluid should not be used in ABS unless specified by the manufacturer of the system.

Silicone fluid can also cause some of the rubber seals to swell as much as 10 percent. In at least one case after a car was switched to silicone fluid, the primary master cylinder cup swelled enough to stick in the bore and not return completely; this caused a serious brake drag because the fluid could not return through the compensating port.

The following precautions can help minimize the amount of water and its resulting corrosive effects in the system:

- The master cylinder should be kept tightly closed, and it should be closed immediately after filling.
- Brake fluid should be purchased in the smallest practical amount. It should not be stored in partially filled containers for long periods of time.
- Brake fluid containers should be kept tightly closed and capped.
- Brake fluid containers should be kept in clean, dry areas.
- Only clean containers should be used for dispensing or transferring fluid.
- Old fluid should never be reused.
- If there is a possibility that fluid is contaminated, it should be thrown away.

Some sources recommend periodic brake fluid changes so that the deterioration of glycol fluids will have no adverse effects; but in today's world, this is often not economically practical. The cost of the periodic changes will probably be greater than that of the parts ruined by corrosion. With the increased use of ABS with its very expensive hydraulic modulator, periodic fluid changes make sense economically and are a wise safety practice. With vehicles that normally operate at very high fluid temperatures, it is probably worthwhile to change brake fluid because of the safety aspect. The cost of a fluid change is minor compared to that of an accident resulting from brake failure. Most road-racing cars change the brake fluid every race. Most informed sources recommend that it be changed at the time of a major brake repair if not more often. At this time, unless the hydraulic system is disassembled and rebuilt, the system should be flushed with new fluid until all the old, dirty, contaminated fluid is removed.

When servicing brakes, the hydraulic system should never be exposed to petroleum-based chemicals, oils, or solvents. These will attack the synthetic rubber compounds used in sealing cups and O-rings, hoses, and the diaphragm in the master cylinder reservoir. Petroleum products will soften and swell the rubber, thereby ruining the sealing ability, the ability to slide in a bore, or the ability to transfer fluid from one place to another. If a petroleum product such as motor oil, gasoline, or solvent enters the hydraulic system, the system should be drained and disassembled and every rubber part should be replaced (Figure 6-90).

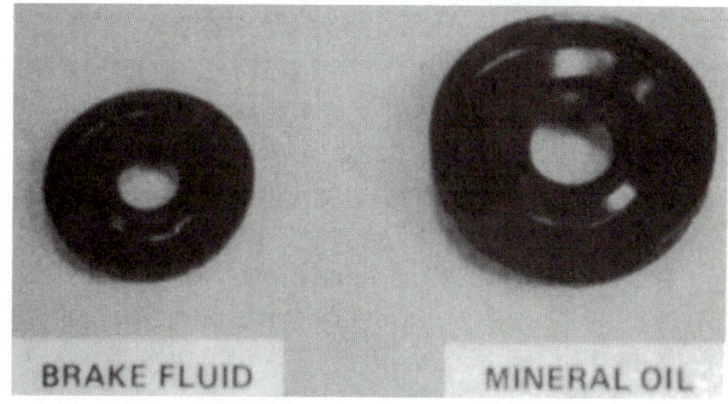

Figure 6-90 These two cups were the same size until the one on the right was placed in mineral (motor) oil. Note how much it has swollen. (Courtesy of General Motors Corporation, Service Technology Group)

6.21 Practice Diagnosis

You are working in a tire shop that also specializes in brake repair and encounter these problems.

CASE 1: The customer has brought in his 1989 Ford Bronco with a complaint of a low brake pedal. He said that his neighbor told him the brakes need to be bled. On checking it, you find the pedal goes almost to the floor fairly easily before becoming solid and firm. What is probably wrong with this vehicle? Is the neighbor's diagnosis correct? What should you do to find the exact cause of this problem?

CASE 2: The customer has brought in a 1983 Honda Prelude with a complaint of a dragging brake at the right front. On checking it you find the wheel is very hot, and when you loosen the bleeder screw, a small amount of fluid squirts out and the brake becomes free. Is this a sign of a faulty master cylinder? A faulty metering or proportioning valve? What should you do to locate the exact cause of this problem?

CASE 3: The customer has brought in his 1991 Chevrolet C1500 pickup with a problem of a grabby left rear brake. Your road test confirms lockup of the tire with only a moderate amount of pedal pressure. What is probably wrong? Could this be a proportioning valve problem? A wheel cylinder problem? What should you do to find the exact cause of this problem?

Terms to Know

anodized
banjo
bleeder screw
brake pedal
bridge bolts
bypass port
changeover
combination valve
compensating port
cup expanders
cups
diagonal split
double flare
dual master cylinder
dust boot
dynamic seal
Equilibrium Reflux Boiling Point (ERBP)
expander washers
front-rear split
heel drag
height-sensing proportioning valve
hydraulic brake system
hydraulics
hygroscopic
International Standards Organization (ISO) flare
inverted flare
master cylinder
metering valve
Pascal's law
pounds per square inch (psi)
primary cup
primary piston
proportioning valve
quick-take-up master cylinder
reservoir
residual pressure check valves
roller burnishing
secondary cup
secondary piston
slope
split point
static seal
stationary seal
stoplight switch
stroking seal
tandem master cylinder
wheel cylinders

Review Questions

1. Hydraulic brake systems are designed to
   a. transmit motion from the driver's foot to the shoes.
   b. transmit force from the driver's foot to the shoes.
   c. multiply the amount of force being transmitted.
   d. all of the above.
2. Hydraulic brakes offer a major advantage in brake systems because they
   a. are self-lubricating.
   b. multiply force.
   c. transmit pressure equally to two or more places.
   d. none of the above.
3. One hundred pounds of force acting on a piston with an area of 1/2 square inch will generate a pressure of
   a. 400 psi.
   b. 200 psi.
   c. 100 psi.
   d. 200 kPa.
4. A fluid pressure of 400 psi acting on a piston with an area of 2 square inches will generate a force of
   a. 400 pounds.
   b. 200 pounds.
   c. 600 pounds.
   d. 800 pounds.

5. Statement A: A hydraulic leak can keep a system from generating high pressures.
   Statement B: Because it is compressible, air in a hydraulic system will cause excessive pressures. Which statement is correct?
   a. A only c. both A and B
   b. B only d. neither A nor B
6. Statement A: The compensating port in a master cylinder allows a fluid flow between the wheel cylinders and the master cylinder reservoir while the brakes are released.
   Statement B: As the brakes are released, fluid will flow from the master cylinder reservoir to the cylinder bore through the bypass port. Which statement is correct?
   a. A only c. both A and B
   b. B only d. neither A nor B
7. Statement A: The master cylinder begins pumping fluid right after the secondary cup passes by the compensating port.
   Statement B: The primary purpose of the primary cup is to keep fluid from leaking out of the end of the cylinder bore. Which statement is true?
   a. A only c. both A and B
   b. B only d. neither A nor B
8. In some brake systems, pressure is kept in the system while the brakes are released by the
   a. wheel cylinder cups.
   b. residual check valve.
   c. replenishing port.
   d. all of the above.
9. In a tandem master cylinder, the primary piston is applied by the pedal pushrod, and the secondary piston is applied by
   a. a second pushrod.
   b. an extension of the primary piston.
   c. hydraulic pressure.
   d. all of the above.
10. In a case of a hydraulic failure in the secondary circuit, the primary piston will operate normally after
    a. the extension of the secondary piston meets the end of the bore.
    b. an extension of the primary piston meets the secondary piston.
    c. both pistons bottom at the end of the bore.
    d. the engine is started.
11. The master cylinder for some diagonal-split systems has
    a. four outlet ports.
    b. a pair of proportioning valves.
    c. an internal differential valve.
    d. all of the above.
12. A quick-take-up master cylinder is designed to
    a. displace more fluid at low pressure and develop the same high pressures normally used.
    b. provide higher pressures during light pedal operation.
    c. operate bigger caliper pistons with less pedal pressure.
    d. all of the above.
13. Statement A: Most wheel cylinders contain two pistons, one spring, and a rubber cup.
    Statement B: Some wheel cylinders use only one piston and cup. Which statement is correct?
    a. A only c. both A and B
    b. B only d. neither A nor B
14. The O-ring around a caliper piston is used to
    a. seal the piston to the bore.
    b. return the piston to a released position.
    c. adjust for lining wear.
    d. all of the above.
15. Statement A: A stroking seal used in a master cylinder moves along with the piston.
    Statement B: A static seal is stationary in the cylinder and allows the piston to move inside it. Which statement is correct?
    a. A only c. both A and B
    b. B only d. neither A nor B
16. All cars use steel brake tubing that has
    A. a double SAE flare at the ends.
    B. an ISO flare at the ends.
    a. A only c. both A and B
    b. B only d. neither A nor B
17. Statement A: Many cars use a proportioning valve(s) in the hydraulic circuit in order to reduce the tendency for rear-wheel lockup during hard stops.
    Statement B: Many cars use a metering valve to reduce the tendency of front-wheel lockup during hard stops. Which statement is true?
    a. A only c. both A and B
    b. B only d. neither A nor B

18. Statement A: A hydraulic pressure failure in one of the hydraulic circuits will center the differential valve and turn on the brake warning light.
Statement B: The differential valve is mounted so that pressure from each hydraulic circuit is at the ends of the valve. Which statement is correct?
    a. A only c. both A and B
    b. B only d. neither A nor B

19. Statement A: The stoplight switch is turned on by brake pedal movement as the brakes are applied.
Statement B: Some cars use a stoplight switch that is turned on by hydraulic pressure. Which statement is correct?
    a. A only c. both A and B
    b. B only d. neither A nor B

20. Statement A: As brake fluid absorbs water, the boiling point increases.
Statement B: Automotive brake fluid standards are controlled by DOT. Which statement is correct?
    a. A only c. both A and B
    b. B only d. neither A nor B

As brake fluid absorbs water, its boiling point decreases.

7 Power Brake Theory

Objectives

After completing this chapter, you should:

- ❑ Be familiar with the different types of power boosters and how they operate.
- ❑ Be familiar with the terms commonly used with power brake units.
- ❑ Understand how each type of power booster operates.

7.1 Introduction

In many driving situations, the driver uses the brakes with frequent regularity, and the brake designs on some vehicles require a large amount of pedal effort. Imagine a four-wheel disc brake system on a full-sized domestic car that is equipped with a full range of accessories and equipment. Without power brakes, very few drivers would have the power to apply the brakes strongly enough to make a hard stop. Power brakes reduce the brake pedal pressure requirement while retaining much of the feel and sensitivity of a nonpower brake. Many vehicles are equipped with power brakes as standard equipment, and they are optional equipment on others.

A power brake system is really a standard brake sytem, sometimes with a different master cylinder but always with the addition of a **booster**. The booster is the source of the power. The booster is usually mounted between the brake pedal and the master cylinder, and it multiplies the force coming from the brake pedal (Figure 7-1). A common term is **power assist**. The booster assists the driver in pushing on the pedal.

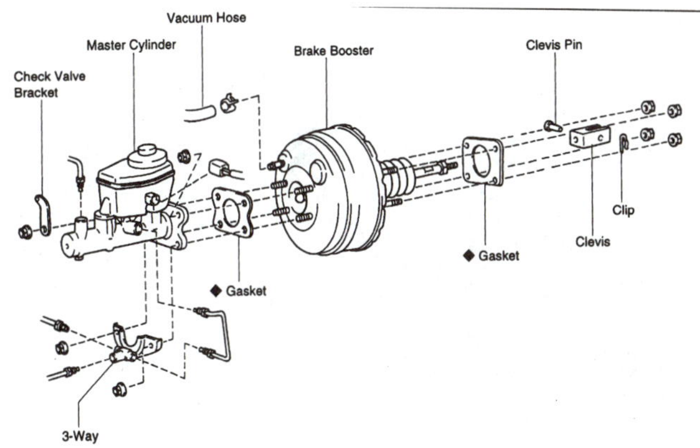

Figure 7-1 A power brake system uses a booster to increase the force from the pushrod to the master cylinder. This system uses a vacuum booster.

In many cases, the power booster is used with a larger-bore master cylinder or a brake pedal with less leverage. The added assist from the booster makes the larger bore or faster leverage practical. Also, the increased displacement from the larger bore helps ensure that there is enough fluid to completely apply the brakes and have a high pedal. This also allows complete brake application with a short stroke of the brake pedal.

Traditionally, power boosters use a vacuum assist. Engine intake manifold vacuum is readily available and almost free. In the early and mid-1970s, hydraulic boosters were introduced. These devices use hydraulic pressure from the power steering system for the assist (Figure 7-2). Hydraulic boosters are smaller in size, more powerful, have a quicker response, and can be easily used in cars with diesel or turbo-charged gasoline engines. Diesel engines do not have a vacuum in the intake manifold; turbo-charged gasoline engines have pressure while under turbo boost. A vacuum pump is required for vacuum booster operation when these engines are used. In the mid-1980s, electrohydraulic boosters were introduced. These units operate similarly to hydraulic boosters and have their own electric-motor-powered hydraulic pump (Figure 7-3). Electrohydraulic boosters run the pump only when necessary for an assist, so they are very efficient and offer improved fuel economy. They are also smaller in size and do not need the hoses between the booster and

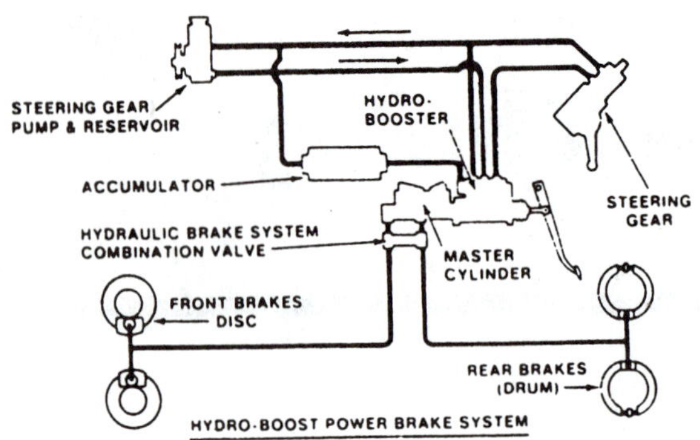

Figure 7-2 Some power booster systems use a Hydro-boost booster. The hydraulic pressure is supplied by the power steering system. (Courtesy of Chrysler Corporation)

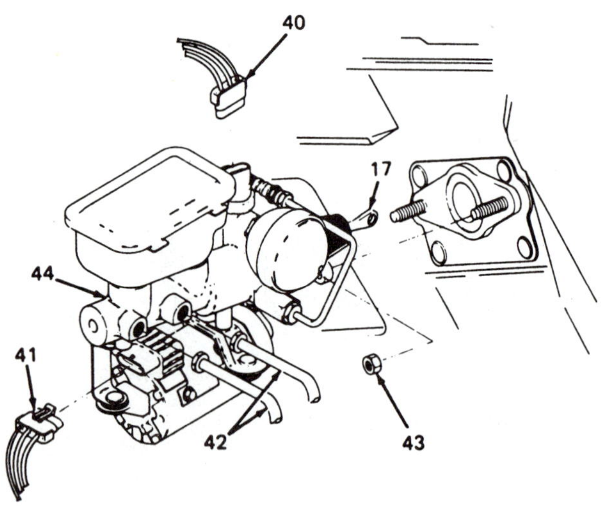

17. PUSHROD
40. ELECTRICAL CONNECTOR
41. ELECTRICAL CONNECTOR
42. BRAKE PIPE
43. NUT
44. POWERMASTER UNIT

Figure 7-3 A Powermaster brake booster combines the master cylinder with an electric motor and hydraulic pump. Note that this booster (44) is being removed for service. (Courtesy of General Motors Corporation, Service Technology Group)

the power steering pump. Electrohydraulic boosters are integrated into the master cylinder and valve assembly of some ABS. Some trucks and busses use a dual power booster. This device is a combination of a vacuum and a hydraulic booster. Another truck booster design combines a hydraulic booster with a reserve electric motor pump. In this case, the electric pump provides hydraulic pressure for stops in case of pressure loss in the power steering system.

7.2 Vacuum Booster Theory

When a gasoline engine operates at a speed less than wide-open throttle, the closing of the throttle plates will cause a **vacuum** in the intake manifold. The action of the pistons pulls air out of the manifold faster than **atmospheric pressure** can push it past the throttle plates.

The term *vacuum* refers to a pressure lower than atmospheric, which is the air pressure created by the weight of the atmosphere around our planet. This pressure is about 15 psi (100 kPa) (1 bar). Pressure less than atmospheric has traditionally been measured in **inches of mercury ("Hg)**, with atmospheric pressure being 0" Hg and a total vacuum being about 30" Hg. An absolute vacuum, with zero air pressure, is equal to 29.92" Hg (Figure 7-4). At idle, many gasoline engines have a manifold vacuum of about 15" to 18" Hg; at cruising speeds it is about 10" Hg; and while decelerating, the vacuum measures about 20" to 21" Hg. The actual measurement for a particular engine will vary depending on the throttle opening, the size of the engine relative to the car, the axle/gear ratio, the mechanical condition of the engine, and the type of emission-control devices. The design of many newer cars has greatly reduced the strength of the manifold vacuum.

In an engine with a manifold vacuum of 10" Hg, there will be a pressure in the manifold of 10 psi (68.9 kPa) absolute. This pressure will be 5 psi (34.5 kPa) less than atmospheric. Imagine a piston with a diameter of 10 inches (25.4 cm) and an area of 78.5 square inches (506.7 cm^2). With atmospheric pressure on one side and a pressure that is 5 psi lower (10" Hg) on the other, the piston

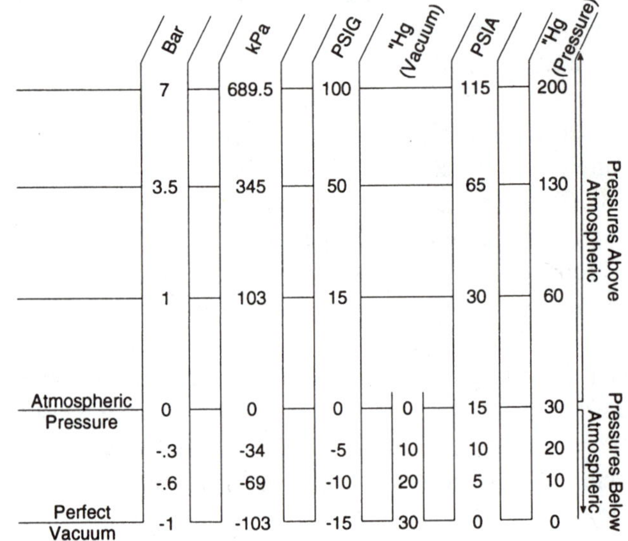

Figure 7-4 Different systems are used to measure air pressure. Pressures below atmospheric are commonly called vacuum.

will generate a force of 5 × 78.5 (392.5) pounds (178.4 kg) toward the low-pressure side. This is approximately the amount of power that power boosters generate. Considering that 50 pounds (22.7 kg) of foot pressure on a brake pedal is fairly high, it is clear that this is a significant assist. The pedal ratio will usually increase the pedal force about six times, to around 300 pounds (136 kg). Booster force will be greater with higher manifold vacuums, larger-diameter boosters, or tandem boosters (Figure 7-5).

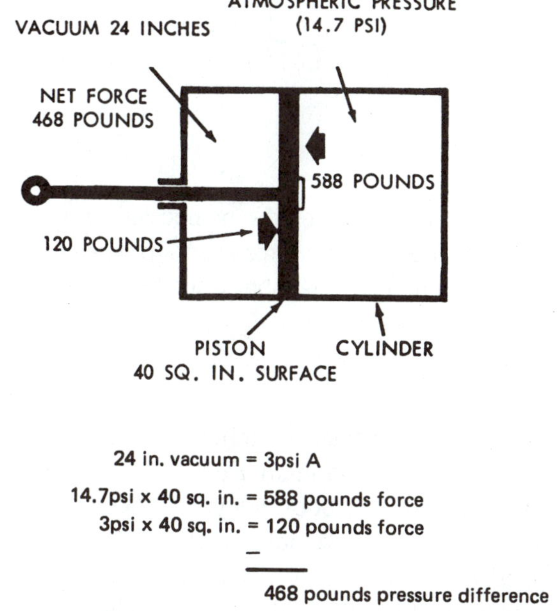

Figure 7-5 If there were a manifold vacuum of 24 inches on one side of a 40-square-inch piston and atmospheric pressure on the other side, there would be a pressure difference of 468 pounds to push the piston toward the left. (Courtesy of John Bean Company)

Diesel engines do not use a throttle plate so they do not have vacuum in the intake manifold. Vehicles with diesel engines and vacuum boosters must use a pump to generate the vacuum for the booster. Vacuum pumps can be powered by an electric motor or run off the engine drive belt, camshaft like a fuel pump, camshaft like a distributor, or mounted on the alternator (Figure 7-6).

Modern boosters are of the **vacuum suspended** type. Manifold vacuum is on both sides of the booster diaphragm when the booster is released. Early booster designs were **atmospheric suspended**. Both sides of the diaphragm were under atmospheric pressure during release. To apply this brake, a valve opened the master cylinder side to the manifold vacuum when the brake pedal was depressed. This system required a vacuum storage chamber to make certain that the necessary vacuum was there when needed. All brake boosters must be able to supply a brake assist if the engine should suddenly stall (Figure 7-7).

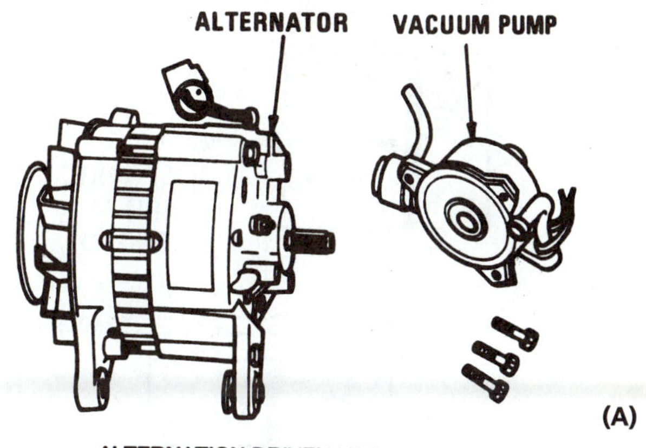

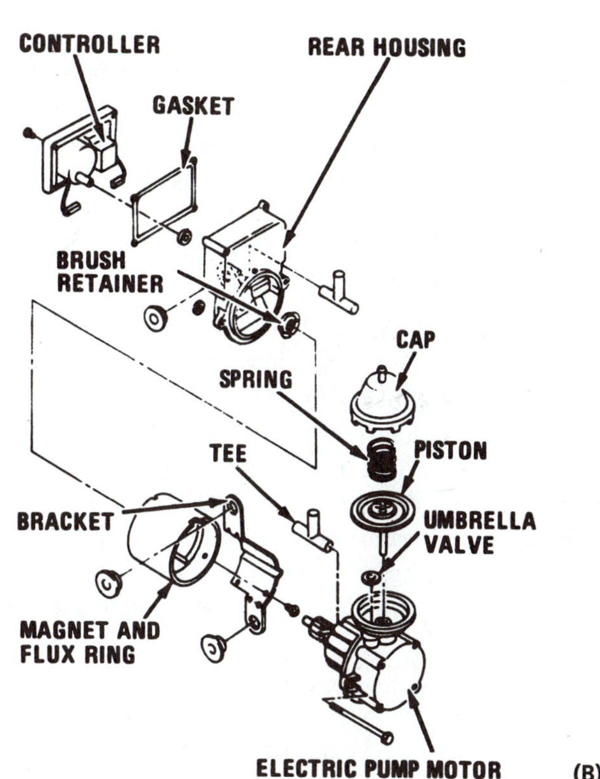

Figure 7-6 Vacuum pumps are used with diesel-engine vehicles to supply vacuum for booster operation. They can be driven by the engine camshaft, drive belt, or alternator (A) or by an electric motor (B). (Courtesy of Brake Parts, Inc.)

Most booster and master cylinder combinations are of the **tandem booster** or **integral booster** design in which the booster is mounted directly behind the master cylinder. Two other booster arrangements are the *linkage*

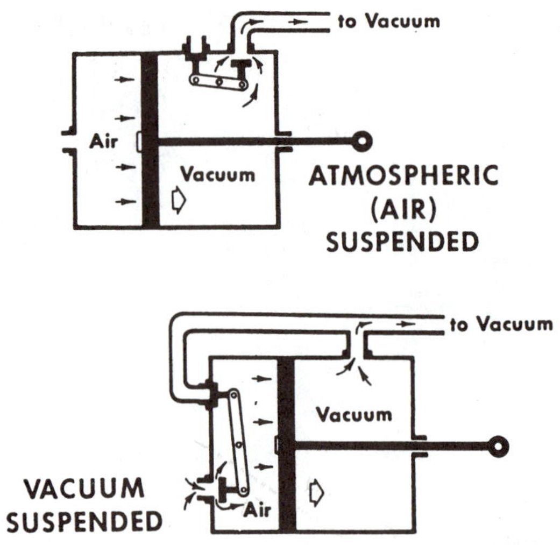

Figure 7-7 Some early boosters were air-suspended; atmospheric pressure was on both sides of the piston while released. All modern boosters are vacuum suspended with vacuum on each side of the piston while released. (Courtesy of Wagner Brake)

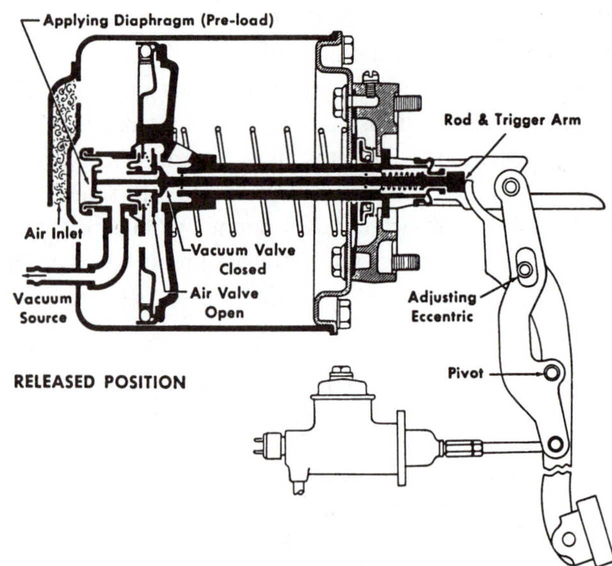

Figure 7-8 In the past, linkage-booster or pedal-assist boosters were used. Booster force was transmitted to the master cylinder through an extension of the brake pedal linkage. (Courtesy of Wagner Brake)

booster style and a remotely mounted unit often called a *pressure multiplier*. The linkage booster unit was mounted above or below the master cylinder and used the pedal linkage to transfer the assist to the master cylinder. These units were used in the 1950s (Figure 7-8). Pressure multiplier units can be mounted anywhere in a vehicle with the only necessary connections being the brake lines and the vacuum hose. Their input is the output pressure from the master cylinder. The booster output pressure is many times greater than the input pressure. The *Bendix Hydrovac* unit, used on many trucks, is a booster of this type (Figure 7-9). This booster style has been successfully retrofitted—added to vehicles that were not originally equipped with power brakes. Ocasionally a pressure multiplier is mounted so that only the front-wheel disc brakes are assisted. On some light trucks and vans, two boosters are used, one for each hydraulic circuit.

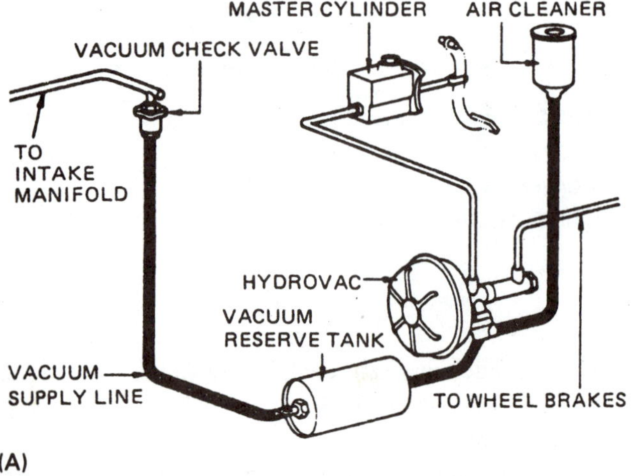

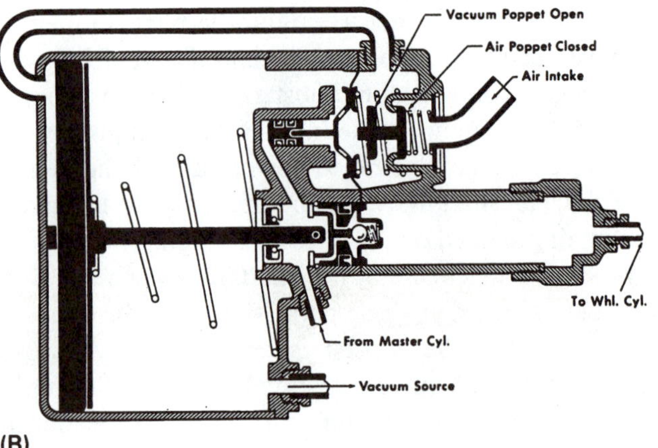

Figure 7-9 A pressure multiplier or Hydrovac booster can be mounted separately from the master cylinder (A). Its internal operation is similar to that of the other styles of vacuum boosters with the addition of its own fluid piston. (Courtesy of Wagner Brake)

7.3 Vacuum Booster Construction

Vacuum boosters are constructed with a rather large—6- to 11-inch (15 to 28 cm) diameter—metal housing that is divided into two sealed chambers. The piston or diaphragm plate is usually sealed to the two chamber halves by a rolling or flexible rubber diaphragm. When the pressure is different on each side, the diaphragm plate will move forward in the housing. In the past, some boosters used a piston with a sliding seal. Others used a booster chamber that was a collapsible bellows (Figure 7-10).

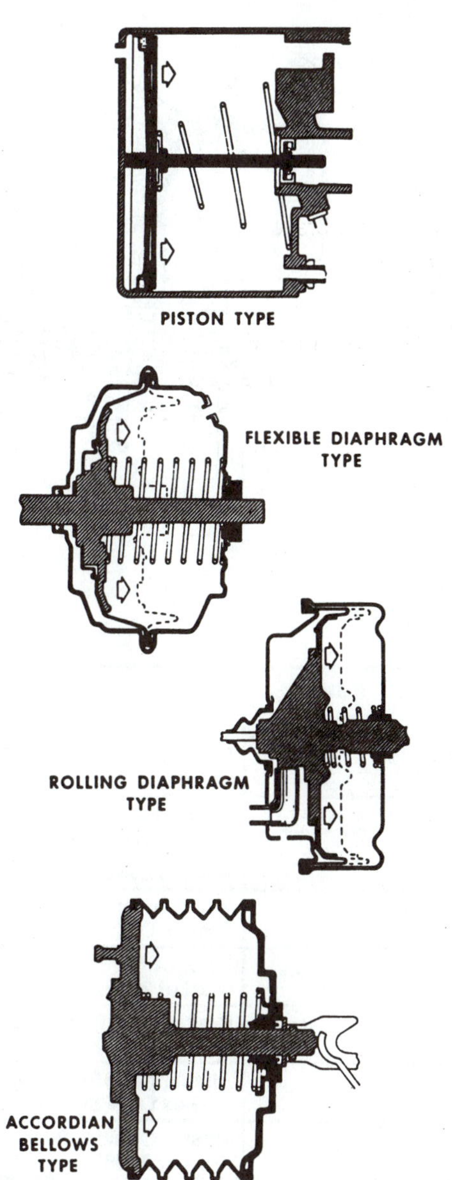

Figure 7-10 Some early booster designs used a flexible bellows or piston as the power unit; modern units use either a rolling or flexible diaphragm. (Courtesy of Wagner Brake)

Most booster chambers are constructed from two interlocking stamped-steel housings. One housing half mounts on the car's engine compartment bulkhead/firewall, and the other half has mounting provisions for the master cylinder (Figure 7-11).

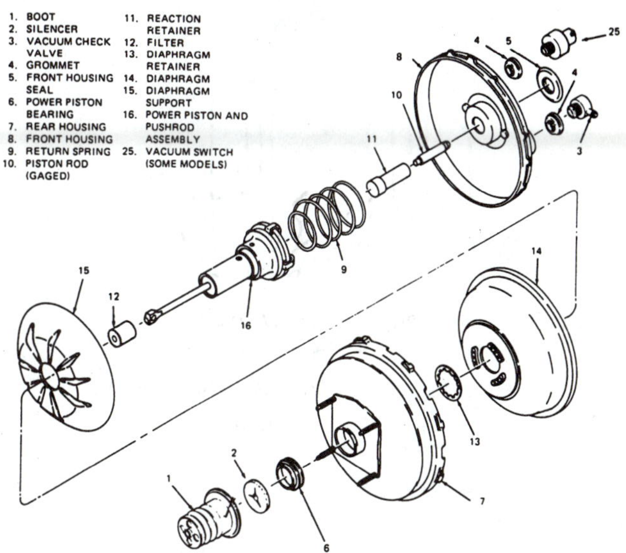

Figure 7-11 An exploded view of a vacuum booster. Note that the front and rear housings (7 and 8) interlock on the outer edge of the diaphragm (14) to form two airtight chambers. (Courtesy of General Motors Corporation, Service Technology Group)

The front or master cylinder section of the booster housing contains the vacuum connection. This connector is often also a check valve. The manifold vacuum pulls air out of the booster, but atmospheric pressure must not push air back into the booster if the engine is not running. The check valve allows air to flow in only one direction—toward the engine. A rubber hose leads from this connector to a fitting on the intake manifold. In some cars, the check valve is in the intake manifold fitting or in a separate unit in the hose (Figure 7-12). The rear or bulkhead side of the booster contains a support bushing and seal for the rear of the diaphragm support plate. An extension of the diaphragm support passes through this opening and slides inward during brake application.

The outer edge of the power piston diaphragm is usually locked between the two booster housing halves. A seal is made between each half as they are locked together. The diaphragm is attached to the diaphragm support plate or piston. This allows it to move the plate and apply force through the plate to the hydraulic pushrod and the primary piston in the master cylinder. The rear portion of the diaphragm support contains a filter element, and is the air intake for the booster. The filter element quiets the

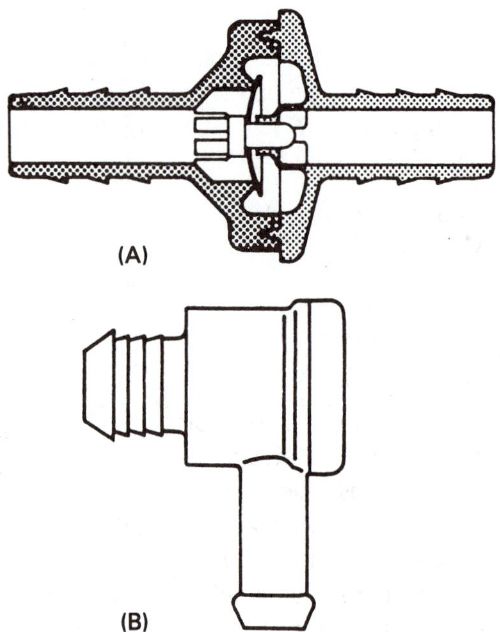

Figure 7-12 Vacuum check valves can be installed in the vacuum hose (top) or into the booster (bottom). Note the upper valve is sectioned to show how air can flow from left to right but not the other way. (Courtesy of ITT Automotive)

airflow and removes dirt particles. Two control valves are built into the center of the support plate, one for vacuum and one for atmospheric pressure or air. The valve plunger slides in the bore in the diaphragm plate. This type of valve is often called a **floating valve** because the valve bore travels with the piston (Figure 7-13).

Some boosters use **tandem diaphragms**. Two separate diaphragms and support plates are tandem mounted so they move together to apply the same hydraulic pushrod. A dividing wall is placed between the two diaphragms so the housing will have two more pressure chambers. A tandem booster will produce twice the assist of a single booster of the same diameter (Figure 7-14). **Runout**, which is the point where a booster is producing its maximum assist, on an 8-inch (200 mm) single booster with a 1-inch (2.54 cm) bore master cylinder, produces a hydraulic pressure of 560 psi (3.861 kPa). A tandem booster with two 8-inch (200 mm) diaphragms and the same master cylinder will produce 1,080 psi (7.447 kPa) at runout. Runout occurs when there is enough brake pedal pressure to completely compress the reaction disc. At this point, there is a full manifold vacuum on one side of the diaphragm(s) and atmospheric pressure on the other (Figure 7-15). A tandem booster is about 20 percent longer because of the added parts.

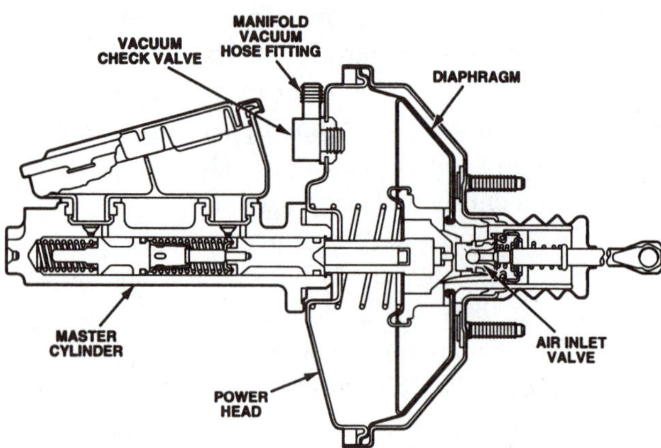

Figure 7-13 A cutaway view of a single diaphragm booster and master cylinder. (Courtesy of General Motors Corporation, Service Technology Group)

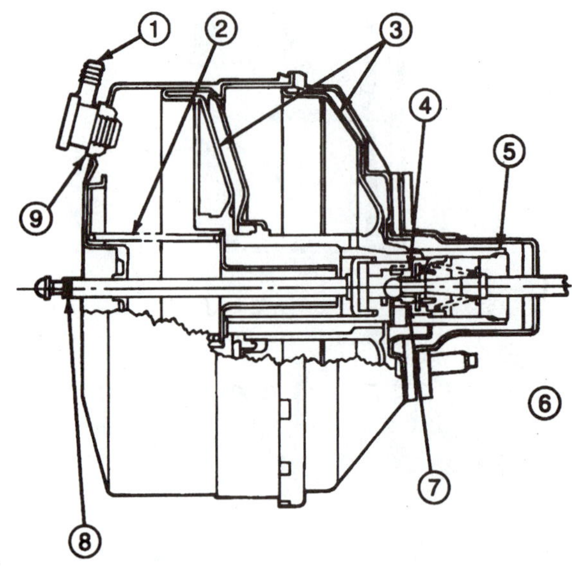

Item	Part Number	Description
1	2365	Power Brake Booster Check Valve
2	—	Return Spring (Part of 2005)
3	—	Tandem Power Diaphragms (Part of 2005)
4	—	Vacuum Port Closed — Brakes On (Part of 2005)
5	—	Filter (Air Inlet) (Part of 2005)
6	—	Brake Pedal Push Rod (Part of 2005)
7	—	Atmospheric Port Open — Brakes On (Part of 2005)
8	—	Master Cylinder Push Rod (Part of 2005)
9	2B176	Check Valve Grommet

Figure 7-14 A cutaway view of a tandem booster showing the two diaphragms (3). (Courtesy of Ford Motor Company)

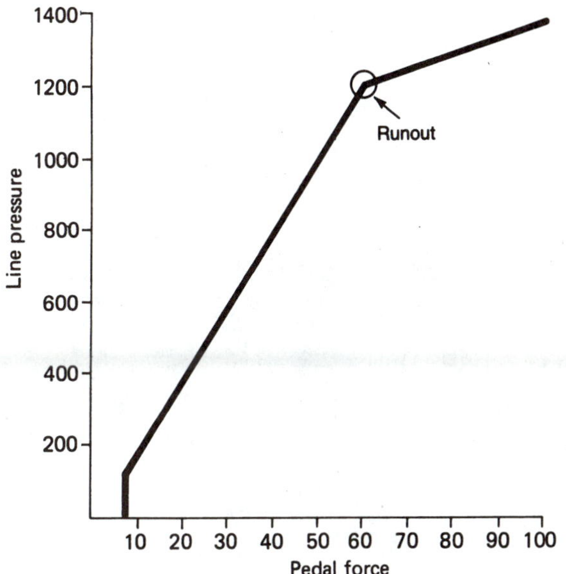

Figure 7-15 This illustration shows the output of a typical vacuum booster. Note that after runout, line pressure increases only from increased pedal pressure; the booster is giving the maximum amount of assist.

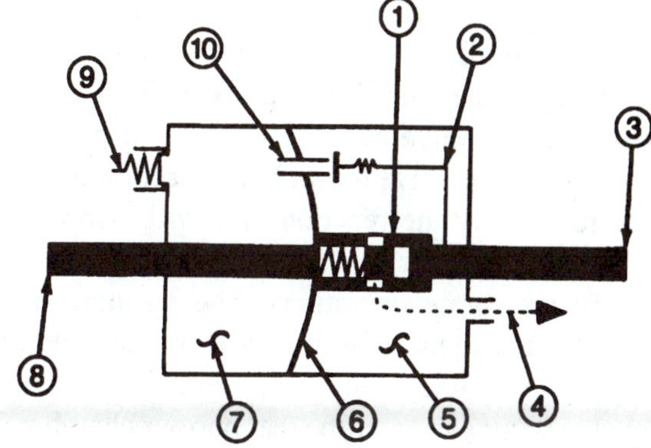

Figure 7-16 This vacuum booster schematic shows the vacuum valve (1), valve equalizer linkage (2), pedal pushrod (3), atmospheric air inlet (4), vacuum chamber B (5), diaphragm (6), vacuum chamber A (7), master cylinder pushrod (8), vacuum check valve (9), and equalizer/vacuum valve (10). During brake application, the pressure in vacuum chamber B increases. (Courtesy of Ford Motor Company)

7.4 Vacuum Booster Operation

Most boosters have three operating positions—released, applying, and holding. These positions are determined by the amount of pressure on the valve pushrod. During release, there is no pressure on the valve rod. During application, there is enough pressure to compress the reaction disc; and during holding, the reaction disc is partially compressed. In the released position, the vacuum valve is open and the air valve is closed. During application, the vacuum valve is closed and the air valve is open. Both valves are closed while holding. These positions are determined by the amount of force on the brake pedal and whether the pedal is moving or stationary (Figure 7-16). The valve is in the application position when there is enough pedal force to compress the reaction disc. Movement of the disc and valves will start the boost, and this in turn will move the diaphragm and support plate. As the support plate moves, it will change the position of the reaction disc and valves, and the unit position will change from applying to holding. The driver can increase or decrease the amount of braking from the holding position by increasing or decreasing pedal pressure. Runout or manual application will occur when there is enough pedal pressure to collapse the reaction disc completely. From this point, pedal pressure will be transmitted through the reaction disc to the hydraulic pushrod along with the booster output force.

When the booster is released, the diaphragm return spring will move the diaphragm plate with the diaphragm and the control valve to the rear of the booster. The control valve spring will then position the control valve so the air valve is closed and the vacuum valve is open. The manifold vacuum will move equally to both sides of the diaphragm (Figure 7-17). This is the only time that air should flow through or into a booster. A booster should have no effect on engine performance other than causing

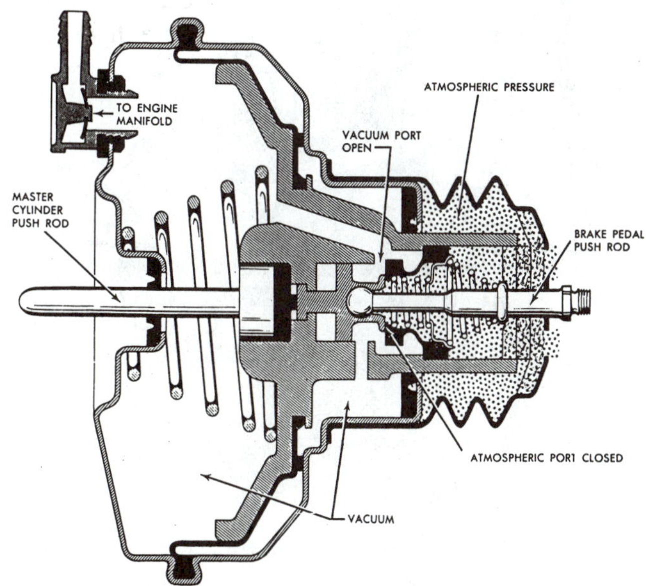

Figure 7-17 In the released position, the control valve opens the vacuum port so manifold vacuum is on both sides of the diaphragm; the atmospheric port is closed. (Courtesy of General Motors Corporation, Service Technology Group)

a slight increase in engine speed for a moment as the brakes are released.

When the brakes are applied, force from the brake pedal through the valve rod or pushrod will push on the inner portion of the valve through the reaction disc to the diaphragm. Usually the reaction disc will compress and allow the valve plunger to change position in its bore to perform two separate operations. The vacuum valve is closed first. This shuts off any flow from one side of the diaphragm to the other. The second operation is to open the air valve so air can enter the rear booster chamber. This provides the pressure differential across the diaphragm for the power assist (Figure 7-18). The diaphragm will then move forward, applying the brakes, and because of the floating valve, it will reset the operation to hold. While in a holding position, the pressure on each side of the diaphragm and the force on the master cylinder will not change (Figure 7-19). During application there is no airflow between the booster and the manifold, so there should be no change in engine operation. A certain amount of air does enter through the air valve into the booster. Sometimes you can hear this airflow, especially if the engine is shut off.

If the engine should stall, the booster will work normally for at least one cycle. It will give an assist for several cycles, but each application will allow more and more air to enter, which reduces the amount of vacuum. Each cycle will provide less and less assist.

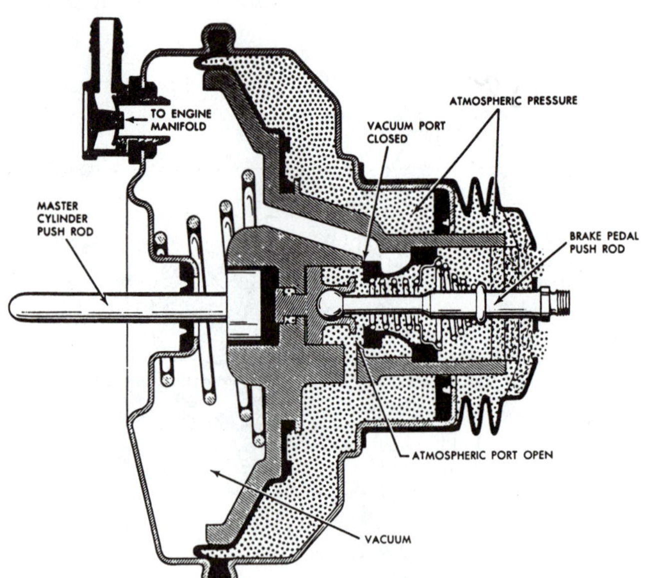

Figure 7-18 As the pedal pushrod moves inward, it will move the control valve so the vacuum port closes and the atmospheric port opens. This raises the pressure on the right side of the diaphragm to produce a power assist. (Courtesy of General Motors Corporation, Service Technology Group)

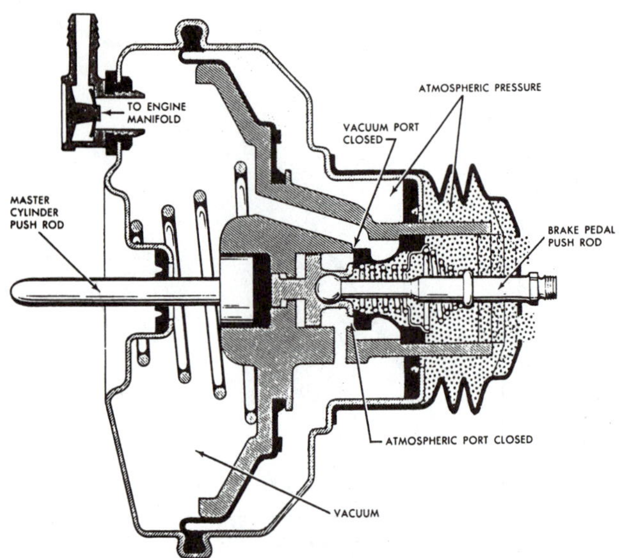

Figure 7-19 In the holding position, both the vacuum and atmospheric ports are closed so the pressure on the right side of the diaphragm will not change. This keeps the same pressure in the brake system. (Courtesy of General Motors Corporation, Service Technology Group)

7.5 Hydraulic Booster Theory

Hydraulic boosters are called **Hydro-boost** by Bendix, the company that first developed them. The external force used to provide the assist with this booster is hydraulic pressure from the power steering pump. This pressure is much higher, about 1,000 to 1,500 psi (6,895 to 10,340 kPa), than the 5- to 10-psi (34- to 69-kPa) pressure differential that can be used with a vacuum booster. Therefore the piston in a hydraulic booster can be made much smaller and still produce a greater assist.

This booster is connected to the power steering pump by two steel and reinforced rubber hoses, a pressure hose, and a return hose. It is also connected to the power steering gear by another pressure hose. Hydraulic flow begins at the pump, goes first to the brake booster, and then continues on to the steering gear (Figure 7-20).

Hydraulic boosters use an **accumulator** to store pressurized fluid, to ensure that the vehicle can stop even if the engine is off. Early accumulators operated by compressing a spring-loaded piston, but newer units use a nitrogen-gas-charged accumulator. An accumulator uses a pressure chamber that is divided by a moveable piston or flexible diaphragm. It has a static charge of nitrogen gas on the side opposite the oil. When the engine runs and the power steering pump develops pressure, pressurized fluid flows

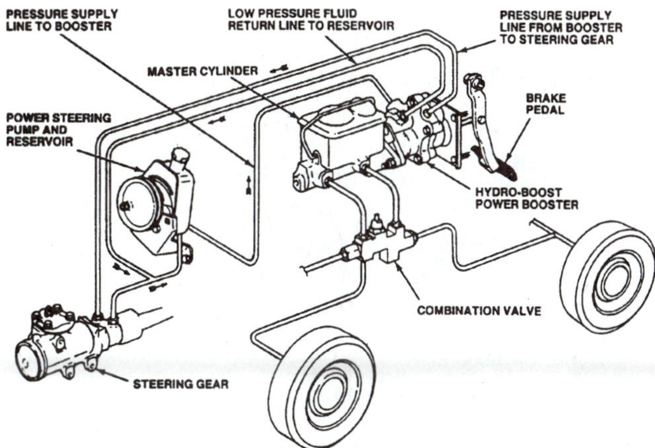

Figure 7-20 A Hydro-boost system showing the arrangement of the hydraulic pressure and return lines. (Courtesy of Hunter Engineering Company)

into the acumulator. Then the pressure of the fluid compresses the spring or gas charge in the accumulator (Figure 7-21). The volume stored in the accumulator is large enough to ensure pressurized fluid for at least one brake application. More than one application is possible, but each will be weaker as the accumulator charge bleeds down. A check valve between the accumulator and the pressure port of the booster prevents a pressure loss back to the power steering pump when the engine stops.

It is important to remember the accumulator when working on a hydraulic booster. The accidental and unexpected discharge of an accumulator that is full of oil under high pressure may cause injury and certainly will cause a mess. The accumulator should be discharged by pumping the brake pedal. Repeat this operation until the feel of the brake pedal indicates a full discharge. The pedal will get harder and harder on each application as the accumulator bleeds down.

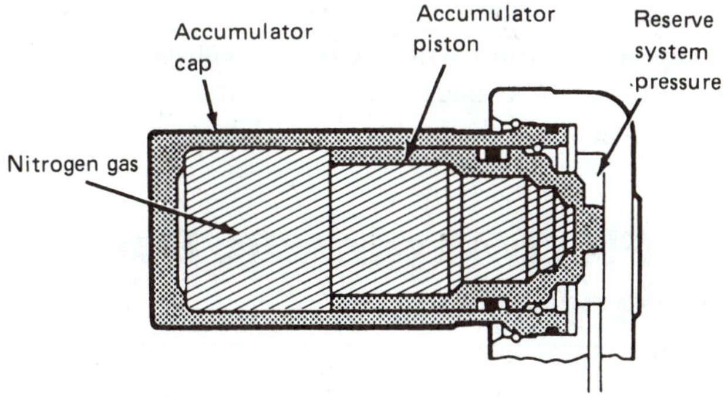

Figure 7-21 Hydraulic boosters store enough pressurized fluid to apply the booster in case the engine stops. The stored oil moves the accumulator piston to the left, compressing the nitrogen gas charge. (Courtesy of Ford Motor Company)

7.6 Hydraulic Booster Construction

This booster uses a cast-iron body or housing which is mounted between the master cylinder and the car's bulkhead. The body section contains the power piston, an open center-spool valve, and a lever assembly, as well as the input and output pushrods. A removable housing cover, sealed by a rubber seal ring, allows access to these parts and helps enclose the high-pressure fluid chamber. The power piston and input rod use lip seals to prevent fluid leaks at these openings (Figure 7-22).

A **spool valve** has precision-sized bands which fit the valve housing bore closely enough to prevent excessive fluid leaks along the bore, yet it is still able to slide in the bore. This valve is used to control the pressure in the booster pressure chamber at the end of the power piston. A spool valve has grooves that allow fluid to flow from one side port to another. Sliding the valve covers and uncovers these side ports. A hollow spool valve is hollow from one end down through the valve to a side passage, allowing another possible fluid passageway.

The position of the spool valve is controlled by a lever. This lever is unique in that the fulcrum can be at either pin A or pin B (Figure 7-23). Pin A is on the power piston, and pin B is on the input rod. In this way, movement of either the input rod or the power piston will change the position of the spool valve and therefore, the pressure in the chamber.

7.7 Hydraulic Booster Operation

In a relaxed or released position, the power piston return spring will position the power piston rearward, toward the brake pedal. With the piston in this position and the brake pedal released, the spool valve will be positioned so that the power cavity is hydraulically connected to the return port. Hydraulic pressure will then return to the power steering pump reservoir (Figure 7-24).

When the driver applies the brakes, the movement of the input rod will move the lever and reposition the spool valve. This action will connect the pressure port to the pressure cavity so that power steering pump pressure will reach the cavity. This pressure at the end of the power piston will cause it to stroke toward the master cylinder and apply the brakes (Figure 7-25). The amount of pressure in the cavity will vary depending on how hard the brake pedal is applied and the pressure setting of the power steering pump. Since the fulcrum is traveling with the piston, if the driver holds

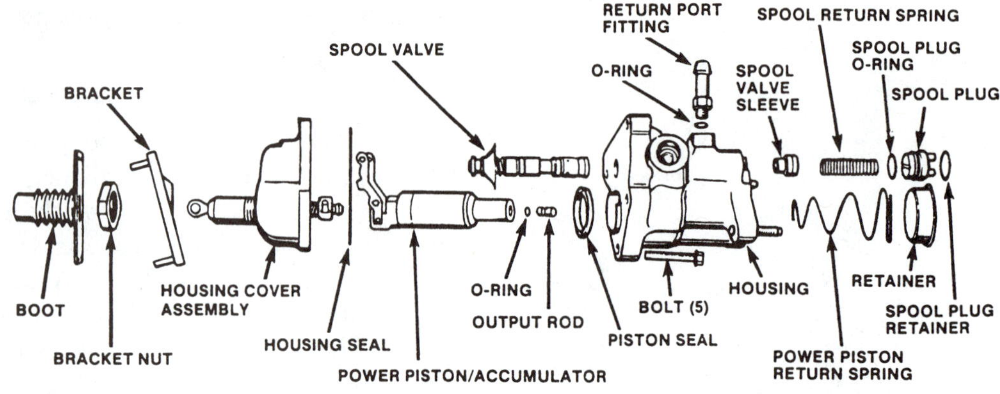

Figure 7-22 An exploded view of a Hydro-boost II unit. (Courtesy of Bendix Brakes, by AlliedSignal)

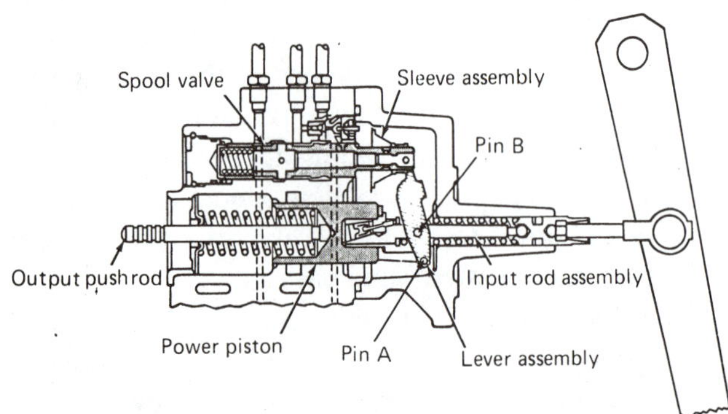

Figure 7-23 A cutaway view of a Hydro-boost unit; the pressure chamber is at the right of the power piston. (Courtesy of General Motors Corporation, Service Technology Group)

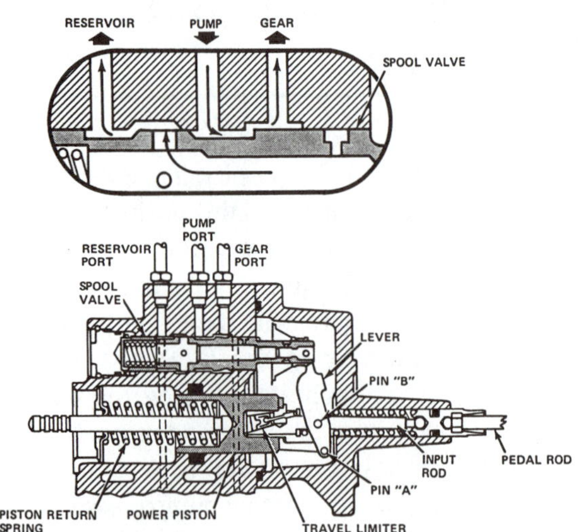

Figure 7-24 When the brakes are released, the spool valve is positioned so pump pressure goes to the steering gear and pressure chamber pressure is released to the pump reservoir. (Courtesy of General Motors Corporation, Service Technology Group)

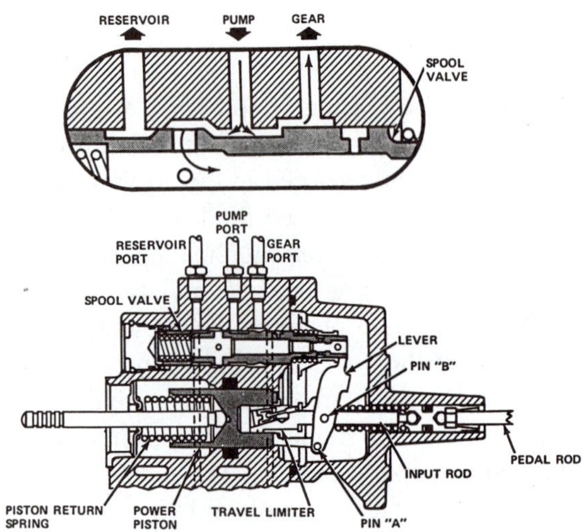

Figure 7-25 Brake pedal pressure (through the pedal rod) moves the spool valve to send pump pressure to the pressure chamber. (Courtesy of General Motors Corporation, Service Technology Group)

the brake pedal steady, the power piston and lever will move the spool valve to the holding position. The pressure in the power chamber will then be held constant. The pressure in this chamber and the amount of braking will change as the driver varies the foot pressure (Figure 7-26).

7.8 Electrohydraulic Booster Theory

This booster, called a *Powermaster*, has its own electric-motor-driven pump to provide the boost pressure. This pump operates only when needed to provide this pressure. A hydraulic booster as just described places a light drag on the power steering pump at all times, which causes a slight

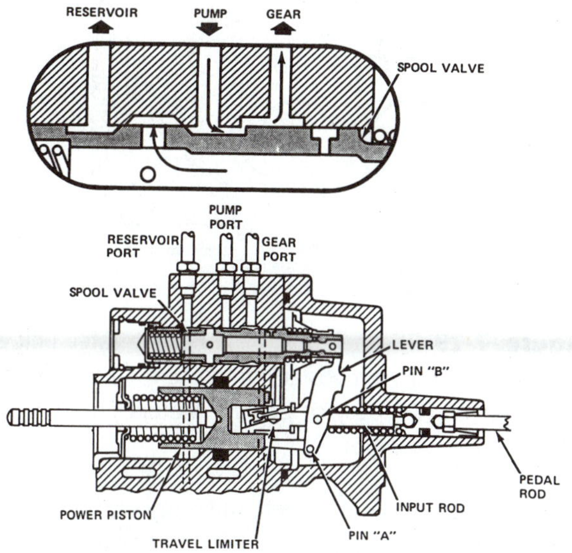

Figure 7-26 In the holding position, the spool valve shuts off the flow to or from the pressure chamber. This keeps the same pressure in the brake system. (Courtesy of General Motors Corporation, Service Technology Group)

loss in fuel economy. Electrohydraulic boosters are more compact (Figure 7-27). They do not require hose connections to the engine, provide a large amount of boost, and can be used with any engine and accessory combination.

The small, 12-volt, electric-motor-driven vane pump charges an accumulator, and the booster operates off this accumulator plus added pump output, if necessary. The operation of the pump is controlled by a pressure switch. When the accumulator pressure falls below 510 psi (3,516 kPa), the switch will close and the pump will run. When the pressure exceeds 675 psi (4,654 kPa), the switch will open, and the pump will stop. The pump normally runs in a short cycle, taking less than 20 seconds to charge the accumulator, and normally will not run again until the brakes are used (Figure 7-28).

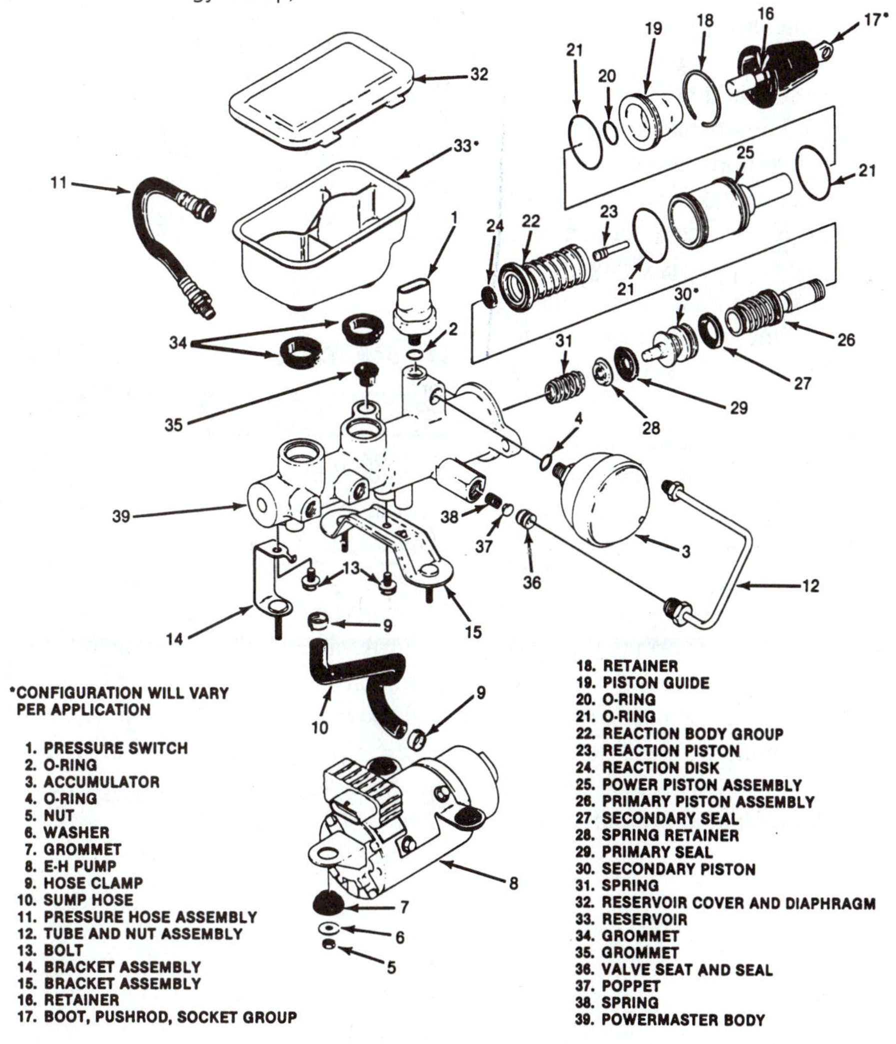

Figure 7-27 An exploded view of a Powermaster unit. (Courtesy of General Motors Corporation, Service Technology Group)

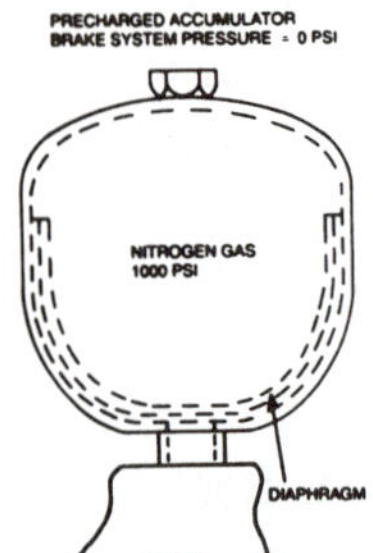

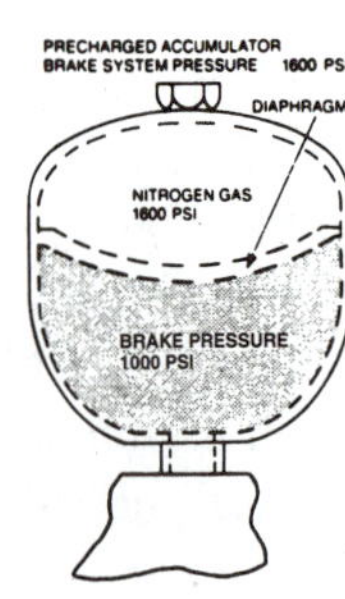

Figure 7-28 An empty accumulator (left) is charged with nitrogen gas at a fairly high pressure. A charged accumulator (right) is partially filled with brake fluid or hydraulic oil at an even higher pressure (Courtesy of Chrysler Corporation)

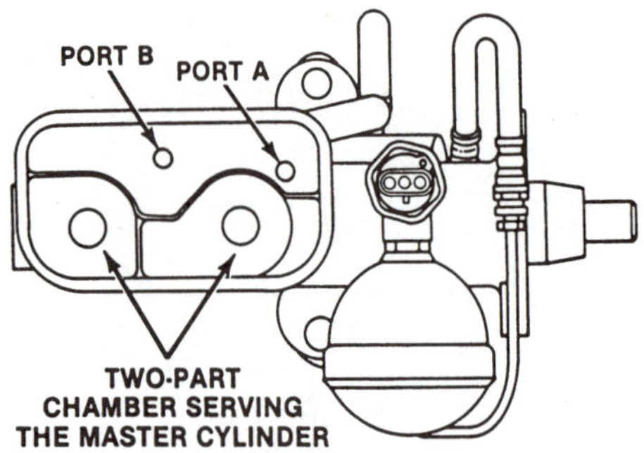

Figure 7-29 The reservoir of a Powermaster unit shows Ports A and B for the booster pump and the two ports for the brake system. (Courtesy of General Motors Corporation, Service Technology Group)

7.9 Electrohydraulic Booster Construction

This booster is combined with the master cylinder in the same cast-aluminum body section. The power piston and the reaction group operate directly against the primary master cylinder piston. The unit is only slightly larger than a nonpower master cylinder.

Brake fluid is used as booster fluid. A three-section reservoir stores this fluid in a section separate from the two master cylinder fluid sections. Two ports at the bottom of the booster section lead to the pump intake and the booster return ports. The fluid level in this reservoir section will drop significantly when the accumulator is charged and should be checked only after discharging the accumulator (Figure 7-29). Otherwise overfilling of the reservoir can occur, which can cause overflow when the accumulator is discharged.

The pump and motor form a single assembly that is secured to a bracket attached to the booster/master cylinder body. The pump intake is connected to the reservoir port by a rubber hose, and the pump outlet is connected to the booster by a reinforced high-pressure hose. Electric power for the pump originates at a 30-amp fuse in the car's fuse block. An electrical connector brings this battery positive (B+) voltage to a relay mounted on the motor. The relay is basically a switch controlled by the pressure switch. When the accumulator pressure drops, the pressure switch will close, the relay will close, and the motor will run. When the accumulator pressure rises, the pressure switch will open, the relay will open, and the motor will stop. The pressure switch has a second set of contacts that will turn on the brake failure warning light if the accumulator pressure drops below 400 psi (2,758 kPa) (Figure 7-30).

The power piston fits into its bore in the body casting and is held in place by the piston guide. Rubber O-rings are used at several points to prevent fluid leaks. A group of parts, called the *reaction group*, fits inside the power piston. These parts include the inner and outer control valves, the reaction disc, and the reaction piston. The chamfered edges at the ends of the inner and outer control valves control the booster pressure in the power piston cavity.

7.10 Electrohydraulic Booster Operation

Even though a different style of valve is used, the electrohydraulic booster operates in essentially the same way as a hydraulic booster. A fluid passage leads from the high-pressure port at the accumulator to the end of the outer control valve, and a return passage leads from the inner control valve to the reservoir return port. The inner and outer valves are positioned by the pushrod, reaction disc, and power piston. They provide release (relaxation), application, and holding of the hydraulic pressure acting on the power piston. These operations are shown in Figures 7-31 through 7-33.

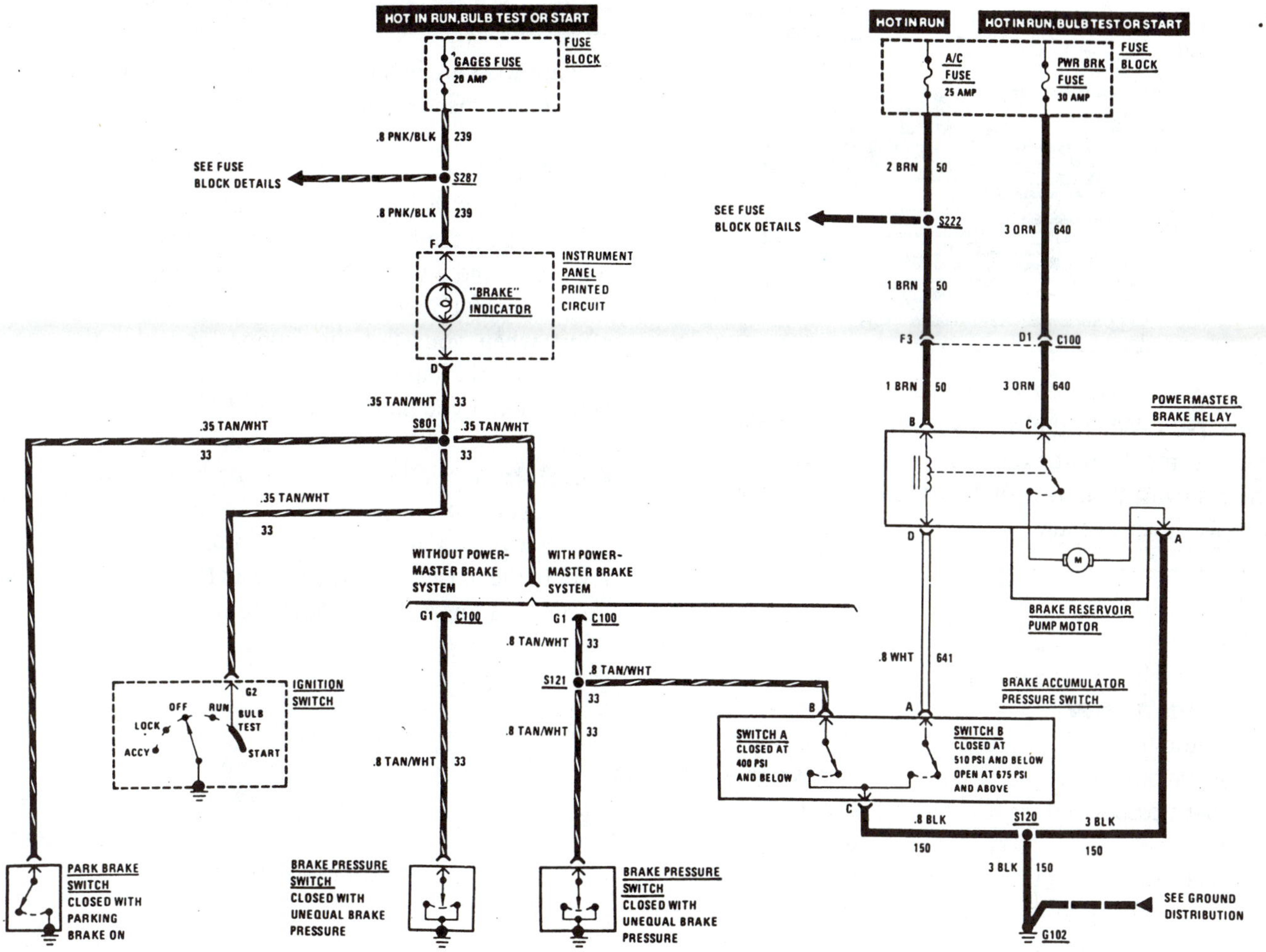

Figure 7-30 The electrical circuit for a Powermaster system shows the brake pressure switch on the accumulator that operates the pump; it can also turn on the brake warning light. (Courtesy of General Motors Corporation, Service Technology Group)

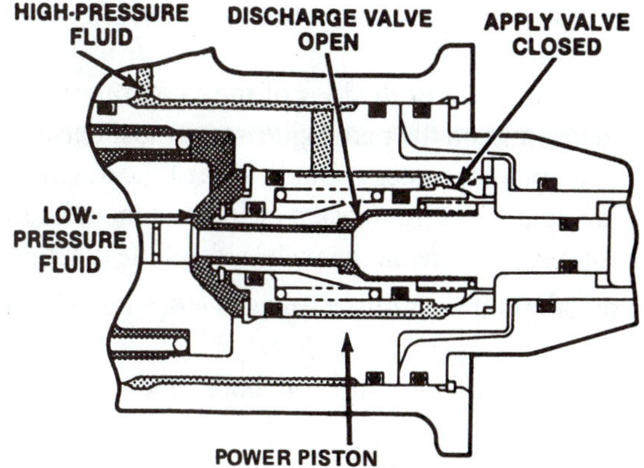

Figure 7-31 With the system at rest, the application valve is closed and the discharge valve is open so there will be no boost pressure on the power piston. (Courtesy of General Motors Corporation, Service Technology Group)

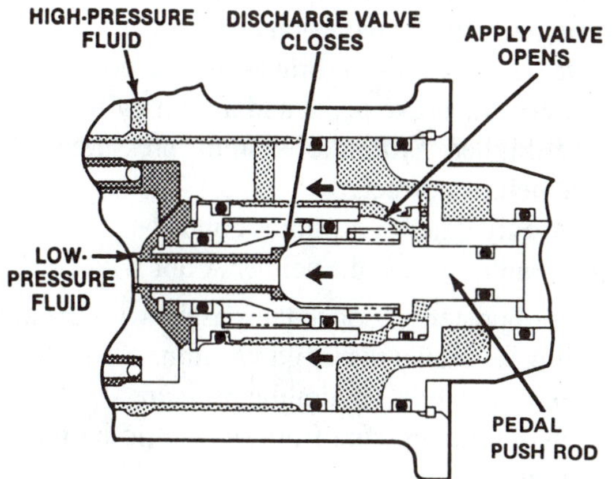

Figure 7-32 As the brakes are applied, the application valve opens to allow fluid pressure to enter the power piston chamber. (Courtesy of General Motors Corporation, Service Technology Group)

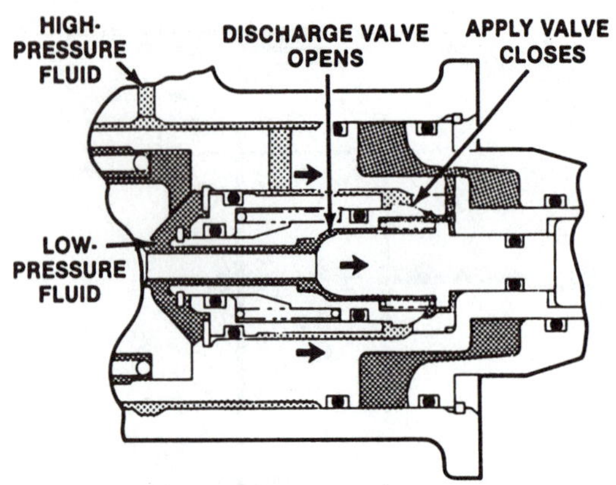

Figure 7-33 When the brakes are released, spring pressure moves the discharge valve to open and release pressure to the reservoir. (Courtesy of General Motors Corporation, Service Technology Group)

7.11 Practice Diagnosis

You are working in a brake and front-end repair shop and these problems are presented to you.

CASE 1: The customer has brought in her 15-year-old Chevrolet Caprice with this complaint: "I really have to push hard on the brake pedal to stop this car." On checking it you find a very hard brake pedal on a high-mileage car; it doesn't seem to make much difference in pedal effort when you start the engine, and also the engine sounds a little rough. What could be wrong with this car? What should you do to check this?

CASE 2: While inspecting the brake system on a 1994 Buick, you notice that the car has an electric motor on the master cylinder instead of a vacuum booster. The fluid level in the master cylinder appears low. What should you do next? Is this a normal or abnormal condition?

Terms to Know

accumulator
atmospheric pressure
atmospheric suspended
booster
floating valve
Hydro-boost
inches of mercury ("Hg)
integral booster
power assist
runout
spool valve
tandem booster
tandem diaphragms
vacuum
vacuum suspended

Review Questions

1. Statement A: Power brakes use larger, more powerful wheel brake assemblies so they are substantially stronger than standard brakes.
   Statement B: Power brakes use a power booster between the brake pedal and master cylinder to produce higher hydraulic system pressures. Which statement is correct?
   a. A only c. both A and B
   b. B only d. neither A nor B
2. Power boosters produce their assist with the aid of
   a. engine intake manifold vacuum.
   b. power steering hydraulic pressure.
   c. hydraulic pressure from an electric-motor-driven pump.
   d. all of the above.
3. In the released position, a vacuum suspended booster has
   a. atmospheric pressure on both sides of the diaphragm.
   b. a vacuum on both sides of the diaphragm.
   c. a vacuum on the pedal side and atmospheric pressure on the master cylinder side of the diaphragm.
   d. a vacuum on the master cylinder side and atmospheric pressure on the pedal side of the diaphragm.
4. In the applied position, a vacuum suspended booster has
   a. atmospheric pressure on both sides of the diaphragm.
   b. a vacuum on both sides of the diaphragm.
   c. a vacuum on the pedal side and atmospheric pressure on the master cylinder side of the diaphragm.
   d. a vacuum on the master cylinder side and atmospheric pressure on the pedal side of the diaphragm.

5. To ensure a power-assisted stop if the engine stalls, vacuum boosters have
   a. a vacuum accumulator in the vacuum supply system.
   b. a check valve in the vacuum hose.
   c. a vacuum pump attached to the booster.
   d. all of the above.
6. Statement A: A linkage booster power assist unit is mounted above the master cylinder.
   Statement B: A hydrovac is a pressure multiplier booster that is mounted separately from the master cylinder. Which statement is correct?
   a. A only c. both A and B
   b. B only d. neither A nor B
7. Statement A: A tandem master cylinder is always used with tandem diaphragm boosters.
   Statement B: Booster runout occurs when there is atmospheric pressure on the pedal side of the diaphragm and the booster is producing maximum assist. Which statement is correct?
   a. A only c. both A and B
   b. B only d. neither A nor B
8. A Hydro-boost unit
   a. develops higher braking pressures than a vacuum booster.
   b. can be easily used with diesel engines.
   c. is always mounted integrally with the master cylinder.
   d. all of the above.
9. To ensure an assisted stop if the engine stalls, a Hydro-boost unit uses
   a. an accumulator in the booster.
   b. a check valve in the hydraulic hose.
   c. an electric-motor-driven hydraulic pump.
   d. all of the above.
10. Statement A: Brake booster valves have three basic operating positions: applying, holding, and releasing.
    Statement B: Booster valve position is determined by the pressure of the driver's foot relative to the braking pressure. Which statement is correct?
    a. A only c. both A and B
    b. B only d. neither A nor B
11. A Powermaster unit
    a. is a self-contained booster and master cylinder combination.
    b. uses brake fluid for the booster and brake hydraulic systems.
    c. uses an accumulator to store pressurized fluid for an assist if the engine stalls.
    d. all of the above.
12. Statement A: The pump of a Powermaster unit is controlled by the booster control valve.
    Statement B: The pump of the Powermaster unit operates continuously when the engine runs. Which statement is correct?
    a. A only c. both A and B
    b. B only d. neither A nor B

Antilock Braking System Theory

Objectives

After completing this chapter, you should:

- ❑ Understand how an antilock braking system operates.
- ❑ Be familiar with the terms commonly used with antilock braking systems.
- ❑ Be familiar with the operating differences in the various types of antilock braking systems.

8.1 Introduction

Antilock braking systems (ABS), sometimes called *anti-skid braking*, have been used to some degree on domestic vehicles since the late 1960s. The early systems were used primarily on luxury cars as an extra-cost option and had limited popularity. Since the early to mid-1980s, ABS has become more popular and is now standard equipment on many vehicle models. Most experts predict almost 100-percent usage of ABS on cars by the year 2000.

Wheel lockup during braking will cause skidding, and this will cause a loss of traction and vehicle control. A tire will generate its greatest amount of traction when it is slipping at a rate of 15 to 25 percent. This will result in longer stopping distances and possible accidents. ABS is designed to prevent wheel lockup and the resulting skid, even under the worst driving conditions, by automatically compensating for changes of traction or tire loading. ABS does not necessarily produce shorter stops, but it greatly improves the driver's ability to control the vehicle when trying to stop quickly (Figure 8-1). On good, dry pavement, an ABS-equipped car will usually stop at about the same rate as a non-ABS-equipped car (Figure 8-2). Both stopping distances will be about equal. Under poor traction conditions, such as on wet or icy pavement, ABS allows the car to stop in a significantly shorter distance with a controlled, steerable stop. On low-traction surfaces, non-ABS systems allow the wheels to lock up easily in most cases, which greatly reduces braking power (Figure 8-3). Remember that braking power is limited by tire traction and driver reaction time and skill (Figure 8-4). A four-wheel ABS system will cycle the braking power on the tire(s) with poor traction while retaining full stopping power in those tires with good traction. In a non-ABS system, if both front or both rear wheels lock up, the ability to steer and maneuver the car will be lost.

Figure 8-1 ABS helps a driver steer the car to a straight stop in all traction conditions (top). Wheel lockup causes skidding and loss of control. (Courtesy of General Motors Corporation, Service Technology Group)

ABS begins with a standard brake system and adds one, two, three, or four modulator or control valves, one to four speed sensors, and an electronic controller for the valves. The modulator valves are used to cycle the hydraulic pressure at the brake assemblies; the speed sensors determine the rotating speed of the wheel; and the electronic controller monitors the speed of the tires and operates the modulator valves to prevent wheel lockup.

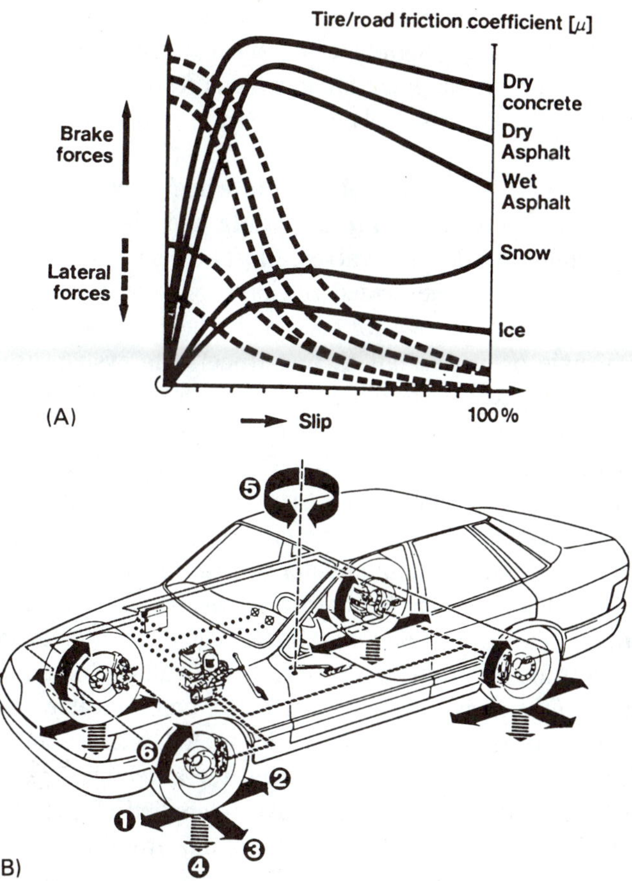

Figure 8-2 By preventing wheel lockup under all traction conditions (A), the tire has maximum traction for acceleration (1), braking (2), cornering (3), normal reaction forces (4), yaw motions (5), and inertia of the tire and wheel (6). (Courtesy of ITT Automotive)

ROAD CONDITION	TIRE CONDITION	RESULTANT COEFFICIENT OF FRICTION
Dry pavement	New tire	1.0 (highest)
Dirt road	New tire	0.9
Dry pavement	Old, worn tire	0.8
Dirt road	Old, worn tire	0.7
Gravel	New tire	0.6
Gravel	Old, worn tire	0.5
Wet road	New tire	0.4
Wet road	Old, worn tire	0.3
Ice	New tire	0.2
Ice	Old, worn tire	0.1 (lowest)

1.0 = Highest coefficient
0.1 = Lowest coefficient

(A)

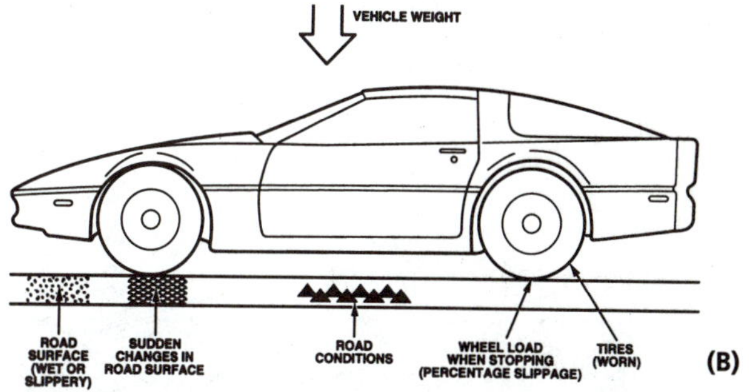

Figure 8-3 The coefficient of friction between the tire and road varies depending on tire and road conditions (A). If a heavily braked wheel exceeds the amount of friction it will lock up (B). (A is courtesy of Chrysler Corporation; B is courtesy of General Motors Corporation, Service Technology Group)

There is a variety in ABS between tandem and diagonal split hydraulic systems and pickups with rear-wheel-only systems (Figure 8-5). Most of these are four wheel systems using either three or four circuits or channels; a circuit is one or more wheels that has a modulator valve(s) controlling the hydraulic pressure. Rear-wheel-only, single-circuit systems are standard equipment on most new pickups (Figure 8-6). One manufacturer names these "RWAL" (rear-wheel antilock) and uses the term "4WAL" (four-wheel antilock) for some four-wheel systems. Most of the early domestic ABS units were two-wheel, rear-wheel-only systems.

As a braked wheel begins to lock up and skid, ABS will cycle the brake off until the wheel is rotating at the correct speed. Then it will cycle the brake back on. If the wheel locks up again, the cycle will be repeated. This cycling is fast, occurring about five to fifteen times a second, and each braking circuit is cycled individually as needed (Figure 8-7).

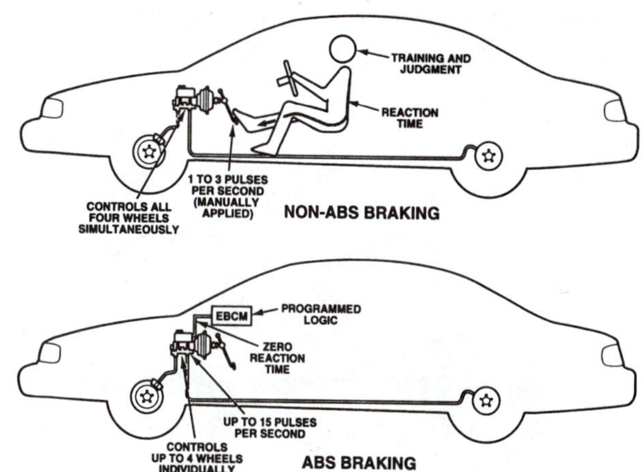

Figure 8-4 Without ABS in a panic stop, the driver must try to prevent wheel lockup by pulsing the brake pedal. With ABS, the system can pulse any or all of the wheels up to fifteen times per second. (Courtesy of General Motors Corporation, Service Technology Group)

Modern electronic systems use three stages of ABS operation. These are (1) normal operation (sometimes called pressure increase or buildup), in which normal braking occurs; (2) pressure holding, in which the pressure

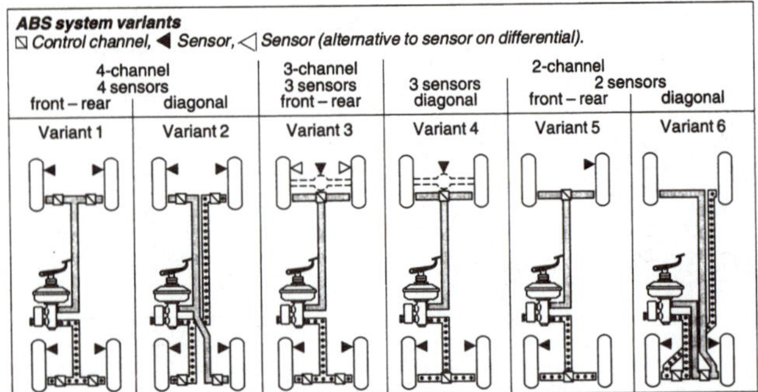

Figure 8-5 ABS systems vary according to the number and location of sensors and the number of conrolled channels or circuits. A rear-wheel-only system will be similar to Variant 4 or 5 without the front control channel. (Reprinted with permission from Robert Bosch Corporation)

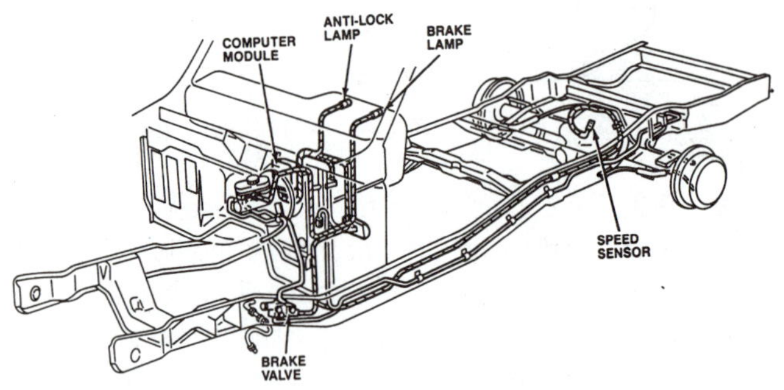

Figure 8-6 A two-wheel (rear only) ABS is used on many pickups as standard equipment. Note the wheel-speed sensor built into the rear axle. (Courtesy of Ford Motor Company)

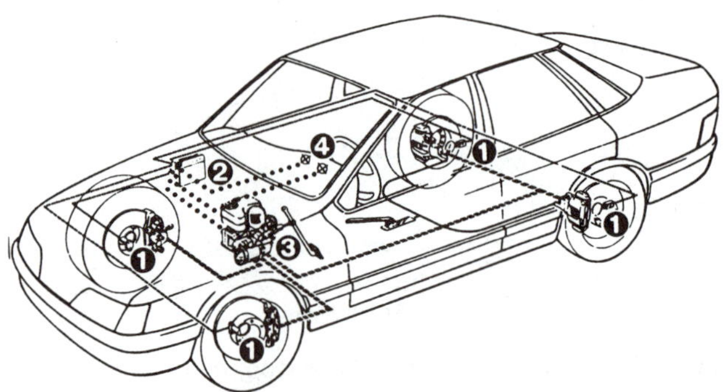

Figure 8-7 In many passenger cars the ABS is a four-wheel system with sensors for each of the wheels. A three-circuit system that combines both rear brakes in the same circuit is shown. The wheel sensors (1) provide the EBCM (2) with the speed of each wheel; if a wheel locks up, the hydraulic control (3) will cycle the brake unit. The warning light (4) informs the driver of a system malfunction. (Courtesy of ITT Automotive)

in the brake assembly is not increased or decreased if wheel lockup should occur; and (3) pressure reduction, decay, or reapplication, in which pressure is released from the brake assembly if necessary to get the wheel turning.

Most ABS units use **electronic wheel-speed sensors** with an **electronic control module** and **electronically controlled modulator valves**. Most current systems use three or four wheel speed sensors—one for each front wheel, and either one for each rear wheel or one for the rear axle or drive shaft. A rear-wheel-only system uses just a rear wheel speed sensor(s). Some systems fit the speed sensor to the speedometer cable or into the transmission or rear axle assembly. Rear-wheel-only systems use a single modulator valve to control both rear wheels at the same time. This is a one-circuit system. Most four-wheel systems use a modulator or control valve for each front wheel and one for both rear wheels. This arrangement is called a *three-circuit* brake. The three circuits are right front, left front, and both rear brake units. Some cars use a *four-circuit* system in which a control valve regulates the hydraulic pressure in each brake assembly (Figure 8-8). There is a tendency for a vehicle to yaw or turn sideways when the brake is applied at one rear wheel and released at the other; modern four-circuit systems use more sophisticated controls to reduce this problem.

A *mechanical ABS*, called a *Stop Control System (SCS)* by its manufacturer, is used on smaller front-wheel-drive cars.

An aftermarket system called Brake-guard ABS is marketed to give "ABS benefits" to all vehicles with hydraulic brakes. This is not a true ABS; it simply connects into each hydraulic circuit to provide a small amount of hydraulic pressure accumulator action to reduce very high pressure spikes.

Most ABS designs will revert to standard braking if the system fails, and normal, non-ABS braking will become available. The electronic systems will turn on an amber antilock warning light if failure should occur. Those systems that use ABS booster pressure to operate the rear brakes will lose the rear brakes and power assist if there is a failure in the boost section (Figure 8-9). The electronic portion of an ABS is designed to check itself each time the car is started. You can sometimes hear or feel the system go through its self-checks as the car is started or as it begins moving. On some cars, if you maintain pressure on the brake pedal during this time, you can feel the effect of the modulators exercising as part of the self-check. Presently, most ABS system problems are the result of poor electrical connections and debris or foreign

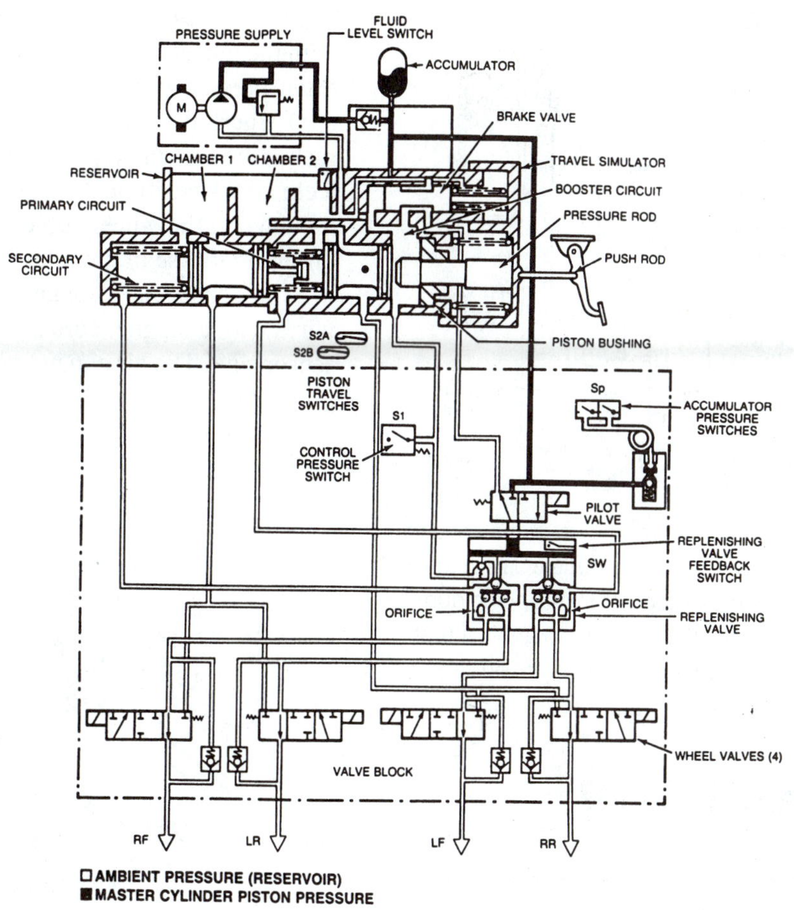

Figure 8-8 A four-circuit ABS. Four wheel valves conrol the pressure to each of the brake calipers to prevent wheel lockup. (Courtesy of General Motors Corporation, Service Technology Group)

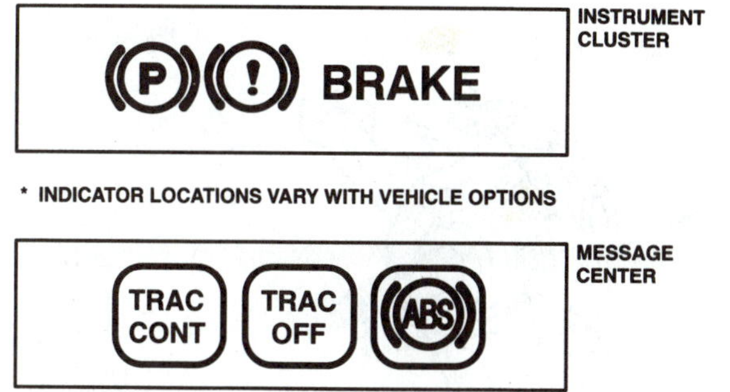

Figure 8-9 Cars equipped with ABS have an amber antilock brake light on the instrument panel in addition to the red brake warning light. This vehicle also has traction control. (Courtesy of Chrysler Corporation)

material in the modulator valve, which causes leakage. Be aware that any system as complex as ABS has quite a few potential problem areas.

Antilock brake systems can also be classified as *integral* or *non-integral*. Non-integral is also called add-on. With some older systems, when ABS is standard equipment, an integral system is used; when ABS is an option, an add-on system is often used. Integral systems have the master cylinder and modulator valve assembly in one unit. Add-on systems mount the modulator between the master cylinder and an existing brake system. Also, some integral systems will use a hydraulic pump and accumulator for power booster assist; one manufacturer calls this a *closed system* and the non-integral ABS an *open system*. An add-on system will have the same master cylinder and power booster as the standard, non-ABS brake, with the modulator valve assembly mounted separately. Some non-integral systems will use a pump for fluid recirculation; this prevents the pedal from moving to the floor as the modulator valves cycle (Figure 8-10).

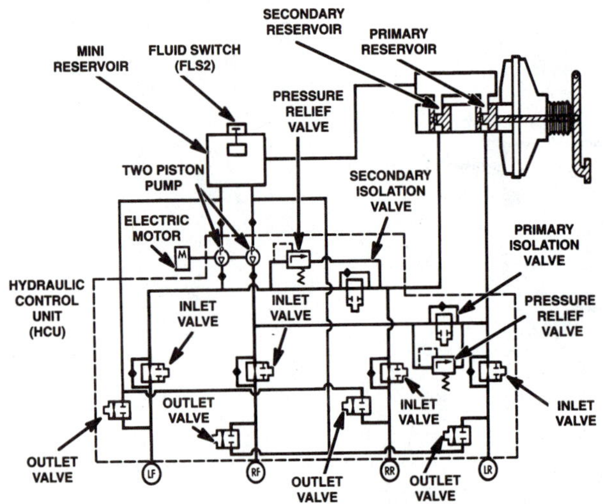

Figure 8-10 This ABS/traction control system has a two-piston pump that will return fluid from the outlet valve(s) to the inlet valve(s) and prevent the brake pedal from dropping during ABS action. This pump also provides pressure to apply a brake to prevent wheel spin and control traction. (Courtesy of Chrysler Corporation)

8.2 Electronic Wheel Sensors

Electronic **wheel-speed sensors** produce an electronic signal as the wheel revolves. These sensors are the "eyes" of the electronic controller, allowing the controller to "see" the rate of deceleration or lockup of the wheel (Figure 8-11). Each sensor uses a gearlike, *toothed rotor*, also called a *sensor ring*, *toothed ring*, *exciter ring*, *tone wheel*, or *reluctor*, on the wheel hub or axle, that rotates with the wheel. Most sensor rings are replaceable, and are often pressed onto the part (CV-joint, rotor, axle shaft, etc.) where they operate. When replacing a sensor ring, make sure that the replacement has the correct number of teeth and is positioned correctly. The wheel sensor can be mounted at the wheel hub or built into the pinion shaft or differential case of a rear axle assembly (Figure 8-12).

The **sensor** is a magnetic induction coil—a coil of wires with a magnet core that is mounted right next to the sensor ring. The air gap between the toothed rotor and the

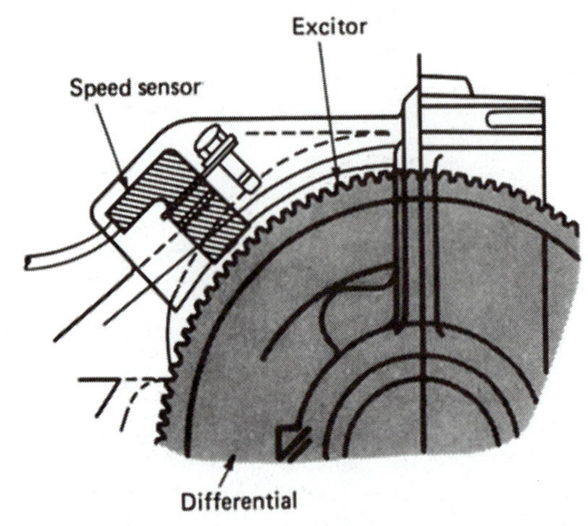

Figure 8-12 Many pickups use an axle-speed sensor mounted in the rear axle. This exciter ring is mounted on the differential case, inside the rear axle housing. (Courtesy of Ford Motor Company)

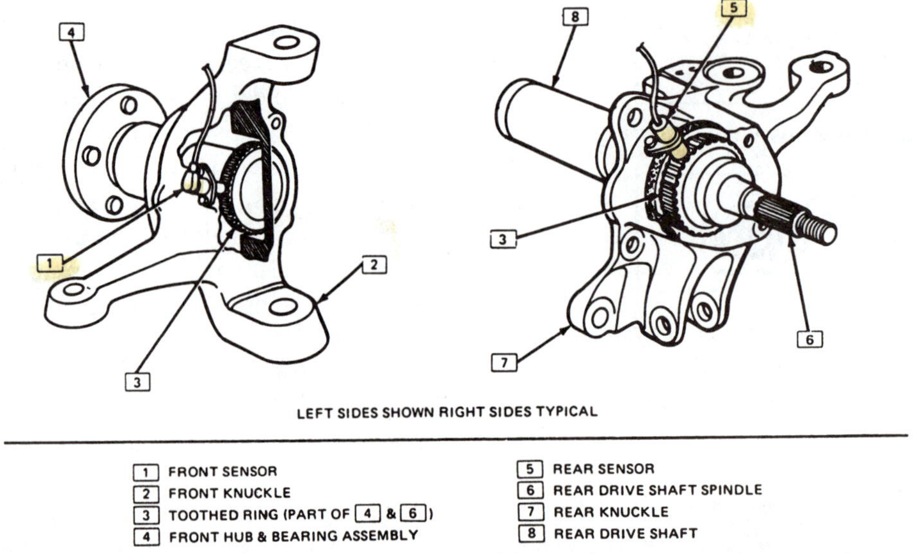

Figure 8-11 On this Corvette, the front wheel sensors (1) are built into the front steering knuckles (2), and the rear wheel sensors (5) are built into the rear wheel support knuckles (7). (Courtesy of General Motors Corporation, Service Technology Group)

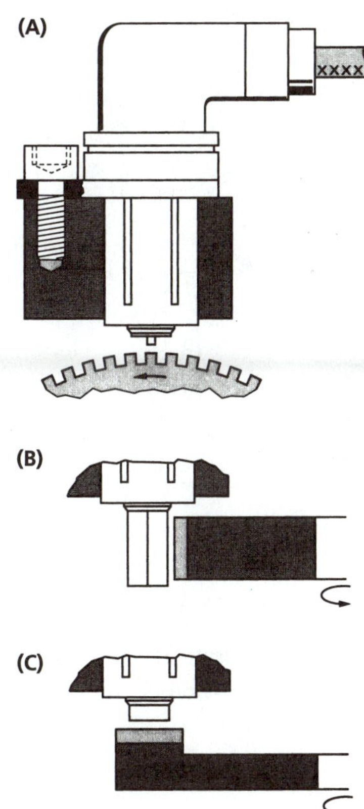

Figure 8-13 A wheel-speed sensor can be mounted so the pole pin is in a radial position outside of the reluctor (A), in an axial position outside the reluctor (B), or alongside the reluctor (C). (Reprinted with permission from Robert Bosch Corporation)

sensor coil is a precise distance to ensure inductance without contact and wear (Figure 8-13). This same type of speed sensor is used in the distributors of some late-model cars and in the speedometer cable for some cruise control systems.

As the wheel rotates, an electrical signal is generated in the induction coil each time one of the teeth of the sensor ring passes the coil. The frequency of this signal is proportional to the speed of the wheel, and this frequency increases and decreases as the wheel speeds up and slows down (Figure 8-14). The electrical signal occurs because the metal teeth of the sensor ring pull the magnetic lines of flux over the coils of wire (Figure 8-15). If a wheel were to lock up under braking, the frequency of the sensor signal for that wheel would drop to zero. Each sensor is connected to the electronic controller by a pair of wires.

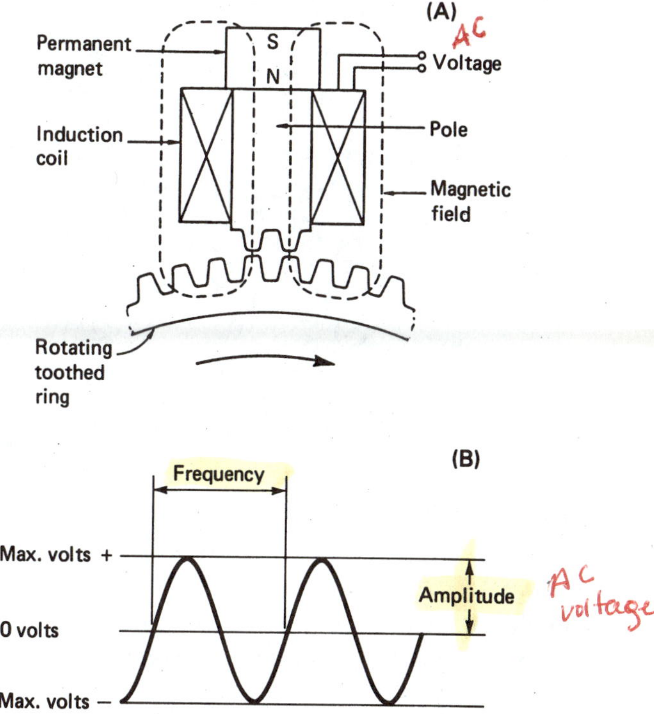

Figure 8-14 As the toothed reluctor ring moves past the induction coil, an AC electrical signal is generated (A). The frequency of the signal will increase as the wheel turns faster (B). (Courtesy of ITT Automotive)

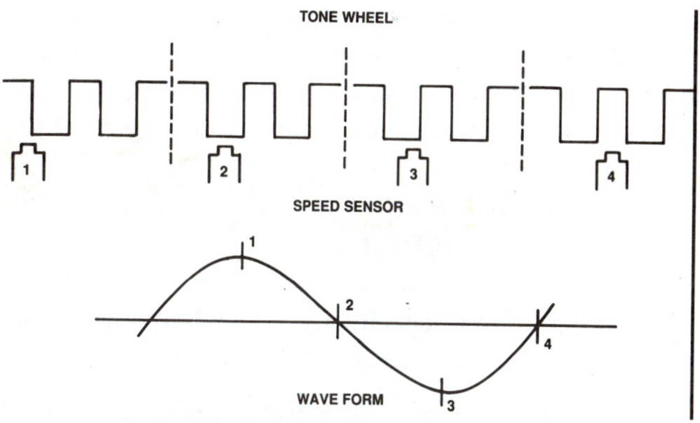

Figure 8-15 The wave form shows a positive voltage signal generated as the front of a tone wheel tooth passes the speed sensor (1). The signal voltage drops to zero as the speed sensor passes the center of the tooth (2) or between two teeth (4). A negative signal is generated as the back of the tooth passes the speed sensor (3). (Courtesy of Chrysler Corporation)

8.3 Electronic Controller

The *electronic controller*, also called an *electronic control module (ECM)*, *electronic brake control module (EBCM)*, or *controller antilock brake (CAB)*, is a microprocessor with about 8K of memory (Figure 8-16). The EBCM receives the signal from each wheel sensor as input and actuates the modulator valves as output. It compares the speed of each wheel (the frequency of the sensor signal) to

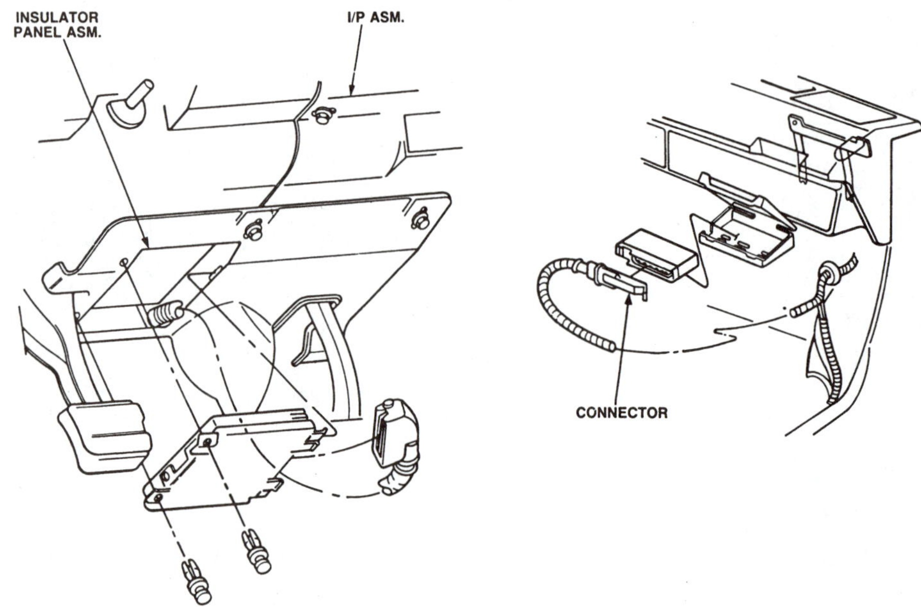

Figure 8-16 The electronic controller (EBCM) is mounted in a relatively cool, clean, well-protected location. In this car, it is mounted under the left or right end of the bottom of the instrument panel. (Courtesy of General Motors Corporation, Service Technology Group)

that of the others and to a deceleration profile programmed and stored in its memory. It determines if wheel lockup is beginning to occur, if the sensor signal frequency is slowing at too fast a rate when compared to the deceleration profile, or if the signal has a significantly lower frequency than the other wheels. The EBCM constantly monitors and compares the wheel speeds. It also checks itself and the system to ensure proper operation. Some systems use two identical controllers, inside the EBCM, which compare each other also. If they do not agree with each other, they shut the system off and turn on a warning light.

When the EBCM determines that a wheel is slowing too quickly or that it has already stopped, it will activate the brake modulator for that wheel (Figure 8-17). The modulator will release the brake enough to allow the wheel speed to increase. As the signal frequency from the wheel comes within the correct profile, the controller will activate the modulator to reapply the brake. Since electrical signals travel at close to the speed of light, these actions can occur very rapidly.

The EBCM is somewhat fragile, in addition to being very expensive. The controller is normally mounted in one of the cleaner, drier, cooler locations where it will be somewhat protected from impact and corrosion. In some cars it is mounted in the trunk area, behind the rear seat, or in one of the side panels (Figure 8-18). In other vehicles it is mounted in the bottom of the instrument panel,

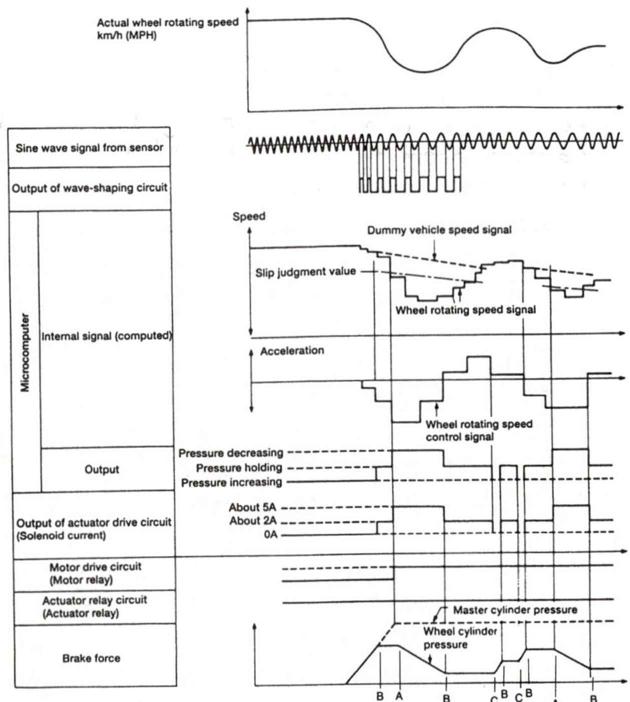

Figure 8-17 A time profile of an ABS-assisted stop. At points A, the EBCM noted the wheel speed was slowing too rapidly and cycled the control valve to hold the brake pressure steady. At points B, the brake pressure was lowered and at points C, the pressure was allowed to increase. (Courtesy of Nissan of North America)

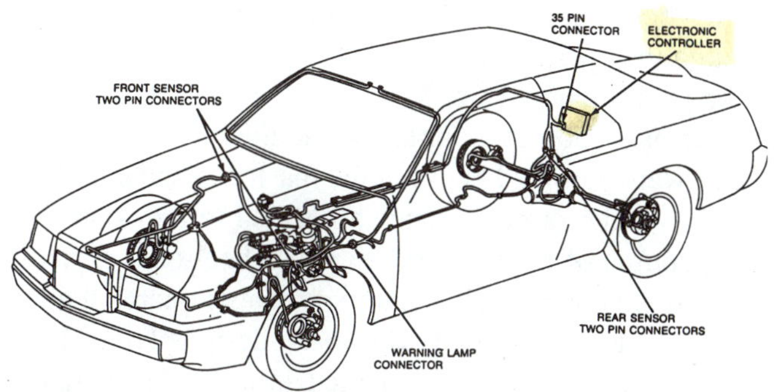

Figure 8-18 This car has the EBCM mounted behind the rear seat, and it is connected to the wiring harness by a 35-pin connector. (Courtesy of Ford Motor Company)

behind or above the glove compartment, in the trunk, or under a seat. It is mounted under the hood in the engine compartment in some cars and pickups.

The electrical circuit for an ABS is somewhat complex and will vary depending on the design of the system and the particular car model. It will always include a power supply for the controller with the ability to power the controller output to the modulators. The circuit will also include connections to the individual wheel sensors and modulators and the brake failure warning-light circuit. A typical circuit is shown in Figure 8-19. This circuit complexity requires advanced electronic troubleshooting skills, well beyond those of the typical brake technician of the 1970s and 1980s.

Normal outputs controlled by the EBCM are the solenoid valves in the modulator, the amber ABS warning light, and the pump motor. The normal inputs are the wheel-speed sensors and in some systems, pump pressure, fluid level, stoplight switch, and a brake pedal travel sensor (Figure 8-20).

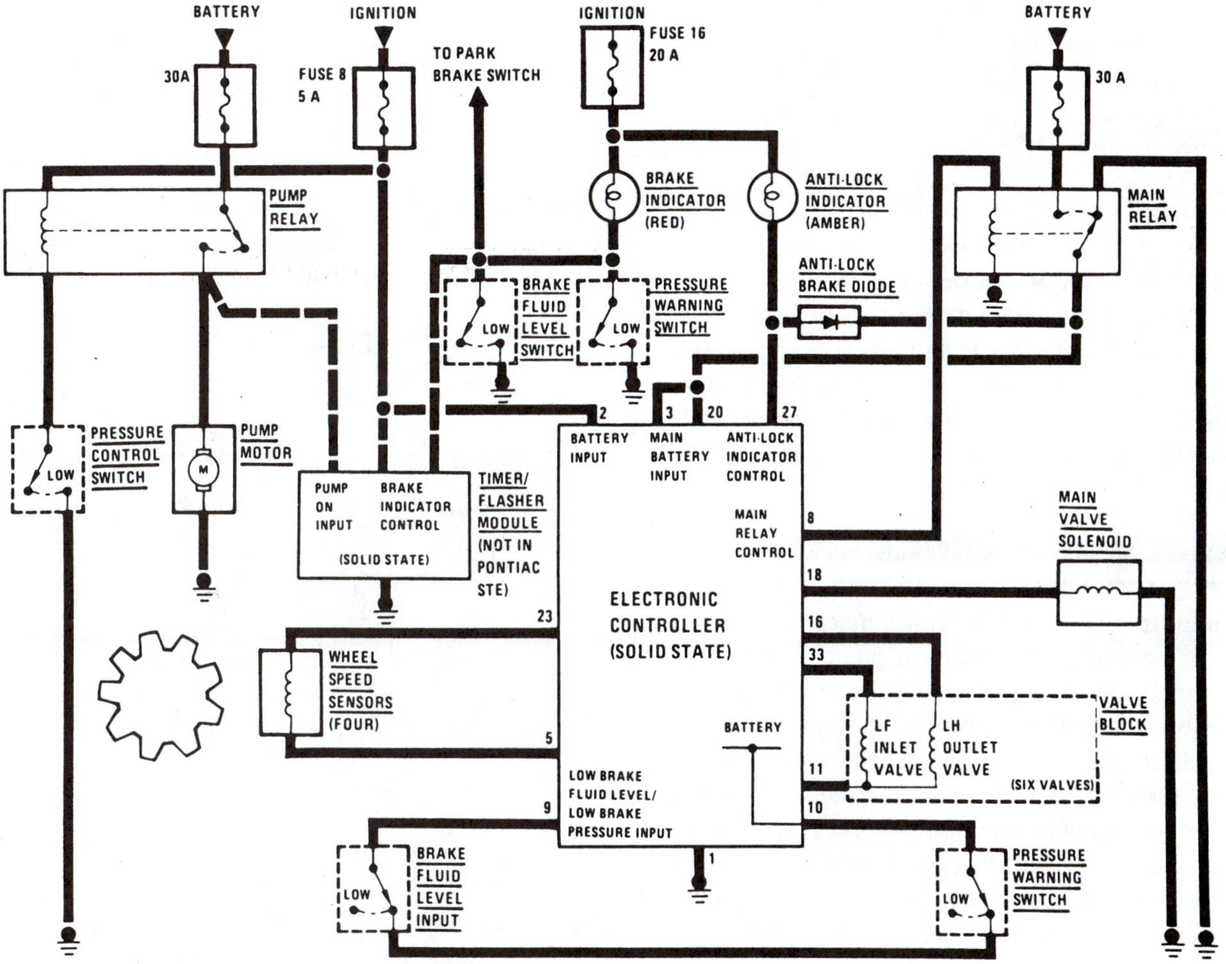

Figure 8-19 An electrical ABS schematic showing the EBCM and most of the connections to it. (Courtesy of General Motors Corporation, Service Technology Group)

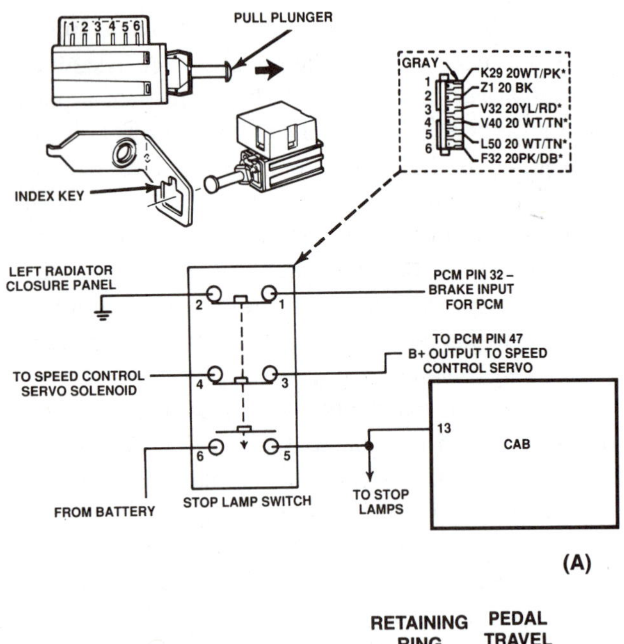

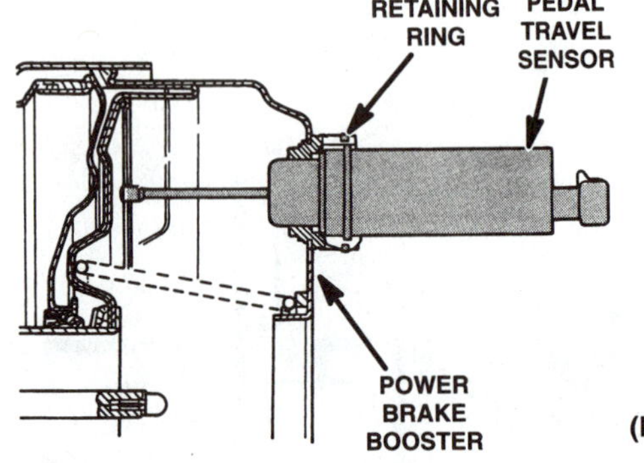

Figure 8-20 This stop lamp switch also provides signals to the Power Control Module (PCM), speed control servo, and Controller Antilock Brake (CAB) to let them know the brakes are applied (A). The Pedal Travel Sensor (PTS) provides a signal so the CAB knows how far the brake pedal has traveled (B). (Courtesy of Chrysler Corporation)

8.4 Electronic Modulator

The **modulator**, also called the *hydraulic control unit* or *actuator*, is the device that cycles or pumps the brakes. During normal stops, the modulator will not change normal brake operation. During very hard stops, the brake pressure will increase in the brake assemblies in a normal manner. If a wheel begins to lock, the modulator must stop any further hydraulic pressure increase at that particular wheel cylinder or caliper. It can hold that pressure in the brake assemblies. If this action is not enough to get the wheel turning at the correct speed, the modulator must reduce the pressure. As soon as the wheel is turning at the correct speed, the modulator must let the pressure in the wheel cylinder or caliper increase again. If reapplication of the brake causes a lockup, the modulator must go through the cycle over again. Most systems can cycle the brake pressure about five to fifteen times a second.

The early domestic units used individual modulators, which were mounted between the master cylinder and the wheel cylinder or caliper and connected to the controller and a source of power. Modern rear-wheel-only systems use a single control valve assembly that is operated electronically (Figure 8-21). Early Bosch-designed units have all the modulator valves mounted in a single hydraulic

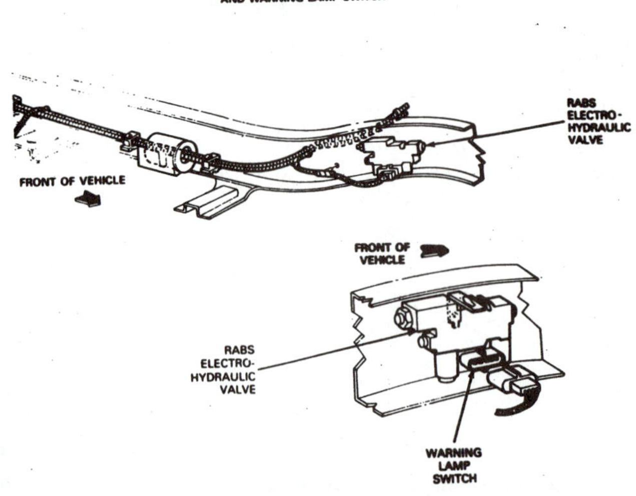

Figure 8-21 This Rear Antilock Brake System (RABS) or RWAL modulator valve is mounted inside the frame rail. (Courtesy of Ford Motor Company)

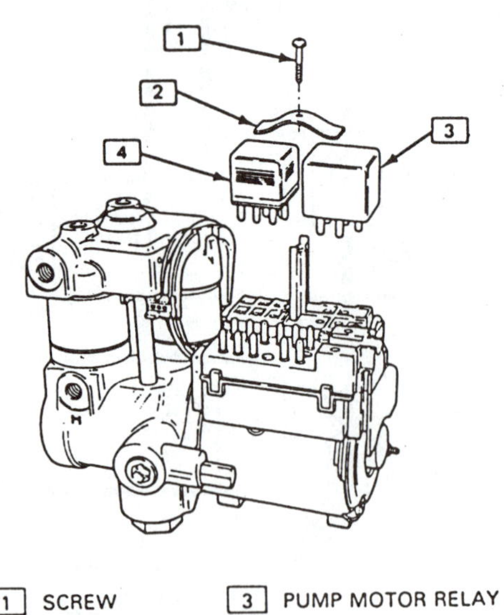

Figure 8-22 A modulator valve from a Bosch-designed system. It has three solenoids that control the three-circuit system. (Courtesy of General Motors Corporation, Service Technology Group)

unit that is connected to the master cylinder by two hydraulic brake lines (Figure 8-22). Early Teves-designed systems combine all the valves into a valve block assembly that is mounted directly on the master cylinder. This unit combines a master cylinder, an electrohydraulic booster, and the valve block (Figure 8-23). The Delco ABS-VI system also combines the master cylinder and the hydraulic modulator; this unit uses a vacuum power booster (Figure 8-24).

The actual valves inside a modulator assembly will vary greatly. Many valves are balls that are normally open or off the seat and can be closed or pushed onto the seat by a solenoid-operated pin. Others can be pushed onto a seat by a spring or fluid pressure and off the seat by the solenoid-operated pin. Some systems use sliding spool valves

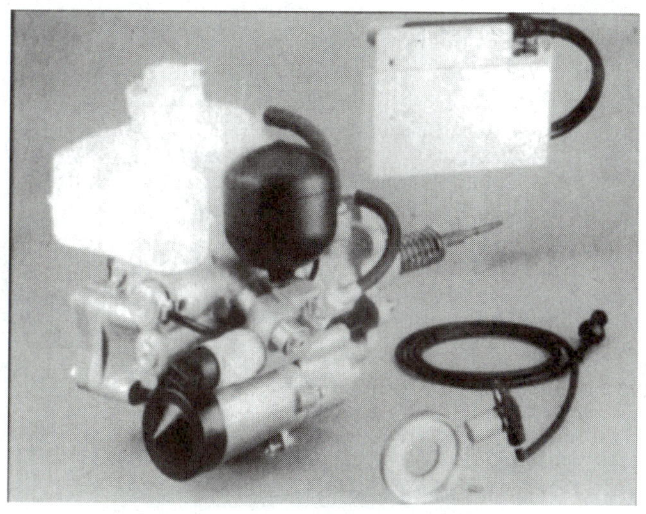

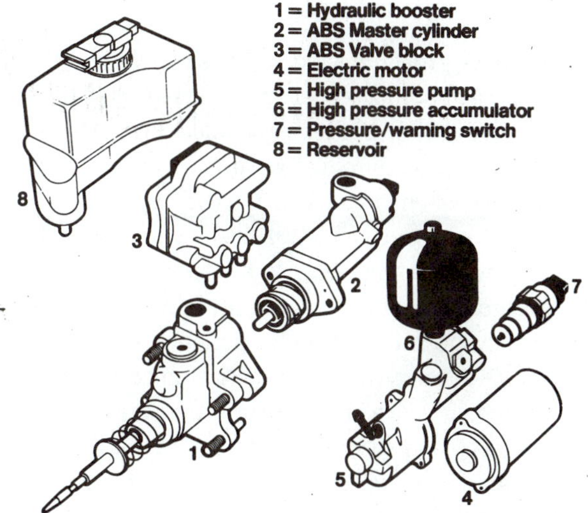

Figure 8-23 The master cylinder–power booster–conrol valve combination used in early Teves, ATE-designed systems. (Courtesy of ITT Automotive)

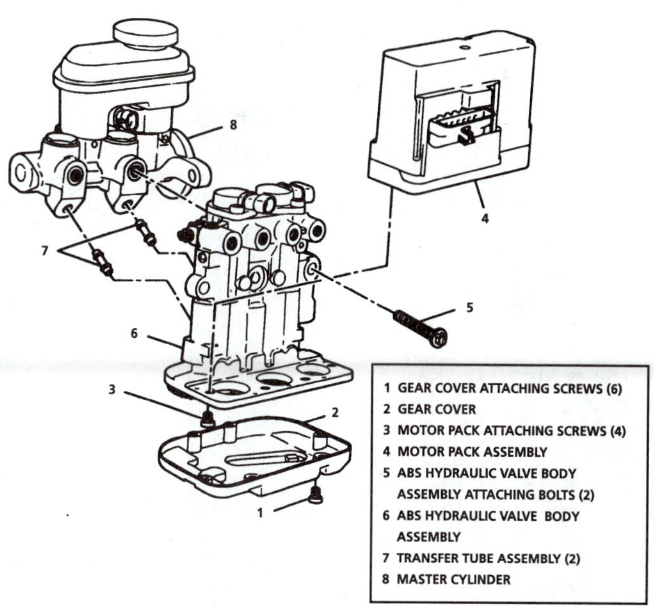

Figure 8-24 This Delco (now Delphi) ABS-VI hydraulic modulator assembly is secured to the master cylinder. Note that the gear cover can be removed to service the gear assemblies and that the motor pack can also be serviced. A special operation must be performed before any service work. (Courtesy of General Motors Corporation, Service Technology Group)

to open or close fluid passages. In some cases, a solenoid will control two ball valves, and this solenoid can have either two or three operating positions (Figure 8-25). In most systems, one solenoid controls pressure increase, inlet, or build, and one will control pressure decrease, outlet, or decay. The Bosch ABS 2S uses one three-position solenoid to allow pressure increase, hold, and decrease. The Delco VI system combines an isolation solenoid that can block pressure increase along with a motor-controlled piston to decrease pressure. Up to ten solenoids and valves are combined into a modulator assembly.

Early modulator valves use a design in which a check valve is normally held open by the end of a piston rod. When closed, this valve will allow a flow only one way—from the wheel cylinder or caliper to the master cylinder. If lockup occurs, the piston rod moves back, and the check valve closes to limit any further increase in brake pressure. Further movement of the piston rod will lower the pressure in the modulator and the wheel cylinder or caliper. Movement of the piston rod, on these early systems, is accomplished by manifold vacuum, hydraulic pressure (power steering pump), or an electric solenoid. When vacuum or hydraulic pressure is used, the vacuum or hydraulic flow is controlled by an electric solenoid. In all cases, the control solenoids are actuated by the EBCM.

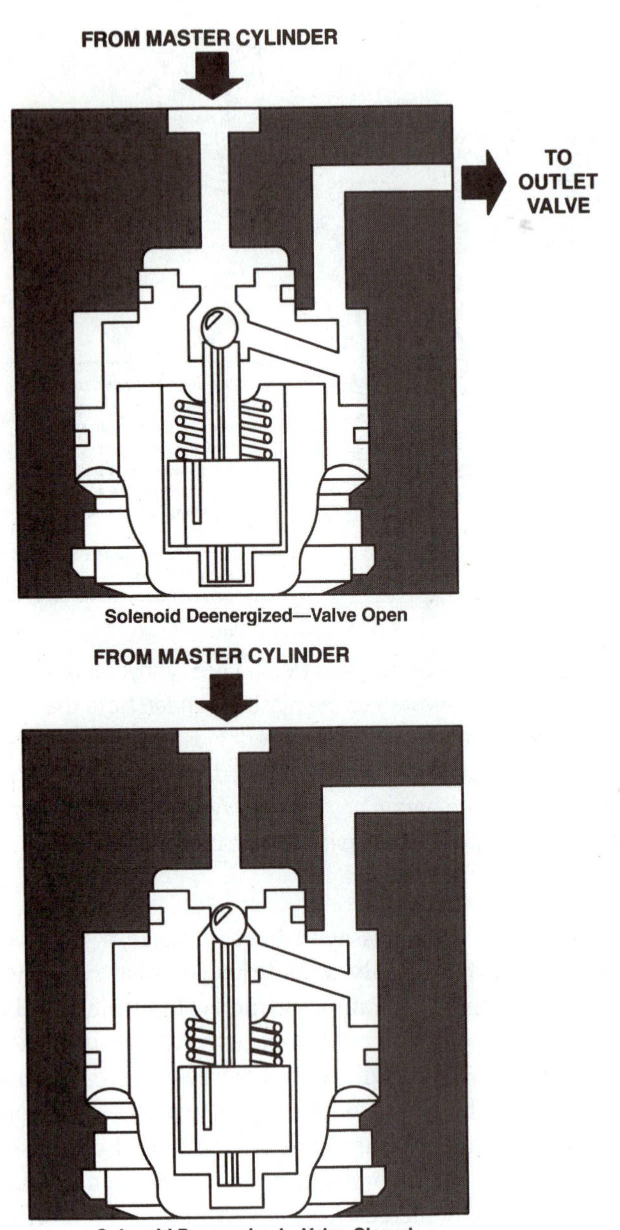

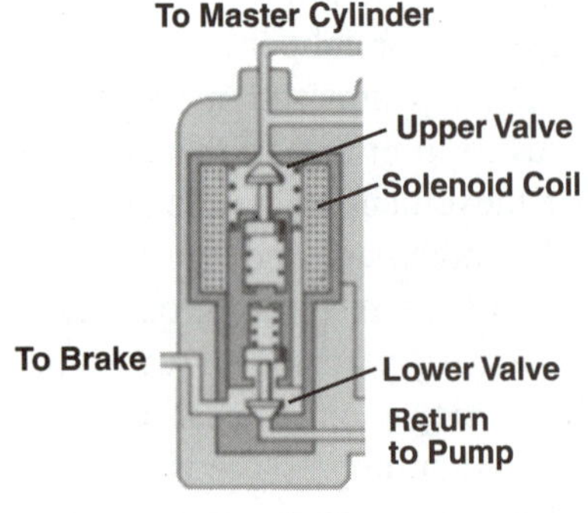

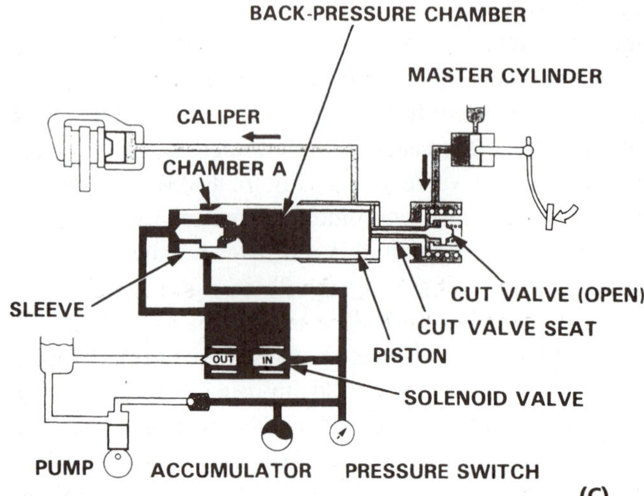

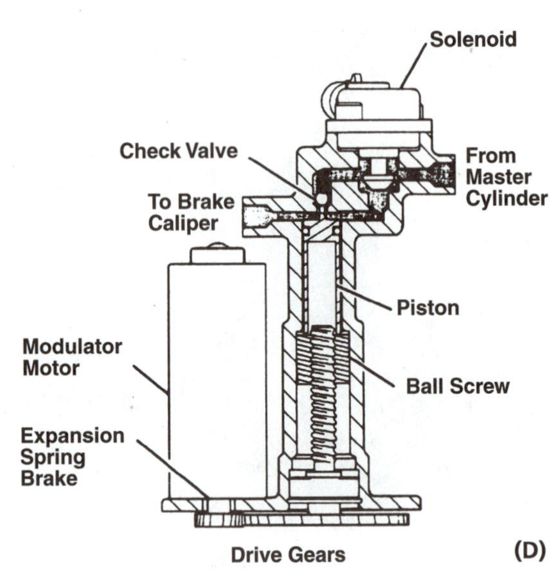

Figure 8-25 A variety of modulator valve types. The inlet/isolation valve (A) is normally open and is closed when the solenoid moves the ball onto the seat. The upper/inlet valve is normally open and the lower/outlet valve is normally closed (B); both are closed when the solenoid moves upward to mid-position, and the outlet valve will open when the solenoid moves to the top position. When the inlet valve is open, the piston moves to the right to open the cut valve, allowing brake pressure to flow to the caliper (C); closing the IN valve and opening the OUT valve will cause the piston to move to the left, closing the cut valve and causing a pressure drop in the caliper. The check and solenoid valves are normally open (D); activating the solenoid will close that valve, and lowering the piston will close the check valve and reduce the pressure in the caliper. (A is courtesy of Chrysler Corporation; B is reprinted with permission from Robert Bosch Corporation; C is courtesy of American Honda Motor Co., Inc.; D is courtesy of General Motors Corporation, Service Technology Group)

8.5 Recirculation Pump

In some systems during an ABS stop, the valve action to hold and release a brake takes fluid from the system, and this will cause the brake pedal to lower, possibly to the floor. This in turn, can cause the circuit to run out of brake pressure. To prevent this from occurring, pumps are used to recirculate or return the fluid from the pressure-reducing or decay valves back into the fluid lines (Figure 8-26). The pump has to be able to develop pressures equal to that in the brake system. The driver should feel a bump(s) from the brake pedal when this occurs. Some systems return this fluid to the master cylinder reservoir.

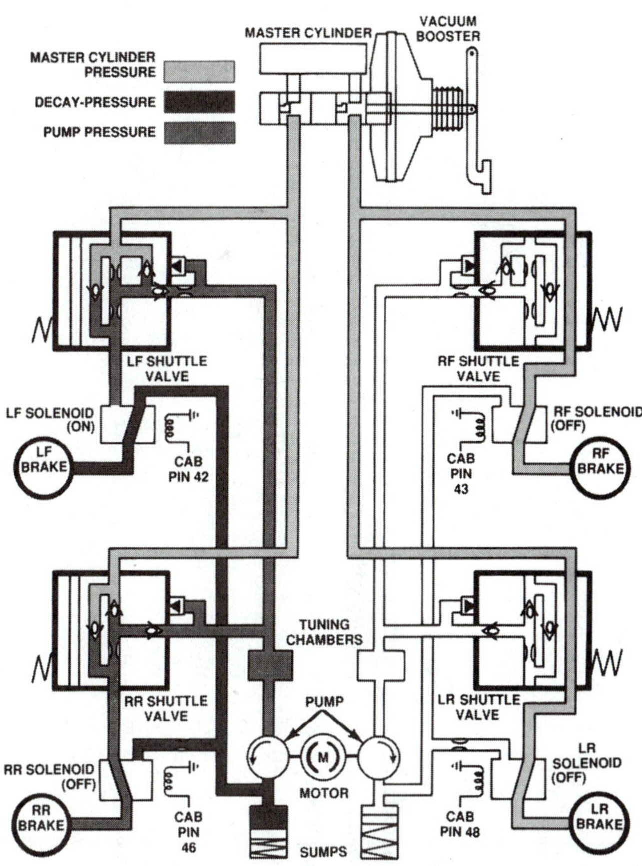

Figure 8-26 This ABS has activated the left front solenoid to move the shuttle valve to reduce (decay) the pressure in the caliper. It has also started the pump to return the fluid to the brake lines for left front and right rear brakes. (Courtesy of Chrysler Corporation)

Decay - In ABS, the action of reducing pressure.

Most systems use a single or double piston pump that is controlled by the EBCM receiving a signal from the brake pedal travel switch or accumulator pressure switch. Some systems store the fluid from the brake valves in an accumulator before starting the pump (Figure 8-27).

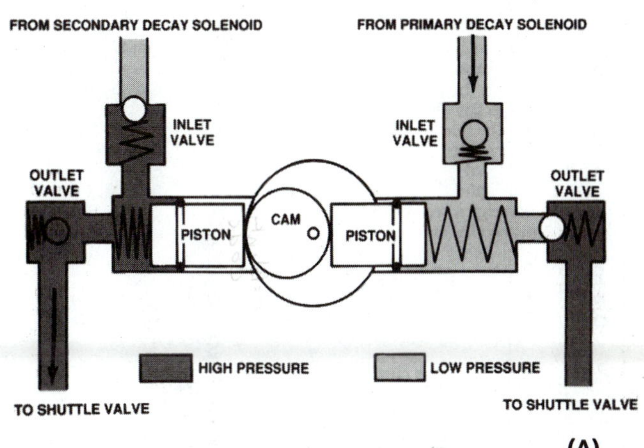

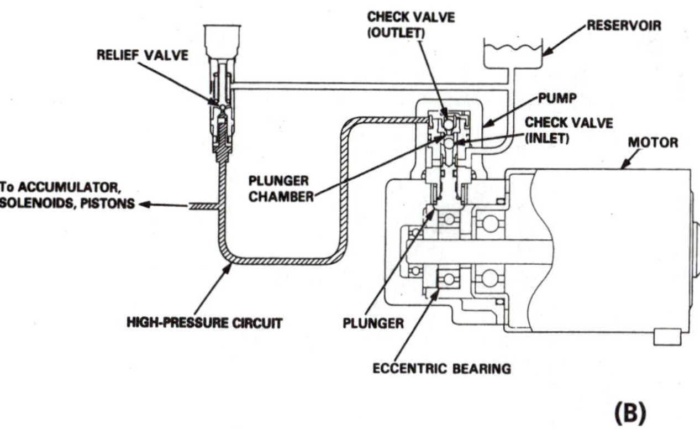

Figure 8-27 This twin-piston pump isolates the primary and secondary circuits (A); the two check valves control the flow in and out of the pump chambers. The pump plunger is operated by an eccentric bearing mounted on the motor shaft (B). (A is courtesy of Chrysler Corporation; B is courtesy of American Honda Motor Co., Inc.)

Recirculator pump pressure is also used in some traction control systems to apply the brake and slow down the spinning wheel.

8.6 ABS Systems

There is a large variety of ABS modulator designs. These systems are manufactured by several domestic (United States), European, and Japanese companies as well as some vehicle manufacturers. Common manufacturers include Bendix, Bosch, Delco Moraine (now Delphi), Kelsey-Hayes, and Teves (now ITT Automotive) (Chart 8-1).

Early Teves systems used electrohydraulic boosters, in which the booster pressure also became the pressure source for the rear brakes, and the modulator valves were integral with the master cylinder. Some versions of this

CHART 8-1

ABS Manufacturer and System	Usage From–to	Integral or Non-integral	Vehicle Make	Car Light Truck Utility Vehicle	Booster Type	Number of Channels	Fluid Return Pump	Diagnostic Trouble Codes	Traction Control Option
Bendix ABX-4	1995–	Non-integral	Chrysler	Car	Vacuum	4	Yes	16	
Bendix 4	1993–95	Integral	Jeep	Car	Elec.-Hyd.	3	Yes	14	
Bendix 6	1991–93	Non-integral	Chrysler	UV	Vacuum	4	Yes	16	
Bendix 9	1989–91	Integral	Jeep	UV	Elec.-Hyd.	3	No	20	
Bendix 10	1990–93	Integral	Chrysler	Car	Elec.-Hyd.	4	No	Yes	
Bendix Mecatronic II	1995–	Non-integral	Ford	Car	Vacuum	4	Yes	Yes	Yes
Bosch 2	1981–90	Non-integral	SEE NEXT LINE		Vacuum	3	Yes	BC=Blink code	Yes
(includes 2E/2S/2U)	*Car models include:* Audi, BMW, Corvette, Ford, G.M., Lexus, Mazda, Mercedes, Mitsubishi, Nissan, Porsche, Rolls Royce, Sterling, Subaru, Toyota; *Utility vehicle models include:* Isuzu, Suzuki								
Bosch III	1987–92	Integral	Chrysler	Car	Elec.-Hyd.	4	No	16	
			G.M.	Car					
Bosch 5	1995–	Non-integral	G.M.	Car	Vacuum	4	Yes	Yes	Yes
(includes ABS/ASR)			Ford	Car					
Bosch 5.3	1997–	Non-integral	G.M.	Car	Vacuum	4	Yes	Yes	Yes
			Toyota	Car					
			Subaru	Car					
Bosch VDC	1996	Non-integral	Mercedes	Car	Vacuum	4	Yes	Yes	Yes
Delco Moraine III	1989–91	Integral	G.M.	Car	Elec.-Hyd.	3	No	63	
Delco Moraine IV	1991–	Non-integral	G.M.	Car	Vacuum	3	No	79	Yes
(Now Delphi)			Geo	Car					
			Saturn						
Honda	1988–	Non-integral	Honda/Acura	Car	Vacuum	3	Yes	Yes	
Kelsey Hayes RWAL	1987–	Non-integral	SEE NEXT LINE		Vacuum	1	No	17	
(includes RABS & EBC2)		*Light truck and utility vehicle models include:* Dodge, Ford, Geo, G.M., Isuzu, Mazda, Nissan							
Kelsey Hayes 4WAL	1988–	Non-integral	Dodge	LT	Vacuum	3	Yes	41	
(includes versions: EBC5, EBC10,			G.M.	LT/UV					
EBC325, & EBC430)			Isuzu	UT					
			Kia	UT					
Nippondenso	1990–	Non-integral	Infiniti	Car	Vacuum	4	Yes	Yes	
			Lexus	Car					

Sumitomo	1987–	Non-integral	Ford Probe	Car	Vacuum	3/4	Yes	Yes	
			Honda	Car					
			Mazda	Car					
Teves Mark II	1985–91	Integral	Ford	Car	Elec.-Hyd.	4	No	42	
			G.M.	Car					
			Saab	Car					
			Volkswagen	Car					
Teves Mark IV	1990–	Non-integral	Chrysler	Car	Vacuum	4	Yes	29	Yes
(now ITT Teves)			Ford	Car					
			G.M.	Car					
ITT Teves Mark 20	1997–	Non-integral	BMW	Car	Vacuum	3/4	Yes	Yes	Yes
			Chrysler	Car					
			Ford	Car					
			Honda	Car					

Note: This chart is printed to give you an idea of which vehicles use the major types of ABS. If servicing a particular system, it is very important to use printed or computerized service information for that particular vehicle. A vehicle manufacturer often uses one ABS type for one vehicle model and other versions from the same system manufacturer or from a different manufacturer for different models. Most newest-version ABS types include TCS as an extra-cost option, and this option will add valves to the hydraulic modulator and possibly an additional channel. Also, some vehicle manufacturers use a system that they have designed and manufactured or a system designed by one of the major ABS manufacturers and manufactured under license of that company. When seeking service information, always use the vehicle make, model, and year and then the ABS type.

ABS Manufacturer and System: indicates the basic design of the system. Many systems have one or more versions. Robert Bosch Corporation now owns the ABS manufacturing assets of Bendix.

Usage: is not necessarily the same for each vehicle manufacturer

Integral or Non-integral: self-explanatory

Vehicle Make: can include any vehicle made by that manufacturer

Car, Light Truck, Utility Vehicle: self-explanatory

Booster Type: most systems will use either a vacuum or electrohydraulic booster

Number of Channels: most systems will have either 3 or 4 braking channels, one system has 1

Fluid Return Pump: this pump returns fluid from the hydraulic modulator to the master cylinder

Diagnostic Trouble Code: indicates if DTCs are available and in most cases how many codes are used, BC indicates Blink or Flash Codes at the MIL

Traction Control Option: indicates if this system can include traction control

design use a different valve arrangement, and also use boost pressure to operate the rear brakes (Figure 8-28). Many modern systems use a master cylinder with a vacuum booster, mount the modulator assembly separately, and connect these together with tubing. Other designs use a separate modulator valve assembly with three (three-circuit system) or four (four-circuit system) pairs of valves (Figure 8-29). The operation of these systems is similar to that just described. During normal braking, fluid flows unchanged through the modulator to the brake assemblies (Figure 8-30). If a wheel starts to lock up, the inlet (or isolation) modulator valve will close to prevent any further increase in brake pressure. If the wheel stays locked up, the outlet or decay modulator valve will open to release pressure. Some systems include an accumulator to store this released fluid, and a pump to return the fluid to the master cylinder reservoir or back into the pressure lines.

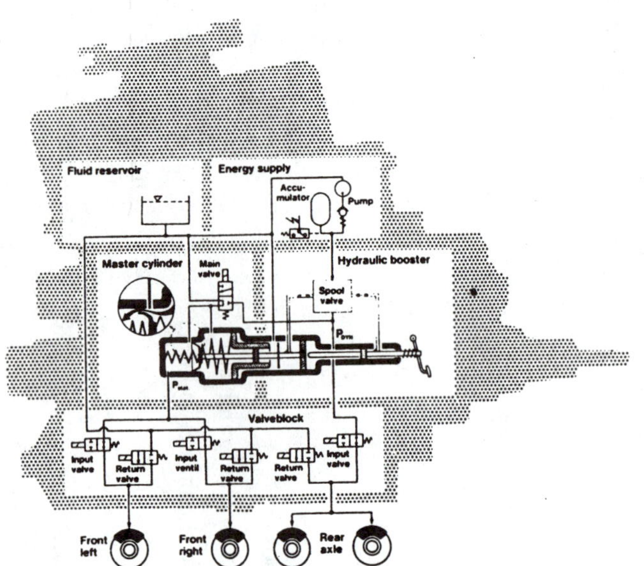

Figure 8-28 A schematic showing the internal flow and the arrangement of the valves in an ABS master cylinder, power booster, and control valve assembly. (Courtesy of ITT Automotive)

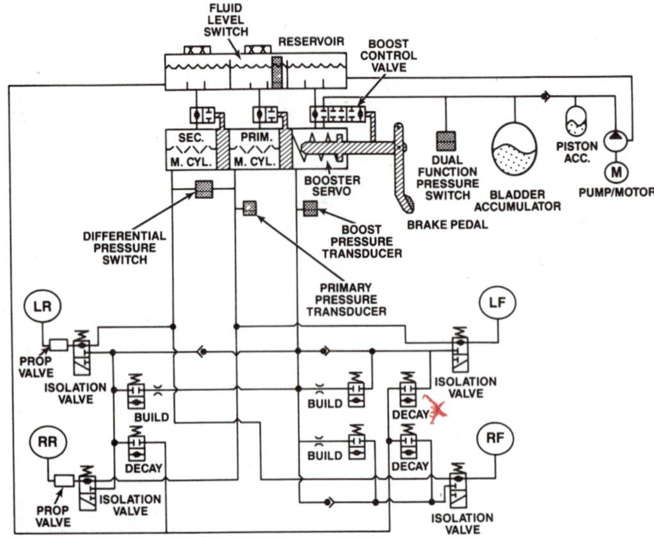

Figure 8-29 This three-circuit system is a closed system and uses a hydraulic pump and accumulator for boost pressure. There are ten valves in the modulator portion that are positioned for normal operation. (Courtesy of Chrysler Corporation)

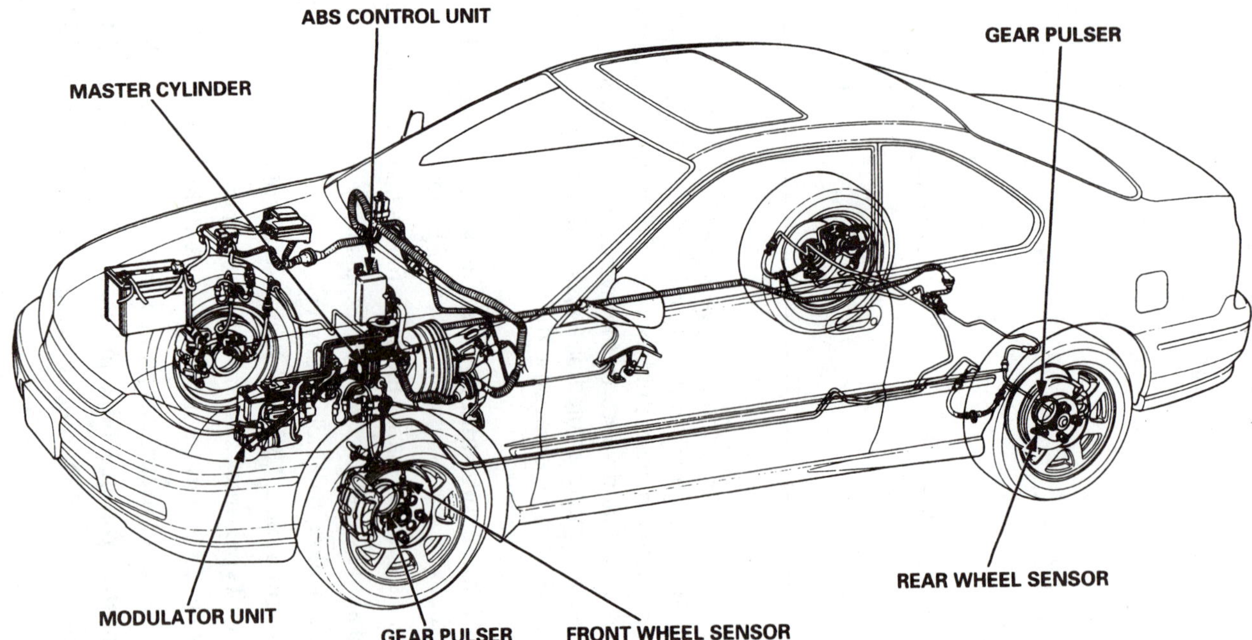

Figure 8-30 Like many others, this ABS system uses a modulator unit that is separate from the master cylinder and vacuum booster. (Courtesy of American Honda Motor Co., Inc.)

Newer modulator designs use a variety of control methods. Several system designs use one or a pair of valves for each hydraulic circuit, and each valve is operated by an electric solenoid. One valve, the inlet, is normally open and normally allows flow between the master cylinder and the wheel cylinder or caliper (Figure 8-31). The other valve, the outlet, is normally closed. When it opens, fluid can flow from the wheel cylinder or caliper to the master cylinder reservoir. If a wheel lockup occurs, the EBCM will close the inlet valve and prevent any further increase in brake application pressure. If the wheel remains locked, the outlet valve is opened to reduce the braking pressure (Figure 8-32).

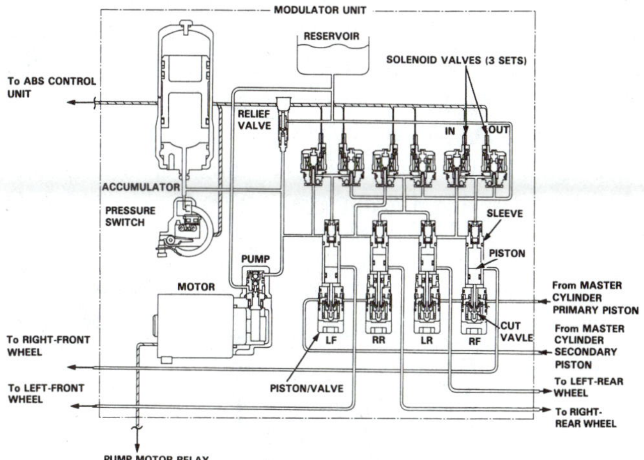

Figure 8-31 This three-circuit system uses a cut valve for each brake unit, but both rear cut valves are controlled by the same pair of in and out valves. These cut valves will block pressure from the master cylinder to the brake circuits. (Courtesy of American Honda Motor Co., Inc.)

The RWAL system uses a single modulator, called the dual solenoid control valve, RABS (Rear Antilock Brake system) valve, and isolation/dump valve, for the rear brakes. This valve is mounted next to the master cylinder or in a remote location closer to the rear. This assembly contains two solenoid valves and an accumulator. During an ABS-assisted stop, the isolation solenoid will close the valve and stop any further pressure increase (Figure 8-33). The other solenoid can open the dump valve if needed, allowing rear pressure to dump into the accumulator, thereby reducing rear brake pressure. Any fluid that enters the accumulator returns to the master cylinder when the brakes are released. A potential problem with this design is that with some vehicles, the brake pedal lowers during each modulator cycle. Approximately eight

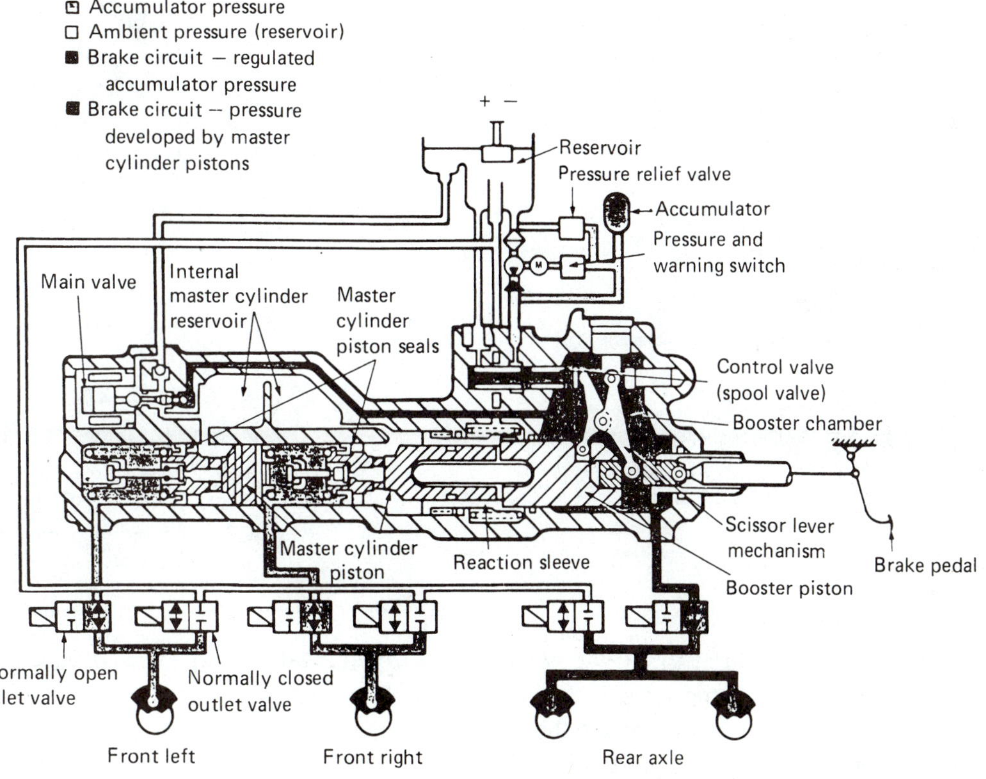

Figure 8-32 During normal braking in this early Teves system, the inlet valves are open and the outlet valves are closed. During the pressure-holding stage, the inlet valve(s) will close, and during the pressure-reducing stage, the outlet valve(s) will open. (Courtesy of General Motors Corporation, Service Technology Group)

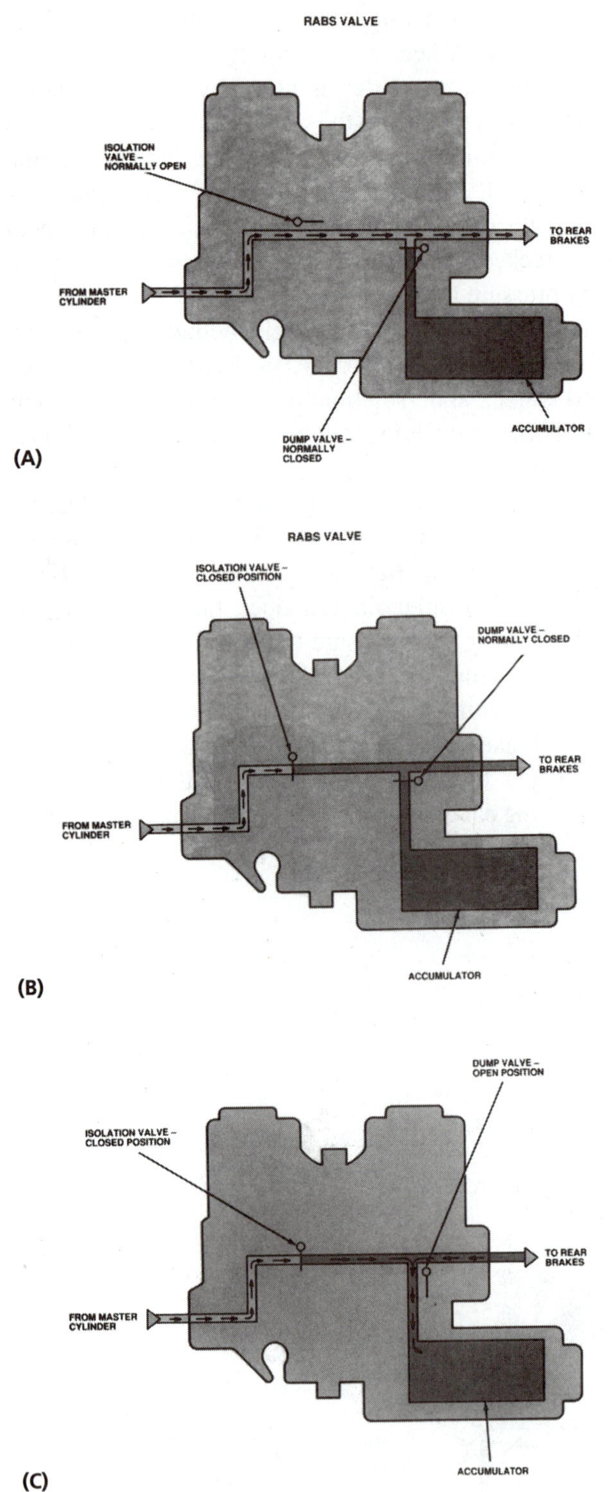

Figure 8-33 During normal braking of this RWAL system, the isolation valve is open and the dump valve is closed so there is a fluid flow between the master cylinder and the rear brakes (A). During the pressure-holding stage, the isolation valve closes (B). During the pressure-reducing stage, the dump valve opens to bleed pressure to the accumulator (C). (Courtesy of Ford Motor Company)

cycles can bring the pedal to the floor, and then the pedal must be pumped to restore braking action.

In some Bosch units, a pump is used instead of an outlet valve. During the pressure-reduction phase, fluid is pumped from the brake cylinder side of the valve back to the master cylinder side of the valve or master cylinder body (Figure 8-34). When the wheel is turning again, the valves are returned to their normal positions.

The Delco/Delphi ABS-VI hydraulic modulator assembly contains two solenoids and three electric-motor-driven pistons. Each front brake circuit uses a single

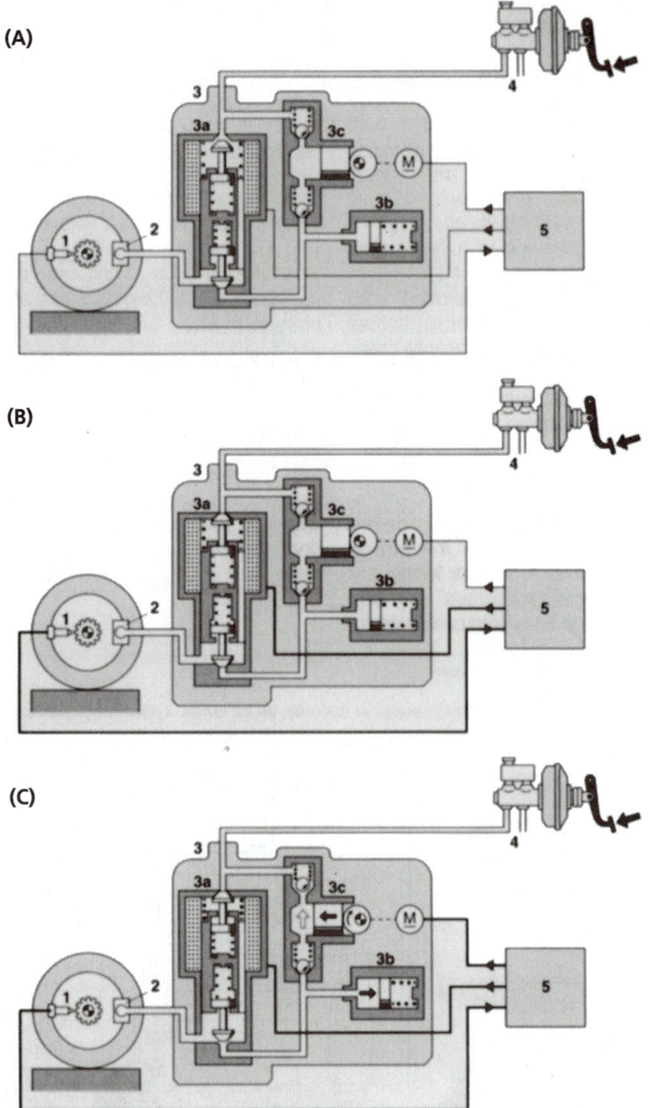

Figure 8-34 During normal braking in this Bosch system (A), braking pressure passes through the control valves. During the pressure-holding stage (B), the solenoid valve is stroked to mid-position. During the pressure-reducing stage (C), the solenoid valve is stroked to full up, opening the outet, and the pump is actuated to return the fluid. (Reprinted with permission from Robert Bosch Corporation)

(A)

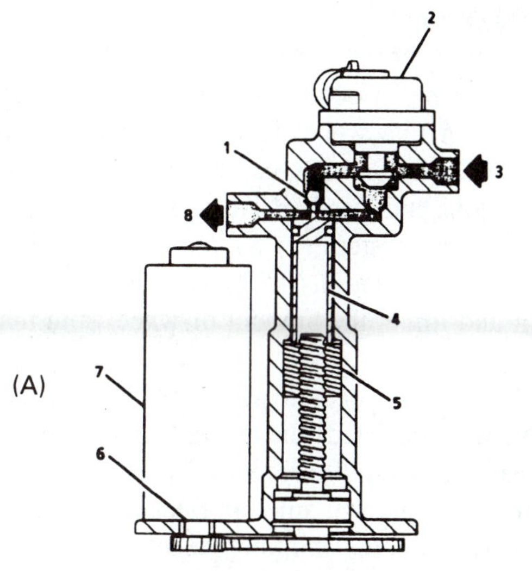

1 CHECK VALVE OPEN
2 NORMALLY OPEN SOLENOID
3 INLET FROM MASTER CYLINDER
4 PISTON FULLY RAISED
5 BALL SCREW IN "HOME" POSITION
6 EXPANSION SPRING BRAKE (ESB)
7 FRONT MODULATOR MOTOR
8 MASTER CYLINDER PRESSURE TO FRONT BRAKE

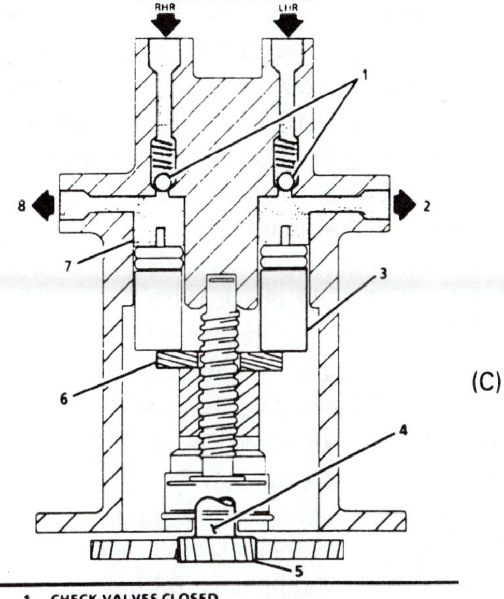

(C)

1 CHECK VALVES CLOSED
2 MODULATED PRESSURE TO LEFT HAND REAR BRAKE
3 PISTON LOWERED TO REDUCE PRESSURE
4 EXPANSION SPRING BRAKE (ESB) LOCATION
5 MOTOR PINION
6 YOKE ON BALL SCREW DRIVES BOTH REAR CIRCUIT PISTONS
7 MODULATION CHAMBER
8 MODULATED PRESSURE TO RIGHT HAND REAR BRAKE

(B)

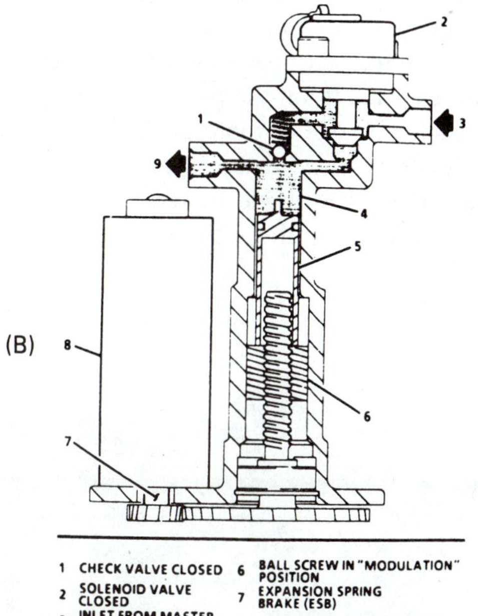

1 CHECK VALVE CLOSED
2 SOLENOID VALVE CLOSED
3 INLET FROM MASTER CYLINDER
4 MODULATION CHAMBER
5 PISTON IN "MODULATION" POSITION
6 BALL SCREW IN "MODULATION" POSITION
7 EXPANSION SPRING BRAKE (ESB)
8 FRONT MODULATOR MOTOR
9 MODULATED ABS PRESSURE TO FRONT BRAKE

Figure 8-35 During normal braking of this Delco ABS-VI system, the piston is in the uppermost position and the solenoid valve is open (A). During the pressure-holding and reducing stages, the solenoid is actuated to isolate the circuit and the piston is lowered to reduce the pressure as needed (B). A piston pair is used to control both rear brakes (C). (Courtesy of General Motors Corporation, Service Technology Group)

piston and motor; the rear brake circuits use a pair of pistons driven by a single motor. Each piston drive mechanism contains a braking mechanism to hold the piston in a fixed position unless driven by the motor. During normal brake operation, the pistons are kept in their uppermost, "home" position where an extension of the piston holds a check valve open (Figure 8-35). During an ABS-assisted stop, one or more of the electric motors is driven to move the piston(s) downward. If the circuit being controlled is a front brake, one or both of the solenoids is also closed. As a piston is driven downward, the check valve is closed to isolate the brake circuit from the master cylinder. From this point, brake circuit pressure is controlled by the piston(s); the lower the piston position, the lower the pressure. The motor(s) will be operated by the EBCM to lower or raise the pressure and obtain the maximum braking pressure without wheel lockup. The solenoids in the front brake circuits provide a redundant circuit so that braking will still be possible should the ABS fail with the pistons lowered.

8.7 ABS Brake Fluid

ABS uses either DOT 3 or DOT 4 fluid; you should use the fluid recommended by the manufacturer. Silicone (DOT 5) fluid should not be used. Air does not move easily through silicone fluid, and when you shock the fluid, any air bubbles tend to become smaller bubbles. During an ABS-controlled stop, the rapid action of the solenoid valves shocks the fluid.

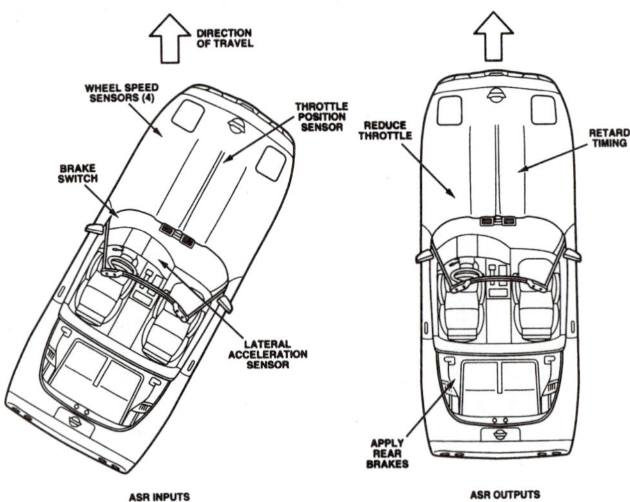

Figure 8-36 This Acceleration Slip Regulation (ASR) uses several sensors to determine if wheel spin is occuring (left); and several methods to control wheel slip (right). (Courtesy of General Motors Corporation, Service Technology Group)

8.8 Automatic Traction Control

Wheel spin can occur while a driver is trying to accelerate, much like wheel lockup occurring during braking. Traction, like braking, can be adversely affected by road conditions and weather. Traction control, also called traction control system (TCS), anti-spin regulation, or automatic slip regulation (ASR), is a system that senses wheel spin and limits the amount of wheel spin that can occur (Figure 8-36).

Wheel spin occurs when the amount of torque at a drive wheel exceeds the available traction, and if a wheel spins more than 20 percent over a normal speed, traction will drop to a lower amount. Because of the differential in the drive axle, the spinning wheel will reduce the amount of torque available to the other drive wheel that is not spinning. Most differentials split torque so an equal amount goes to each drive wheel. Wheel spin will cause a slower acceleration because of the reduced traction and possible side slip and spinout if it should occur while on a curve. Wheel spin also causes unnecessary and excessive tire and differential wear.

TCS shares the ABS wheel-speed sensors and microprocessor. The microprocessor compares the speeds of the two drive wheels with each other and to the non-drive wheels. Excessive speed at a drive wheel indicates wheel spin. Depending on the particular system, wheel spin is controlled by one or more methods: applying the brakes on the drive wheel that is spinning (on an RWD car) or reducing engine torque by retarding the timing, closing the throttle, or shutting off one or more fuel injectors (Figure 8-37).

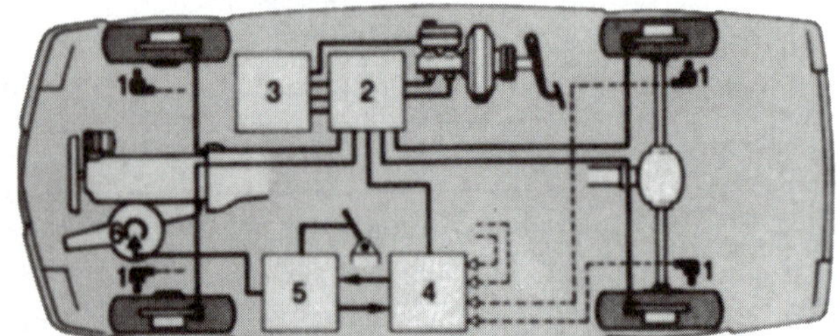

Figure 8-37 This TCS system shares the wheel-speed sensors (1), hydraulic modulator (2), and control unit (4) with the ABS. Added are a hydraulic modulator that can apply the rear brakes (3), control unit (5), and an engine throttle control (6). Some systems also include an engine timing or fuel-injector control. (Reprinted with permission of Robert Bosch Corporation)

8.9 Practice Diagnosis

You are working in a new-car dealership and encounter these problems.

CASE 1: You are prepping a new car for delivery to the customer, and you notice the car has ABS. The ABS warning light operates as you start the engine and goes out a few seconds after startup. But, as you start on your road test, the light comes back on. What is this light indicating? Is this a normal or abnormal condition? If abnormal, what is probably wrong with the car?

CASE 2: One of your top-of-the-line cars that was sold about 11 months ago has come back with a complaint of a low brake pedal. Along with most of the possible accessories, this car has a big engine, automatic transmission, four-wheel disc brakes, and ABS. Is this an ABS problem? What could be causing the low pedal? What should you do next?

Terms to Know

antilock braking system (ABS)
electronic control module
electronically controlled modulator valves
electronic wheel-speed sensors
modulator
modulator valves
sensor
wheel-speed sensor

Review Questions

1. Antilock brake systems are designed
   a. with more powerful brakes to provide quicker stops.
   b. to prevent wheel lockup during stops.
   c. to provide a warning system informing the driver that a skid is occurring.
   d. all of the above.
2. Statement A: Antilock brake systems measure wheel speed using electronic sensors.
   Statement B: Rear-wheel speed sensors can be mounted in the rear axle housing. Which statement is correct?
   a. A only
   b. B only
   c. both A and B
   d. neither A nor B
3. Statement A: If wheel lockup occurs, an ABS will cycle the braking pressure in all the wheel brake assemblies.
   Statement B: A control valve is used to allow pressure buildup, stop pressure increase, or reduce pressure in the wheel brake assemblies. Which statement is correct?
   a. A only
   b. B only
   c. both A and B
   d. neither A nor B
4. An antilock brake system can use
   a. one braking control circuit.
   b. two braking control circuits.
   c. three braking control circuits.
   d. any of the above.
5. Statement A: The brake warning light for cars with ABS is changed from red to an amber color.
   Statement B: If an electrical failure occurs in the ABS controls, the system will revert to a standard brake system and turn on an amber warning light. Which statement is correct?
   a. A only
   b. B only
   c. both A and B
   d. neither A nor B
6. As a wheel rotates, the wheel-speed sensor will send out an electrical signal with a
   a. frequency that will vary with the speed of the wheel.
   b. voltage that will decrease as the wheel speed changes.
   c. current that will increase as the wheel speed changes.
   d. all of the above.
7. The electronic brake control module
   a. monitors the speed signals from the wheel-speed sensors.
   b. operates the hydraulic control unit(s).
   c. checks the electrical circuits constantly to make sure they are operating correctly.
   d. all of the above.

8. Statement A: If the speed of one wheel drops faster than that of the others, an ABS will operate to cycle that brake.
   Statement B: If the speed of a wheel drops faster than that in the profile programmed into the EBCM, an ABS will operate to cycle that brake. Which statement is correct?
   a. A only c. both A and B
   b. B only d. neither A nor B
9. Statement A: The first stage of ABS control is to stop an increase in the hydraulic pressure in a caliper or wheel cylinder.
   Statement B: The second stage of ABS control is to reduce the hydraulic pressure in the brake assembly. Which statement is correct?
   a. A only c. both A and B
   b. B only d. neither A nor B
10. Two technicians are discussing TCS. Technician A says that a TCS system can reduce engine power if wheel spin occurs. Technician B says you can tell if TCS is working if the accelerator pedal pulses during wide-open throttle acceleration on wet roads. Who is right?
   a. A only c. both A and B
   b. B only d. neither A nor B

9 Brake System Service

Objectives

After completing this chapter, you should:

- ❑ Be able to perform a detailed brake inspection.
- ❑ Be able to check brake pedal operation for excessive side motion, sticking, or binding, while checking the pedal for proper free travel and reserve.
- ❑ Be able to check brake pedal feel for indications of fluid leakage or sponginess.
- ❑ Be able to inspect a master cylinder for correct fluid level and fluid leaks.
- ❑ Be able to inspect a master cylinder reservoir for proper fluid motions during application and release.
- ❑ Be able to inspect rigid and flexible brake lines and fittings for leaks, wear, and damage.
- ❑ Be able to remove a brake drum, inspect for lining or drum wear, and determine the cause of poor stopping, noise, pull, or other problems which can be visually located.
- ❑ Be able to remove a brake caliper, inspect for lining or rotor wear, and determine the cause of poor stopping, noise, pull, or other problems which can be visually located.
- ❑ Be able to remove and replace a wheel and torque-tighten the lug nuts.
- ❑ Be able to road-test brakes to test for correct operation and identify any problems.
- ❑ Be able to use a diagnostic chart to determine the possible cause of a brake problem.
- ❑ Be familiar with metal surface finishes.
- ❑ Be familiar with basic electrical troubleshooting procedures.
- ❑ Be able to perform the ASE Tasks relating to brake system inspection and problem diagnosis (see Appendix 1).

9.1 Introduction

After determining that a brake component is faulty, the next operation is to repair that particular component or the whole system. Brake systems can be repaired one component at a time, but the majority of the repairs are made as a brake job. Specific problems such as brake pull, pedal pulsation, or a failing power booster can often be corrected by repairing the faulty component. A good technician will not fix anything that does not need fixing. However, a car that is five years old or older, has gone 50,000 miles (80,450 km) or more, and has a lining that is close to wearing out is becoming a candidate for a brake job—a total brake overhaul. As mentioned earlier, it should be the intent of the brake technician to produce a car with "new-car" brake performance. This is the same goal as that of a technician working on an engine overhaul, a tune-up, a front-end repair and alignment, and so on. Because of the safety ramifications, the brake technician must be even more concerned about top-quality performance once the job is completed. In fact, the brake repair must not only produce a car that "stops well," but one that stops well enough to meet unforeseen emergencies for a reasonable length of time. A quality brake job—like the brakes of a new car—will allow a car to stop well for the life of the new friction material.

Unfortunately, there is no simple and inexpensive way to thoroughly test a brake system, especially the hydraulic system, other than to dismantle it and make visual checks.

It would be nice to be able to plug a meter into the system to measure its present and future effectiveness, but no such instrument is available at this time. As we will see, the eyes and other senses of the brake technician are the major means for determining if the brake system and its parts are good or bad.

Remember that customers have different perceptions of how a car stops. We know that a vehicle is capable of stopping at a certain rate before wheel lockup; that a certain pedal pressure is required; and that the pedal feel can vary between car models. Competent technicians should become familiar with the proper braking action of the vehicle models they work on, so that valid recommendations can be made to the motorist (Figure 9-1).

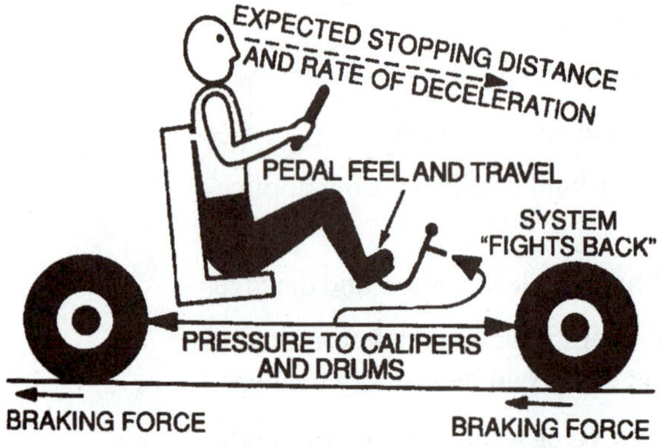

Figure 9-1 When a driver applies the brakes, a combination of feelings and perceptions indicates brake operation. If things don't feel or seem right, there may be a problem. (Courtesy of Ford Motor Company)

At this time, there is no trade standard for a brake job. However, most experts recommend that it include the following:

- Replacement of all friction material.
- Careful checking, reconditioning, and/or resurfacing of friction surfaces on the rotors and drums.
- Replacement of brake shoe return springs.
- Cleaning, inspection, and lubrication of backing plates.
- Cleaning, inspection, and replacement of certain caliper-mounting hardware parts, and lubrication of caliper-mounting hardware.
- Disassembly, cleaning, and inspection of all major hydraulic units—wheel cylinders, calipers, and master cylinder. Note that some master cylinders are nonrebuildable.
- Replacement of all old brake fluid and bleeding of all air from the system.
- Repacking and adjusting serviceable wheel bearings.
- Inspection and/or replacement of hub and axle grease seals.
- Inspection of all steel and rubber brake lines.
- Adjustment of the parking brake.
- Checking and adjustment, if necessary, of brake pedal free travel and the stoplight switch.
- Checking the operation of the hydraulic valves and switches.

A coalition group of the Brake Manufacturer's Council (BMC) has developed a set of uniform guidelines to be followed during a brake inspection as part of the **Motorist Assurance Program (MAP)**. These guidelines help ensure honest and proper recommendations to the motorist as to the condition of the braking system. They also help define the fine line that dictates whether the part *must* be replaced or *should* be replaced. According to MAP, a part must be replaced or repaired if it no longer performs its intended purpose, does not meet its design specification, or is missing. You should suggest that a part should be replaced if it is close to the end of its useful life; if the motorist wants better performance, less noise, no worry; or to comply with the vehicle manufacturer's maintenance recommendation. The MAP guidelines for brake service and repair can be obtained from one of the member companies or the Motorist Assurance Program, Washington, D.C.

Most brake service operations begin with an inspection and/or a set procedure for diagnosing a particular problem so that the best, quickest, and least expensive repairs can be performed. This chapter will describe these procedures.

SAFETY TIP: Brake repair must produce proper safe brake operation. To produce brake operation and to protect yourself from injury, the following general precautions should always be followed:

- Wear eye protection.
- Follow the repair procedure that is recommended by the vehicle manufacturer.
- Support the car in a safe and secure manner before working under it.
- Do not stir up brake dust; protect yourself so you do not breathe this dust. The use of compressed air to blow brake assemblies clean is strictly prohibited. The asbestos fibers it might contain could be harmful to you. Use proper asbestos dust control methods and equipment.
- When using an air hose for cleaning up or drying parts, be careful of the air blast and the particles it might contain.
- Use the proper tool for the job and use that tool in the correct manner.

- Do not allow a strain to be placed on a brake hose.
- Do not allow grease, oil, solvent, or brake fluid to get on the brake lining or the braking surfaces of the rotors or drums.
- Do not let oil or solvents enter the hydraulic system.
- Do not allow brake fluid to be sprayed or squirted into your face or eyes.
- Do not allow brake fluid to be spilled on painted surfaces.
- Replace all damaged, worn, or bent parts.
- Check replacement parts against the old parts to ensure exact or better-quality replacement.
- Make certain that all replacement bolts and nuts are of the same size, type, and grade as the OEM parts.
- Tighten all bolts, nuts, fittings, and bleeder screws to the correct torque and lock them in place by the correct method.
- Do not move the car after working on it until a firm brake pedal is obtained.
- Carefully road-test the car after working on it to make certain that it is operating safely and correctly.

9.2 Brake Inspection

A brake inspection is performed to determine the condition of the brake system. The inspection can determine the cause of a complaint or serve as preventative maintenance—to determine when and if service is necessary (Figure 9-2).

At the same time that the brakes are being inspected, the competent technician will also note the condition of the car's tires, wheel bearings, shock absorbers, wheel alignment (by noting tire wear), suspension system, and drive train. The average car owner has little knowledge of what is happening under his or her car. In most cases, this person is very appreciative of any information that will help ensure safe and efficient operation of the car. Most technicians feel a strong obligation to inform the motorist that a part is failing or will probably fail in the near future.

Most of the operations performed during a brake inspection will be described more completely in the chapters dealing with drum, disc, hydraulic system, or parking brake service. While making a brake inspection, it is a good practice to follow a set procedure and mark your findings on an inspection form or checklist. This helps ensure that no important steps are forgotten, and it also provides a record for the car owner (Figure 9-3).

To perform a brake inspection, you should:

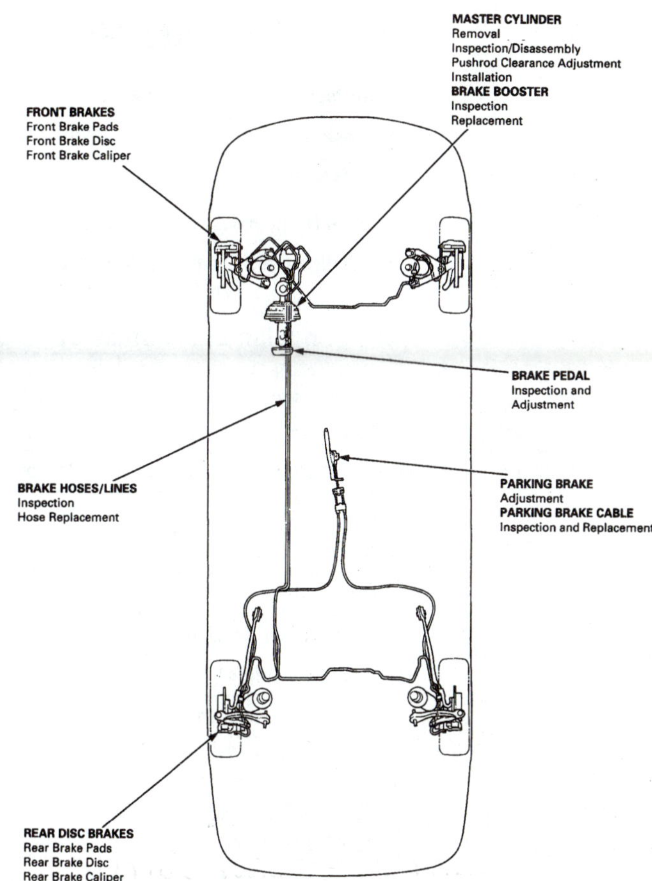

Figure 9-2 The entire brake system should be checked during an inspection. (Courtesy of American Honda Motor Co., Inc.)

1. Depress and release the brake pedal several times (the engine should be running on cars with power boosters). The pedal should have 1/16 to 1/8 inches (1.5 to 3 mm) of free travel before engaging the master cylinder piston. Then it should move smoothly and quietly to a firm pedal, and there should be no excessive side-to-side motion (Figure 9-4).
2. Depress the pedal heavily. On older cars, there should be no sponginess, and the pedal should stop with at least one-half of the available pedal travel left in reserve. It should be noted that many new cars have a soft, almost spongy pedal instead of the rock-hard pedal of the past and with only about 20-percent reserve (Figure 9-5).
3. Depress the pedal moderately, about 25 to 35 pounds (11 to 16 kg), for about fifteen seconds, making sure it does not sink to the floor. Relax the pedal pressure to a light pressure of about 5 to 10 pounds (2 to 4 kg) and hold this pressure for a little longer to make certain that the pedal does not sink under light pressure. The first check is for an external leak or an internal master cylinder leak. The second step checks for an internal master cylinder leak, often called bypassing (Figure 9-6).

Complete Brake Job Checklist

Customer Name ______________________________

Address ____________ Telephone Number: Work ____________ Home ____________

Vehicle Make ____________________ Year ____________ Mileage ____________

License Number ______________________________

Approximate time required for service ______________________________

Inspected by ______________________________ Date ____________

The condition of the brake system on this vehicle is as follows:

Brake System Components	Required Services	$ Estimate Parts	Labor
Disc Pads*	☐Semi-metallic ☐Conventional		
Disc Caliper*	☐Recondition ☐Replace		
Disc Hardware*	☐Replace		
Disc Rotor*	☐Recondition ☐Replace		
Grease Seals*	☐Replace		
Front Wheel Bearings*	☐Repack ☐Replace		
Wheel Cylinders*	☐Recondition ☐Replace		
Rear Brake Hardware*	☐Replace		
Brake Shoes*	☐Replace		
Brake Drums*	☐Recondition ☐Replace		
Parking Brakes	☐Adjust ☐Lubricate ☐Replace		
Power Brake Booster	☐Service ☐Replace		
Master Cylinder	☐Recondition ☐Replace		
Brake Fluid*	☐Drain old, and add new fluid		
Lines, Hoses, Combination Valve	☐Replace		
Stop Light	☐Replace Bulb ☐Replace Switch ☐Adjust ☐Replace		
Other ____________			
	Total		

***Wagner's Complete Brake Job includes reconditioning or replacement of these items**

McGRAW-EDISON

Figure 9-3 A checklist to use while inspecting a brake system. (Courtesy of Wagner Brake)

4. Depress and release the pedal several times under varying amounts of pressure as you watch the warning light on the dash. Also, have an assistant watch the stoplights. The warning light should not come on, but the stoplights should come on each time the pedal is depressed and go off each time it is released.
5. Check the brake warning light operation by cranking the engine. The light should come on as the engine is cranking. On cars equipped with ABS and/or an airbag(s), these warning lights should also come on during cranking and remain on for a few seconds after the engine starts.
6. Apply the parking brake. The lever should not travel more than two-thirds of the available distance and should provide enough braking power to hold the car in place. Some technicians test this power by trying to drive the car with the brake applied. Check the parking brake warning light; it should come on as the parking brake is applied.
7. On power-brake-equipped cars, with the engine off, depress the brake pedal several times to exhaust the booster reserve. Hold the pedal down with a light pressure and start the engine. As the engine starts, the pedal should drop slightly but noticeably. With

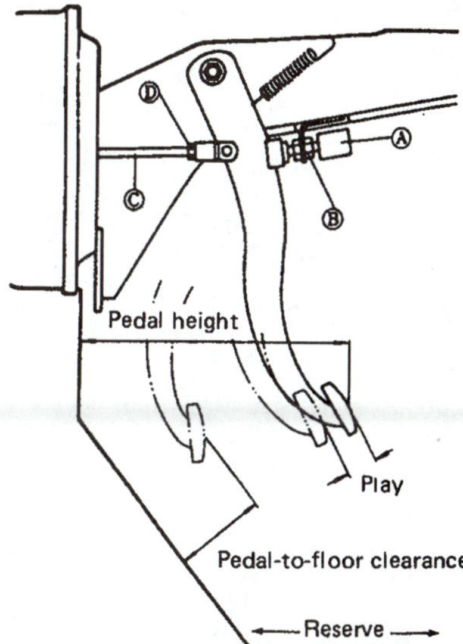

Figure 9-4 A brake pedal should have 1/16 to 1/8 inches of free play, only a small amount of side play, and a firm, hard pedal before it gets halfway to the floor. In this unit, free travel is adjusted by loosening the lock nut (D) and turning the pushrod (C). Stoplight switch (A) adjustment should be checked after adjusting free travel. (Courtesy of Mazda Corporation)

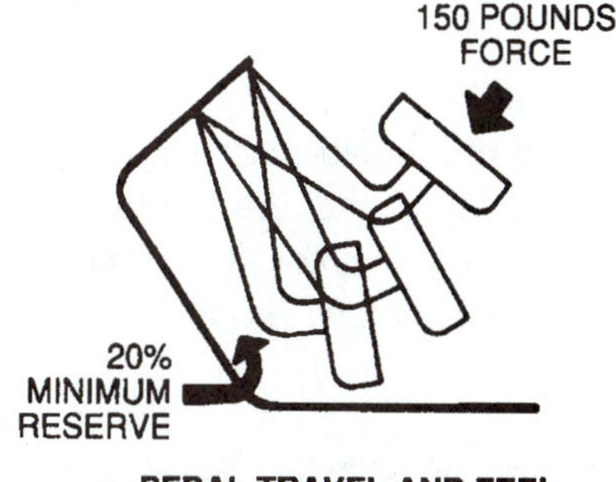

Figure 9-5 There should be a minimum of 20-percent reserve with a pedal pressure of 150 pounds on this late-model vehicle. (Courtesy of Ford Motor Company)

hydraulic boosters, the pedal should drop and then rise back up (Figure 9-7).

After running the engine several moments more, shut the engine off, wait ninety seconds, and depress the pedal lightly. One or more assisted brake applications should occur.

8. Check the master cylinder for external leaks at the line fittings, mounting end, or reservoir cover. Leaks at the line fittings can often be repaired by loosening and retightening the tubing nut. Fluid leaks at the booster where the master cylinder is mounted indicate internal failure of a master cylinder cup.

Figure 9-6 A brake-pedal pressure gauge can be used to measure the force on the brake pedal. (Courtesy of SPX Kent-Moore, Part # J-28662)

9. Remove the reservoir cover, note the condition of the diaphragm, and make sure the vent hole in the cover is open.
10. Check the fluid level. It should be 1/4 to 1/2 inches (6 to 13 mm) from the upper edge of the reservoir or at the level indicated by the marks on the reservoir (Figure 9-8). On cars with electrohydraulic boosters, pump the brake pedal about 20 to 25 times with the key off (up to 40 times on some systems) to return the fluid from the accumulator to the reservoir before checking the fluid level (Figure 9-9). On master cylinders equipped with fluid level switches, make sure the wires are in good condition with sound connections at the switch.
11. Run a clean finger around the bottom of the reservoir, and check for rust, dirt, or other contamination. Clean the fluid from your finger immediately. If the fluid appears contaminated, place a sample in a clean, glass jar so it can be inspected more completely. If test equipment is available, test the fluid for contamination and lowering of boiling point (Figure 9-10).

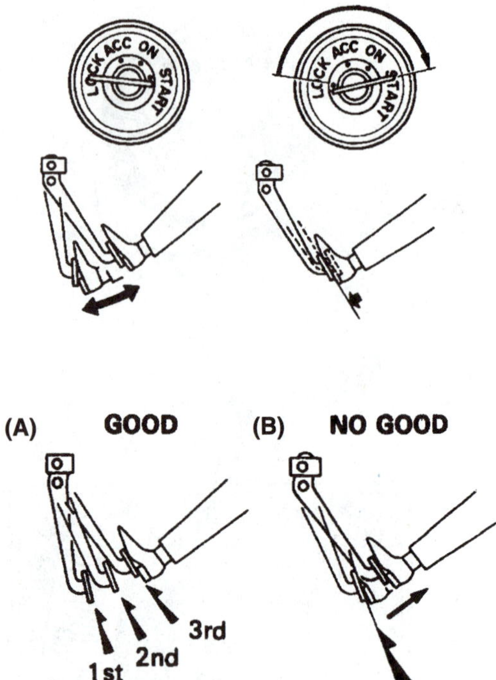

Figure 9-7 When checking a power brake system, pump the brake pedal and note the change in reserve (A). Then with your foot on the pedal, start the engine (B); the pedal should drop slightly. After a few minutes, there should be a pedal reserve.

12. Watch the fluid in the reservoirs as the pedal is depressed. A slight spurt or swirl should occur over the compensating ports during application, and also in the drum brake section of a tandem system during release. Then, observe the fluid levels in both master cylinder reservoirs during several hard pedal depressions to make certain that one does not rise as the other one falls.

CAUTION *Do not lean directly over a master cylinder during brake application or release. Fluid can spurt or spray high enough to get in your face or eyes. It is a good practice to cover the reservoir with clear plastic wrap to contain the fluid so you can observe the movement.*

NOTE: Because of the fluid bypassing through the quick-take-up valve, the fluid level in the reservoir of a quick-take-up master cylinder will normally rise as the brakes are applied and fall during their release.

13. On power-brake-equipped cars, inspect the vacuum hose and check valves, hydraulic lines, and electrical connections to make certain they are in good condition.

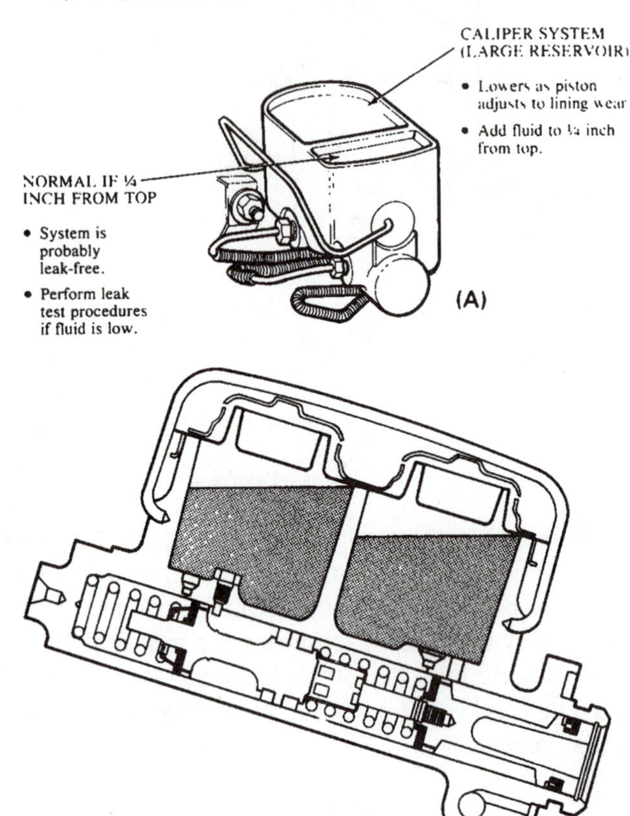

Figure 9-8 Normal brake fluid is 1/4 inches below the top of the reservoir (A); on angled master cylinders, the fluid level is almost at the top of the reservoir (B). (A is courtesy of Ford Motor Company; B is courtesy of General Motors Corporation, Service Technology Group)

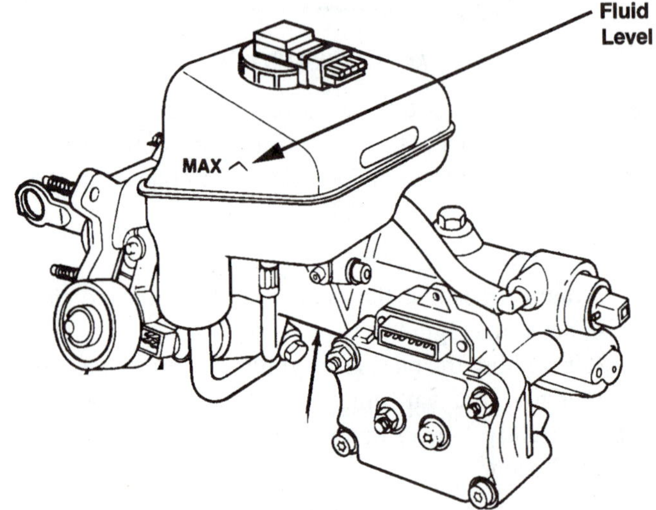

Figure 9-9 With the accumulator charged, fluid level on this plastic ABS-unit reservoir should be at the MAX level. Some units should have the accumulator discharged before checking fluid level. (Courtesy of Ford Motor Company)

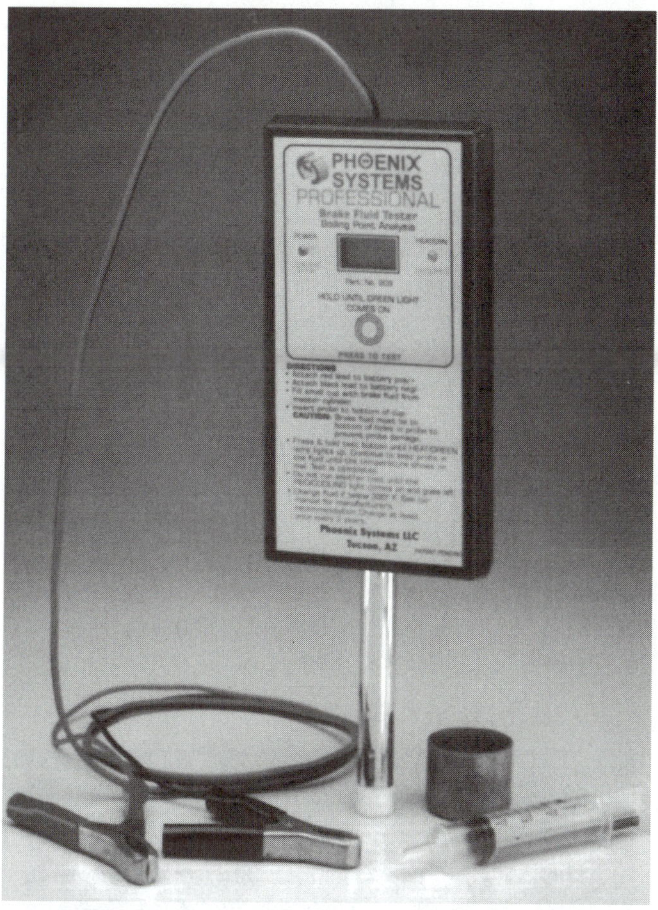

Figure 9-10 This tester checks the boiling point of the brake fluid, and this gives a quick and accurate check for moisture contamination. (Courtesy of Phoenix Systems, L.L.C.)

14. Raise and securely support the car. Place a jack or the support arms of a hoist under some part of the car that is strong enough to support the weight. Some cars have a jacking pad at the bumper or on the side of the car. Other places that are commonly used are the frame, a frame cross-member, a reinforced unibody box section, an axle, or a suspension arm. Lift the tire and wheel off the ground and, if using a jack, place a jackstand under some strong portion of the car (Figure 9-11).

NOTE: The rear axle of some FWD cars will bend if lifted by the center of the axle. These axles should be lifted one side at a time, with the jack placed near the spring.

NOTE: If you are unsure of where to lift the car, consult the "Lifting and Jacking" section of the car manufacturer's service manual or a technician's service manual (Figure 9-12).

Figure 9-11 Two different types of jackstands that should be placed under a car for safe support. Never work under a vehicle supported only by a jack.

15. Rotate and shake or rock each of the wheels as you check for excessive looseness and free rotation of the wheel or axle bearings. A rough, harsh feeling accompanied by a growling, grinding noise indicates a faulty wheel bearing (Figure 9-13).

16. Remove at least one front and one rear wheel using the following procedure. Some technicians prefer to inspect all four brake assemblies.

    a. Remove any wheel covers (often locked).

    b. Put an indexing chalk-mark on the end of the lug bolt that is closest to the valve stem. This is done to ensure replacement of the wheel assembly in the same location so that no change in the balance or the runout of the tire and wheel will occur. If the valve core is positioned between two studs, mark both of them or draw a line on both the wheel and the hub (Figure 9-14).

    c. Remove the lug nuts using a six-point lug nut socket or lug wrench. If you are using an air-impact wrench, excessive speed can cause lug bolt or nut galling. Adjust the wrench speed or lubricate the threads to ensure that galling does not occur.

    SAFETY TIP: The exhaust of an air-impact wrench can cause dust hazards by blowing asbestos fibers into the air.

    d. Remove the tire and wheel assembly.

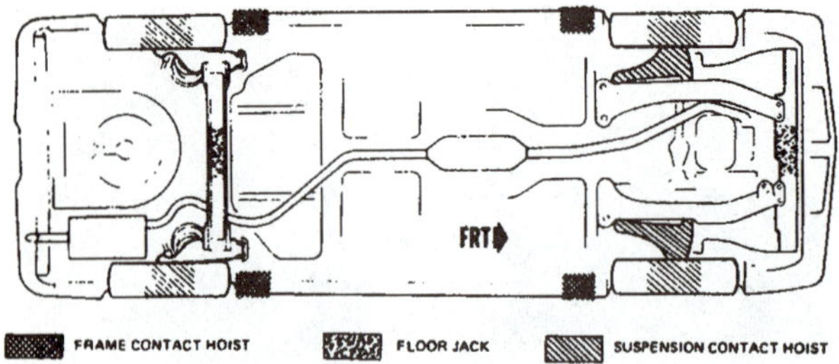

Figure 9-12 The jacking and lifting points are shown in the manufacturer's or mechanic's service manual. These points are strong enough to support the car. (Courtesy of General Motors Corporation, Service Technology Group)

Figure 9-13 Wheel bearing is quickly checked by rocking the tire as shown here. Push in at the top while pulling out at the bottom. (Courtesy of Hunter Engineering Company)

NOTE: Steps 17 through 22 will be described in more detail in Chapter 10.

17. Mark the drum next to the previously marked stud(s), and remove the drum (Figure 9-15).

SAFETY TIP: OSHA requirements state: "There should be no visible dust during brake inspection and repair." Removal of a brake drum can release dust and asbestos fibers. The recommended method of preventing this is to flood the brake assembly using a brake washer; rotate the drum as you thoroughly wet the inside components. Aerosol sprays and vacuum enclosures can also be used.

NOTE: On brake drums that are mounted at the front of rear-wheel-drive cars or at the rear of front-wheel-drive cars, the drum can often be removed at the wheel bearing as a wheel, hub, and drum assembly.

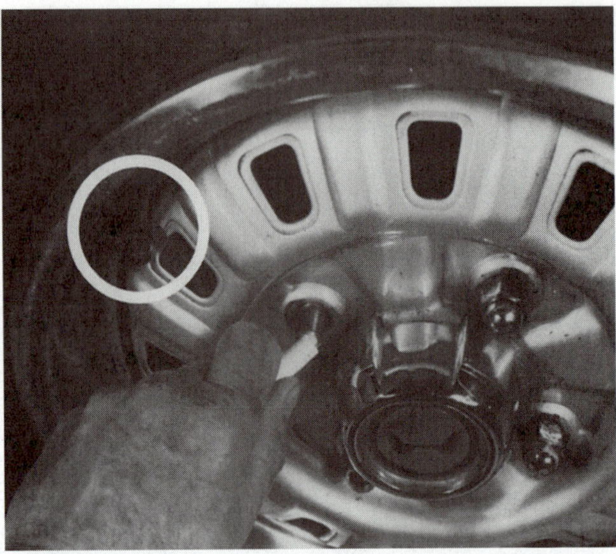

Figure 9-14 Placing a chalk mark at the end of the wheel stud closest to the valve stem (circled) will help you replace the wheel in the correct position.

Figure 9-15 Before removing a brake drum, place a chalk mark next to the marked wheel stud. This index mark helps you replace the drum in the same position.

NOTE: Follow the manufacturer's recommendations on drum removal if you are not sure of how to do it.

18. Check the brake lining for wear—the amount and pattern. There should be a minimum of 1/32 to 1/16 inches (0.8 to 1.5 mm) of lining at the thinnest point above the rivets on riveted shoes, or above the shoe

rim on bonded shoes. The wear should be even and equal across the shoe (Figures 9-16 and 9-17).

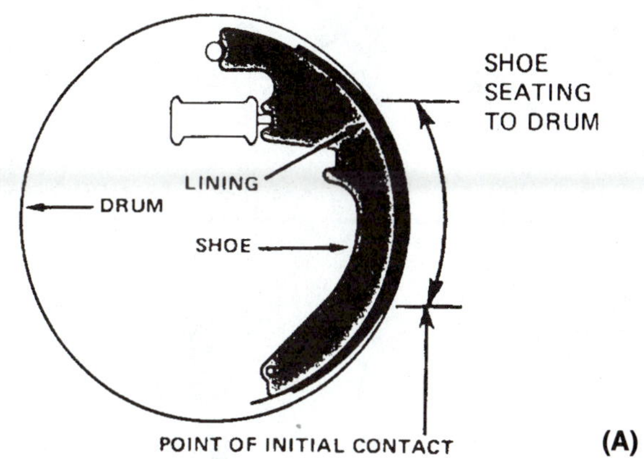

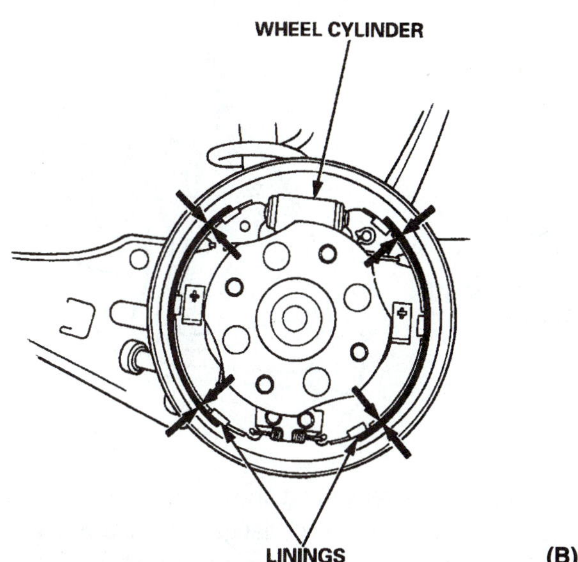

Figure 9-16 Normal brake lining wear occurs with the greatest amount of wear at the center of the lining (A). You should also measure lining thickness near the ends of the lining (B). (A is courtesy of Ford Motor Company; B is courtesy of American Honda Motor Co., Inc.)

Also check the lining for contamination, glazing, or cracking (Figures 9-18 and 9-19).

19. Check the brake springs for distortion or stretched or collapsed coils, twisted or nicked shanks, or severe discoloration (Figure 9-20). Check the self-adjuster and parking brake linkage for distorted or stretched parts and correct operation. Check the self-adjuster operation by prying the secondary shoe away from the anchor or pulling the cable. The adjuster mechanism should operate and turn the adjuster (Figure 9-21). Also check that the released shoe returns to its anchor.
20. Check the wheel cylinder for leakage. On units with external boots, pull a boot back to look inside. A small amount of fluid dampening is normal. Actual wetness is not.
21. Check the drum friction surface for cracks, unusual wear, or a worn or distorted surface. Measure the drum in several locations for size and roundness. Each measurement should be smaller than the maximum diameter indicated on the drum and within 0.010 inches (0.25 mm) of each other (Figure 9-22).

NOTE: Steps 22 through 27 will be described in more detail in Chapter 11.

22. Remove the caliper in the manner recommended by the manufacturer. On some cars, it is possible to visually check the lining thickness with the caliper in place (Figure 9-23). Many technicians prefer to remove the caliper to allow a more complete check of the lining, rotor, and caliper. During removal of floating calipers, check the amount of clearance at the caliper-to-mount abutments (Figure 9-24).
23. Inspect the lining for wear, noting the amount and pattern. There should be a minimum of 1/32 to 1/16 inches (0.8 to 1.5 mm) of lining at the thinnest point above the rivets on riveted lining or above the backing on bonded lining (Figure 9-25). The pad wear should be even and equal from the inner edge to the outer edge, with no more than 1/16 inches (1.5 mm) more wear at one end than the other. Check the lining for cracks, contamination, or glazing.
24. Check the caliper mounting hardware for wear or distortion.
25. Check the caliper piston boot for cracks or tears and leakage. No fluid seepage is considered acceptable.
26. Check the friction surfaces of the rotor for unusual wear. Measure the rotor in several locations at the center of the friction surface. Make sure that its width is greater than the minimum width dimension indicated on the rotor (Figure 9-26). If there is a complaint of pedal pulsation or brake lockup, or if it is probable that a lining replacement will be needed, measure the amount of rotor runout and parallelism. Then compare these values to the manufacturer's specifications (Figure 9-27).
27. Replace the caliper using the manufacturer's recommended procedure, being sure to tighten the caliper mounting bolts or guide pins to the correct torque.

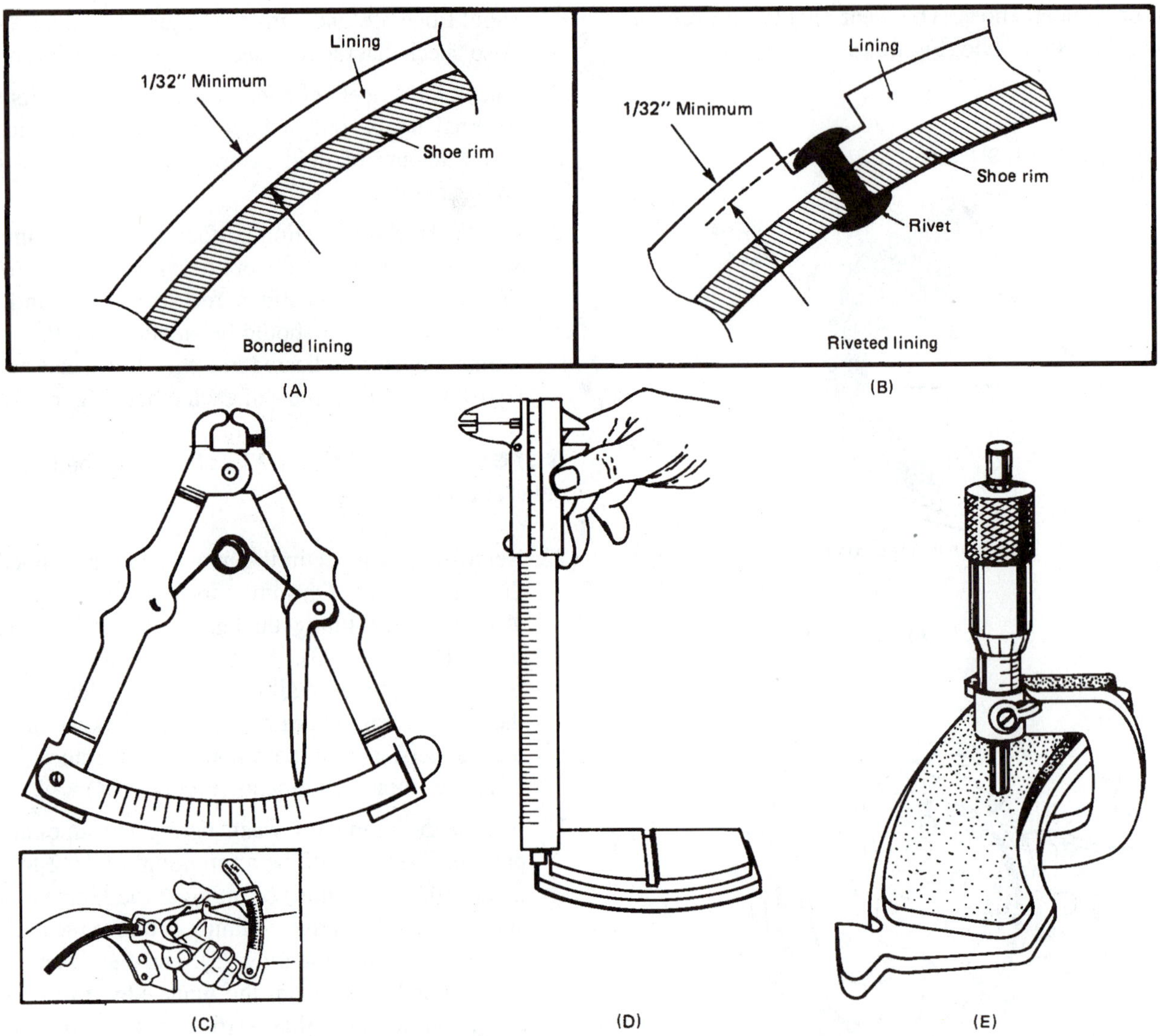

Figure 9-17 There should be a minimum of 1/32 (0.031) inches of lining material remaining above the shoe rim on bonded shoes (A) or above the rivets on riveted shoes (B). The thickness can be measured using a special gauge (C), a vernier caliper (D), or a micrometer (E). If measuring the overall width, don't forget to subtract the thickness of the metal backing. (C is courtesy of KD Tools; D is courtesy of Mazda Corporation)

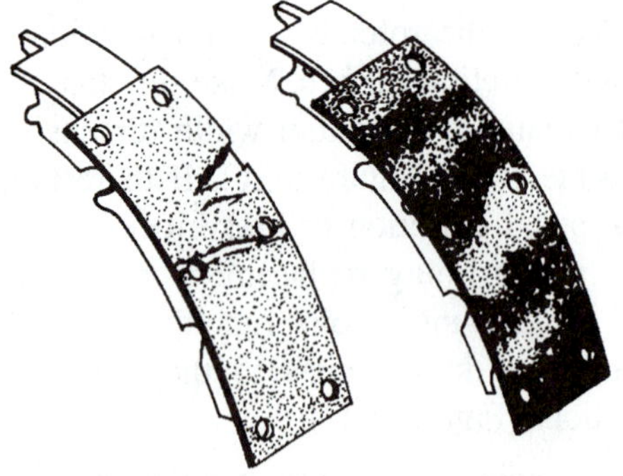

Figure 9-18 Check the lining to make sure it is not chipped, cracked, or contaminated with oil, grease, or brake fluid.

NOTE: In cases where brake drag or abnormally fast pad wear is the complaint, it is a good practice to check for excessive brake drag. Do this by measuring the amount of force required to turn the rotor (with the caliper removed) using a scale as shown in Figure 9-28. Next install the caliper, apply the brakes a few times, and rotate the rotor at least ten revolutions. Then remeasure the force required to turn the rotor. Subtracting the first measurement from the second will give the amount of drag caused by the brakes. If it is excessive—over 20 to 25 pounds—the caliper or caliper mounts should be serviced.

28. If one or more problems were located, or there was a complaint of a specific problem and the cause has not

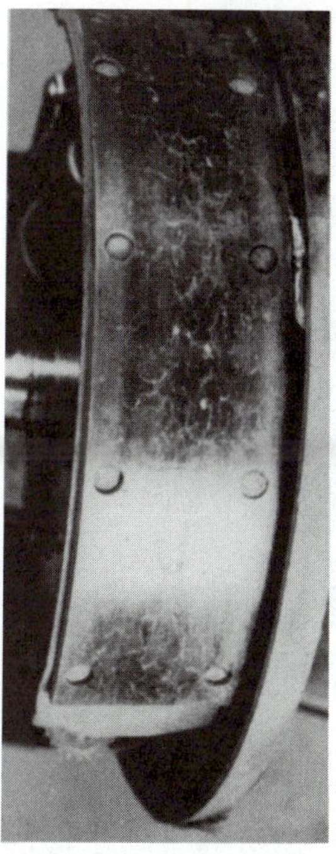

Figure 9-19 This lining is badly glazed. Note the shiny appearance, the fine cracks, and the darker color toward the center.

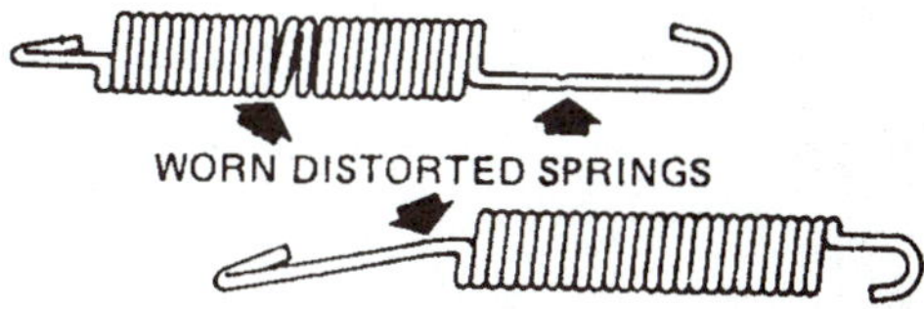

Figure 20 Badly bent, nicked, or stretched springs can cause dragging brakes or pull during stops. These should be replaced. (Courtesy of Wagner Brake)

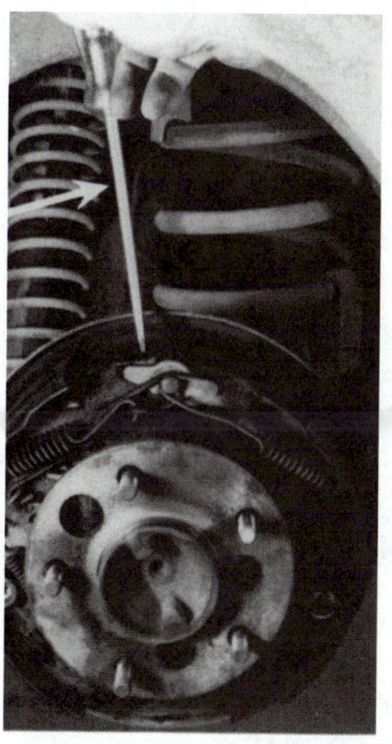

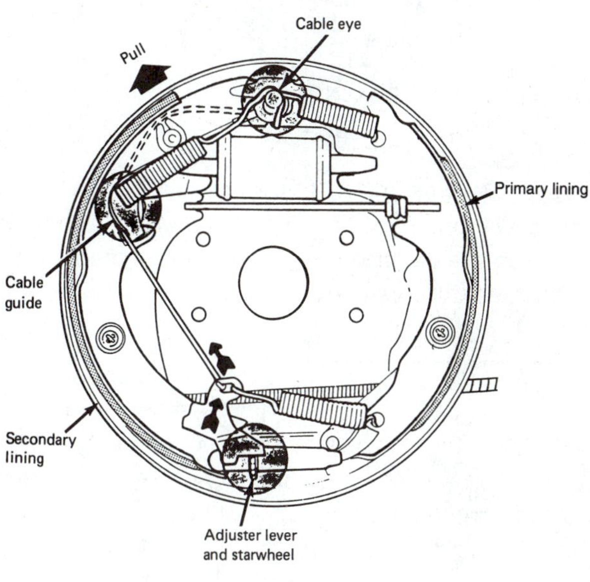

Figure 9-21 Operation of the self-adjuster on a duo-servo brake can be checked by prying the secondary shoe away from the anchor (A) or pulling on the adjuster cable (B). The motion should cause the lever to turn the starwheel. (B is courtesy of Ford Motor Company)

been found, one or both of the other wheels will need to be removed and the brake assemblies inspected. If a preventative maintenance inspection is being made or the cause of the complaint has been determined, replace the wheels. The following procedure should be used:

a. Check the lug bolts and the wheel nut bosses for worn or elongated holes. Damaged wheels should be replaced. Place the wheel over the lug bolts with the valve core next to the previously marked stud(s).

b. "Snug" down the lug nuts, making sure the tapered portion of the lug nut enters the tapered opening of the wheel nut bosses. While the tire and wheel are in the air, it is difficult to apply enough torque to completely tighten the nuts because the wheel will spin. However, they can and should be tightened enough to hold the wheel in the correct position.

c. Lower the car onto the ground and immediately complete step d.

CAUTION *Don't allow yourself to forget step d.*

d. Tighten the lug nuts to the correct torque using a tightening pattern which moves back and forth

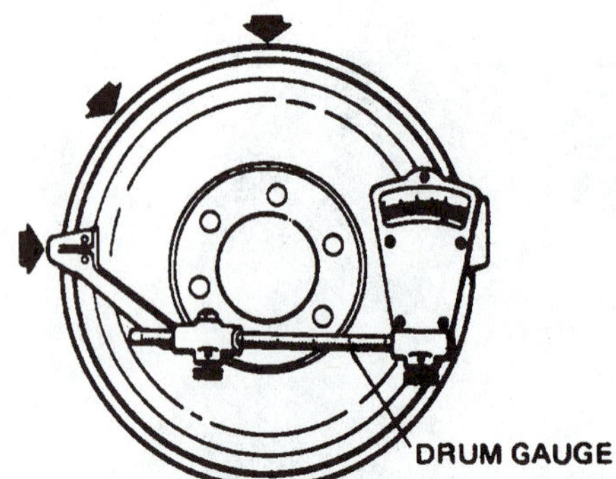

Figure 9-22 A brake drum can be measured using a drum gauge/micrometer to make sure it is not too large. Measuring at several locations tells if it is round. (Courtesy of Wagner Brake)

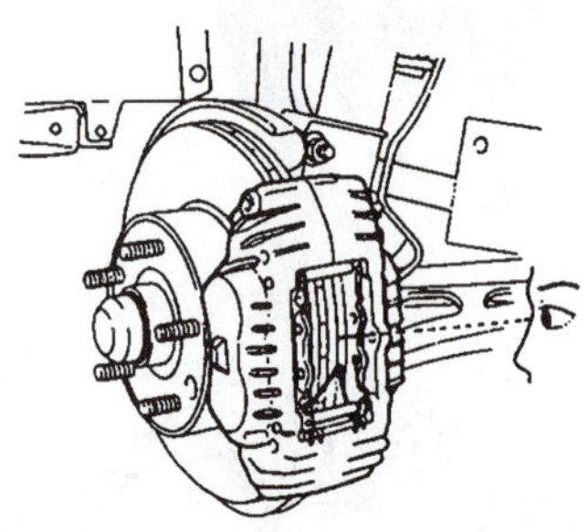

Figure 9-23 On some cars, it is possible to check lining thickness without removing the shoe. Many technicians prefer to remove the shoe for a more accurate check. (Courtesy of Mazda Corporation)

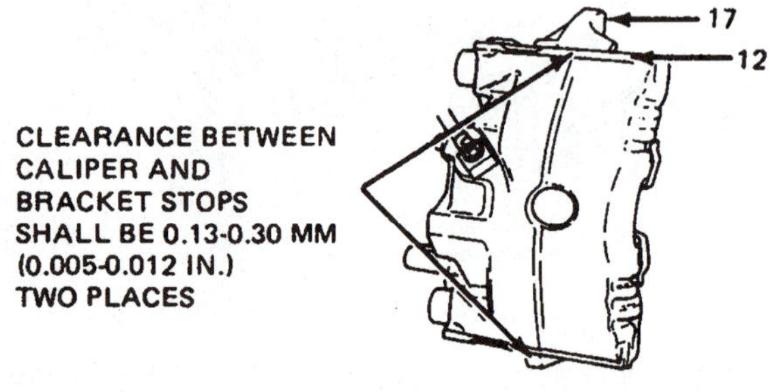

Figure 9-24 As a caliper is being removed, the clearance between the caliper and mounting bracket should be checked. (Courtesy of General Motors Corporation, Service Technology Group)

across (not around) the wheel (Figure 9-29). A brake drum or rotor can be distorted by overtightening the lug bolts or using the wrong order. Torque sticks used with an air-impact wrench have become a popular method for many shops to tighten lug nuts quickly (Figure 9-30).

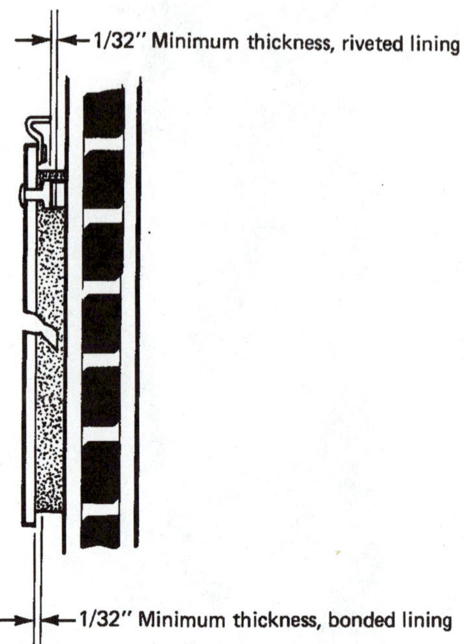

Figure 9-25 Disc brake pads should have a minimum of 1/32 (0.031) inches of lining above the pad backing on bonded shoes or above the rivet on riveted lining. The wear should be even across the lining. (Courtesy of Wagner Brake)

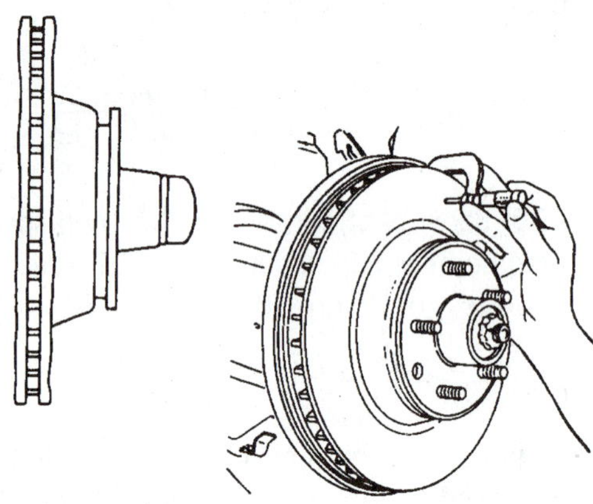

Figure 9-26 Rotor thickness is measured using a micrometer. The measurements should be taken at the center of the friction surface. (Reprinted from Mitchell Anti-Lock Brake Systems, with permission of Mitchell Repair Information, LLC)

e. Replace the wheel cover or hubcap using a rubber hammer or the hammer portion of the hubcap tool.

NOTE: Uneven or excessive lug nut torque can cause braking problems. As a rotor is heated and then cools through normal driving, parts of the rotor will cool at different rates because of heat transfer to the wheel. This can cause the rotor to distort, which results in runout or thickness variation.

Figure 9-27 Rotor runout is measured using a dial indicator; any wobble will show up as the rotor is rotated. (Reprinted from Mitchell Anti-Lock Brake Systems, with permission of Mitchell Repair Information, LLC)

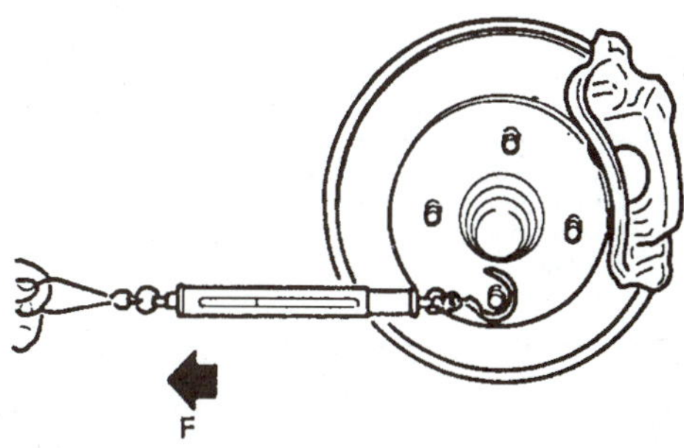

Figure 9-28 Drag caused by the caliper can be measured using a spring scale. If there seems to be too much drag, it can be measured with the caliper removed. (Courtesy of Nissan North America, Inc.)

Lug nut torque will vary depending on size of the lug bolt and the wheel center section material. Aluminum wheels sometimes use a slightly higher torque because of the solid center section. However, a slightly lower torque is sometimes recommended to reduce nut to wheel galling and compression of the wheel nut bosses. If the manufacturer's torque specifications are not available, use the following—based on the lug bolt size:

Lug Bolt Size	Torque	
⅜ × 24	35–45 ft-lb	(48–61 N-m)
7/16 × 20	55–65 ft-lb	(75–88 N-m)
½ × 20	75–85 ft-lb	(102–115 N-m)
M10 × 1.5	40–45 ft-lb	(54–61) N-m)
M12 × 1.5	70–80 ft-lb	(95–108 N-m)
M14 × 1.5	85–95 ft-lb	(115–129 N-m)

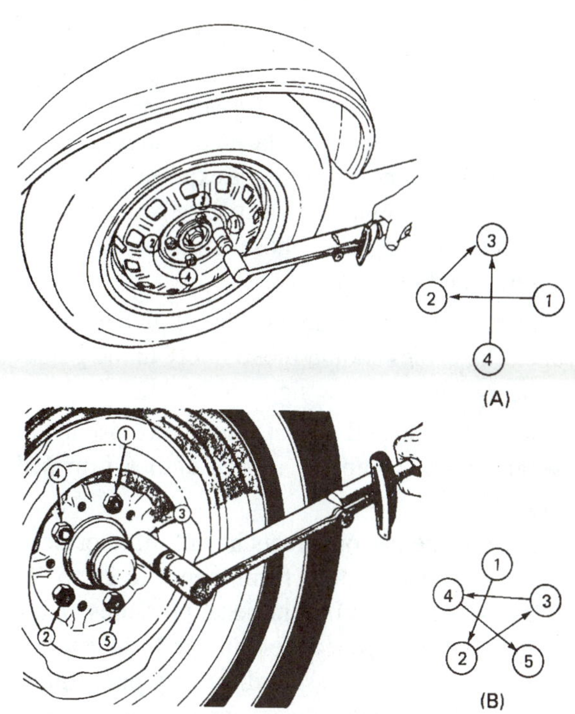

Figure 9-29 Wheel studs or lug bolts should be tightened to the correct torque using a crisscross or star-shaped pattern. (Courtesy of Chrysler Corporation)

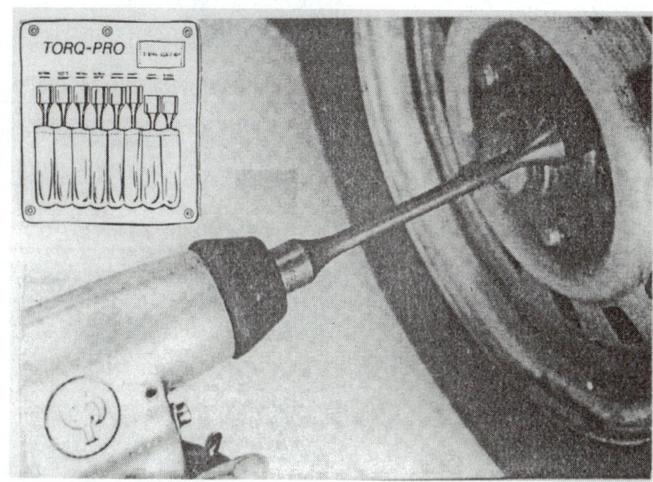

Figure 9-30 Torque sticks (inset) resemble a torsion bar with a socket built onto it, and they are used with an air-impact wrench to tighten lug nuts to the correct torque.

SAFETY TIP It is a good practice to retighten the lug nuts after driving 10 to 20 miles (16 to 30 km). With alloy-center wheels, it is a good idea to check lug nut tightness again after another 100 miles (161 km).

29. Check all visible steel lines for kinks or collapsed sections which might cause fluid restriction or leaks

(Figure 9-31). Check all flexible hoses for leaks, deterioration, cuts, chafing, rubbing, excessive cracks, bulges, or loose supports. Surface cracks in the outer cover are normal, but replacement should be recommended if the inner cording is exposed (Figure 9-32). If there is a complaint of a spongy pedal, it is a good practice to check for hose swelling as a partner applies the brakes.

30. Check the parking brake cables, equalizer, and linkage. The cables should move freely in the guides and housing and not show signs of fraying.
31. Lower the car and operate the brake pedal through several slow, complete strokes until a firm pedal is obtained.
32. Road-test the car on streets having little or no traffic and make several stops from speeds of 20 to 25 mph (32 to 40 kph) at different pedal pressures. While the stop is occurring, check for pull, grab, squeal, or other unusual noises; excessive dive; or a pulsating pedal. Any faults indicate a need for further inspection to determine their cause.

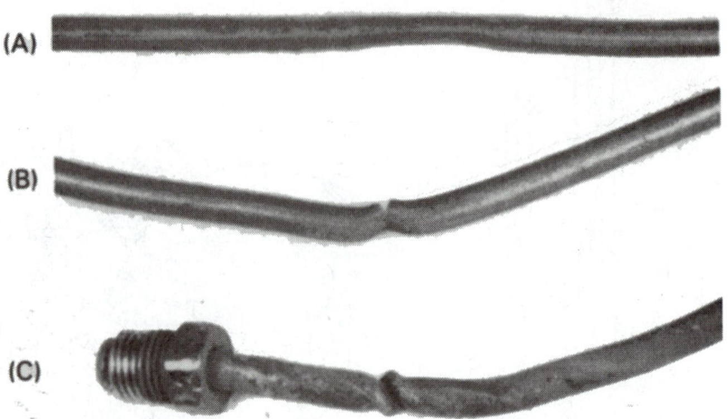

Figure 9-31 Some of the brake tube problems encountered are dents that can cause restriction (A), kinks that can cause leaks (B), and twists that can cause both problems (C).

CAUTION *Be sure to conduct the road test in such a way that there is no danger to yourself or other motorists and no violation of any applicable traffic regulations.*

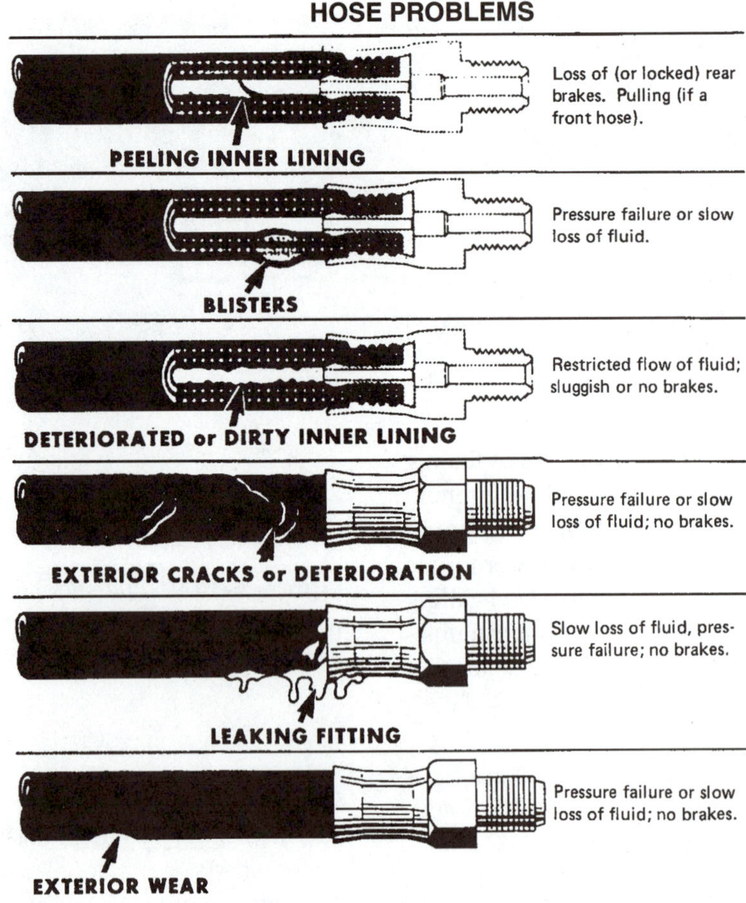

Figure 9-32 The most common causes of hose failure are shown here. (Reprinted from Mitchell Anti-Lock Brake Systems, with permission of Mitchell Repair Information, LLC)

33. Evaluate the findings of your inspection and make your recommendations on the inspection report.

9.3 Troubleshooting Brake Problems

Proper diagnosis is essential in locating the cause of a brake problem. Once the cause has been determined, it is usually fairly easy to make the repair. The repair procedures for the various systems will be described in the following chapters.

When a particular brake problem shows up, an experienced brake technician will often know the cause based on his or her experience. However, beginners will usually need to refer to a troubleshooting guide. These guides are published by the various vehicle manufacturers and by aftermarket brake-part manufacturers. They generally list common problems that are encountered, along with the probable causes of each problem, with the most likely causes listed first. The following troubleshooting guide is divided into sections dealing with different types of problems. It is possible for a particular problem to involve two or three of these sections.

It should be noted that some problems, such as pull, chatter, and pedal pulsations, can be caused by faults in the car's suspension system, tires, or wheel bearings, and that pedal pulsation during hard braking can be the result of normal ABS operation. Table 9-1 on page 163 provides definitions for the common terms referring to brake problems.

SAFETY TIP: When making a decision concerning a repair, remember that lining replacement must always be done in pairs. Any change at one end of an axle must be accompanied by the same change at the other end. Rotor or drum machining must also be done in pairs. The friction surfaces of both rotors on an axle must have the same finish. Both drums on an axle must have the same finish and diameter. If one end of the axle has a difference in friction material or friction surface, the car will probably pull during braking; this is an unsafe condition which must be avoided.

9.4 Brake Repair Recommendations

After completing a brake inspection or following a troubleshooting guide to determine the cause of a problem, it is time to make a decision about the repair. The competent

BRAKE SYSTEM DIAGNOSIS CHARTS

RED BRAKE WARNING LAMP FUNCTION

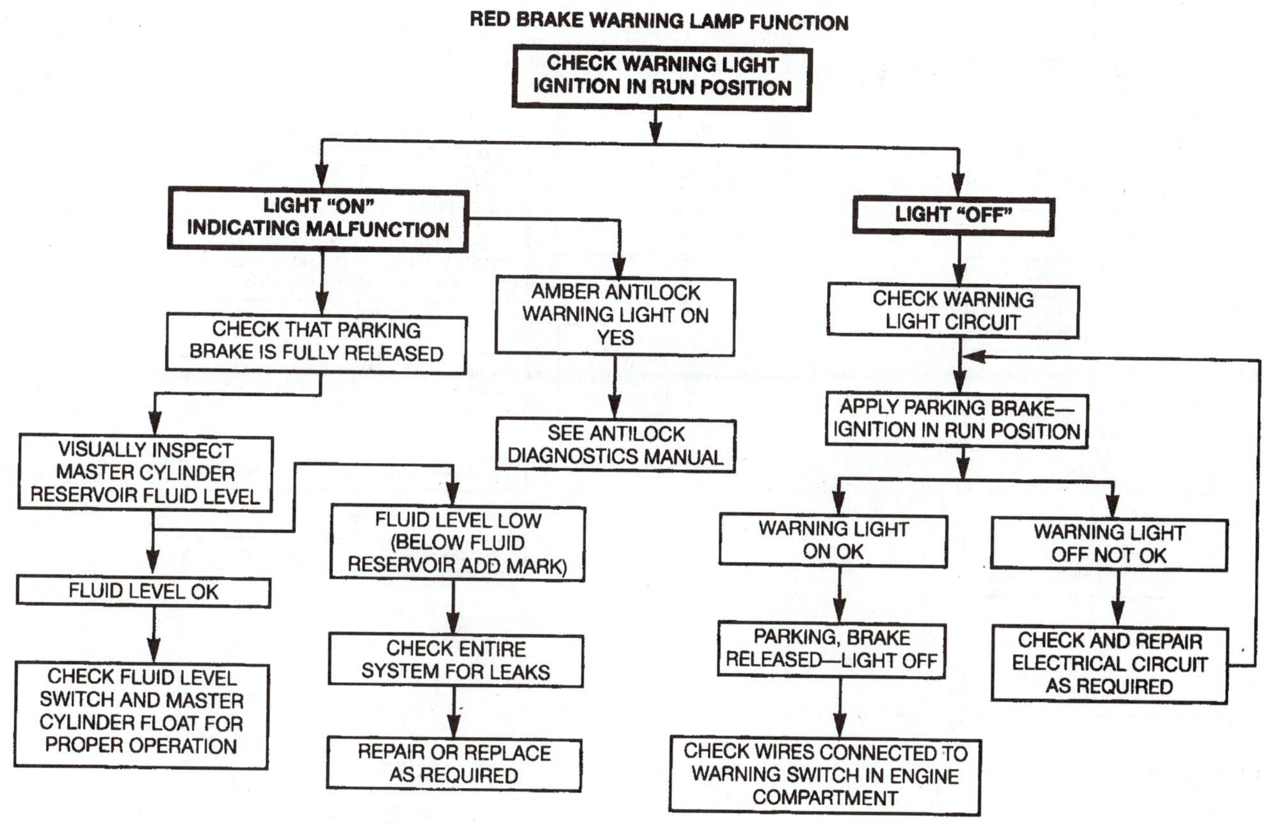

MISCELLANEOUS BRAKE SYSTEM CONDITIONS

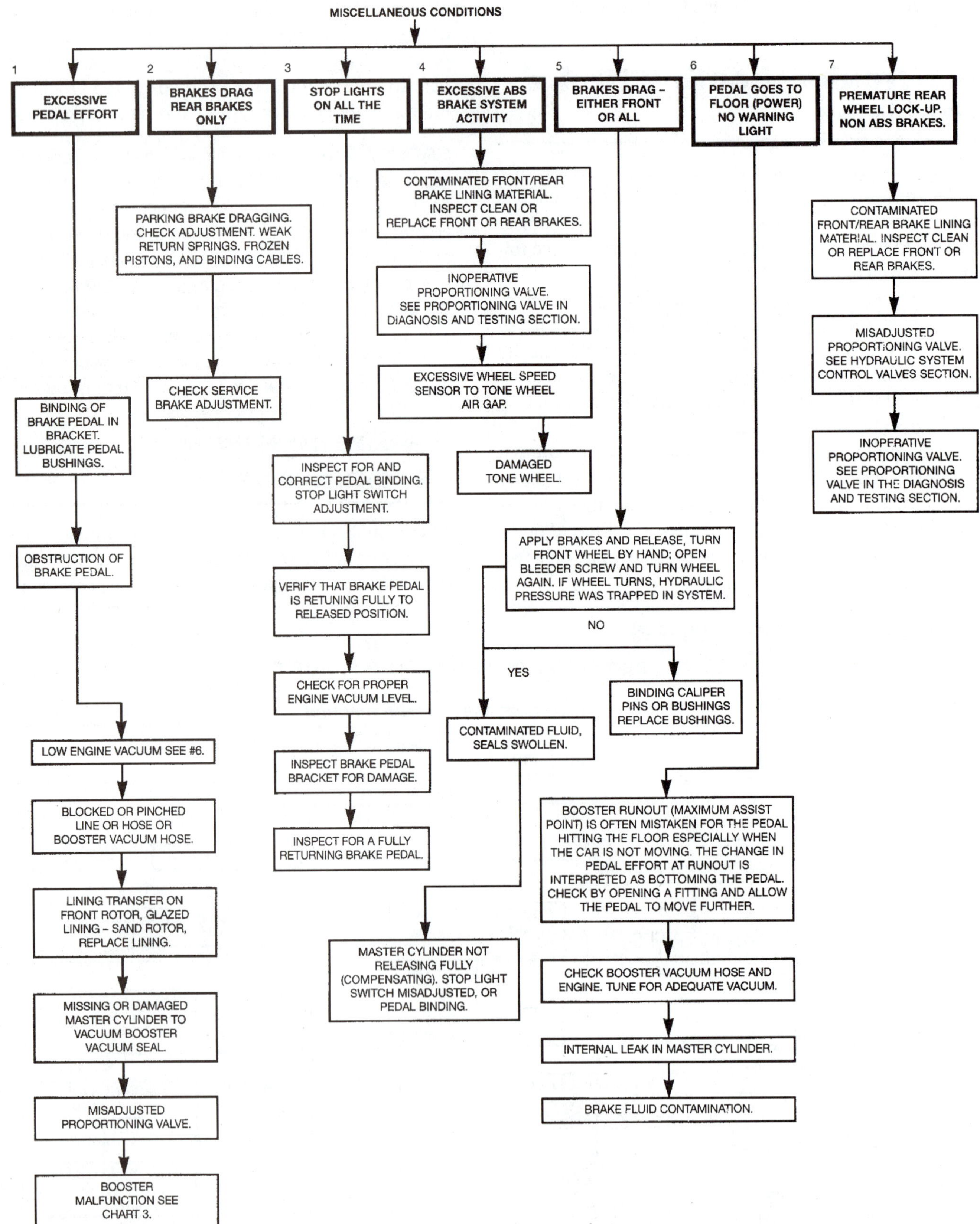

POWER BRAKE SYSTEM DIAGNOSTICS

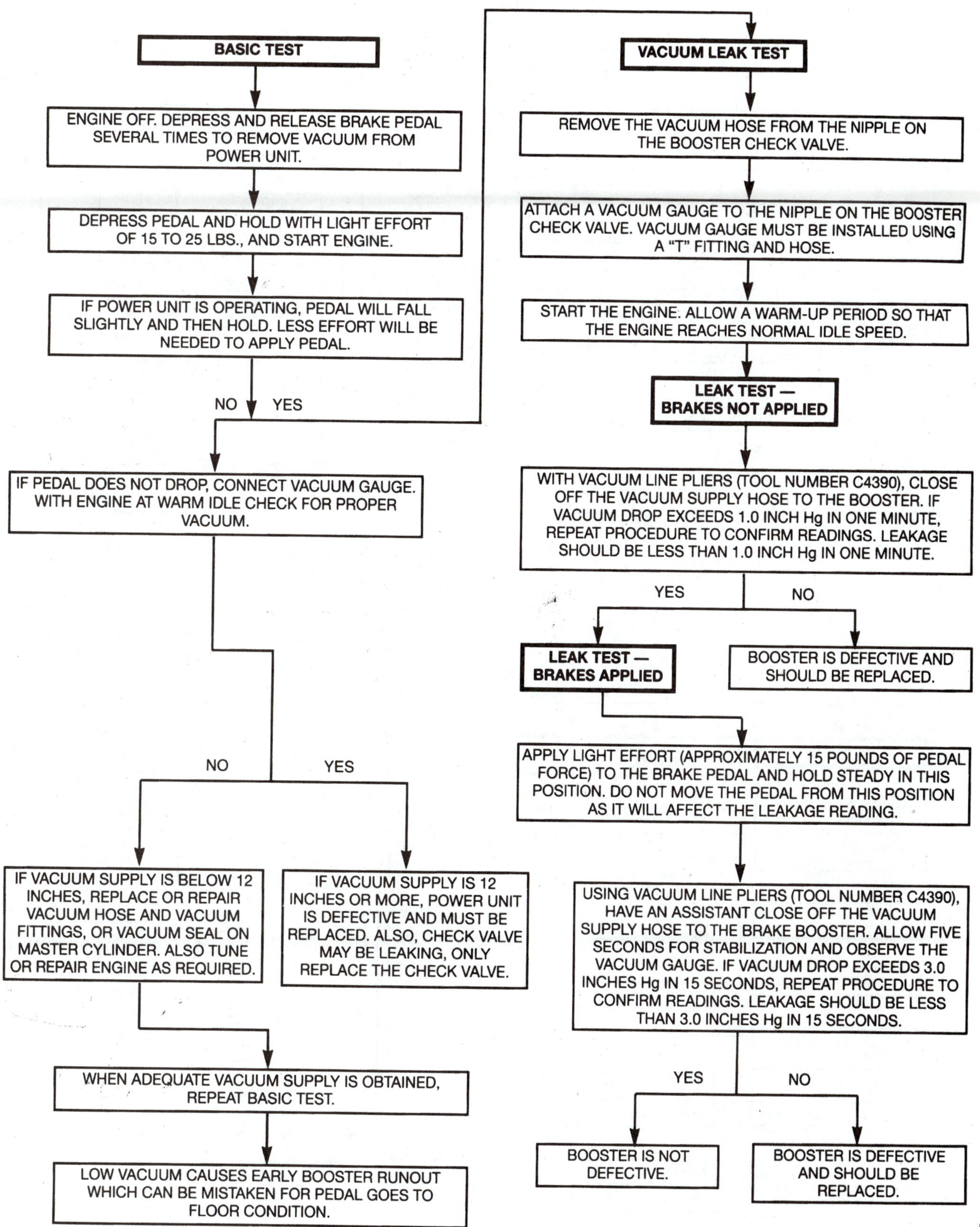

BRAKE NOISE

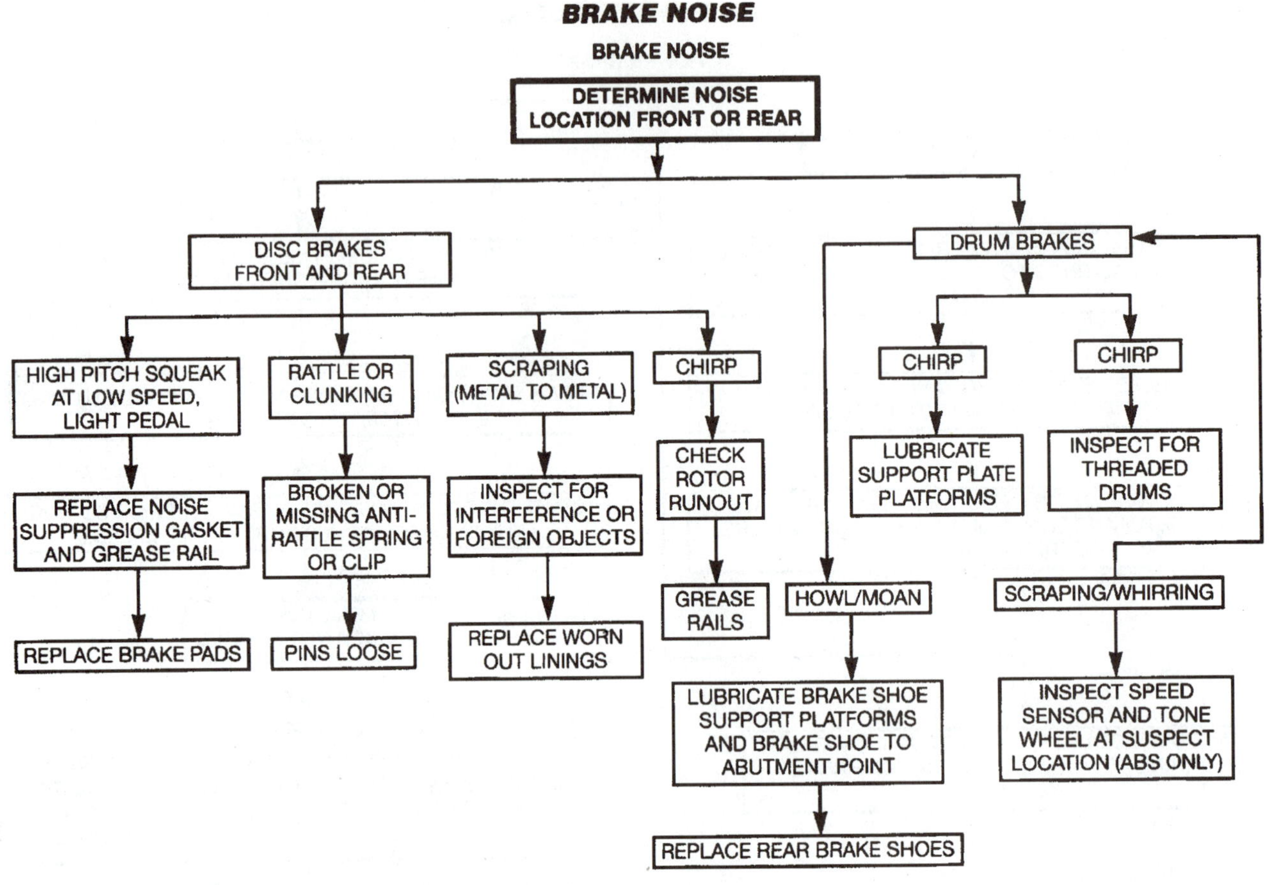

VEHICLE ROAD TEST

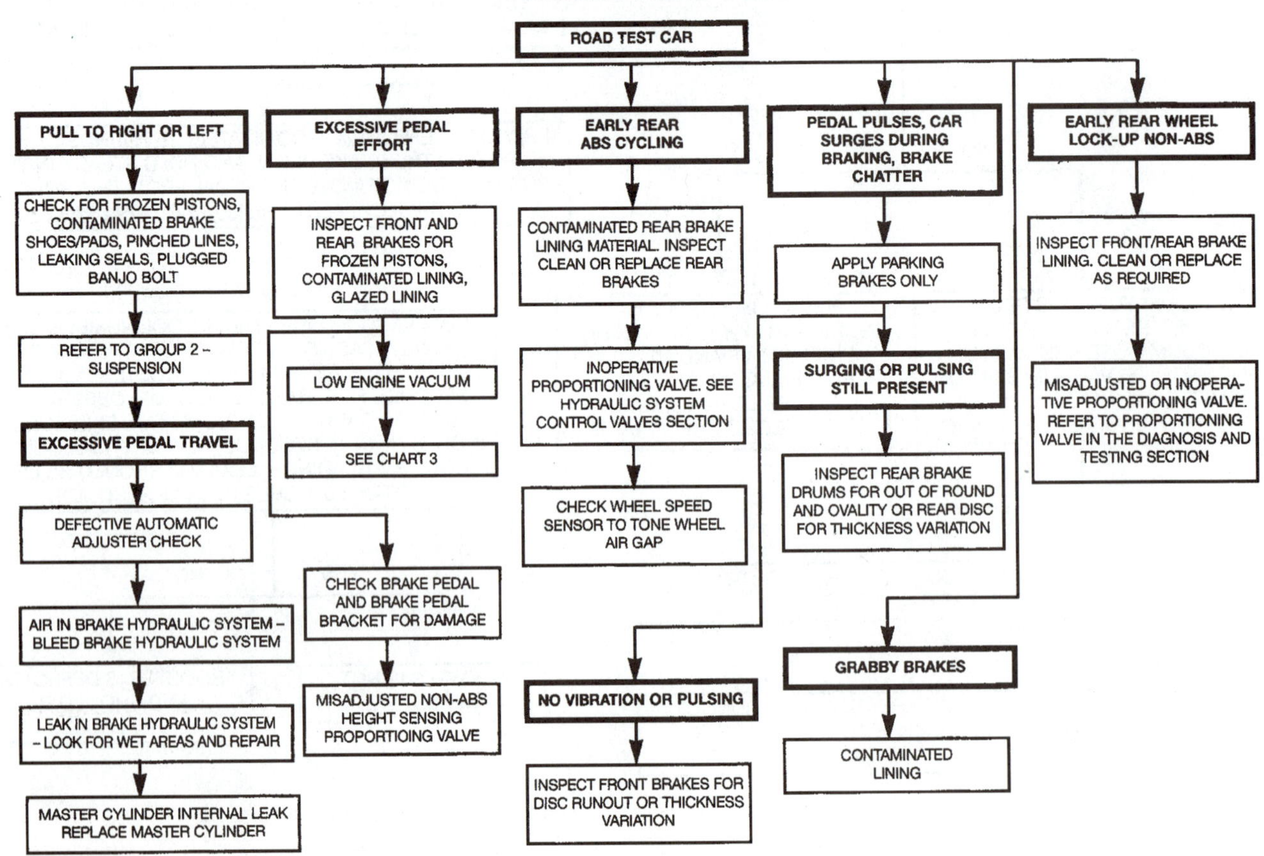

Table 9-1
There is a variety of brake problems, and the terms to describe them vary somewhat among technicians. The following should help you understand these terms and improve your communication with other technicians and future customers.

Symptom	What it is
Chatter	One or more brakes do not apply smoothly but in an abrupt, apply-and-release manner.
Dive	The front of the vehicle lowers during a stop.
Drag	One or more brakes do not release completely but remain partially or completely applied, also called "binding."
Fade	Braking power reduces or disappears during a stop.
Grab	One or more brakes locks up too easily.
Grind	Harsh, abrasive sound like someone grinding metal with a low-speed grinder.
Hard pedal	The pedal is very firm and it takes more-than-normal pressure for braking.
Knock	A harsh "knock" sound as the brake is applied, also called "thump" or "clunk"; this problem can be caused by worn suspension parts.
Lock	One or more brakes apply completely so the wheel does not revolve.
Low pedal	The pedal travels farther than normal.
Moan	A low-frequency noise during a stop, also called "groan."
Noise	A noise that is different from normal braking.
No pedal	The brake pedal goes all the way down with no brake application.
Pedal squawk	A chirp or squawk as the pedal is applied.
Pull	The vehicle tends to turn right or left as the brakes are applied.
Pulsating pedal	The pedal moves up and down as the car is stopping, also called "pedal surge."
Sinking pedal	The pedal slowly lowers under a steady pedal pressure, also called "fading pedal."
Soft pedal	The pedal goes down too easily, also called "pedal falls away."
Spongy pedal	The pedal action is the same as pushing on a spring, also called "springy pedal."
Squeal	A high-pitched noise during a stop; this usually can be varied by changing pedal pressure, also called "squawk."
Surge	A pulsating, not smooth, stopping action.

technician presents the findings to the car owner, along with the various choices to be made and the cost and ramifications of each choice. The car owner always makes the final decision as what repairs are to be made; never make a repair that is unsafe or not industry-approved; and don't forget the MAP Guidelines.

When there is a specific problem that can be corrected by an adjustment or replacement of one or a few parts, and the system is sound and has an adequate amount of good brake lining and a good hydraulic system, it is best to simply repair the defect. However, with older cars with a tired hydraulic system, worn lining, or both, it is often better to recommend a complete brake job. After a certain point, it is more economical to make all the necessary repairs at one time. This should also give the car owner more security. Remember that a brake job should give safe, "new-car" operation for several years and many thousands of miles.

9.5 Special Notes on Surface Finish

At times, a brake technician is concerned with the surface finish or texture of a metal surface. This concerns the flatness or roundness—also called the waviness—as well as the roughness of the metal surface. The finish of the friction surfaces of the brake drums and rotors is important because these surfaces must mate, that is, cause the new lining to wear to an exact fit. This is called **lining break-in** or **burnishing**. The surface finish of the bore of a wheel cylinder or master cylinder is important because a rough or faulty bore can cause a poor seal or fast rubber cup wear.

The actual surface finish is determined by the following factors: roughness—the finely spaced irregularities caused by the machining process; waviness—an irregular surface height caused by unequal wear during operation or deflection during machining; scoring—a severe form of waviness; and flaws—very irregular imperfections caused by rust, impact, or defects in the metal (Figure 9-33). The pattern left by the machining operation is often

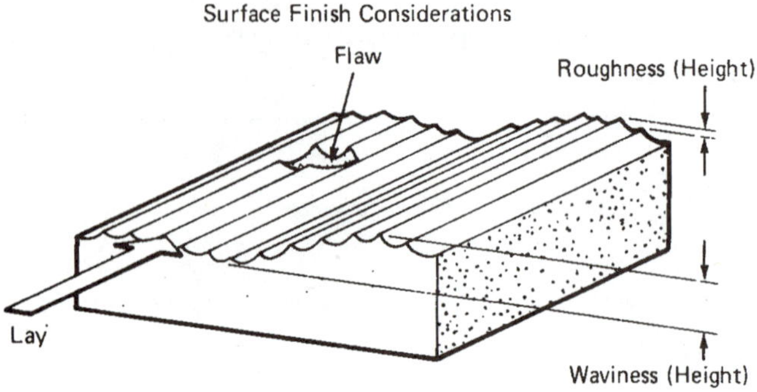

Roughness: the finely-spaced surface irregularities usually in a consistent pattern produced by the machining processes.

Waviness: an irregular surface condition of larger size than roughness caused by wear.

Flaws: irregularities or damage that do not appear in a consistent pattern.

Lay: the predominant direction of surface pattern. (A directional pattern is shown.)

Figure 9-33 Metal surface finish is a combination of roughness, waviness, and any flaws.

called a **directional finish**. It runs in a predominately parallel path. It is called a **nondirectional finish** if there is no regular pattern (Figure 9-34). The direction of the pattern machine marks is often called **lay** by machinists.

The surface finish is measured as the average depth of the irregularities. It is affected by the four factors involved. A certain surface finish is often referred to as 20 to 100 rms, meaning the average scratch depth is between 20 to 100 microinches. The term *rms* refers to *"root mean square"* which is a method of averaging scratch depth. A microinch is very small, 1/1,000,000 or 0.000001 inch (0.0254 mm).

The cylinder bore surface is fairly easy to deal with. It needs to be a very smooth cylinder shape, which is an almost perfect circle with straight sides. Normal piston operation in a hydraulic cylinder does not cause bore wear. The most common problem is corrosion, which causes pitting and other flaws. Normally, a corroded bore is not repaired, only cleaned, and if pitted excessively, the cylinder is replaced.

More problems are encountered with the surface finish of a rotor or drum because of the wear caused by friction material contact during stops. Problems involving the surface of a drum are even worse because of its round, cylindrical shape. As it wears, the friction surface of a drum or rotor will normally burnish to an extremely smooth surface (Figure 9-35). A certain amount of waviness often accompanies this burnishing. Any waviness in the surface will require that the replacement lining—which is essentially flat—wear to make full contact with the drum or rotor surface. If the worn surface of the drum or rotor is too smooth, this wear during lining break-in will be very

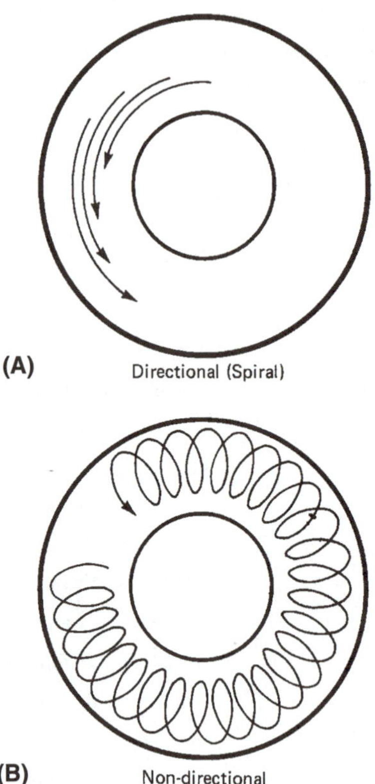

Figure 9-34 The finish (machine marks) on rotor A are in a circular, directional pattern. Rotor B has a nondirectional finish with the machine marks going in all directions.

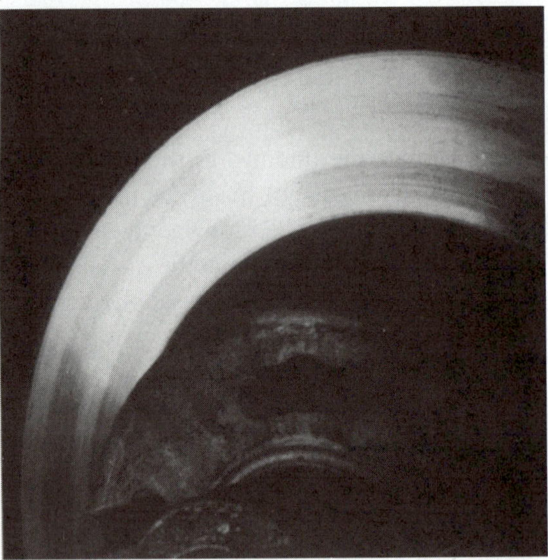

Figure 9-35 This used rotor has a very smooth, shiny surface that is common because of the burnishing of the lining.

slow. There is a good possibility that the new lining might be used in a hard stop before it is broken in. An improperly fitted lining—one that is not broken in—will not develop a full braking force and will also tend to overheat in those areas that are in contact.

The correct surface finish for lining break-in is about 40 to 80 microinches of roughness with no waviness. A smoother finish will cause a very slow break-in. A rougher finish will probably cause faster break-in but also rapid lining wear and possibly grabbiness. A nondirectional pattern is also desirable because scratches which run across the direction of rotation promote rapid lining break-in.

9.6 Electrical Diagnosis and Repair

Occasionally a brake technician needs to repair a fault in the electrical portion of a stoplight, brake warning light, or antilock braking system circuit. In the past, diagnosing and repairing these circuits could often be accomplished with a limited knowledge of electrical circuits and a simple test light. With the introduction of modern, solid-state computerized electronics in the ABS, there is a greater need to understand the system and electrical test procedures. The ability to measure voltage and resistance and interpret these measurements has become very important.

Solid-state electronics is at the heart of computerized circuits. It includes transistors, diodes, and integrated microchip devices. These are control and sensing devices which are quite fragile compared to other automotive electrical devices.

A course in basic automotive electronics is necessary to thoroughly understand electricity and how to measure it. The description that follows is merely a brief review.

A technician is normally concerned with three measurable aspects of electricity: **volts**, **amperes** and **ohms** (Figure 9-36). Voltage, also called electrical pressure, is the push that forces electricity to flow through a wire or component. This flow is measured in amps or amperes. In a car, voltage is supplied by a source of electrical power—either the battery or the alternator. Amperage multiplied by voltage gives **watts**, which measure the amount of electrical power.

A **circuit** is a complete electrical path that allows amps to flow from the power source, through the electrical components, and back to the power source (Figure 9-37). This path is composed of wires (commonly called conductors), safety devices such as a fuse or circuit breaker, usually one or more switches, and the electrical component(s). In a car, a **ground circuit** is used to conduct electricity from a component back to the source of power. The ground circuit uses the metal of the car body, frame, engine block, and so on, as an electrical conductor. In modern cars, the negative (-) terminal of the battery and

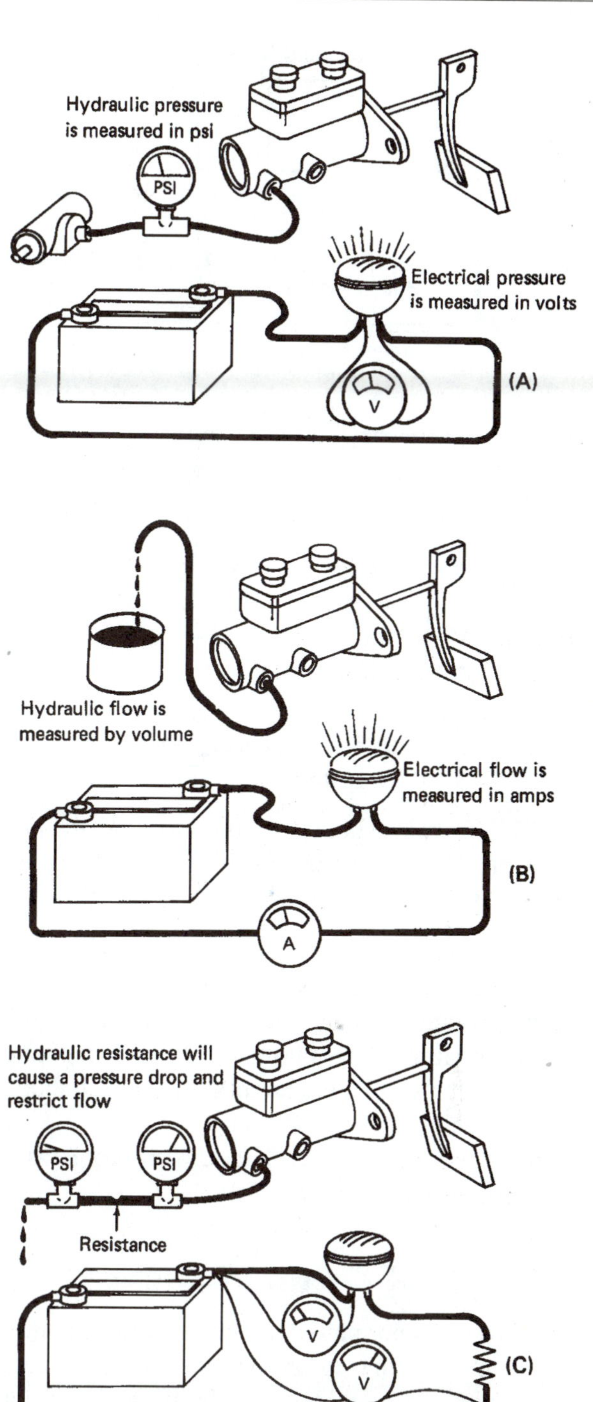

Figure 9-36 If we compare hydraulics with electricity, pressure and voltage (A), fluid flow and current flow (B), and resistance to flow (C) are very similar.

the alternator are connected to ground, and the positive (+) side is insulated.

Ohms are units of electrical resistance. The amount of current flow in a circuit is determined by the resistance

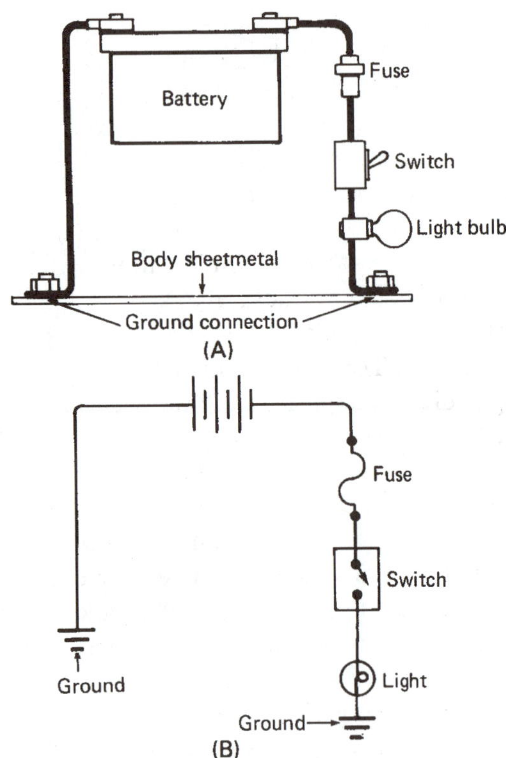

Figure 9-37 Electric diagrams are used to show the path through a circuit. They are usually drawn in a schematic form (B) using symbols in place of pictures of the components.

of the components. A large amount of resistance will stop or severely limit current flow, and a small amount of resistance will allow a large current flow. The symbol Ω (omega) is often used to signify ohms.

Diagnosing solid-state ABS electrical problems is described in Chapter 15.

9.6.1 Electrical Circuit Problems

Electrical problems normally fall into the following three categories: **open**, **short**, or **grounded**. Except in some solid-state units, these problems are fairly easy to check. An open circuit is a broken, incomplete circuit through which no current will flow (Figure 9-38). Open circuits are usually caused by a broken wire or a burned-out fuse or fusible link. A loose or dirty connection can cause a high-resistance, partially open circuit. A switch opens a circuit intentionally when it is turned off. An open circuit will have power source voltage in it up to the point where the circuit is open.

A grounded circuit occurs when a current-carrying wire or component touches ground, the bare metal of the car body, and so on. Normally, insulation on the wire prevents a grounded circuit, but it can wear through and allow the metal conductor to touch the ground metal (Figure 9-39). A grounded circuit provides a low-resistance path for current flow, and the rate of current flow increases because of this drop in resistance. This will usually burn out a fuse, open a circuit breaker, or burn up a wire. A grounded circuit will often produce an open circuit when it burns out a fuse, fusible link, or wire. A grounded circuit is also called a **short to ground.**

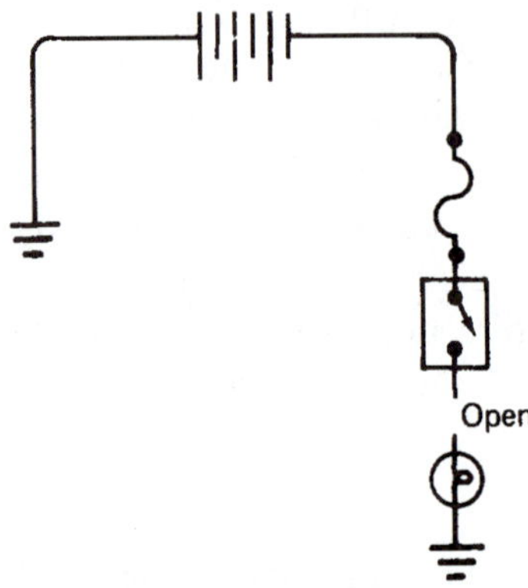

Figure 9-38 An open circuit; there is a break in the circuit that will stop current flow.

A **short circuit** is sometimes found in electrical units that use coils of wire. If the coils short and make electrical contact with each other, a shorter-than-normal electrical path is created. A short will lower the resistance of a component, which will increase the current flow. It will also reduce the operating efficiency of the unit (Figure 9-40). A short can also occur between two wires if they both lose their insulation and the conductors make contact.

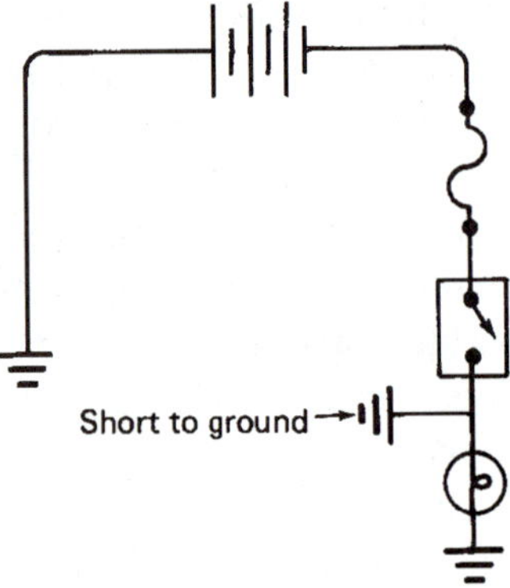

Figure 9-39 A grounded circuit or short-to-ground occurs if the insulation wears and lets the conductor touch ground.

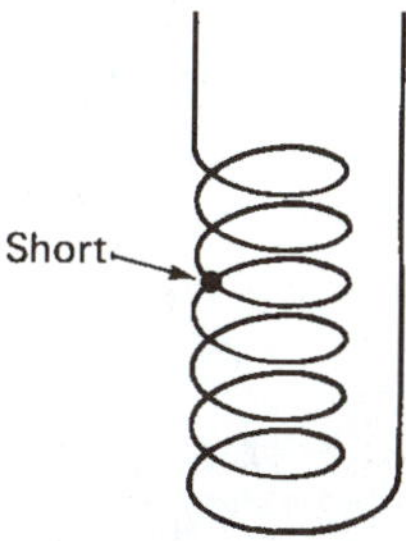

Figure 9-40 A short circuit occurs if the insulation wears so the conductor of two wires touches. Current flow will take the shorter path and bypass that coil.

9.6.2 Measuring Electrical Values

A technician often uses a test light or a volt-ohmmeter when troubleshooting an electrical circuit. A test light is a simple, inexpensive device which can quickly indicate if a circuit has voltage (Figure 9-41). The brightness of the light can also give an indication of the amount of voltage. A test light is a handy device for checking simple circuits, but it should not be used on the sensor portion of solid-state circuits. The current flow through the test light can damage the relatively fragile transistors and integrated circuits in some solid-state electronic equipment. A test light which uses a light-emitting diode (LED) can be used with these circuits. The very high resistance of an LED will allow it to draw very little current. High-impedance test lights, which draw very little current, are now available for checking computerized circuits.

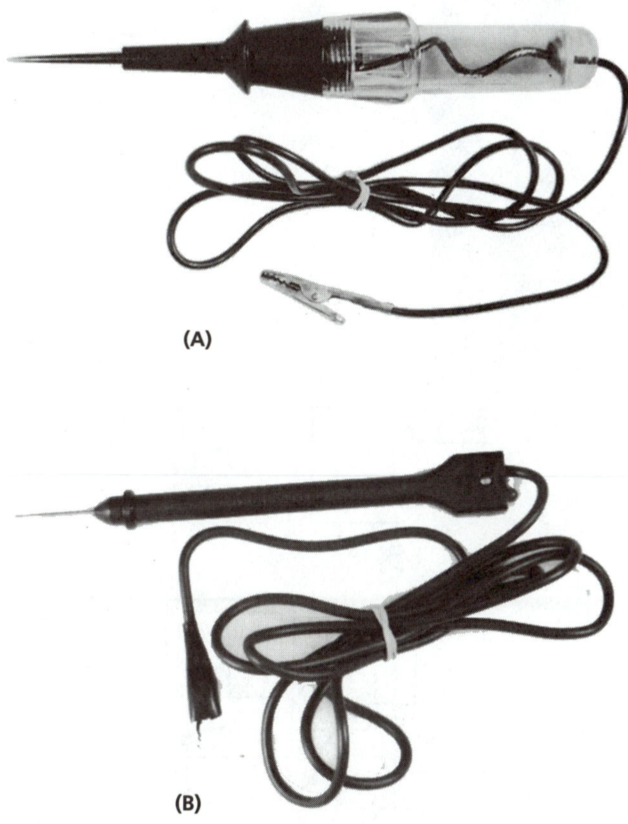

Figure 9-41 A standard test light (A) and a high-impedance test light (B). The wire clip is connected to a good ground, and the light in the handle will light up when the probe contacts voltage. The standard light should not be used on computer-controlled circuits.

Volt-ohmmeters allow accurate measurement of voltage and resistance. These meters are commonly available in analog or digital form (Figure 9-42). An analog meter has a needle that sweeps over a scale to give a reading. A digital meter displays a number that is the actual reading. Analog meters are simple and relatively inexpensive, but some of them should not be used with certain solid-state electronic units. Like the test light, some draw enough current from the circuit to damage the electronic components. The current draw of the meter depends on its internal resistance. If the internal resistance is at least 10 megohms (10,000,000Ω), the meter can be safely used on solid-state circuits. A digital volt-ohmmeter has over 10 megohms of internal resistance. Only digital meters or a high-impedance test light should be used for checking solid-state circuits.

Voltage is measured by connecting the negative (-) lead of a voltmeter to a good, clean body ground and making contact between the positive (+) lead and various connection points in the insulated portion of the circuit. The voltage of the circuit at that point will be read on the meter. On many meters, it is important to select and set the meter to the correct voltage range before taking a reading. Some meters can be damaged if you measure a voltage higher than the meter setting. Always select a range higher than the value you expect to read. After measuring the voltage, the meter can be reset to a lower voltage scale if desired—as long as the range of the lower scale exceeds the voltage being read (Figure 9-43).

Resistance is measured by connecting both leads of an ohmmeter to the two connections of a component or to both ends of a wire. A reading indicates whether the circuit is complete and the amount of resistance. An ohmmeter is self-powered. It will cause a small amount of current flow and measures the amount of flow to determine the resistance. An ohmmeter should never be connected to a circuit that contains voltage or is connected to a battery. The added voltage from the circuit can damage the meter.

When using an ohmmeter, the range of the ohmmeter is selected and set to the value that you expect to read on the meter. Many meters have a range selector switch with ranges like × 1, × 1,000 (1k), × 10,000 (10k), and so on. The reading on the meter should be multiplied by the range

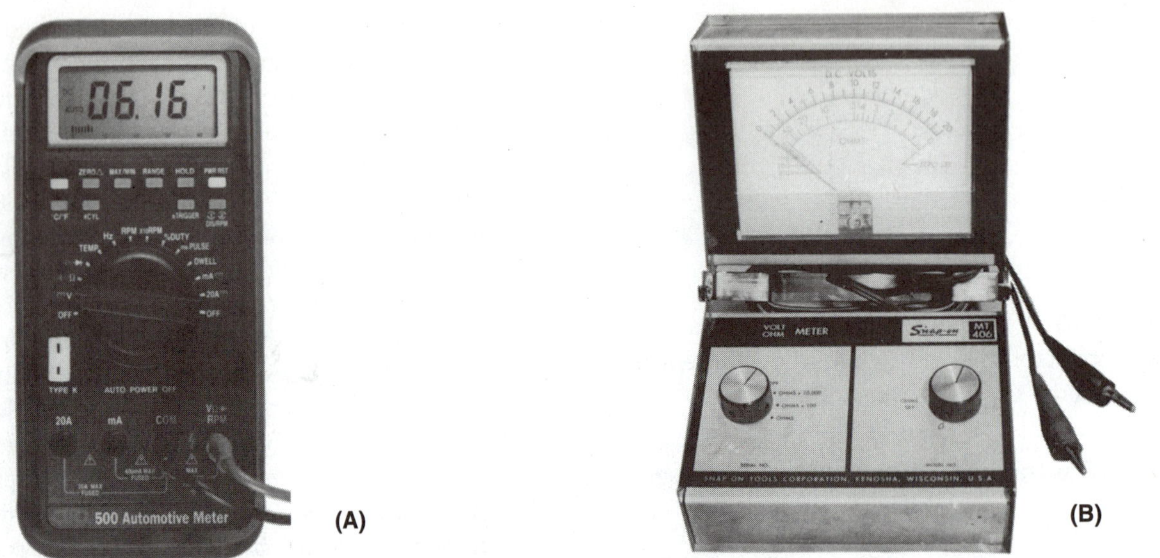

Figure 9-42 A digital (A) and an analog (B) volt-ohmmeter. Note that the digital meter is reading 06.16 DC volts on both the display and bar graph. The digital meter is capable of many other functions (A is courtesy of SPX/OTC)

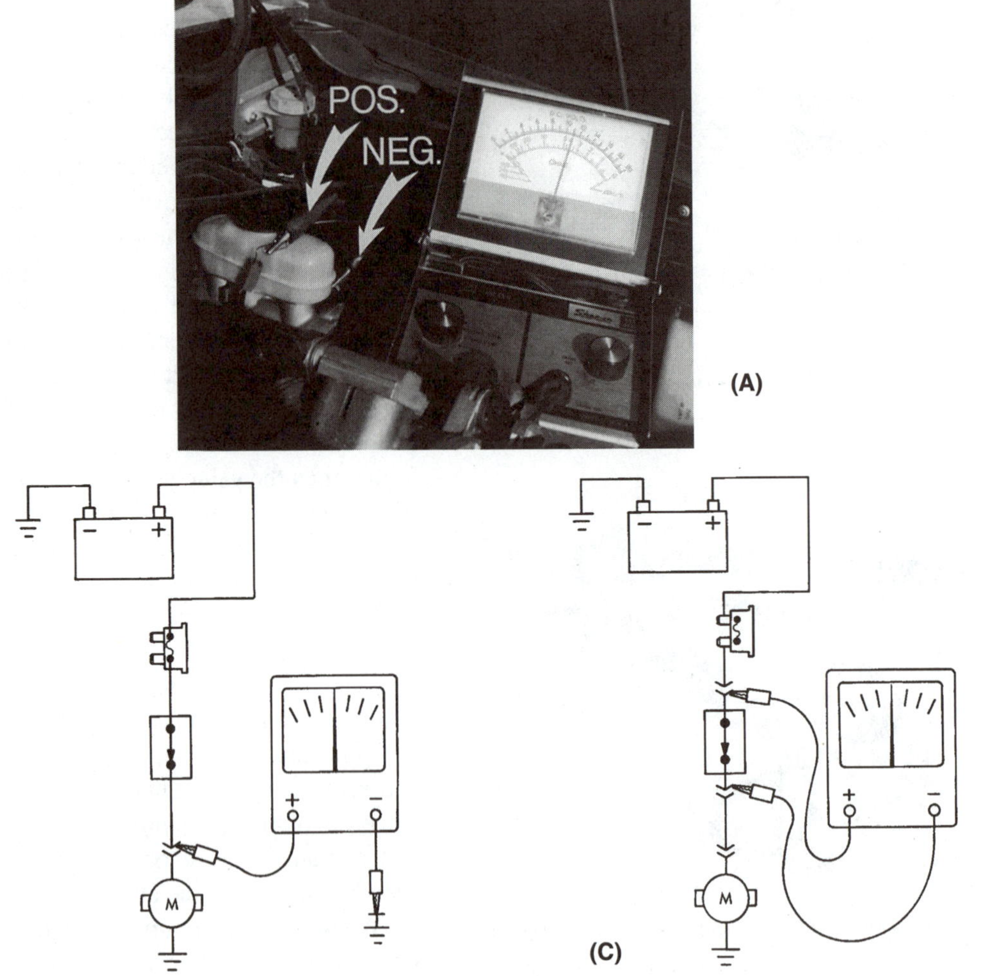

Figure 9-43 Voltage is being measured at the fluid-level switch connector using paper clips to make contact inside the connector (A). A schematic view of a similar check is shown in (B); connecting the meter lead to both sides of the switch will check for voltage drop across the switch (C). (A is courtesy of SPX/OTC; B and C are courtesy of Chrysler Corporation)

setting to determine the amount of resistance. After selecting a range, the meter leads should be connected together and the meter read. It should read zero, because there is no resistance between the leads. Many meters have a knob that is used to calibrate the meter to zero at this time. When the leads are separated, the meter should read at the top of the scale, often marked as infinity. A digital meter will read "OL" for out of limits. The leads of the meter are then connected to the terminals of a component or to the ends of a wire. A zero reading indicates a complete circuit with no resistance. A high reading indicates a large amount of resistance or a possible open circuit. When this occurs, it is a good practice to switch the meter to a higher scale, recheck the calibration, and remeasure the resistance (Figure 9-44).

9.6.3 Interpreting Readings

Meter readings help only the technician who knows what the readings should be. Readings taken from different locations in a circuit or from different circuits often vary. Experienced technicians are familiar with many of the simple circuits, such as the one for a brake warning light. They usually know what the voltage or resistance should be at the different points in the circuit. When a different or new and more complex circuit is encountered, a wiring diagram is often used.

A wiring diagram shows the electrical path through the various switches and components of a system. It usually indicates the color coding of the wires, the size of the bulbs and fuses, and the locations of fuses, fusible links, and connectors in the circuit. A technician uses this diagram to locate points where voltage or resistance checks can be made and to determine what kind of reading should be taken at these locations.

The diagram of the rear turn and stoplight circuit shown in Figure 9-45 is for a late-model domestic vehicle that uses combination stoplights and turn lights along with a high-mounted stoplight at the center of the rear window. A system with brake lights that are separate from turn indicators will be simpler. This circuit, including stoplight and turn indicator switches, can be diagnosed using a test light or a voltmeter. If they are disconnected from the circuit, the switches can be checked using an ohmmeter. The stoplight portion of this circuit should have power to it at all times. Referring to Figure 9-45, power leaves the 15-amp fuse at the lower connection and should also be present at the top (or IN) connection to the brake on/off switch (stoplight switch). When the brakes are applied, electricity should flow through the brake switch, through wires 511 and 810 to the high mount stop lamp and the multi-function switch. Wire 511 is color-coded light green, and wire 810 is red and light green (red with a light green stripe or tracer). The multi-function switch includes the turn light switch with connections 9 to the left stop lamp and 5 to the right stop lamp; with the turn switch centered, electricity should be at both of these connections when the brake switch is on. Electricity should flow through wire 9 (light green and orange) to the left stop lamp and through wire 5 (orange

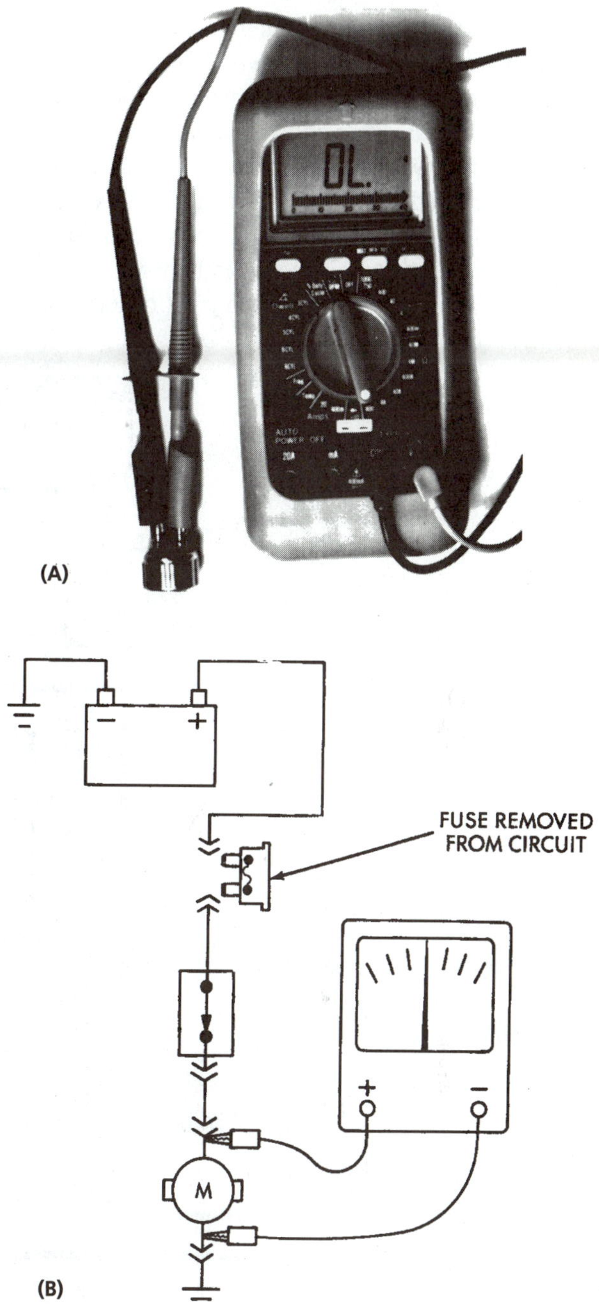

Figure 9-44 The stoplight switch (A) has a resistance that is out-of-limis (OL)/infinite showing an open circuit; the resistance should drop to almost zero when the brakes are applied. A schematic view of a similar check is shown in (B). (A is courtesy of SPX/OTC; B is courtesy of Chrysler Corporation)

and light green) to the right stop lamp. All three stop lamps are connected to ground through wire 57 (black).

In this circuit, a technician can often determine the problem by observing how it operates or fails to operate because it is divided into two parallel circuits. A total failure will probably be caused by a power loss at the fuse or brake switch. A failure of both side lights can be caused by a faulty turn-hazard switch or the wires between this switch and the fuse. A partial failure of the right or left side will probably be caused by a faulty turn-hazard switch or an open circuit in the wires from the switch to the rear junction of the lights. A failure of a single light will probably be caused by a faulty bulb, a faulty light ground, or a faulty wire between the bulb and the junction. Knowing the voltage at a given point or the resistance between two given points will let the technician determine exactly where the fault is. At that point, correcting the problem is usually easy.

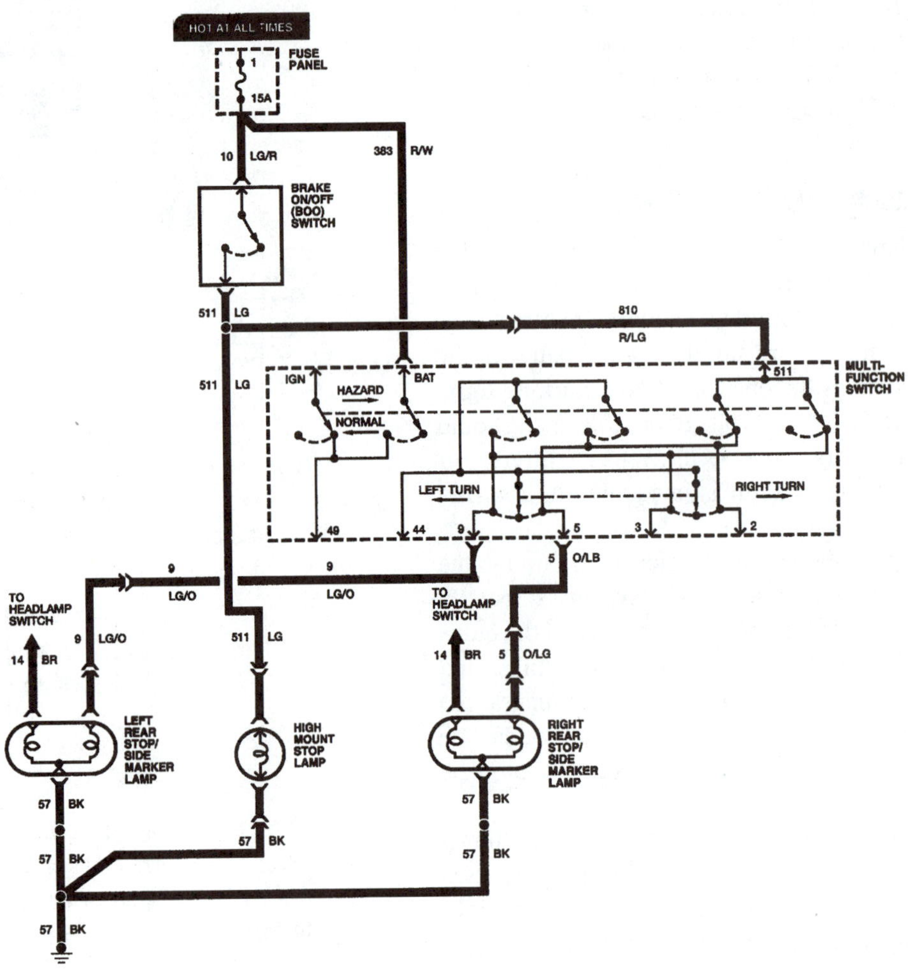

Figure 9-45 A schematic showing the stoplight for a late-model vehicle; note how the circuit from the brake on/off switch is affected by the turn-indicator switch. (Courtesy of Ford Motor Company)

9.7 Practice Diagnosis

You are working in a brake and front-end shop and encounter these problems.

CASE 1: While inspecting the brakes on a three-year-old Toyota, you find that the center stoplight does not light when the pedal is depressed; the other two lights do come on. What is probably wrong? If you check out this possibility, and you find it is good, what should you do next?

CASE 2: The complaint on the eight-year-old pickup is a very dim stop light on the right side. You measure the voltage at the contacts for the bulb, and find that one of them has 6 volts, the other one has 0 volts, and the bulb socket also has 0 volts. What is probably wrong? Have you found the cause of the dim bulb? What should you do next?

Terms to Know

amperes	lay	short
burnishing	lining break-in	short circuit
circuit	Motorist Assurance Program (MAP)	short to ground
directional finish	nondirectional finish	volts
ground circuit	ohms	watts
grounded	open	

Review Questions

1. Two technicians are discussing brake inspection procedures. Technician A says that a brake pedal must have a slight amount of free movement before it starts to move the master cylinder piston. Technician B says that the pedal should become firm before it travels halfway to the floor board. Who is right?
   a. A only
   b. B only
   c. both A and B
   d. neither A nor B
2. A sinking brake pedal is an indication that
   a. there is air in the system.
   b. the shoes require adjustment.
   c. there is a leak in the system.
   d. the lining is worn.
3. Technician A says that the brake warning light should come on as the engine cranks. Technician B says that this light usually comes on under a very hard pedal. Who is right?
   a. A only
   b. B only
   c. both A and B
   d. neither A nor B
4. Technician A says that a low brake pedal is a definite indication that there is air in the hydraulic system. Technician B says that a low brake pedal can usually be corrected by bleeding the brakes. Who is right?
   a. A only
   b. B only
   c. both A and B
   d. neither A nor B
5. Technician A says the best test for a power booster is to apply the brakes hard during a stop. Technician B says that a good booster will cause the brake pedal to lower if the engine is started while the brakes are being applied. Who is right?
   a. A only
   b. B only
   c. both A and B
   d. neither A nor B
6. A small swirl or spurt in the fluid of the master cylinder reservoir as the brakes are applied indicates
   a. normal operation.
   b. a faulty piston seal.
   c. a low brake-fluid level.
   d. all of the above.
7. Technician A says that a brake lining should be replaced when it wears to a thickness of 0.005 to 0.010 inch. Technician B says that lining thickness is always measured from the shoe rim or backing to the outer edge of the lining. Who is right?
   a. A only
   b. B only
   c. both A and B
   d. neither A nor B
8. During an inspection, a brake drum should be checked to make certain that
   a. there are no cracks.
   b. there is a smooth friction surface.
   c. the diameter is less than the maximum specified diameter.
   d. all of the above.
9. A brake rotor being checked should always be
   a. thicker than the thickness dimension on the rotor.
   b. thinner than the maximum size dimension on the rotor.
   c. measured with a ruler.
   d. all of the above.
10. Technician A says that rotor runout is checked using a dial indicator. Technician B says that a pedal pulsation problem can be caused by nonparallel friction surfaces on the rotor. Who is right?
   a. A only
   b. B only
   c. both A and B
   d. neither A nor B
11. Technician A says that improper tightening of the wheels can distort the shape of the rotor or drum. Technician B says that the lug bolts can come loose if they are not tightened correctly. Who is right?
   a. A only
   b. B only
   c. both A and B
   d. neither A nor B
12. A brake hose should be replaced if
   a. it leaks.
   b. it has bulges or bubbles in the outer cover.
   c. it has rub marks extending into the outer layer of the cord.
   d. any of the above.

13. Technician A says new brake springs should be installed if an overly stretched spring is discovered. Technician B says that if oil-soaked lining is found on one wheel, the lining at both ends of the axle should be replaced. Who is right?
    a. A only
    b. B only
    c. both A and B
    d. neither A nor B
14. Technician A says that a nondirectional surface finish has no scratch marks on it. Technician B says that the actual surface finish is a combination of waviness, irregular flaws, and machining scratches on the metal surface. Who is right?
    a. A only
    b. B only
    c. both A and B
    d. neither A nor B
15. Technician A says that a worn drum or rotor surface is usually too smooth to properly wear in new brake lining. Technician B says the correct surface finish for a rotor friction surface is about 60 microinches. Who is right?
    a. A only
    b. B only
    c. both A and B
    d. neither A nor B
16. Which of the following is not normally used to determine a brake system electrical problem?
    a. a test light.
    b. a wiring diagram.
    c. an oscilloscope.
    d. a voltmeter.

10 Drum Brake Service

Objectives

After completing this chapter, you should:

- ❑ Be able to remove, clean, measure, and inspect brake drums for wear or damage, and determine if they should be machined or replaced.
- ❑ Be able to mount a drum on a lathe and machine it according to the manufacturer's procedures and specifications.
- ❑ Be able to clean and remove brake shoes, springs and any related hardware and determine the necessary repair.
- ❑ Be able to clean, inspect, and lubricate a backing plate.
- ❑ Be able to lubricate and install brake shoes, springs, and related hardware.
- ❑ Be able to adjust brake shoes and install brake drums.
- ❑ Be able to perform the ASE Tasks for drum brake diagnosis and repair (See Appendix 1).

10.1 Introduction

A typical brake job on a drum brake assembly consists of the following: drum removal, drum machining, shoe removal, rebuilding or replacement of the wheel cylinder, thorough cleanup of the backing plate and all small parts, careful reassembly of the parts, lubrication of these parts as they are assembled, adjustment of the shoe-to-drum clearance, and adjustment of the parking brake. The parts should be carefully inspected at least three times—before disassembly, during cleanup, and during reassembly—to locate any faulty or damaged components that might affect the final job. All damaged parts should be replaced. A brake system cannot work as it should if one or more parts are faulty.

After the drum has been removed, the disassembly inspection serves three major purposes—to locate damaged parts, to determine the relationship of all of the parts, and (if there are unusual wear patterns) to determine problems involving the backing plate alignment or drum surface. A bell-mouthed, concave, or convex drum surface or a distorted backing plate is indicated by the abnormal lining wear it causes. Some newer brake designs have subtle differences in spring or linkage positioning (Figure 10-1). These differences should be noted so that mistakes will not be made during reassembly.

Competent brake technicians realize that in disassembling a unit, they become the authority on how that unit

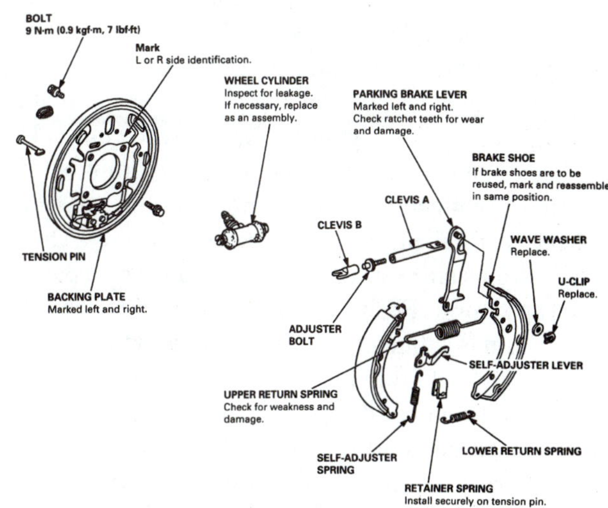

Figure 10-1 An exploded brake assembly showing various checks. (Courtesy of American Honda Motor Co., Inc.)

goes together. A service manual should be followed as a guide to the disassembly and reassembly steps. These manuals, however, tend to be somewhat general and will sometimes fail to show all the important details. As you inspect a unit in order to locate damaged parts, you should check the hold-down and return spring attachment positions, how and where the self-adjusters are attached, and how and where the parking brake mechanism is positioned (Figure 10-2). Determine any obvious differences which can help you, because you are the one who is going to put the unit back together. For some of the more complicated assemblies, it might be necessary to draw a simple sketch to remember the details. As a last resort with complex units, work on one side at a time. Most cars have a right-side brake that is a mirror image of the left one. If one side is left intact, it can serve as a pattern until the other one is assembled.

Figure 10-2 This return spring was installed upside down and was nearly worn in two from rubbing on the hub.

SAFETY TIP The following sections will describe operations on a single brake assembly. All brake operations are normally done in pairs. The same operation is done at each end of the axle to ensure even, straight braking. Don't forget the safety tips that were mentioned in Chapter 9.

10.2 Drum Removal

Several different methods are used to secure a drum onto an axle or hub. For the drive axle of rear-wheel-drive cars, the most common style is a "floating drum." It is slid over the axle pilot, which centers it and the wheel studs (commonly called lug bolts). It is held in place with speed nuts and the wheel lug nuts (Figure 10-3). The speed nuts, also called Tinnerman nuts, are used to keep the drum on the axle during manufacturing and assembly of the car. After the wheel is installed, they become unnecessary. Other drive axle drums are secured to the axle flange by a pair of bolts or cap screws (Figure 10-4). In a few cases, the center of the drum is splined for the axle and used to drive the wheel. This style of drum is usually secured to the axle by a large nut at the end of the axle (Figure 10-5).

Drums used on nondrive axles—the front drum on a rear-wheel-drive car and the rear drum on a front-wheel-drive car—use several different drum-to-hub attachment methods also. The hub and drum are normally treated as an assembly. If a faulty drum or hub requires replacement,

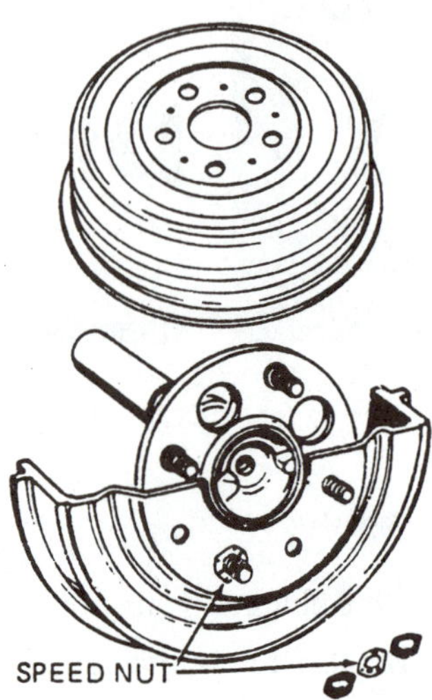

Figure 10-3 A floating brake drum is centered on the axle by the pilot hole in the center and is often held on the axle by speed nuts. (Courtesy of Wagner Brake)

Figure 10-4 This drum is held on the axle flange by two Phillips-head screws. Removing these screws allows drum removal. Another screw under the hole (arrow) can be turned to push the drum off if it is stuck.

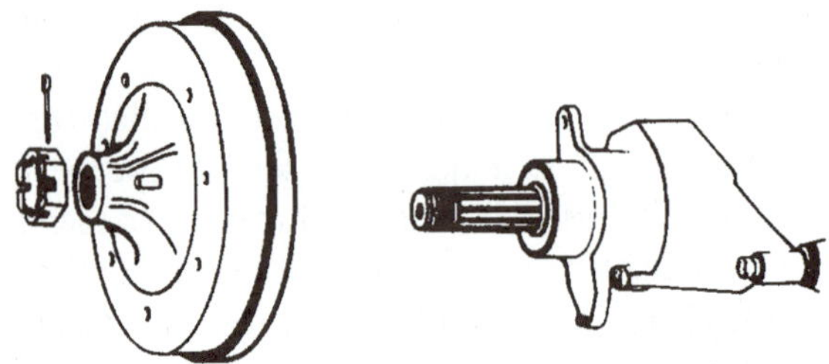

Figure 10-5 The brake drums on some cars are splined onto the axle and held in place by a nut. (Courtesy of Bendix Brakes, by AlliedSignal)

it is usually less expensive to separate them so the faulty portion can be replaced. The most common method of securing the drum to the hub is to use shouldered wheel studs that are swaged into place (Figure 10-6). **Swaging** is also called *upsetting*, *peening*, or *riveting*. The drum is normally placed over the hub, the wheel studs are pressed through the hub and drum, and the shoulder of the stud is swaged to lock it in place (Figure 10-7). Some drums are positioned on the inner side of the hub, some on the outer side. Other methods of drum-to-hub attachment use rivets to lock them together or hold the drum in place with speed nuts, much like the floating drum on a rear axle.

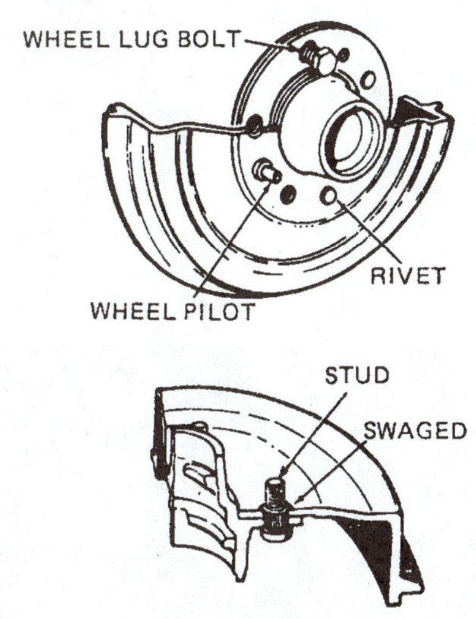

Figure 10-6 Most front drums are secured to the hub by swaging a portion of the wheel stud shank. Other front drums are riveted to the hub, and a few float on the hub like a rear drum. (Courtesy of Wagner Brake)

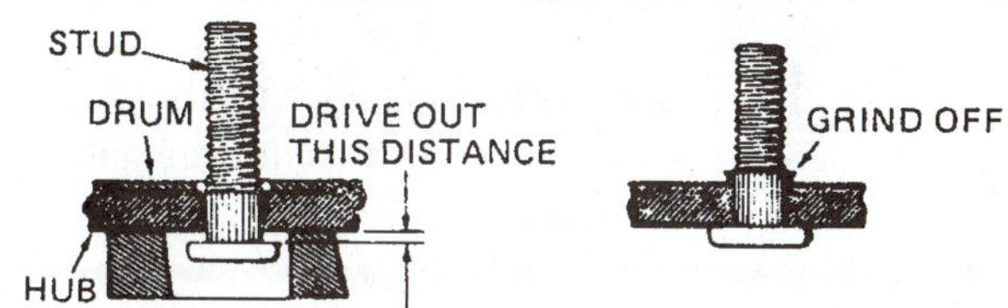

Figure 10-7 On some cars, the wheel stud is pressed through the hub and drum and then swaged. The swaging should be removed before driving or pressing the stud out. (Courtesy of Wagner Brake)

Removal of a drum can be made difficult by three different problems: failure to release the parking brake, freezing of a drum onto the axle flange or pilot, and shoes which have adjusted to a worn drum surface. The solution for the first problem is to simply release the parking brake, but solving the other two problems can be more difficult. With frozen drums, apply penetrating oil around the axle pilot and wheel studs and then strike the area between and inside the wheel studs using a hammer and punch (Figure 10-8). If this does not work, use a plastic hammer and strike the outer edge of the drum with a blow directed toward the axle, repeating this procedure several times. A drum's stamped-steel center section occasionally catches and digs into the axle pilot, and the hammer blows tend to remove the interference (Figure 10-9). If this method works, be sure to check the drums for cracks that might result. If this technique does not work, apply heat on the drum face around the axle pilot in the area inside

Figure 10-8 Occasionally a frozen drum can be loosened by squirting penetrating oil around the axle flange and wheel studs and then striking the area between the wheel studs with a punch and hammer.

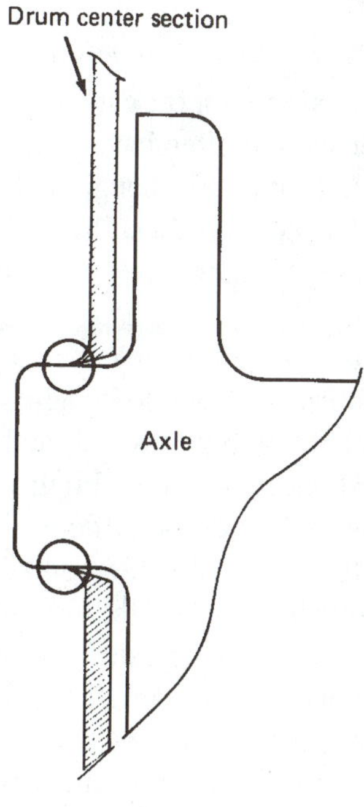

Figure 10-9 The center hole of some drums can catch and grip the axle pilot tighter as you pull the drum off.

the wheel studs. This method is effective only on drums having stamped-steel inner sections (Figure 10-10). Move the torch flame evenly around the axle flange until the center section expands and loosens its grip. A puller is also available to pull floating drums off the axle (Figure 10-11). It should be noted that the use of heat and pullers are last-resort methods of drum removal because they can damage the drum. If you use these severe drum-removal methods, it is always a good practice to check the drum for straightness. Place the drum—open-side down—on a flat surface and lay a straightedge across the center section. The straightedge should be parallel to the top of the surface (Figure 10-12).

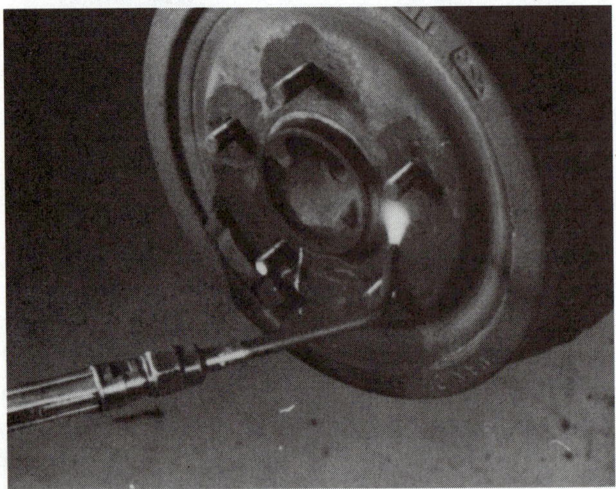

Figure 10-10 A drum with a tight pilot hole can usually be easily removed by heating the center area—inside the wheel studs.

The drums on some import cars and light trucks that have a cast inner section often have a pair of holes drilled and topped close to the axle flange. Machine screws or bolts can be threaded into these holes where they will press against the axle flange and push the drum off the axle.

If the drum is loose on the axle but will not pass over the shoes, the shoes will have to be adjusted to a smaller size. The drum is probably grooved, and the shoes have adjusted outward into the groove (Figure 10-13). On cars with self-adjusters, this can be difficult. It will be necessary to locate the adjuster screw, obtain access to the screw and adjuster lever, move the self-adjuster lever out of the way, and turn the adjuster screw in the correct direction. Most cars will have an access plug positioned in the backing plate or the face of the drum that can be removed so you can reach the mechanism. This plug can be an easily removed rubber or metal plug, or it can be a stamped knockout slug. Usually you can reach the lever

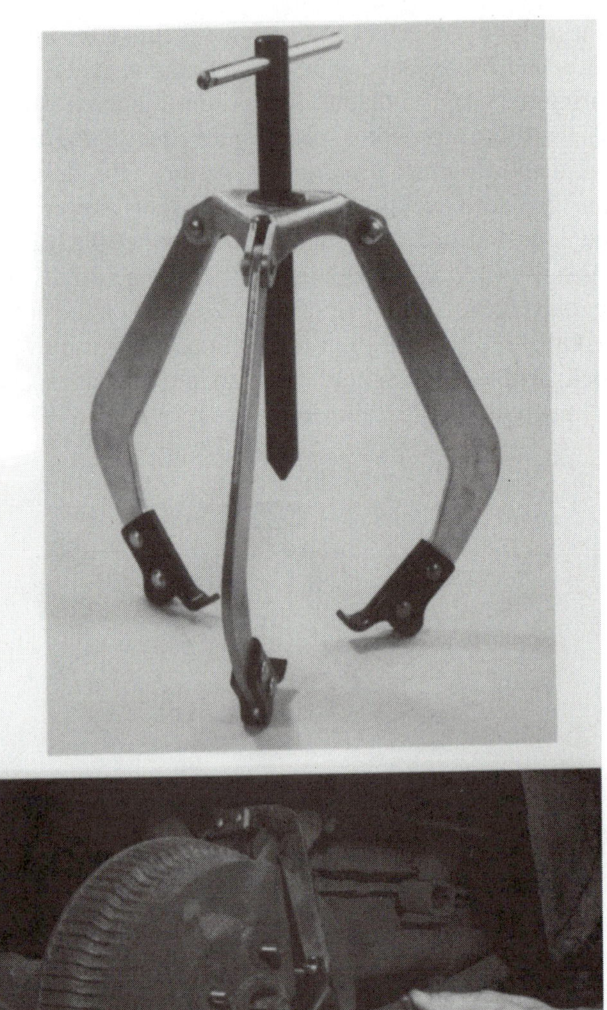

Figure 10-11 This puller is used to remove tight drums. If much force is used, be sure to check the drum for straightness after removal. (Courtesy of KD Tools)

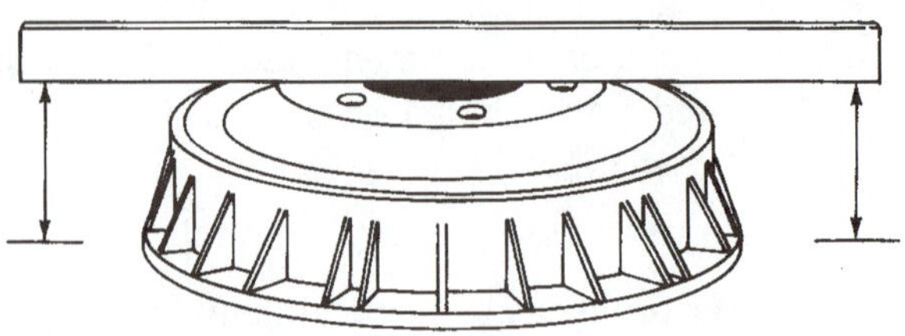

Figure 10-12 A drum is checked for straightness by placing it on a flat surface and placing a straightedge across the mounting surface. The straightedge should be parallel to the flat surface.

using a long, thin screwdriver or an ice pick and carefully push it away; it needs to move only about 1/16 inch (1.5 mm). With the lever pushed back, use a brake adjuster tool or an old screwdriver to rotate the adjuster screw (Figure 10-14). It is a good idea to preplan how you will release the self-adjuster each time you work on a different brake design. As a last resort, if you cannot back off the self-adjuster, cut off the ends of the shoe hold-down pins where they extend through the backing plate. Side- or diagonal-cutting pliers ("dikes") can be used for this. This usually will allow you to pull the drum outward far enough to reach in and release the adjuster. The pins are relatively inexpensive, and the time you would spend fighting a drum would cost much more (Figure 10-15).

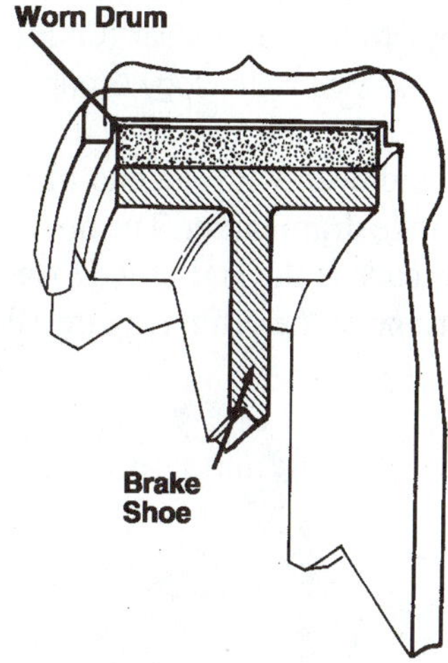

Figure 10-13 If the shoes have been adjusted to fit a badly worn drum, the drum will catch on the lining as you try to remove it.

SAFETY TIP It is recommended that the brake drum be wet down thoroughly or enclosed in the chamber of an OSHA-approved brake cleaning system during the removal steps (Figure 10-16).

To remove a brake drum, you should use the following procedure:

1. Raise and support the car in a secure manner.
2. Remove the wheel as described in Chapter 9.
3. Release the parking brake.
4. Prevent the release of brake dust by thoroughly wetting down the exterior of the drum using a low-pressure brake washer. Rotate the drum and try to get the solution inside of the drum to wet the interior. An aerosol spray or vacuum enclosure can also be used.

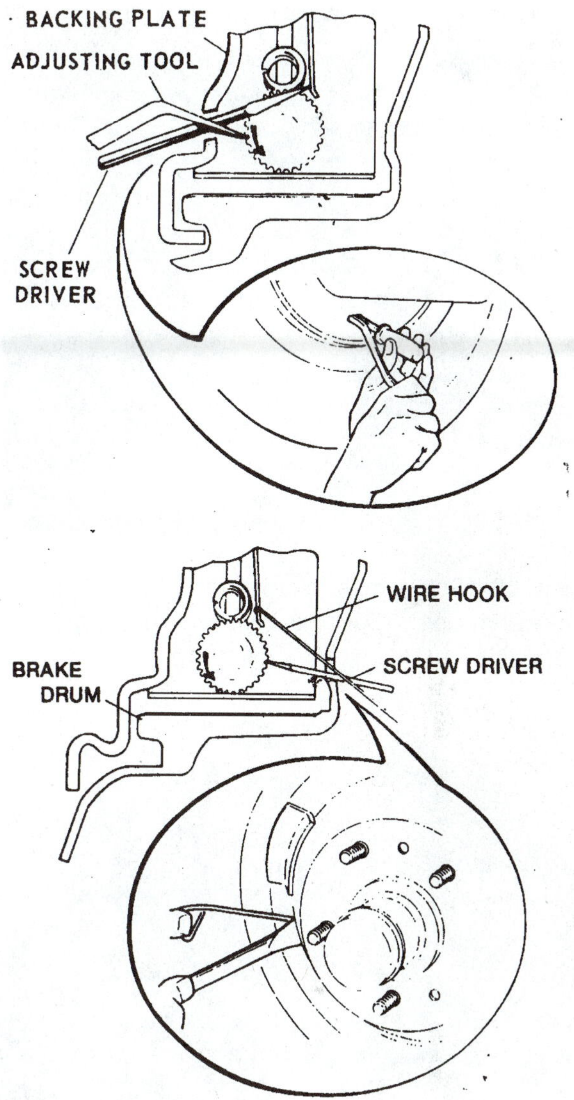

Figure 10-14 To back off a self-adjuster, it is necessary to push or pull the lever away from the starwheel so you can rotate the starwheel. (Courtesy of General Motors Corporation, Service Technology Group)

5. Determine the method of drum attachment and remove the drum.
   a. On nondrive axle drums, remove the dust cap, wheel-bearing cotter pin, and adjusting nut. Then slide the drum and hub along with the wheel bearings off the spindle (Figure 10-17). This process will be described more completely in Chapter 16.
   b. If the drum is a floating drum, remove the speed nuts and slide it off the axle flange and lug bolts.
   c. If the drum is secured with bolts, remove the bolts and slide it off the axle. If the drum is stuck, check the drum for two holes threaded into the face. The bolts used to retain the drum or any

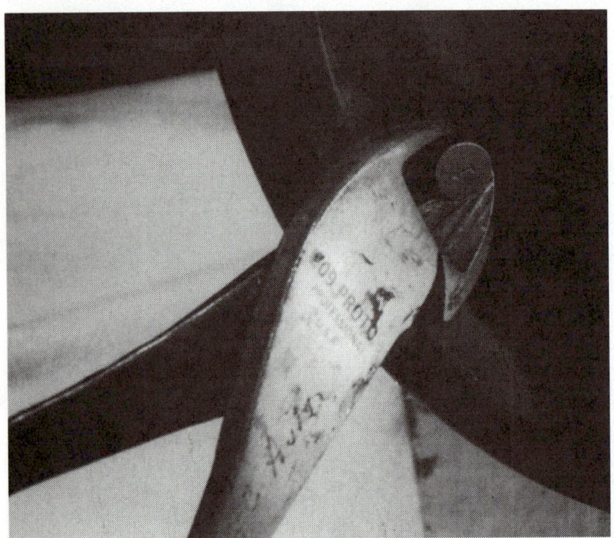

Figure 10-15 If all else fails, cut the heads off the hold-down pins. This lets you pull the drum outward and gain access to the adjuster.

Figure 10-16 An enclosure has been placed over the brake assembly, and it has been connected to a HEPA-filter-equipped vacuum cleaner. Any asbestos fibers released during drum removal will be safely caught in the filter. (Courtesy of NILFISK OF AMERICA, INC.)

bolts of the correct size can be threaded into these holes and used as pullers to remove the drum.

d. If the drum is secured by a single nut on the axle, remove the nut and slide the drum off the axle. If the drum is stuck, replace the nut so it is free of

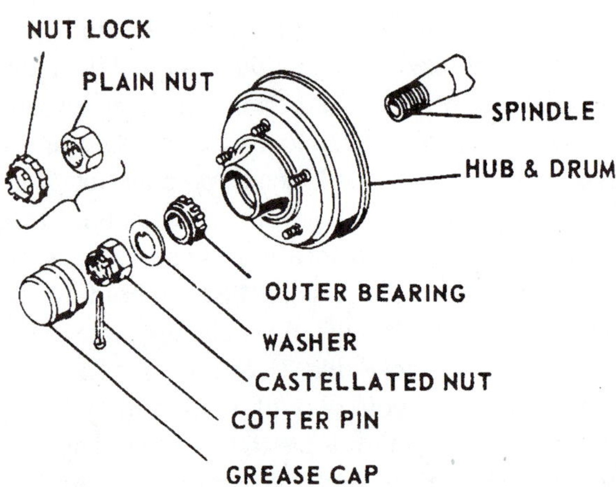

Figure 10-17 The front drums on light trucks and older RWD cars are normally retained by the front wheel bearings. Rear drums on many FWD cars are similar. (Courtesy of General Motors Corporation, Service Technology Group)

the drum and even with the end of the axle; then install a brake drum puller. The nut is used to protect the threads at the end of the axle. Tighten the puller to remove the drum (Figure 10-18).

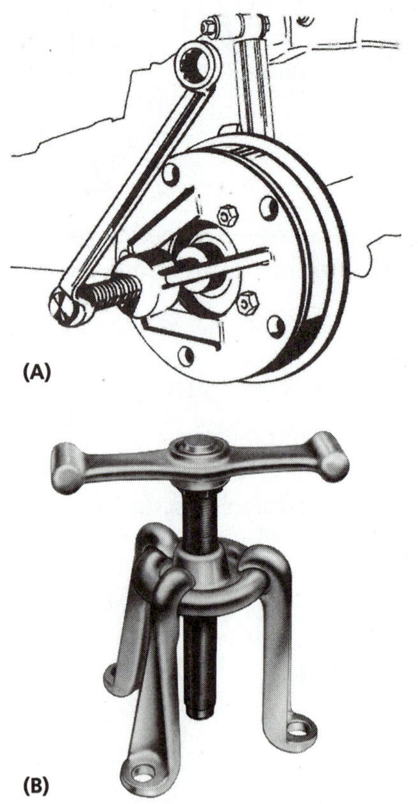

Figure 10-18 If the rear drum is splined to the axle, it is usually necessary to use a puller to remove it. Many technicians prefer to loosen the nut a few turns and leave it on the threads to protect them while pulling the drum. (A is courtesy of Volkswagen of America; B is courtesy of SPX/OTC)

6. Thoroughly vacuum any dust or dirt from the inside of the drum or wash the drum in soapy water or solvent and air-dry it. If the drum is washed in a petroleum-based solvent, the friction surface must be reconditioned or cleaned using denatured alcohol or a commercial brake friction surface cleaner to remove all traces of the solvent residue (Figure 10-19).

Figure 10-19 After removal, it is a good practice to thoroughly clean the drum to remove any traces of asbestos and to make inspection easier.

10.3 Drum Inspection

Most technicians check the brake drum as it is being removed. Brake drum inspection should include: checks for cracks; a scored friction surface; a bell-mouthed, concave, or convex friction surface; hard spots and heat checks; an oversize diameter; and an out-of-round diameter. Faulty drums must be replaced or remachined. Drum **machining** is also called *turning*, *truing*, or *remachining*.

A check for a cracked drum includes a visual inspection and an auditory inspection. Look for cracks across the friction surface and at the lug bolt holes. For the auditory check, lightly support the drum by the inner hole and strike the outer edge lightly with a steel hammer. The drum should make a ringing, bell-like sound. If it makes a dull thud or the sound of plain metal striking metal, it is cracked (Figure 10-20). A cracked drum should be replaced.

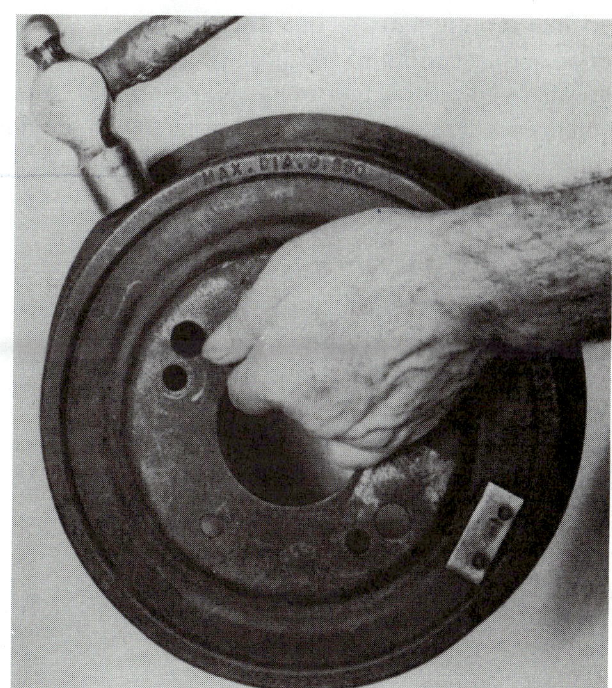

Figure 10-20 Support the drum loosely as shown and strike it lightly with a hammer. You should hear a bell-like sound. A dull thud indicates a cracked drum.

A scored friction surface is caused by grit on the lining or contact with a rivet or shoe rim. A scored drum will have a lining with ridges or grooves (Figure 10-21). Any scores that are deeper than 0.010 inches (0.25 mm) require that the drum be turned. An experienced technician can usually guess the depth of the grooves; the most practical way to measure them is to use a brake lathe to see how deep a cut is needed to remove them.

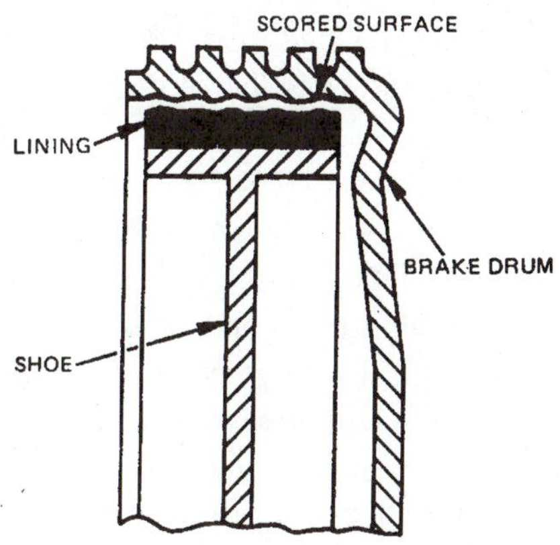

Figure 10-21 A badly scored drum. Note how the lining wears to match the scored surface. Such drums should be machined or replaced. (Courtesy of Wagner Brake)

A **bell-mouthed drum**, one with a concave or convex friction surface, will also be accompanied by shoes that have uneven wear on the lining (Figure 10-22). These problems produce a drum friction surface with varying diameters. If the diameters vary more than 0.010 inch, the drum should be turned.

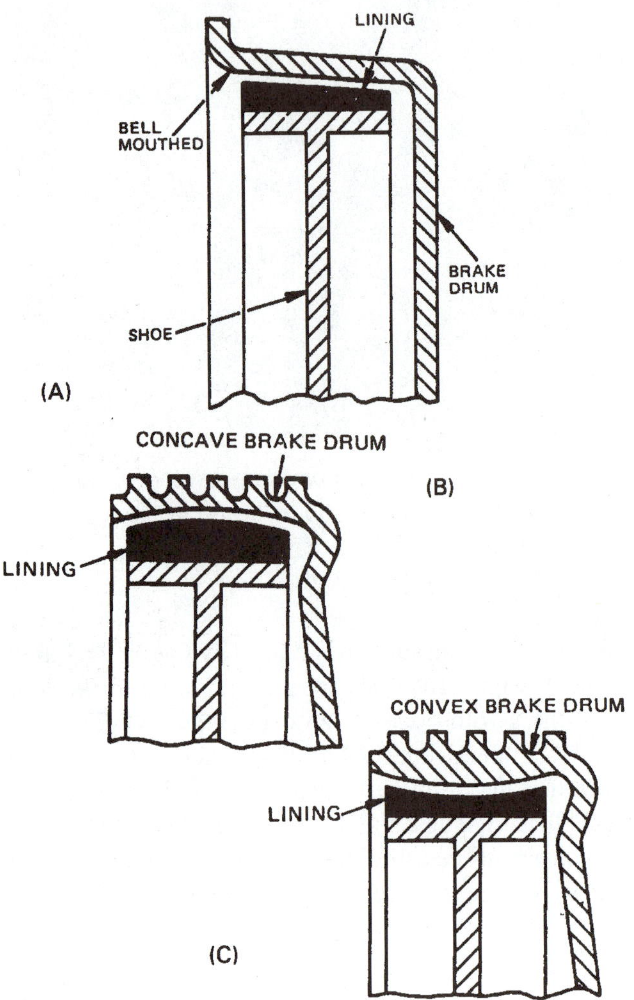

Figure 10-22 A bell-mouthed drum (A), a concave drum (B), and a convex drum (C). Note that in each case the lining wears to match the drum surface. (Courtesy of Wagner Brake)

Hard spots, also called *chill spots* and *heat checks*, are the result of very high-heat conditions. Hard spots are caused by a metallurgical change from cast iron to steel, and an overheated drum will often show a bluish or golden tint. Heat checks are visually evident in the worn drum surface, and excessive checking will require drum replacement (Figure 10-23). As the drum is turned, hard spots will appear as raised, hard areas or islands in the drum's surface. These spots also cause a series of "clicking" noises when they contact the cutter bit as the drum is being turned. The raised portion of a hard spot can often be removed with a carbide cutter bit (although the raised area is often a result of drum deflection) or by grinding. Drum grinding is a slower, more expensive operation that requires a grinder attachment for the lathe (Figure 10-24). Some sources consider grinding a hard spot a temporary repair and recommend replacement of any drum with hard spots.

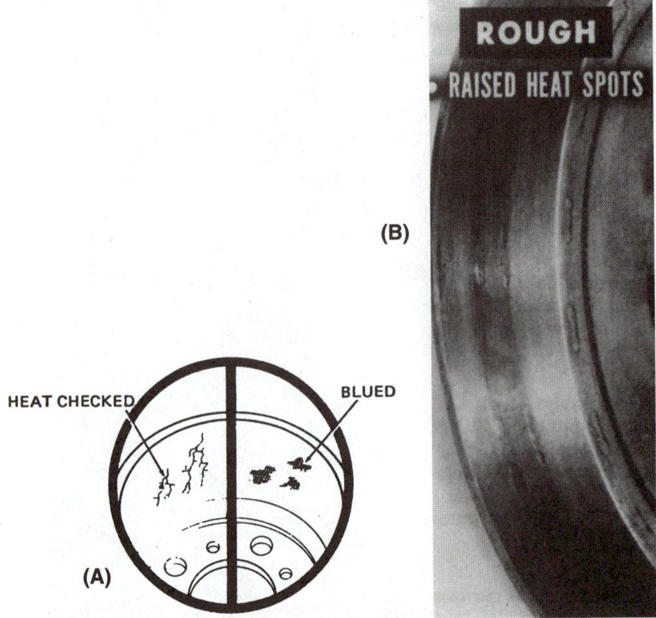

Figure 10-23 Heat checks or blued spots on the friction surface are signs of an overheated drum with hard spots (A). The hard spots will show up as the drum is turned (B) (A is courtesy of Wagner Brake; B is courtesy of General Motors Corporation, Service Technology Group)

Drum diameter is measured using a **brake drum micrometer** which is commonly called a **drum mike**. Measuring the diameter in three or four directions allows the drum to be checked for out-of-roundness (Figure 10-25). A drum with a diameter greater than the maximum allowable should be replaced. A drum with diameter differences greater than 0.005 inches (0.12 mm) should be turned.

Since 1971, all brake drums have been made with the maximum diameter dimension indicated on them (Figure 10-26). This is the wear or discard diameter of the drum, and a drum larger than this must not be used. There are three commonly used drum dimensions. These are the *original diameter*, the *maximum machining* or *rebore diameter* (which is usually 0.060 inches [1.5 mm] larger than the original), and the *wear diameter* (which is about 0.090 inches [2.3 mm] larger than the original). These

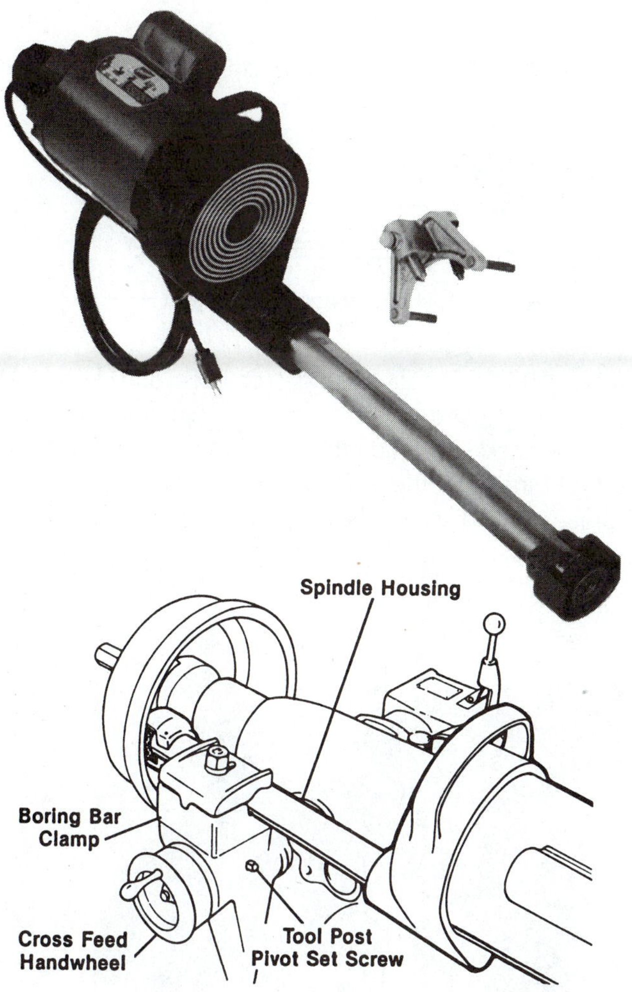

Figure 10-24 A brake drum grinder is secured to the tool post of a brake drum lathe and is used to grind hard spots so they are flush with the drum surface. (Courtesy of Ammco Tools, Inc.)

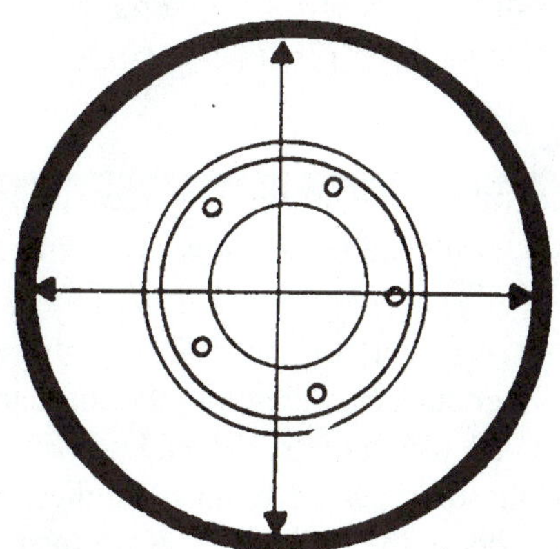

Figure 10-25 Drum diameter should be measured in two or more directions. If the measurements are different, the drum is out of round. (Courtesy of Bendix Brakes, by AlliedSignal)

sizes can vary; always check the manufacturer's specifications. A drum should never be turned to a size larger than the maximum machining diameter. When a drum is marked with a "Max. Dia." (maximum diameter) or "Discard Dia.," this size can be considered the maximum size that the drum can be machined to. On the drums used in domestic cars before 1971, the original diameters were usually even inches or half-inches, such as 9, 9½, 10 inches, and the maximum allowable diameters were 0.060 inches greater than this or 9.060, 9.560, 10.060. However, a few manufacturers used drums with original diameters that were 0.030, 0.060, and 0.090 inches larger than an 11- or 12-inch diameter. The maximum diameters for these drums are listed in service manuals.

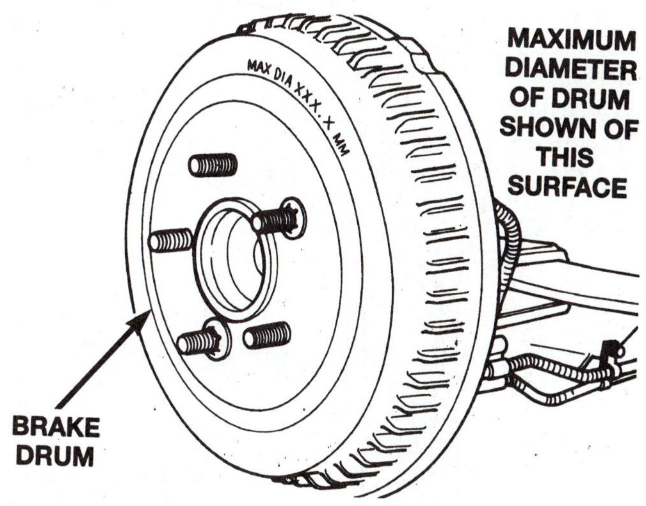

Figure 10-26 All modern brake drums have a maximum allowable dimension indicated on them. (Courtesy of Chrysler Corporation)

Any drum that is too large must be replaced. When a drum is too thin, it loses its ability to absorb heat, and the operating temperatures can rise excessively. A drum that is too thin also loses its structural strength. An excessive amount of deflection can occur during braking, and excessive deflection will cause a spongy brake pedal (Figure 10-27).

To measure drum diameter, you should:

1. Adjust the drum micrometer or "mike" to the original diameter of the drum (Figure 10-28).

NOTE: A mike can be checked against a standard or a large outside micrometer to determine its accuracy. Inaccurate drum mikes should be recalibrated (Figure 10-29).

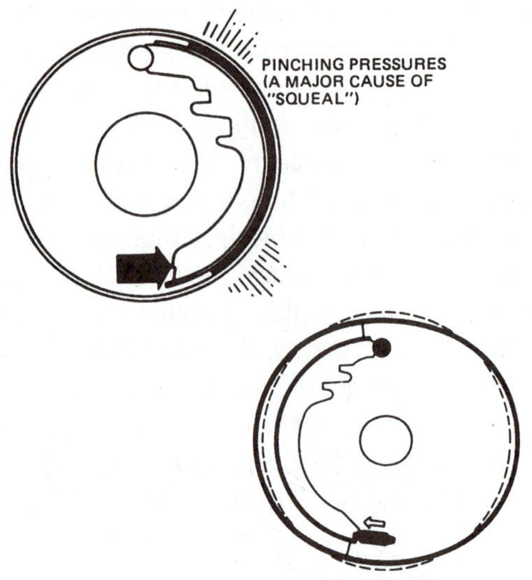

Figure 10-27 A brake drum will deflect under high brake pressures, and thinner drums will deflect more. Deflection can cause a spongy pedal or a pinching of the lining ends, which can cause squeal. (Courtesy of Wagner Brake)

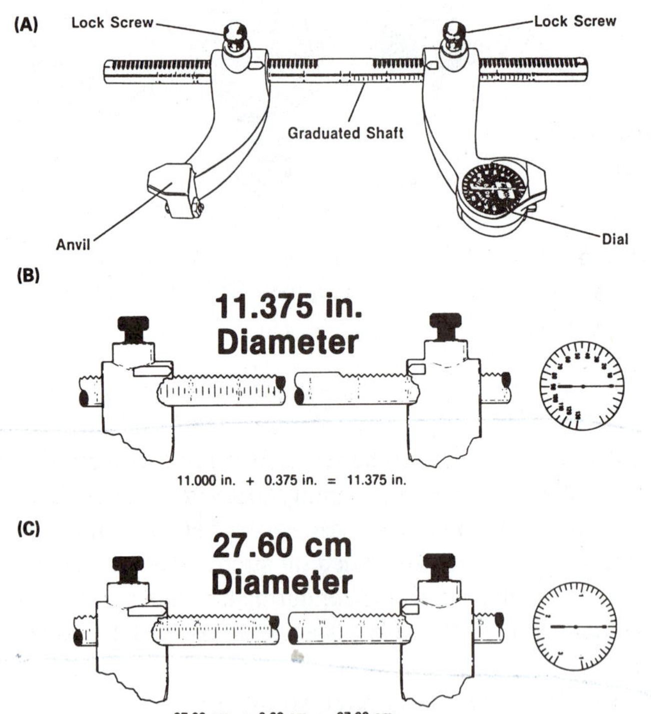

Figure 10-28 A brake drum micrometer or mike must be set to the drum diameter by moving the anvil and dial assemblies to the proper location on the graduated shaft. The mike at [B] has been set to 11.375 inches; the metric mike at [C] has been set to 27.60 cm, or 276 mm. (Courtesy of Ammco Tools, Inc.)

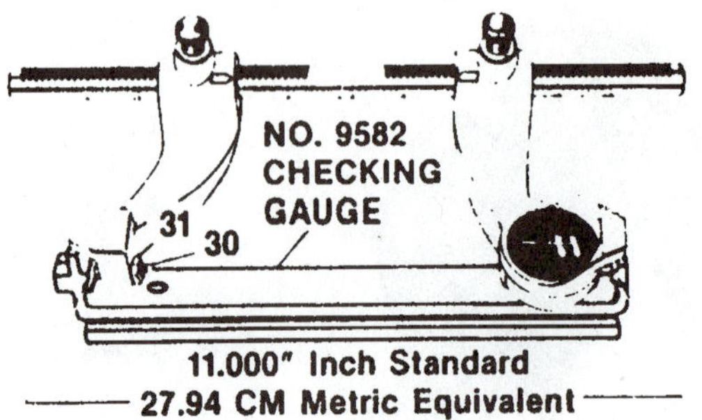

Figure 10-29 Checking gauges or large outside micrometers can be used to determine the accuracy of a drum mike. The checking standard shown has an inside width of exactly 11 inches. If the reading differs, the mike should be calibrated by adjusting the set screw (30). (Courtesy of Ammco Tools, Inc.)

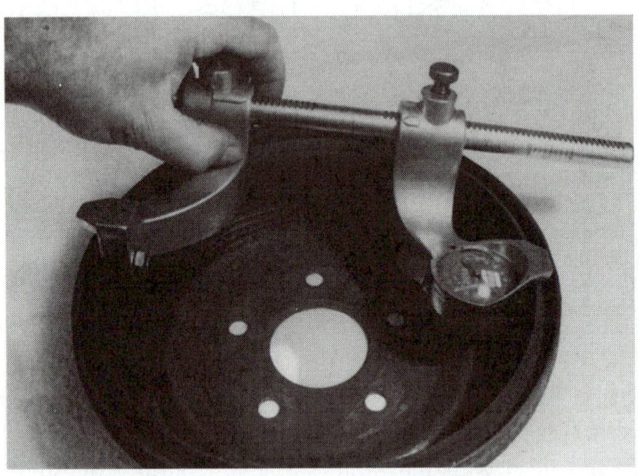

Figure 10-30 To prevent damage, a drum mike is always inserted into the drum with the dial end entering first. It is always removed with the dial end leaving last.

2. Place the mike in the drum so the measuring end, next to the dial, enters first and the rigid end enters last (Figure 10-30).
3. Hold the rigid end of the mike firmly against the drum's inner surface and swing the measuring end in an arc as you watch the scale reading (Figure 10-31). Position the mike at the point of highest reading on the dial. This is the amount of drum oversize. When added to the size setting of the mike, it is the actual drum diameter. For example, if the mike is set to 10 inches and the dial reads 0.020, the drum diameter is 10.020 inches (Figure 10-32).

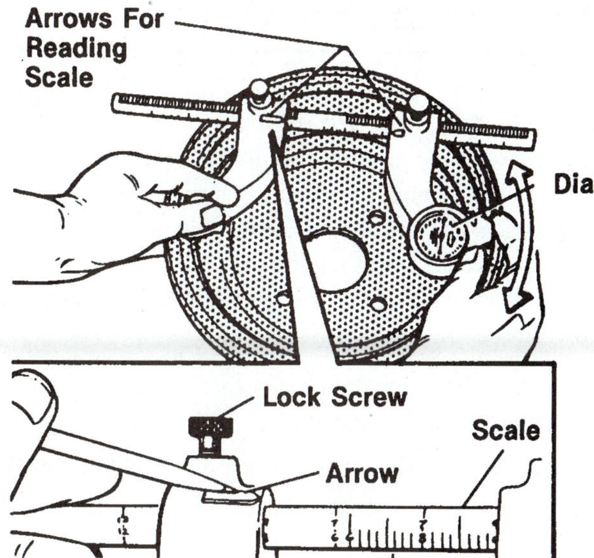

Figure 10-31 After the mike has been placed into the drum, the anvil end is held tightly against the drum while the dial end is moved to the highest reading on the dial. (Courtesy of Ammco Tools, Inc.)

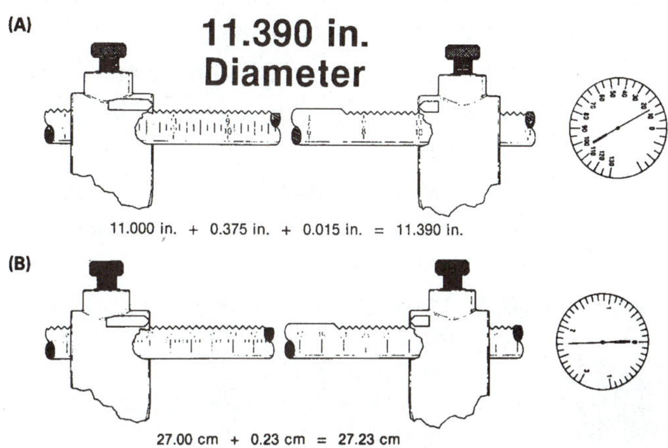

Figure 10-32 A drum mike is read by adding the dial reading to the graduated shaft setting. In (A), the size is 11.390 inches (11.00 + 0.375 [3/8] + 0.015). In (B), the size is 27.23 cm or 272.3 mm (27.00 + 0.23). If the mike was set to the actual drum diameter, the reading on the dial would be the amount of drum oversize. (Courtesy of Ammco Tools, Inc.)

4. Lift the mike out of the drum so the rigid end comes out first. The mike can be damaged if the measuring end leaves first and snaps out.
5. Remeasure the drum in two or three more locations and compare the measurements. A difference in readings indicates drum out-of-roundness.

A brake drum can also be checked for *radial runout* which can cause a pulsating brake pedal and possible brake grab. Remount the drum onto the hub backwards, and position a dial indicator on the drum's inner surface (Figure 10-33). Then rotate the drum while watching the dial indicator. More than 0.006 inches (0.15 mm) of runout can cause a problem. This check can be effected by hub runout. If the amount of runout is excessive, reindex the drum on the hub, and repeat your check; if it is still excessive, the drum should be turned or replaced.

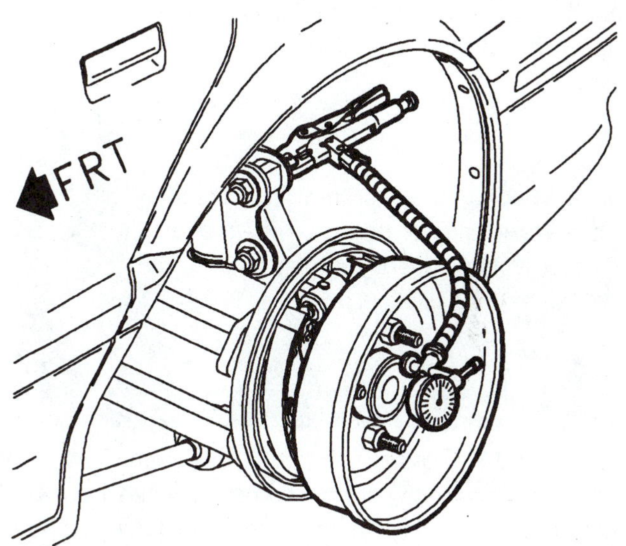

Figure 10-33 Mounting the drum onto the hub backwards allows a dial indicator to be used to measure a drum for runout or out-of-round. More than 0.011 inches (0.28 mm) of runout can cause problems. (© Saturn Corporation, used with permission)

After the drum checks, the technician makes a decision on what to do. Any faulty drums must be replaced, and it is recommended that drums be replaced in pairs. Besides the problem of turning the new drum to match the old one at the other end of the axle, the structural differences between the new and old drums can cause stopping differences during a hard stop. Excessively worn or out-of-round drums should be turned, and drums should always be turned in pairs. The right and left drums on an axle should have the same surface finish and almost the same diameter. A size difference of 0.010 inches (2.5 mm) is allowable.

In the past, most sources recommended always turning a drum, as long as it stays within a safe diameter. The very smooth surface we see on a used drum is not always as flat as it appears. When a new lining is installed, it has to wear or be broken in to exactly match the drums' friction surface. There is only partial lining-to-drum contact until the lining wears in, and a worn, burnished drum surface is usually too smooth to wear in the shoes within a reasonable length of time. Many people want to test their brakes with a really hard stop as soon as the job is done. A hard stop with a lining that has poor contact can severely overheat those portions of the lining that contact the drum. It is recommended that a drum have a friction surface finish of about 40 to 80 microinches. This surface is very flat, having an average scratch depth of 40 to 80 microinches. If we disregard the slight waviness on a worn drum surface, the burnished surface of an average smooth, worn drum is probably smoother than 10 microinches (Figure 10-34).

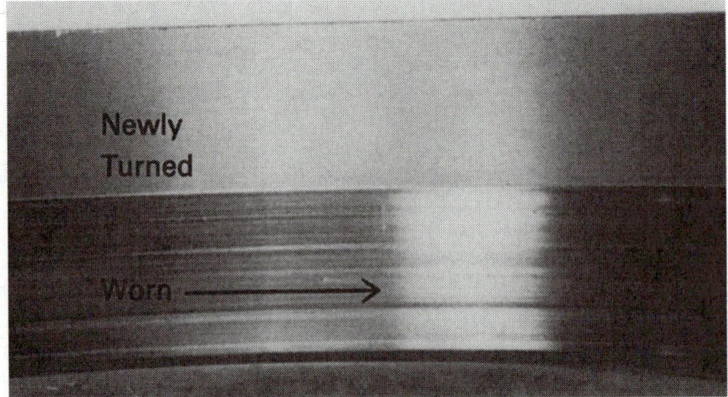

Figure 10-34 The inner portion of this drum shows light scoring, but note how the friction surface has been burnished to an extremely smooth finish. The outer portion of this drum has been turned to a smooth surface with the correct surface finish for breaking in the new lining.

10.4 Drum Machining

A brake drum lathe is used to turn a drum. There are several different popular makes of drum lathes, and their operation is fairly similar (Figure 10-35). The drum is mounted on an **arbor**, also called a *mandrel* or *spindle*, which rotates the drum as a **cutter bit** or tool is moved straight across the friction surface.

When turning a drum, it is a good practice to use the following recommendations:

- Use only sharp cutter bits that are sharpened to a slightly curved or rounded cutting point.
- Always use the correct drum-mounting adapters.
- A vibration, silencer, or chatter band must always be installed snugly around the drum.
- Be aware that the surface finish of the drum will be affected by the sharpness of the cutter bit, the use of the vibration band, the depth of cut, the feed rate, the speed of the arbor, and the mechanical condition of the brake lathe.
- Turn the worst drum first and then turn the other drum to the same finished diameter.
- Front drums should be turned with the hub. Drums that are loose or easily removed should be secured to the hub using lug nuts.
- Drums with wheel bearings should be checked to make sure that the bearing races or cups are in

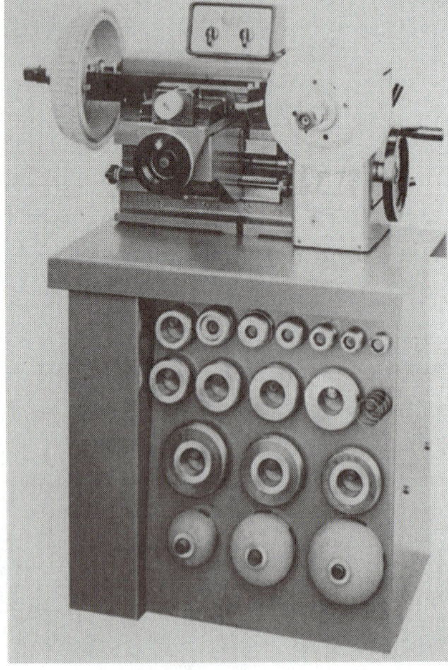

Figure 10-35 Three different drum and disc combination lathes with adapters. (A) is a standard lathe, (B) is a computerized version, and (C) is a dual spindle lathe. (A and B are courtesy of Ammco Tools, Inc; C is courtesy of Hunter Engineering Company)

good condition and tight in the hub. Damaged races should be replaced before turning the drum.

- Never bump the cutter bit or push it against a drum that is not revolving. The hard carbide bits will chip or break rather easily.
- The inner and outer wear ridges should be removed as a beginning step.
- Always remove the smallest amount of metal possible except when matching drum diameters.
- It is a good practice to deburr the outer edge of the surface after machining with 80-grit sandpaper held lightly against the edge of the rotating drum surface in the hand or on a very small disc.
- A turned drum should be cleaned using a commercial braking surface cleaner or denatured alcohol. Wash the turned surface with a clean shop cloth and the liquid until the cloth shows no more contamination.
- After cleaning, do not touch the new friction surface. Pick up the drum by the outer diameter or by the center hole.

When a drum is turned, minute particles of metal and carbon from the cast iron will become embedded in the surface scratches. Unless they are washed out, these particles can become embedded in the new lining surface. Once there, they can cause abnormal braking, squeal, or excessive future drum wear.

NOTE: Most brake shops do not have an occasion to store brake drums, but if they do, brake drums should be stored flat. A drum stored on its edge will tend to go out-of-round. For this reason, new drums should always be checked for roundness.

SAFETY TIP: When machining a drum use the caution that should always be exercised around moving machinery. Be particularly concerned about the revolving drum, the cutter bit, the revolving vibration band, and the revolving lug bolts on some drums.

The following machining procedure is very general. You should always follow the method recommended by the manufacturer of the brake drum lathe you are using. To turn a brake drum, you should:

1. Clean the drum or drum and hub. All oil and grease in the drum and hub should be washed off with a solvent or soap and water and then air-dried.
2. Select the correct mounting adapters. Always make sure there is no grit or dirt between the mounting adapters and the drum, so that the drum will be centered as it is locked securely to the arbor. Make sure the cutter bit is out of the way while the drum is being mounted on the arbor.
   a. Floating rear drums are centered onto the arbor using a properly sized cone that should fit partially through the center hole. This cone is usually spring-loaded to ensure a snug fit into the drum. A pair of bell-shaped adapters fit on each side of the drum's face to prevent drum runout as they secure the drum to the arbor (Figure 10-36).
   b. Drums with hubs and wheel bearings are normally centered and held in place by a pair of cones that enter partially into the wheel-bearing cups. On drums that are attached to the hub with speed nuts, several lug nuts should be installed and torque-tightened to ensure a tight fit. It is usually necessary to also use a few flat washers to ensure complete tightening.

NOTE: Some sources recommend attaching a wheel center to the drum with the lug bolts tightened to the correct torque. This is done to introduce any distortion that will be caused by the lug bolts so that this distortion can be machined out as the drum is turned.

3. Wrap the silencer band around the drum so it is snug, but not overly tight, and interlock the end clasp under the band. The band should be wrapped so that the free end comes over the top of the drum toward you as you are installing it from the side of the lathe with the cutter (Figure 10-37). A loose silencer band will allow the drum to vibrate and "ring" as it is being turned. This will leave a rough drum surface with a herringbone pattern (Figure 10-38). A loose band will also tend to loosen further and fly off. A too-tight silencer band will stress the belt and can distort the drum.
4. Select the correct spindle speed for the drum diameter. Then adjust the speed mechanism, usually a V-belt and various-sized pulleys, to produce the correct speed (Figure 10-39).
5. Adjust the position of the cutter bit tool holder so there is enough cutter or drum travel to machine the drum. Keep the extension of the tool holder or drum as short as practical. This reduces flexing and usually produces a better drum finish.
6. Make sure the automatic feed(s) is disengaged and then turn on the lathe. Adjust the feed handwheel to position the cutter and drum lengthwise so the cutter is midway across the friction surface. Carefully adjust the cross-feed-diameter-handwheel so the cutter bit

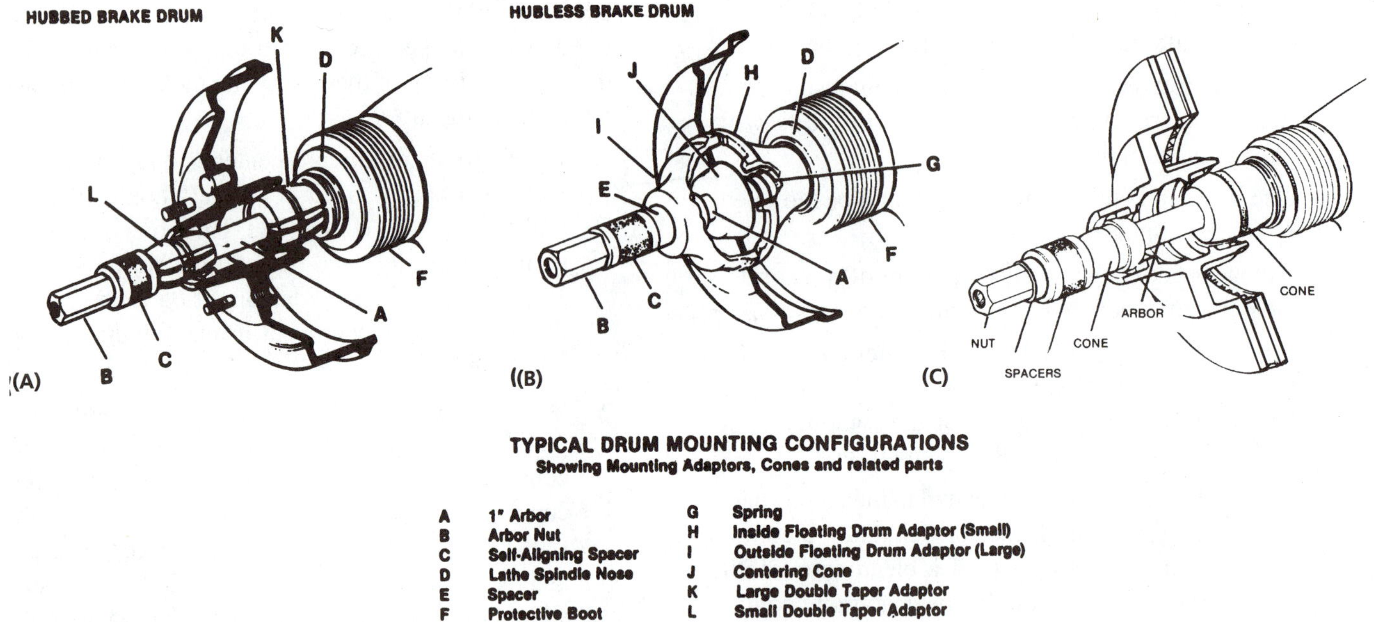

Figure 10-36 Brake drums with hubs are normally mounted on the lathe arbor as shown in (A). Floating rear drum mounting is shown in (B), and rear drum with a small center hole is mounted as shown in (C). (Courtesy of Ammco Tools, Inc.)

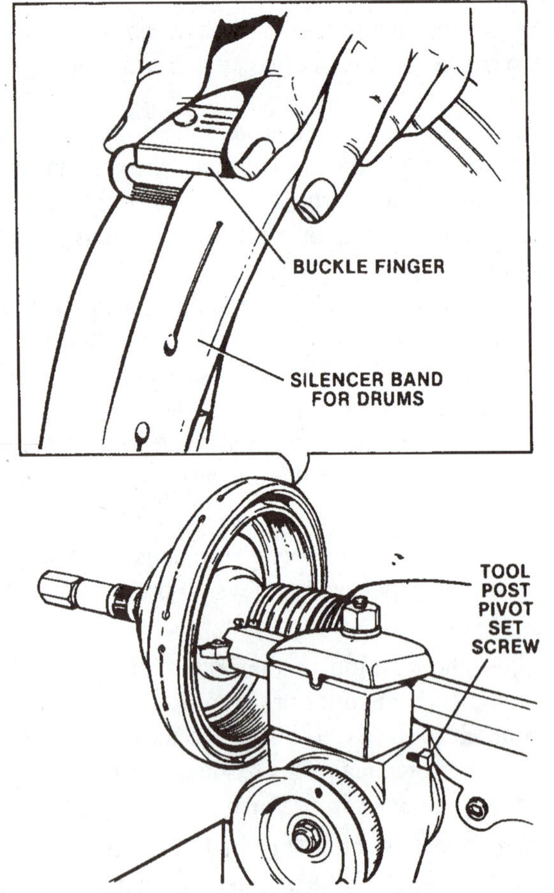

Figure 10-37 A silencer band should be wrapped snugly around the drum with the buckle coming toward you over the top. The buckle finger is then slid under the band to secure it in place. (Courtesy of Ammco Tools, Inc.)

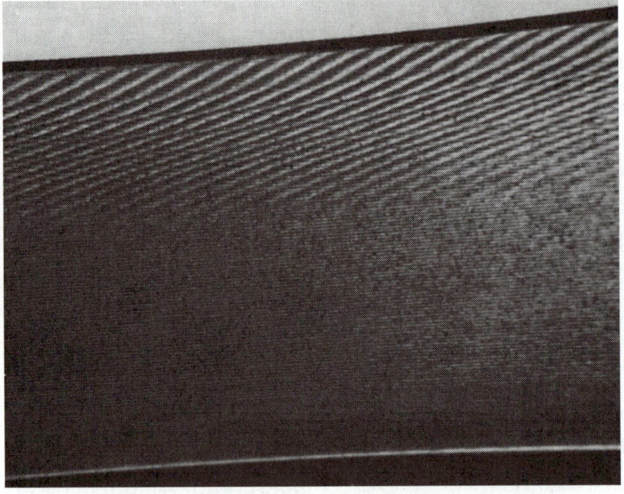

Figure 10-38 Note the odd, angled pattern on the outer portion of this drum. It resulted from the drum surface vibrating as it was being turned. This is not an acceptable surface finish.

makes light contact with the drum. At this point, reset the cross-feed dial to zero or note the diameter indicated on the dial (Figure 10-40).

7. Turn the cross-feed dial inward about 0.010 or 0.020 inches and turn off the machine so you can check the scratch mark (Figure 10-41). If the scratch mark made by the cutter bit extends all the way around the drum, the drum is round and mounted properly. If there is a short scratch partway around the drum, the

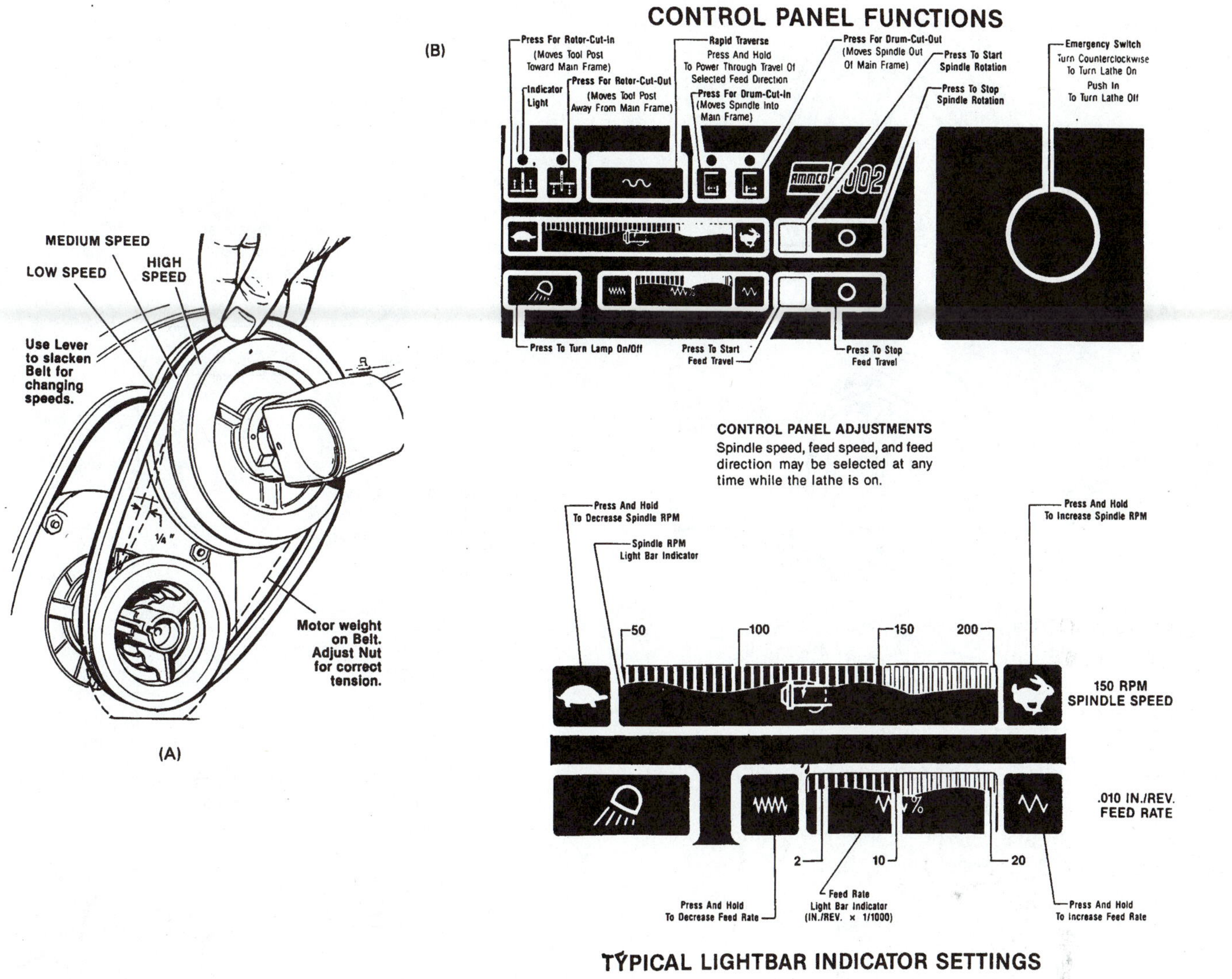

Figure 10-39 The arbor speed is adjusted by changing the drive belt position on some lathes (A) or by pressing the correct button on a computerized lathe (B). (Courtesy of Ammco Tools, Inc.)

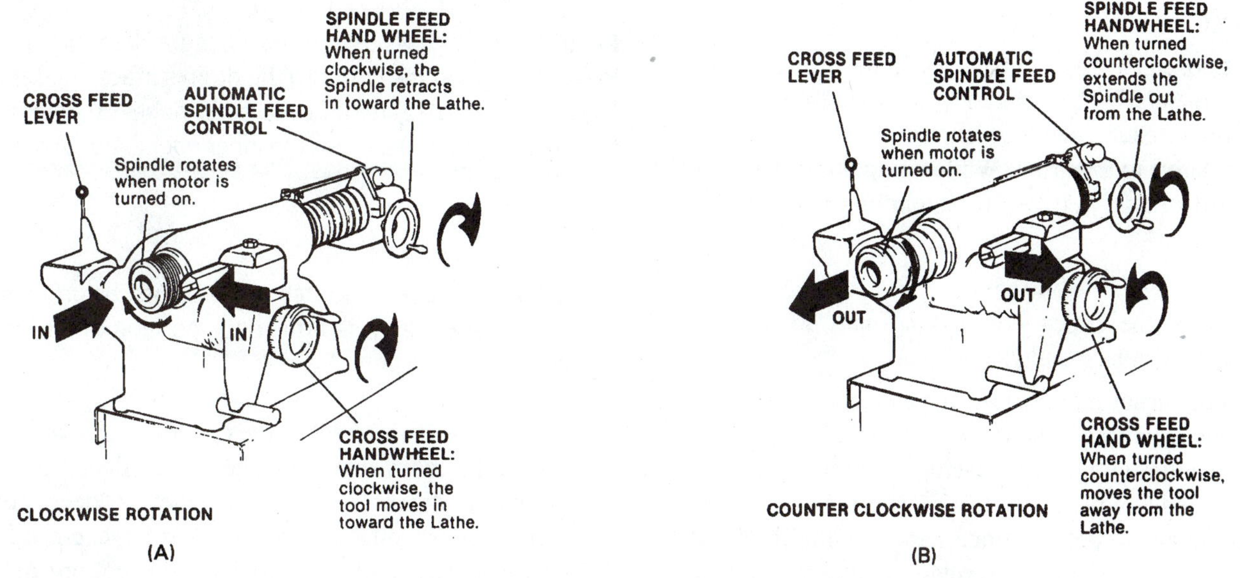

Figure 10-40 On most lathes, the spindle-feed handwheel controls the position of the cutter in and out of the drum, while the cross-feed handwheel controls the depth of cut. (Courtesy of Ammco Tools, Inc.)

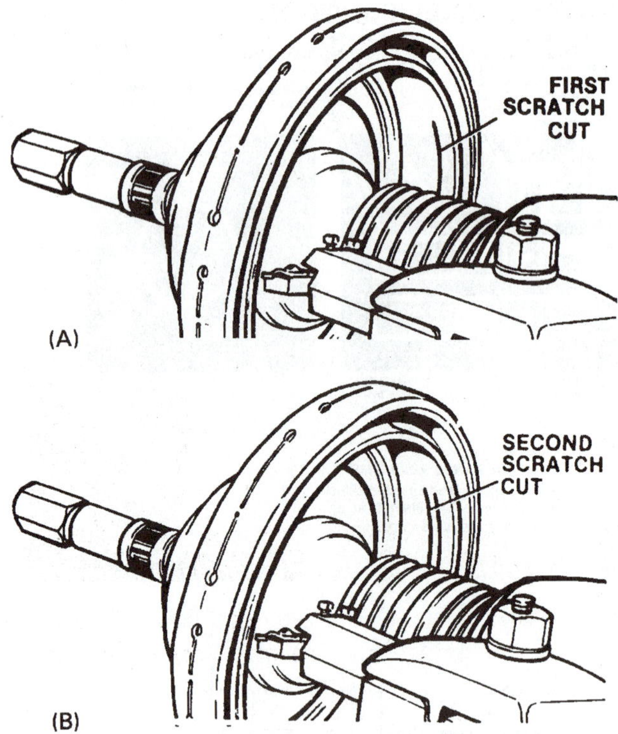

Figure 10-41 After the drum is mounted, make a scratch cut to see how true the mounting is. A short cut indicates an off-center drum (A). In this case, loosen the drum, rotate it half a turn, and make another scratch cut alongside the first one (B). If the cuts are side by side, the drum has worn off-center. (Courtesy of Ammco Tools, Inc.)

drum is out-of-round or mounted off-center. An off-center drum will force you to remove more metal than necessary to clean it up. If the drum is off-center, you should

a. Loosen the arbor nut and rotate the drum one-half turn relative to the center mounting cone. Also, rotate each of the mounting bells one-half turn relative to the drum. While doing this, rub off any dirt or rust between the mounting bells and the drum.

b. Retighten the arbor nut.

c. Move the cutter sideways a small amount and repeat steps 6 and 7.

8. With the drum mounted as true as practical and with the lathe running, adjust the cross-feed back to zero so the cutter is lightly contacting the drum. Move the cross-feed handwheel so that the cutter bit is brought to the outer ridge. Continue hand-feeding the handwheel slowly so that this ridge is cut away (Figure 10-42).

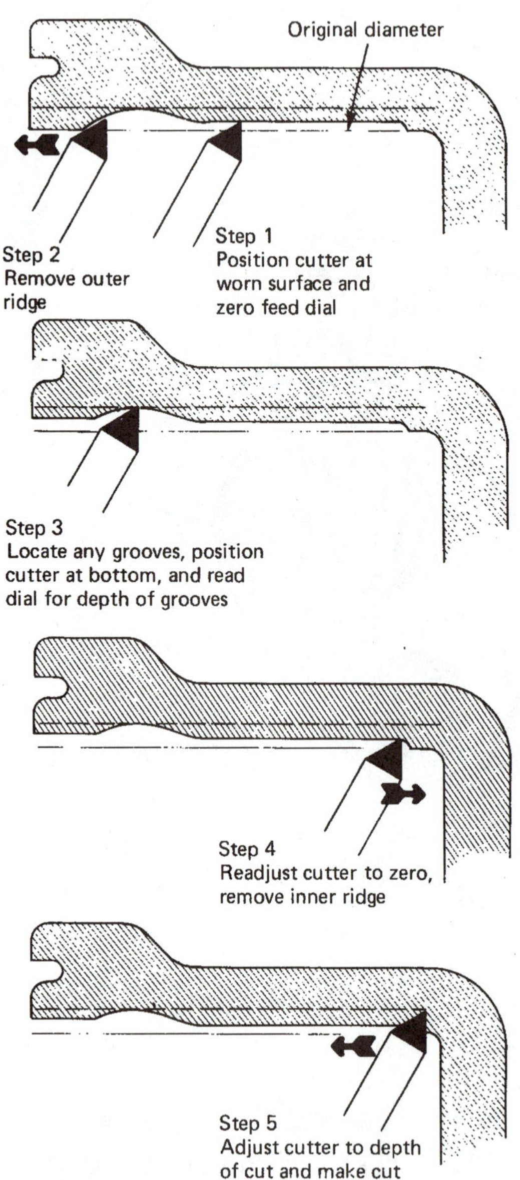

Figure 10-42 The steps in machining a drum are as follows: (1) position the cutter to the drum surface and set the feed dial to zero; (2) remove the outer ridge; (3) determine the depth of cut; (4) remove the inner ridge; and (5) make a cut across the friction surface.

NOTE: Watch the scratch marks across the drum from the cutter to easily determine the position of the cutter bit.

9. Reverse the handwheel so the cutter bit moves inward across the drum surface to the inner ridge. Note any grooves in the friction surface. If they are deeper than a few thousandths of an inch, turn the cross-feed handwheel so the cutter touches the bottom of the deepest groove. Note the depth of the groove indi-

cated on the dial (Figure 10-43). Return to a zero setting, move the cross-feed handwheel so the cutter bit is next to the inner ridge, and continue hand-feeding the feed handwheel carefully to cut away the inner ridge (Figure 10-44).

Figure 10-43 The depth of a groove in a drum can be determined by adjusting the cutter bit to the bottom of the groove and noting the reading on the cross-feed dial. Note that the outer ridge has been removed.

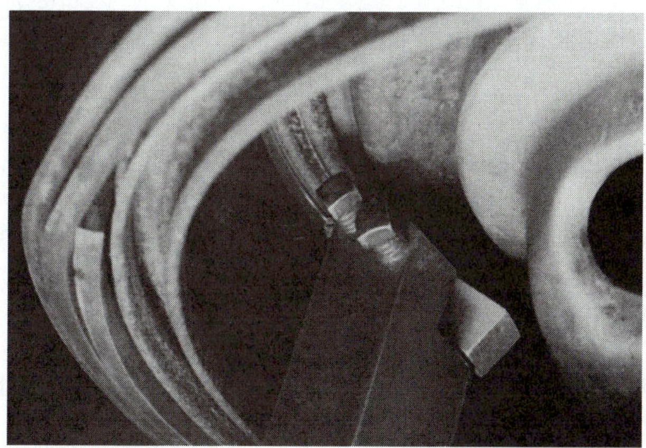

Figure 10-44 This cutter has been positioned next to the inner wear ridge; it will be hand-fed inward to remove this ridge before making the first cut across the drum surface.

10. Decide on the depth of cut that will be necessary to clean up the drum. The last finish cut should ideally be about 0.004 inches (0.1 mm) deep and at a feed rate of 0.002 to 0.006 inches per revolution. The slower rate will produce a smoother finish but take three times as long as a rough cut. Too fast or too deep a cut will leave a rough finish (Figure 10-45). A good, smooth, properly mounted drum can usually be cleaned up with one finish cut. A grooved, rough, out-of-round, or off-center drum will require one or more cuts. You should have a good idea as to the size of the cut from the trial cut you made in step 9. The maximum depth of a rough cut is determined by the strength or capacity of the lathe. A rate of 0.020 inches per revolution is usually used on a fast, roughing cut.

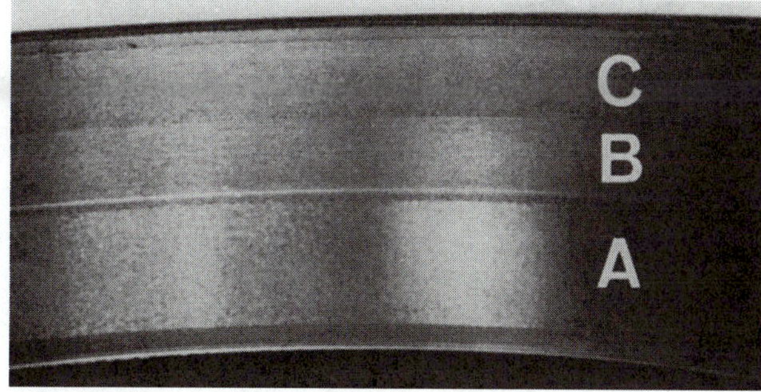

Figure 10-45 A drum with a ground finish (A), a finish cut (B), and a roughing cut (C) allows a comparison of the surface finishes.

11. With the machine running, adjust the feed rate to the desired rate of feed and tighten the feed rate lock to secure that setting. Adjust the cross-feed handwheel to the desired depth of cut and tighten the cross-feed lock to secure that setting. Engage the automatic feed and, as the cut begins, observe the machine, cutter, and drum for any unusual noises or movement. If something appears wrong, stop the machine and recheck the settings.
12. After the cut is finished and the cutter bit moves past the outer edge of the drum, disengage the automatic feed. If you have made the final cut and the friction surface of the drum is cleaned up, shut off the lathe. If you have only made a rough pass or part of the surface was skipped, repeat steps 10 through 12.
13. Remove the silencer band from the drum and remove the drum from the arbor.
14. Remeasure the drum to be sure that it is not oversized. One new brake lathe has a digital readout that shows the actual diameter as you are machining.
15. Mount the second drum from the same axle on the arbor and turn it. Be sure to use the same feed-rate and cross-feed dial setting on the final finish pass.
16. Wash both newly machined drum surfaces with denatured alcohol (Figure 10-46).

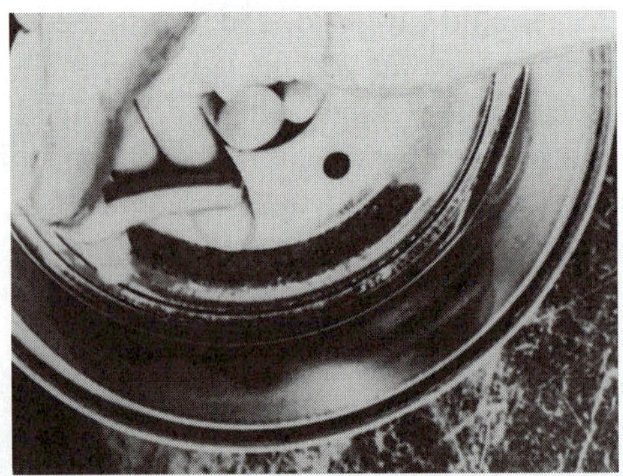

Figure 10-46 After the drum has been turned, it should be washed to remove small metal chips and carbon particles. Use denatured alcohol, soapy water, or brake cleaner, and air-dry it.

Figure 10-47 This technician is using an air nozzle to knock loose all dirt and dust so these particles can be drawn into the HEPA filter on the vacuum attachment. (Courtesy of NILFISK OF AMERICA, INC.)

10.5 Shoe Assembly—Predisassembly Cleanup

The brake assemblies should be given a preliminary cleanup before disassembly to remove the brake and road dust. This cleanup prevents this dust from being stirred up and allows a better view of the parts of their relationship.

A straight air blast *must not* be used because it will drive the dust into the air where you or others might breathe it. Two methods are approved for this cleaning operation: vacuum and wet cleaning.

Several companies market a vacuum cleaner attachment which mounts onto the backing plate over the brake shoe assembly. These attachments include an air-hose port and, in some cases, a viewing window. The attachment is connected to the backing plate and to a HEPA filter-equipped vacuum cleaner. Then the vacuum cleaner is started, and the air hose is operated. The air blast knocks the residue loose from the various brake parts, and this dust is pulled into the vacuum cleaner where it is collected for disposal. A HEPA filter–equipped vacuum cleaner must be used (Figure 10-47). Small asbestos fibers can pass through the filter of a standard shop vacuum cleaner and be blown into the air.

The wet method uses an air-powered suction nozzle, or a hand spray bottle, to spray soapy water or an aerosol can of brake cleaner. A pan is placed under the brake assembly, and the soapy water or brake cleaner is sprayed over the parts (Figure 10-48). The first spraying action should be soft enough to wet the dust particles and not blow them into the air. Dust and asbestos particles are now trapped in the wet solution and washed down into the pan. Asbestos has a good tendency to wet, and to mat when wet, and can be collected by this method. The pan is often left under the backing plate so that, during removal, the parts can be dropped into it for soaking and further cleaning. The used cleaning solution and any cleaning rags which are contaminated with asbestos dust must be disposed of in an approved manner.

SAFETY TIP: Many commercial brake cleaners contain Perchlorethylene (or Perc) which has been identified as a possible human carcinogen. It is thought to cause an increase in cancer rate. This chlorinated solvent is being studied by various air-quality groups to have its usage limited or reduced.

10.6 Brake Spring Removal and Replacement

Most brake shoes are held in place on the backing plate and anchors by the return springs and the hold-down springs. Removal of these two sets of springs allows removal of the shoes. Before the shoes come loose from the backing plate, it is very important to note the relationship of the parts. Some brake assemblies almost fall apart

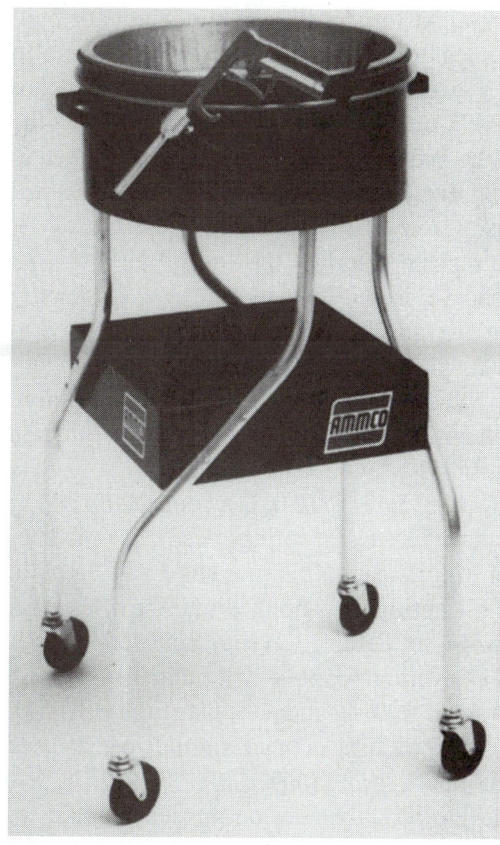

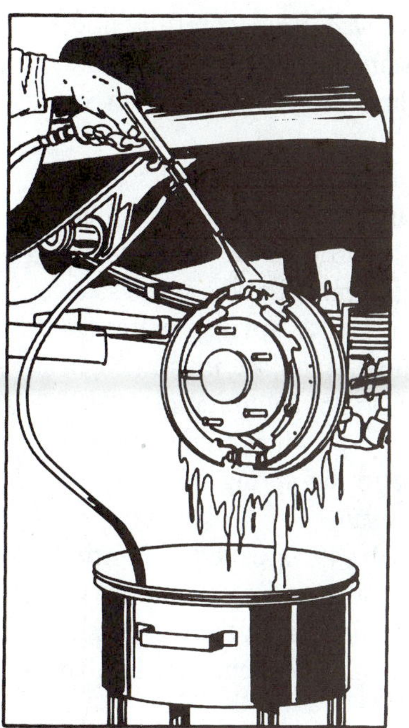

(A)

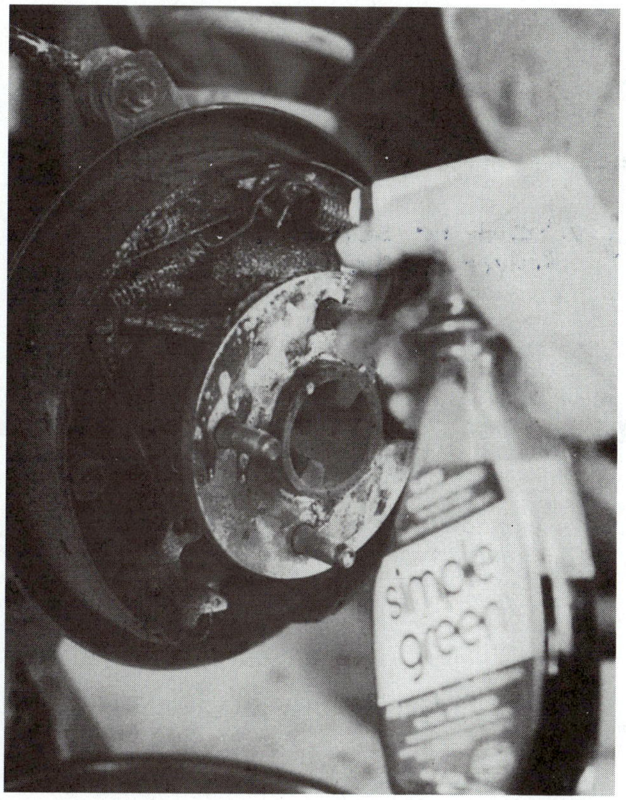

(B)

(C)

Figure 10-48 A brake assembly can be cleaned with a soapy-water mist (A). The asbestos particles along with other dirt are trapped in the solution and washed into the pan. Another method is to use a household cleaner and spray bottle (B) or brake cleaner (C). (A is courtesy of Ammco Tools, Inc.; C is courtesy of Locktite Corp.)

after the springs have been removed. Also, carefully note the way that the springs are connected or hooked into the shoe and if they have a right and left, top and bottom, or in and out.

Several styles of return springs are used, and several styles of brake spring tools are available to aid in their removal and installation. Each particular style of spring requires a tool of a particular design. All of these tools are called **brake spring tools** or **brake spring remover-installers**. These tools and their usage are illustrated in Figures 10-49 through 10-53. Do not try to remove these springs with ordinary pliers, except when recommended by the manufacturer. Damage to the spring or injury to yourself will probably be the result.

The tool shown in Figure 10-49 is called a **Bendix brake spring tool**, and it works on most duo-servo brakes. It is available in three basic shapes, the most common being an offset form having two bends. It is also available in a straight form with a plastic handle or built into one of the legs of a pair of brake spring pliers. To remove a return spring, the tool is placed over the anchor pin with the tool's projection inside the spring eye—the curved tang at the end of the spring. The tool is rotated to move the spring away from the anchor and then leaned toward the spring to lever the spring over the end of the anchor pin. To replace a spring, the end of the tool with the slightly concave area is first placed over the anchor with the spring eye around it. Then the tool is levered to stretch the spring far enough to slide it over the anchor pin. Do not overstretch the spring, as this will weaken it. A good technician will push the spring over onto the anchor as soon as it is stretched far enough.

A tool, shown in Figure 10-50, has been developed to remove and replace the return springs used on some late-model General Motors cars. This tool has a wire hook to catch and pull the spring. It is hooked into the spring eye and placed over a rivet holding the shoe guide and anchor block in place. Levering the tool will stretch the spring enough to remove it from its anchor post (Figure 10-51). The spring is replaced in a manner similar to that used with the Bendix brake spring tool.

Brake spring pliers are used on nonservo brakes of both past and present designs. This tool, shown in Figure 10-52, has two pointed jaws. One of them is pointed inward to hook into a rivet hold or into the lining, and the other is hook-shaped for catching the spring eye. The tool is placed over the lining, with the hooked jaw engaging the end of the spring, and squeezed to stretch the spring enough for removal or replacement. Many technicians place a thin piece of plywood between the plier jaw and the brake lining to keep from gouging the lining.

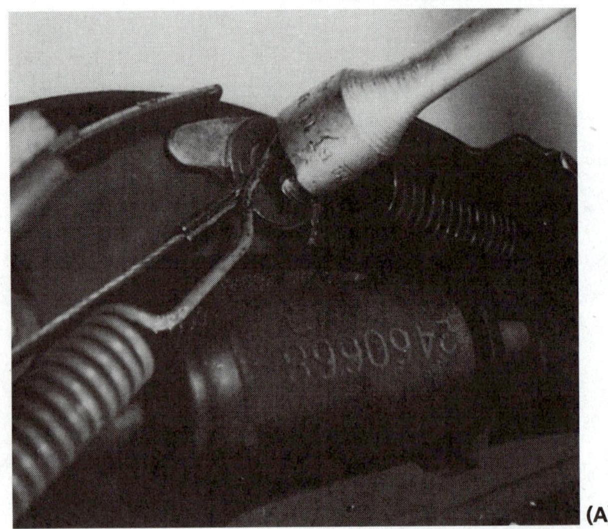

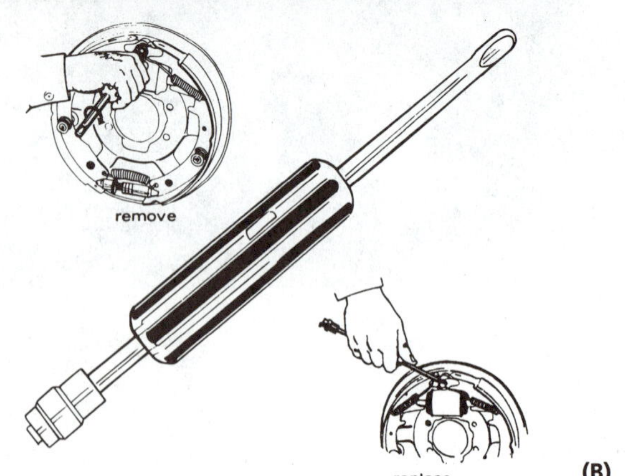

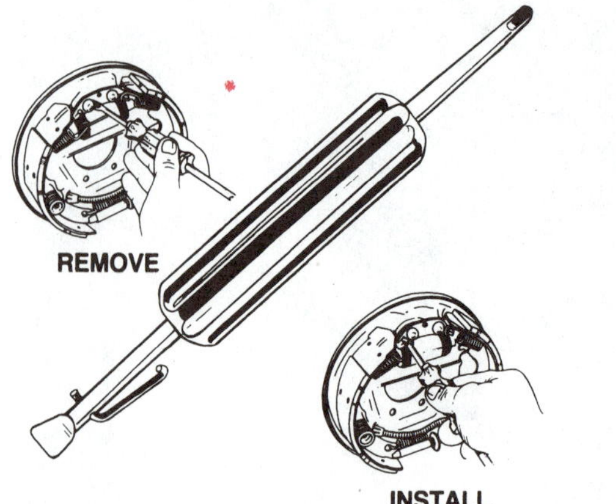

Figure 10-49 A brake spring tool is placed over the anchor pin and rotated to lift the spring over the anchor and remove it (A). Two different styles are shown in (B) and (C). (A is courtesy of Ford Motor Company; B and C are courtesy of Lisle Corporation)

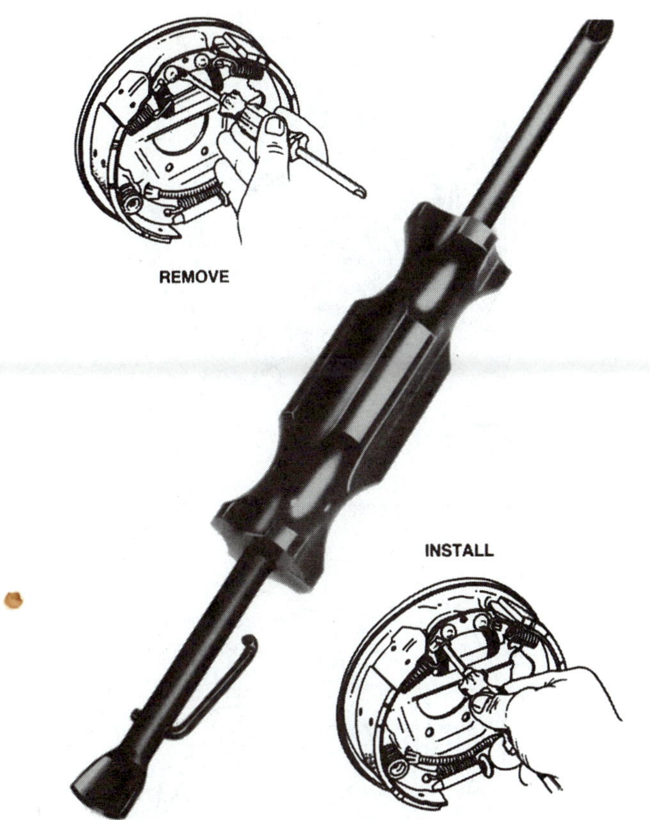

Figure 10-50 This special tool is used to remove or replace the return spring on some late-model General Motors brake assemblies. (Courtesy of SPX Kent-Moore, Part # J-29840)

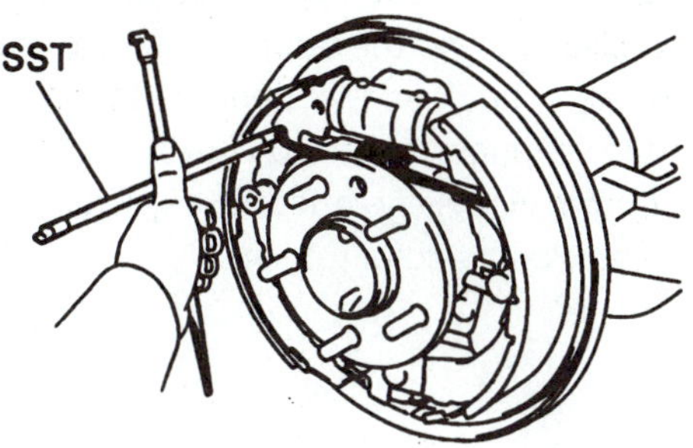

Figure 10-51 This special service tool is used to remove the return spring on this nonservo brake.

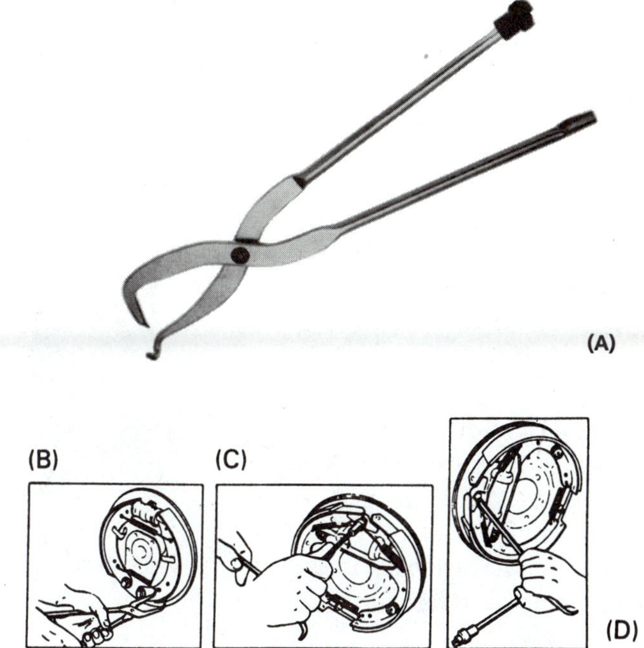

Figure 10-52 Brake spring pliers (A) are used in different ways to remove or replace return springs. (Courtesy of KD Tools)

Another brake spring tool from the past is shown in Figure 10-53 and sometimes called a **Lockheed brake spring tool**. This somewhat L-shaped tool has one pointed end and a washerlike eccentric attachment at the other end. To remove a spring, the tool end is inserted into the hole in the brake shoe web with the end of the spring in the notch in the eccentric. The tool is rotated so the spring is away from the shoe web and then leaned to lever the spring end out of the shoe. To replace the spring, the pointed end is passed through the spring eye and into the hole in the shoe web. Then the spring end is levered into the hole.

In a very few cases, the use of pliers is recommended by a manufacturer to grip the shank of the spring and stretch it to allow removal and replacement. Normally, the use of pliers is avoided because of their tendency to nick or scratch the spring. Pliers usually cannot grip stronger return springs tight enough to prevent slippage without nicking or bending the spring (Figure 10-54).

The return springs on some nonservo designs are removed by lifting the shoe over the shoe guide at the anchor or sliding it off the end of the anchor. This allows the shoe to be moved over so the spring can be unhooked. The springs are replaced by placing one shoe in position, connecting the springs, and pulling the second shoe far enough to allow it to be placed back on the anchor. With this design, the hold-down springs are removed first and installed last (Figure 10-55).

There are essentially five different styles of hold-down springs and a few different special tools to aid in their removal and installation (Figure 10-56).

The most common style of hold-down spring has a short coil, one or two concave washers with a slot in the center, and a pin, sometimes called a nail, which has a

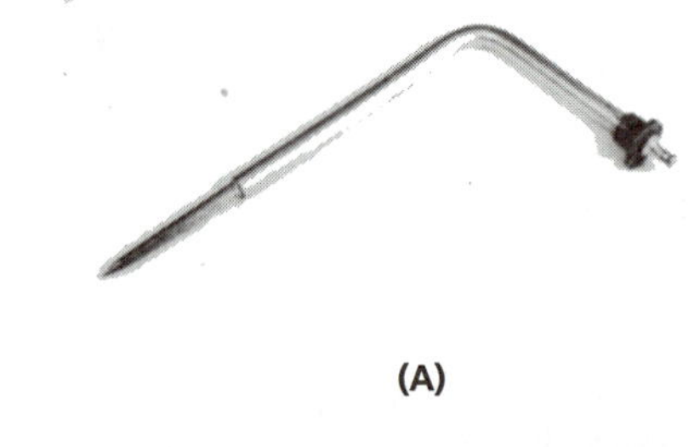
(A)

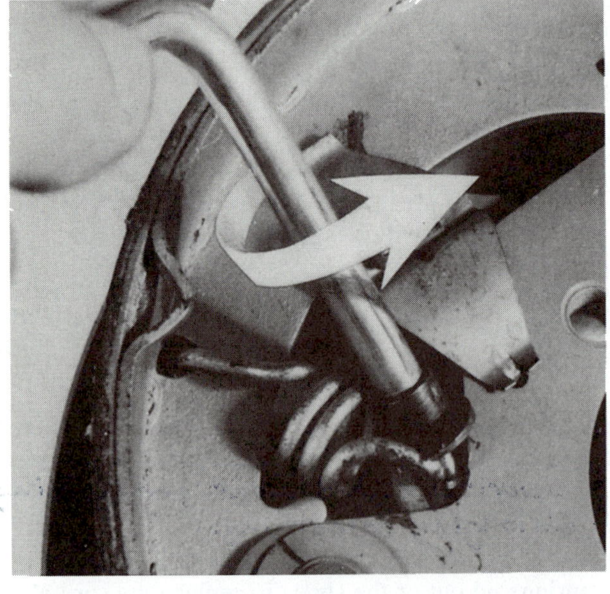
(B)

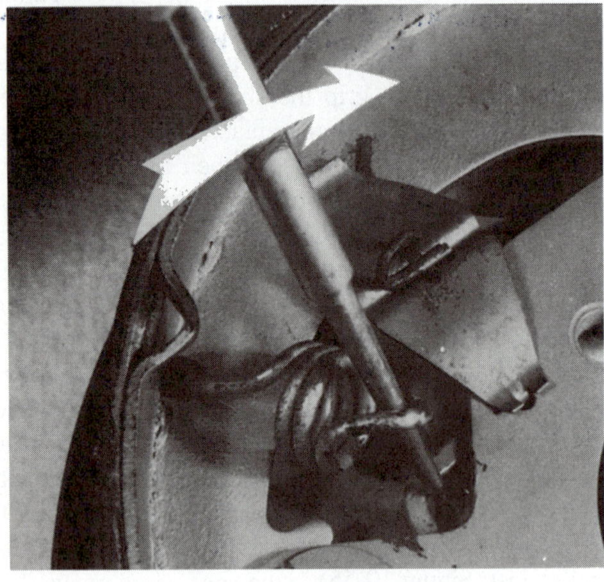
(C)

Figure 10-53 This Lockheed brake spring tool is placed so the notch engages the spring (B) and is rotated to remove the spring. The pointed end is used to replace the spring (C).

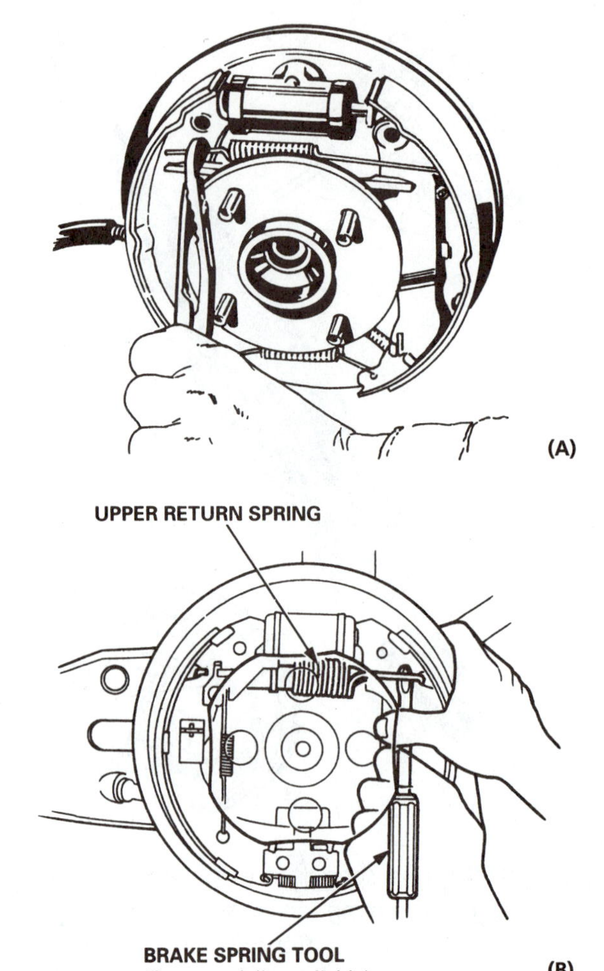

Figure 10-54 On some brakes, the manufacturer recommends using pliers to grip the return spring (A) or sliding a brake spring tool to hook on the spring (B) so the spring can be stretched far enough for removal. (A is courtesy of General Motors Corporation, Service Technology Group; B is courtesy of American Honda Motor Co., Inc.)

flattened, somewhat arrow-shaped end (Figure 10-57). For removal, the tubular end of the brake shoe retaining spring tool is first placed over the washer. Then the washer is pushed inward, turned to align the slot with the flats of the pin, and released. Installation is the reverse of removal, but it is important to make sure that the flats on the pin align with the depressions in the washer. When working on many General Motors cars with self-adjusting duo-servo brakes, note that the bottom spring cup has an extension that serves as a pivot for the self-adjuster lever.

Some hold-downs use a flat, U-shaped spring that engages the flattened end of the hold-down pin. It can be gripped with a pair of common pliers and pushed inward to allow release or installation (Figure 10-58).

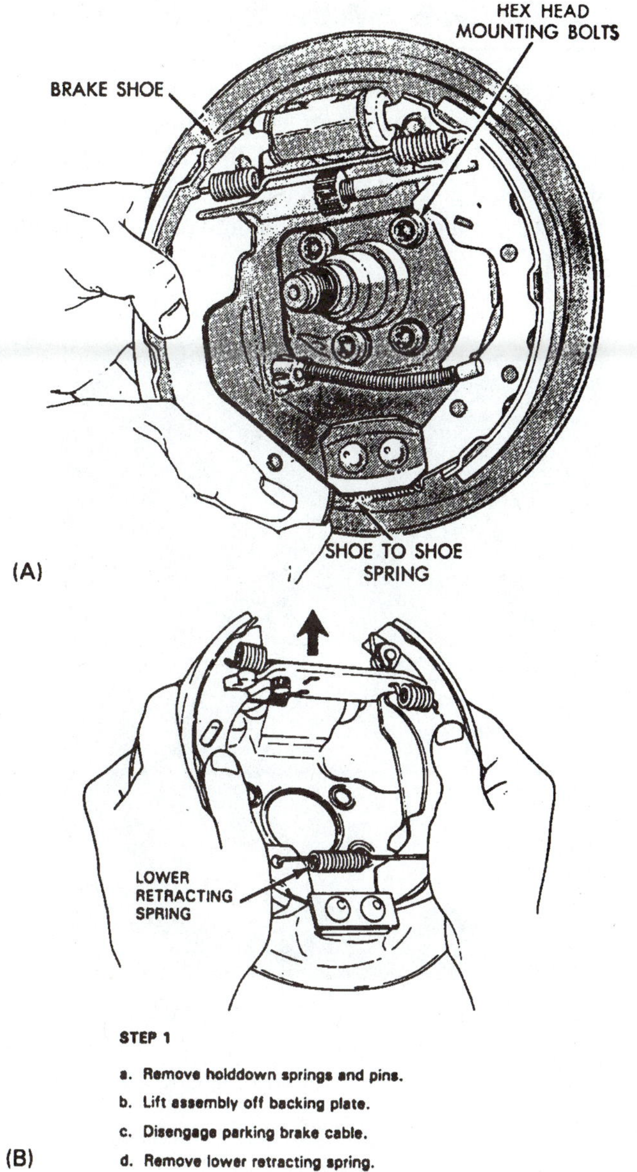

Figure 10-55 After removing the hold-down springs on some brakes, the brake shoes (with return springs still attached) are pulled off the anchor block. (A is courtesy of Chrysler Corporation; B is courtesy of Ford Motor Company)

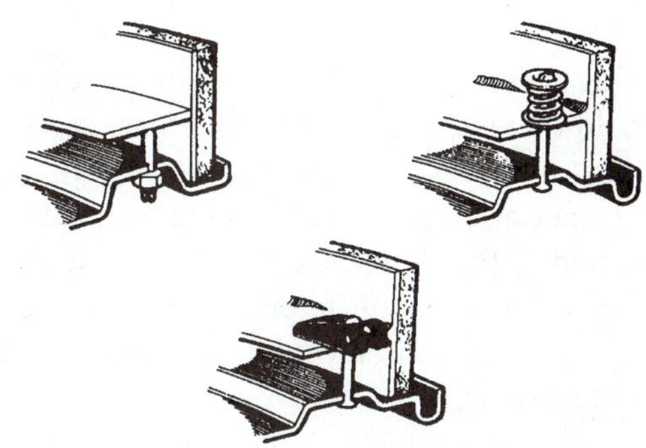

Figure 10-56 Several styles of hold-down springs are used. (Courtesy of LucasVarity Automotive)

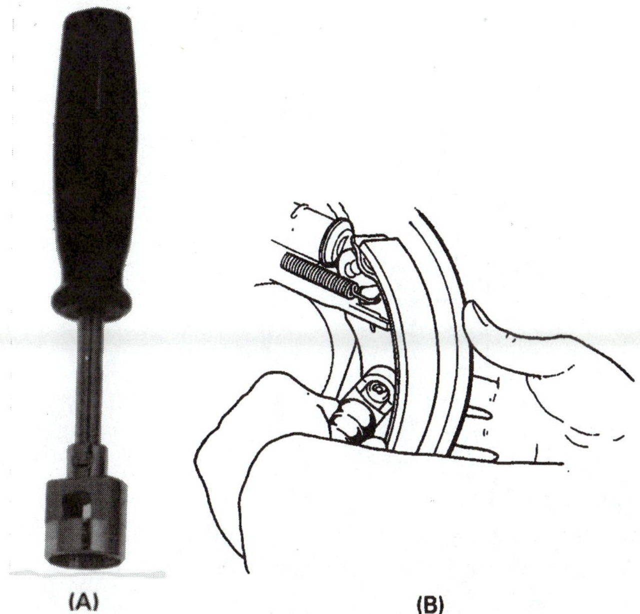

Figure 10-57 This hold-down tool (A) is used to push the retainer inward and rotate it during spring removal and installation. It is usually necessary to hold the retainer pin with one hand while the other hand uses the tool (B). (A is courtesy of KD Tools; B is courtesy of LucasVarity Automotive)

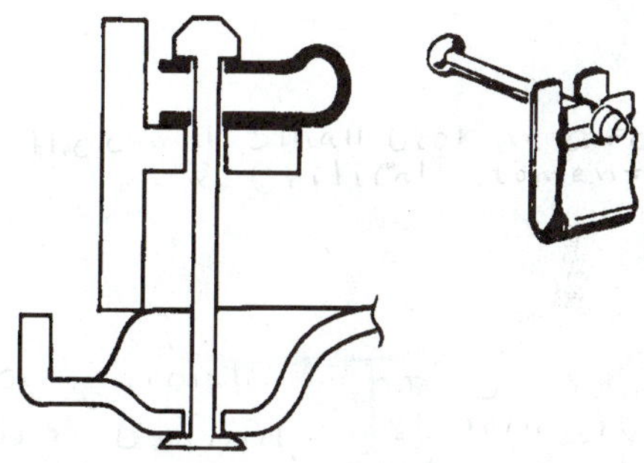

Figure 10-58 This hold-down spring is compressed while the pin is rotated to disengage it.

In the past, the hold-down spring used on some domestic cars was a coil spring that passed partially through the hole in the shoe web and hooked onto a tang that passed through the backing plate. Because of its shape, it was sometimes called a *beehive spring*. A special tool was used that passed into the spring and pushed on the last coil so the spring could be stretched enough to be hooked or unhooked (Figure 10-59). If this special tool is not available, a small screwdriver that will enter, but not pass through the spring, will work.

Some pickups and light trucks use a mousetrap-shaped style of hold-down spring on the shoe which hooks into a curved wire extending from the backing plate. The odd-

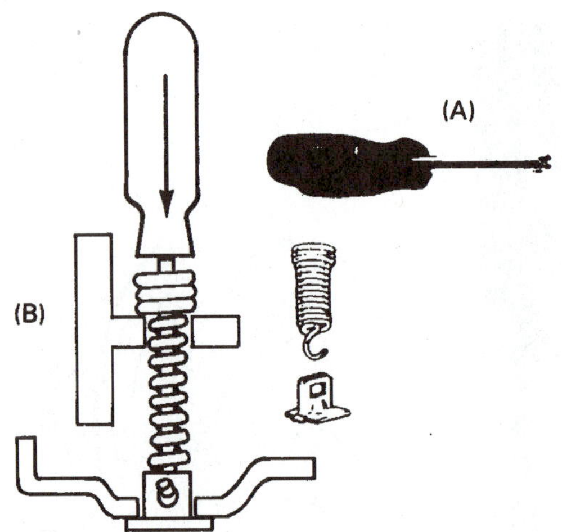

Figure 10-59 A special tool (A) is used to push a beehive-style hold-down spring inward so it can be unhooked (B). (Courtesy of Snap-on Tools Company, Copyright Owner)

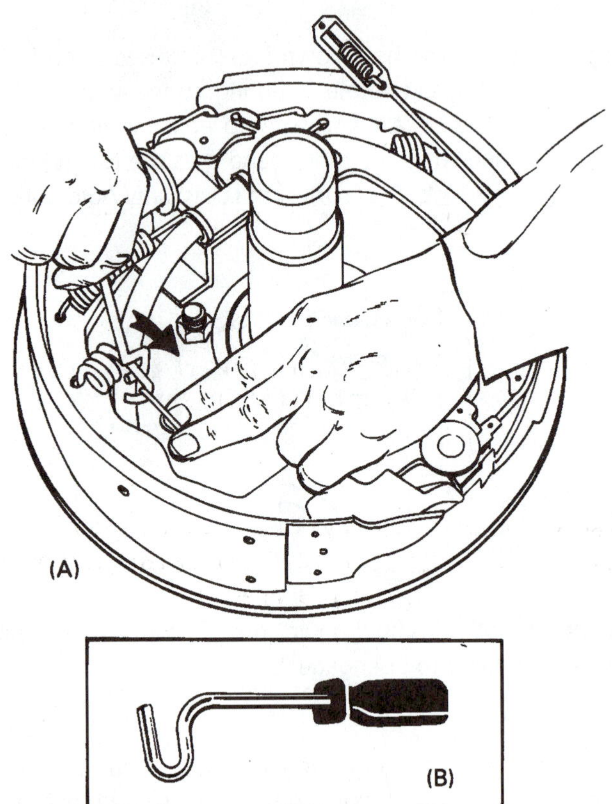

Figure 10-60 This coil-type hold-down spring must be pulled toward the brake shoe so the pin can be disconnected (A). A special tool (B) makes this easy. (A is courtesy of Chrysler Corporation; B is courtesy of Snap-on Tools Company, Copyright Owner)

shaped tool shown in Figure 10-60 is required to bend the spring far enough to hook or unhook it.

Some brake designs use a U-shaped, flat spring that hooks into the backing plate and rests on the shoe web. The brake shoe is simply slid out from under or slid back under this spring during removal or during installation.

10.7 Brake Shoe Removal

During shoe-removal operations, many technicians position a pan of soapy water or solvent under the brake assembly. As the parts are removed, they are placed in the liquid for soaking. This allows a quick and easy cleanup and reduces the amount of dust being stirred up (Figure 10-61).

Figure 10-61 The cleaning pan was left under the brake assembly and the parts were dropped into it as they were removed. Cleanup of the parts and backing plate is quick and easy.

When removing and replacing brake shoes, the manufacturer's recommended procedure for the particular car should be followed. The following procedure is very general. To remove a pair of brake shoes, you should:

1. Carefully check the shoes to note the types of return and hold-down springs and how they are connected to the shoes and backing plate. Also, carefully note how the adjuster mechanism and parking brake linkage are connected to the shoes, usually the secondary shoes. You should note that different colors are used on the springs, as they can help you remember the correct spring positions.
2. If you are not planning to service the wheel cylinder, install a wheel cylinder clamp to prevent "popping" of the wheel cylinder pistons out of their bores (Figure 10-62).

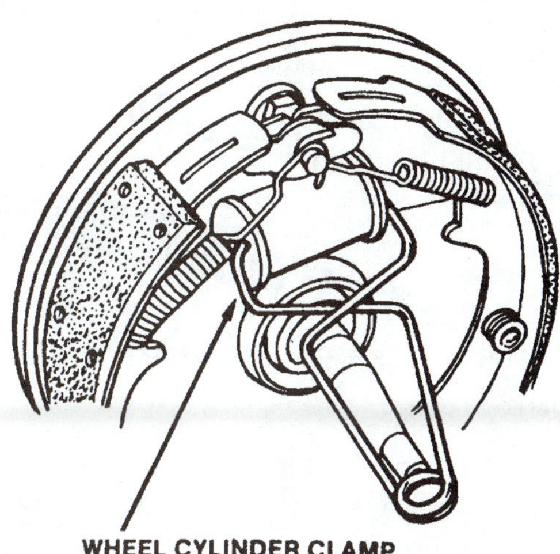

Figure 10-62 A wheel cylinder clamp can be put on a wheel cylinder to keep the pistons from coming out when the shoes are removed. (Courtesy of Bendix Brakes, by AlliedSignal)

3. Remove the return springs. Also, remove any self-adjuster mechanism which has been released.

   NOTE: On some duo-servo brakes, the shape of the anchor will cause the outer return spring to be stretched a little farther than the inner. It is a good practice to note which spring is removed first so it can be installed last.

4. Remove the hold-down springs and any self-adjuster mechanism that has been released.

   NOTE: On those brakes where the shoe is pulled or slid off the anchor block to release the return spring, the hold-down springs are normally removed first.

5. If necessary, disconnect any remaining parking brake or self-adjuster linkage from the shoes. Some parking brake levers merely hook into the shoe, and others are held in place by an E-clip or a C-clip. An E-clip is removed by prying it off with a small screwdriver (Figure 10-63). A C-clip is most easily removed with pliers or a sharp punch as shown in Figure 10-64. If it is necessary to disconnect the parking brake lever from the cable, place the jaws of a pair of side- or diagonal-cutting pliers over the cable. Then slide the pliers down the cable to compress the spring, and then tighten the pliers to lightly grip the cable, holding the spring compressed while you disconnect the cable from the lever (Figure 10-65).
6. If the wheel cylinder is to be serviced, remove it. This procedure will be described in Chapter 12.

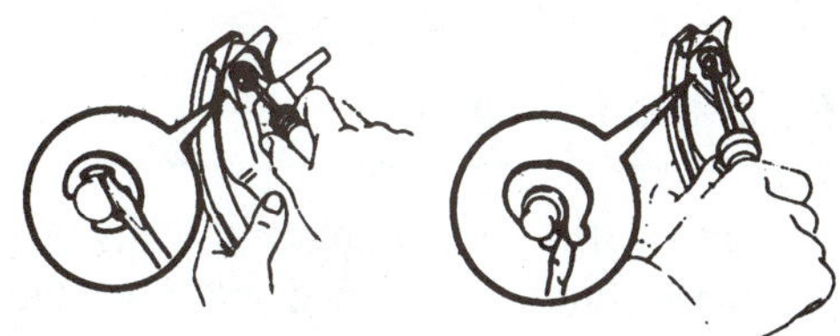

Figure 10-63 An E-clip (left) or C-clip (right) retainer can be removed using a small screwdriver.

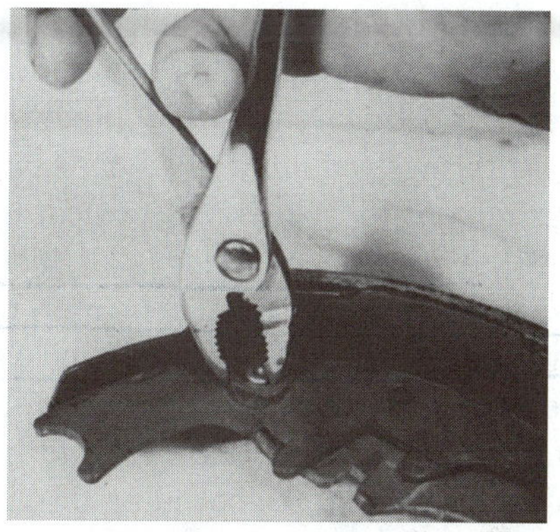

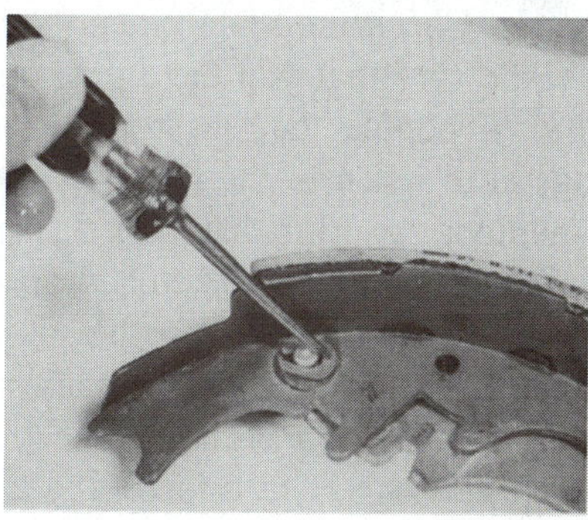

Figure 10-64 A C-clip retainer can be removed by squeezing it with a pair of pliers (A) or by tapping it out of its groove using an awl and hammer (B).

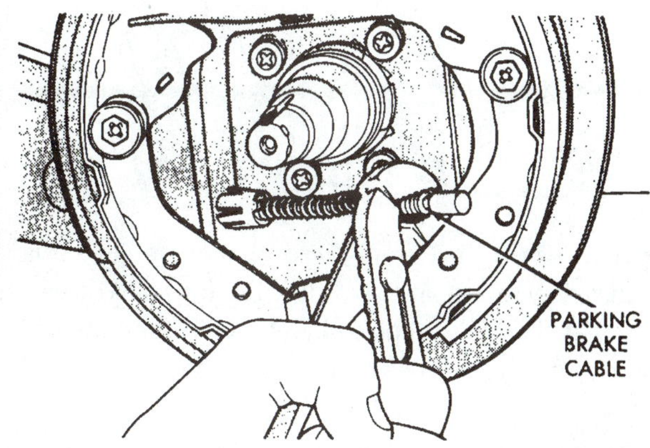

Figure 10-65 Gripping the parking brake spring with a pair of pliers allows you to hold the spring compressed so the brake cable can be removed or replaced on the lever. (Courtesy of Chrysler Corporation)

10.8 Component Cleaning and Inspection

After shoe removal, the backing plate and any part that is going to be reused should be cleaned and inspected. These parts should be washed in soapy water or solvent and air-dried. On brake drums mounted on hubs with wheel bearings, the wheel bearings should also be cleaned and repacked. This operation will be described in Chapter 16.

As mentioned earlier, many technicians will normally install new return springs and hold-down springs. Several companies market a kit, often called a "hardware set," that includes these springs and other small parts that are often worn or damaged or for which replacement is advisable (Figure 10-66). A shoe return spring is a critical part. If it is weak or stretched, early shoe application or drag can occur, and this can cause a pull or rapid shoe wear (Figure 10-67). There are no strength or length specifications for return springs available to the brake technician, and their low cost does not warrant the spending of much time to check them. Return springs are made in many different sizes and shapes. Check the replacement spring against the old one to be sure that you have the correct spring (Figure 10-68). A popular but questionable method for testing springs is to listen to the sound as the spring is dropped on the floor. A spring that makes a ringing noise is faulty because the coils are stretched apart. Always replace the return springs in sets at both ends of the axle. The springs at each end must be the same strength to reduce the possibility of a brake pull. Check any spring that is to be reused for severe discoloration, rusting, or stretch coils. Any damaged or doubtful springs must be replaced.

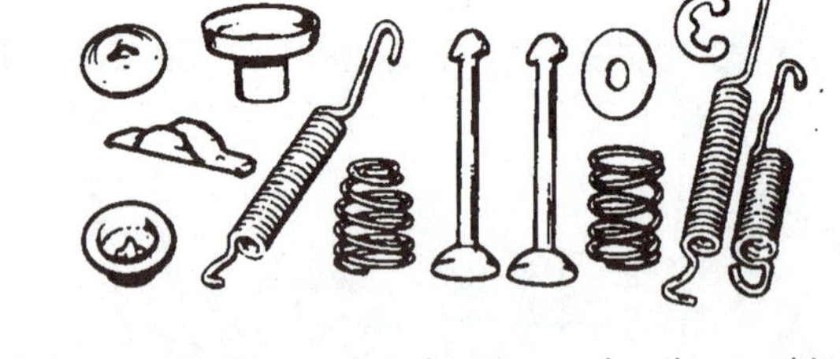

Figure 10-66 A Combi-Kit or hardware kit. This set includes new return and hold-down springs plus other small parts that should be replaced along with the brake shoes. (Courtesy of Rolero-Omega, Division of Cooper Industries)

On cars with self-adjusters, the adjusting lever should be checked for wear. With cable-type self-adjusters, be sure to check the lever for incorrect bends or cracks in the pivot area, and the cable for signs of fraying or stretching

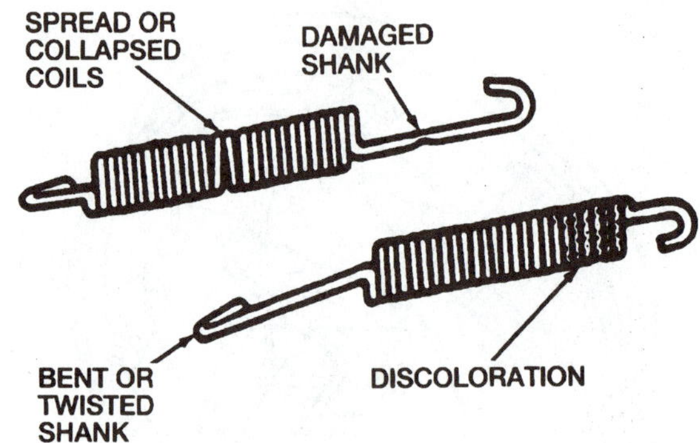

Figure 10-67 Damaged return springs should be replaced to ensure proper brake shoe application and release. (Courtesy of Ford Motor Company)

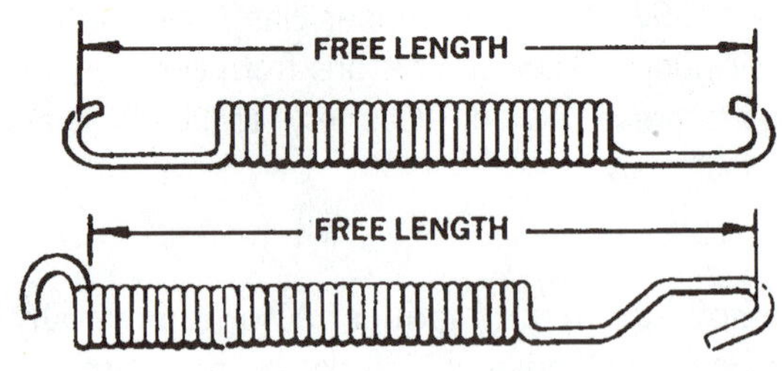

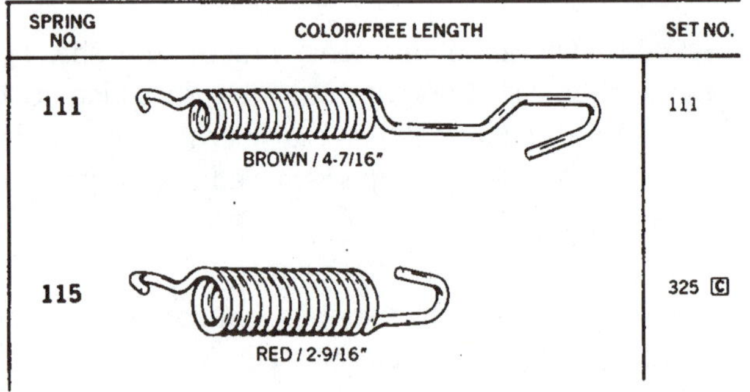

Figure 10-68 There are many shapes and sizes of return springs. Replacements should be carefully checked against the old ones to make sure they are correct. (Courtesy of Rolero-Omega, Division of Cooper Industries)

(Figure 10-69). Each style of self-adjuster has its own wear points. Any damaged parts should be replaced. For example, on 1971 to 1975 General Motors cars using leading-trailing shoes and nonservo brakes, special pliers are necessary to disassemble or shorten the strut and adjuster assembly. This strut lengthens to adjust for lining wear, and it must be shortened when new lining is installed. Failure to use the correct tool will result in broken adjuster locks.

Note that adjuster screws and self-adjuster levers are right- and left-hand and cannot be switched. Many adjuster screws will be stamped with an "L" or an "R," indicating the thread direction. Usually, on duo-servo brakes, the left-hand brake uses right-hand threads, and

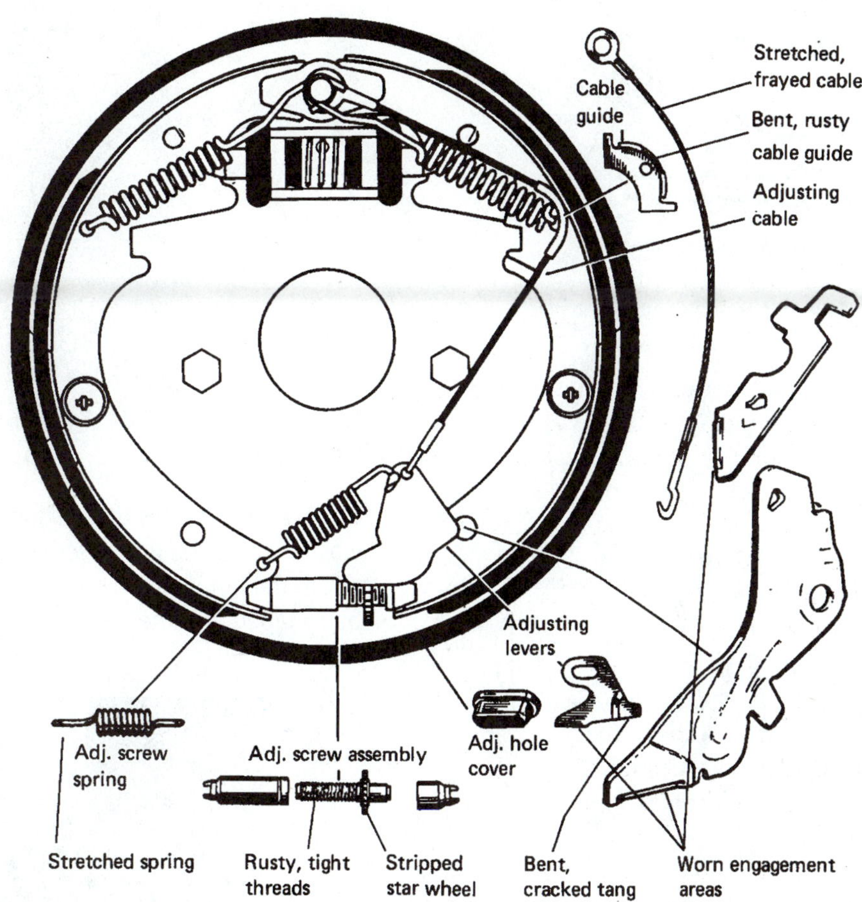

Figure 10-69 The various parts of a self-adjuster should be checked for wear or damage.

vice versa (Figure 10-70). When cleaning adjuster screws, they should be unthreaded to allow thorough cleaning of all dirt or rust from the threads and also to allow proper lubrication. Any adjuster with damaged threads or teeth should be replaced.

While checking a backing plate, make sure that the platforms are clean and smooth and not grooved badly enough to catch a shoe (Figure 10-71). Small imperfections can be cleaned up with a fine file or emery cloth.

If the shoe shows a lining wear pattern indicating a cocked shoe, the backing plate should be checked to see if the platforms are worn or distorted (Figure 10-72). On brakes with an axle flange, the relative height of the platforms can be easily checked by using this procedure: Attach a dial indicator to the axle flange, adjust the indicator stylus and dial to zero on one of the platforms, and rotate the axle so the remaining platforms can be measured (Figure 10-73). At one time, a simple tool was positioned over the spindle to allow platform height checks. This tool is shown in Figure 10-74. A badly bent backing plate or one with badly worn platforms should be replaced. When replacing a backing plate, be sure to tighten the mounting bolts to the correct torque.

10.9 Component Lubrication

Drum brake components are carefully lubricated before and during assembly to reduce future wear and noise (Figure 10-75). The lubricant used must be an approved grease. It must remain where it is applied and not run onto the friction surfaces. Several of the aftermarket brake component manufacturers market a brake assembly grease. Note that this product should not be confused with hydraulic brake assembly fluid which is used inside the hydraulic system. Some brake technicians prefer to use an anti-seize compound for assembly grease because it has superior heat-tolerance characteristics and an ability to retain excellent lubricating quality over a long period of time.

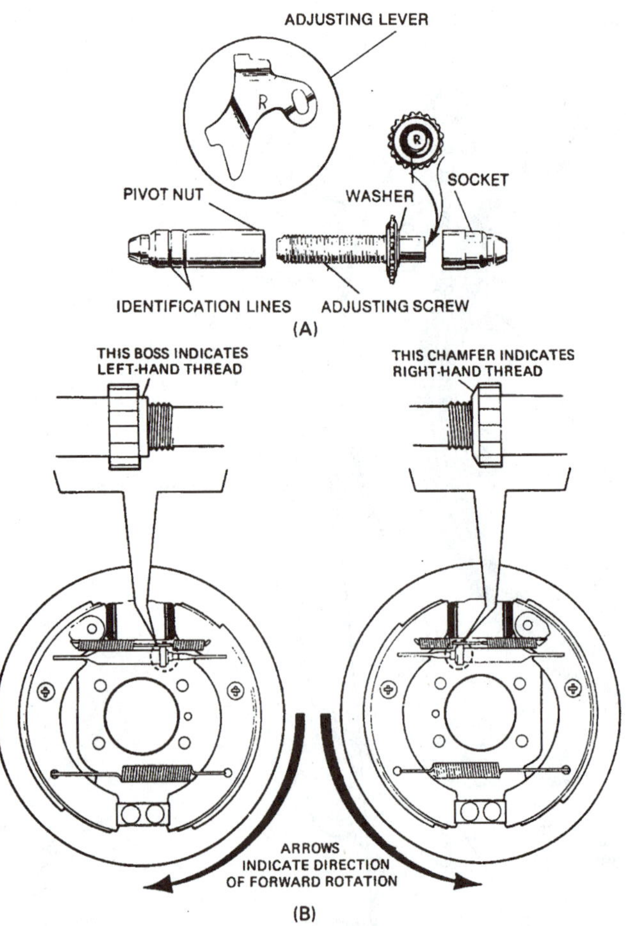

Figure 10-70 The right and left adjuster mechanisms are identified by various markings. (A is courtesy of General Motors Corporation, Service Technology Group; B is courtesy of LucasVarity Automotive)

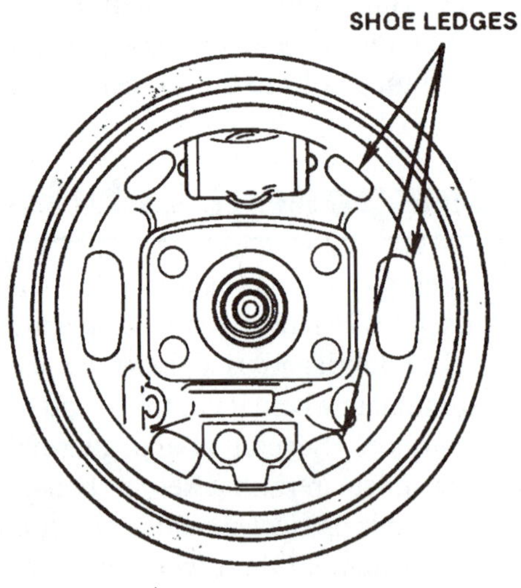

Figure 10-71 After cleaning the backing plate, it should be checked for worn shoe ledges/platforms and loose mounting bolts and anchors. (Courtesy of Chrysler Corporation)

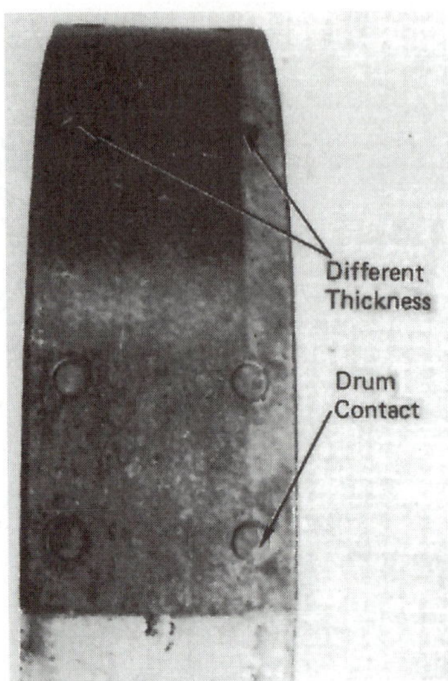

Figure 10-72 This brake shoe has worn so the right rivet was touching the drum, yet there was 1/32 inches of lining above the left rivet. This indicates that either the drum was badly tapered or the backing plate was cocked.

Figure 10-73 The dial indicator is mounted on the axle flange to check the platform heights; if they are different, the shoe will be crooked with the drum.

Normally a thin film of grease is applied to each surface where there will be a rubbing metal-to-metal contact. The smallest amount of grease should always be used, to prevent the possibility of grease falling onto the friction surfaces. Grease will contaminate the lining and will probably cause grabbing. A thin layer of grease should be

Figure 10-74 This simple gauge is used to check the relative height of the brake platforms; after adjusting it to fit one of the platforms, the axle is rotated to allow checking of the others.

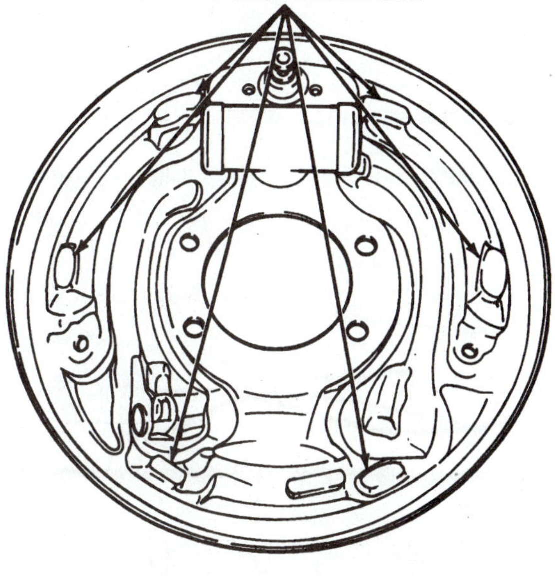

LUBRICATE LEDGES WITH
DISC BRAKE CALIPER SLIDE
GREASE, D7AZ-19590-A
(ESA-MIC172-A) OR EQUIVALENT

Figure 10-75 The shoe ledges/platforms should be lubricated with a thin film of lubricant before installing the shoes. (Courtesy of Ford Motor Company)

applied to each of the backing plate platforms. Most technicians will grease the threads of the adjuster screws before threading them together. This grease will seal the threads to prevent water or dirt from entering (Figure 10-76). A thin film of grease is also applied between the adjuster screw and its socket. During reassembly of the shoes, most brake technicians will apply a thin grease film at the parking brake lever pivot, at each end of the parking brake strut, at each pivot of the self-adjuster mechanism, at the point where the adjuster screw contacts the shoes, at the anchor pin or blocks, and at every metal-to-metal contact (Figure 10-77).

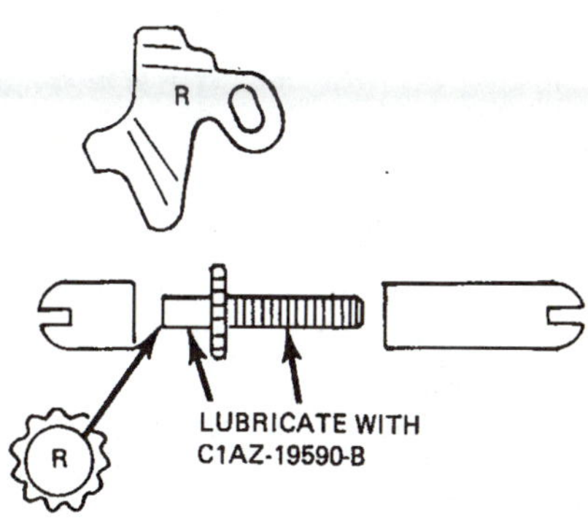

Figure 10-76 The adjusting screw should be lubricated before installation. (Courtesy of Ford Motor Company)

10.10 Brake Shoe Preinstallation Checks

Before installation on the backing plate, the brake shoes should be checked for proper lining attachment and proper curvature. Use care to keep the new lining clean. Clean your hands before handling the new shoes and develop the habit of handling the shoes by the web or the edges of the rim—not by the lining. Lay the shoes out on a clean bench in the position in which they will be installed and determine if there are shoe differences; for example, a primary and a secondary shoe or a leading and a trailing shoe. Compare the new shoes with the old shoes if there are any doubts about the correct lining length or lining or hole placement.

After determining the correct position for a particular shoe, check the lining-to-shoe attachment to be sure that the lining is positioned straight and does not overhang the edges of the shoe rim. Any overhanging lining will require that the shoe be returned to the supplier or that it be removed with a file or coarse sandpaper (Figure 10-78). If it is necessary to remove overhanging lining, be careful to remove it in an OSHA-approved manner and not to breathe any of the sanding dust.

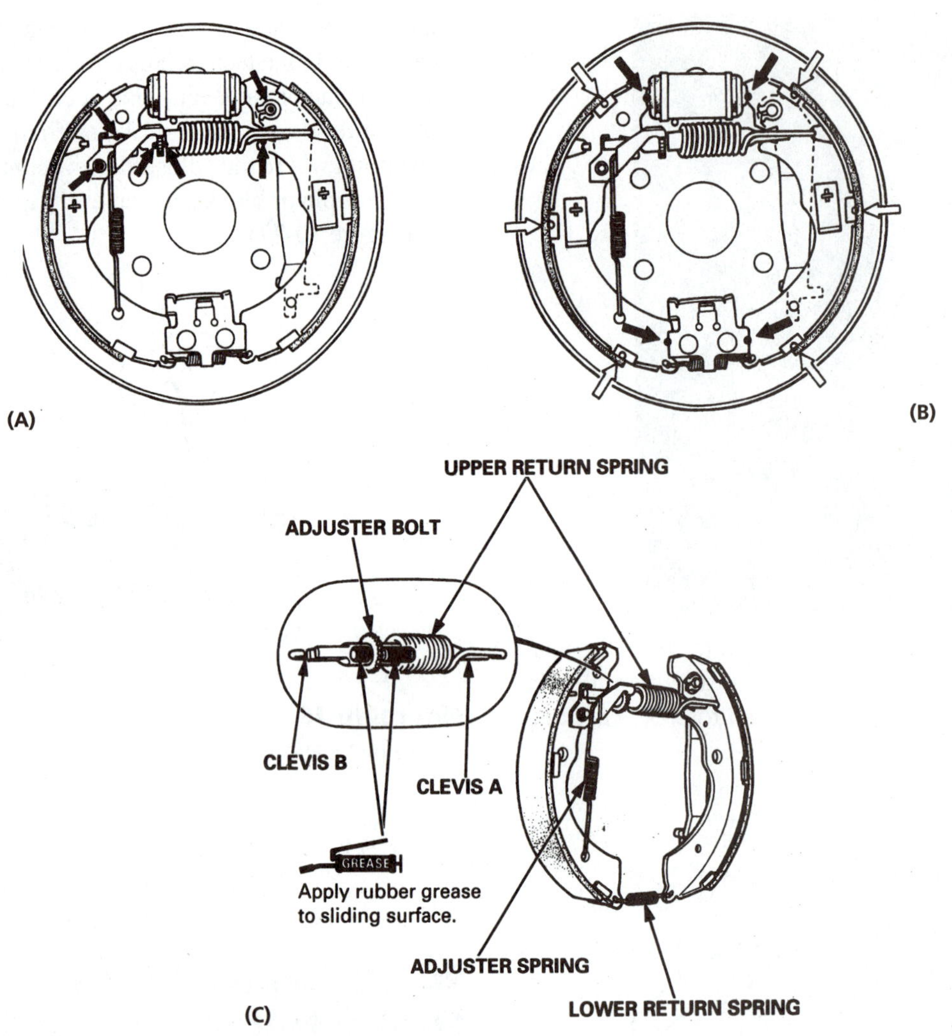

Figure 10-77 This manufacturer recommends lubricating the parts with brake cylinder grease (A), Molykote 44 MA (B), and rubber grease (C). (Courtesy of American Honda Motor Co., Inc.)

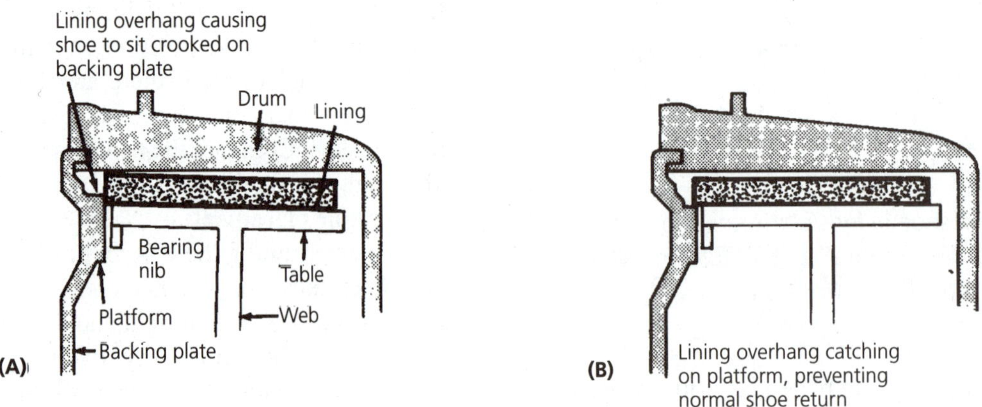

Figure 10-78 If the lining overlaps the shoe table/rim, it can move the shoe out from the platform and cause it to sit crookedly (A). This overhanging lining can also catch on the platform (B) and prevent the shoe from returning properly.

It is a good practice to always check the lining-to-drum fit. This is sometimes called a heel and toe clearance check. The lining should almost fit the curvature of the drum, to ensure a quick break-in of the lining. There should be a slight clearance at the ends of the shoe, and this slight clearance should allow the shoe to rock in the drum (Figure 10-79). There should never be a clearance between the center of the shoe and the drum, as this would cause erratic, grabby brake action. Heel and toe clearance helps ensure that the ends of the lining are not pinched because of drum deflection during a hard stop. This clearance is checked by placing the shoe in the drum and squeezing the lining into the drum. One hand is placed so it grips the web of the shoe and the outside of the drum. With the other hand, gauge the clearance at the ends of the lining. A 0.005-inch (0.13 mm) feeler gauge should fit between the lining and the drum at each end, and a 0.010-inch (0.25 mm) gauge should not (Figure 10-80).

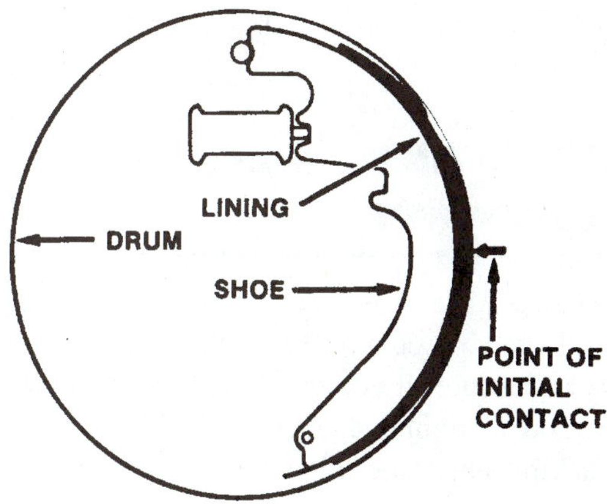

Figure 10-79 The new lining should contact the drum at the center of the lining. (Courtesy of Bendix Brakes, by AlliedSignal)

In the past, brake shoes were commonly available with a standard or an oversize lining. As a drum is turned to a larger diameter, the curvature or arc of the friction surface will become flatter, and the heel and toe clearance will increase. An oversize lining was slightly thicker, and it was ground with a slightly flatter arc. The lining on brake shoes is normally ground on a radius that is one-half of the brake diameter minus 0.030 inches (0.76 mm). This is sometimes called an eccentric or cam grind. A standard lining was ground to fit drums that were standard size to 0.030 inches oversize. Oversize lining was ground to fit drums that were more than 0.030 inches oversize. Today, the lining on most relined shoes is ground to a standard size.

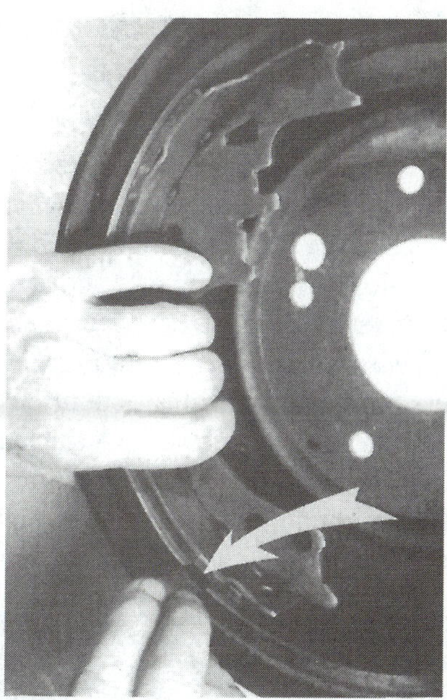

Figure 10-80 Lining clearance is checked by squeezing the shoe against the drum and trying to fit a feeler gauge at each end.

After finding that the shoe is the correct one, that there is no lining overhang, and that the heel and toe clearance is acceptable, some brake technicians will cover the lining with one or two strips of loosely applied masking tape. This tape keeps the lining clean while the shoe is being installed on the backing plate. The tape is removed just before the drum is installed (Figure 10-81).

Figure 10-81 After checking the lining, some technicians cover it with masking tape to keep it clean until the drum is installed.

10.11 Regrinding Brake Shoes

In the past, most brake shops would regrind or re-arc a lining if the shoe did not have the correct heel and toe clearance. This ensured the best fit and shortest break-in period for the lining. Most shops had a brake shoe grinder as well as a brake drum lathe. Because of concerns about asbestos fibers, very few shops nowadays grind brake shoes. Most shoe grinders are designed with vacuum dust-collection systems, some of which meet OSHA standards. As the lining is being re-arced, the dust-collection system collects the dust in a container so it can be disposed of easily and safely (Figure 10-82).

Figure 10-82 A brake shoe grinder is used to regrind brake lining to an arc that matches the drum. (Courtesy of Ammco Tools, Inc.)

SAFETY TIP At this time, because of the hazards connected with asbestos dust, many shops are no longer regrinding brake shoes in the field (outside manufacturing plants). However, there are some shoe grinders that meet OSHA requirements and can be safely used. It is important that these units be maintained properly to ensure proper compliance with the law.

There are two styles of shoe regrinding: a standard or plain grind and a fixed-anchor grind. Standard arc grinding creates a lining that is curved to match the drum diameter minus 0.030 inches and is used for sliding anchor or adjustable anchor shoes. A fixed-anchor grind positions the shoe so the anchor eye is in the same position relative to the shoe clamp pivot that the brake's anchor pin is in relative to the center of the axle or spindle. A good fixed-anchor grind will produce a shoe that has the correct arc and will be centered on the drum when it is installed (Figure 10-83). When using this equipment, be sure to follow the directions of the manufacturer.

10.12 Brake Shoe Installation

In most cases, the assembly of the shoes on the backing plate follows a procedure that is the reverse of the disassembly procedure. It is a good practice to always follow the method recommended by the vehicle manufacturer. The backing plate and adjusting screw should be lubri-

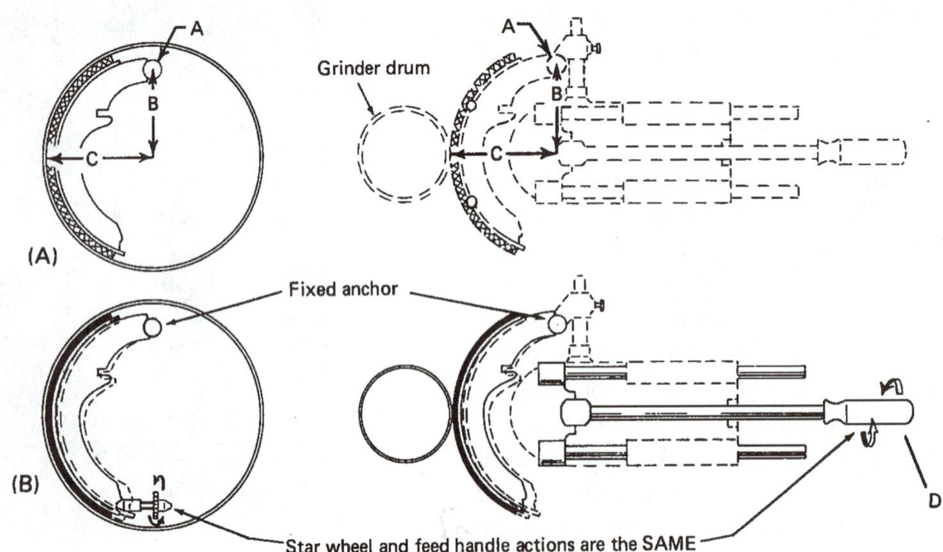

Figure 10-83 The shoe grinder positions the lining in the correct position, relative to the grinder drum, so it can be ground to the correct arc. (Courtesy of Ammco Tools, Inc.)

cated, and a slight amount of lubricant should be used on the other parts as they are installed.

To install brake shoes on a backing plate, you should:

1. Check the shoes to determine the correct placement—primary or secondary, leading or trailing—and install the wheel cylinder if it has been removed.
2. If necessary, install the parking brake lever on the secondary or trailing shoe and connect the parking brake cable to the lever.

On vehicles with manually adjusted duo-servo brakes, overlap the anchor ends of the shoes as you hook the adjuster spring between the two shoes. Then place the adjuster screw in position and spread the shoes to their normal position to retain the spring and adjuster screw. Check that both the spring and the screw are in the correct position relative to the shoes and backing plate (Figure 10-84).

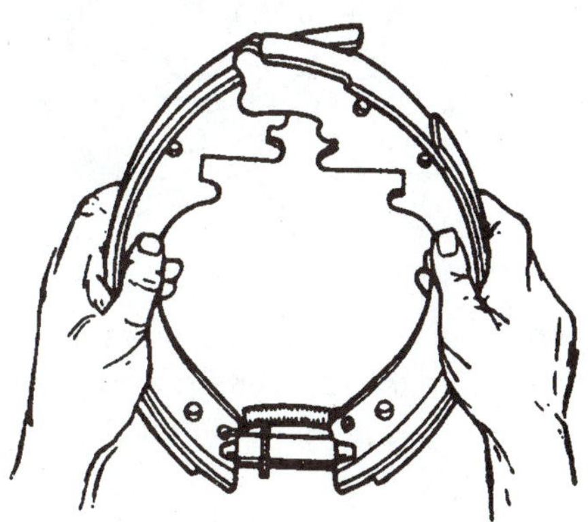

Figure 10-84 The first replacement step with manually adjusted, duo-servo brakes is to connect the adjuster spring and screw. (Courtesy of Bendix Brakes, by AlliedSignal)

Figure 10-85 When installing this style of hold-down spring, make sure the flats at the end of the pin are completely in the washer recesses.

3. Install the hold-down spring and parts on the primary shoe. Make sure that the hold-down pin is completely in place in the detents of the washer (Figure 10-85).
4. Install the hold-down spring and parts on the secondary shoe. As this is being done, attach the self-adjuster mechanism and parking brake strut, if necessary.
5. Install the return springs. The normal order of installation on a duo-servo brake is to install the primary shoe return spring first and then the secondary shoe return spring. If necessary, attach the self-adjuster mechanism as the return springs are being installed. The spring should be hooked into the shoe and then installed on the anchor or second shoe.

NOTE: On cable-type self-adjusters, make sure the cable guide stays against the shoe web and does not lift and allow the cable to go under the guide.

NOTE: Some technicians recommend using a pair of pliers to close the spring eye to the point where the spring end is parallel to the shank (Figure 10-86). This prevents the spring eye from ever opening up (Figure 10-87)

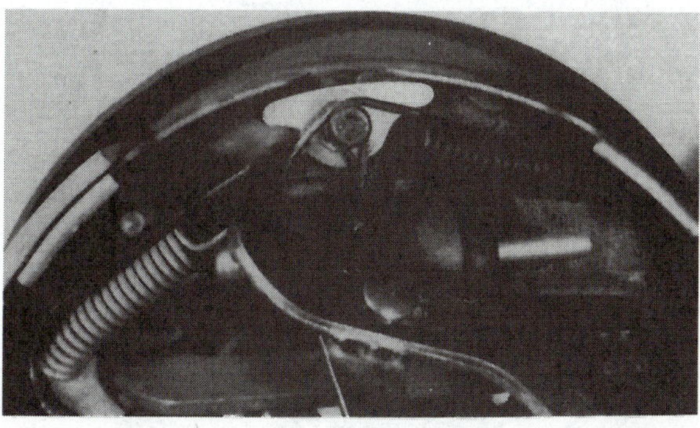

Figure 10-86 After installing return springs, some technicians use pliers to squeeze the tang of the spring until it is parallel with the shank.

6. Inspect installation to be sure that everything is correctly installed. On most types of brakes, the shoes should be returned to the anchors, and there should be a slight clearance at the parking brake strut. Test the self-adjuster to check its operation and then remove the masking tape, if used (Figure 10-88).
7. On cars that use adjuster screws, preadjust the brake shoe clearance. This can be done easily with a **brake shoe gauge**. First, adjust the gauge to fit inside the drum (Figure 10-89). Then adjust the shoes to a slight gauge clearance (Figure 10-90). If a gauge is not

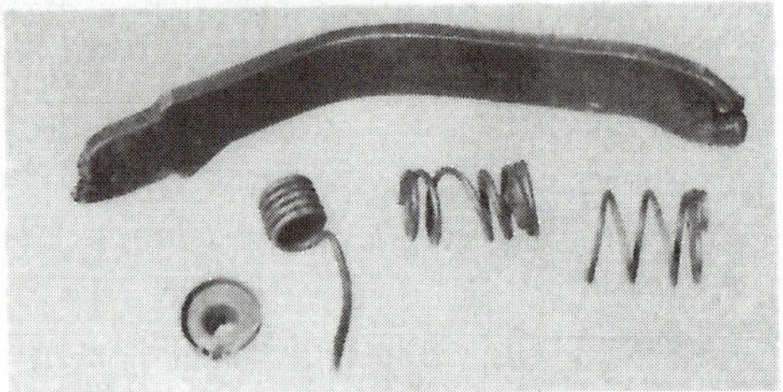

Figure 10-87 This severely worn drum is probably the result of a broken return spring.

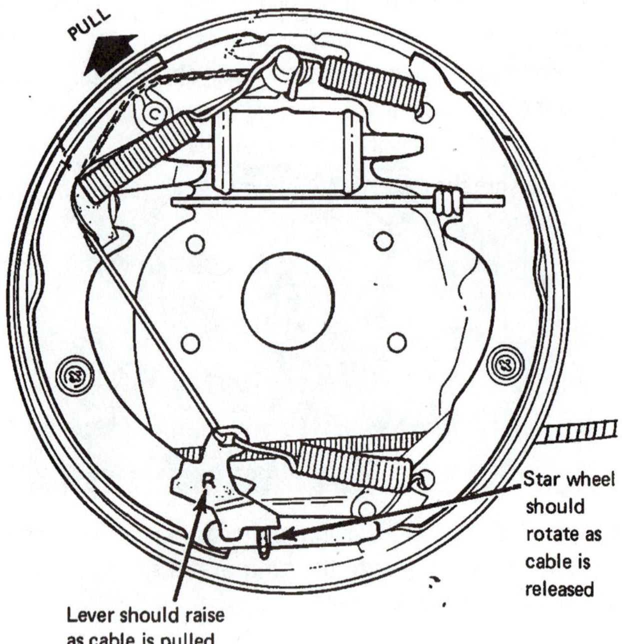

Figure 10-88 It is a good practice to check self-adjuster operation before installing the drum. (Courtesy of Ford Motor Corporation)

available, expand the shoes until a drag is felt when trying to slide the drum in place. Then remove the drum and turn the adjuster inward about one-half turn to increase the clearance.

8. Install the drum.

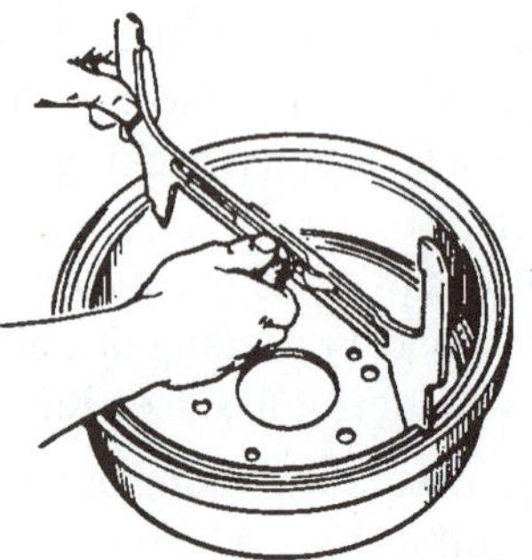

Figure 10-89 A brake shoe gauge speeds up the procedure to preset lining clearance before installing the drum. The first step is to adjust the gauge to the drum diameter. (Courtesy of Ford Motor Company)

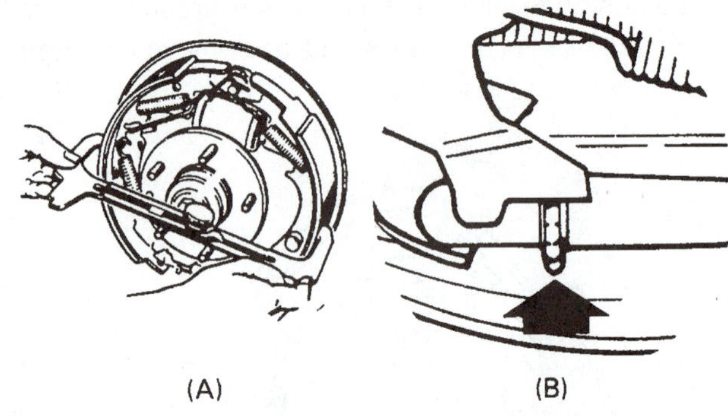

Figure 10-90 After adjusting the gauge size, place it over the lining (A) and turn the adjuster screw to bring the lining to almost the size of the gauge (B). (Courtesy of Ford Motor Company)

NOTE: On drums with a hub and wheel bearings, the wheel bearings should be cleaned, repacked, and adjusted. These operations will be described in Chapter 16.

NOTE: On a few older cars, adjust the anchor to perform the major brake adjustment required by some manufacturers. This procedure is described in older shop manuals.

9. If the wheel cylinder has been rebuilt or replaced, bleed the air from the hydraulic system.
10. Install the wheel and tighten the lug bolts to the correct torque using the proper tightening pattern. Note that overtightening the lug bolts can distort the brake drum.
11. Complete the brake shoe adjustment.

a. On duo-servo brakes with self-adjusters, slowly and carefully back up the car while applying and releasing the brakes. You should feel the brake pedal height increase as the brakes self-adjust.
b. On nonservo brakes with self-adjusters, slowly apply and release the parking brake. Some designs completely adjust in one application, but some require several applications. As you apply and release the service brake, you should feel an increase in the brake pedal height as the adjustment occurs.
c. On manually adjusted duo-servo brakes, turn the adjuster screw using a brake-adjusting tool or spoon as you slowly rotate the wheel. Use the tool to pry the teeth of the adjuster screw downward (tool handle upward) to expand the screw in most cases. Expand the shoes until the wheel is almost locked and then back off the adjustment until the brake is just free of drag. You should be able to hear a slight rubbing as the wheel is turned (Figure 10-91).

NOTE: After completing a brake adjustment, some sources recommend that you tap the brake drum with a hammer. It should make a ringing noise; a dull noise indicates that the shoes are in contact with the drum surface. This is also useful to help diagnose the problem of an overheating, dragging brake. It is not a good practice to adjust the lining to the point of a definite drag.

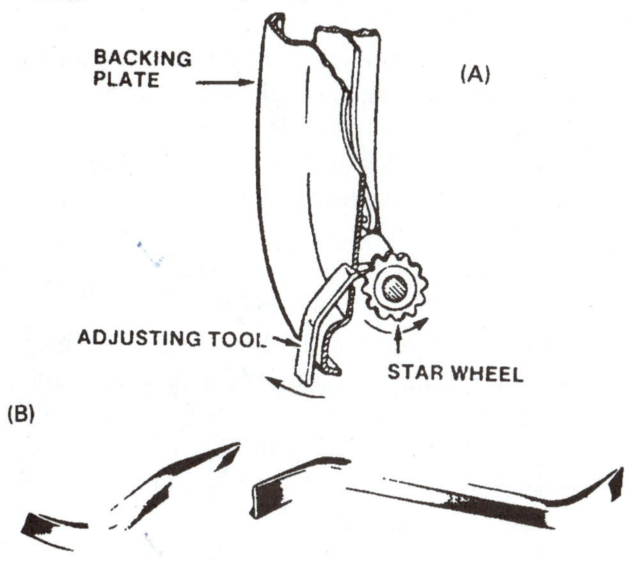

Figure 10-91 On vehicles with manual adjustment, clearance is adjusted after drum installation. Turn the starwheel to lock up the drum, and then back it off until there is a very slight drag (A). An adjustment tool or "spoon" should be used (B). (A is courtesy of Bendix Brakes, by AlliedSignal; B is courtesy of Snap-on Tools Company, Copyright Owner)

If the brake pedal is still low after the final adjustment, remove the drums, one at a time, and use the brake shoe gauge as described in step 7 to determine if there is excessive clearance in any of the brake assemblies. Once the faulty brake is located, a careful inspection will usually show the cause of the excess clearance. On cars with four-wheel drum brakes, save time by using the parking brake as a diagnostic tool. If the brake pedal height increases after the parking brake has been applied, there is excess clearance in one or both of the assemblies where the parking brake is.

10.13 Completion

Several checks should always be made after completing a brake job or brake service operation. They are as follows:

- Be sure that all nuts and bolts are properly tightened to the manufacturer's specifications and all locking devices properly installed.
- Be sure that there is no incorrect rubbing or contact between parts or brake hoses.
- Be sure that the master cylinder has the correct fluid level.
- Be sure that there are no fluid leaks.
- Be sure that there is a firm, high brake pedal.
- Clean any fingerprints or grease from the vehicle.
- Perform a careful road test to make sure that the brakes operate correctly and safely.

New friction material must be properly broken in. Unless it is specifically stated otherwise by the lining manufacturer, new lining must never be "burned in." New lining can be ruined if it is overheated. Light or moderate stops should be made until the lining and drum surfaces are fully burnished and seated in. This varies with the type of driving. Some sources recommend about 250 to 2,000 miles (400 to 3,000 km) of driving. The brake pedal will firm up and become more solid as the lining wears to exactly match the drum radius, and drum and shoe deflection will disappear. Poorly fitted lining will require an even longer, more gradual break-in period.

10.14 Practice Diagnosis

You are working in a brake and front-end shop and encounter these problems.

CASE 1: The ten-year-old car has high mileage and shows a lot of abuse. It has been brought in with a complaint of noisy brakes, and on the road test, you hear metal-to-metal contact at the rear when the brakes are applied. With the car back in the shop, you remove the rear tires, but when you try to pull the drums, you find that they come out only about an eighth of an inch. What is happening? What should you do next? Is this going to be an inexpensive repair?

CASE 2: The car has 80,000 miles on it and is in pretty good condition. When you pull the rear drums, you find riveted lining that is worn down to the rivets on the outer edge, but there is about 1/16 inches of lining above the inner rivets. What is wrong? How can you check this?

CASE 3: The 1990 passenger car has been brought in with a complaint of overheating rear brakes. Your road test confirms a dragging condition, and when you check them, you find the right rear is very hot, the right drum makes a dull sound when you tap it with a hammer, and the left rear makes a ringing sound. What is wrong? What should you do to locate the cause of this problem? Could this be a bad proportioning valve?

Terms to Know

arbor
bell-mouthed drum
Bendix brake spring tool
brake drum micrometer
brake shoe gauge
brake spring pliers
brake spring remover-installers
brake spring tools
cutter bit
drum mike
hard spots
Lockheed brake spring tool
machining
swaging

Review Questions

1. Two technicians are discussing a brake job. Technician A says that brake operations should be done in pairs, performing the same operation at each end of the axle. Technician B says that usually the only operation that is really necessary is to remove and replace the shoes. Who is right?
   a. A only
   b. B only
   c. both A and B
   d. neither A nor B
2. The rear drums of most front-wheel-drive cars are usually removed
   a. by sliding them off the axle flange after the speed nuts have been removed.
   b. along with the hub and wheel bearings.
   c. from the axle flange after removing two bolts.
   d. none of the above.
3. Technician A says that care should be exercised to avoid breathing the dust released while removing a drum and shoes. Technician B says that brake dust can be safely handled by using a HEPA filter-equipped vacuum cleaner or a soap-and-water spray system. Who is right?
   a. A only
   b. B only
   c. both A and B
   d. neither A nor B
4. Technician A says that the drum must be catching on the axle flange in cases where it slides outward about 1/4 to 1/2 inches and then seizes. Technician B says that it is possible to back off and loosen a self-adjuster using a brake spoon and ice pick. Who is right?
   a. A only
   b. B only
   c. both A and B
   d. neither A nor B
5. Technician A says that a hammer can be used to check a brake drum for cracks. Technician B says that a standard dial indicator can be used to check for an out-of-round drum. Who is right?
   a. A only
   b. B only
   c. both A and B
   d. neither A nor B

6. There will not be good contact between the new lining and the drum's friction surface if the drum is
   a. scored.
   b. bell-mouthed.
   c. barrel-shaped.
   d. any of the above.
7. Technician A says that the diameter of a brake drum is measured using a drum mike. Technician B says that the maximum diameter for a drum is usually indicated on the drum. Who is right?
   a. A only
   b. B only
   c. both A and B
   d. neither A nor B
8. The diameter of a 10-inch drum has been measured at three locations, and the readings are 10.005, 10.010, and 10.015 inches. The drum is
   a. good, and it can be safely used.
   b. oversize, and it should be replaced.
   c. out-of-round, and it should be turned.
   d. none of the above.
9. Two rear drums measure 9.535 and 9.505 inches, and both are slightly scored. The maximum allowable size for the car is 9.560 inches. The technician should
   a. replace both rear drums with new ones.
   b. turn both drums to 9.560 inches.
   c. turn the largest drum to the smallest possible diameter and the smaller drum to the same size.
   d. turn each drum to the smallest possible diameter.
10. Technician A says that the lining can overheat if the drum is turned too thin. Technician B says that a spongy brake pedal can be the result if the drums are turned to too large a diameter. Who is right?
    a. A only
    b. B only
    c. both A and B
    d. neither A nor B
11. Technician A says that a rubber belt should be wrapped around a brake drum that is being turned to deflect the metal chips. Technician B says that a herringbone pattern in a drum that was just turned is the result of taking too deep a cut. Who is right?
    a. A only
    b. B only
    c. both A and B
    d. neither A nor B
12. Brake shoe return springs are normally removed using
    a. brake spring pliers.
    b. a Bendix brake tool.
    c. a pair of combination pliers.
    d. any of the above depending on the brake design.
13. Before removing the brake shoes and springs, you should
    a. clean off the dirt and dust using an airgun.
    b. purchase new shoes and drums.
    c. clean off the dirt and dust using an OSHA-approved method.
    d. bleed the fluid out of the wheel cylinder.
14. A car has a faulty rear-axle seal, and the lining has been soaked with grease. Technician A says that the grease can be washed off using the correct friction surface cleaner. Technician B says that the bad axle seal and the lining on both rear brakes must be replaced. Who is right?
    a. A only
    b. B only
    c. both A and B
    d. neither A nor B
15. Technician A says that a weak brake shoe return spring can cause a pull during a stop. Technician B says that a weak return spring can cause rapid lining wear. Who is right?
    a. A only
    b. B only
    c. both A and B
    d. neither A nor B
16. Technician A says that a common shoe hold-down spring with a coil spring, pin, and washer can be removed by pushing in on the washer and turning it one-quarter turn. Technician B says that the hold-down spring can be used to secure the self-adjuster cable guide in place. Who is right?
    a. A only
    b. B only
    c. both A and B
    d. neither A nor B
17. A self-adjusting, duo-servo rear brake is being assembled. Technician A says that the shoe with the shortest lining should be placed in the position closest to the front of the car. Technician B says that the adjuster screw threads and the backing plate platforms should be coated with a thin film of lubricant. Who is right?
    a. A only
    b. B only
    c. both A and B
    d. neither A nor B
18. When a brake shoe is placed in the drum so the lining is against the inner friction surface, there should be
    a. about 0.010 inches of clearance at the center of the lining.
    b. about 0.007 inches of clearance at each end of the lining.
    c. a perfect fit so there is no clearance anywhere between the lining and the drum.
    d. a 0.005-inch clearance between the lining and the drum at the center and each end.

19. A self-adjusting, duo-servo brake is being assembled. Technician A says that the parking brake lever should be attached to the secondary shoe. Technician B says that the self-adjuster cable guide or actuator lever should be attached to the primary shoe. Who is right?
    a. A only
    b. B only ✓
    c. both A and B
    d. neither A nor B

20. After new brake lining has been installed, it is a good practice to
    a. burn in the lining with about ten hard stops from 50 mph.
    b. adjust the lining to a definite drag so it will wear in.
    c. make a series of slow stops from moderate speeds ✓ so as to not overheat the lining as it wears in.
    d. all of the above.

11 Disc Brake Service

Objectives

After completing this chapter, you should:

- ❑ Be able to remove, clean, measure, and inspect rotors for wear or damage and determine if they should be machined or replaced.
- ❑ Be able to mount a rotor on a lathe and machine it according to the manufacturer's procedures and specifications.
- ❑ Be able to remove a caliper assembly and clean and inspect it for leaks and damage.
- ❑ Be able to clean and inspect caliper mounts and slides for wear and damage and lubricate them as required.
- ❑ Be able to remove, clean, and inspect pads and retaining hardware to determine needed repairs, adjustments, and replacements.
- ❑ Be able to lubricate and install pads, calipers, and related hardware.
- ❑ Be able to perform the ASE Tasks for disc brake diagnosis and repair (See Appendix 1).

11.1 Introduction

A typical brake job on a disc brake assembly consists of the following: caliper removal, rotor resurfacing or reconditioning, caliper rebuilding, lining replacement, caliper hardware replacement, and careful reassembly with lubrication of each part. The parts should be carefully inspected at least three times—before disassembly, during cleanup, and during reassembly—to locate any faulty or damaged components that should be replaced.

The disassembly inspection begins as soon as the wheel is removed and serves two major purposes: to locate damaged parts and to determine the relationship of the various parts. After the caliper is removed, the inspection continues on to the lining and the rotor. Some caliper designs provide visual access to check lining wear (Figure 11-1). But with this limited view, you often cannot tell whether the lining is riveted or if it is developing tapered wear.

Some newer brake designs have subtle differences in the caliper mounts and hardware items. These differences often provide opportunities for mistakes to be made during reassembly, which can change the performance of the system. Again, competent brake technicians realize that when

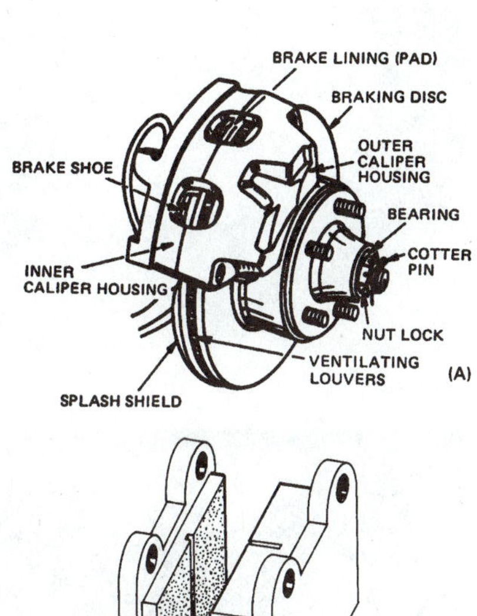

Figure 11-1 Some calipers allow the thickness of the lining to be quickly checked by looking through the openings (A). When the lining is worn to1/32 to 1/16 inches above the rivets or pad backing (B), it should be replaced. (A is courtesy of Wagner Brake; B is courtesy of ITT Automotive)

Figure 11-2 These two rotors were ruined because the outboard lining wore out completely. In both cases, the inboard lining and rotor surfaces are still good. This was caused by calipers that did not release completely and dragged.

they disassemble a unit, they become the authority on how the unit goes together. A service manual should be followed for guidance through the disassembly and reassembly steps, but many manuals fail to show the exact positioning of each and every small part. As in the case of drum brakes, when a technicians encounter a new and usually more complicated design, they will often work on one caliper at a time, saving the other to use as a pattern if needed.

When one pad wears faster than the other, a simple rule can be followed. Excessive wear of the outboard pad indicates a sticking, dragging caliper, and excessive wear of the inboard pad is caused by a sticking piston (Figure 11-2). During brake application, the amount of hydraulic force available is enough to move a partially stuck piston or caliper to apply the shoes. But since disc brakes have very weak release mechanisms, the piston or caliper will tend to stay applied and will drag. Rubber parts in the caliper and caliper hardware will harden and lose their resiliency with age and from the heat of braking. As they harden, they can no longer cushion or reposition the caliper or piston at release as they were intended to do.

In some cases, disc brake pads can be simply removed and replaced—the old ones are slid out of the caliper, and the new ones are slid in. This simple procedure will not be described in this text because, in the author's opinion, this pad replacement is not a brake job. This operation cannot guarantee "new-car" braking performance for the life of the friction material. As with drum brakes, the entire braking assembly should be reconditioned when the linings are replaced. A thorough and fast disc brake repair has been made very easy with the increased availability of **loaded calipers** (Figure 11-3). These are rebuilt calipers that come equipped with new hardware and shoes installed in them. Simple pad replacement should be done only in instances where the brake technician knows the rest of the system is in good, safe, operating condition.

Figure 11-3 A loaded caliper is a rebuilt caliper with new hardware and shoes. (Courtesy of Nuturn Corporation)

SAFETY TIP: The following sections will describe service operations on a single brake. But remember that brake operations are performed in pairs. The same operation is done at each end of the axle to ensure even, straight braking. Also, don't forget the safety points which were mentioned in Chapter 9.

11.2 Removing a Caliper

Before the caliper is removed, it is a good practice to make sure that the master cylinder reservoir is no more than one-half to two-thirds full. You should remove some of the fluid if necessary (Figure 11-4). One of the first

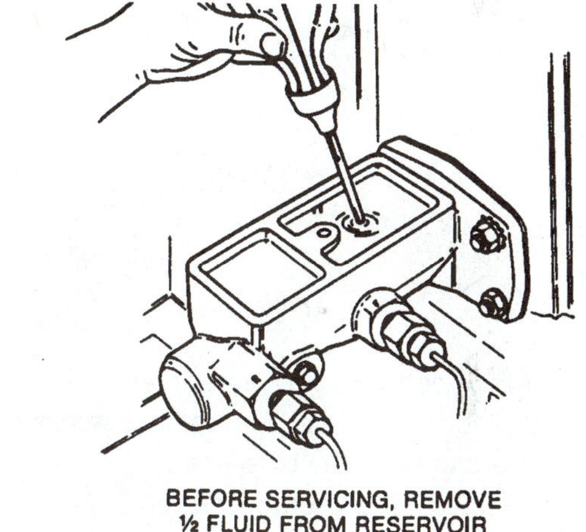

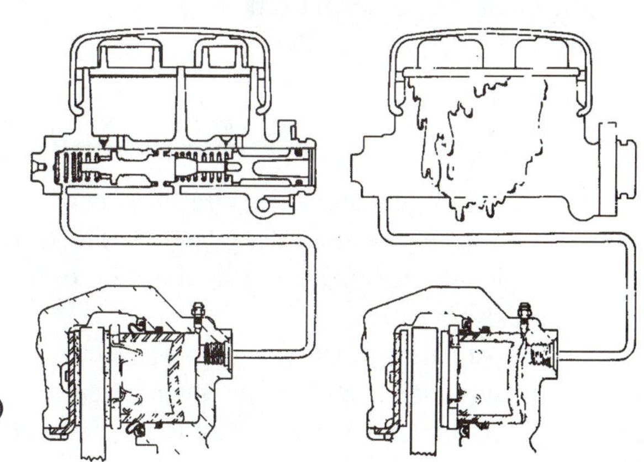

Figure 11-4 Before compressing caliper pistons, remove fluid from the reservoir (A). Fluid pushed back from the calipers will cause the reservoir to overflow (B). (A is courtesy of Ford Motor Company)

steps in actual caliper removal is to retract the pistons—move them partway back into the caliper. This is necessary to get enough pad clearance to move them past the wear or rust ridge at the outer edge of the rotor. As the pistons are retracted, the fluid returning to the reservoir can cause it to overflow, resulting in spilled brake fluid. Spilled brake fluid will create a mess or damage the painted surfaces under the master cylinder. A much better way of solving this problem is to attach a small hose between the bleeder screw and a container. Then open the bleeder screw while the pistons are being retracted and allow the excess fluid to flow into the container (Figure 11-5). This is the preferred procedure on all vehicles

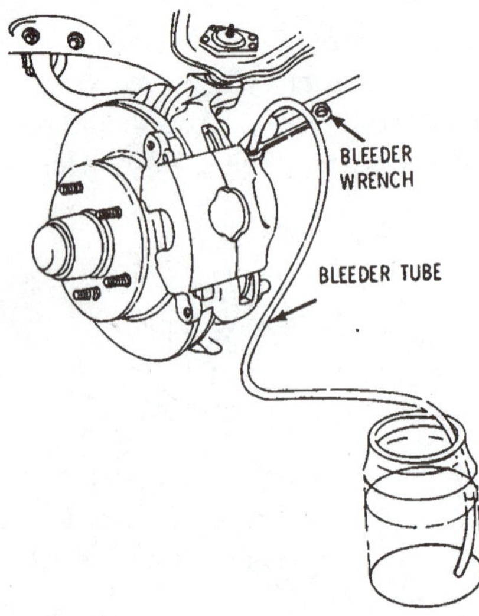

Figure 11-5 A better way to prevent reservoir overflow is to connect a bleeder hose and catch container to the bleeder screw and open the bleeder screw while the piston is being retracted. This method prevents dirty fluid and debris from being pushed into ABS modulator valves. (Courtesy of General Motors Corporation, Service Technology Group)

because it does not force dirty, contaminated fluid back into the ABS hydraulic modulator. Debris being forced back into an ABS modulator can cause sticking or leaking problems of the modulator valves, and if forced into a master cylinder, the debris can cause seal wear or leaking of a quick-take-up valve.

For the most part, calipers that include a parking brake can be serviced in the same way as other calipers, except that the parking brake cable will have to be disconnected and reconnected. Another difference is that the piston must be retracted by methods other than those mentioned in the next section; these procedures are described in most service manuals. The internal hydraulic rebuilding services for calipers will be described in Chapter 12.

It is a good practice to remove heavy rust deposits from the outer edge of the rotor before removing the pads or caliper. This can be done fairly easily and quickly by resting a scraper against the caliper as the rotor is turned by hand. The scraping can be followed by rubbing with coarse sandpaper if a cleaner surface is desired (Figure 11-6).

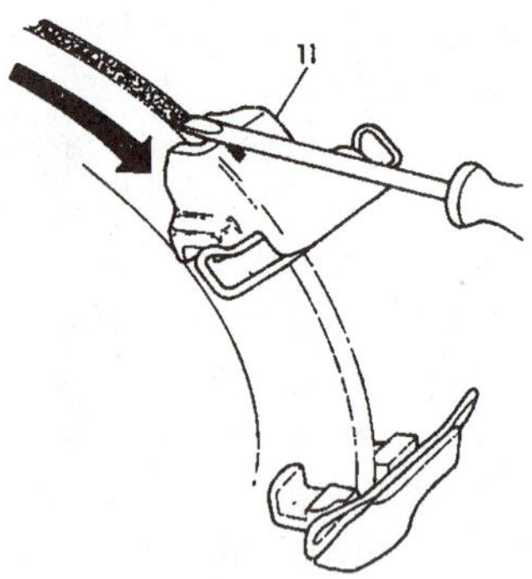

Figure 11-6 A heavy rust deposit at the outer rotor edge can be removed by scraping it with a chisel or old screwdriver as you turn the rotor by hand. (Courtesy of LucasVarity Automotive)

11.2.1 Retracting Caliper Pistons

There are several commonly used methods for retracting caliper pistons. On calipers with two or more pistons, note the amount of effort that is required to retract each piston. A piston that is really hard to retract is probably dragging. On fixed calipers, a large pair of slip-joint pliers can grip the edge of the pad and the outside of the caliper. Slowly squeezing the pliers will retract the pads and pistons into the caliper body (Figure 11-7). Another way to retract the

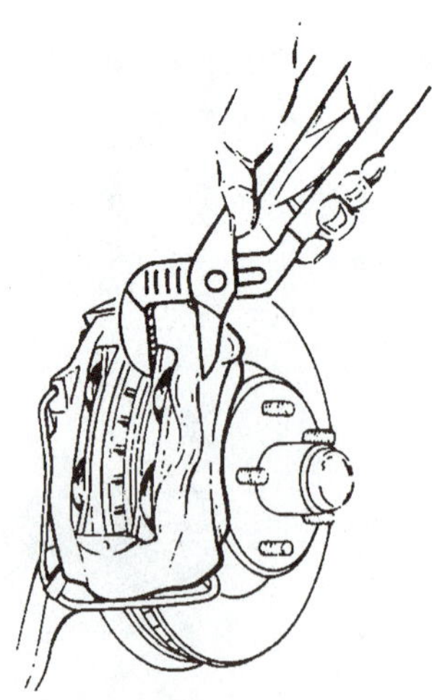

Figure 11-7 Large pliers can be used to squeeze the brake pad and force the piston to retract. (Courtesy of General Motors Corporation, Service Technology Group)

pistons on a fixed caliper is to use a large screwdriver or a special tool (Figure 11-8). Slide the tool between the pad and the piston or between the pad and the rotor. Then slowly and carefully push the piston back into its bore. Be

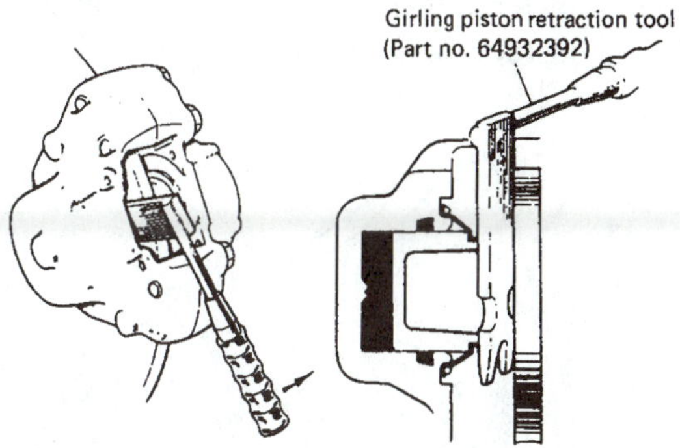

Figure 11-8 This special tool is slid between the rotor and piston and twisted to force the piston back. (Courtesy of LucasVarity Automotive)

careful not to scratch the rotor, piston boot, or lining. If the pads have been removed, there are several types of spreader tools that can be slid between the two pistons to retract them further.

Several different methods can be used to retract the piston in a floating caliper. A large screwdriver or pry bar can be hooked onto the cooling fins of a vented rotor or onto the edge of a solid rotor, and the caliper can be pried outward. This action will push the piston into the caliper (Figure 11-9). Another method is to place a large C-clamp over the caliper body and outboard pad. Tightening the C-clamp will force the caliper outward and retract the piston (Figure 11-10). In some cases, it is possible to position large slip-joint pliers over the edge of the inner brake shoe or caliper support and the inboard side of the caliper body. Slowly squeezing the pliers will retract the piston (Figure 11-11). On calipers with an integral parking brake mechanism, the piston is retracted by rotating it with a special tool (Figure 11-12).

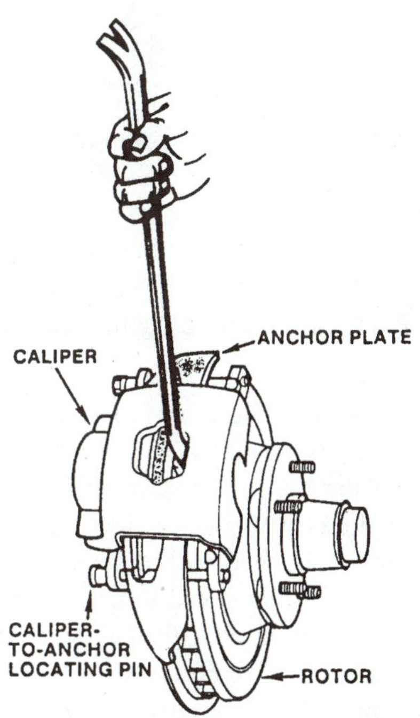

Figure 11-9 A pry bar or large screwdriver can be hooked to the rotor and pulled outward to retract the piston.

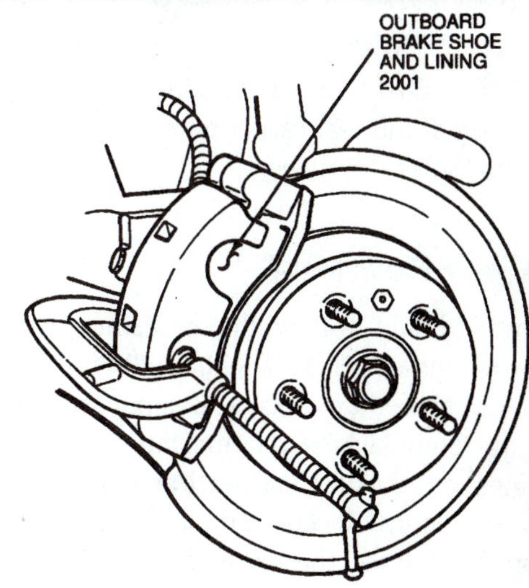

Figure 11-10 The C-clamp positioned over the caliper and outboard pad can be tightened to force the piston into its bore. (Courtesy of Ford Motor Company)

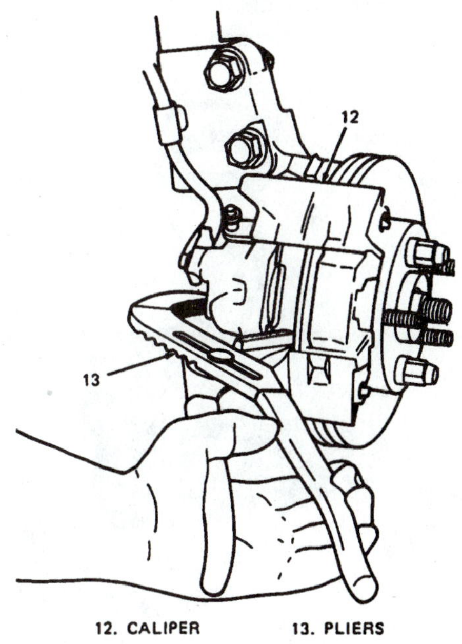

Figure 11-11 These large pliers are gripping the pad and caliper; squeezing the pliers will retract the piston. (Courtesy of Delco Moraine Division, General Motors Corporation, Service Technology Group)

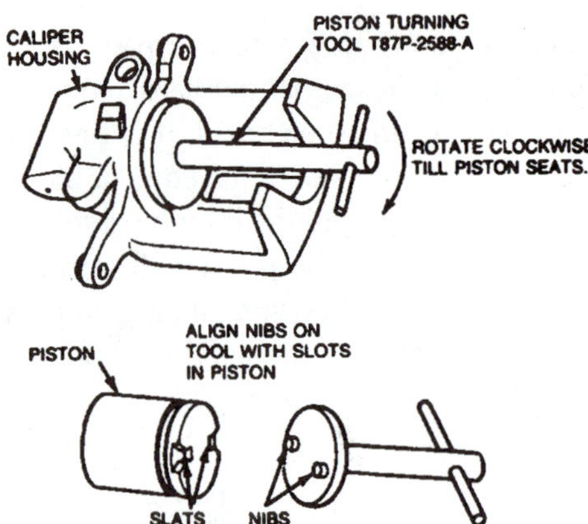

Figure 11-12 A special tool (T87P-2588-A) can be used to rotate the piston of this rear caliper to retract the piston. (Courtesy of Ford Motor Company)

11.2.2 Caliper Removal

Caliper removal should follow the procedure recommended by the manufacturer. This will usually follow one of three general procedures. As a floating or sliding caliper is being removed, the relationship of the small clips and springs, often called **caliper hardware**, should be noted so that correct replacement can be made. Depending on the caliper, the hardware will consist of shims, bushings, support keys, and antirattle springs. In a few cases, you will find a pair of components such as antirattle springs or support keys, that are almost alike but not interchangeable (Figure 11-13).

The brake hose presents another problem. This hose will be damaged if it is twisted or kinked too tightly. The caliper should never be allowed to hang on this hose. A piece of wire or bungee cord should be kept handy so the caliper can be suspended from some convenient location inside the fender well after it has been removed from the mounts (Figure 11-14). Some technicians use a piece of 1/8- or 3/16-inch rod bent into an "S" shape about 3 or 4 inches long, just for this purpose.

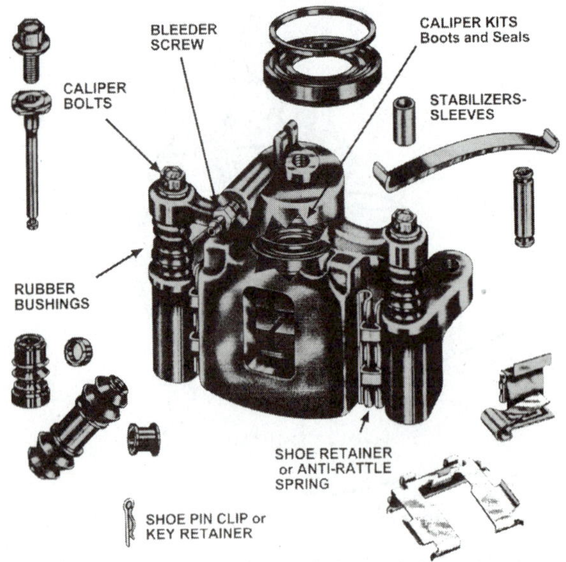

Figure 11-13 A variety of hardware is used with disc brakes. (Courtesy of Rolero-Omega, Division of Cooper Industries)

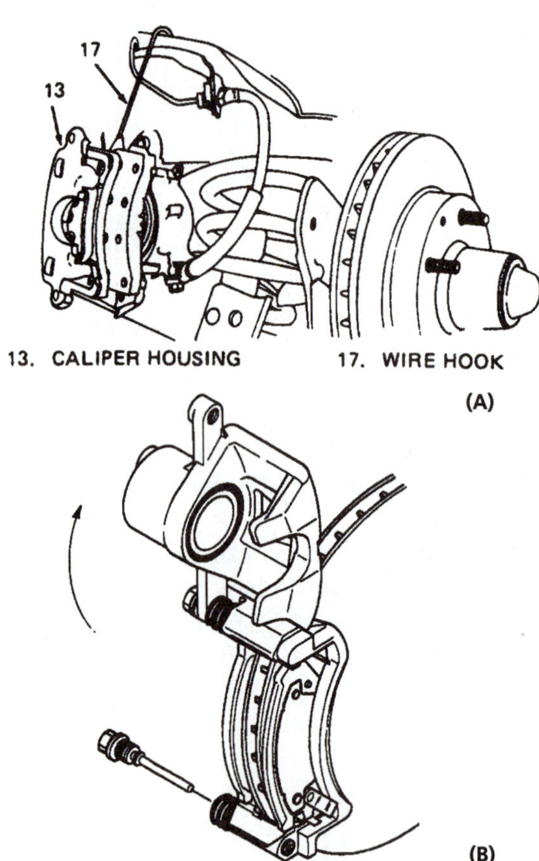

Figure 11-14 When a caliper is removed from its mounts, it should be suspended by a wire (17) and not allowed to hang on the hose (A). Some calipers can be pivoted to a vertical position and left on one of their mounting bolts (B). (A is courtesy of General Motors Corporation, Service Technology group; B is © Saturn Corporation, used with permission)

If the caliper is to be removed completely, the hose will need to be disconnected. This presents a choice as to which end of the hose to remove (Figure 11-15). If the hose threads directly into the caliper, there is no choice. The other end—attached to the steel brake line at the frame—must be disconnected to prevent twisting of the hose. If the hose is secured to the caliper using a banjo fitting and a drilled bolt, there is a choice. The inlet fitting or banjo fitting bolt can be unscrewed, releasing the fitting and the hose from the caliper. If this is done, two new copper washers should be used when this fitting is

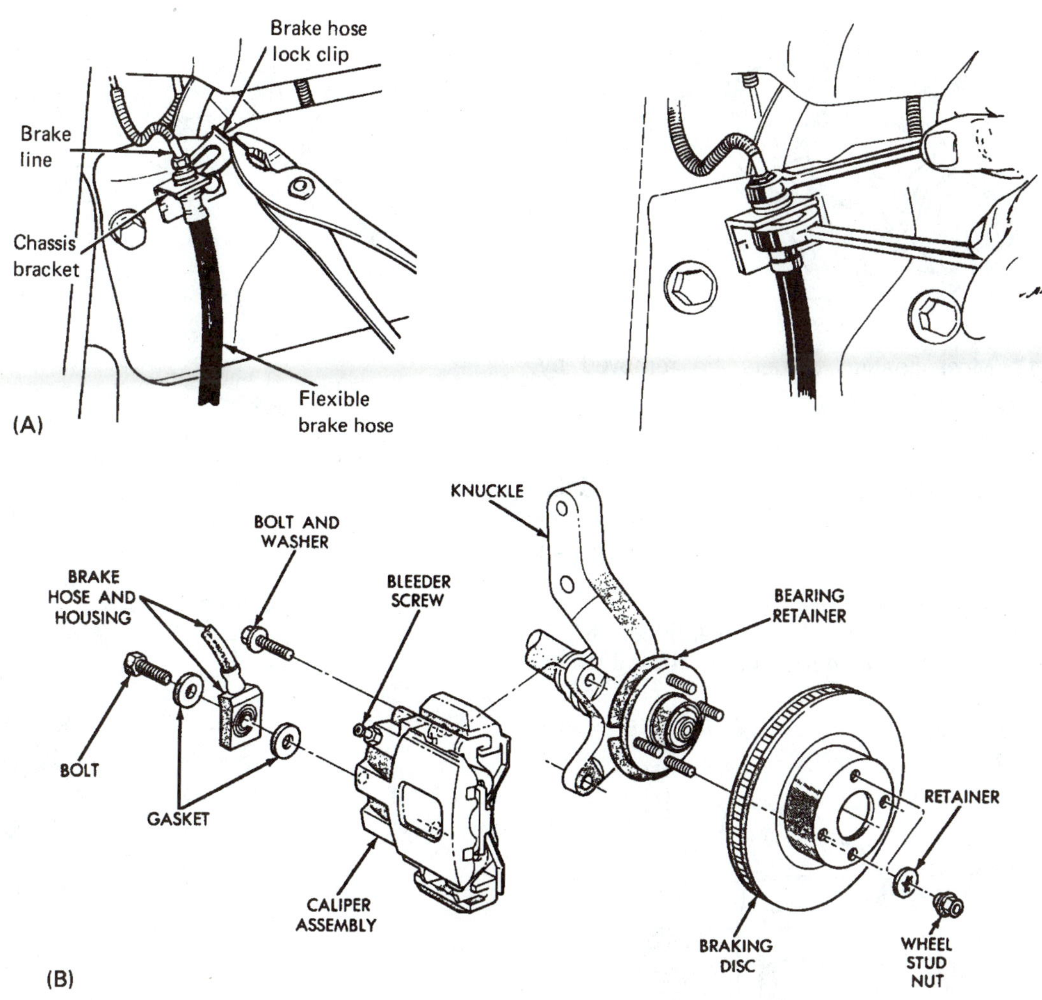

Figure 11-15 When a caliper is removed, the brake hose can be disconnected at the frame (A) or the caliper (B). The copper gaskets/washers should be replaced if the hose is disconnected from the caliper. (A is courtesy of Brake Parts, Inc.; B is courtesy of Chrysler Corporation)

replaced. In some areas, these washers are not easily available at local parts houses. Many brake technicians normally disconnect this hose at the frame end. This is usually a simple operation of unscrewing the steel tube nut from the hose end, sliding the hose retaining clip from the hose, and slipping the hose end out of the frame bracket.

Fixed calipers are normally removed by unscrewing the caliper mounting bolts at the steering knuckle or axle. The removal of these bolts will allow the caliper assembly and the brake hose to be lifted off the rotor (Figure 11-16).

Some floating calipers are removed by unscrewing the **guide pins**, also called *guide bolts* or *locating pins* (Figure 11-17). The removal of these pins will allow the caliper to be lifted out of the caliper mount and off the rotor. Caliper abutment clearance should be checked on floating and sliding calipers as they are removed. This check will be described later in this chapter.

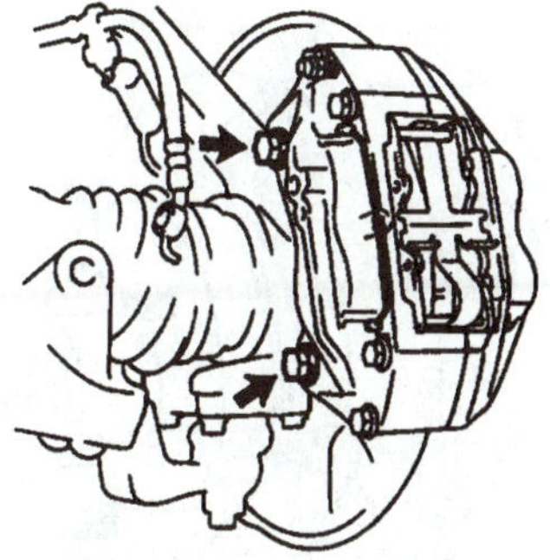

Figure 11-16 The caliper mounting bolts are removed to allow removal of a fixed caliper.

Several types of sliding calipers are held in place by one or two *support keys* or *guide plates*. These keys or plates are secured by a retaining screw or by cotter or split pins. After removing the retaining screw or pins, the support

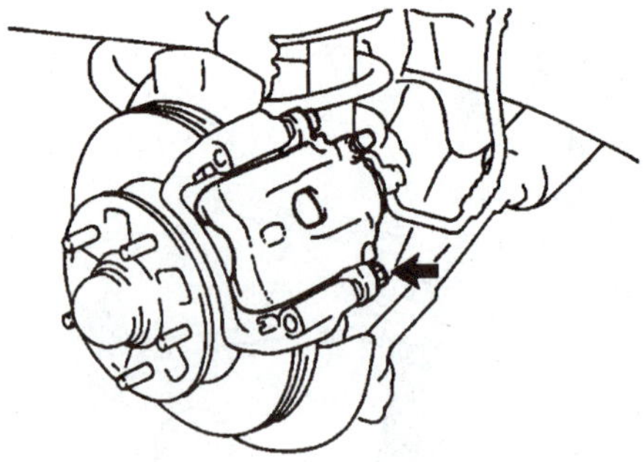

Figure 11-17 Most floating calipers are removed by removing the mounting bolts or guide pins.

key(s) can be driven out from between the caliper and the mounting bracket. Then the caliper can be lifted out of the mounting bracket and off the rotor (Figures 11-18 and 11-19).

Several import calipers use a large support yoke or mounting adapter or bracket. In most cases, the caliper or

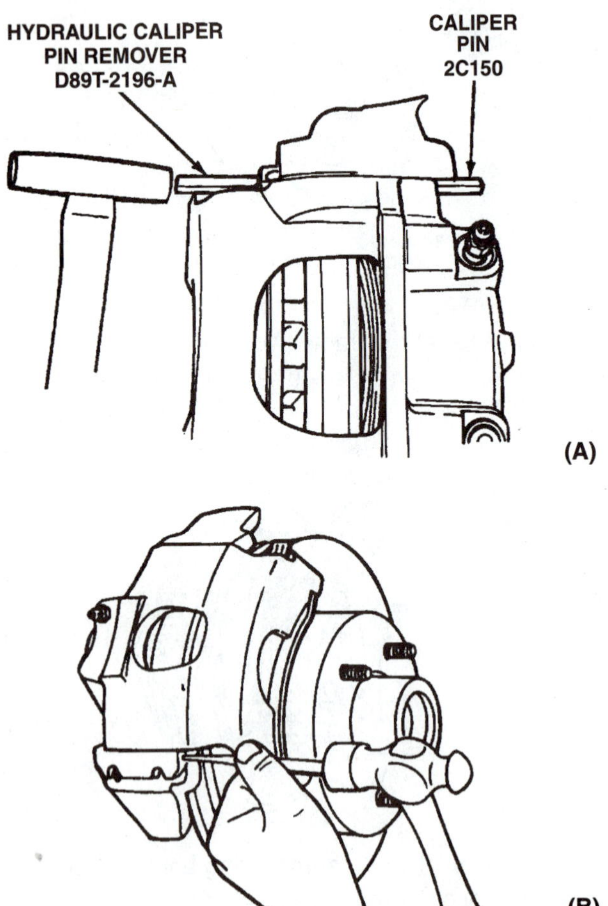

Figure 11-18 The caliper pin or support key is driven out using a special tool (A) or punch (B) so the caliper can be removed from the mounting bracket. (Courtesy of Ford Motor Company.)

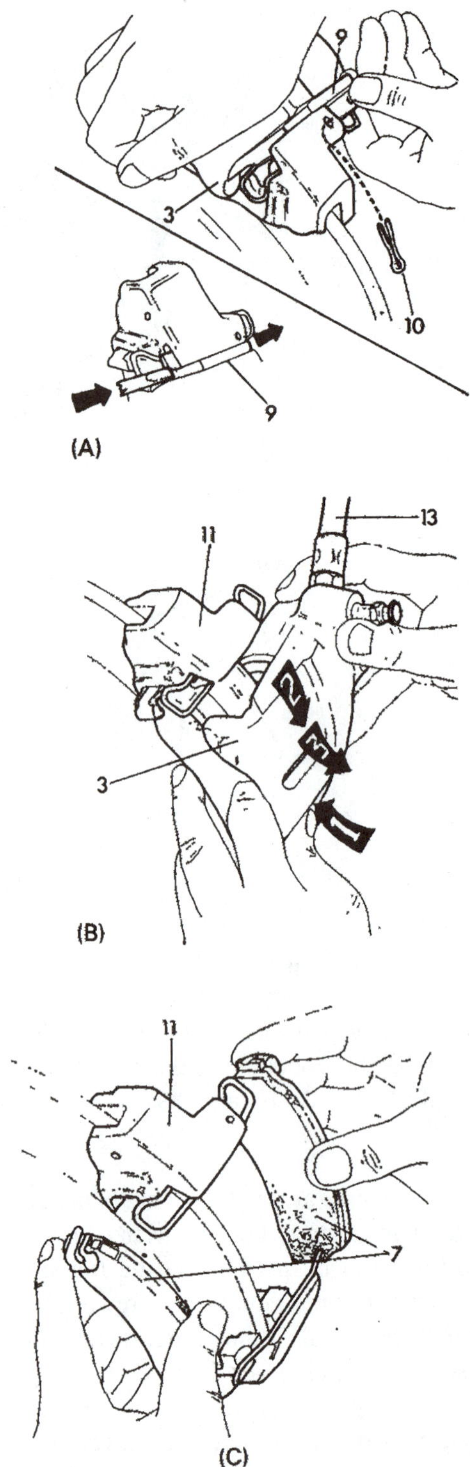

Figure 11-19 This caliper is removed (A) by the split pins (10) and guides (9), and then (B) by lifting the caliper (3) out of the bracket. (Courtesy of LucasVarity Automotive)

support yoke is removed similarly to a floating or sliding caliper. Occasionally it is necessary to remove the mounting support to remove the rotor. A service manual should always be checked to determine the exact procedure (Figure 11-20).

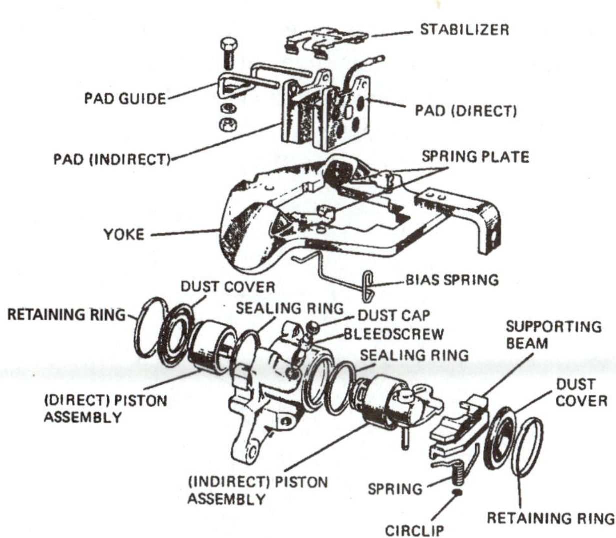

Figure 11-20 A yoke caliper like this one should be removed according to the manufacturer's procedure. (Courtesy of LucasVarity Automotive)

11.3 Rotor Inspection

The rotor should be carefully inspected as part of the routine when doing a brake job and when trying to determine the cause of a particular complaint. Problems such as a pulsating brake pedal, pulsating or vibrating brake action, or a grabby brake are often caused by a rotor that has an excessive runout or parallelism problem. Excessive runout can also knock the pads back and cause a low brake pedal.

In the past, most sources recommended that when the pads are replaced, the rotor should be *reconditioned* or *resurfaced*. A rotor is reconditioned by **turning** it on a rotor lathe; this is also called *truing* or *remachining*. As in the case of used drums, the smooth, burnished surface of the rotor is too smooth for good lining break-in. The ideal surface finish for breaking in organic linings is about 30 to 60 microinches (0.76 to 1.5 μm). For semimetallic linings it is about 40 to 80 micro inches (1 to 2 μm). Surface finishes are not normally measured in the field. They are discussed here only to give you an idea of how smooth or rough a rotor surface should be. When you install a new rotor, check the surface finish and use that as a guide.

In deciding whether a particular rotor should be resurfaced, reconditioned, or replaced, a brake technician considers the following: the type of rotor (vented or solid), the amount of scoring, the width of the rotor, the amount of thickness variation (parallelism), the amount of runout, how the rotor is mounted, the type of repair equipment available, and the manufacturers' recommendations. One manufacturer recommends that the rotors should be replaced in pairs; if one needs replacement, replace both. Also, it is not recommended to mix rotor types; if a car with composite rotors needs a replacement, replace that rotor with a composite type or replace both rotors with cast types. A rotor cannot be turned if the machining process will make it too thin.

Some manufacturers do not recommend turning the rotors on certain vehicles; these should be replaced if there is excessive scoring or thickness variation. Runout problems can often be cured by reindexing the rotor on the hub. Some manufacturers require using an on-car lathe if turning a rotor.

One domestic manufacturer has stated that there are only three reasons that justify turning a rotor: excessive runout, excessive thickness variation, and excessive grooving (0.006" or deeper), and even then, resurfacing can only be done if there will be sufficient thickness after turning. They also state that if more than 0.015" has worn from the rotor, it should be replaced.

Begin with a visual inspection. Check both the inner and outer friction surfaces for scoring and wear. Normal wear will usually produce a smooth but wavy rotor surface. Scores are abnormal wear caused by a metal or abrasive object digging into the rotor surface (Figure 11-21). If the scores or waves are deeper than 0.010 to 0.015 inches (0.25 to 0.38 mm), the rotor should be turned to produce a flat surface for breaking in the new lining. Many technicians treat the inner and outer rust ridges as scoring because they can interfere with the contact between the new lining and the rotor. Some older General Motors cars were produced with a rotor that had a deep groove machined midway around the friction surface, and this groove should not be confused with scoring (Figure 11-22).

Figure 11-21 This rotor is ruined because the scoring is too deep; machining cuts to true the surface will produce a rotor that is too thin.

Figure 11-22 The groove in this rotor was produced by the manufacturer; it is unworn.

Also, check visually for rotor glaze (a highly glassy surface which is somewhat normal); blueing; contamination from foreign materials such as oil, paint, or silicone sprays; and heat checks. The smooth, glassy surface should be roughened up by resurfacing the rotor to ensure proper pad break-in. If the rotor shows severe heat checking, it should be turned or replaced (Figure 11-23). A contaminated rotor should be thoroughly cleaned by using a friction surface cleaner or by machining.

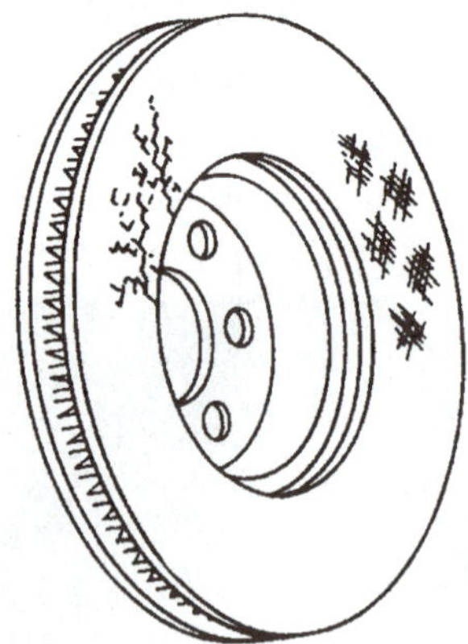

Figure 11-23 An overheated rotor will have small cracks or heat checks on the rotor surface. (Reprinted from Mitchell Anti-Lock Brake Systems, with permission of Mitchell Repair Information, LLC)

11.3.1 Rotor Measurements

Before a rotor is reused, it should be measured for size. This is normally done by carefully measuring the thickness of the friction surface with an outside micrometer or caliper. This measurement is normally taken midway between the inner and outer edges of the friction surface (Figure 11-24). The minimum thickness for a rotor is specified by the manufacturer and, since 1971, this dimension has been indicated on the rotor (Figure 11-25).

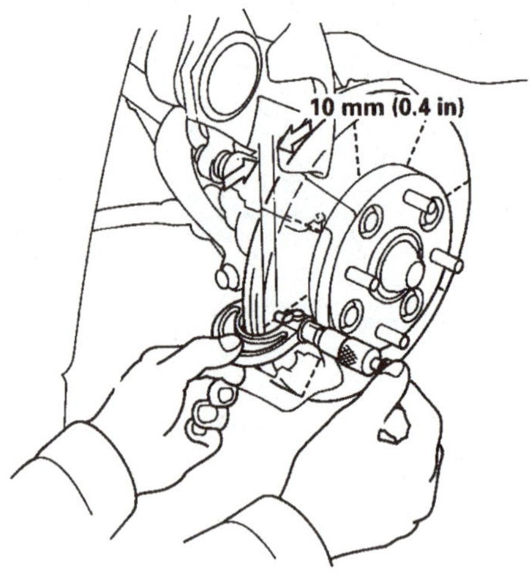

Figure 11-24 Rotor thickness is measured using a micrometer positioned in the middle of the friction surface. Measuring at eight points helps ensure that you find the thinnest spot and also checks for thickness variation. (Courtesy of American Honda Motor Co., Inc.)

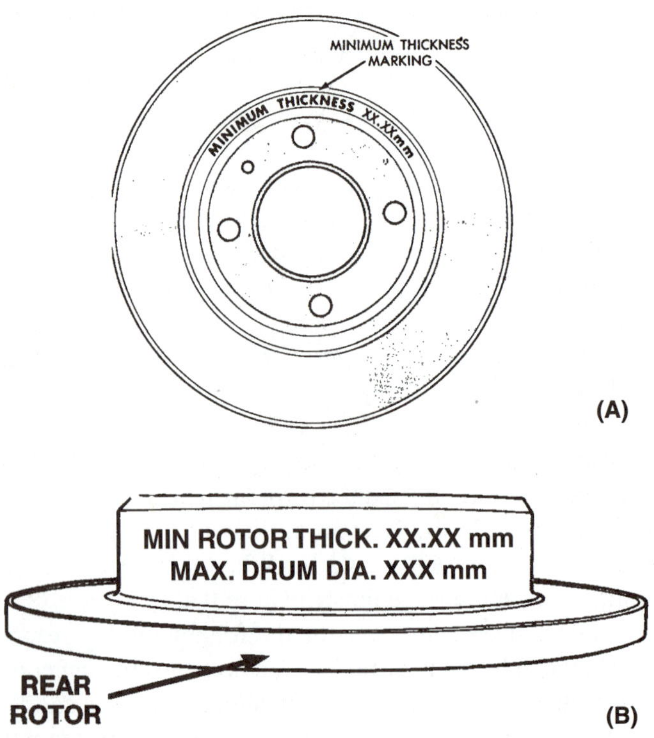

Figure 11-25 All modern rotors have their minimum thickness indicated on them (A). Rotor B is a rear rotor with a drum parking brake; it also bears the maximum drum diameter. (Courtesy of Chrysler Corporation)

A rotor that is too thin should not be used. As in the case of a drum, a rotor that is too thin will not have sufficient heat-sink capability and overheating of the lining will result. This is currently a potential problem with non-vented rotors used on front-wheel-drive cars. Another problem is that too thin a rotor will position the caliper piston too far out of the bore.

While the rotor is being measured for thickness, it can also be checked for **parallelism**. This is sometimes called **thickness variation** because it is a variation in the thickness across the rotor's friction surface, which causes a lack of parallelism. If the two friction surfaces of a rotor are not parallel to each other, the pads must move together and apart during stops as they try to follow the changing friction surfaces. This action can be felt as a pulsation—a rising and falling—of the brake pedal, or as a vibrating, grabby action of the brakes (Figure 11-26). To check parallelism, the rotor is carefully measured at eight to twelve different locations around the rotor as the technician tries to find the thickest and thinnest rotor widths. The measurements are normally taken in the middle of the rotor surface. Subtracting the thin measurement from the thick measurement will give the amount of thickness variation or parallelism (Figure 11-27). Compare these findings with the manufacturer's specifications (Figure 11-28). As a general rule, if the thickness varies more than 0.0005 to 0.001 inches (0.013 to 0.025 mm), the rotor should be turned or replaced. Some technicians prefer to check for parallelism problems using one or two dial indicators. This procedure will be described later in this chapter.

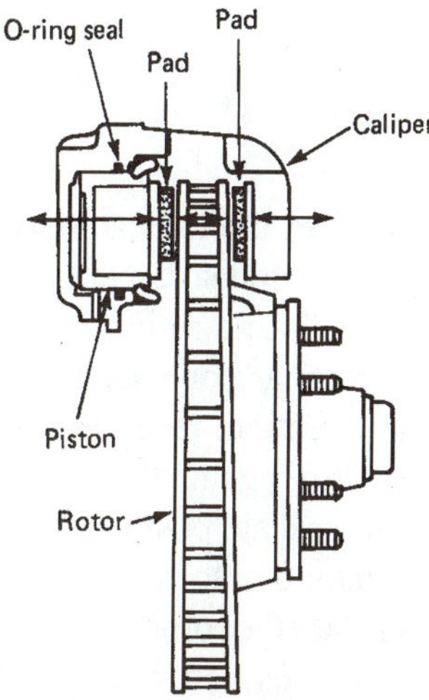

Figure 11-26 If the rotor surfaces are not parallel, the pads will move inward at the thin area and then back out when the thick area passes between them. This causes a pulsation of the brake pedal and surging brake action.

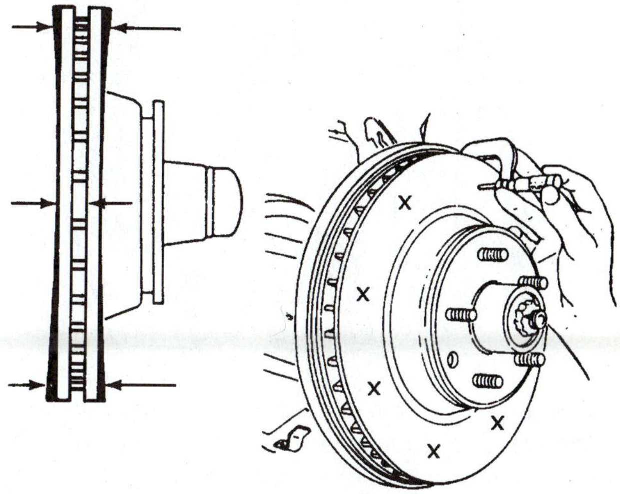

Figure 11-27 Rotor thickness variation is checked by measuring the rotor at eight to twelve places and comparing the measurements. (Reprinted from Mitchell Anti-Lock Brake Systems, with permission of Mitchell Repair Information, LLC)

BRAKE SPECIFICATIONS

Disc Thickness (New)	0.7500 in.
Disc Thickness (Minimum)	0.6850 in.
Disc Parallelism (Thickness variation, max)	0.0007 in.
Disc Runout (Maximum TIR)	0.0030 in.
Surface Finish	15–80 micro inches

Figure 11-28 Typical rotor specifications. (Courtesy of General Motors Corporation, Service Technology Group)

Also, some manufacturers recommend checking the rotor for wedge-shaped wear while measuring the rotor thickness. Measure the rotor as close to the center of the rotor's friction surface as the micrometer frame will allow and then measure it again at the outer edge. A difference in measurements indicates wedge wear (Figures 11-29 and 11-30). If it is excessive, the rotor should be turned or replaced.

Rotor runout is a wobbly, side-to-side motion of the friction surfaces which occurs as the rotor revolves. It will knock or kick the pads away from the rotor as the car is driven. The result will be too much pad clearance and a low brake pedal the next time the brakes are applied. It can also cause side-to-side motion of the caliper during a stop, and if the caliper doesn't slide easily, can cause a pulsating, vibrating stopping action (Figure 11-31).

Rotor runout, often called lateral runout, is measured using a dial indicator (Figure 11-32). The dial indicator is attached to the steering knuckle, spindle, caliper mounting

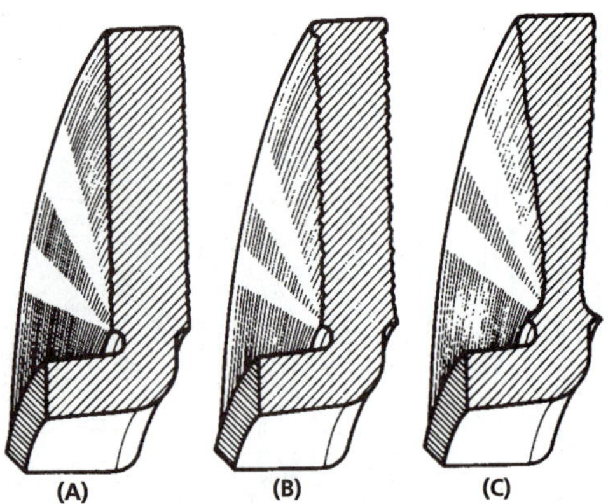

Figure 11-29 Rotor (A) is slightly worn; rotor (B) shows severe scoring and roughness; and rotor (C) is worn to a wedge shape. (Courtesy of ITT Automotive)

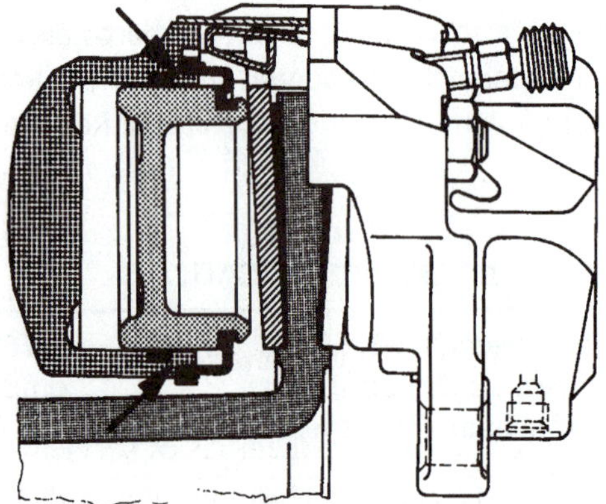

Figure 11-30 A wedge-shaped rotor can cause the piston to cock in the bore, a caliper to cock on its mounts, or a caliper to fracture. (Courtesy of ITT Automotive)

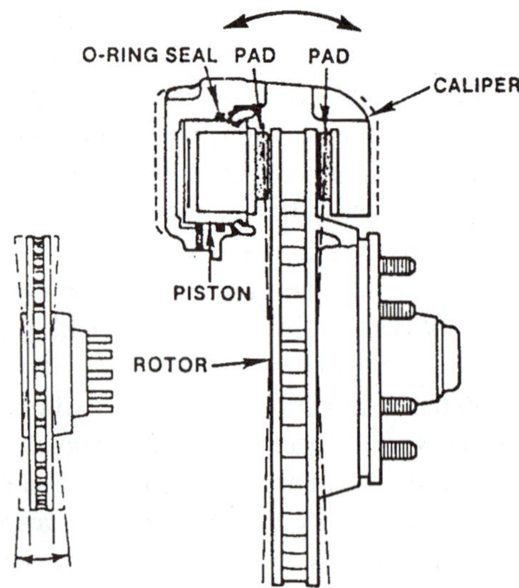

Figure 11-31 Excessive rotor runout forces the pads and caliper to move back and forth sideways when the brakes are applied. This can cause surging of the brakes, vibration, and steering-wheel oscillation.

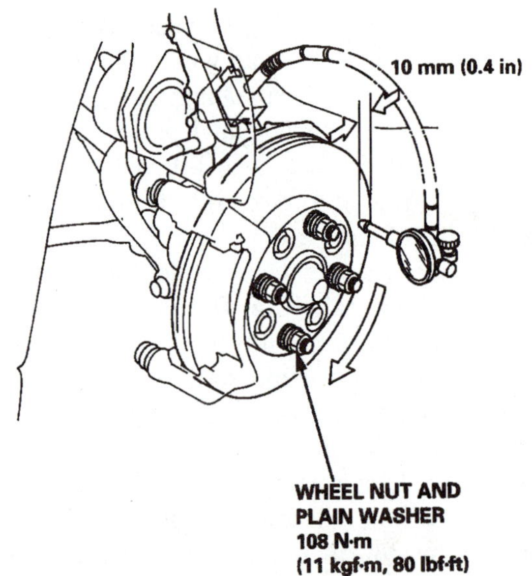

Figure 11-32 Rotor runout is checked by mounting a dial indicator on the rotor friction surface and rotating the rotor. Note the wheel nuts holding the rotor tight; on some cars the wheel bearings should be snugged up. (Courtesy of American Honda Motor Co., Inc.

bracket, or caliper. Then the indicator stylus is positioned in the middle of the friction surface. Be sure to mount the indicator so the stylus is at a right (90°) angle to the rotor surface. Rotate the rotor as you watch the indicator needle. If it moves in one direction, reverses, moves in the other direction, and reverses again, there is runout. Stop the rotor at the most downscale position of the indicator needle, adjust the indicator dial to zero, and mark this position on the rotor. Now rotate the rotor until the needle is reversing at its most upscale position, note the reading, and mark the rotor again. This is the highest, most outward part of the rotor, and the indicator reading is the amount of runout. Compare the measured amount of runout with the manufacturer's specifications. If the measured amount is greater, the rotor must be turned or replaced. As a general rule, runout should not exceed 0.005 inches (0.13 mm).

As you are measuring runout, note the location of the high and low spots. If they are on opposite sides of the rotor, the problem is truly one of runout, and the rotor is wobbling. If the indicator needle dips for only a portion of the rotor and most or part of the rotor is flat (no indicator motion), there is a hollow, concave area in the rotor's surface. This is more a problem of parallelism than of runout. Note that loose or rough wheel bearings will affect runout readings. It is recommended that you snug up the wheel bearings to zero clearance before measuring runout. Be sure to readjust the bearings to the correct clearance before the car is driven (Figure 11-33).

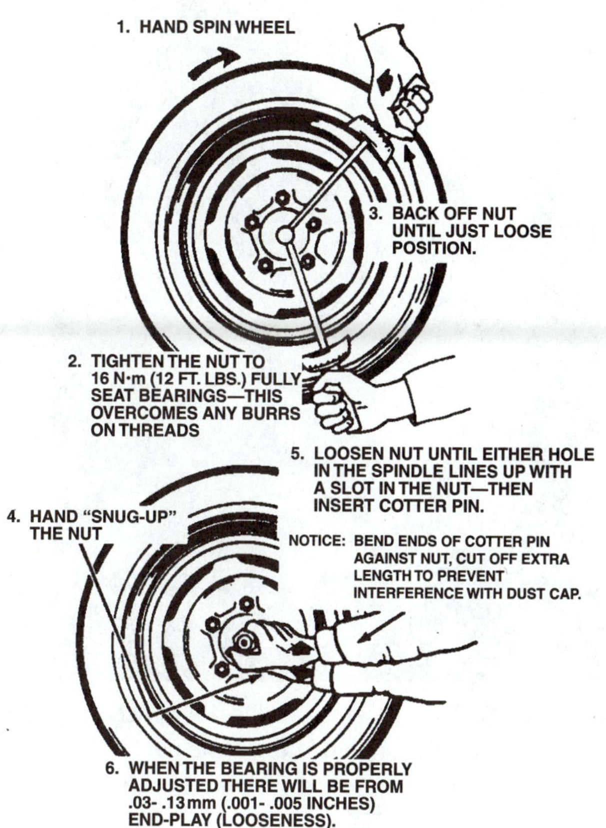

Figure 11-33 If wheel bearings are tightened for a rotor runout check, they must be readjusted before the car is driven. (Courtesy of General Motors Corporation, Service Technology Group)

Tapered shims are available to remove runout at the rear rotors of older Corvettes (Figure 11-34). On this car, as well as on some others, a distorted axle flange can produce rotor runout. This runout can be eliminated only by turning the rotors on the car or on the axle, which usually requires removal of the axle shaft from the car. This particular rotor has a new thickness of 1.250 inches (31.75 mm) with a machining limit of 1.230 inches (31.24 mm). Only 0.020 inches (0.5 mm) can be machined off in this rotor to true it. A shim, which is available in increments of 0.001 inches (0.25 mm), is placed between the rotor and the axle flange with the thick part of the shim in the correct position to reduce the amount of runout to the allowable specifications. The rotor and shim are indexed as needed to find a location where the amount of runout is acceptable.

Don't forget that you are really trying to determine what needs to be done about the rotor. Any rotor that exceeds the runout or parallelism limits must be turned or replaced, and any rotor that will become too thin must be replaced.

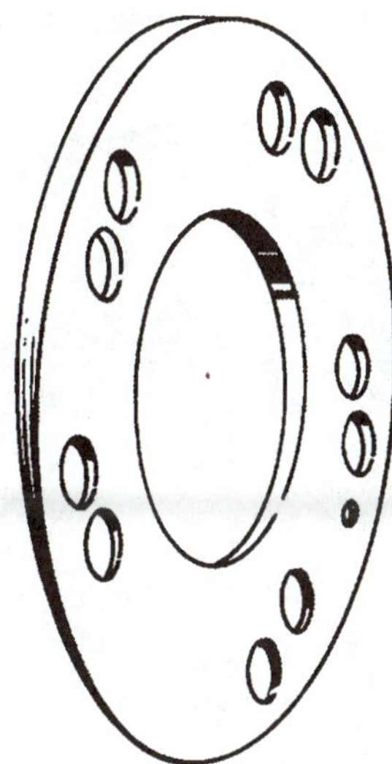

Figure 11-34 This tapered shim can be placed between the rotor and axle flange to correct runout; it is available in varying amounts of taper. (Courtesy of Stainless Steel Brakes Corp.)

11.4 Rotor Refinishing

A brake technician has four methods of servicing a rotor: resurfacing by hand, off-car turning, on-car turning, and replacing. When choosing a method, the brake technician must remember that both rotors on an axle must have the same surface finish.

The ideal rotor surface is a flat surface that has very shallow scratches in it. These scratches should be about 40 to 80 microinches (0.76 to 1 μm) deep, and they should run in every direction. This is called a **nondirectional** or **swirl finish**. A rotating sander will produce a series of more desirable circular but irregular scratches running in every direction, and the overlapping scratches produce the nondirectional, swirl finish (Figure 11-35). When a lathe cuts a rotor, it produces a scratch that slowly spirals from the center to the outer edge—constantly flowing in the same circular direction (Figure 11-36).

Rotor replacement becomes mandatory if the rotor is too thin or will become too thin if turned to remove defects like deep scoring or excessive runout. In other cases, where the rotor is worn thin but still not undersize and a new rotor is inexpensive and easily replaceable, it is a good idea to replace the rotor. If the rotor has worn a large amount (more than halfway to the discard thickness) on the original pads, it is a good idea to replace it because it will likely wear past the minimum thickness during the next set of pads. A new rotor with its greater heat-sink capability will usually result in a longer lining life. In

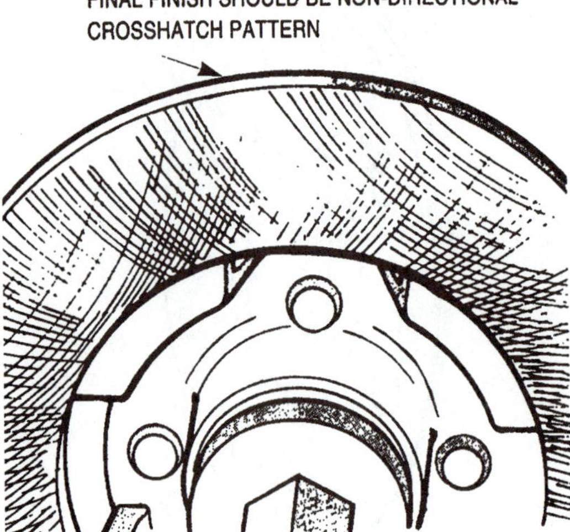

Figure 11-35 A nondirectional/crosshatch surface has a series of circular scratches that run in all directions. (Courtesy of Chrysler Corporation)

some cases, the replacement rotor can be very expensive, or rotor replacement can be a difficult, time-consuming operation (Figure 11-37). In this case, the added expense does not warrant replacement, and it is usually better to refinish the old rotor.

Figure 11-36 A drum lathe produces a directional finish with all the scratches running in the same direction.

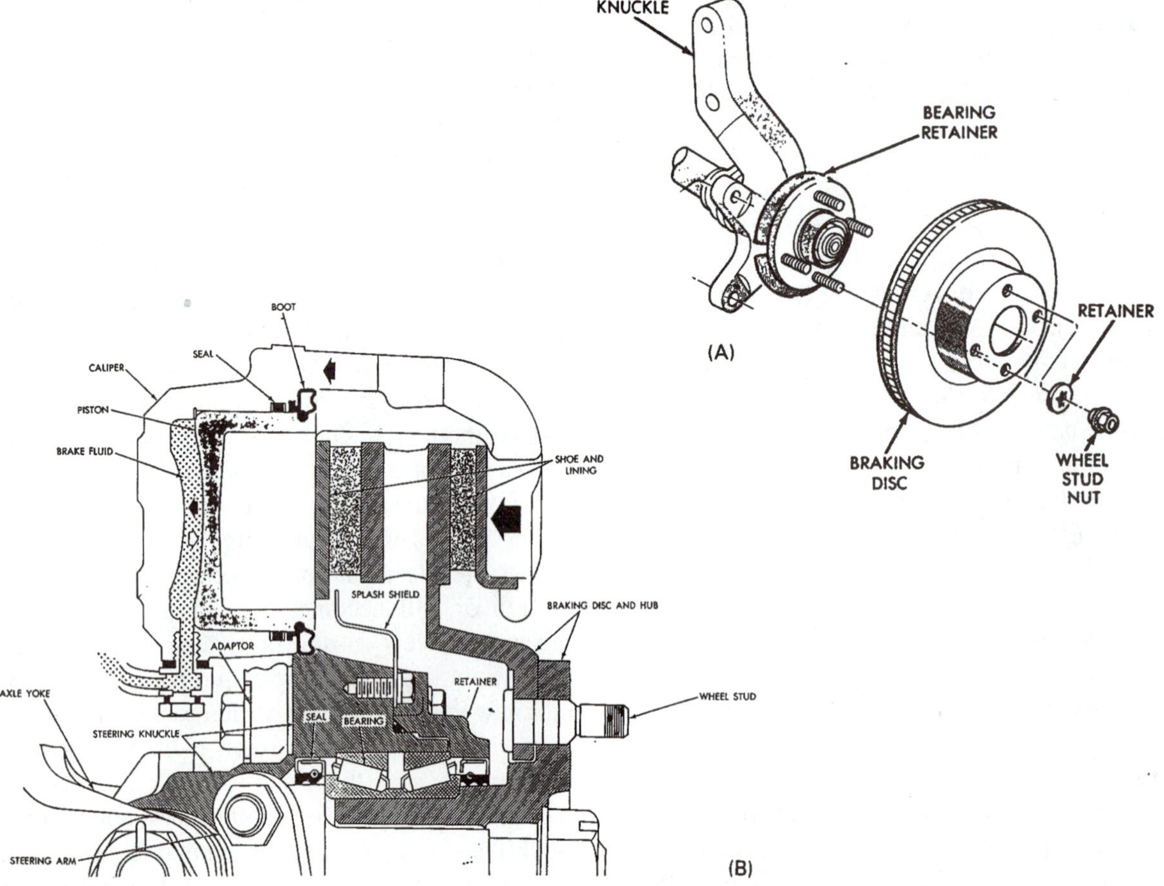

Figure 11-37 The rotor on many FWD cars (A) is held in place by the wheel and lug nuts so it is easy to R&R. Some rotors are held captive behind the hub flange (B); they are much harder to R&R. (Courtesy of Chrysler Corporation)

Resurfacing a rotor by hand is not recommended by many manufacturers because of the possibility of producing a different surface finish on the two rotors. Some technicians resurface a rotor in order to break the glaze on an otherwise good rotor while removing as little metal as possible. Some front-wheel-drive cars experience very rapid lining wear after the rotors have been turned, and this can be blamed on a loss of the rotor's heat-sink capability. As mentioned earlier, the car will have a brake pull toward one side if the two rotors do not have the same surface finish. This pull will probably disappear after a few stops, however, because the pads will quickly burnish out the small difference in the scratches. Resurfacing is desirable to remove glaze, old lining residue, or light rusting on true rotors, because it removes a minimum amount of metal. Resurfacing will not true up a rotor. If there is excessive runout or thickness variation, the rotor must be turned or replaced. Resurfacing a rotor by hand sanding or using a disc sander normally produces a nondirectional finish. This procedure will be described later in this chapter.

A rotor is turned on a brake lathe (Figure 11-38). At one time, turning a rotor was an off-car operation only.

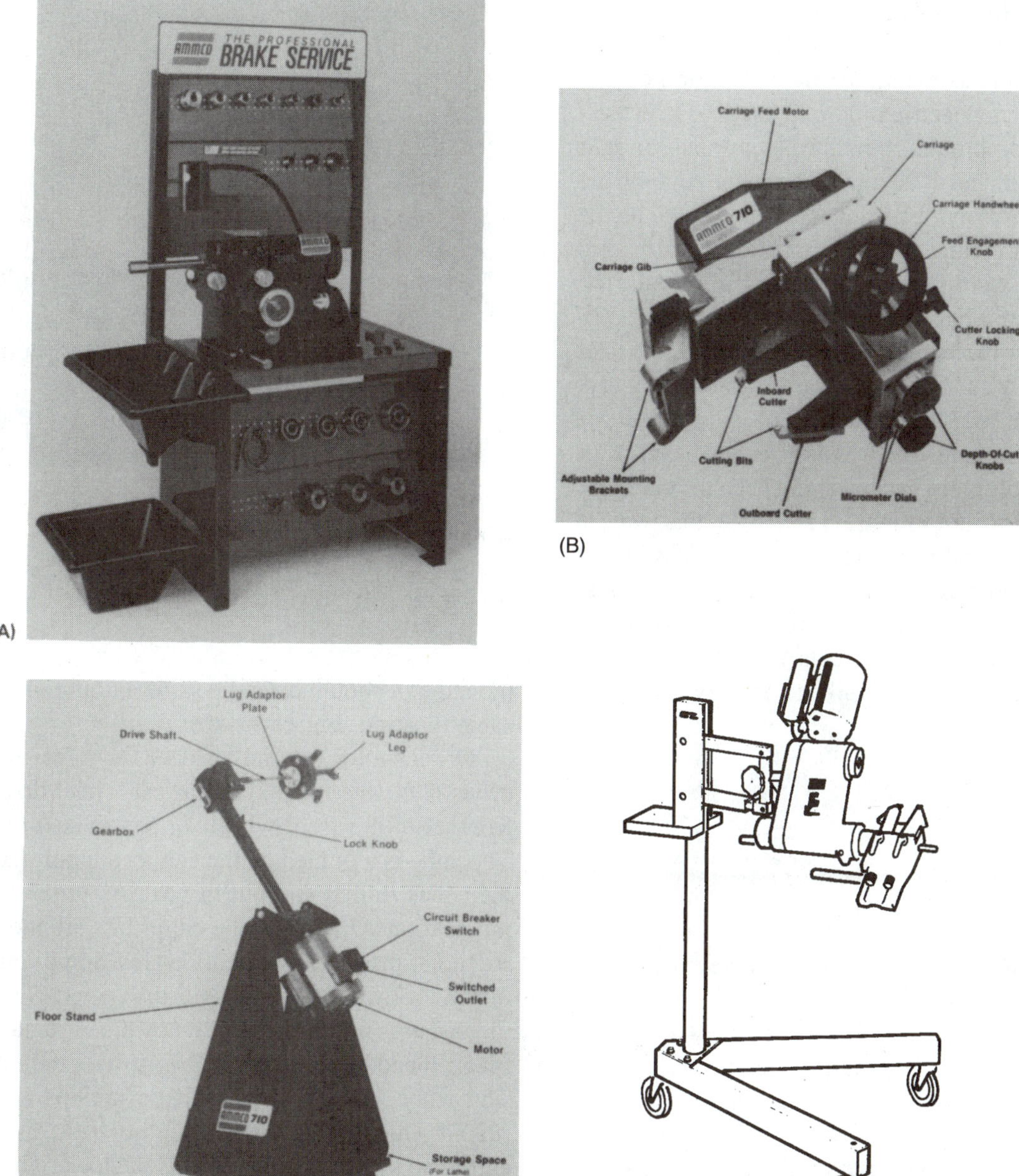

Figure 11-38 There is a variety of rotor lathes. (A) is a standard off-car lathe. (B) is an on-car lathe that is mounted onto the caliper brackets. The rotor is driven by the car's engine or a rotor-driving unit (C). (D) is an on-car lathe that is mounted onto the hub and includes a drive motor. (A is courtesy of Ammco Tools, Inc.; D is courtesy of Hunter Engineering Company)

Today, on-car lathes and turning devices are common. When an off-car lathe is used, the rotor is removed from the car and mounted on the lathe mandrel or arbor. As the rotor is rotated, a pair of cutter bits is moved straight across the friction surfaces. Some of the earlier lathes used a single cutter that required two cutting operations, one for each side of the disc. All new machines use twin cutters which cut both sides at the same time. This method is faster and produces truer friction surfaces. As the worn friction surface is removed by the cutters, truly parallel but thinner friction surfaces are produced.

When an on-car lathe is used, the rotor is left-mounted in its normal manner, and the lathe mechanism is either attached to the caliper mount or to the wheel-mounting bolts. Depending on the unit, the rotor is rotated by the vehicle's engine, an electric motor that is part of the lathe unit, or by a separate electric motor.

When a rotor is turned, the thinnest cut made should be at least 0.004 inches (0.1 mm). The outer edge of the rotor surface tends to harden from use, which could be the result of work-hardening from the pads or heat-treating that occurs during hard stops. With very shallow cuts, the point of the cutter bit has trouble penetrating the hard surface and will wear rapidly from the pressure and the heat generated. Shallow cuts tend to smear the rotor surface. A slightly deeper cut will use the side of the cutter to cut the harder surface skin, and the increased cutter-bit-to-rotor contact will transfer more heat and help keep the cutter bit from becoming too hot (Figure 11-39). In this way, a turned rotor will be reduced a minimum of 0.008 inches (0.2 mm)—0.004 inches on each side. If there is thickness variation, rotor runout, or mounting runout, the rotor will be even thinner because deeper cuts will be required. If the rotor is not mounted on the arbor correctly, mounting runout will occur. This runout will require a deeper cut on each side and also cause runout when the rotor is remounted on the car. When turning rotors that have excessive runout on the car, it is a good practice to mark the high spot of the runout. After the rotor is mounted on the lathe, it should have the same amount of runout in the same location (Figure 11-40).

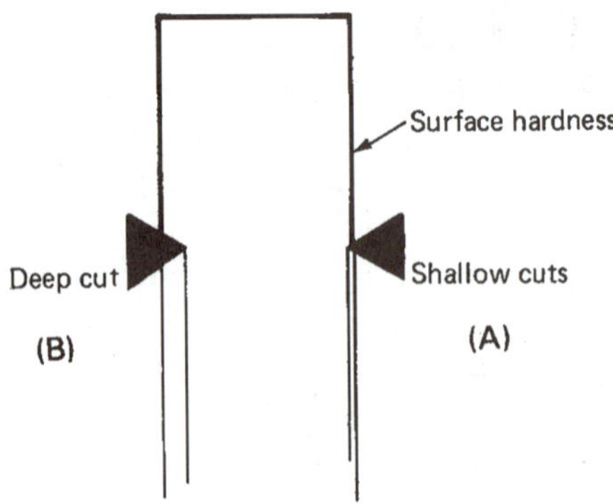

Figure 11-39 A shallow cut uses the very tip of the cutter bit (A), and this places a great load on a small area. A deeper cut (B) uses more of the stronger part of the lathe bit.

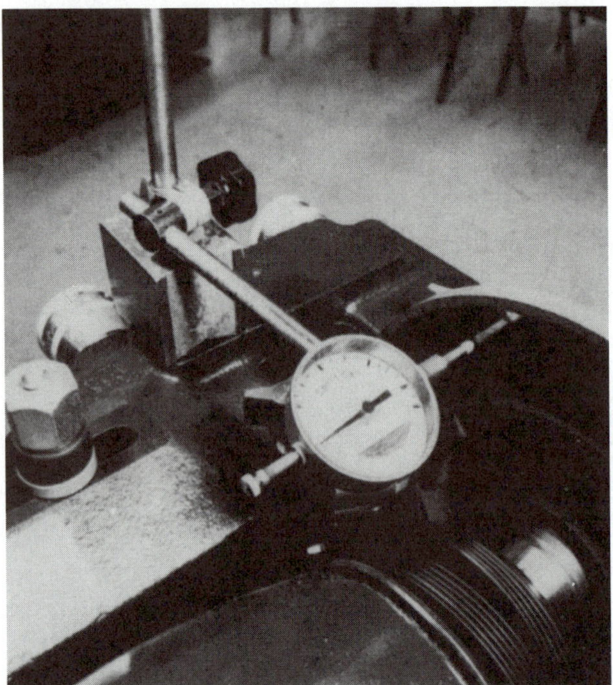

Figure 11-40 This dial indicator is set up to measure the amount of runout with the rotor mounted on the lathe. It should indicate the same high and low spots as when it was on the car.

On-car rotor turning is ideal for rotors that are on a drive axle—the rear axle of a rear-wheel-drive car or the front axle of a front-wheel- or four-wheel-drive vehicle. The lathe is mounted on the caliper support (adapter), and the car is started and put in gear. As the engine or drive motor rotates the rotor, the cutter bits are moved past the friction surfaces (Figure 11-41). Some on-car lathes include a drive motor to spin the motor. Since the rotor mounting is not disturbed, removal and replacement time is saved and there are no problems with mounting runout. The rotor surfaces will be true and parallel, with no runout. One turning device uses abrasive discs instead of cutters to smooth the rotor. This device produces a nondirectional finish. Several manufacturers market an on-car lathe with a portable drive motor that allows any rotor to be turned on the car.

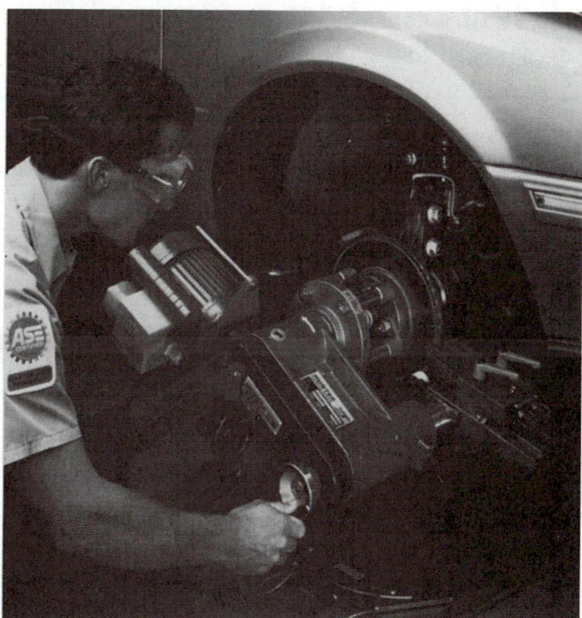

Figure 11-41 This front rotor is being turned on-car with the rotor in its original, undisturbed position. (Courtesy of Hunter Engineering Company)

When turning a rotor, use the following recommendations:

- Use only sharp cutter bits.
- Always use the correct rotor-mounting adapters.
- A vibration, silencer, or antichatter device must always be installed snugly around or against the rotor.
- Be aware that the surface finish of the rotor will be affected by the sharpness of the cutter bit(s), the use of the vibration band, the depth of cut, the feed rate, the speed of the lathe arbor, and the mechanical condition of the brake lathe.
- Front rotors (rear-wheel-drive cars) should be turned with the hub. Rotors that are loose or easily removed should be secured with lug nuts.
- Composite rotors must be mounted using clamping adapters that simulate the clamping action of a wheel. Composite rotors have a greater tendency to vibrate, and this produces a poor finish.
- Rotors with wheel bearings should be checked to make certain that the bearing races are in good condition and tight in the hub. Damaged races should be replaced before turning the rotor.
- Never bump the cutter bit or push it against a rotor that is not revolving. The hard carbide bits that are commonly used chip and break easily.
- Always check and remove mounting runout before turning a rotor.
- The inner and outer ridges should be removed as a beginning step.
- Always remove the smallest amount of metal possible, but don't make too shallow of a cut (less than 0.004 inches).
- Always cut both sides of the rotor before removing it from the lathe.
- Rotors used with fixed calipers should have the same amount of metal machined from each friction surface.
- To achieve an acceptable final finish, hold a swirl tool with 120-grit paper or a sanding block with 150-grit paper against the rotating friction surfaces for 60 seconds using a medium pressure (Figure 11-42).

ROTOR STOPPING EFFICIENCY

- new OEM rotor 100%
- standard turning procedure 60%
- standard turning procedure
  - * plus sanding block for 60 sec. 110%
  - * plus swirl tool for 60 sec. 90%
  - * plus swirl tool for 20 sec. 60%
- rotor-good condition- no resurfacing 96%

Figure 11-42 A comparison of the stopping efficiency of various rotor friction surfaces. (Courtesy of Hunter Engineering Company)

- Don't forget that sanding will further reduce the rotor thickness.
- It is a good practice to deburr the outer edge after machining using a file or 80-grit sandpaper—either handheld or on a small disc—placed lightly against the rotating rotor surface.
- A turned rotor should be cleaned using a commercial brake surface cleaner or denatured alcohol (Figure 11-43). Wash the turned surfaces with a clean shop cloth and the liquid until the cloth shows no more contamination.
- After cleaning, do not touch the new friction surfaces. Pick up the rotor by the outer diameter or by the center hole.

As in the case of a drum, when a rotor is turned, minute particles of metal and carbon from the cast iron

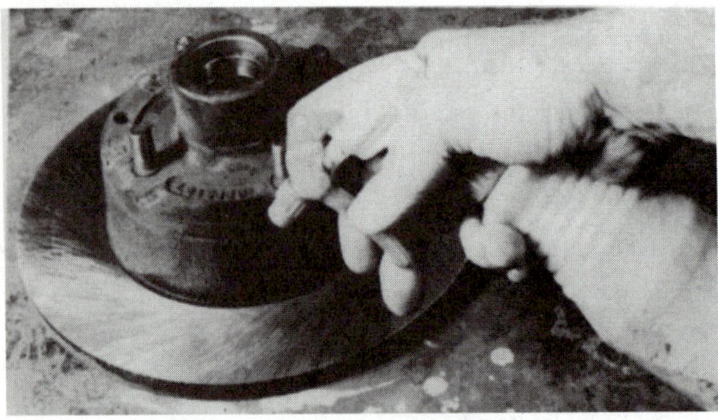

Figure 11-43 After a rotor has been turned, the friction surfaces should be flushed clean to remove all traces of carbon and small metal chips.

will become embedded in the surface scratches. Unless they are washed out, these particles will become embedded in the new lining surface. Once there, they can cause abnormal braking, squeal, and/or excessive future rotor wear.

11.5 Machining a Rotor Off-Car

SAFETY TIP: When machining a rotor, use the caution that should always be exercised around moving machinery. Be particularly concerned about the revolving drum, the cutter bit, the revolving vibration band, and the revolving lug bolts on some drums.

When a rotor is machined, follow the directions furnished by the manufacturer of the machine you are using. The description which follows is very general. To turn a rotor, you should:

1. Clean the rotor and hub. All oil and grease in the hub or on the rotor should be washed off with solvent and air-dried.
2. Select the correct mounting adapters. Always make sure there is no grit between the adapters and the rotor so the rotor will be exactly centered as it is mounted with no runout. Make sure the cutter(s) is out of the way while the rotor is being mounted.

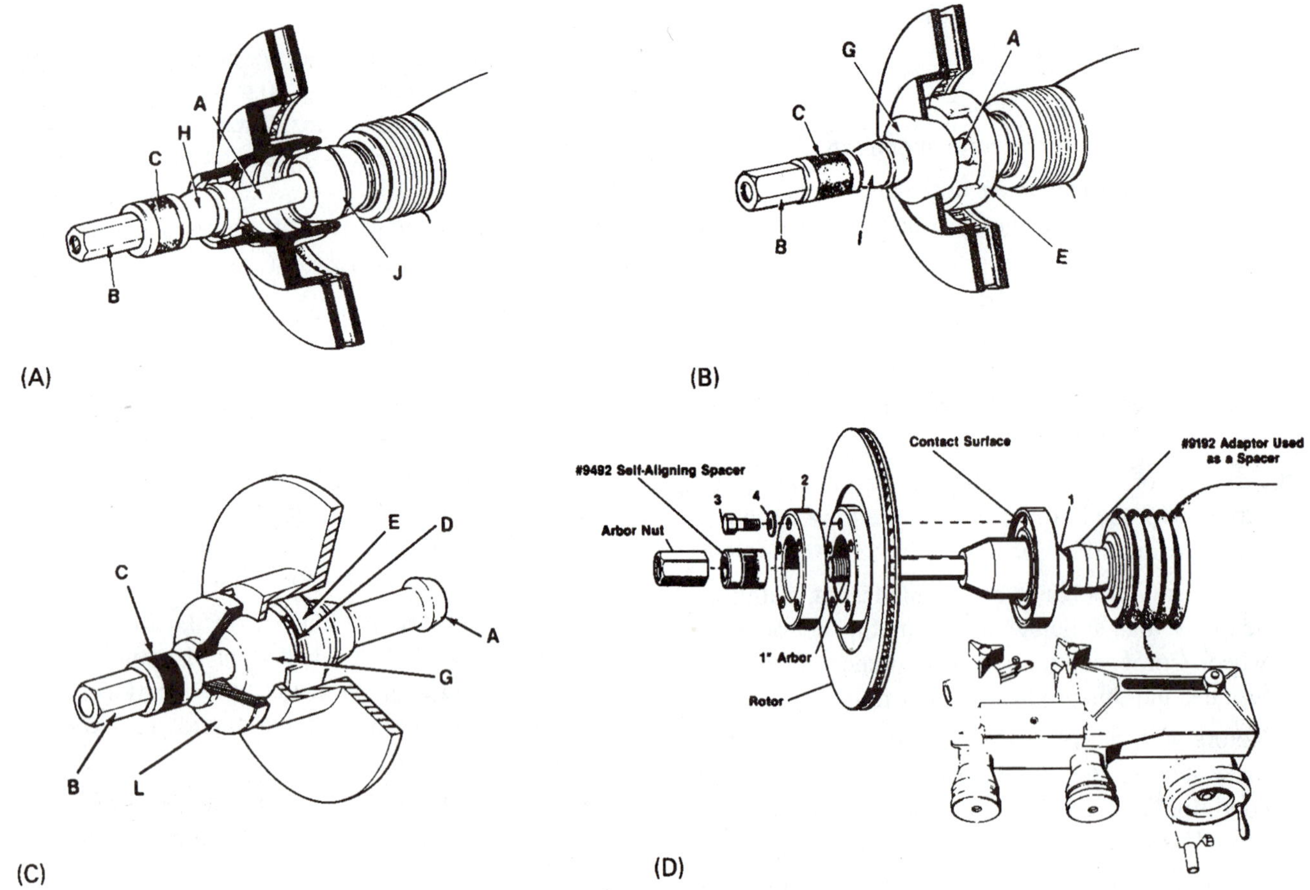

(A) (B) (C) (D)

Figure 11-44 Rotors with hubs are normally mounted on the lathe using an adapter entering the wheel-bearing races (A). Hubless rotors are centered using a cone (G) and kept from wobbling by one or two hubless adapters (B; C.) Composite rotors require special mounting adapters to secure the center section (D). (Courtesy of Ammco Tools, Inc.)

a. Rotors with hubs are normally centered and held in place by a pair of tapered cones that enter partway into the wheel-bearing cups (Figure 11-44).

b. Rotors without hubs are normally squeezed between a single cone passing partway through the center hole and a hubless adapter. The cone centers the rotor, while the hubless adapter eliminates lateral runout.

c. Composite rotors are normally clamped between two adapters that also center the rotor.

3. Mount rotors with the proper adapters, and tighten the mounting nut to the correct torque. DO NOT OVER-TIGHTEN.

4. Select the correct arbor speed for the rotor diameter. Then adjust the speed mechanism, usually a V-belt and various-sized pulleys, to produce the correct speed. Too fast a speed will shorten the life of the cutter bit(s), and too slow a speed wastes time (Figure 11-45).

	Rough Cut	Finish Cut
Spindle Speed		
10" & Under	150-170 RPM	150-170 RPM
11"-16"	100 RPM	100 RPM
17" & larger	60 RPM	60 RPM
Depth of cut		
(Per Side)	0.005"-0.010"	0.002"
Tool Cross Feed		
(Per Rev.)	0.006"-0.010"	0.002" max
Vibration Dampener	Yes	Yes
Sand Rotors Final Finish	No	Yes

Figure 11-45 Before turning a rotor, select the correct drive/spindle speed depending on the rotor diameter. Also note the other recommendations. (Courtesy of Ammco Tools, Inc.)

5. Adjust the position of the cutter bit holder so the cutter bits are centered or even across the rotor and there is enough travel to completely true the rotor surfaces (Figure 11-46). The extension of the tool holder should be kept as short as possible to produce the best finish.

6. Make sure the automatic feed is disengaged and turn on the lathe. Adjust the cross-feed mechanism to position the cutter(s) in the center of the friction surface. Slowly and carefully feed one of the cutter bits

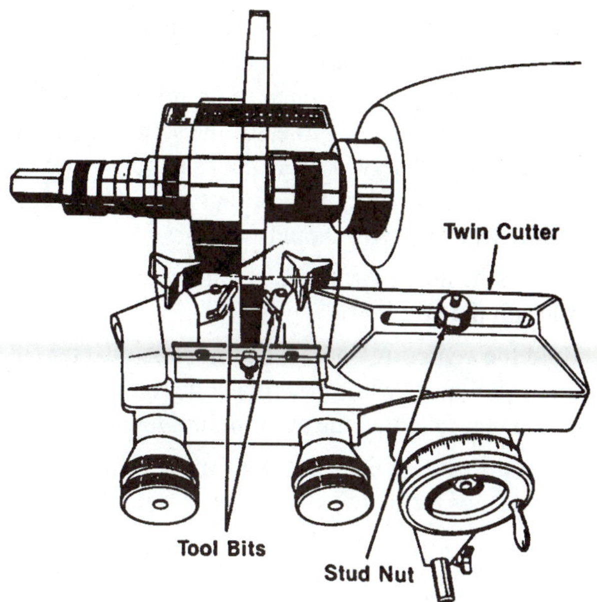

Figure 11-46 After mounting a rotor, loosen the tool holder stud nut and adjust the tool holder so the cutter bits are exactly opposite each other. (Courtesy of Ammco Tools, Inc.)

inward until it just begins to cut the rotor surface. It should make a shallow scratch. As you are doing this, observe the mirror image of the cutter bit in the rotor surface (Figure 11-47). You hope it will stay relatively still and not move in and out. A lot of in-and-out movement indicates runout. This can also be observed by watching the clearance between the cutter bit and the rotor.

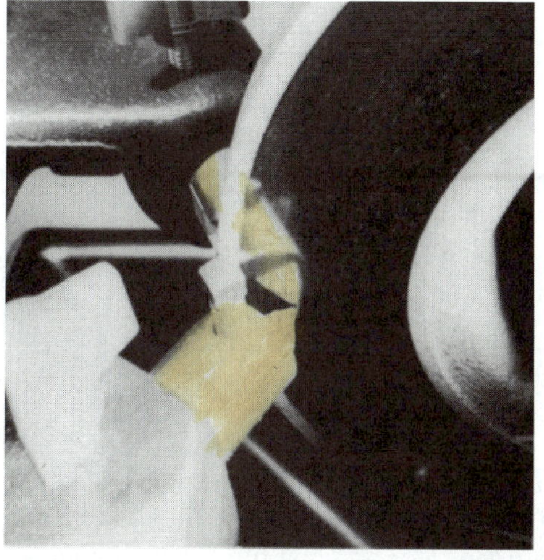

Figure 11-47 Note the mirror image of the cutter bit in the smooth rotor surface; if it moves in and out when the lathe is turned on, the rotor has runout.

7. Stop the lathe so you can inspect the scratch (Figure 11-48). If little runout was observed and the scratch runs almost all the way around the rotor, the rotor is mounted true and can be turned. If the scratch is short and runout was observed, the rotor is either mounted crooked or has substantial runout. To check for mounting trueness, you should:
   a. Loosen the nut securing the mounting adapters and rotor on the arbor.
   b. Hold the centering cone(s) and hubless adapter (if used) stationary while you rotate the rotor one-half turn. Then retighten the arbor mounting nut.
   c. Move the cross-feed handwheel one revolution and repeat steps 6 and 7.
   d. If the new scratch indicates an acceptable mounting, the rotor can be turned. If the new scratch is alongside the first scratch, the runout is being caused by the rotor. The rotor can be turned. If the new scratch is one-half turn from the first scratch and there is still an excessive amount of runout, there is a mounting problem. This can be caused by a bent, loose, or improperly indexed arbor; the wrong adapters; grit between the adapters and the rotor; or loose wheel-bearing cones in the hub.
8. Attach the silencer or vibration device to the rotor. It is usually a band wrapped around the outer edge of the rotor or a pair of pads resting against the rotor surfaces (Figure 11-49).

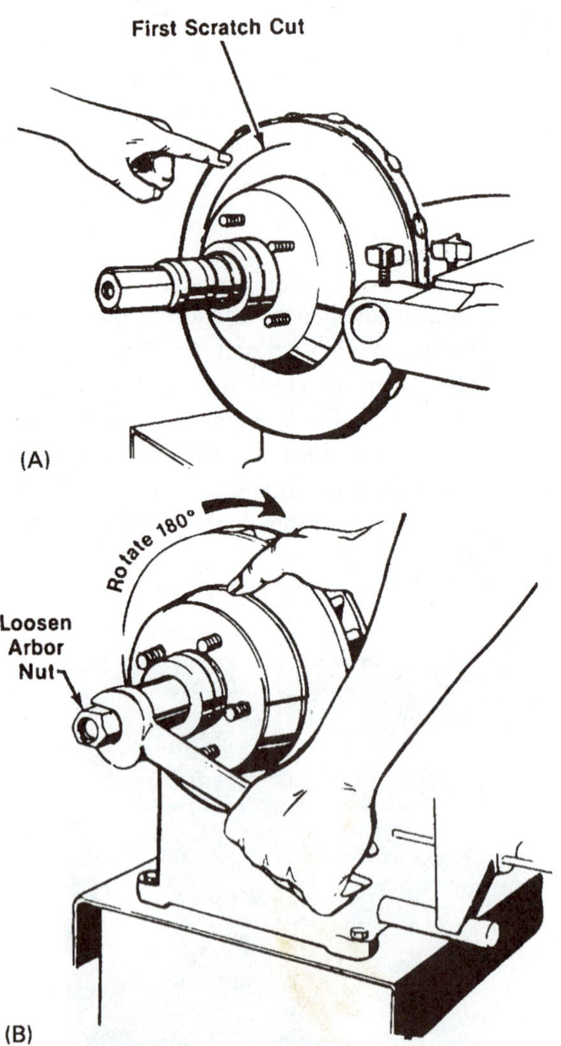

Figure 11-48 Runout can be checked by making a light scratch cut (A); a very short cut indicates too much runout. Next, rotate the rotor one-half turn on the lathe, and make a second scratch cut. The relative position of the second scrach cut gives a good indication of what is causing the runout. (Courtesy of Ammco Tools, Inc.)

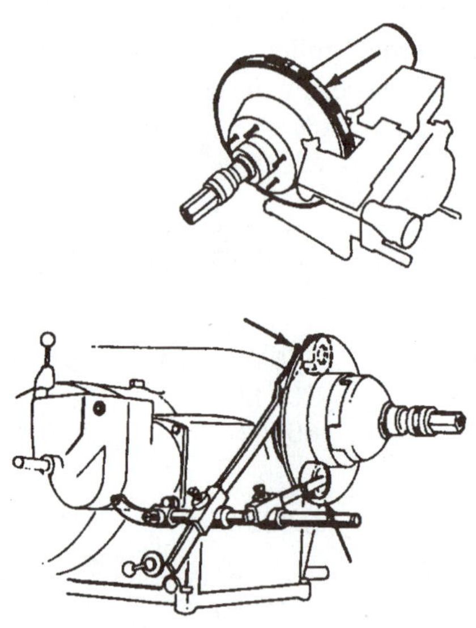

Figure 11-49 The most common vibration dampeners or silencers (arrows) are a large rubber band that wraps around the edge of the rotor or a pair of pads that drag on the friction surfaces. A vibrating rotor causes a rough surface finish. (Courtesy of Ammco Tools, Inc.)

9. Start the lathe and adjust the cutter bit(s) to just contact the rotor surfaces(s). Adjust the cutter feed dial(s) to zero or note the reading(s) (Figure 11-50).
10. Make sure the safety shield is correctly positioned over the cutters and safety glasses are worn (Figure 11-51).
11. Turn the cross-feed handwheel to bring the cutter(s) to the outer rust ridge and slowly hand-feed the machine to remove this ridge (Figure 11-52).
12. Turn the cross-feed handwheel to move the cutter(s) inward—toward the rotor's hub. As you do this, watch for any grooves. If you locate any, feed the cutter bit to

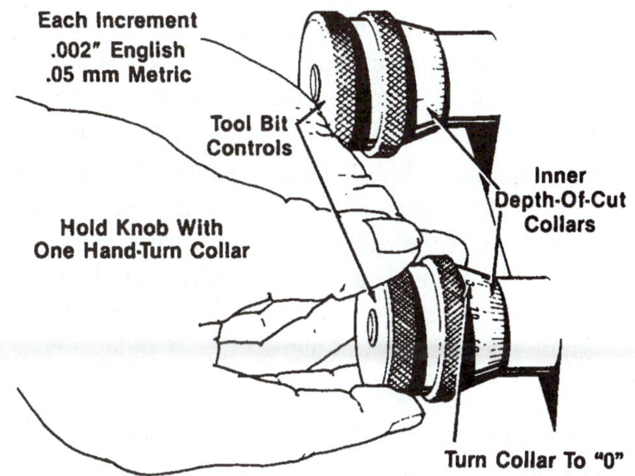

Figure 11-50 Before starting a cut, the cutter bits are adjusted to lightly touch the rotor, and the cutter feed dials should be set to zero. (Courtesy of Ammco Tools, Inc.)

Figure 11-52 With the cutter bits adjusted to a very light cut, they are hand-fed outward to remove the outer wear/rust ridge before cutting the friction surface.

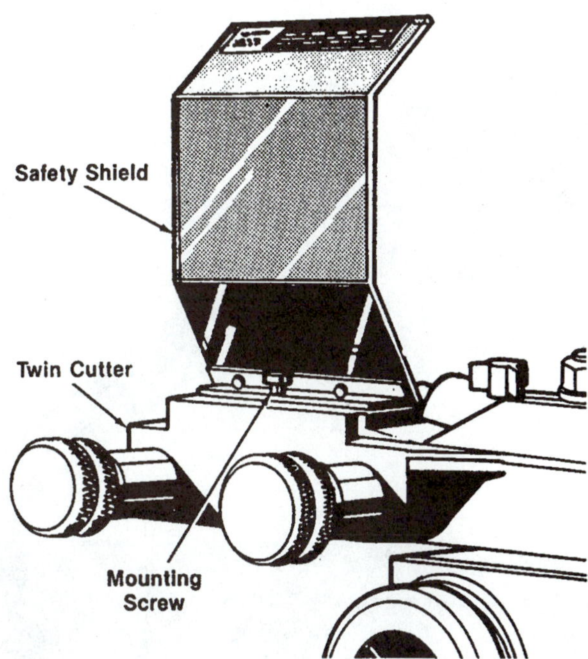

Figure 11-51 A safety shield should be positioned between you and the cutters as the rotor is machined. (Courtesy of Ammco Tools, Inc.)

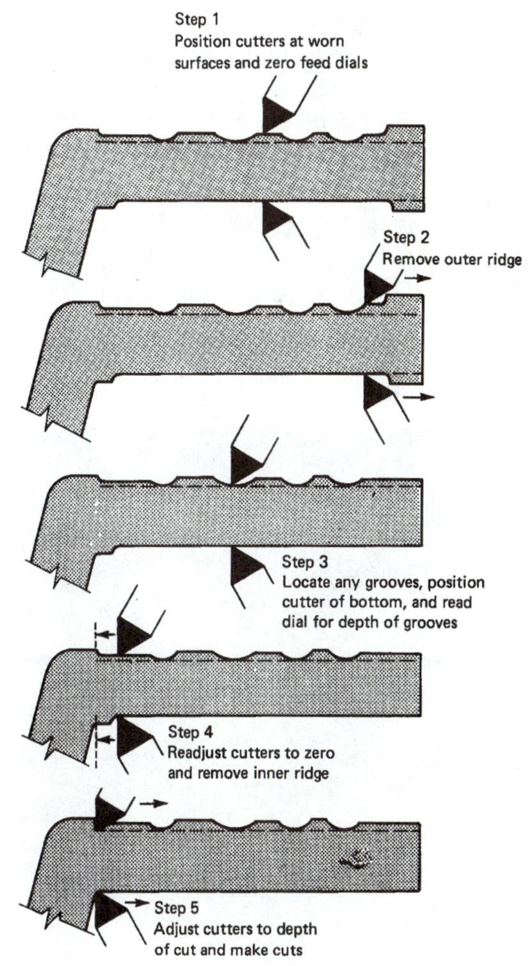

Figure 11-53 The steps used in machining a rotor are: (1) set the feed dials to zero; (2) remove the outer ridges; (3) determine the depth of cuts; (4) remove the inner ridges; and (5) cut the friction surfaces.

the bottom of the groove to determine how deep a cut is necessary to remove it (Figure 11-53). Return the cutter bit to zero and continue turning the cross-feed to bring the cutter(s) to the inner wear ridge. Hand-feed the machine to remove the inner wear ridge (Figure 11-54).

13. Select the depth of cut. If no grooves were found, the rotor can often be turned in one finish pass. The ideal **finish cut** is between 0.004 and 0.005 inches (0.1 and

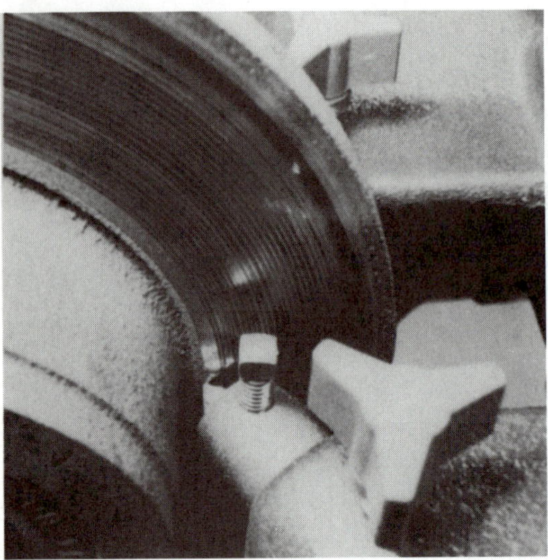

Figure 11-54 Following the cut for the outer ridges, hand-feed the cutters inward to remove the inner wear ridges.

0.15 mm) deep—on each side of the rotor—with a feed rate of 0.002 inches (0.05 mm) per revolution. If the rotor requires a cut deeper than 0.006 inches, it is recommended that two or more cuts be made. The maximum depth of the cut and the correct feed rate are determined by the machine being used. The depth is generally about 0.010 inches (0.25 mm) per side with a feed rate of about 0.006 inches (0.15 mm) per revolution. This is called a **roughing cut** and should always be followed by a finish cut (Figure 11-55).

14. With the machine running, adjust the cutter(s) to the desired depth of cut and the automatic feed to the rate of cut. Engage the automatic feed and, as the cut begins, observe the machine, cutter bit(s), and rotor for any unusual noise or movement. If something appears wrong, stop the machine and recheck your settings.
15. After the cut is finished and the cutter bit(s) moves past the outer edge of the rotor, disengage the automatic feed. If the cut was the finish cut and the friction surfaces are cleaned up, go to step 16. If the cut was a rough pass or part of the friction surface was skipped, repeat steps 14 and 15.
16. Remove the silencer device.
17. Many technicians will hold a disc sander (with a 120-grit disc on it) against the rotating rotor for a few moments to produce a nondirectional finish. This should be repeated on both sides of both rotors using the same time interval and sander pressure on each surface (Figure 11-56). A simple, handheld refinisher

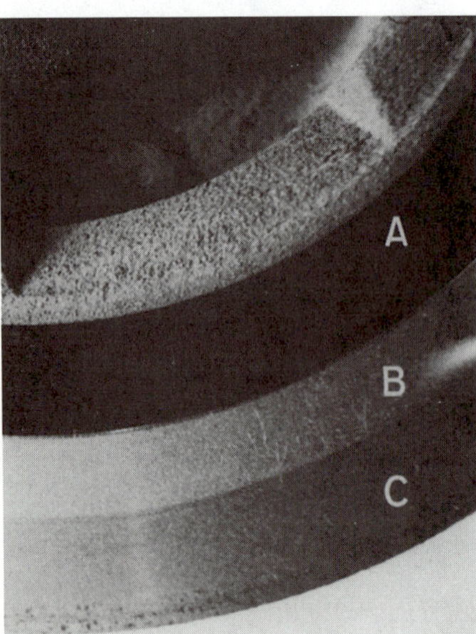

Figure 11-55 This rotor has an uncut, burnished surface (A), a section cut at a fast feed rate for a rough cut (B), and a section cut at a slow rate for a finish cut (C).

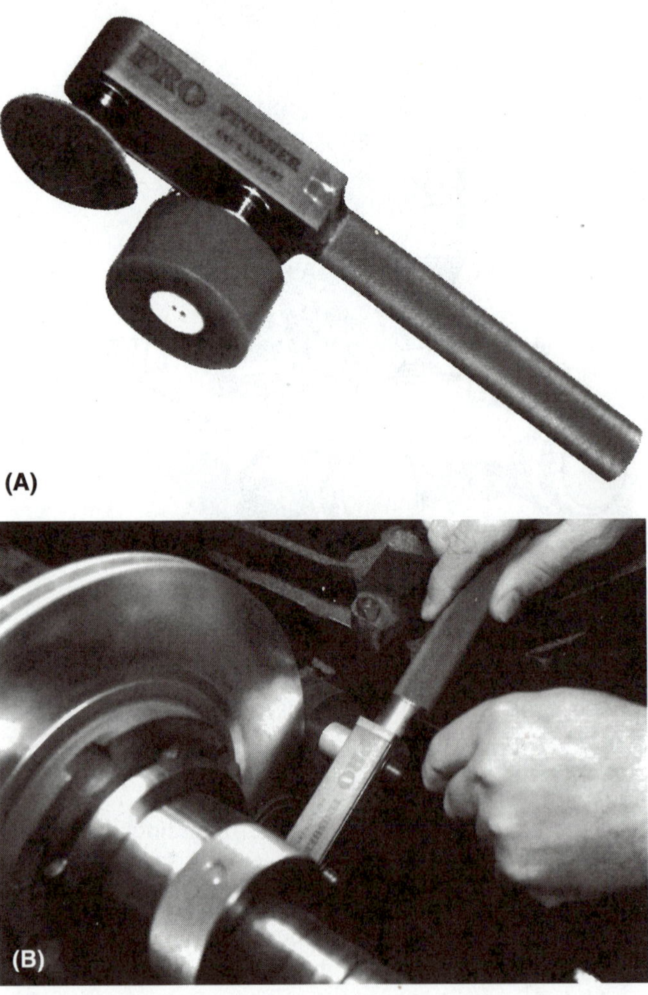

Figure 11-56 This Pro Finisher (A) is handheld and driven by the drive wheel against the outer edge of the turning rotor. About 30 seconds on each side produces a smooth, nondirectional finish. (Courtesy of RSR Enterprises LTD)

can be positioned on each rotor surface for about 30 seconds on each side to produce a nondirectional finish (Figure 11-57).

18. It is a good practice to stroke a smooth file along the outer edge of the rotor to deburr or chamfer the sharp corner. This will prevent shaving or chipping of the brake lining as the caliper is installed (Figure 11-58).

19. Remove the rotor from the lathe.

20. Remeasure the rotor to make sure that it is not undersize. One new lathe model includes a digital readout that displays the actual rotor width.

21. Mount the second rotor on the arbor and turn it. Be sure to use the same feed rate on the finish pass.

22. Wash the newly machined friction surfaces using denatured alcohol or brake surface cleaner.

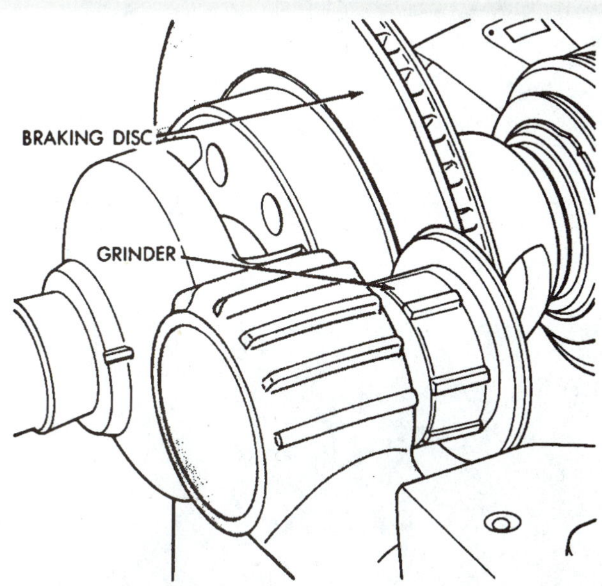

Figure 11-57 This grinder, mounted in the lathe, is adjusted to run lightly against the turning rotor to produce a nondirectional finish. (Courtesy of Chrysler Corporation)

Figure 11-58 It is a good practice to stroke a file along the outer edges of the rotor to remove the sharp corners (inset). While doing this, avoid the spinning rotor and lug bolts.

11.6 Machining a Rotor On-Car

At the present time, there are several different methods of machining rotors on the car. One of these uses twin cutter tools, and the rotor is turned by the vehicle's engine. Another uses twin abrasive pads, and the rotor is turned by the vehicle's engine. A third uses twin cutter tools, and the rotor is turned by an electric motor contained in the unit. This self-powered unit is different in that it is connected to the rotor using the wheel lugs, with a stabilizing link to the caliper support bracket. The other two are attached to the caliper support bracket. When using any of these units, be sure to follow the operating procedure provided by the manufacturer.

SAFETY TIP: If performing this operation, you will be working next to a turning rotor in a confined area. Be careful, especially with vehicle-powered units. It is difficult to shut them off quickly in cases of emergency.

To turn a rotor on the car, you should:

1. Raise and securely support the car.
2. Remove the wheels and the calipers.
3. Attach the lathe unit to the steering knuckle.
   a. On caliper-bracket-mounted units, bolt the mounting adapters to the caliper support bracket or to the caliper mounting bosses on the steering knuckle. Then bolt the lathe assembly to the mounting adapters (Figure 11-59).
   b. On hub-mounted units, mount the adapter to the hub, and adjust the adapter to compensate for any runout caused by mounting errors. Mount the lathe onto the adapter (Figure 11-60).
4. Adjust the cutter units or abrasive discs so they are centered on the rotor.
5. If applicable, attach the antivibration band around the rotor and the protection band around the exposed wheel lug bolts (Figure 11-61).

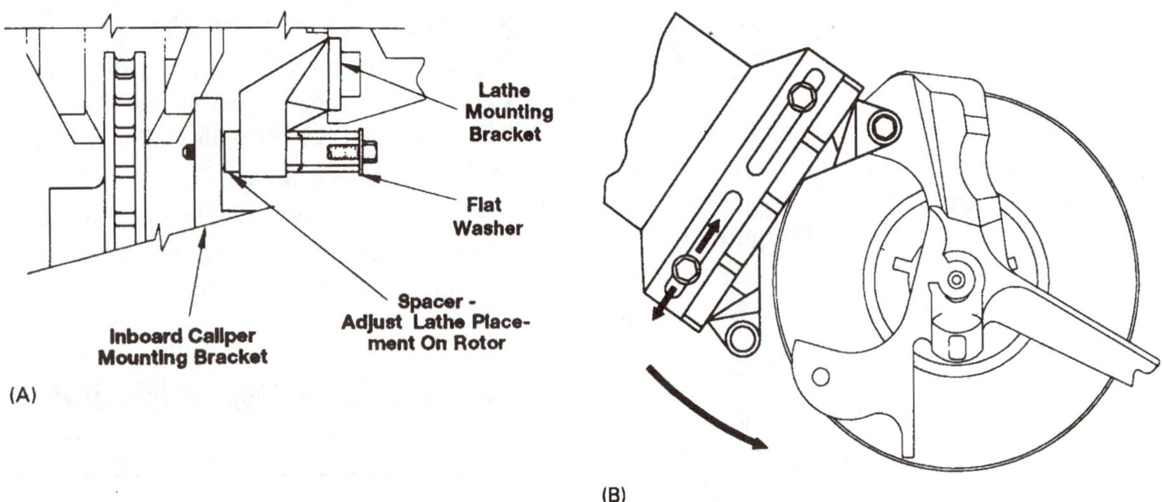

Figure 11-59 This on-car lathe is attached to the caliper mounting points by adjusting the attaching bracket and using the proper spacers. (Courtesy of Ammco Tools, Inc.)

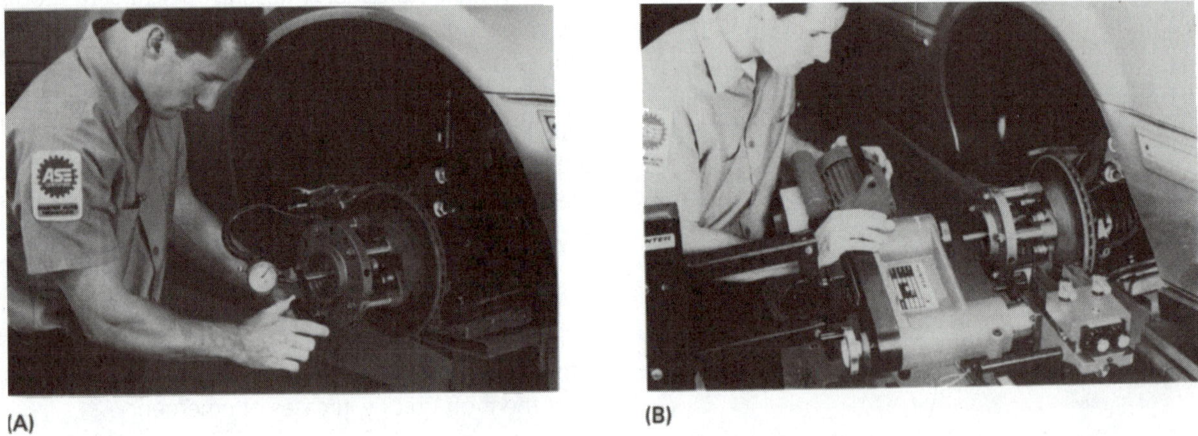

Figure 11-60 This on-car lathe is mounted onto the hub by first attaching an adapter and adjusting it to compensate for any runout (A). The lathe is then connected to the adapter (B). (Courtesy of Hunter Engineering Company)

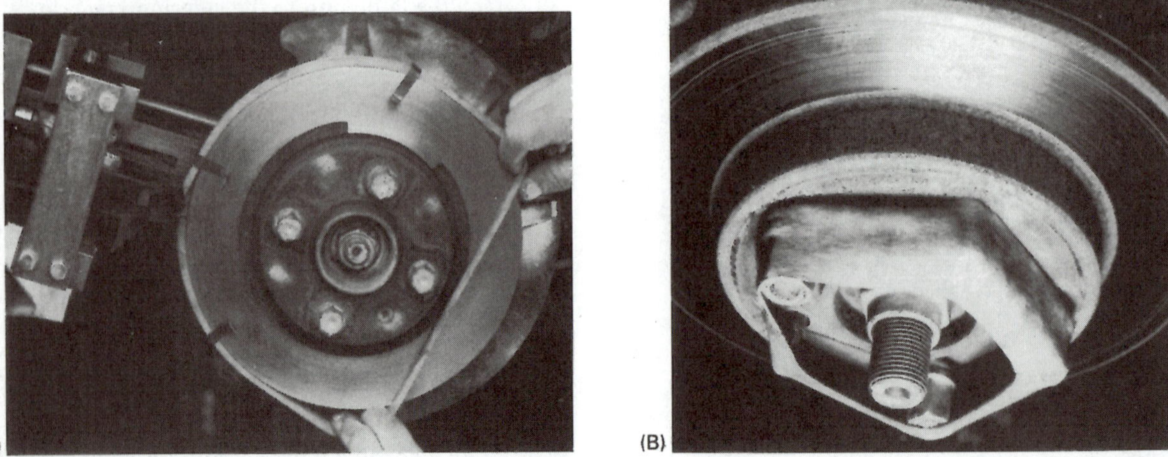

Figure 11-61 A vibration band is being positioned around the rotor with the aid of metal clips (A). The protector band placed over the lug bolts (B) helps keep you from being caught on them. (Courtesy of Kwik-Way Mfg. Co.)

On abrasive disc units, select and install the disc—fine, medium, or coarse grit—for the desired cutting rate. This rate is determined by how much material must be removed from the disc.

6. On vehicle-powered units, attach a clamp to the rotor on the other side of the car. The clamp should be positioned against the caliper abutments to prevent rotation of the rotor (Figure 11-62).
7. On vehicle-powered units, start the engine and carefully put the car in gear, first or reverse, as required to turn the rotor in the correct direction.

   On self-powered units, turn on the switch to start the lathe.
8. On units using cutter bits, turn the cutter bit cross-feed to position the cutters midway on the rotor and turn the feed knobs so the cutters just touch the rotor. From this point, the remainder of the cutting procedure is quite similar to that with an off-car lathe (Figure 11-63).

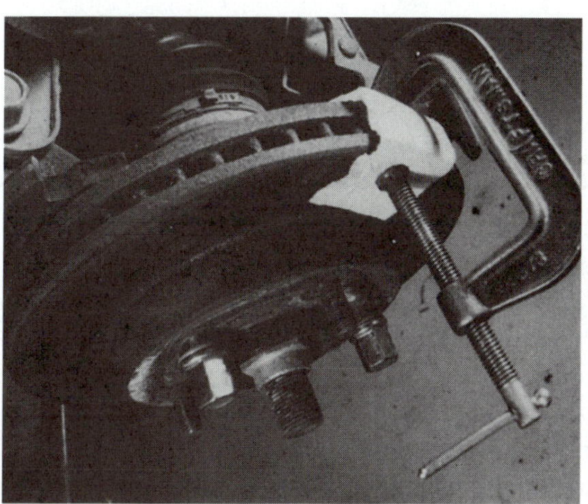

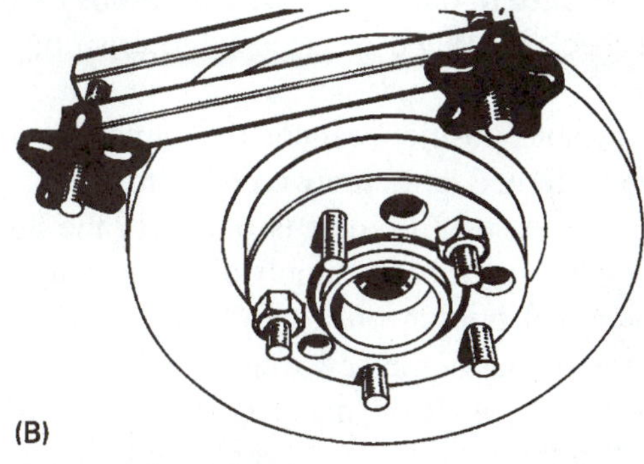

(B)

Figure 11-62 If using the car's engine to turn the rotor, a C-clamp (A) or rotor clamp (B) will prevent the opposite rotor from turning. Note the cardboard to keep the clamp jaws from marking the rotor. (Courtesy of Can Am)

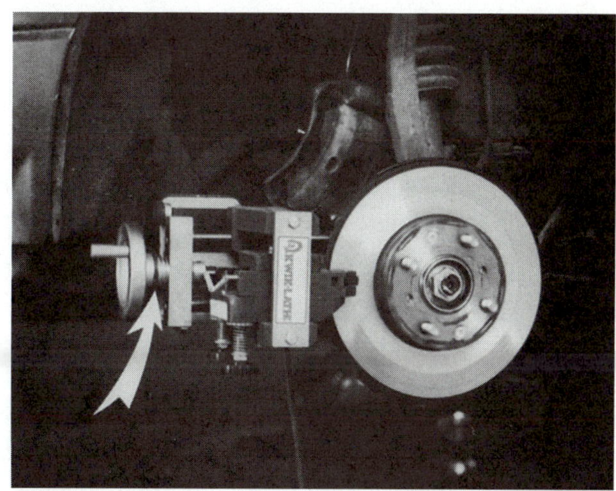

Figure 11-63 This rotor is being machined using a Kwik-Lathe. The cutter bits are fed outward by an electric motor turning the drive belt (arrow).

On abrasive disc units, turn the knobs to apply pressure between the abrasive pads and the rotor. Continue applying feed pressure until the rotor is cleaned up. If the operation was begun with coarse discs, stop cutting when the rotor is almost clean and switch to fine pads (Figure 11-64).

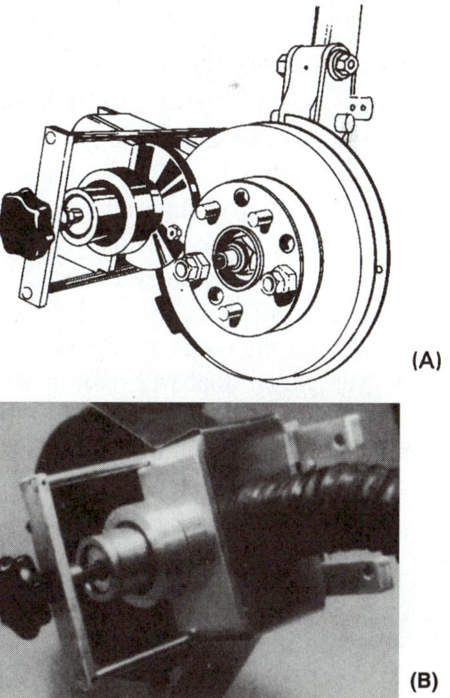

Figure 11-64 A special grinder with an abrasive pad on each side is being used to machine the rotor while the rotor is turned using the car's engine (A). A vacuum cleaner attachment (B) can be used to remove dust and debris. (Courtesy of Can Am)

9. When the cutting or grinding operation is complete, stop the rotor and remove the machining unit.
10. Machine the second rotor in the same manner.
11. Wash the metal and carbon residue from the newly machined rotor surfaces using denatured alcohol or brake surface cleaner.

11.7 Resurfacing a Rotor

Resurfacing a rotor by hand is not recommended by all manufacturers, but it has been used successfully by many brake technicians. It has the advantage of being quick, does not require removal of the rotor, leaves a nondirectional finish, and removes a minimum amount of metal from the rotor. This method can be used only on good rotors because it will not true a rotor. If there is excessive runout, thickness variation, or grooving, the rotor must be turned or replaced.

SAFETY TIP: With this method, you will be using a disc sander in confined quarters. You should wear eye protection and, in some cases, breathing and ear protection. Also, be careful of the rapidly rotating abrasive disc in the limited area.

To resurface a rotor, you should:

1. Raise and securely support the car.
2. Remove the wheels and the calipers.
3. Attach a sharp, medium-grit disc (120- to 160-grit) to a drill sanding pad or a small disc grinder; a 3- or 4-inch disc is ideal. With the drill or disc sander running and the abrasive disc held almost flat against the rotor to prevent gouges, move the disc back and forth around the rotor to obtain an even scratch pattern on the friction surfaces. Rotate the rotor as necessary so that you can sand the entire friction surface. Try to obtain an even pattern that will remove all or nearly all of the glazed surface. Remove the rust ridges at the inner and outer edges as you work (Figure 11-65).

Figure 11-65 This rotor is being refinished by sanding with a drill-powered sanding disc.

4. Repeat this operation on both sides of both rotors. The caliper mounting area is used for access to the inner friction surface (Figure 11-66).
5. Wash the metal and carbon grit from the newly ground friction surfaces using denatured alcohol or brake surface cleaner.

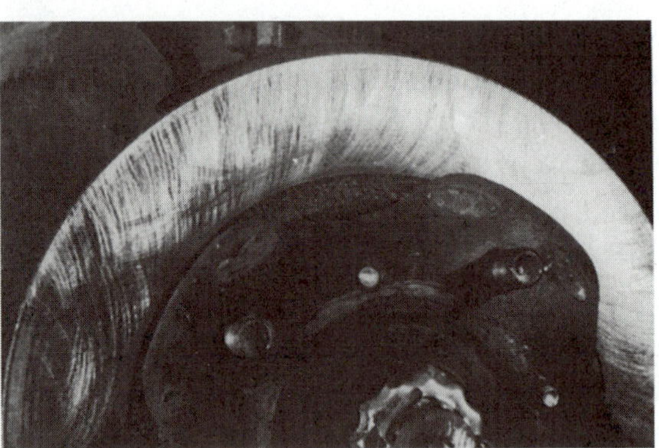

Figure 11-66 This rotor was refinished with a disc sander; note the shiny surface has been roughened by the nondirectional sanding scratches.

11.8 Rotor Replacement

If a rotor is faulty, it must be replaced. On many cars this is simply a matter of lifting the rotor off the hub and setting a new one in its place. As this is done, always make sure there is no dirt or grit between the rotor and hub, which can cause runout. Rotor runout should be checked after mounting by securing the rotor to the hub with two or three lug nuts and washers properly tightened. Measure the amount of runout using a dial indicator as previously described (Figure 11-67).

If the runout is not within specifications, **index** it—move the rotor to one of the other positions on the hub—and recheck it. When an acceptable position is found, mark one of the lug bolts and, right next to it, the rotor. If an acceptable position cannot be found, the rotor will need to be turned, preferably on the hub.

In some cases, the rotor is bolted to the hub. In this case also, replacement is simply a matter of unbolting the old rotor and bolting on a new one. Again, be sure to check the runout of the new rotor.

The front rotors on many rear-wheel-drive cars are secured to the hubs using swaged lug bolts, as in many brake-drum-and-hub combinations. To separate the rotor from the hub, it is recommended that the first step be removal of this swaged area using a special cutter. Next, the lug bolts should be pressed out, one at a time, using a special fixture or hub anvil so the rotor or hub flange is not

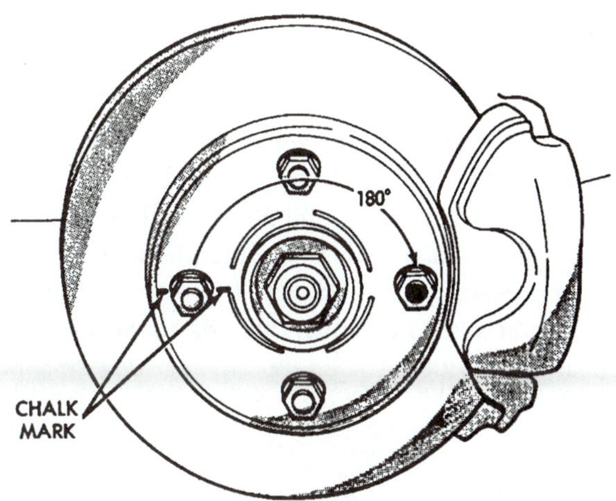

Figure 11-67 Occasionally, excessive rotor runout can be reduced to tolerable limits by reindexing the rotor. A four-bolt rotor can be turned 180° or 90° in either direction. (Courtesy of Chrysler Corporation)

distorted (Figure 11-68). The new rotor is placed on the hub, and new lug bolts are pressed in place, one at a time, again using the special fixture or support anvil. Be careful that no pressing force is placed on the rotor, as this could crack or distort the rotor. The new lugs should be locked in place by using a center punch to upset the metal at the shoulder of the new lug bolts (Figure 11-69). When selecting the new lug bolts, be aware that they have several important dimensions that should be considered (Figure 11-70). The new rotor will often have to be turned to remove runout.

Some rotors are riveted onto the hub or axle flange. These rivets will need to be removed in order to replace the rotor (Figure 11-71). In some cases, the rivet head can be shaved off with a sharp chisel. In many cases, it is helpful to drill a hole first, slightly smaller than the rivet's

Figure 11-69 A series of punch marks has been made around this new wheel stud to lock it in place.

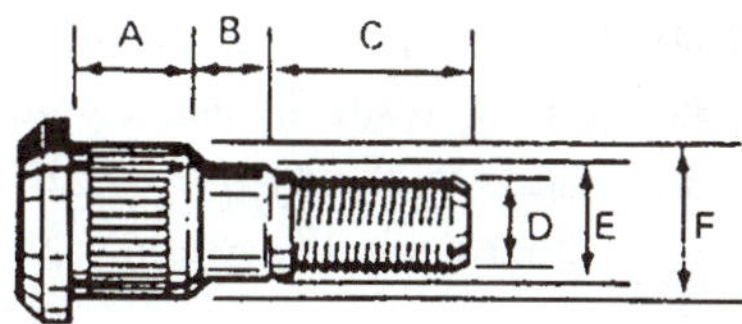

Figure 11-70 The important dimensions when selecting a replacement wheel stud are serration length (A), shoulder length (B), thread length (C), thread diameter (D), shoulder diameter (E), and serration diameter (F) plus thread pitch and direction. Replacement wheel studs should be an exact match of the original. (Courtesy of Dorman® Products, Division of R&B Inc.)

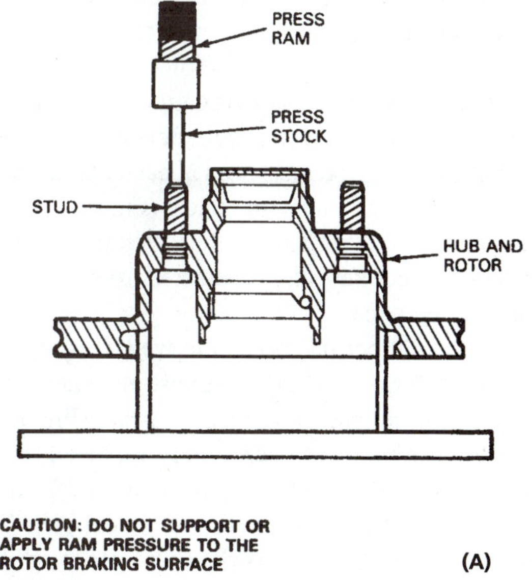

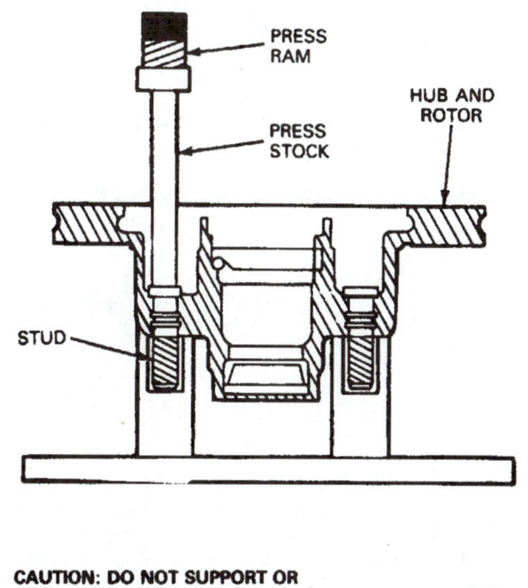

Figure 11-68 Some rotors are secured to the hub by the wheel studs that are pressed in place. If replacing a rotor, hub, or wheel stud, support the rotor properly while pressing the studs out (A) or back in (B). (Courtesy of Ford Motor Company)

Figure 11-71 An older Corvette rotor is secured to the axle by a set of rivets. (Courtesy of Stainless Steel Brakes Corp.)

diameter, partially through the rivet first. After cutting the rivet heads, the rest of the rivet can be driven out using a punch (Figure 11-72). The rotor can then be worked off the hub or axle. A new rotor is usually not reriveted because it will be held in place by the wheel lug bolts. It is a good practice to check the runout as previously described in this section. As mentioned earlier, tapered shims that can be used to reduce runout are available for some cars.

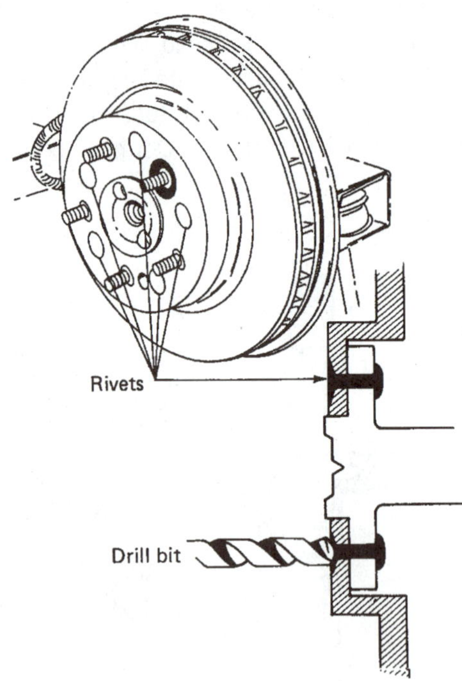

Figure 11-72 When removing a rotor that is secured to the axle or hub by rivets, drill through the head of the rivet with a drill bit about the same size as the rivet shank. Stop drilling after passing through the head, and slice the rivet head off with a chisel. The remaining rivet shank can be driven out with a punch.

The rotors on some front-wheel-drive cars are mounted on the inside of the wheel hubs. On these cars, it is necessary to pull the front hubs, and this procedure will disturb the front hub bearings. The service procedure for these bearings is described in Chapter 16. Be sure to follow the manufacturer's recommendations for rotor replacement and bearing service.

11.9 Caliper Service

As a caliper is being removed, the abutment areas where the floating and sliding calipers meet the caliper support should be checked to make sure that they are clean and free of rust so the caliper will not bind during application and release. With fixed calipers, the areas where the pad ends press against the caliper should be checked. These areas must allow pad or caliper movement during application and release, but they should not allow excess clearance. A sloppy fit will cause a knock when forward and then reverse braking occurs (Figure 11-73).

The caliper abutment clearance is checked after the caliper guide pins or mounting bolts have been removed. At this time, pry the caliper up against one of the stops or abutments and measure the clearance at the other abutment (Figure 11-74). This clearance is specified by most manufacturers. If this specification cannot be located, a rule of thumb is 0.005 to 0.020 inches (0.1 to 0.5 mm). Excessively tight calipers will cause a bind and probably drag. The flats at the abutments can be finely filed to remove rust or dirt or increase the clearance. Excessive clearance will probably cause a knock. The normal solution is to replace the caliper or caliper support. On some General Motors cars, it is recommended that the caliper abutment clearance be measured at each end of the caliper with the mounting bolts in place (Figure 11-75). One caliper, used on light trucks, has five different support keys. The correct key size is selected after measuring the abutment clearance.

After the caliper has been removed, it is a good practice to remove the sleeves, bushings, insulators, antirattle springs, and other pad or caliper hardware (Figure 11-76). Some of these items are commonly replaced during pad replacement. The bushings and insulators can usually be pried out using a small screwdriver or seal pick. The new parts should be carefully worked into place so they will not be cut or torn. Water can be used as a lubricant to aid in the installation if necessary. Make sure that the new

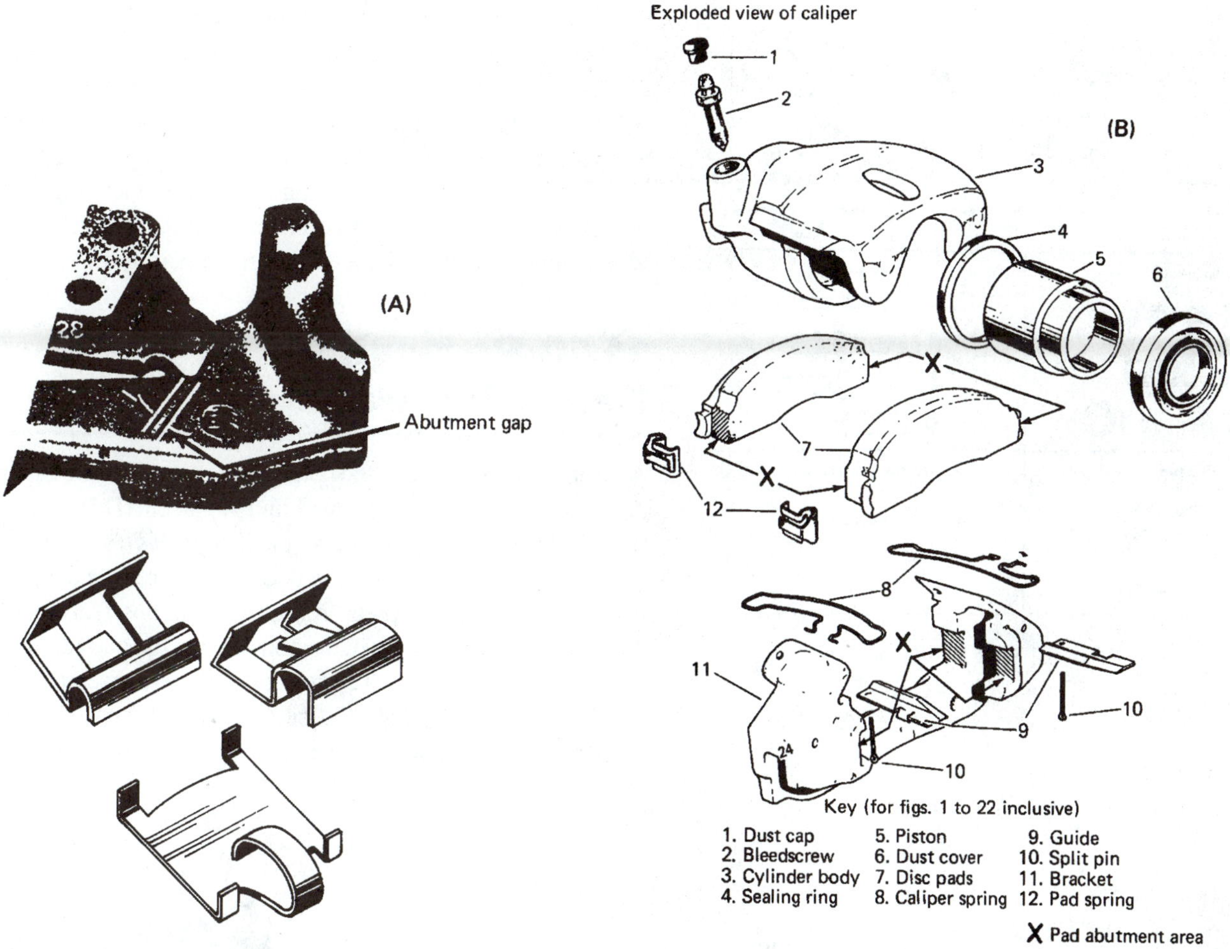

(A)

Abutment gap

Exploded view of caliper

(B)

Key (for figs. 1 to 22 inclusive)

1. Dust cap
2. Bleedscrew
3. Cylinder body
4. Sealing ring
5. Piston
6. Dust cover
7. Disc pads
8. Caliper spring
9. Guide
10. Split pin
11. Bracket
12. Pad spring

X Pad abutment area

Figure 11-73 Clearance between the inboard pad and caliper mount (A) ensures the pad is free to move during application and release. An antirattle clip or caliper pad spring (B) is used to prevent rattles. (A is courtesy of Ford Motor Company; B is courtesy of LucasVarity Automotive)

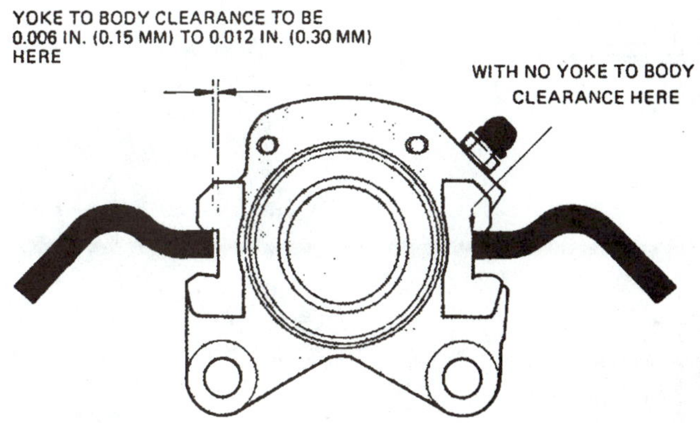

Figure 11-74 Manufacturers recommend checking abutment clearance when replacing brake pads. (Courtesy of LucasVarity Automotive)

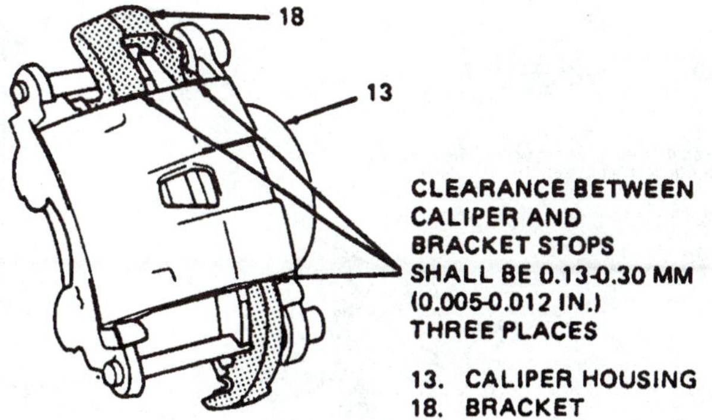

Figure 11-75 This manufacturer recommends checking abutment clearance between the caliper and the bracket stops with the guide pins installed. (Courtesy of General Motors Corporation, Service Technology Group)

bushings and insulators are correctly seated and not twisted (Figure 11-77).

The caliper should be wire-brushed to remove any dirt, rust, or grease deposits. It is recommended that the caliper be rebuilt at this time to ensure proper caliper operation and no future leakage for a reasonable period of time. This will require removing the piston, cleaning the

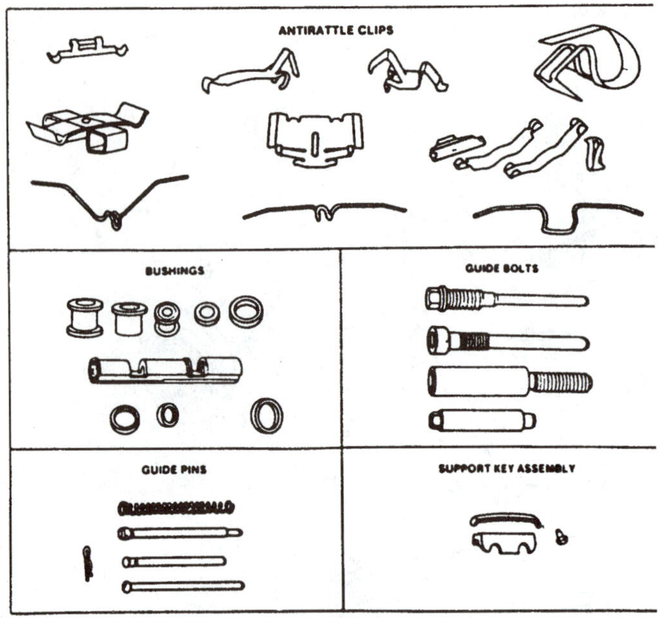

Figure 11-76 Caliper hardware varies among different brake assemblies; some of the different types of hardware are shown. (Courtesy of Bendix Brakes, by AlliedSignal)

piston and bore, and reassembling the caliper using a new piston seal and boot. This procedure will be described in Chapter 12. If the caliper is not going to be rebuilt, carefully inspect the boot to make sure it is not cracked, cut, or torn. Remember that it will have to last as long as the new lining. Also, compress the piston farther into the caliper to test its action. A sticky piston, any fluid leakage, or a faulty boot indicates that a caliper must be rebuilt.

11.10 Removing and Replacing Pads

As mentioned earlier, the pads on some fixed-caliper designs can be easily removed and replaced. This operation in its simplest and most basic form is as follows: remove the wheel, remove the pad retainer, retract the pistons, lift out the old pads, retract the pistons (further if needed), slide in the new pads, and replace the pad retainer and wheel. However, this simple procedure is discouraged for the average car. It should be done only when you can guarantee that the rotor and caliper are in excellent operat-

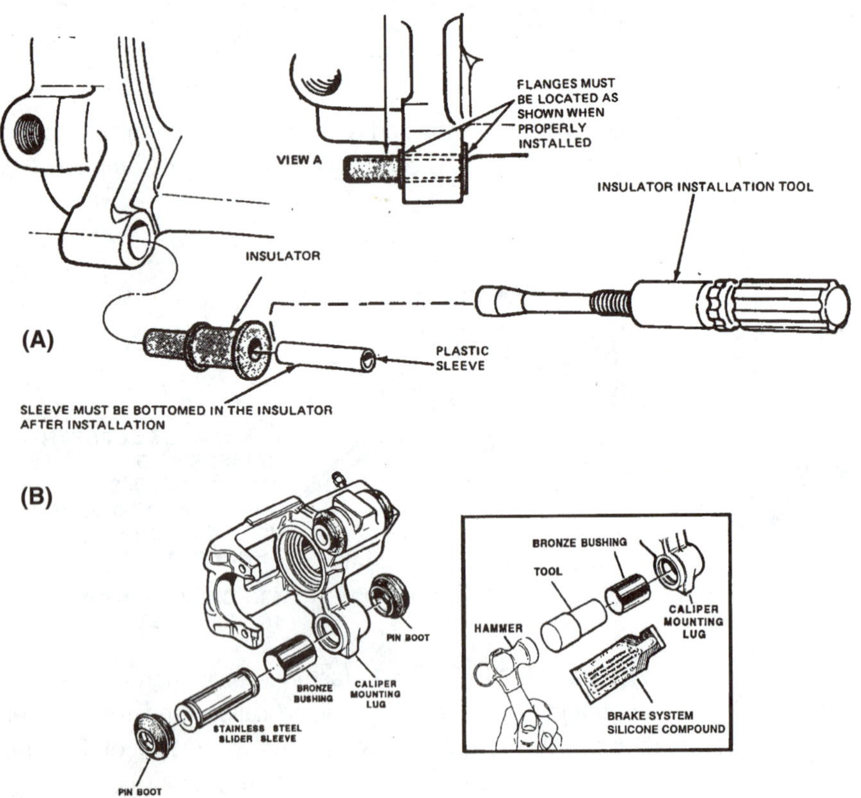

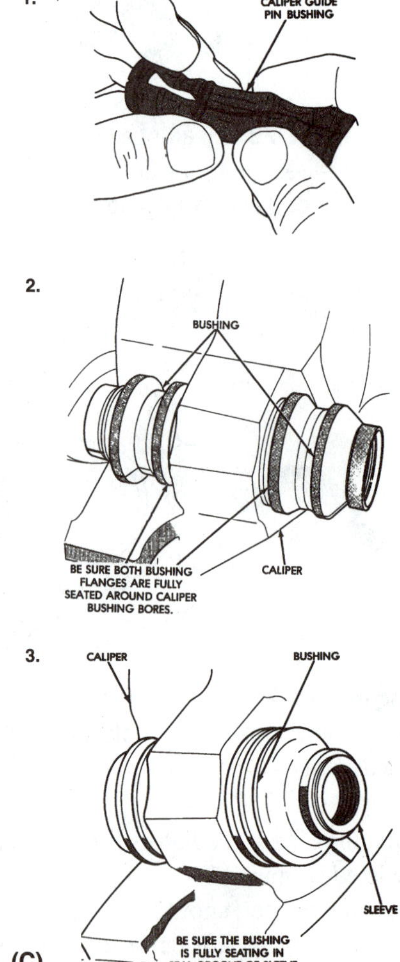

Figure 11-77 Caliper guide bushings are installed in different ways. Caliper (A) requires a special tool (can be shop-made); (B) can be repaired using an aftermarket kit; (C) is installed in several steps. (A is courtesy of Ford Motor Company; B is courtesy of Rolero-Omega, Division of Cooper Industries; C is courtesy of Chrysler Corporation)

ing condition. Most manufacturers and competent brake technicians recommend following a procedure similar to that described in this chapter. The extra service steps really don't take that much more time, and they give you the assurance that the job will result in "new-car" performance for the life of the new lining.

Some fixed-caliper designs and all floating- or sliding-caliper designs require removal of the caliper to replace the pads. The pads are retained in the caliper by slightly different methods in each design, but there are some general similarities.

Pads used with fixed calipers and the inboard pad on floating and sliding calipers must both move relative to the caliper during application and release. These pads must be able to slide freely relative to the caliper but must not move too far or rattle while released. An antirattle clip or antirattle spring is sometimes used with such pads to prevent rattles (Figure 11-78). The inboard pad on many rear calipers has a tab that must be aligned with a piston cutout (Figure 11-79). Some designs use a retainer to attach the pad to the caliper piston (Figure 11-80). This serves two purposes: stopping rattles and pulling the pad back with the piston to provide a slight clearance between the pad and the rotor when the piston retracts.

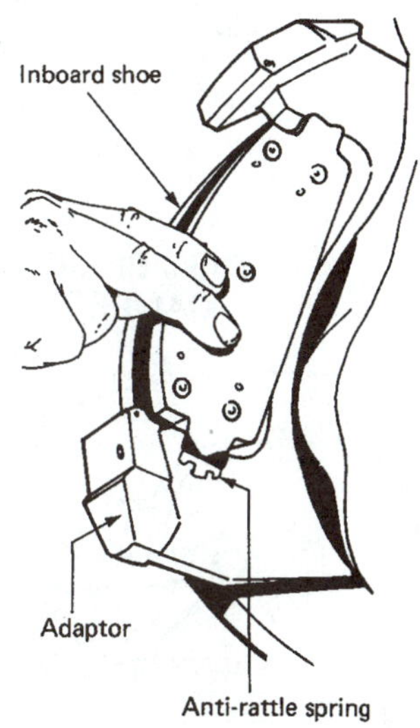

Figure 11-78 The antirattle spring is installed along with the inboard shoe/pad. (Courtesy of Chrysler Corporation)

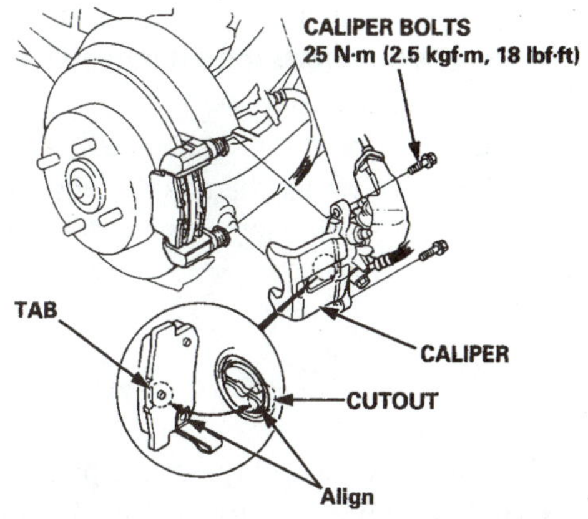

Figure 11-79 The piston on many rear calipers must be turned to index with the inboard pad. (Courtesy of General Motors Corporation, Service Technology Group)

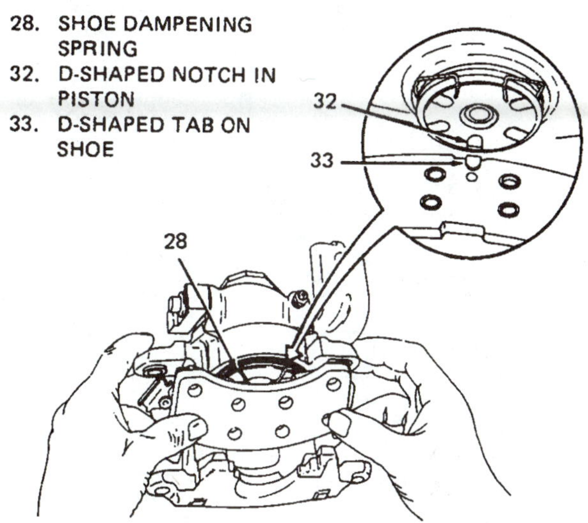

Figure 11-80 Many calipers use a retainer spring to hold the inboard pad to the piston. (Courtesy of General Motors Corporation, Service Technology Group)

The outboard pad on a floating or sliding caliper is nearly always secured tightly to the caliper. This stops pad rattle, helps pull the pad away from the rotor during release, and helps reduce squeal. Many cases of brake squeal can be attributed to vibration between this pad and the caliper. Several methods are used to stop this pad vibration. The most common involve tabs or spring clips on the pad, which fit over lugs on the caliper (Figures 11-81 and 11-82). Occasionally, brake squeal can be eliminated by coating the pad backing with a special compound or by making it fit the caliper more tightly (Figure 11-83).

Some brake pad sets have four identical shoes. A few fixed-caliper sets have shoes that are almost the same except that the lining is offset on the pad. The leading edge of the pad is the end that has the largest lining gap (Figure 11-84). Many floating- and sliding-caliper lining sets have two different shoes: two inboard shoes and two outboard shoes. Many sets include wear sensors, which are usually placed at the trailing edge of the outboard pad. A few pad sets have four different pads—each pad has a specific location such as inboard right hand, and

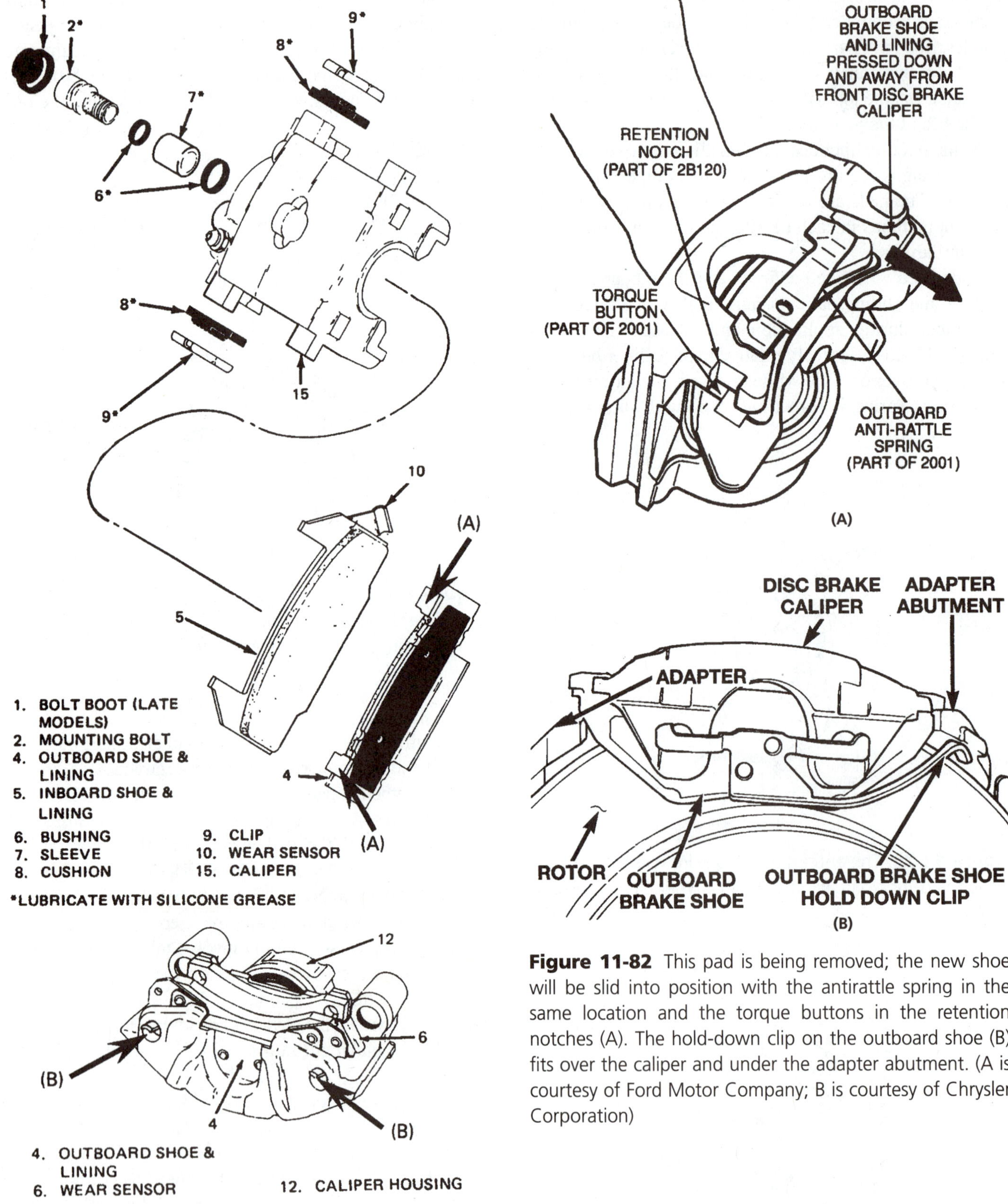

Figure 11-82 This pad is being removed; the new shoe will be slid into position with the antirattle spring in the same location and the torque buttons in the retention notches (A). The hold-down clip on the outboard shoe (B) fits over the caliper and under the adapter abutment. (A is courtesy of Ford Motor Company; B is courtesy of Chrysler Corporation)

Figure 11-81 Some outboard pads have tabs that are bent to secure the pad to the caliper. (Courtesy of General Motors Corporation, Service Technology Group)

so on. It is a good practice to lay out the pads from a set on a clean surface and check for pad backing or lining differences to ensure that each of the pads will be positioned correctly.

Figure 11-83 A center punch was used to tighten the tabs on this outboard pad in order to stop a squeal.

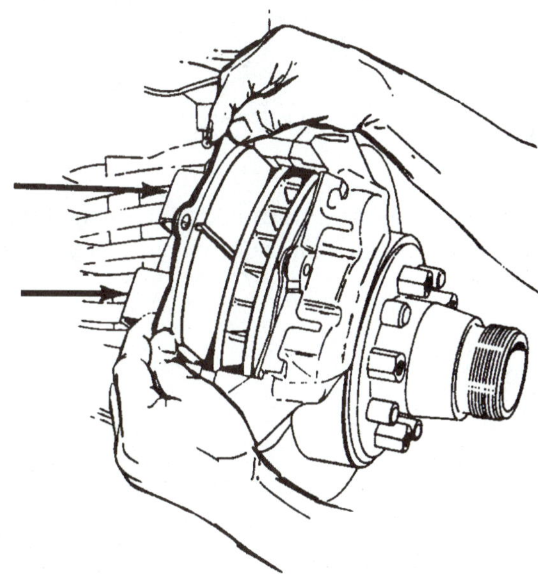

Figure 11-84 The pads on this older Corvette are slid into the gap between the caliper and rotor. Note the two clips (arrows) that are being used to keep the pistons retracted in their bores. (Courtesy of General Motors Corporation, Service Technology Group)

It is a common trade practice to use a *noise suppressant* with disc brake pads. The most popular noise suppressants are (1) a compound that is applied to the back of the pad and (2) a plastic or fiber shim (Figure 11-85). Do not use more than one at a time as the suppressant can cause the shim to pull or creep away, reducing its effectiveness.

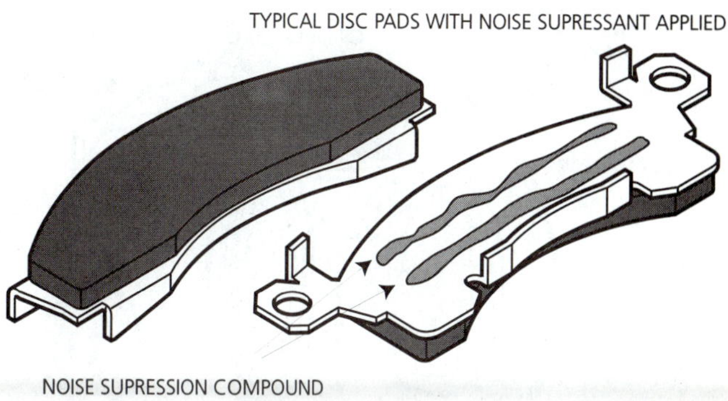

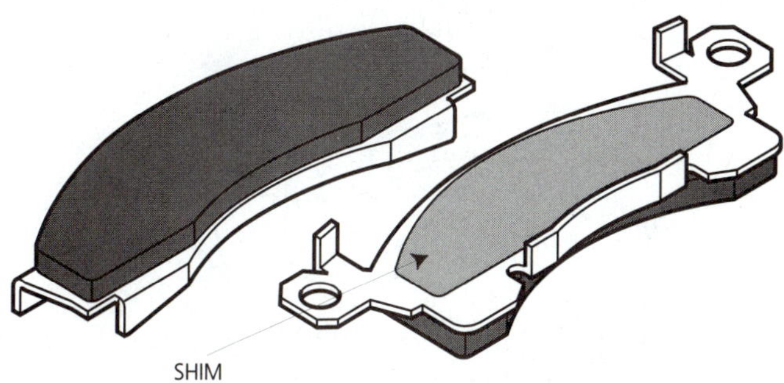

Figure 11-85 Noise-suppression compound or a pliable shim is often used on the back side of disc brake pads to prevent squeal/noise; do not use both on the same pad. (Courtesy of Wagner Brake)

When replacing pads on a caliper, it is wise to follow the manufacturer's recommendations to ensure proper and complete installation. To remove and replace brake pads in a caliper, you should:

1. Remove the inboard pad and antirattle spring from the caliper, the caliper support, or piston.
2. Remove the outboard pad from the caliper.
3. Check the replacement pads to make sure they are the correct pads and to determine where they will be installed.
4. Make sure the piston is bottomed in the caliper bore.
5. Install the outboard pad. It is important that this pad fit tightly against the caliper and be held securely in this position. Some pads use retainer tabs that should be bent into position after the caliper is installed and the brakes are applied (Figure 11-86).
6. Install the inboard pad and antirattle spring or retainer clip. Make sure that this pad fits squarely against the piston.

Note that on some fixed-caliper designs, a piston or pad shim must be installed and aligned during pad replacement (Figure 11-87).

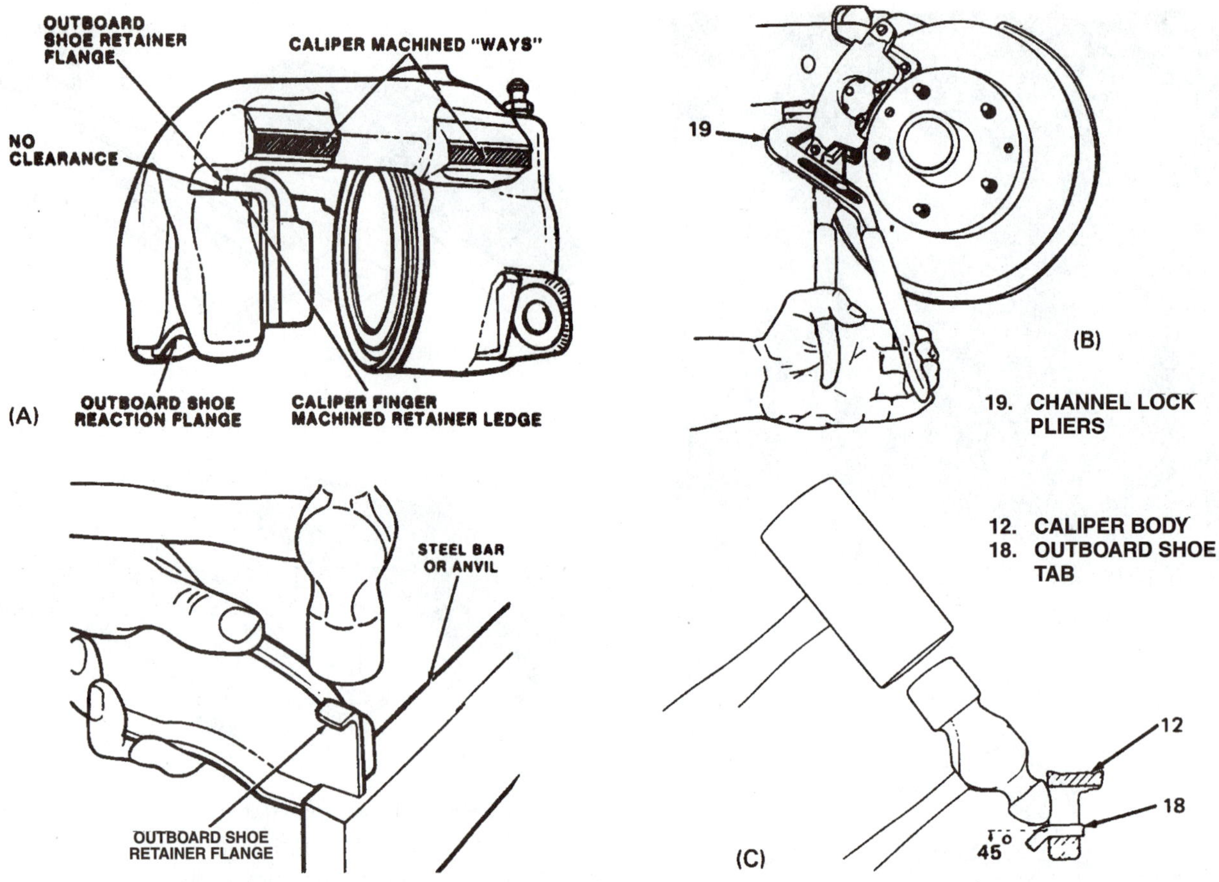

Figure 11-86 On many calipers, the outboard pad should be locked onto the caliper (A). This can be done using pliers (B) or a hammer(s) (C). (A is courtesy of Chrysler Corporation; B and C are courtesy of General Motors Corporation, Service Technology Group)

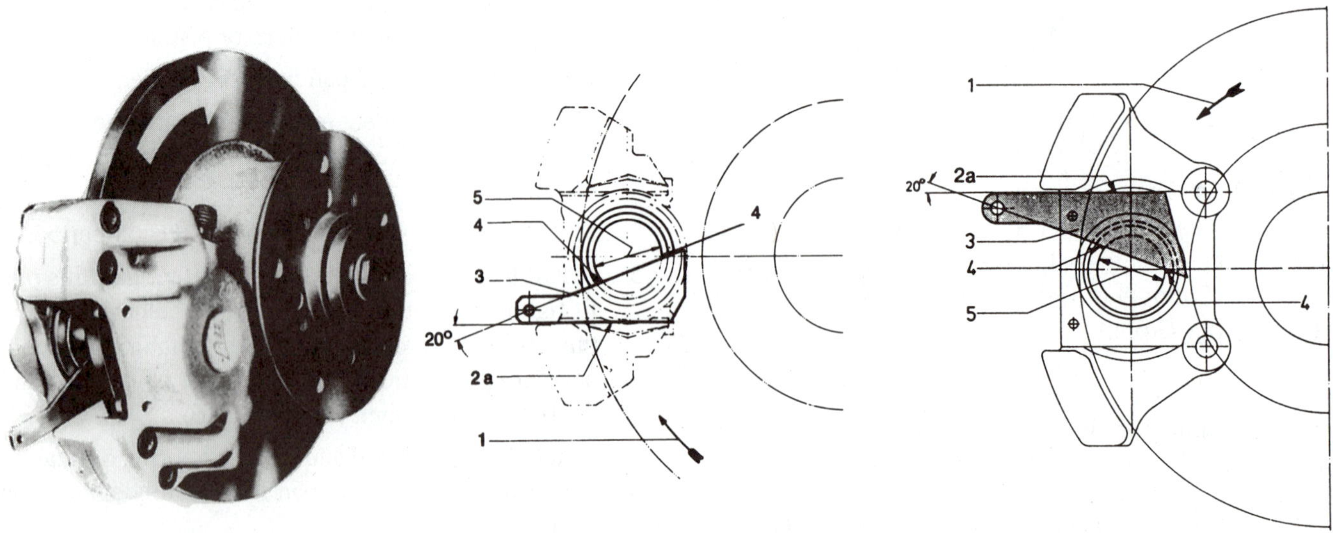

Figure 11-87 This manufacturer recommends using a gauge to position the piston properly before installing the pads. (Courtesy of ITT Automotive)

11.11 Caliper Installation

At this point, the rotor has been checked and serviced, the caliper has been cleaned and rebuilt, the caliper bushings, insulators, or other hardware parts have been replaced, and new pads have been installed. All that is left is to lubricate the unit in the proper areas and to remount it (Figure 11-88). Molylube and antiseize compound are popular lubricants for metal to metal contact areas, but only silicone-based, high-temperature lubricant should be used at rubber-to-metal contacts. Again, follow the manufacturer's installation procedure.

To replace a caliper, you should:

1. Apply the correct amount of the proper lubricant to the bushings, sleeves, or slides as required by the manufacturer. When lubricating external areas where grease might fall onto the lining or friction surfaces, use it sparingly (Figure 11-89).

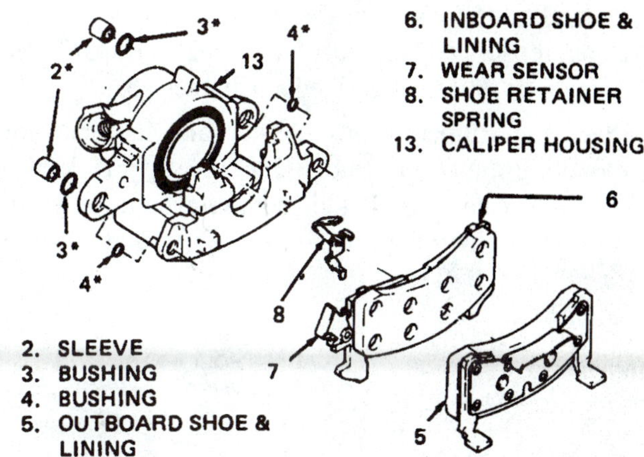

Figure 11-89 When this caliper is installed, a film of silicone grease should be applied to the sleeves (2) and bushings (3 and 4). (Courtesy of General Motors Corporation, Service Technology Group)

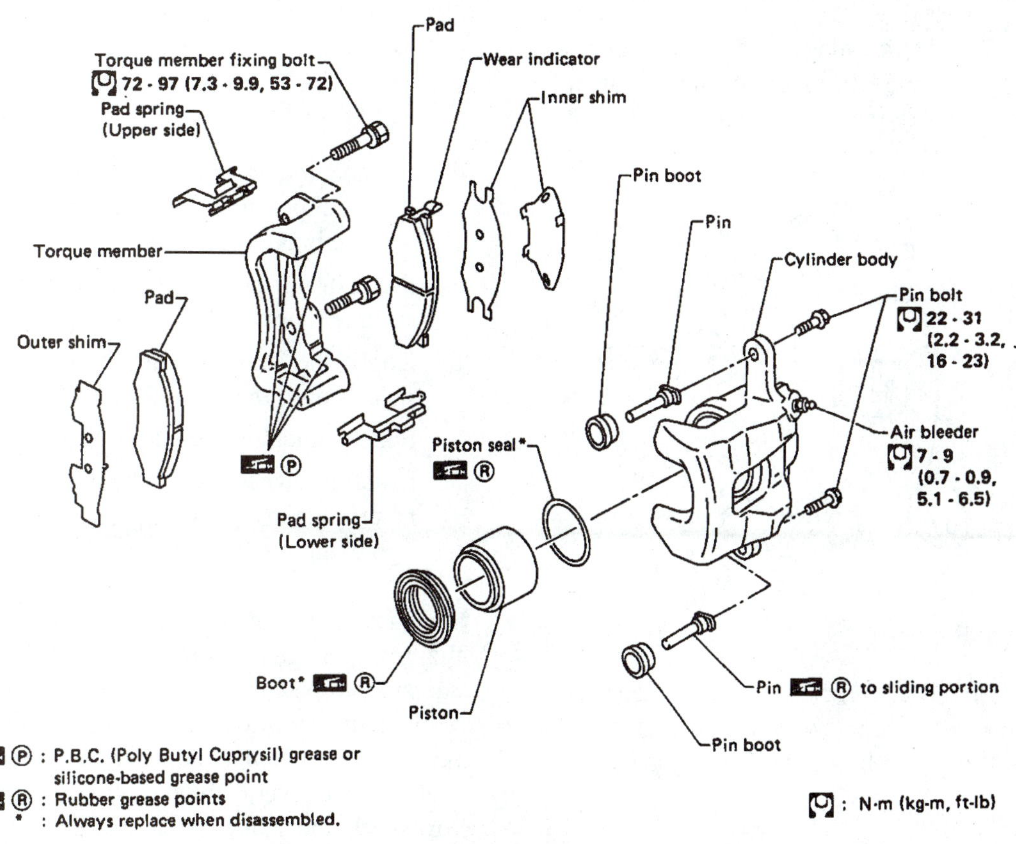

Figure 11-88 As a caliper is assembled, all of the bolts and nuts, along with the bleeder screw, must be tightened to the correct torque, and lubricant must be applied to the proper locations. (Courtesy of Nissan North America, Inc.)

2. On brakes where serviceable wheel bearings are used, clean, repack, and readjust the wheel bearings. These operations will be described in Chapter 16.
3. Place the caliper over the rotor and lubricate the mounting bolts or guide pins or support key if required. Then install the mounting bolts or guide pins and tighten them to the correct torque (Figures 11-90 and 11-91).

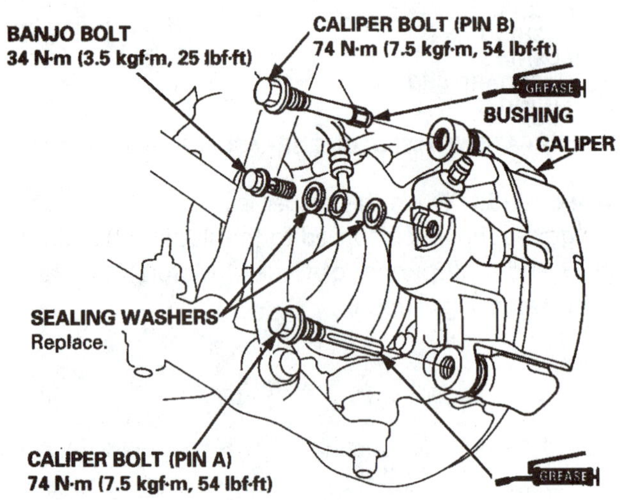

Figure 11-90 The torque ratings for the caliper bolts and banjo bolt along with the lubrication recommendation for this caliper. (Courtesy of American Honda Motor Co., Inc.)

Figure 11-91 On this caliper, the support key is driven back in place and the retaining screw tightened to the correct torque. (Courtesy of Bendix Brakes, by AlliedSignal)

On calipers using a support key(s), drive the key(s) into position and replace the retaining bolt or pins (Figure 11-92).

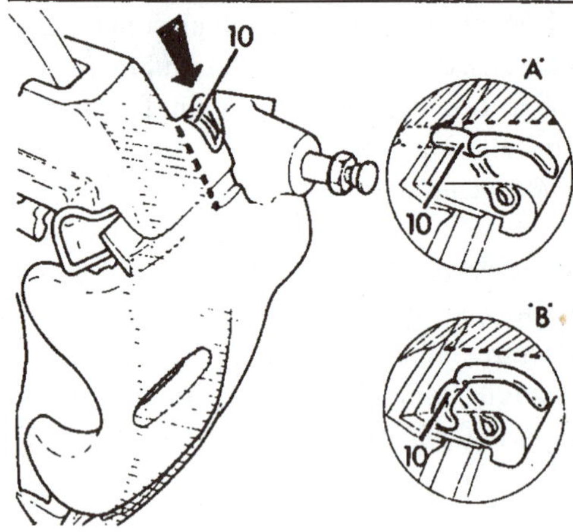

Figure 11-92 During installation, all retaining pins and clips must be properly installed to ensure that they stay in place and do not interfere with caliper operation. (Courtesy of LucasVarity Automotive)

4. Connect the brake hose and refill the master cylinder. If the caliper includes a parking brake, reconnect the parking brake cable.
5. Bleed the air out of the caliper cylinder(s). This operation will be described in Chapter 12.
6. Apply the brakes using slow and complete pedal strokes until there is a firm, high brake pedal.
7. Finish clinching the outboard pads if necessary.
8. Install the wheel and tighten the lug bolts to the correct torque using the correct tightening pattern. Overtightening the lug bolts can distort the rotor.

On most disc brake assemblies, it is very easy to adjust the lining clearance: simply apply the brake, using slow and smooth pedal applications until there is a firm pedal. With many rear brake calipers using a mechanical parking brake, the mechanical adjuster has to reset. Applying the brake pedal several times with slow, smooth, complete strokes should make the adjustment. If the shoes fail to adjust on General Motors units, hold the inboard shoe in the applied position using a screwdriver while an assistant applies and releases the brakes (Figure 11-93).

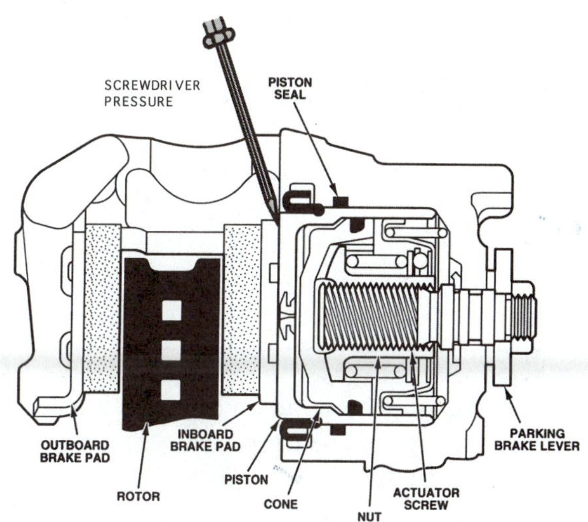

Figure 11-93 Some rear disc brakes self-adjust easily, others do not; self-adjustment can be hastened by placing a screwdriver (as shown) to hold the inboard pad in the applied position while the brake pedal is applied and released.

11.12 Completion

Several checks should always be made after completing a brake job or brake service operation. They are as follows:

- Be sure that all nuts and bolts are properly tightened and that all locking devices are properly installed.
- Be sure there is no incorrect rubbing or contact between parts or brake hoses.
- Be sure that the master cylinder has the correct fluid level.
- Be sure that there are no fluid leaks.
- Be sure that there is a firm, high brake pedal.
- Perform a careful road test to ensure that the brakes are operating safely and correctly.
- Tighten the lug nuts to the correct torque using the proper tightening pattern. Although the road test will show no problems, uneven wheel nut torque can cause later brake problems.

New friction material must be properly worn in. This is especially important if the rotors have not been turned. Unless it is specifically stated otherwise by the lining manufacturer, new lining must never be burned in. New lining can be ruined if it is overheated. Light or moderate stops should be made until the lining and rotor surfaces are fully burnished and seated in. This varies with the type of driving. Some sources recommend about 250 miles (400 km) of driving.

11.3 Practice Diagnosis

You are working in a brake and front-end shop and encounter these problems.

CASE 1: You did a brake job on this five-year-old Toyota last week with new rotors and pads along with rebuilt calipers at the front, and new lining, turned drums, and new wheel cylinders at the rear, but it has come back with a complaint of pedal pulsations. What could have gone wrong? What checks should you make?

CASE 2: The old 1982 Chevrolet has a problem of a grabby right front brake. When you pull the right front wheel, you find worn but still good lining and no fluid or grease contamination. What else could be causing this problem? What should you do next?

CASE 3: The customer is complaining of brake squeal on the five-year-old Escort, and your road test confirms a squeal that changes in level as you change the pedal pressure. What can be causing this? What should you do to find the cause of this problem? What methods can you use to correct it?

CASE 4: The five-year-old car has been brought in with a complaint of a rattle. Your road test confirms a light rattle noise, and it seems to be coming from the left front. Could this be a brake problem? What might be causing it? What should you do to locate the exact cause?

Terms to Know

caliper hardware
finish cut
guide pins
index
loaded calipers
nondirectional finish
parallelism
roughing cut
swirl finish
thickness variation
turning

Review Questions

1. Two technicians are discussing disc brake service procedures. Technician A says that rapid wear of the outboard pad is an indication that the caliper mounts are not properly lubricated. Technician B says that a sticky caliper piston can cause abnormally fast wear of the inner pad. Who is right?
   a. A only c. both A and B
   b. B only d. neither A nor B
2. Before removing a caliper, you should
   a. loosen the bleeder screw.
   b. remove a portion of the fluid from the master cylinder.
   c. tighten the wheel bearings.
   d. all of the above.
3. To make removal of the caliper from the rotor easier, you should retract the piston into the caliper using a large
   a. C-clamp.
   b. pair of pliers.
   c. screwdriver.
   d. any of the above.
4. Technician A says that too much clearance between the caliper and the mounting bracket can cause a knock as the brakes are applied. Technician B says that if this fit is too tight, the only solution is to replace the caliper or the mounting bracket. Who is right?
   a. A only c. both A and B
   b. B only d. neither A nor B
5. Technician A says that as a caliper is being removed, it should be lowered down gently so it will not damage the hose as it hangs. Technician B says the caliper should be removed completely or suspended from a wire. Who is right?
   a. A only c. both A and B
   b. B only d. neither A nor B
6. A floating caliper is normally removed from a car by disconnecting the brake hose and removing the
   a. caliper guide pins.
   b. mounting bracket from the steering knuckle.
   c. caliper bridge bolts.
   d. brake pads.
7. As you are doing a brake job, you discover a cracked caliper. You should
   a. weld up the crack.
   b. fill the crack with epoxy sealant.
   c. replace the caliper.
   d. just rebuild the caliper and not worry about the crack.
8. Technician A says that it is a good practice to turn a rotor during lining replacement to ensure a flat surface for the new pads to run against. Technician B says that the smooth, shiny surface of the average worn rotor is ideal for new lining break-in. Who is right?
   a. A only c. both A and B
   b. B only d. neither A nor B
9. The thickness of a rotor has been measured in six places, with the largest measurement being 0.9345 inches and the smallest 0.9333 inches. The specifications for this rotor are nominal thickness, 0.882 inches; allowable runout, 0.003 inches; finish, 16 to 79 rms; thickness variation, 0.0005 inches. Technician A says that this rotor should be replaced. Technician B says that the rotor should be turned before it is reused. Who is right?
   a. A only c. both A and B
   b. B only d. neither A nor B
10. The correct tool for measuring rotor thickness is a
    a. dial indicator.
    b. outside micrometer.
    c. brake drum micrometer.
    d. ruler graduated in millimeters.

11. When servicing disc brakes, brake technicians normally do not check a rotor for
    a. runout.
    b. scoring.
    c. parallelism.
    d. bell mouth.

12. Technician A says that the ideal rotor friction surface finish has a series of somewhat circular scratches and is called a nondirectional finish. Technician B says that this finish is a result of the normal cutting action of most rotor lathes. Who is right?
    a. A only
    b. B only
    c. both A and B
    d. neither A nor B

13. You are doing a brake job on a car with one slightly scored rotor (about 0.015 inches deep); the other rotor is smooth. Both rotors measure close to the new rotor thickness. You should
    a. turn both rotors to the minimum allowable thickness.
    b. turn both rotors to the largest size possible that will clean up the friction surface on the scored rotor.
    c. turn each rotor the minimum amount that will clean up the surfaces on that particular rotor.
    d. replace the scored rotor with a new one and run the smooth rotor.

14. A car has excessive vibration during a stop, and the rotor has been found to have an excessive amount of runout. Technician A says that runout can sometimes be corrected by reindexing the rotor on the hub. Technician B says that runout can be corrected on some cars by installing a shim between the hub and the rotor. Who is right?
    a. A only
    b. B only
    c. both A and B
    d. neither A nor B

15. Technician A says that the thickness of the rotor has no effect on brake lining temperature. Technician B says that a semimetallic brake lining is often used in OEM installations where lining temperatures are higher. Who is right?
    a. A only
    b. B only
    c. both A and B
    d. neither A nor B

16. Technician A says that if replacement is necessary, new bearing cones should be installed before turning a rotor. Technician B says that it is a good practice to wash a rotor using alcohol after it has been turned. Who is right?
    a. A only
    b. B only
    c. both A and B
    d. neither A nor B

17. When new pads are installed in a floating caliper, the outboard pad is usually
    a. fastened tightly to the caliper.
    b. lubricated using a high-temperature silicone grease.
    c. attached to the caliper with new antirattle clips.
    d. all of the above.

18. As a caliper is being replaced, be sure to
    a. lubricate the required contact points with the correct lubricant.
    b. tighten the mounting bolts to the correct torque.
    c. bottom the caliper piston before trying to place the caliper over the rotor.
    d. all of the above.

19. Technician A says that a caliper can be distorted if the lug bolts are overtightened. Technician B says that a rule of thumb is to tighten all lug bolts to 45 foot-pounds of torque. Who is right?
    a. A only
    b. B only
    c. both A and B
    d. neither A nor B

20. After new lining has been installed, you should
    a. check it out with a good, hard stop from 55 mph.
    b. make sure you have a firm brake pedal before moving the car.
    c. adjust the lining clearance by turning the starwheel with the correct brake spoon.
    d. all of the above.

12 Hydraulic System Service

Objectives

After completing this chapter, you should:

- ❑ Be able to diagnose stopping problems that are caused by improper operation of a master cylinder, wheel cylinder, caliper, brake line or hose, or brake valve.
- ❑ Be able to remove and replace a master cylinder and adjust the pedal pushrod length.
- ❑ Be able to disassemble, clean, inspect, and measure a master cylinder bore and other internal parts to determine if it should be rebuilt or replaced.
- ❑ Be able to reassemble a master cylinder.
- ❑ Be able to hone a master cylinder, wheel cylinder, or caliper bore as recommended by the manufacturer.
- ❑ Be able to remove and replace a wheel cylinder.
- ❑ Be able to disassemble and clean a wheel cylinder, inspect it for wear or damage, and reassemble it.
- ❑ Be able to disassemble and clean a caliper, inspect it for wear or damage, and reassemble it.
- ❑ Be able to remove and replace brake lines, hoses, fittings, and supports.
- ❑ Be able to select, handle, store, and install brake fluids.
- ❑ Be able to bleed or flush a brake hydraulic system by any of the approved methods.
- ❑ Be able to test, inspect, and replace brake valves.
- ❑ Be able to inspect, test, and replace brake light switches, wiring, and bulbs.
- ❑ Be able to adjust load- or height-sensing proportioning valves.
- ❑ Be able to reset a brake pressure differential valve.
- ❑ Be able to pressure-test a brake hydraulic system.
- ❑ Be able to perform the ASE Tasks for hydraulic system diagnosis, testing, and adjustment (See Appendix 1).

12.1 Introduction

The hydraulic components of a brake system can be serviced on an individual, as-needed basis or as part of a brake job. A major brake job normally includes a careful inspection and rebuilding of the hydraulic components. Many service shops prefer to remove and replace hydraulic components like master cylinders and wheel cylinders rather than rebuild them. The hydraulic system is arguably the most critical portion of a brake system. A failure in this system can produce a total, or at least a substantial, loss in braking ability.

The hydraulic components begin to deteriorate as the car is being assembled. The hygroscopic nature of brake fluid causes it to absorb moisture from the atmosphere. This moisture starts the chemical reactions that produce corrosion of the metal parts of the system, especially the bottoms of the cylinder bores. Rubber parts are vulcanized to form various shapes. Before vulcanizing, rubber is very pliable. Vulcanizing, primarily a heating-under-pressure process, changes rubber to the familiar, stable, but elastic substance. Heat, in any form, can cause rubber parts to self-vulcanize. They will continue to harden and lose their elastic nature. As this happens, cracks will begin to appear in the surface of the rubber, indicating that hardening is occurring. A certain amount of mechanical wear will also occur on dynamic, moving seals, especially on those in the master cylinder and somewhat on those in the wheel cylinders, because these particular seals undergo a lot of movement. If a seal has to move over rough bores or debris, the critical areas at the edges of the seal will wear that much faster or possibly be cut. A cut seal will leak.

After any hydraulic component is serviced, it becomes necessary to bleed all the air from the system. As mentioned earlier, air is compressible and will cause a low, spongy brake pedal. Some components, especially master cylinders, are bench-bled before they are installed. **Bench-bleeding** involves filling the component with fluid and removing the air before mounting it on the car. Some wheel cylinders are also bench-bled because they are mounted in such a way that they cannot be completely bled when installed on the car.

SAFETY NOTE: Hydraulic system repair must produce safe brake operation. To obtain proper brake operation and to protect yourself from injury, the following general precautions should be observed:

- Wear face or eye protection.
- Do not allow brake fluid to be sprayed on or splashed in your face or eyes. Do not hold your face over the master cylinder reservoir while the pedal is being pumped.
- Do not use petroleum products such as solvent, gasoline, kerosene, and so on, for cleaning hydraulic components. Only denatured alcohol or commercial hydraulic brake system cleaners should be used. Immersion in hot water can be used to remove traces of petroleum products if necessary. The part should then be dried immediately.
- Handle hydraulic components and parts with clean hands. The smallest trace of petroleum from greasy hands can damage rubber parts.
- Never soak parts, especially rubber ones, in alcohol. They should be cleaned and dried immediately.
- Use caution when using compressed air for cleaning. The air blast sometimes contains dirt or rust particles and can drive these particles, plus other debris from the parts you are cleaning, into your skin or eyes.
- Make sure that the air you are using to clean hydraulic brake parts comes from a clean source and does not contain oil.
- Do not hone aluminum bores or use abrasive methods to clean them. Any scratch in an anodized bore can allow rapid corrosion and future failure.
- Do not leave brake fluid containers, the master cylinder reservoir included, open to the atmosphere. Close them as soon as practical.
- Never reuse brake fluid.
- Store and handle brake fluid in clean containers.
- Do not allow brake fluid to spill, especially onto painted surfaces. If it spills, wipe it up immediately and flush the area with clean water.

12.2 Working with Tubing

When removing or replacing a hydraulic brake component, it is often necessary to disconnect or connect a steel brake line. A tubing or flare-nut wrench should always be used for this purpose (Figure 12-1A). Using an open-end wrench can round out corners of the tube nut and make removal extremely difficult. A brake technician will usually have flare-nut wrenches in both a 6- and a 12-point style in metric and fractional sizes. The greater gripping power of 6-point wrench allows maximum torque for breaking the nut loose, and the 12-point style allows working in tighter locations. It is also a good practice to use two wrenches when removing a line fitting from a union or a component that tends to turn with the fitting nut. This will often cause the line to twist. One wrench is used to prevent the twisting while the other unscrews the nut.

Many technicians make a practice of finger-tightening a connection at least two revolutions before using a wrench. A tube nut that is slightly misaligned will tend to cross-thread. This can ruin the unit to which the brake line is being connected or the tube nut. Most of us are not strong enough to cross-thread a fitting with our fingers, and two revolutions of the tube nut assures us the nut is started truly straight.

Like any other fastener, nut, or bolt, a tube nut should be tightened to the correct torque. If it is too loose, the connection will probably leak under pressure. If it is too tight, the threads might strip (especially brass ones) or the

(A)
FLARE NUT WRENCH
BRAKE LINE
BRAKE HOSE
(B)

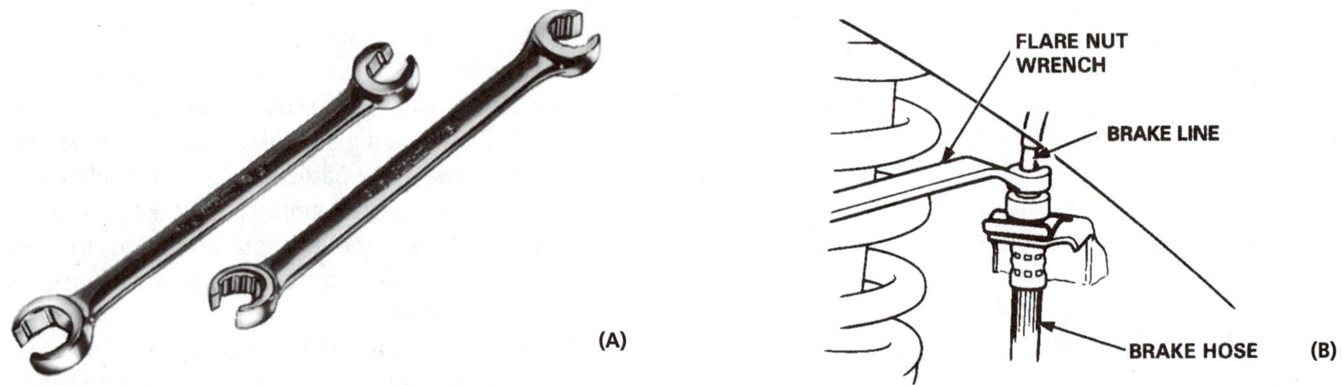

Figure 12-1 Either a 6- or 12-point flare nut wrench should be used when loosening or tightening a tube nut (A). The wrench design allows it to fit over the nut with the best gripping power (B). (A is courtesy of Snap-on Tools Company, Copyright Owner; B is courtesy of American Honda Motor Co., Inc.)

ABS MODULATOR UNIT-to-BRAKE LINE
19 N·m (1.9 kgf·m, 14 lbf·ft)
BRAKE HOSE-to-CALIPER
(BANJO BOLT)
34 N·m (3.5 kgf·m, 25 lbf·ft)
BRAKE LINE-to-DUAL PROPORTIONING VALVE
19 N·m (1.9 kgf·m, 14 lbf·ft)
MASTER CYLINDER-to-BRAKE LINE
19 N·m (1.9 kgf·m, 14 lbf·ft)
BRAKE LINE-to-BRAKE HOSE
15 N·m (1.5 kgf·m, 11 lbf·ft)
REAR DISC BRAKE:
BLEED SCREW
9 N·m (0.9 kgf·m, 6.5 lbf·ft)
REAR DRUM BRAKE:
7 N·m (0.7 kgf·m, 5 lbf·ft)
BRAKE HOSE-to-CALIPER
(BANJO BOLT)
34 N·m (3.5 kgf·m, 25 lbf·ft)
BLEED SCREW
9 N·m (0.9 kgf·m, 6.5 lbf·ft)
ABS MODULATOR UNIT-to-BRAKE LINE
19 N·m (1.9 kgf·m, 14 lbf·ft)

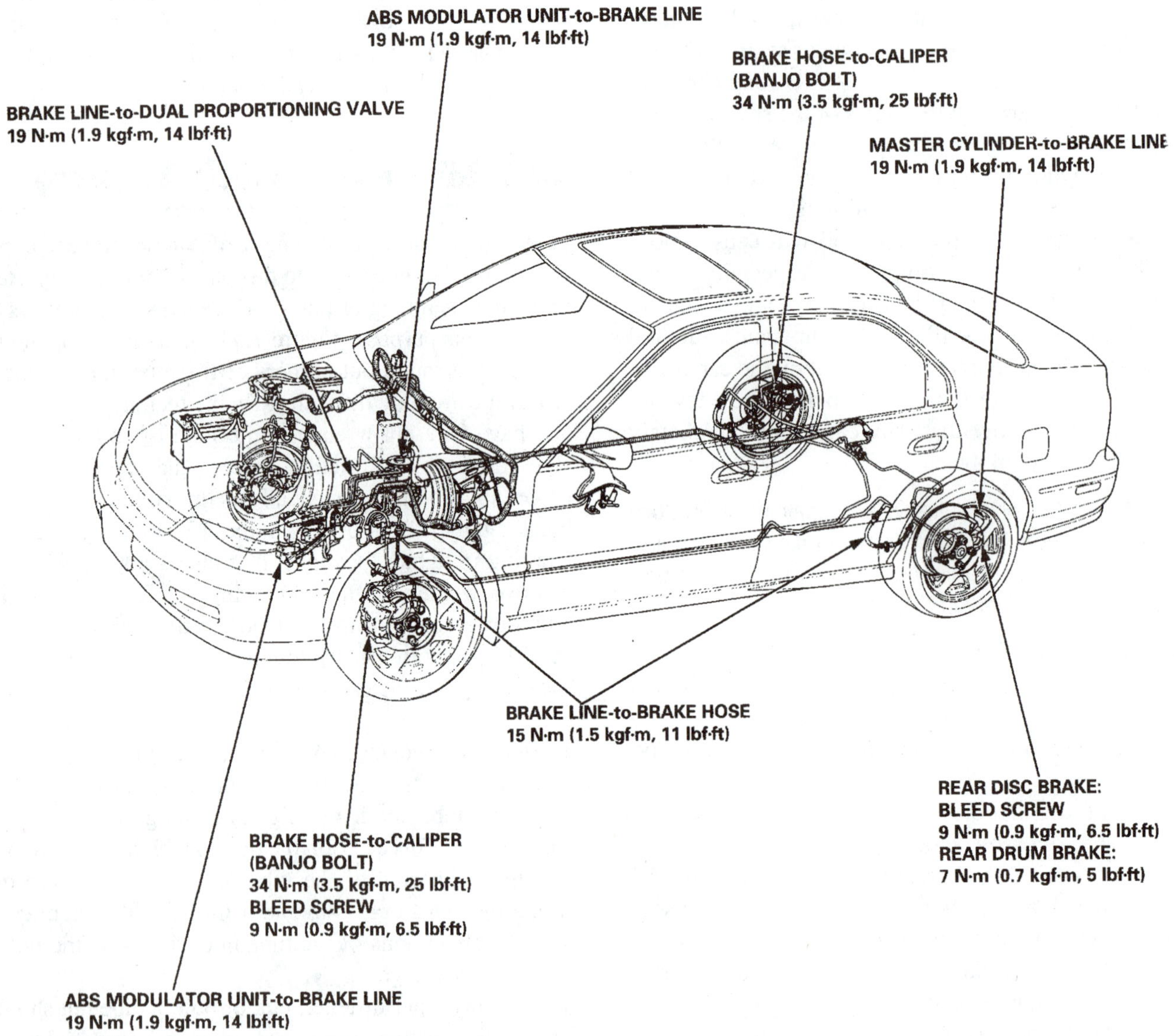

Figure 12-2 All tubing and bleed screws (bleeder valves) should be tightened to the correct torque when a component or tube is replaced. (Courtesy of American Honda Motor Co., Inc.)

tube seat or the tubing flare might be damaged. Also, over time, an overtightened connection becomes nearly impossible to take apart. A torque wrench should be used as the connection is being tightened, but at many connections this is extremely difficult if not impossible. Most brake technicians use their experience and their ability to judge correct tightness as a guide when tightening line fittings. The correct torque for tube nuts and other screwed connections is provided in Appendix 4 (Figure 12-2).

12.2.1 Tubing Replacement

Two flare styles are used for brake tubing: double flare and ISO (Figure 12-3). Double-flare fittings use fractional sizes, and ISO fittings are metrically sized. Sometimes these will both be used on a vehicle (Figure 12-4).

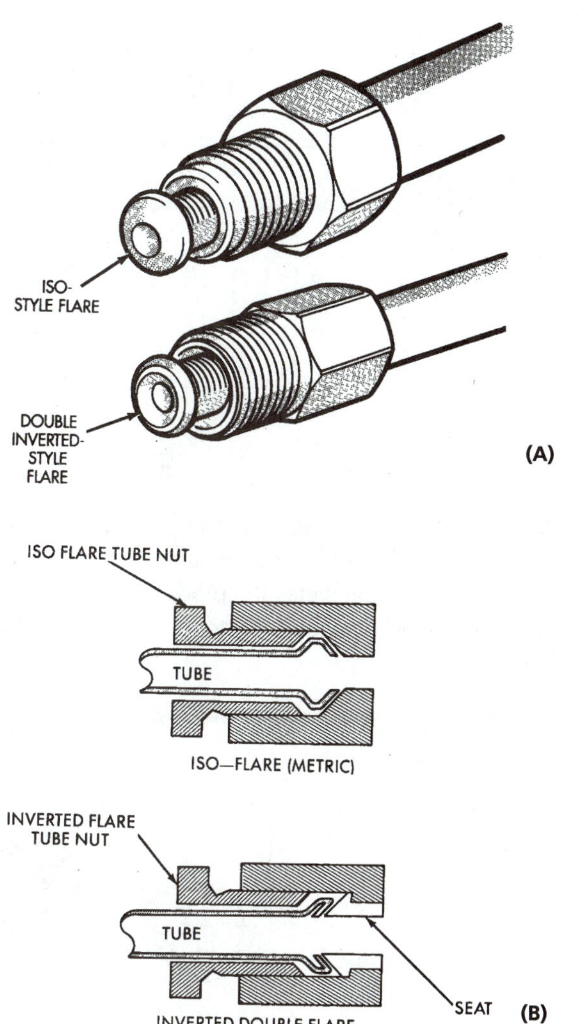

Figure 12-3 ISO and inverted flares can be identified by the shape of the flare (A). The fitting seat and tube nut also differ (B). (Courtesy of Chrysler Corporation)

Normally a faulty steel tube or one with a faulty tube nut should be replaced. Replacement tubes are available in various lengths with an end type and tube nut to match those on the car. The new tube should be carefully bent to the right shape using the old tube as a guide. A tubing bender should be used to prevent kinks while making sharp bends. If a tube is kinked, it will probably fracture at the point of the kink and develop a leak. If a proper tubing bender is not available, some technicians place the tubing in the groove of a pulley (alternator or water pump) and use both hands and thumbs to carefully hand-bend tubing (Figure 12-5). If a tube is too long for a particular location, it can be bent into a coil to use up some of the extra length. Be sure to lay the coil loops in a horizontal position to avoid air traps which will cause bleeding problems later. Tubing that is too short can be joined to another section of tubing by using a union to make it longer.

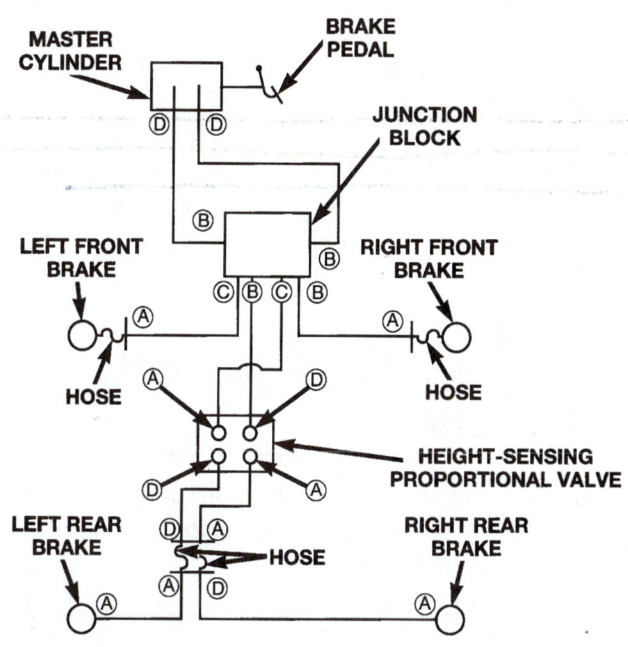

Figure 12-4 A modern vehicle might use both ISO and inverted flare fittings; note that ISO fittings use metric threads, and inverted flare fittings use fractional-inch-sized fittings. (Courtesy of Chrysler Corporation)

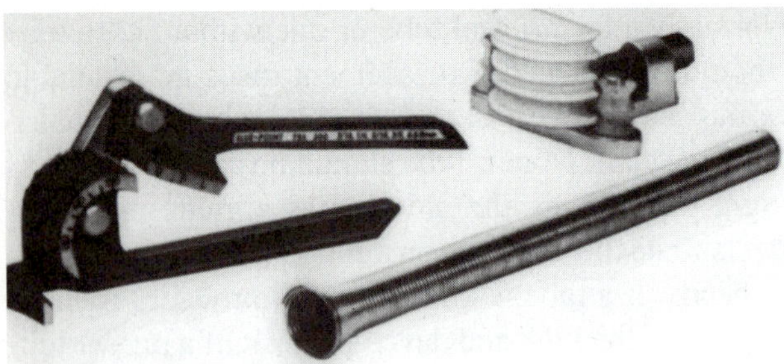

Figure 12-5 Three different styles of tubing benders. These tools allow tubing to be bent to a short radius without kinking. (Courtesy of Snap-on Tools Company, Copyright Owner)

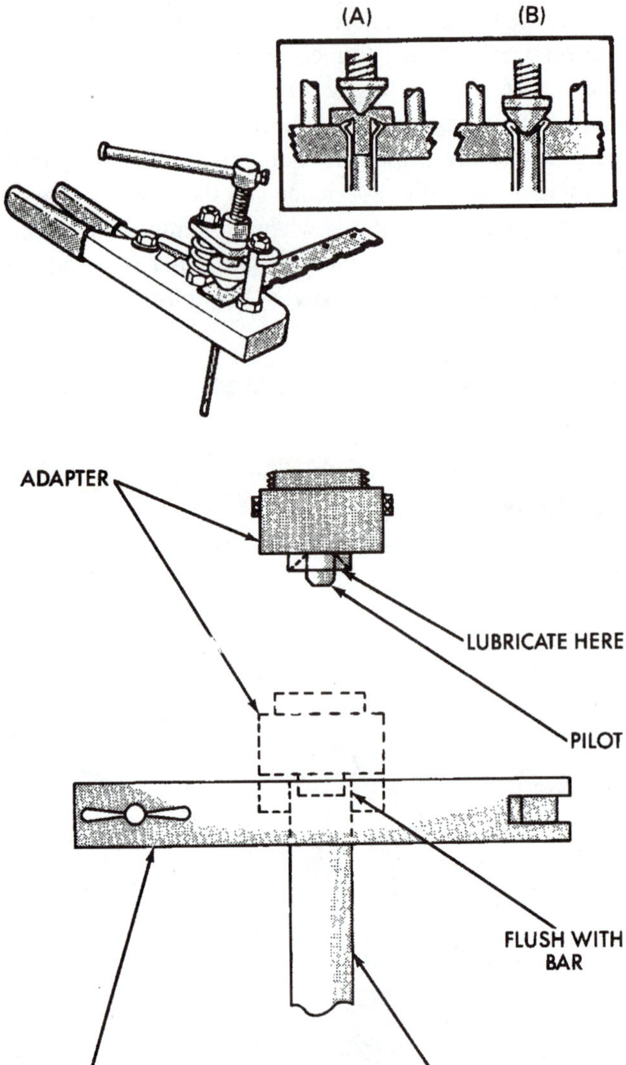

Figure 12-6 Making a double flare (top) uses a two-step process. The first step (A) begins the flare using an adapter, and the second step (B) completes the process. An ISO flare is also made using a special adapter (bottom). (Courtesy of Chrysler Corporation)

If necessary, a new flare can be put on a tube to allow it to be shortened, repaired, or have a new tube nut installed. Flaring is often avoided because steel tubing is strong and difficult to flare and special flaring tools are required. Steel tubing must be double-flared when using an SAE flare (Figure 12-6). This requires an extra step and the use of a special adapter. ISO flares can also be made in the field if the special flaring tool is available. Compression fittings, sometimes used on fuel lines, must never be used on brake lines. Also remember that copper tubing must never be used for hydraulic brake lines. When repairing tubing, always use a tubing cutter to avoid any small metal chips and to ensure a straight, smooth cut (Figure 12-7).

12.3 Servicing Hydraulic Cylinders

A stroking rubber seal must have a straight bore of a precise size. A caliper that uses a stationary, square-cut O-ring seal around the piston has different bore requirements. It can tolerate some imperfections in the bore. Rust or corrosion deposits, dirt, and other residue must be removed from a bore used with a stroking seal. On cast-iron bores, small scratches, pits, or deposits can sometimes be removed by honing. On aluminum bores, they cannot. Most technicians will not take the gamble of rebuilding a unit with a questionable bore. A leaky wheel cylinder can easily ruin a brake lining set. A leaky master cylinder can cause a loss of braking power. Many sources recommend against rebuilding aluminum wheel cylinders or master cylinders.

NOTE: Cylinders with aluminum bores cannot be honed. They can be cleaned using a fiber or nylon brush.

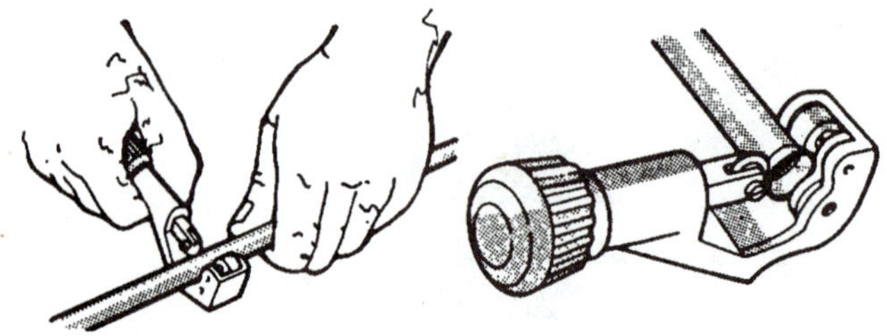

Figure 12-7 A tubing cutter is the best tool for cutting tubing because it makes a straight, neat cut with no metal chips. Note the groove in the cutter rollers so a cut can be made close to the flare (right). (Courtesy of Chrysler Corporation)

A cylinder is slightly larger than the piston, but if it is too large, heel drag can occur as a result of the rubber cup being forced between the piston and the bore. Remember that heel drag can cause sticking or nibbling of the cup by the piston. Bore size is normally measured using a narrowed feeler gauge or a strip of shim stock. You can cut a feeler gauge to a narrow width, about 1/8 to 1/4 inches (3 to 6mm), using sharp tin snips. After cutting, you should smooth any burrs with a wet stone or fine sandpaper. The gauge thickness should be as follows:

Cylinder Size	Gauge Thickness
¾ to 1³⁄₁₆ in.	0.006 in.
19 to 30 mm	0.15 mm
1¼ to 1⁷⁄₁₆ in.	0.007 in.
32 to 37 mm	0.18 mm
Over 1½ in.	0.008 in.
Over 38 mm	0.20 mm

To measure bore clearance, place the gauge strip lengthwise in the cylinder and try to slide the piston over it (Figure 12-8). If the piston will not enter past the gauge strip, the cylinder is good. If the piston slides past, the cylinder is oversize and should be replaced. It should be noted that the primary face of some master cylinder pistons has an undersize land and therefore should not be used when gauging bore sizes. This particular face does not have replenishing holes, and during master cylinder release, replenishing fluid flows around this piston instead of through it.

The size of a cylinder can be gauged by using a hole gauge that is adjusted to the size of the piston plus the allowable clearance (Figure 12-9). In the past, a special gauge set was marketed for brake cylinders. A gauge was included for each common cylinder size, and each gauge was the maximum size for the cylinder. If the cylinder is the same size as the gauge or larger, it is oversize and should not be rebuilt (Figure 12-10).

To service a bore that uses a stroking seal in a master cylinder, wheel cylinder, or some calipers:

1. Wash the bore with denatured alcohol.
2. Inspect the cylinder bore under a strong light. Move the cylinder around so that the light is concentrated on the area you are checking. Then turn the cylinder so you can check the entire inner surface where the seals stroke. Pay particular attention to the bottom side of the bore (Figure 12-11).

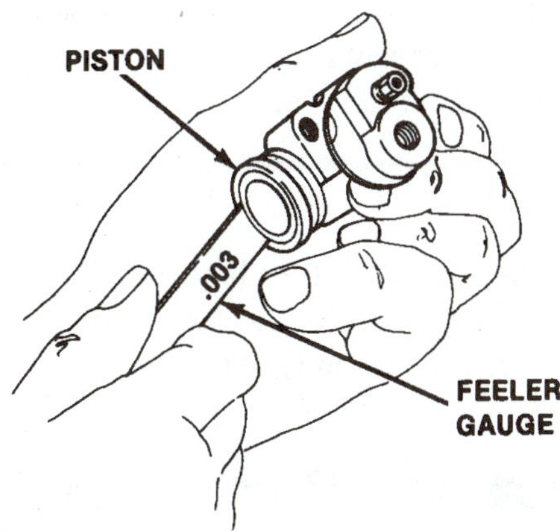

Figure 12-8 A cylinder bore size can be checked by placing a feeler gauge in the bore and then trying to slide the piston in also. If the piston enters past the feeler gauge, the bore is too large. (Courtesy of Brake Parts, Inc.)

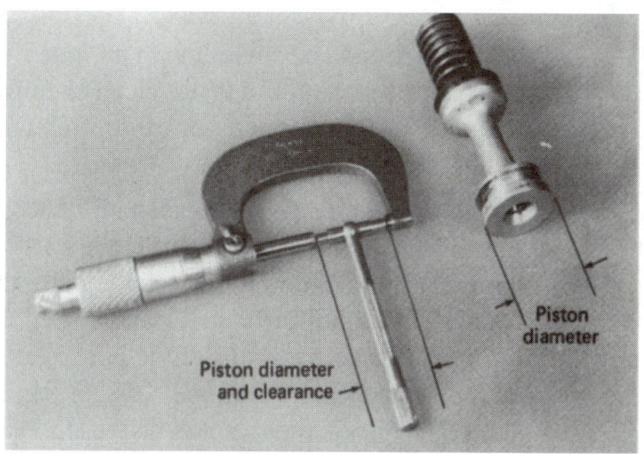

Figure 12-9 The micrometer is set to the piston diameter plus the maximum clearance, and the bore gauge is set to the micrometer. If the bore gauge is smaller than the cylinder, the cylinder is too large.

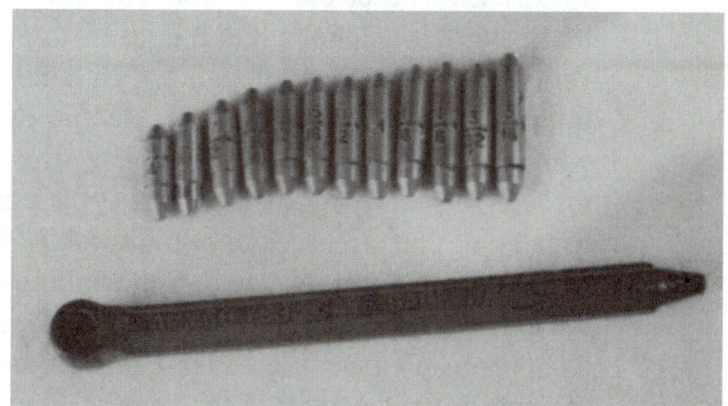

Figure 12-10 At one time a cylinder bore gauge set was available for measuring cylinders. Each gauge rod was sized to the maximum allowable for each bore. If it passed through the bore, the bore was too large.

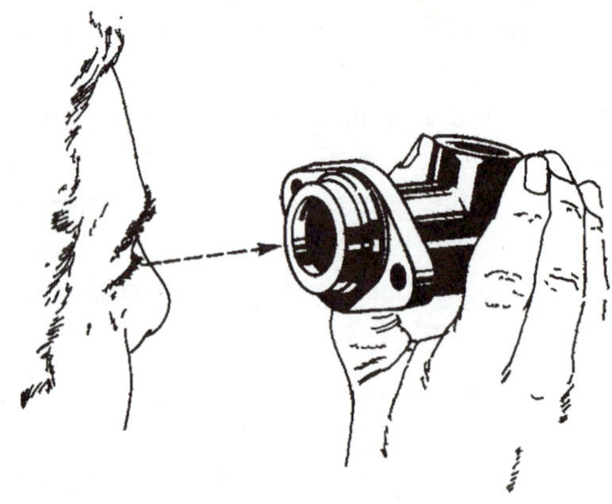

Figure 12-11 After cleaning, a bore should be carefully inspected to determine if there are any pits or corrosion. A strong light source is necessary for a thorough inspection. (Courtesy of ITT Automotive)

3. If the bore is clean and smooth with no imperfections, it can be reused. If it is not clean or imperfections are found, one of the following procedures should be used:
   a. *On aluminum bores*: Use a small, fiber or nylon brush or a clean piece of cloth wrapped around a wooden dowel to swab the bore with denatured alcohol or brake cleaner. Then air-dry it and inspect it again. If the bore is clean and has no imperfections (discolorations should be disregarded), it can be reused.
   b. *On cast-iron bores*: Use either crocus cloth wet with alcohol and moved in a circular direction or a brake cylinder hone (Figure 12-12). A carbide-tipped brush sold under the name Flex-Hone can also be used (Figure 12-13). Either type of hone should be wet with brake fluid or alcohol while it is used, to prevent the abrasive from clogging. The hone or brush should be turned with an electric drill while moving it with slow, complete strokes through the entire cylinder. Several strokes are usually all that are necessary or allowable. Be careful not to overstroke against the bottom of the cylinder or out the end of the bore. Many technicians prefer to use a Flex-Hone because the cutting stones tend to stay cleaner and it removes debris faster.

A hone can fly apart if it is removed from the bore while spinning.

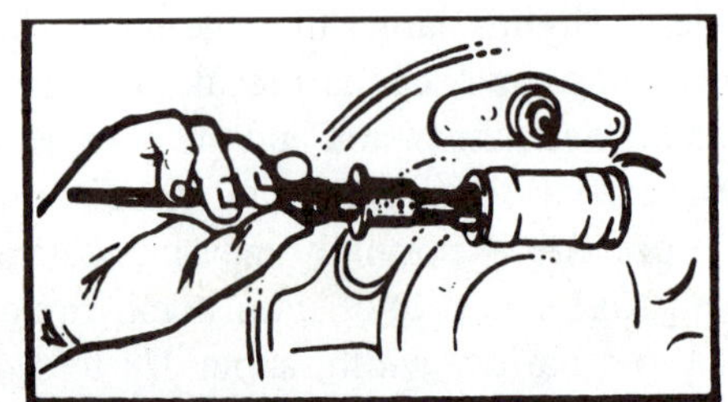

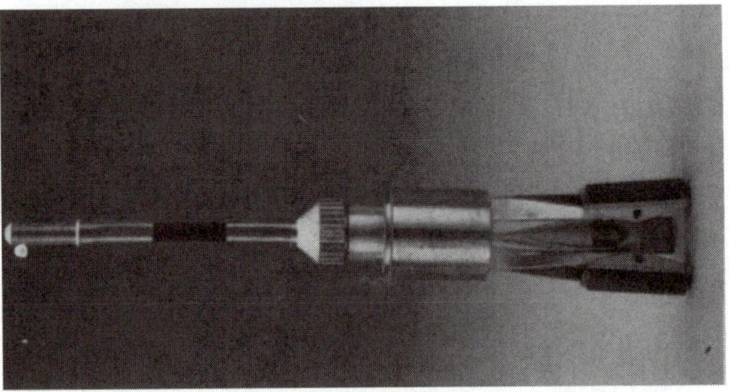

Figure 12-12 A wheel cylinder hone is normally turned using a drill. Be sure to keep the stones lubricated and clean, and do not remove the hone from a cylinder while it is spinning. (Courtesy of KD Tools)

Figure 12-13 A carbide-tipped brush or Flex-Hone. This device is also spun by a drill in a bore to clean the cylinder. It should be lubricated during use. (Courtesy of KD Tools)

4. Repeat steps 1 and 2. If the bore is still not clean, it should be replaced. Any further honing will make it oversize and unusable.

12.4 Master Cylinder Service

Normal master cylinder service includes removal, rebuilding, and replacement. Many service shops replace the master cylinder with a new rebuilt unit rather than rebuilding the one from the vehicle. If the master cylinder is faulty, a new or commercially rebuilt unit can be installed. A rebuilt master cylinder usually offers substantial price savings over a new unit, but rebuilt units should be carefully checked to ensure adequate quality control.

Normally, any master cylinder with a good bore can be rebuilt in a repair shop if a parts kit can be obtained. Many technicians prefer to rebuild the master cylinder that is on a car because they are sure that it will fit and have line fittings of the correct size. Some replacement master cylinders will have outlets of a different size or at slightly different locations. Outlets of different size require step-up or step-down fittings (Figure 12-14). The tubing can usually be bent slightly to fit a different location.

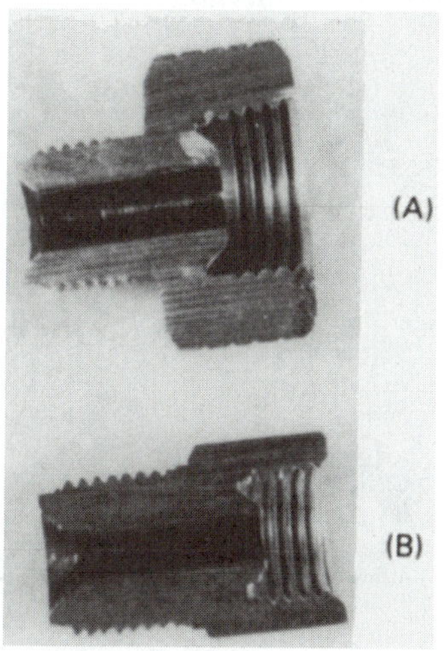

Figure 12-14 A step-down (A) and a step-up (B) adapter. These fittings allow an inverted fitting to be connected to a different-sized opening.

When installing a new master cylinder, it is a good practice to flush the cylinder with clean brake fluid. This is done to remove any debris which might be left over from the manufacturing process or chemical coatings that were used to protect the cylinder from corrosion. To flush a master cylinder, simply fill the reservoir(s) and the cylinder bore(s) about one-third full with clean brake fluid, install the reservoir cover, plug the line port(s), shake the cylinder to work the fluid all around, and drain out all the fluid. Finally, bench-bleed and install the cylinder in a normal manner.

12.4.1 Master Cylinder Removal

Removal and replacement procedures for most power brake (vacuum or hydraulic booster) units are very similar. Manual brake (no power assist) master cylinder removal often requires that the pushrod be disconnected from the brake pedal. If you are not sure of the exact procedure, check a service manual (Figure 12-15).

To remove a master cylinder:

1. Take off the reservoir cover and remove the fluid from the reservoir.
2. Disconnect any wires connected to the reservoir or master cylinder body.
3. Disconnect the brake tube(s). Depending on the system, there will be one, two, or four tubes. Be sure to use a tubing wrench for this operation.
4. Remove the nuts or bolts attaching the master cylinder to the power booster or vehicle bulkhead.
5. Slide the master cylinder off the booster or bulkhead. If it will move only a short distance but no farther, replace one of the nuts or bolts (finger-tight) to support the master cylinder, and then disconnect the pushrod from the brake pedal. After disconnecting the clip or pin and clip holding the pushrod to the pedal, repeat steps 4 and 5 (Figure 12-16).

Remove

1. Disconnect electrical lead and four hydraulic lines.
2. Remove two attaching nuts.
3. Remove master cylinder as shown.

Install

Notice: See notice at the beginning of this section.

1. Install master cylinder as shown and torque attaching nuts to 30-40 N·m (22-30 ft. lbs.)
2. Attach electrical lead and four hydraulic lines. Torque tube nuts to 13.6-20.3 N m (120-180 in. lbs.).

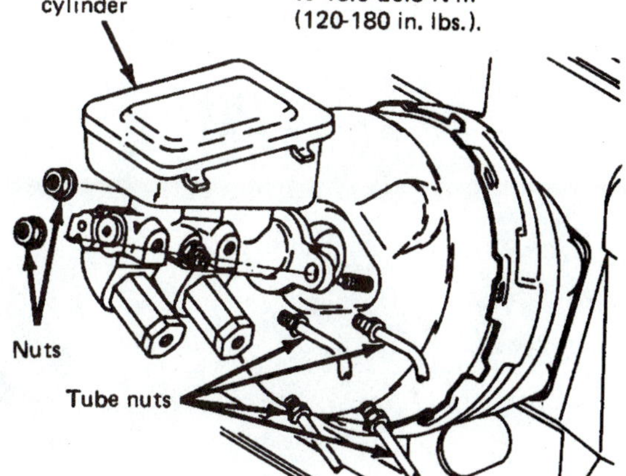

Figure 12-15 After the tubing and electrical connections have been disconnected, most master cylinders can be removed by unscrewing the two attaching nuts and lifting the master cylinder off the booster. (Courtesy of General Motors Corporation, Service Technology Group)

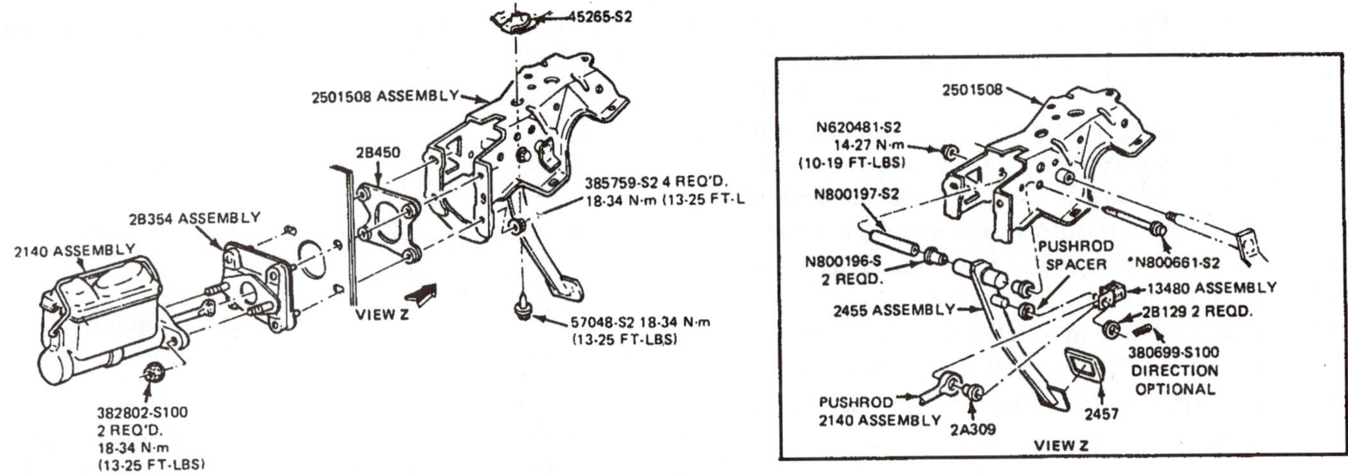

Figure 12-16 It is usually necessary to disconnect the pushrod from the brake pedal when removing a nonpower brake master cylinder. (Courtesy of Ford Motor Company)

NOTE: With some power boosters, the pushrod will be free; test this by trying to pull it outward. If it comes out, tape the pushrod to the front of the windshield to keep it from getting lost or forgotten.

12.4.2 Rebuilding a Master Cylinder

Master cylinder rebuilding is a process of disassembly, cleaning, servicing and checking the bore, and reassembly using new rubber parts. A master cylinder rebuilding kit is required (Figure 12-17). The kit for a tandem master cylinder usually contains a complete primary piston assembly, primary and secondary seals for the secondary piston, and a primary piston retaining ring. In some cases, it will also contain, as necessary, a new residual check valve(s) and tube seat(s) or reservoir grommets. Some reservoirs can be removed by simply loosening the clamp and sliding the reservoir off the master cylinder (Figure 12-18).

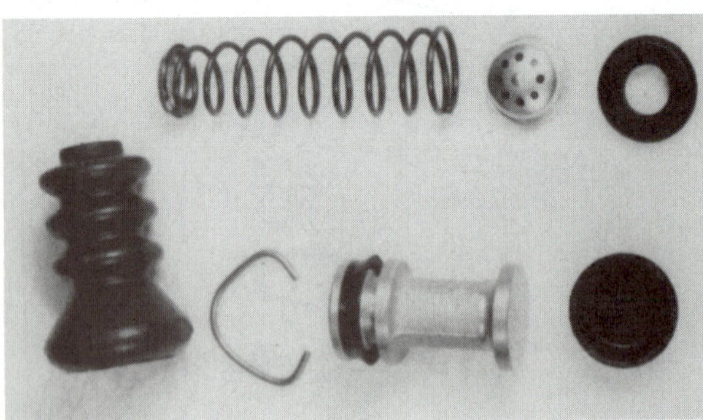

Figure 12-17 An overhaul kit for a single master cylinder. It includes (clockwise from the top) a new return spring, residual valve, grommet, primary cup, piston with secondary cup, retaining ring, and dust boot. A kit for a tandem master cylinder will also include primary and secondary seals for the secondary piston.

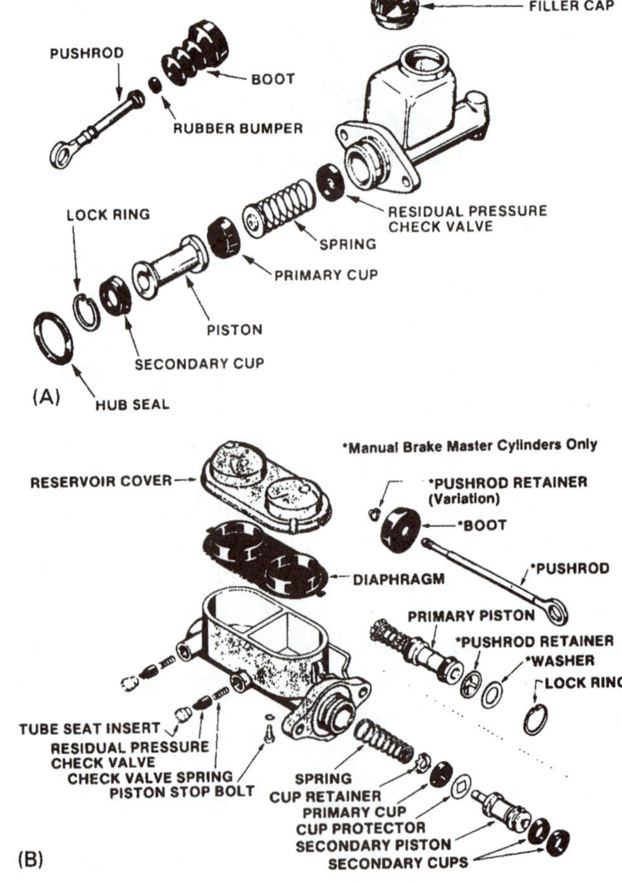

Figure 12-18 A disassembled single (A) and tandem master cylinder (B). Their service procedures are very similar. (Courtesy of Bendix Brakes, by AlliedSignal)

Master cylinder disassembly and reassembly procedures will vary slightly depending on whether the reservoir is removable; whether residual check valves are used; whether it is a single, tandem, or quick-take-up unit; and whether it contains internal valves or external switches (Figure 12-19). Again, it is a good practice to follow the procedure recommended by the manufacturer.

To rebuild a master cylinder:

1. Some reservoirs have a vacuum seal that can be pried from the master cylinder assembly (Figure 12-20).
2. Remove the reservoir cover and pour out any fluid that remains. Hold the unit over a container and stroke the pistons a few times to pump any fluid out of the bore. Use a rounded wooden dowel or metal rod for a pushrod if necessary.
3. To remove the reservoir on units with plastic reservoirs:
   a. Clamp the master cylinder body in a vise by gripping a master cylinder mounting ear.
   b. Remove any retaining pins or clips.
   c. Insert a large screwdriver or pry bar between the cylinder body and the reservoir and pry the reservoir off the body. Use care, as it can break (Figure 12-21). Note that a retaining pin is used on some reservoirs; this pin must be removed first.
   d. Remove the rubber grommets from the cylinder body (Figure 12-22).
4. On some tandem units, you will need to locate and remove the secondary piston stop bolt or pin

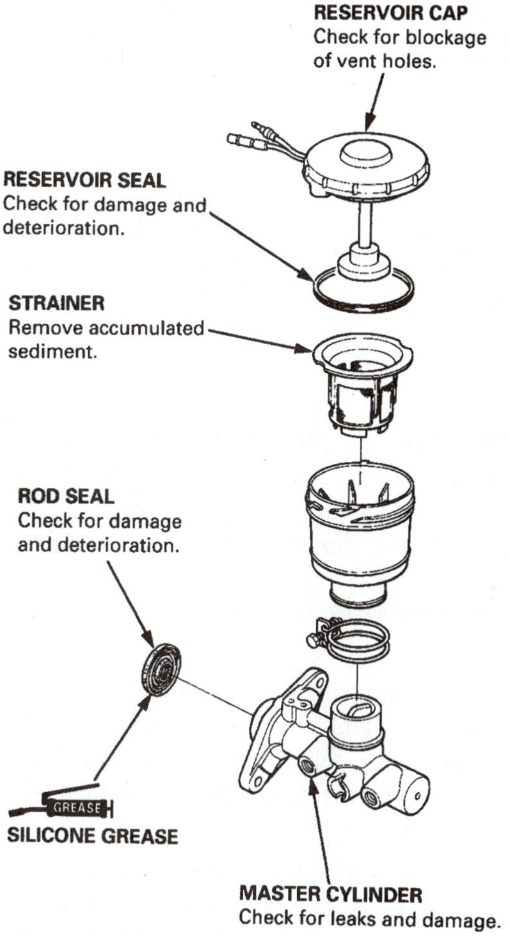

RESERVOIR CAP
Check for blockage of vent holes.
RESERVOIR SEAL
Check for damage and deterioration.
STRAINER
Remove accumulated sediment.
ROD SEAL
Check for damage and deterioration.
GREASE
SILICONE GREASE
MASTER CYLINDER
Check for leaks and damage.

Figure 12-19 Loosening the clamp allows removal of the reservoir with its strainer, seal, and cap from the master cylinder body. Note the fluid level switch is contained in the cap. (Courtesy of American Honda Motor Co., Inc.)

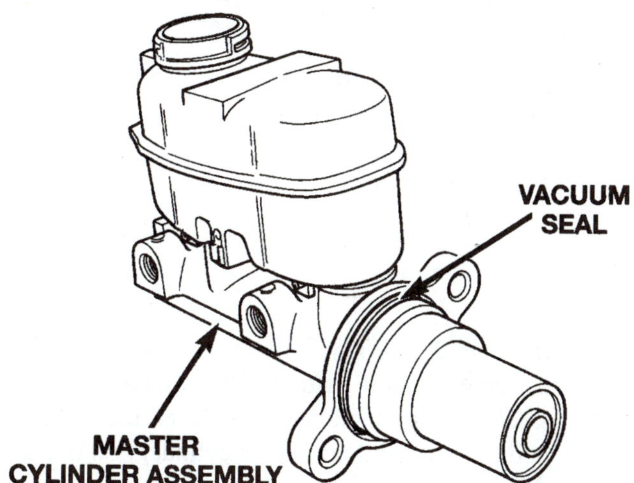

Figure 12-20 This master cylinder has a vacuum seal that can be removed by prying it off. (Courtesy of Chrysler Corporation)

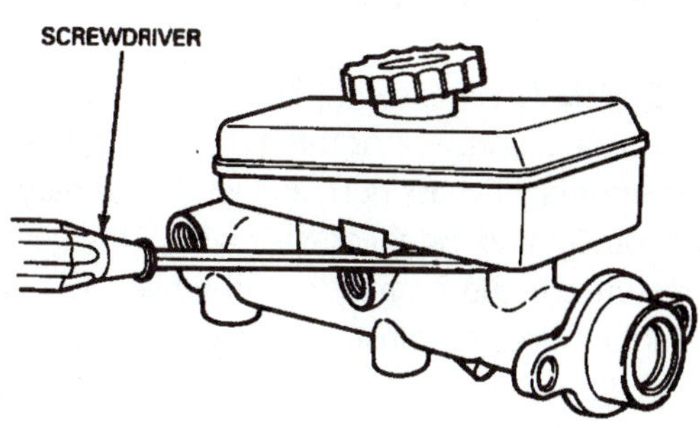

Figure 12-21 The reservoir of many composite master cylinders is removed by prying it off the cylinder body. Some use a retaining pin(s) that must be removed first. (Courtesy of Ford Motor Company)

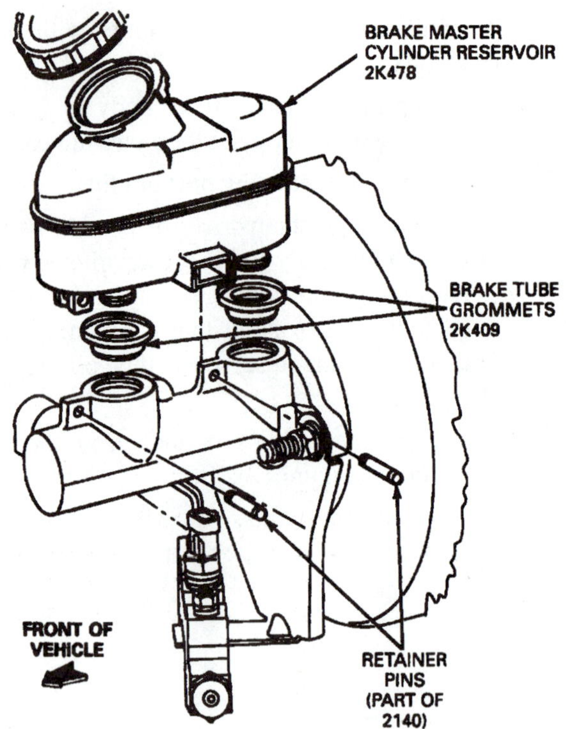

Figure 12-22 Reservoir grommets are pulled or pried out. New grommets should be lubricated as they are installed. (Courtesy of Ford Motor Company)

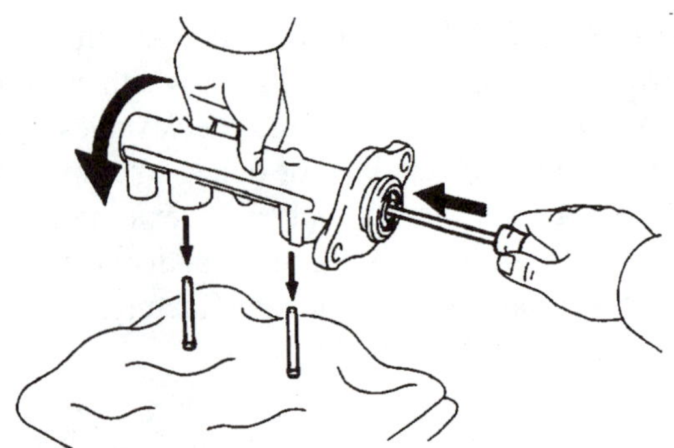

Figure 12-23 Some master cylinders use a stop pin for one or both pistons; they are removed by turning the cylinder body upside down and pushing the piston inward.

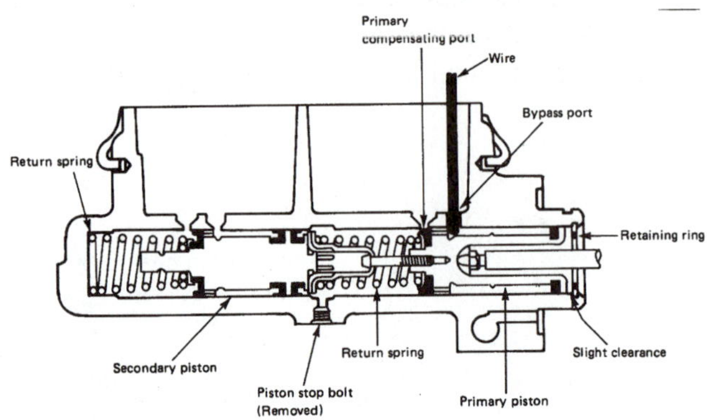

Figure 12-24 If the piston is slid inward, the shank of a small drill bit can be placed through the primary bypass port to hold the piston and return spring slightly compressed as the retaining ring is removed.

(Figure 12-23). This bolt or pin enters the cylinder bore from the bottom of the reservoir or at the side or bottom on the outside of the cylinder body. Many tandem master cylinders do not use this stop bolt.

5. Clamp the cylinder body in a vise by gripping a mounting ear as described in step 3, push inward slightly on the primary piston, and remove the primary piston retaining ring. Then remove the primary piston and spring.

HELPFUL HINT: Some technicians push the primary piston inward, slide a iron wire or the shank of a small drill bit through the primary bypass port, and let the primary piston slide back slowly and catch on the wire. This keeps the spring pressure off the primary piston while the retaining ring is being removed or reinstalled (Figure 12-24).

6. Slide the secondary piston out of the bore. If it is stuck, either grip it with needlenose pliers and pull it out, slam the cylinder body onto a block of wood (bore opening down), or use air pressure.

CAUTION *If air pressure is used, be careful that the piston does not fly out. It can be contained by wrapping a shop cloth around the cylinder body, covering the bore end, and holding the cloth securely while applying air pressure to the secondary outlet port (Figure 12-25).*

7. Check the placement and direction of the seals and disassemble the secondary piston. Do not disassemble the primary piston or disturb the position of the screw unless so directed by the manufacturer.
8. If the rebuilding kit includes a replacement check valve and a tube seat, probe the outlet ports with a small wire or a straightened paper clip to determine if residual valves are used in the master cylinder. If the outlet port contains a residual valve, you should be able to feel the rubber valve

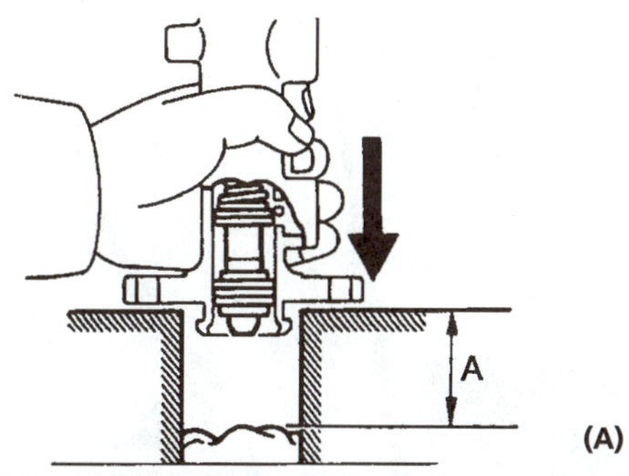

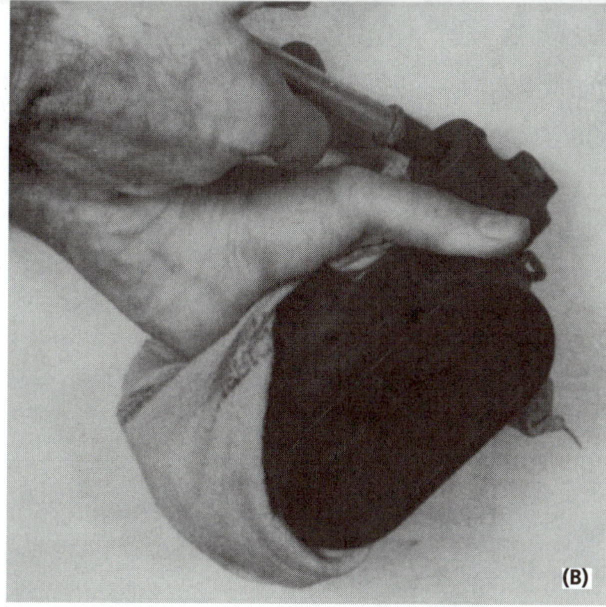

Figure 12-25 A stuck master cylinder piston can be removed by tapping the cylinder down onto two wooden blocks (A) or using air pressure to blow it out (B). Note the shop cloth to protect the piston as it drops (A) and the shop cloth to catch the piston (B).

with the wire. It will be about ¼ inches (6.3 mm) past the tube seat. Do not remove the tube seat unless the outlet contains a check valve and you have a replacement. If there is a check valve and you have a replacement, remove the tube seat according to the following procedure:

a. Thread a #6-32 or a #8-32 self-tapping machine screw through the outlet port. It should thread in about ¼ inches.
b. Using two screwdrivers placed as shown in Figure 12-26, pry the screw upward to lift out the tube seat.
c. Remove the tube seat, check valve, and spring.

NOTE: Never install a check valve in a master cylinder port if it did not originally have one.

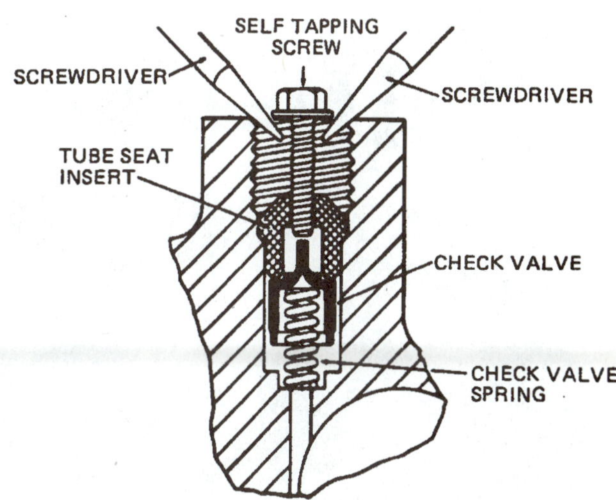

Figure 12-26 A tube seat and check valve assembly can be removed by threading a self-tapping screw into the insert and prying upward as shown here. (Courtesy of Wagner Brake)

9. Remove any other valves or switches as directed.
10. Using denatured alcohol or brake clean, thoroughly clean the reservoir, cylinder body, and any other parts that will be reused. Dry these parts with compressed air, making sure that all ports and passages are clean and open. Inspect all the parts to be reused to make sure they are in good condition. If necessary, service the cylinder bore as described earlier in this chapter.

If the bore is acceptable and the other parts are in good condition, the master cylinder can be assembled using the following procedure:

1. If a check valve has been removed, place the new spring, valve, and seat in position and lightly tap the new seat downward using a flat punch. Make sure the seat remains straight as you start it into the bore. The tube seat will be completely seated as the brake tube is tightened in place.
2. Carefully install the new seals on the secondary piston in the same position as the original ones. This installation is easier if they are wet with brake assembly fluid or brake fluid (Figure 12-27).

NOTE: A dry rubber cup should never be slid into a dry bore. Assembly fluid is a rather thick fluid that is used to lubricate the rubber cups. It also helps reduce future rusting in dry areas of the bore. Silicone (preferred) or glycol brake fluid can also be used.

(A)

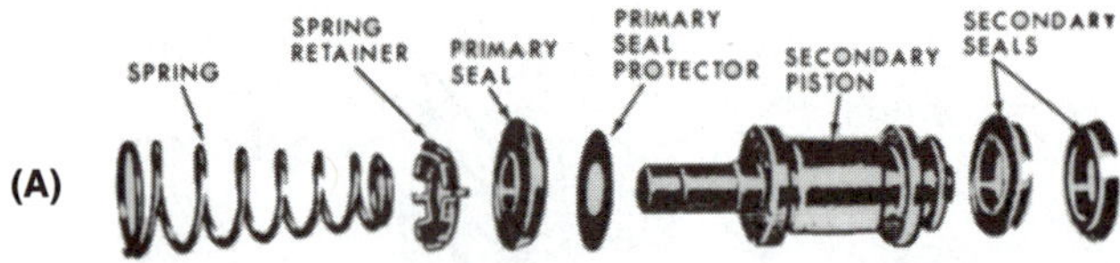

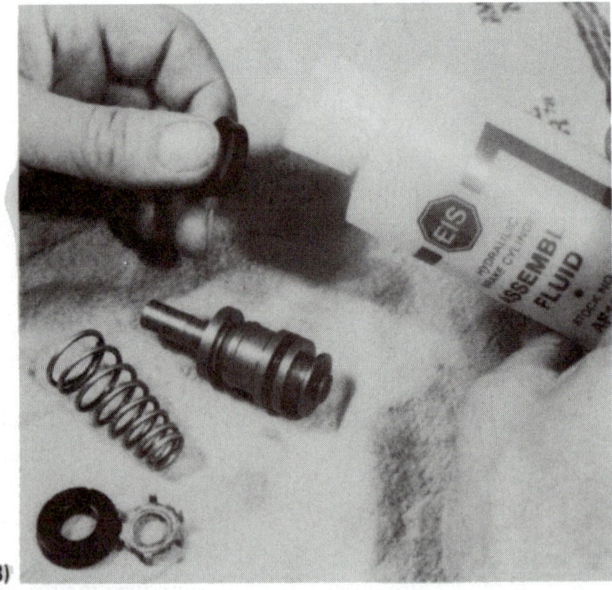

(B)

Figure 12-27 During assembly, the seals must be positioned correctly on the piston (A) and lubricated with assembly fluid (B) or brake fluid as they are slid over the bands on the piston. (A is courtesy of General Motors Corporation, Service Technology Group)

3. Coat the new secondary seals and the cylinder bore with assembly fluid, and slide the secondary piston and its spring into the bore. Make sure the lips of the cups do not catch on the edges of the bore as they enter.
4. Coat the seals on the new primary piston assembly with assembly fluid and slide it into the bore (Figure 12-28). Note that on some manual brake master cylinders, the pushrod is attached to the primary piston using a retainer inside the piston. To separate the piston, clamp the pushrod in a vise and pry upward on the piston (Figure 12-29). To install the pushrod in the new piston, insert the pushrod into the new piston an press it inward until the retainer snaps into place.
5. Push the primary piston into the bore, compressing the spring so the retaining ring can be installed. Some technicians use a holding device for the primary piston as described in step 4 of the disassembly procedure.
6. If the master cylinder uses a secondary piston stop screw, install it. In a few cases, it will be nec-

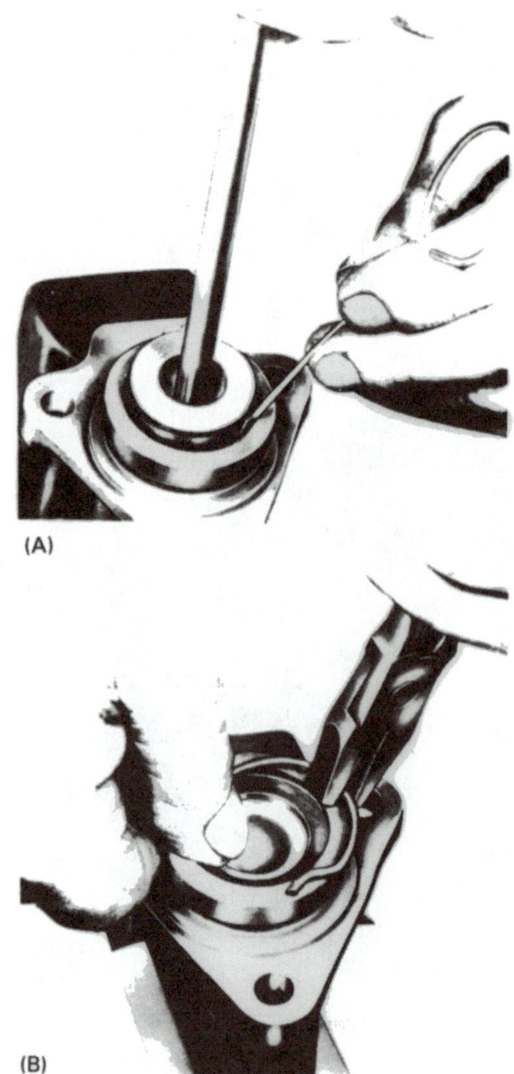

(A)

(B)

Figure 12-28 As a piston is slid into the bore, it is often necessary to coax the lip of the seal past the end of the bore with a smooth instrument (A). With the piston in place, the retaining ring can be replaced (B). (Courtesy of ITT Automotive)

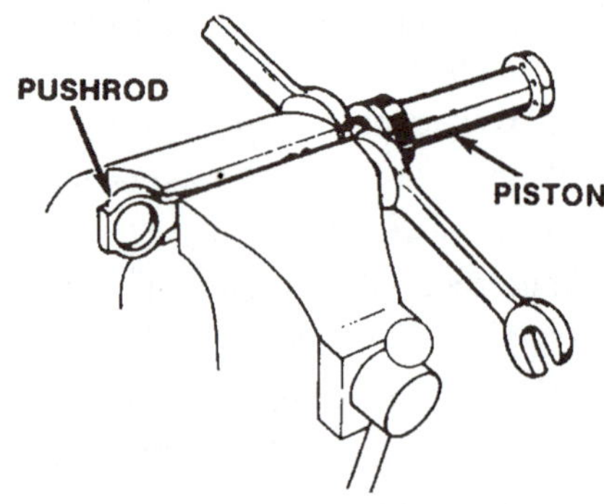

Figure 12-29 On many manual-brake master cylinders, the pushrod is attached to the piston by a retainer. Some can be removed by prying with two wrenches as shown here. (Courtesy of Bendix Brakes, by AlliedSignal)

essary to push inward on the primary piston so the secondary piston will move inward far enough to allow the screw to enter.

7. If the reservoir has been removed, wet the new grommets with assembly fluid and install them in the cylinder body. Wet the extensions of the reservoir with assembly fluid. With the cylinder body held in a vise, push the reservoir onto the body using a rocking motion. The bottom of the reservoir should contact the top of the grommets (Figure 12-30). Some technicians find it easier to place the reservoir upside down on a bench and push the cylinder body downward onto it with a rocking motion.
8. Replace all other switches or valves.

The master cylinder is now ready for bench-bleeding and installation.

12.4.3 Bench-Bleeding a Master Cylinder

Master cylinders have several areas that can easily trap air. This is especially true of master cylinders that are mounted at an angle. Also, step bore master cylinders with quick-take-up valves offer additional places to trap air. Bench-bleeding easily removes the air from these pockets. In some cases, it is almost impossible to completely bleed a master cylinder after it is installed on the car. When bench-bleeding master cylinders that have four outlet ports, plug the lower port of each section and bleed both sections using the two upper ports. Some replacement master cylinders include bleeder screws which make bleeding a fairly easy, on-car operation. There are two commonly used methods of bench-bleeding master cylinders.

To bench-bleed a master cylinder using tubes:

1. Secure the cylinder body in a vise by gripping a mounting ear.
2. Select a tube nut or adapter of the correct size and install a tube in each outlet port. The tubes should curve up and over into the reservoir (Figure 12-31).
3. Fill the reservoir about one-half to three-quarters full with brake fluid. The fluid level should be above the ends of the tubes.
4. Push the primary piston inward using slow, complete strokes and allow the piston to return slowly. You should observe air bubbles leaving the tubes on each pumping stroke. Bleeding will go faster if you close off the tubing during the return stroke. This is done by pinching the tubing (if it is plastic or rubber) or by putting your finger over the end of a metal tube (Figure 12-32).
5. Continue step 4 until no more air bubbles are expelled on the pumping stroke. Occasionally, it will help to tilt the master cylinder bore up or down during the pumping stroke. On quick-take-up master cylinders, you need to generate 75 to 100 psi of pressure to open the quick-take-up valve and bleed it.

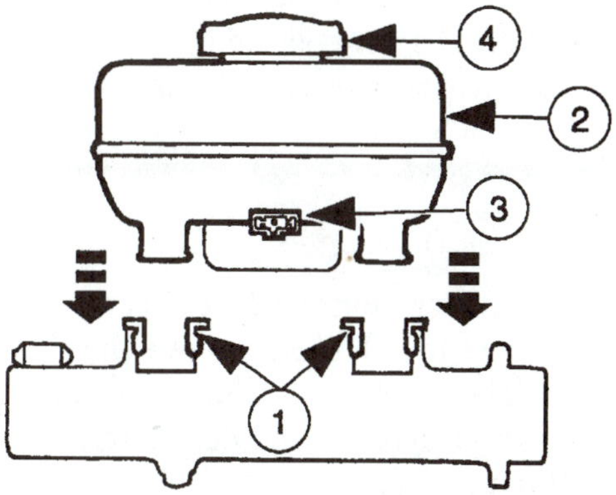

Figure 12-30 With the new grommets (1) installed and wet with assembly fluid or brake fluid, the reservoir (2) is pushed onto the body until it snaps securely into the grommets. (3) indicates the connection for the fluid level warning switch. (Courtesy of Ford Motor Company)

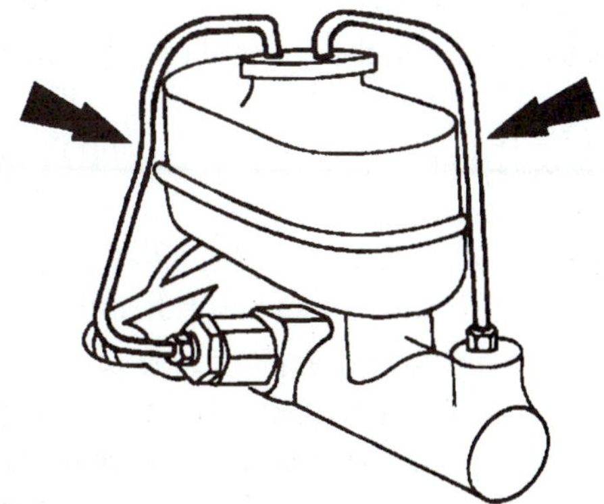

Figure 12-31 The two bleeding tubes (arrows) run from the outlet ports to below the brake fluid level in the reservoir. As the master cylinder piston is stroked, air will be pumped form the cylinder through the tubes. (Courtesy of Ford Motor Company)

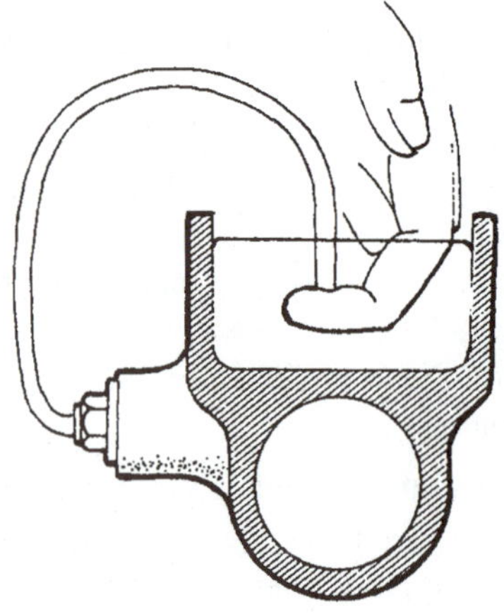

Figure 12-32 When bench-bleeding using tubes, the process will often go faster if the return flow is stopped during the piston return stroke as shown here. (Reprinted from Mitchell Anti-lock Brake Systems, with permission of Mitchell Repair Information, LLC.)

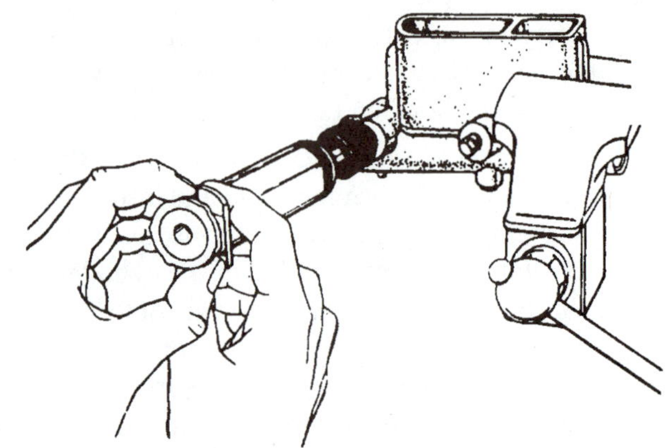

Figure 12-33 A special syringe is available for bench-bleeding master cylinders. The first step is to partially fill the reservoir and then suck fluid out of the outlet port. Air will come out with the fluid. (Reprinted from Mitchell Anti-lock Brake Systems, with permission of Mitchell Repair Information, LLC.)

6. After bleeding, remove the bleeding tubes and plug the ports. Test your bleeding operation by applying pressure to your push rod; it should not move inward. Any movement of the pushrod indicates that air is remaining in the master cylinder pressure cylinder or that there is some other fault in the master cylinder.

The master cylinder is then ready to be installed. The bleeder tubes can usually be left in place until the brake lines are attached.

To bench-bleed a master cylinder using an EIS Sur-Bleed syringe:

1. Secure the master cylinder in a vise with the bore tilted upward at the pushrod end. When clamping the sides of the master cylinder, do not clamp by the cylinder bore and do not clamp the reservoir too tightly.
2. Install a plug in each of the outlet ports.
3. Fill the reservoir about half full with brake fluid.
4. Remove one of the plugs, depress the plunger of the syringe completely, press the syringe firmly against the port to make a seal, and slowly pull outward on the plunger. You should observe fluid and air entering the syringe through the cylinder port (Figure 12-33).
5. Remove the syringe and, while holding it vertically, depress the plunger until all the air is removed (Figure 12-34).

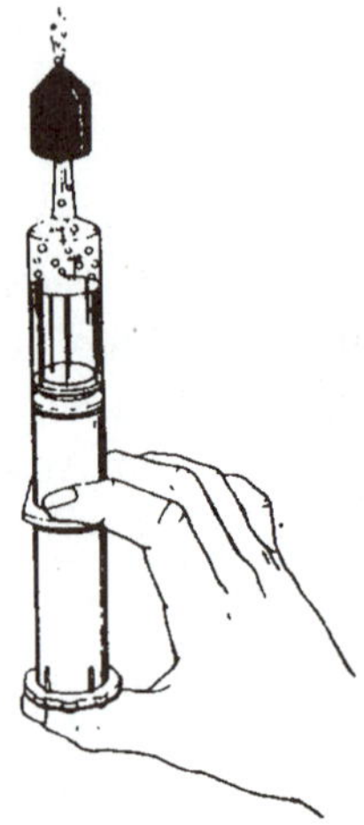

Figure 12-34 After sucking air and fluid out of the outlet port with the syringe, the air should be expelled from the syringe as shown here. (Reprinted by from Mitchell Anti-lock Brake Systems, with permission of Mitchell Repair Information, LLC.)

6. Place the syringe back against the outlet port and depress the plunger, pushing the fluid left in the syringe back into the cylinder. You should observe some air and fluid entering the reservoir (Figure 12-35).
7. Repeat steps 4 through 6 until there are no more air bubbles and then replace the plug in the outlet port.
8. Repeat this operation on the other cylinder port. On quick-take-up master cylinders, use the

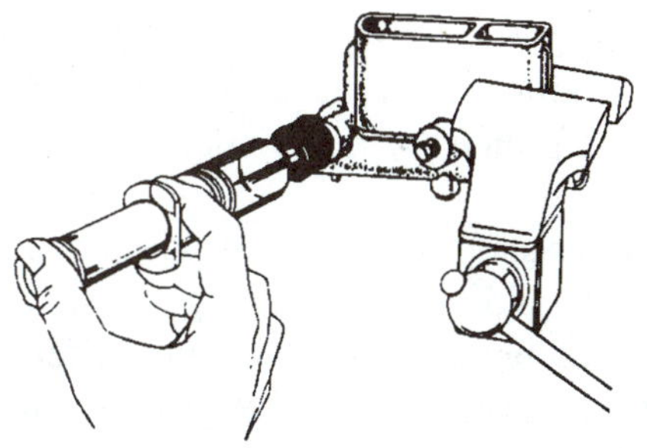

Figure 12-35 With the air removed from the syringe, the fluid is pushed back into the outlet port. The fluid flow should flush any remaining air back through the compensating port into the reservoir. (Reprinted from Mitchell Anti-lock Brake Systems, with permission of Mitchell Repair Information, LLC.)

syringe to pull fluid and any trapped air through the valve.

The master cylinder is then ready to be installed. The plugs should be left in place until the master cylinder is mounted and the lines are being connected. You can test your bleeding operation using step 6 of the tubing bleeding method.

To bench-bleed a master cylinder using a Phoenix reverse injector:

1. Secure the master cylinder in a vise, and install plugs for the outlet ports if they are not already plugged.
2. Remove the plug(s) from the secondary outlet port(s).
3. Inject fluid into the secondary outlet port until no more bubbles are seen in the fluid stream entering the front reservoir (Figure 12-36). Or, on master cylinders with four outlet ports, inject fluid into the lower outlet port until fluid drips from the upper port. Then, plug the lower port and inject fluid into the upper outlet port until the fluid stream entering the front reservoir is free of air bubbles.
4. Replace the plug in the secondary outlet(s).
5. Repeat steps 2 through 4 on the primary section.
6. Test your operation by trying to push the primary piston inward; it should not move.

12.4.4 Master Cylinder Replacement

Master cylinder replacement is essentially the reverse of removal. If the only repair of the hydraulic system was the master cylinder, a careful bleeding of the lines as they are

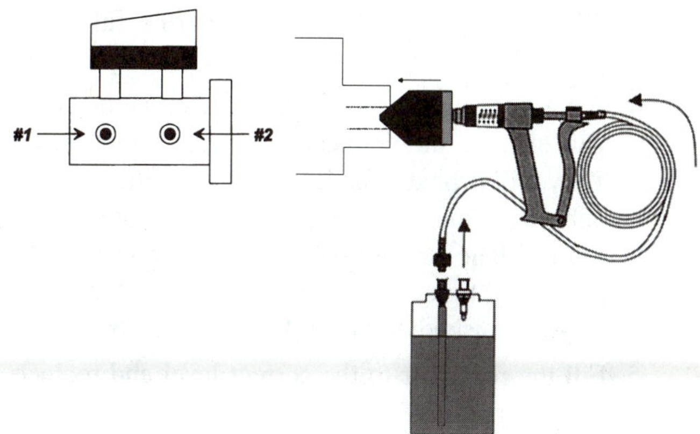

Figure 12-36 The Phoenix reverse injector can be used to pump fluid from a fluid source (right). To bleed a master cylinder, a port adapter allows it to inject fluid through the outlet ports and remove all of the air. (Courtesy of Phoenix Systems, L.L.C.)

being connected will usually save bleeding the whole system. If other parts of the system were worked on, they will need to be bled also.

If the old master cylinder used a boot or hub seal, a new one should be installed as the master cylinder is being replaced (Figure 12-37).

To replace a master cylinder:

1. Place the master cylinder in position on the booster or bulkhead, replace the mounting bolts or nuts, and tighten them to the correct torque.

   On manual brake cars, reconnect the pushrod to the brake pedal as necessary.
2. Remove the plugs or bleeder tubes from the outlet ports as you connect the brake lines. Do not tighten the lines yet. Place a shop cloth under each line fitting to catch any fluid that may leak out.

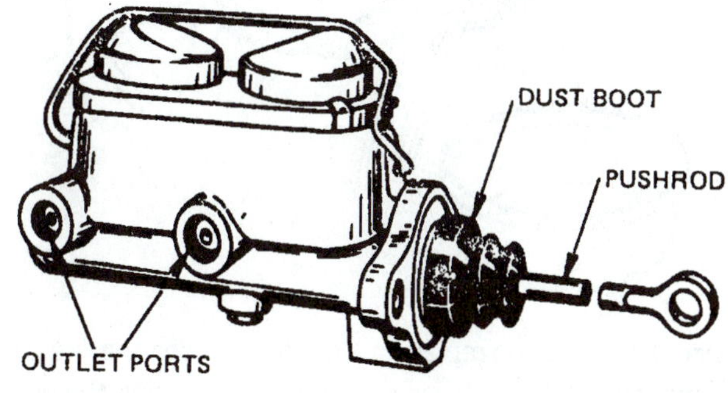

Figure 12-37 Many manual-brake master cylinders use a rubber dust boot to keep the end of the cylinder bore clean. (Courtesy of Wagner Brake)

3. Fill the reservoir about three-fourths full with brake fluid.
4. Have an assistant slowly push the brake pedal as you observe the connections at the outlet ports. They will probably be leaking some fluid with air bubbles. Continue the pedal strokes until only fluid with no air bubbles leaves the connection. At this point, tighten the connection while the pedal is being pushed downward (Figure 12-38).
5. Fill the reservoir to the correct level and replace the cover.
6. Reconnect any wires that were disconnected.
7. Check the brake pedal free travel and adjust it if necessary. There should be 1/16 to 1/8 inches (1.6 to 3.1 mm) of free travel before the pushrod engages the primary piston in the master cylinder.

12.5 Wheel Cylinder Service

Wheel cylinder service includes removal, rebuilding, and replacement of rubber parts. If the wheel cylinder is faulty, a new one should be installed. Wheel cylinders are normally serviced during lining replacement. The installation of new lining will disturb the position of the wheel cylinder pistons and cups and move them inward over the rust, corrosion, or other debris in the center of the cylinder. An otherwise sound-appearing wheel cylinder will often start leaking soon after a lining replacement, and the fluid leaking onto the new lining will ruin it (Figure 12-39).

Most service shops install a new wheel cylinder rather than rebuild the old one. If installing a new wheel cylinder, it is a good practice to flush the cylinder with clean brake fluid in a manner similar to that recommended in Section 12.4. Flushing the new cylinder ensures that small metal particles remaining from the machining process will not damage the hydraulic system.

A wheel cylinder with a good bore can usually be rebuilt. In many cases this can be done with the cylinder body fastened to the backing plate. In some cases, the wheel cylinder is mounted on a backing plate having "piston stops." These stops prevent the pistons from sliding out of the end of the cylinder bore (Figure 12-40). If the mounting prevents disassembly of the unit, the wheel cylinder must be removed from the backing plate, either entirely or at least an inch or so. Some sources recommend always moving the wheel cylinder to a bench for service, because you will be able to clean and inspect it better.

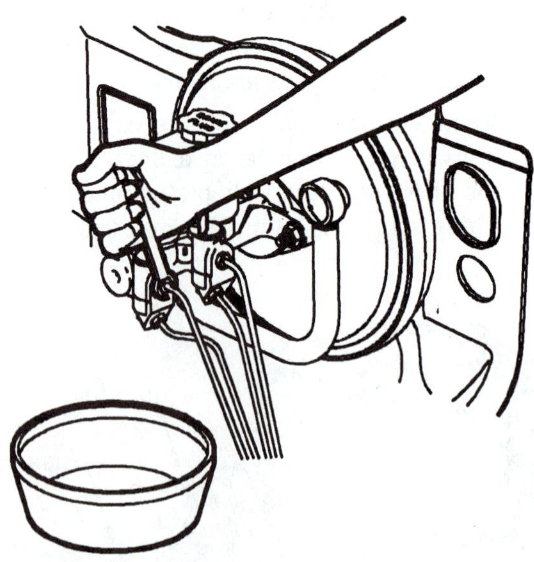

Figure 12-38 When replacing a master cylinder, the line connections can be bled by having someone lightly apply the brake pedal as you tighten the tube nut. After fluid leaks out and the air bubbles stop, tighten the nut before pedal is lifted. (© Saturn Corporation, used with permission)

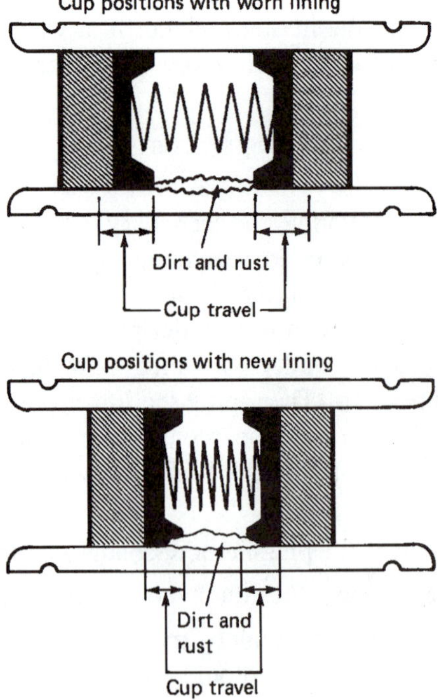

Figure 12-39 As brake lining wears, the piston and cups will often move outward in the bore, and dirt and rust deposits will form in the bottom. When new lining is installed, the cups will be positioned back toward the center, and leaks will often result because the cups are pushed over the dirt and rust.

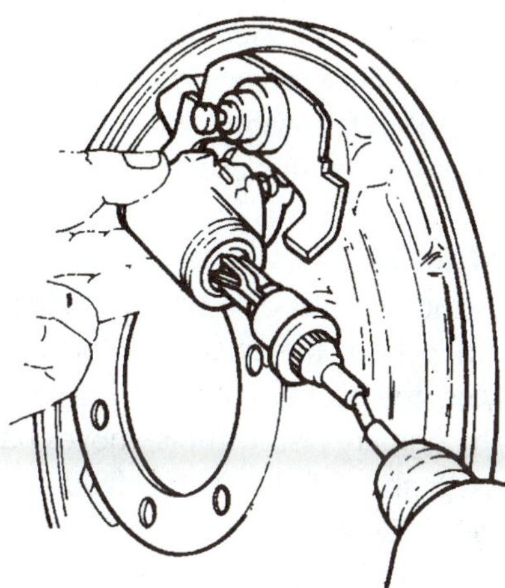

Figure 12-40 Many wheel cylinders can be rebuilt while mounted on the backing plate. This unit uses piston stops, so the mounting bolts were removed for piston and boot removal and access to the ends of the bore. (Courtesy of Brake Parts, Inc.)

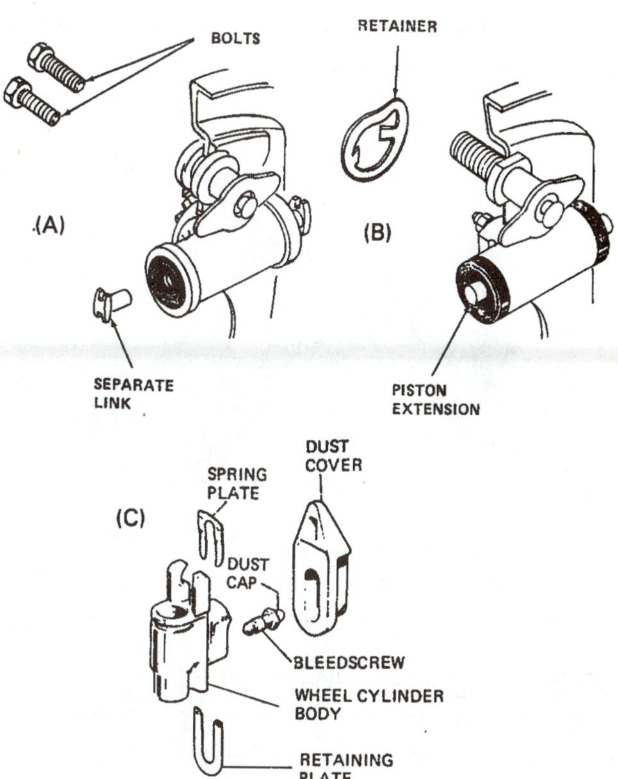

Figure 12-41 The most common methods of attaching wheel cylinders are bolts (A), a ringlike retainer (B), and a U-shaped retainer plate (C). (A and B are courtesy of General Motors, Service Technology Group; C is courtesy of LucasVarity Automotive)

12.5.1 Wheel Cylinder Removal

The following three methods are commonly used to attach a wheel cylinder to the backing plate: nuts or bolts, a spring lock retainer, and a U-shaped lock plate with shims (Figure 12-41). On a few older cars, the anchor pin passed through a boss on the wheel cylinder and was threaded into the steering knuckle. The hydraulic tube or hose is also attached to the cylinder body. Wheel cylinder removal is normally done during a lining replacement. On a single-fault repair of a leaky wheel cylinder, it is usually necessary to remove and replace the contaminated lining.

To remove a wheel cylinder:

1. Remove the brake shoes as described in Chapter 10. If you are servicing a faulty wheel cylinder and the lining is still good, sometimes it is possible to merely remove the return spring(s) and slide the shoes out of the way.
2. Disconnect the brake line from the wheel cylinder. Be sure to use a flare wrench and not to bend the metal tubing any more than necessary. Bending the brake line will make its replacement more difficult (Figure 12-42).
3. Disconnect the wheel cylinder mounting.
   a. *If bolts are used*, remove the bolts or nuts.
   b. *If a spring lock retainer is used*, remove the retainer using a pair of awls or a special tool. Insert the points of the awls or tool into the recess slots between the retainer tabs and the

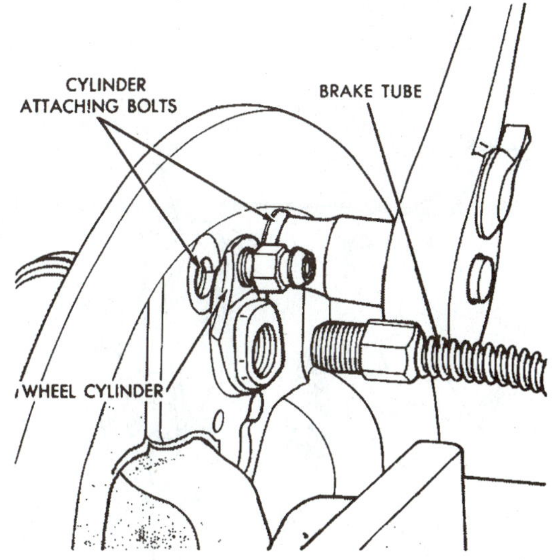

Figure 12-42 After removing the brake shoes, a wheel cylinder can be removed by unscrewing the attaching bolts. (Courtesy of Chrysler Corporation)

wheel cylinder pilot. Then bend both tabs outward and over the pilot at the same time (Figure 12-43 and 12-44).

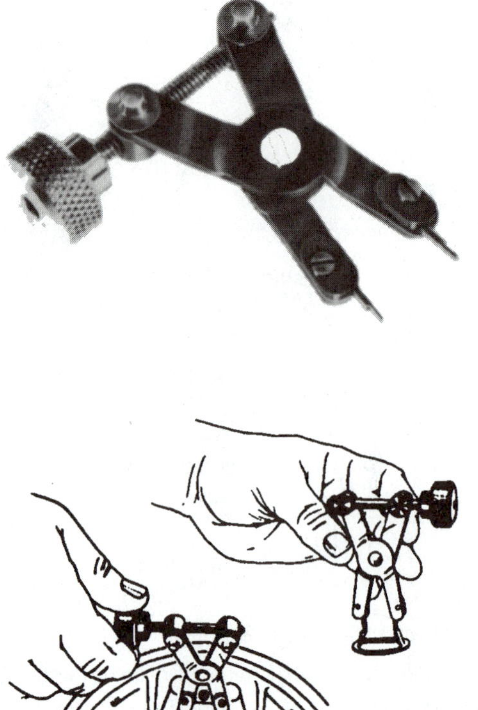

Figure 12-43 This special tool is used to expand the retaining ring so it can be removed or replaced. (Courtesy of SPX Kent-Moore, Part # J-29839)

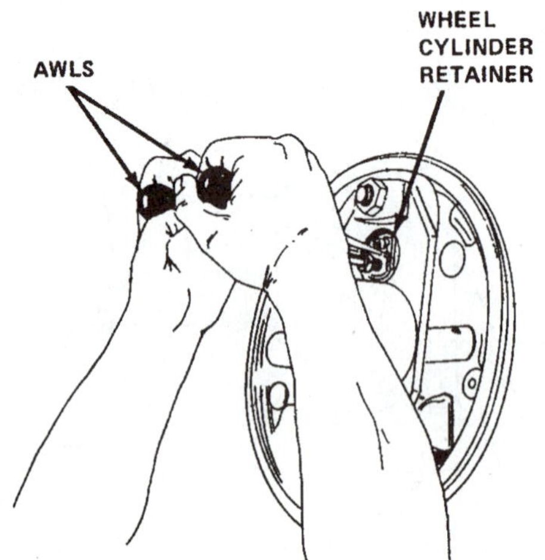

Figure 12-44 A wheel cylinder retaining ring can be removed using a pair of awls to pry the retainer tabs outward. (Courtesy of General Motors, Service Technology Group)

c. *If a lock plate is used*, tap out the retainer plate using a screwdriver and hammer. Be sure to note the relationship and position of the lock tab and any shims used (Figure 12-45).

HELPFUL HINTS: When a service operation is done which requires the removal of a wheel cylinder, caliper, or line, the fluid drip from the open line can be stopped by partially applying the brake. Prop the brake pedal to the point where the primary piston cup(s) moves past the compensating port(s); this will block the fluid flow from the reservoir(s) (Figure 12-46)

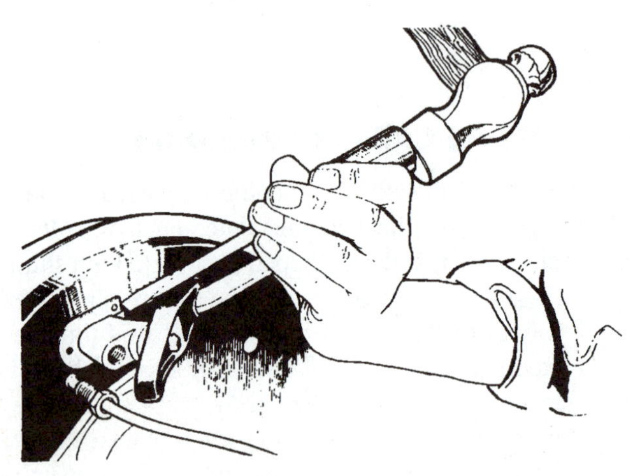

Figure 12-45 This U-shaped retainer is slid out of its groove using a punch or screwdriver and hammer. (Courtesy of Lucas-Varity Automotive)

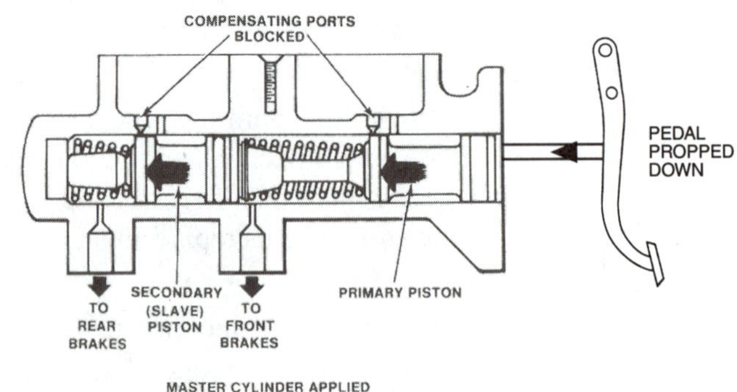

Figure 12-46 Propping the brake pedal downward positions the primary cups to shut off flow from the compensating ports and reservoir. This stops flow toward the master cylinder or drips from a disconnected line.

12.5.2 Reconditioning a Wheel Cylinder

Wheel cylinder rebuilding is normally a process of disassembly, cleaning, servicing and checking the bore, and reassembly using new rubber parts. Some technicians feel that a wheel cylinder is not worth rebuilding, especially if it has a stuck or broken bleeder screw or a stuck piston. Others feel that it is an easy-enough operation to check out a wheel cylinder and then go ahead and rebuild it if it is good enough. A wheel cylinder rebuilding kit will be required. This kit normally contains new cups and boots and sometimes a new spring or cup expanders (Figure 12-47). Some larger shops purchase cups and boots in bulk rather than stocking individual kits.

To rebuild a wheel cylinder:

1. Loosen the bleeder screw. If it snaps off, discard the wheel cylinder and replace it with a new one. The procedure to loosen a stuck bleeder screw is described later in this chapter.

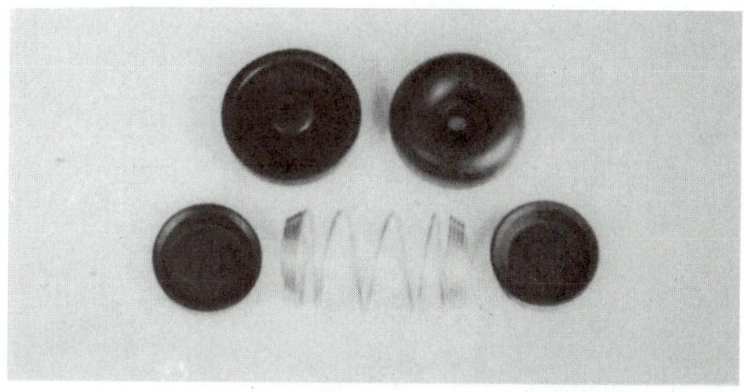

Figure 12-47 A wheel cylinder rebuilding kit includes new cups, boots, and a spring. Note that the spring is shaped to place an outward pressure on the lips of the cups.

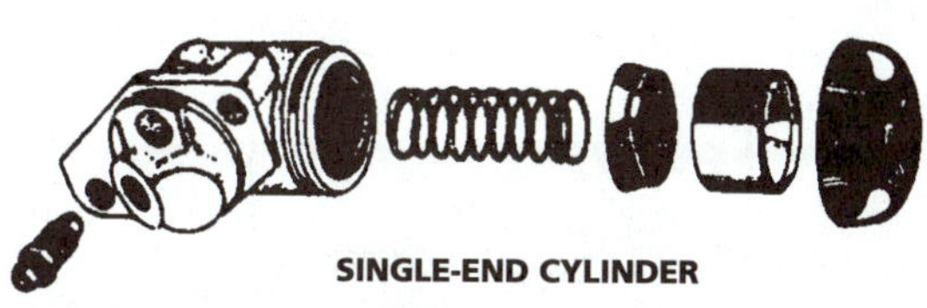

Figure 12-48 An exploded view of a single- and a double-piston wheel cylinder. Note the position of the rubber cups relative to the pistons. (Courtesy of Wagner Brake)

2. Remove the brake shoe links (Figure 12-48).
3. *If external boots are used*, pull them off. *If internal boots are used*, insert a screwdriver through the center opening in the boot to the edge of the boot and wheel cylinder and pry the boot loose. Be careful not to damage the cylinder bore (Figure 12-49).
4. Slide the pistons out of the bore. If one of them is stuck, insert a wooden dowel through the bore and tap it out. If both pistons are stuck, wrap a shop cloth around the cylinder body so it covers both ends of the bore. Then grip the cloth tightly around the cylinder body and use air pressure through the cylinder port to blow the pistons loose (Figure 12-50).

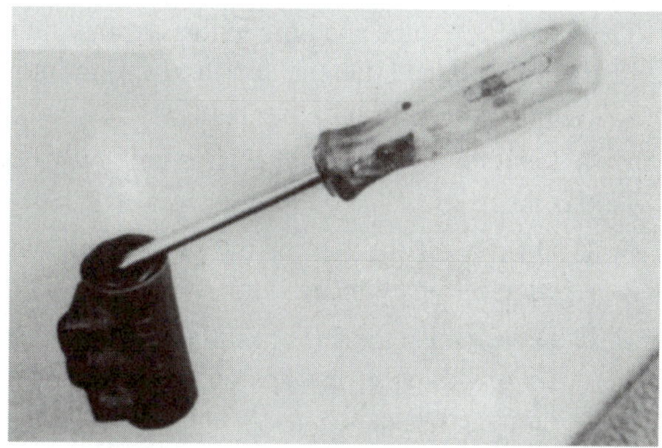

Figure 12-49 This boot, which uses an internal retainer, is removed by inserting a screwdriver under the edge of the boot and twisting the screwdriver to pry it upward.

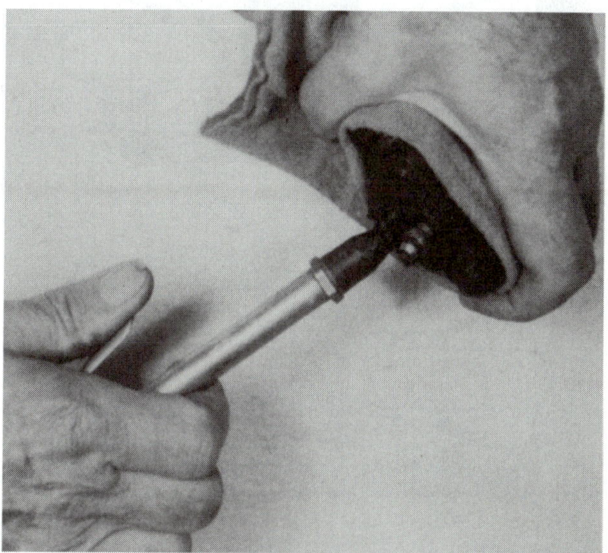

Figure 12-50 If the pistons are stuck in a wheel cylinder, they can be blown loose. Note the shop cloth wrapped around the wheel cylinder to prevent the cups from flying out.

5. Remove the piston cups and spring.
6. Thoroughly clean the cylinder body, piston, spring, bleeder screw, an any other small parts that are to be reused. Air-dry the parts, making sure that all the ports and passages are clean and open. Inspect the parts to make certain they are in good condition. If necessary, service the bore as described in Section 12.3. Some technicians have successfully rebuilt wheel cylinders (duo-servo brakes) that have small pits in the center of the bore, away from the area where the piston cup strokes.

If the bore is acceptable and the other parts are in good condition, the wheel cylinder can be assembled using the following procedure:

1. Wet the new cups and bore with assembly fluid or brake fluid and slide the cups and spring into the cylinder bore. Be sure that you:
   - Position the cups so their lips are toward each other.
   - Do not let the lips of the cups catch on the edge of the cylinder.
   - Do not push the cups past the ports in the center of the bore; the lips of the cups might be damaged because they tend to catch in the ports.
   - Do not push the cups inward in the cylinder so far that they cover the fluid inlet or bleeder ports (Figure 12-51).
   - Fit the expander in position if required.
2. Wet the piston with assembly fluid and slide it into position, making sure the flat side of the piston is next to the flat side of the cup (Figure 12-52).
3. Replace the boots.
   a. *If they are external boots*, slide them into place, making sure they fit snugly into the grooves on the outside of the cylinder ends (Figure 12-53).

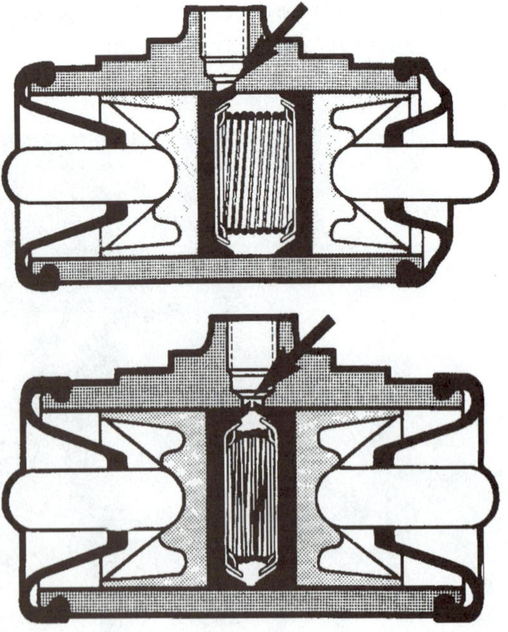

Figure 12-51 If the cups are not in the right positions, they will leak or the lips of the cups can be damaged. (Courtesy of ITT Automotive)

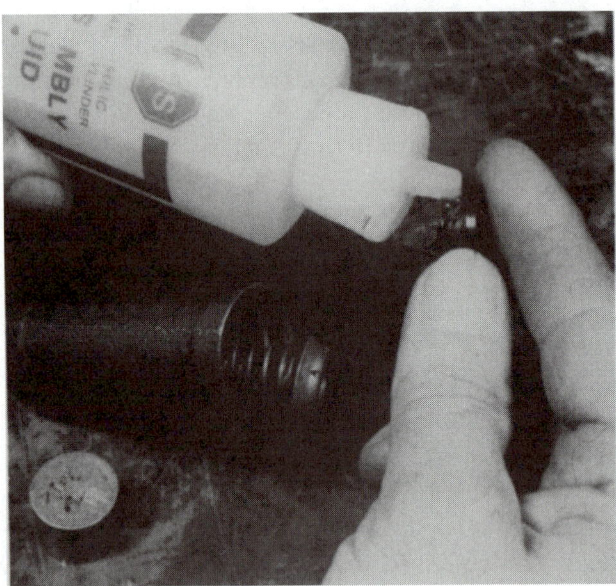

Figure 12-52 As a wheel cylinder is assembled, the bore, pistons, and cups should be lubricated with assembly fluid or brake fluid.

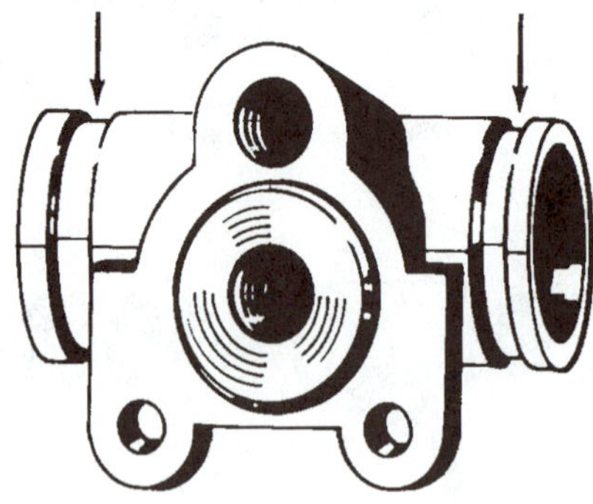

Figure 12-53 The boot-retaining grooves (arrows) must be clean to ensure a tight seal between the dust boots and cylinder body. (Courtesy of ITT Automotive)

b. *If they are internal boots*, press them into place by hand. They should work completely into position.

4. Install the bleeder screw finger-tight.
5. Install the brake shoe links if required.

12.5.3 Wheel Cylinder Replacement

Wheel cylinder replacement is essentially the reverse of the removal procedure. Many technicians prefer to connect the brake line first. This allows freedom to move the cylinder slightly to align it with the tube. Some wheel cylinders are in very tight locations, and a little misalignment makes it extremely difficult to connect this line.

Some wheel cylinders are difficult to bleed while mounted on the backing plate. These cylinders should be bench-bled or filled with fluid before they are installed. A vertically mounted wheel cylinder and one with the bleeder screw mounted in the line fitting are examples of this. Any portion of the wheel cylinder above the bleeder valve seat cannot be bled. To bench-bleed a wheel cylinder, simply position it with the port in the uppermost position and fill it with fluid. Next, place a plastic or rubber plug or cap over the port until the brake tube or fitting is ready to be attached (Figure 12-54).

To replace a wheel cylinder:

1. Place the wheel cylinder in position on the backing plate and thread the tube nut or brake hose into the wheel cylinder. Finger-tighten.

   Note that on a wheel cylinder using a spring lock retainer, the retaining ring cannot be installed using a socket with the brake line in place.
2. Replace the mounting attachment.

   a. *If bolts are used*, replace the nuts or bolts and tighten them to the correct torque (Figure 12-55).

   b. *If a spring lock retainer is used*, it is recommended that a new retaining ring be used. Two methods can be used to replace this retainer. The simplest is to use the special tool shown in Figure 12-43 to expand the retainer tabs enough for them to be placed in position. Then remove the tool. Another way is to wedge a block of wood between the axle flange and the wheel cylinder to hold the wheel cylinder in position. Then press in on the retainer ring until the tabs snap into place. A 1⅛-inch, 12-point socket is the right size to use as a pushing tool (Figure 12-56).

   c. *If a lock plate is used*, lubricate the wheel cylinder, backing plate, and shims as required. Then place the wheel cylinder in position and slide the shims and lock plate into place.

NOTE: It is possible for a spring-lock-retainer-style wheel cylinder to rotate in the backing plate so the ends of the pistons will pass the brake shoes, and this can allow the wheel cylinder pistons to pop out of the bore. Make sure that the wheel cylinder is fully seated in the backing plate and the retainer is fully seated in wheel cylinder cavity.

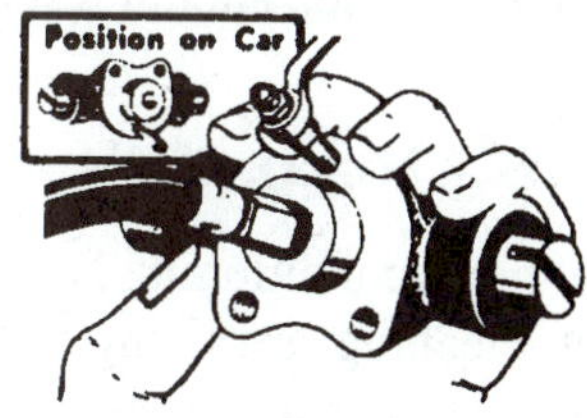

Figure 12-54 Some wheel cylinders are mounted with the bleeder screw in a position other than the top, and they must be bled before mounting on the backing plate. Vertical wheel cylinders can be bled by sliding a thin feeler gauge past the cup to allow the air to leak out. (Reprinted from Mitchell Anti-lock Brake Systems, with permission of Mitchell Repair Information, LLC.)

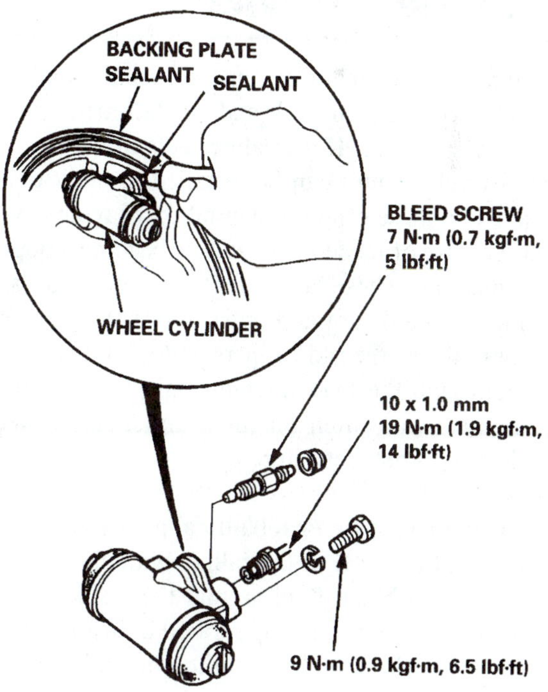

Figure 12-55 Some manufacturers recommend using a sealant between the wheel cylinder and backing plate as the wheel cylinder is replaced, to help keep water and dirt out of the brake assembly. (Courtesy of American Honda Motor Co., Inc.)

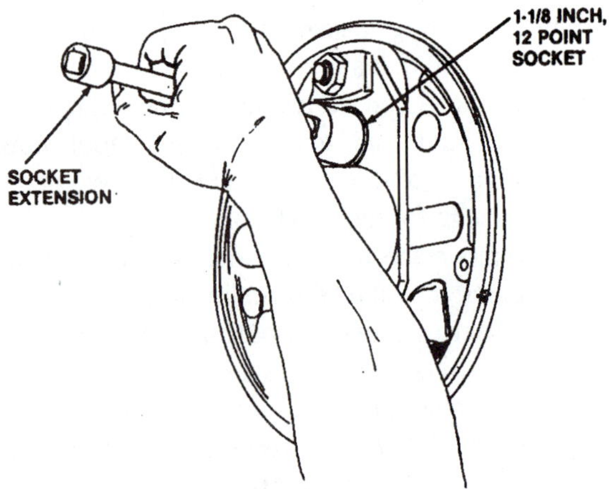

Figure 12-56 A wheel cylinder retaining ring can be installed by driving it into place as shown here. The wheel cylinder must be blocked in place as this is done. (Courtesy of General Motors Corporation, Service Technology Group)

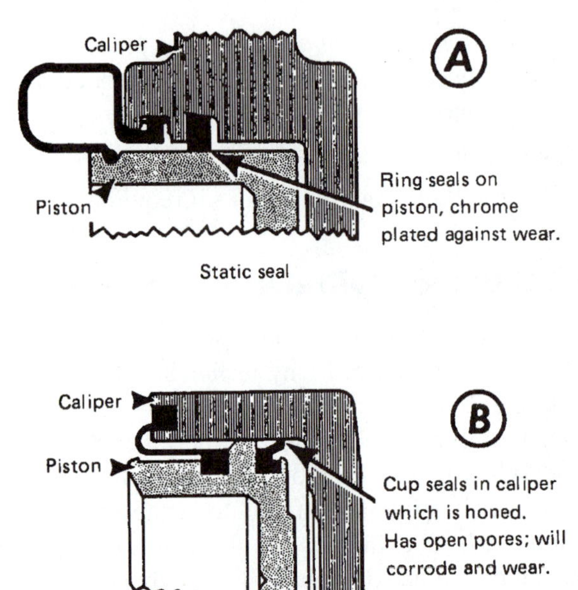

Figure 12-57 Most calipers use a static seal—a square O-ring seal in the caliper body (A). Some older calipers use a stroking seal—a lip seal on the piston (B). A stroking seal demands a good, straight, and smooth bore. (Reprinted from Mitchell Anti-lock Brake Systems, with permission of Mitchell Repair Information, LLC.)

3. Tighten the brake line to the correct torque.
4. Bleed the air from the brake line and wheel cylinder.

12.6 Caliper Service

Normal caliper service includes removal, rebuilding, and replacement. Caliper removal and replacement were described in Chapter 11. If the caliper is faulty, a new or commercially rebuilt unit can be installed. Rebuilt units are available and these offer substantial cost savings over new units. The use of loaded calipers has become a popular repair method. These are rebuilt calipers that are loaded with new hardware and brake shoes. The technician simply removes the old calipers and worn parts and replaces them with the loaded/rebuilt unit. Calipers that are stroking seals are often rebuilt using stainless-steel cylinder sleeves to prevent future bore corrosion (Figure 12-57).

When installing a new or rebuilt caliper, it is a good practice to flush the inside of the caliper with clean brake fluid. Simply slosh brake fluid around the inside of the unit in a manner similar to that described in Section 12.4 to remove any debris which might contaminate the hydraulic system.

Normally, a caliper with a good bore can be rebuilt in a repair shop if a caliper kit can be obtained. Many technicians prefer to rebuild the caliper that is on a car because it is a relatively quick and easy operation and an exact replacement is ensured. Don't forget that the bleeder screw positions on the right- and left-side calipers are different; they are mirror images of each other.

Caliper rebuilding during lining replacement is strongly recommended. Most calipers use a square-cut O-ring seal. You should remember that this O-ring not only seals in hydraulic pressure, but also returns the piston to a released position. Time and heat tend to harden this rubber seal, and when it loses its ability to return the piston, lining drag will result. Time and heat also cause the boot rubber to crack and break. This will allow bore corrosion which will also cause piston sticking and drag. Remember that a brake job should last quite a few years.

12.6.1 Reconditioning a Caliper

Caliper reconditioning is normally a process of disassembly, cleaning, servicing and checking the cylinder bore and piston, and reassembly using new rubber parts. A caliper rebuilding kit is required, and this usually includes a new boot and piston seal.

Many newer calipers use phenolic pistons. These pistons should be treated gently because they can crack, chip, or break. Some cracks and chips are acceptable as long as they do not extend completely across the piston face, are not in the area of the piston seal, and do not enter the dust boot groove (Figure 12-58). A phenolic piston will swell if it contacts a petroleum product, grease, oil, or solvent. If this should occur, wash it in denatured alcohol, and wipe it dry with a clean paper towel.

After removal of the pistons, some fixed calipers, mostly domestic, can be split in half. Removal of the bridge bolts will separate the two halves, giving better access to the pistons and cylinders for service, cleaning, and inspection (Figure 12-59). Some manufacturers, mostly foreign, do not recommend separating the caliper halves (Figure 12-60).

Removing the pistons is sometimes a difficult operation in caliper rebuilding. There are several off-car methods of piston removal. They are easier if the piston is partially removed while still on the car. When removing the caliper, first remove the brake pads and place the caliper back over the rotor. Next, pump the brake pedal. The hydraulic pressure will move the pistons outward in their bores (Figure 12-61). Moving each of the pistons about one-half inch or so is usually sufficient for easier removal later. When the caliper is off the car, the pistons are commonly removed using air pressure. This is not recommended for phenolic pistons by some manufacturers. The on-car method just described is recommended instead. A phenolic piston that is really stuck can be removed by breaking it into several pieces using a hammer and chisel. When doing this, be sure to wear eye protection. Other off-car methods that can be used with metal pistons employ several types of piston pullers. These tools all grip the piston so that it can be pulled out (Figure 12-62). For really tight pistons, some shops will adapt a grease or zerk fitting to the caliper port and, using a grease gun, force the piston out. If this method is used, be sure to thoroughly clean out all traces of grease.

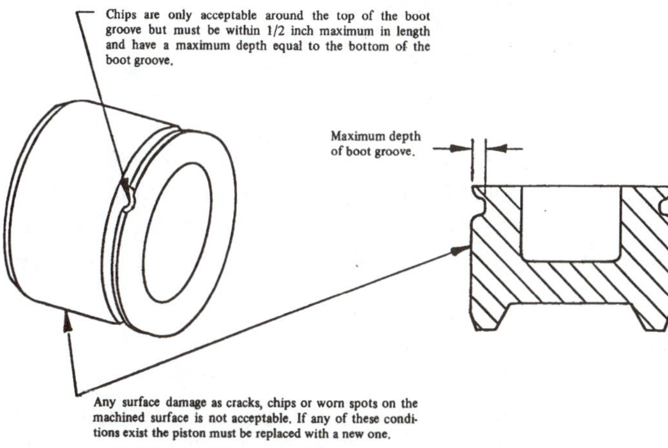

Figure 12-58 A phenolic piston with small chips at the end is acceptable for reuse, but any damage in the sealing area requires replacement. (Courtesy of Brake Parts, Inc.)

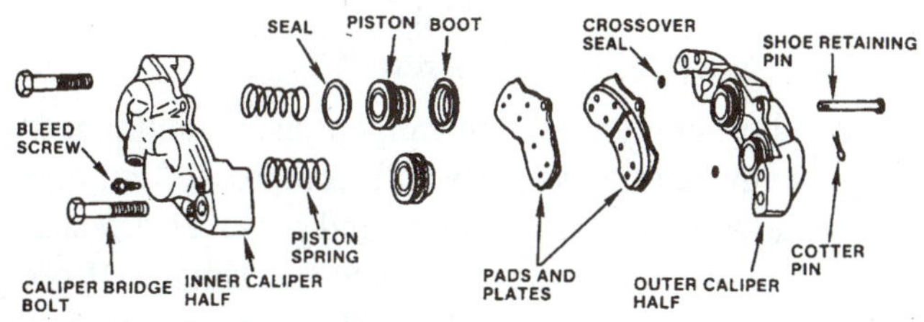

Figure 12-59 When servicing domestic fixed calipers, it is usually recommended that they be completely disassembled. (Courtesy of Bendix Brakes, by AlliedSignal)

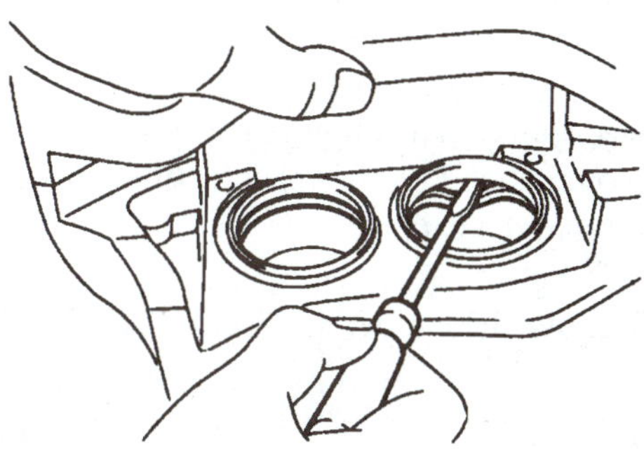

Figure 12-60 Some manufacturers recommend rebuilding a fixed caliper while the two caliper halves are still together.

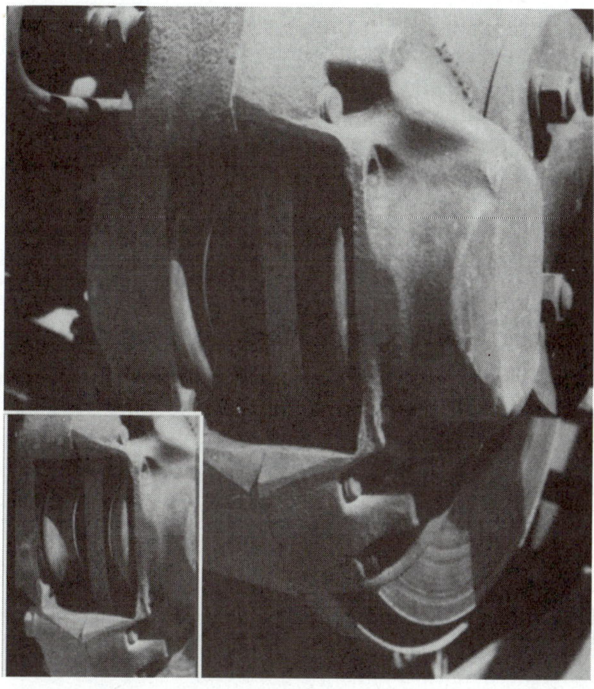

Figure 12-61 Caliper pistons can be moved outward in their bores by removing the pads and applying the brake pedal (inset). This operation must be done before removing the caliper.

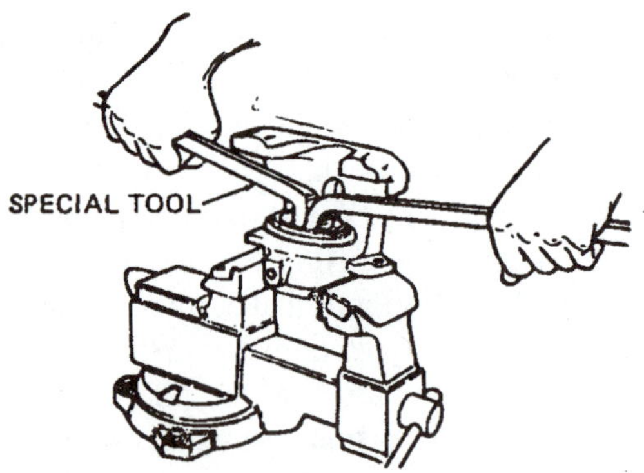

Figure 12-62 Several styles of caliper piston removal tools are available to help pull a piston out of the bore. (Courtesy of Wagner Brake)

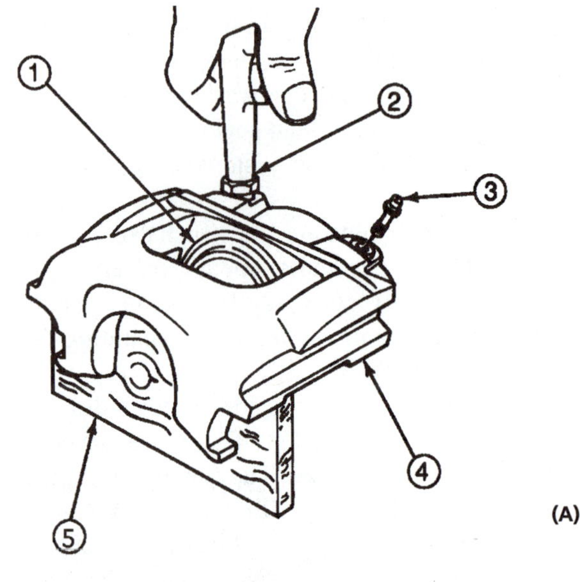

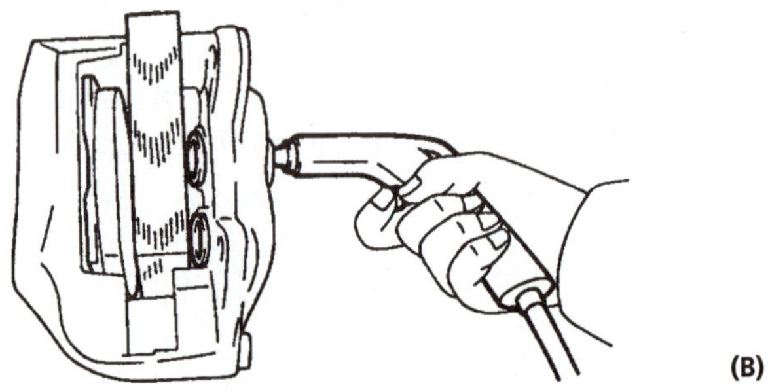

Figure 12-63 An air gun (2) can be used to help remove the piston (1); the wood block (5) keeps the piston from traveling too far (A). This also works on fixed calipers with multiple pistons (B); note the wood block and brake pad to limit piston travel. (A is courtesy of Ford Motor Company)

The description that follows is very general. The procedure recommended by the manufacturer should be followed. To rebuild a caliper:

1. Remove the caliper as described in Chapter 11.
2. Use a wire brush to remove dirt and grease from the outside of the caliper. Light grease deposits and oil can be washed off with denatured alcohol.
3. Loosen the bleeder screw and drain the old fluid out of the caliper. If the bleeder screw breaks off, the caliper can be saved by installing a replacement from a bleeder screw kit. This requires that the broken bleeder screw be drilled out and threads be cut into the caliper so the bleeder screw fitting can be installed. Each kit contains directions for installation.

   If a bleeder screw is so tight that you feel breakage is probable, follow the steps given in the Section, "Bleeding Brakes" 12.7 to loosen it.
4. Place a wood block (a 1×4 or 2×4, about 6 inches long) in the caliper opening so the piston cannot fall or fly out of the bore. Then carefully apply air pressure to the caliper port (Figure 12-63). It is recommended that the air pressure be limited to about 30 psi (207 kPa). Try to move the piston slowly out of the bore in a controlled fashion. Note that the bleeder screw should be closed.

CAUTION *Do not place your fingers in the way of the piston or try to catch it.*

NOTE: On multipiston calipers the loosest piston(s) will move out first leaving the tightest piston(s) to be removed by another method. Observe the pistons. After one of them moves a little way, wedge it in place or hold it with a clamp so the other one(s) is forced to move.

5. Remove the boot. If it is retained by a metal ring, lift the ring off or use a screwdriver to pry the boot off the caliper (Figure 12-64).
6. Remove the piston seal or O-ring using a sharpened wooden dowel or plastic rod to pry it out of the groove in the caliper or piston (Figure 12-65). If a metal device is used, be careful to not damage the bore or O-ring groove.
7. Remove the bleeder screw.
8. Thoroughly wash the caliper, piston, and bleeder screw. Dry these parts with compressed air, making sure that all the ports and passages are clean and open. Then inspect them for damage. If necessary, service the cylinder bore on calipers with stroking seals using the procedure described in Section 12.3. For calipers with an O-ring seal, the primary sealing surface is on the piston, and it must be clean and smooth. A rough or otherwise

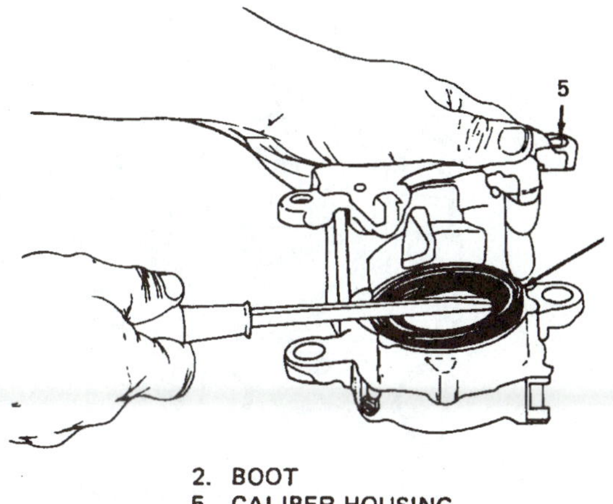

Figure 12-64 To remove a boot held in place by a metal retainer ring, slide a screwdriver under it and twist the screwdriver. (Courtesy of General Motors Corporation, Service Technology Group)

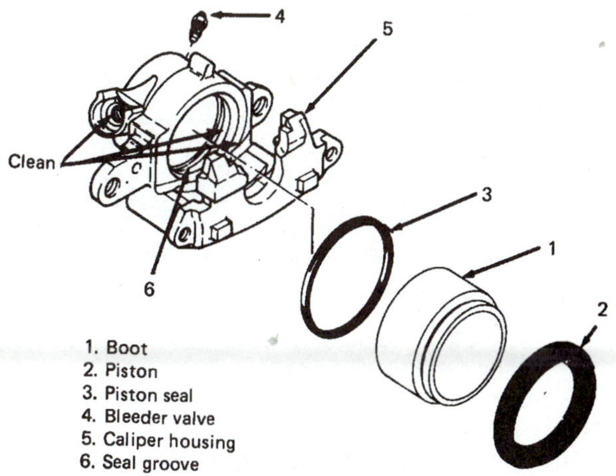

Figure 12-66 As a caliper is cleaned, be sure that the O-ring groove, boot retainer groove, and passages into the caliper and to the bleeder screw are clean. (Courtesy of General Motors Corporation, Service Technology Group)

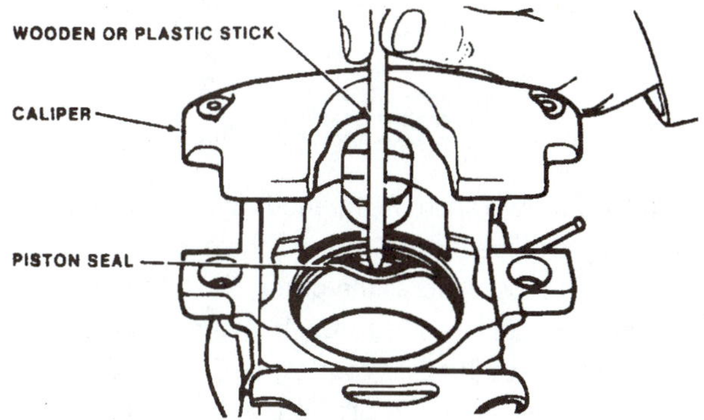

Figure 12-65 With the piston out, the O-ring can be removed from its groove using a pointed wooden or plastic stick. (Courtesy of Chrysler Corporation)

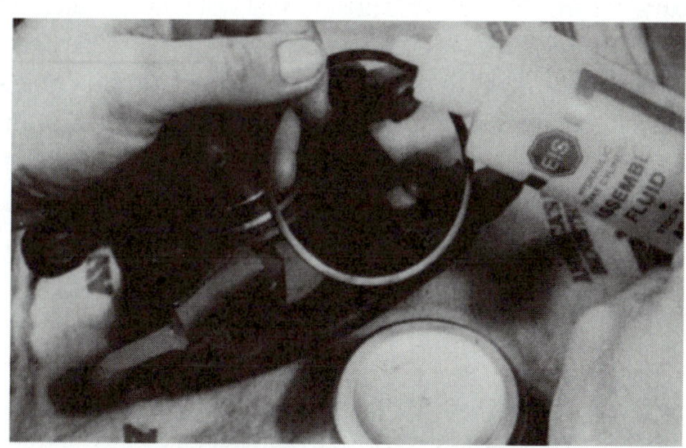

Figure 12-67 The O-ring seal and piston should be lubricated with assembly fluid or brake fluid as they are being installed.

damaged piston must be replaced. Also, the O-ring and boot grooves in the caliper must be clean. Remove any rust or dirt from these grooves (Figure 12-66).

If the caliper bore and piston are acceptable, the caliper can be reassembled using the following procedure:

1. *If an O-ring seal is used*, place the new O-ring in its groove, making sure that it is properly seated and not twisted. Lubricate the caliper bore, O-ring, and piston with assembly fluid or brake fluid and slide the piston partially into the bore (Figure 12-67).

   If a stroking seal is used, lubricate the piston and seal with assembly fluid or brake fluid and work the seal into its groove in the piston, making sure it is properly seated. Lubricate the cylinder bore and carefully slide the piston and spring, if required, into the bore. Be careful that the seal lip does not catch on the edge of the bore (Figure 12-68). A seal installer will facilitate this job. A 0.005-inch (0.10-mm) feeler gauge with smooth edges can also be used to guide the seal lip into the bore.
2. Install the boot.
   a. *If the outer edge of the boot is secured by the bore*, the boot should be installed on the piston and worked into its groove before the piston is slid into the bore. As the piston enters the bore, the boot will be locked in place (Figure 12-69). An alternate method of installing this type of boot is to place it into its bore groove, insert

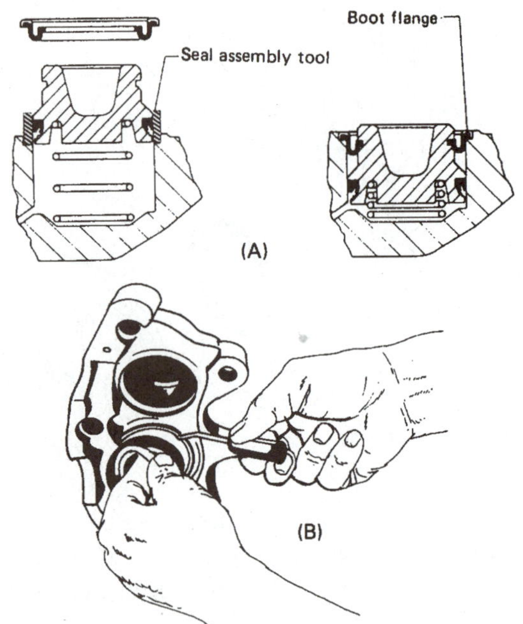

Figure 12-68 The lip of a stroking seal will tend to catch on the edge of the cylinder bore and be damaged. A seal assembly tool (A) or a smooth seal installer (B) can be used to guide the seal into the bore. (A is courtesy of General Motors Corporation, Service Technology Group; B is courtesy of Brake Parts, Inc.)

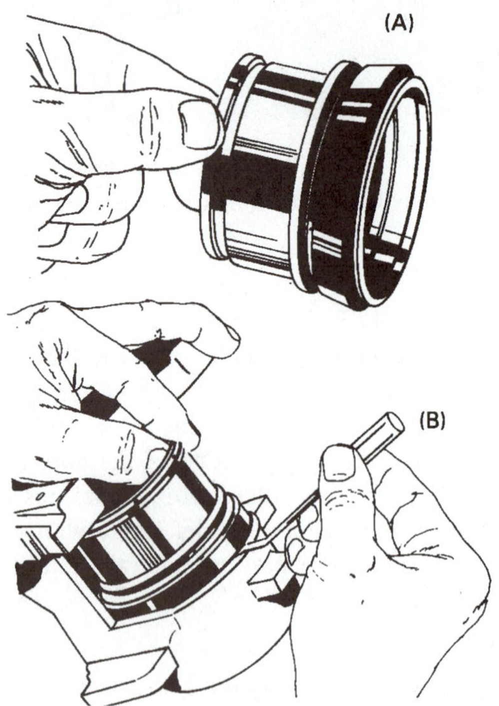

Figure 12-69 The lower edge of the boot shown here must be fitted into its groove before the piston is installed. The recommended way of doing this is to slide the boot over the piston (A), work the boot into the groove (B), and then slide the piston into the bore. (Courtesy of Brake Parts, Inc.)

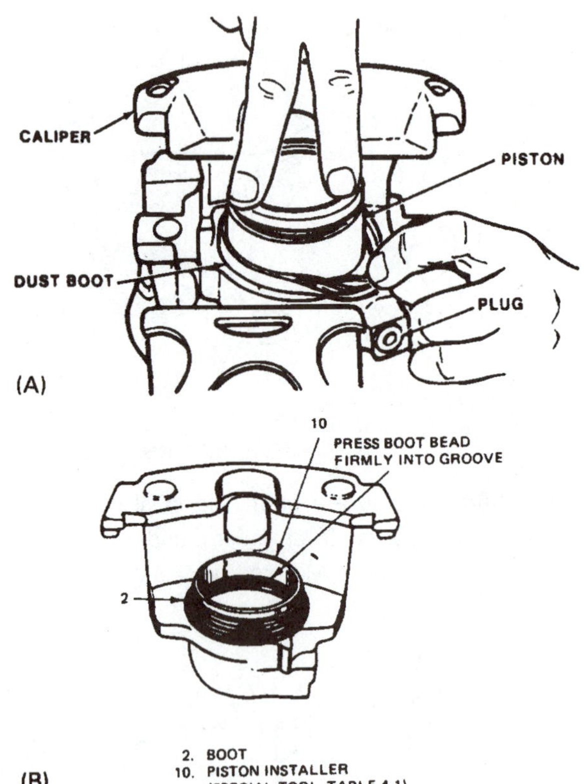

Figure 12-70 Two alternate methods of installing a boot into a Kelsey Hayes caliper are positioning the boot into the groove and stretching it outward as the piston is slid through it (A) or using a piston installer to keep the boot expanded as the piston is slid into place (B). (A is courtesy of Chrysler Corporation; B is courtesy of General Motors Corporation, Service Technology Group)

a special piston installer, and slide the piston through the installer (Figure 12-70).

b. *If the outer edge of the boot is retained by a metal ring molded into the boot*, with the piston partially installed in the bore, slide the boot over the piston. Then position the boot in the counter bore in the caliper and use a boot installer and hammer to seat it (Figure 12-71).

c. *If the outer edge of the boot is secured by a separate metal ring*, with the piston partially in the bore, slide the boot over the piston. Then position the boot in the groove in the caliper, and work the retaining ring into position.

Note that some boots are locked into the caliper before the piston is installed. Note also that the boot on Delco Moraine fixed-caliper pistons should be sealed to the piston and caliper using thin beads of silicone sealant. This will prevent moisture from entering and causing bore corrosion (Figure 12-72).

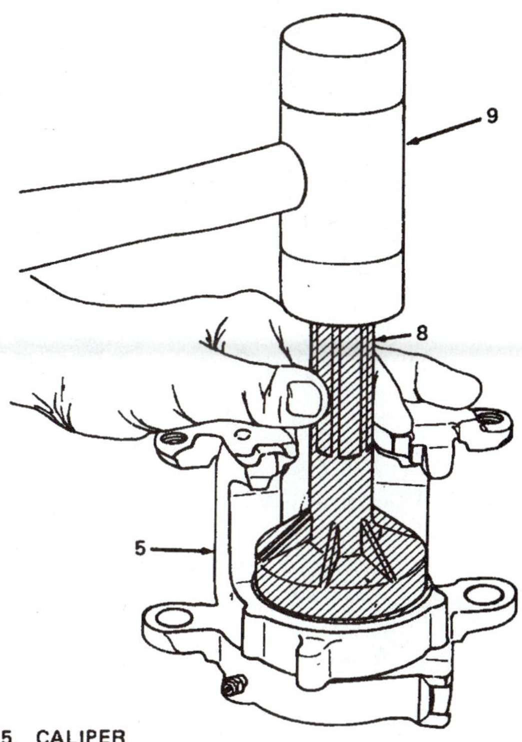

Figure 12-71 Boots that have metal support rings built into them should be installed in the caliper using the proper-sized seating tool. (Courtesy of General Motors Corporation, Service Technology Group)

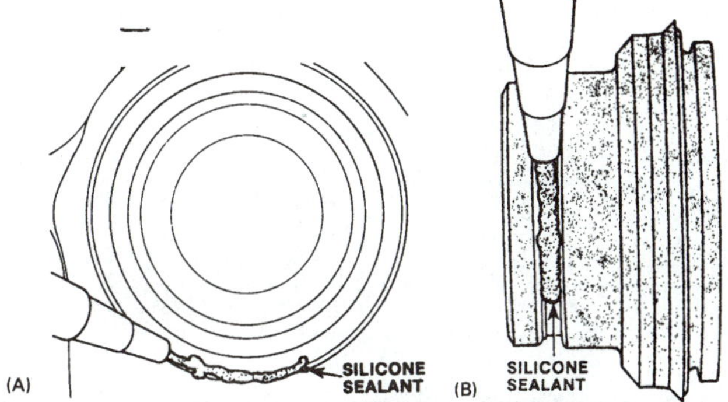

Figure 12-72 It is recommended that the boot of a fixed caliper with stroking seals be sealed to the caliper (A) and piston (B); this prevents moisture from entering the bore and causing corrosion. (Courtesy of Bendix Brakes, by AlliedSignal)

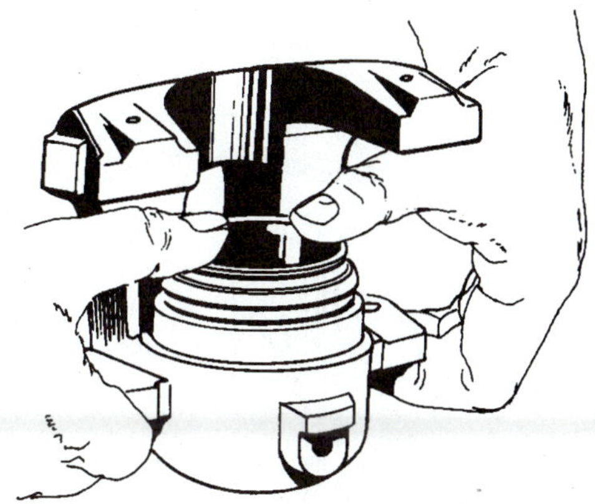

Figure 12-73 Once the piston enters the O-ring, it can usually be worked to the bottom of the bore using both thumbs as shown here. (Courtesy of Brake Parts, Inc.)

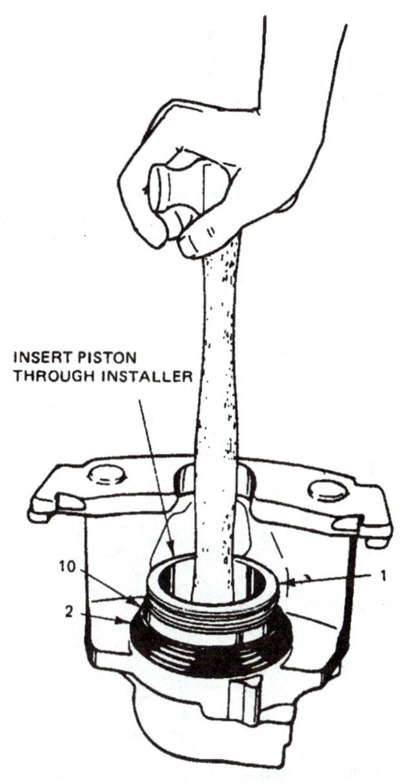

Figure 12-74 If the piston is too tight to be worked in using your thumbs, a hammer handle can be used to tap it into place. (Courtesy of General Motors Corporation, Service Technology Group)

3. Make sure the boot seats properly and retract the piston completely into the bore. This can be done by rocking it inward using both thumbs or by pushing it in with a hammer handle (Figures 12-73 and 12-74).
4. For fixed calipers, if the caliper halves were separated, place new O-ring seals between them if required. Then install and tighten the bridge bolts to the correct torque (Figure 12-75).
5. Install and finger-tighten the bleeder screw.
6. Install the lining in the caliper and the caliper on the car as described in Chapter 11.
7. Bleed the brake line and caliper.

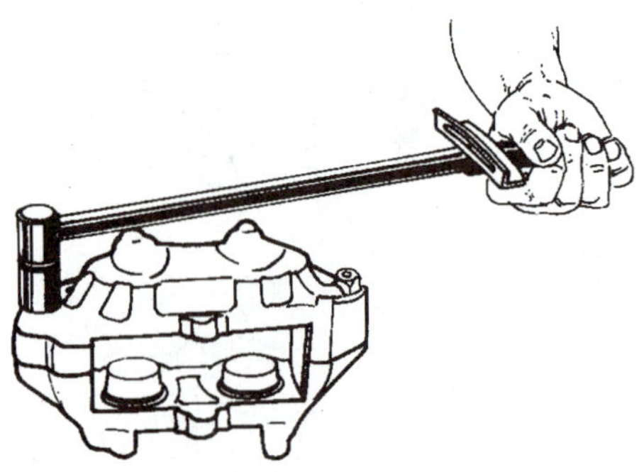

Figure 12-75 If a fixed caliper has been disassembled, new O-rings should be installed and the bridge bolts tightened to the correct torque. (Courtesy of Brake Parts, Inc.)

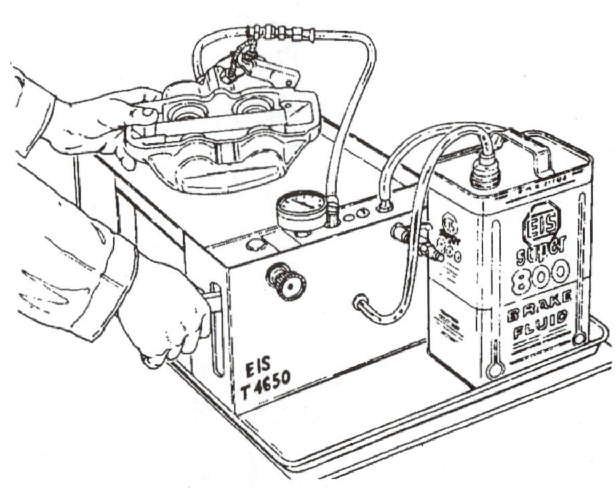

Figure 12-76 A caliper test bench is used to apply pressure to a caliper to make sure there are no leaks or a sticky piston(s). (Reprinted from Mitchell Anti-lock Brake Systems, with permission of Mitchell Repair Information, LLC.)

A caliper test bench is used by some rebuilders and service shops to pressurize a rebuilt caliper and check for leaks and proper piston movement (Figure 12-76).

12.6.2 Reconditioning a Caliper with a Mechanical Parking Brake

At the present time, the following two basic caliper designs are used on domestic cars with a mechanical parking brake mechanism: Delco Moraine (used on General Motors cars) (Figure 12-77) and Kelsey-Hayes (used on Ford products) (Figure 12-78). Several different designs are used on import cars. The hydraulic portion of these calipers uses a square-cut O-ring seal and a boot that are essentially the same as those used on the standard caliper; a mechanical mechanism has been added. The additional mechanism requires different disassembly, checking, and reassembly procedures, and the various calipers differ in the procedure required. On cars that use rear disc brakes with drum-type parking brakes, the

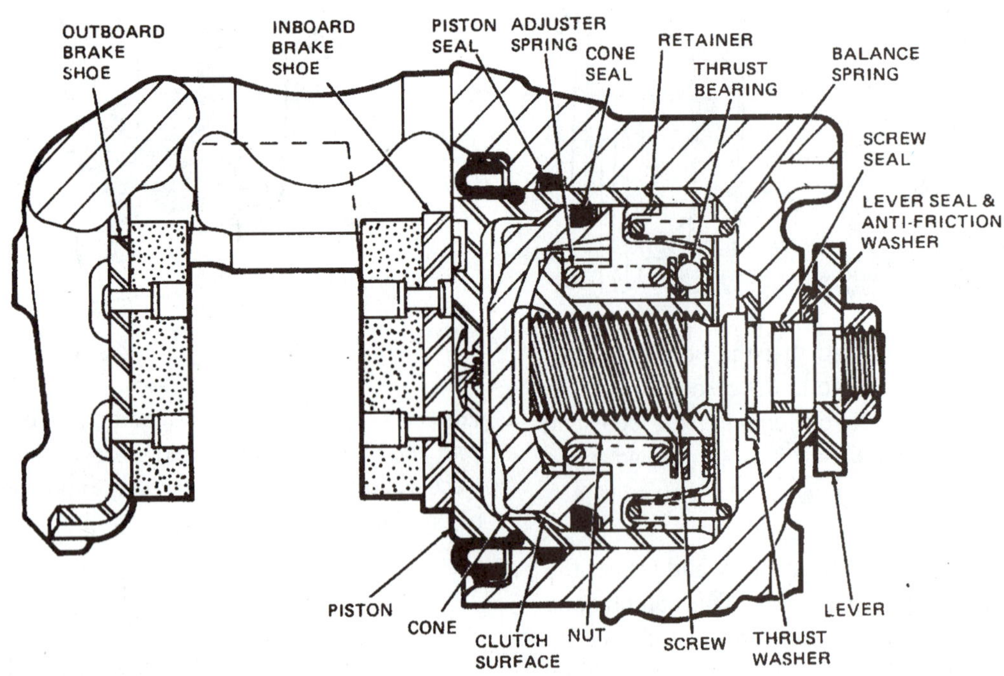

Figure 12-77 A cutaway Delco Moraine rear caliper. The boot, piston seal, and caliper seal can be serviced. (Courtesy of General Motors Corporation, Service Technology Group)

caliper is serviced in the same manner as a front caliper; the service of this parking brake is described in Section 14.2 (Figure 12-79).

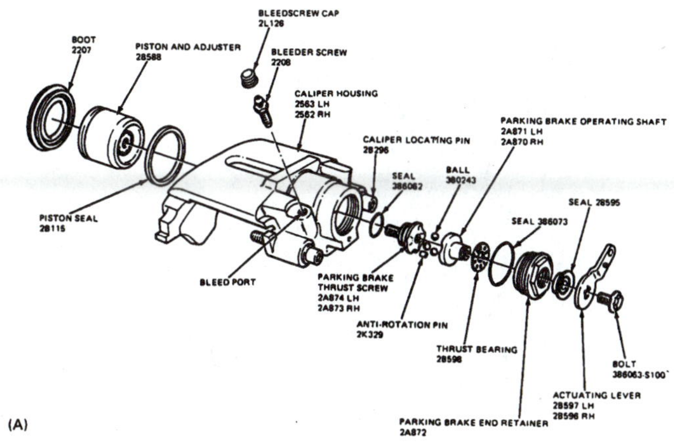

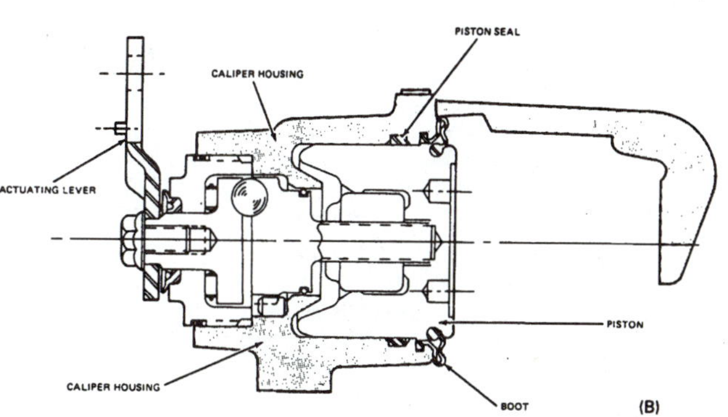

Figure 12-78 An exploded view (A) and cutaway view (B) of a Kelsey Hayes rear caliper. Note the parking brake mechanism is removed from the rear of the caliper. (Courtesy of Ford Motor Company)

The manufacturer's overhaul procedure should be followed when rebuilding these calipers. In general, to rebuild a caliper that includes a mechanical parking brake mechanism:

1. Remove the caliper from the car.
2. Clean the outside of the caliper using a wire brush. Grease and oil can be washed off using denatured alcohol or brake cleaner.
3. Loosen the bleeder screw and drain the old fluid.
4. Remove the piston.

On Delco Moraine units, place a wooden block or folded shop cloth in the caliper opening to protect the piston and then remove the nut and parking brake lever. Attach a wrench to the hex of the actuator screw and rotate the screw to move the piston out of the bore. The actuator screw is turned clockwise on right-side calipers and counterclockwise on left-side calipers (Figure 12-80).

On Kelsey-Hayes units use the following procedure:

1. Remove the parking brake lever.
2. Unscrew the retainer and lift it off with the operating shaft and thrust bearing.
3. Remove the three metal balls and, using a magnet, take out the antirotation pin. If the pin catches, tightly rotate the thrust screw (Figure 12-81).
4. Remove the thrust screw by unscrewing it.
5. Install the special tool and push the piston out of the caliper.

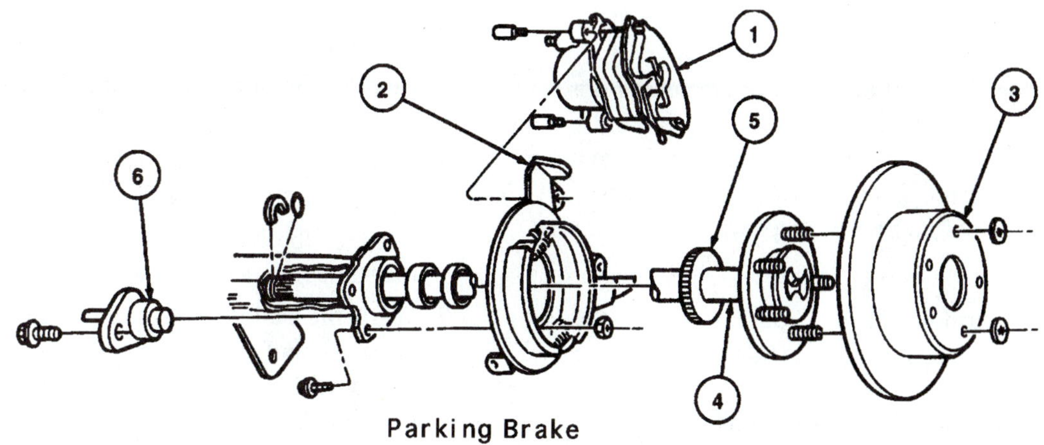

Figure 12-79 This rear disc brake assembly uses a caliper (1) that is similar to a front caliper. The drum-type parking brake uses the inner portion of the rotor (3). (Courtesy of Ford Motor Company)

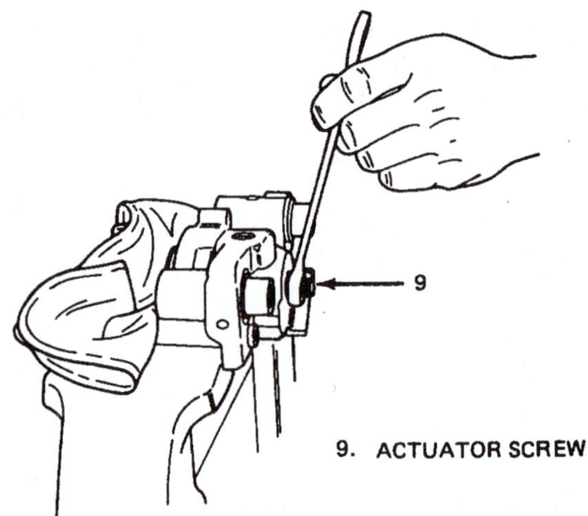

Figure 12-80 On this caliper, using a wrench to turn the actuator screw in the application direction will move the piston out of the caliper bore. (Courtesy of General Motors Corporation, Service Technology Group)

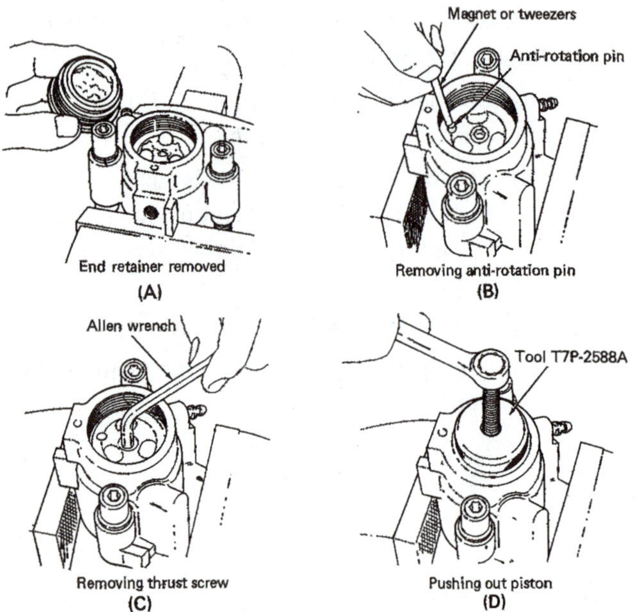

Figure 12-81 The parking brake mechanism is removed from this caliper by unscrewing the end retainer (A), lifting out the balls and antirotation pin (B), and unscrewing the thrust screw (C). The caliper piston can then be pushed out using the special tool (D). (Courtesy of Ford Motor Company)

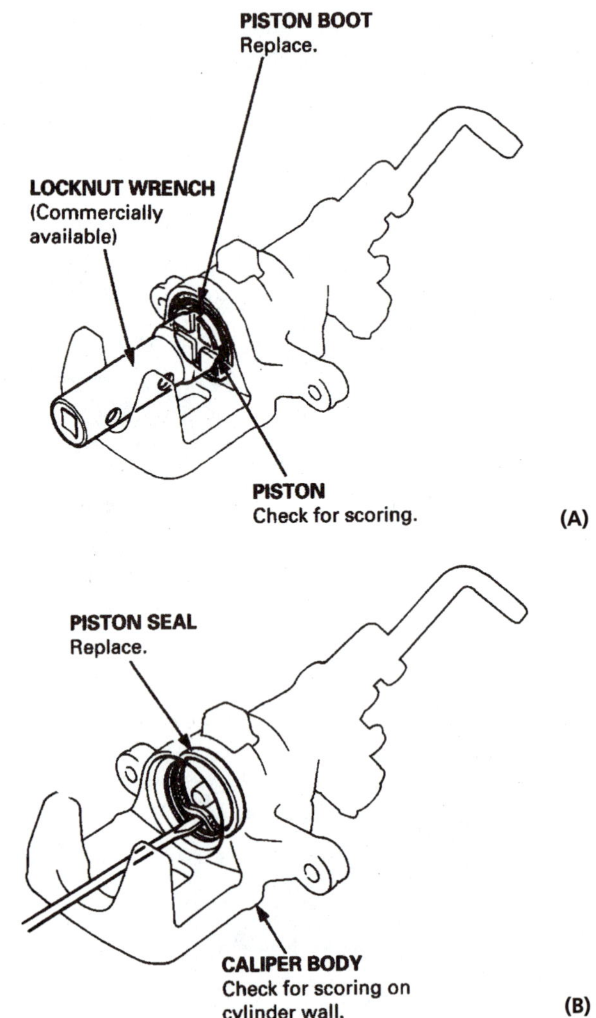

Figure 12-82 The locknut wrench is used to unscrew the piston and remove it from this rear caliper (A). With the piston removed, the piston seal can be removed using a sharp wood or plastic tool (B). (Courtesy of American Honda Motor Co., Inc.)

On other units, follow the procedure given by the manufacturer; it will be somewhat like this:

1. Remove the piston by rotating it using a special tool, and remove the piston seal (Figure 12-82).
2. Using the special tool, compress the adjuster spring and remove the retaining ring/circlip (Figure 12-83).
3. Remove the adjusting bolt and spring parts.
4. Remove the sleeve piston and O-ring seal (Figure 12-84).
5. Remove the lever and cam assembly with seal (Figure 12-85).
6. Remove the boot, O-ring, and bleeder screw in the same manner as for a standard caliper.
7. Wash the caliper and internal parts with denatured alcohol, air-dry them, and inspect them as you would a standard caliper, with the following additional steps.
   a. *On Delco Moraine units*, remove the rubber check valve from the center of the piston face and inspect the inner area. If there are signs of brake fluid, moisture, pitting, or corrosion, the

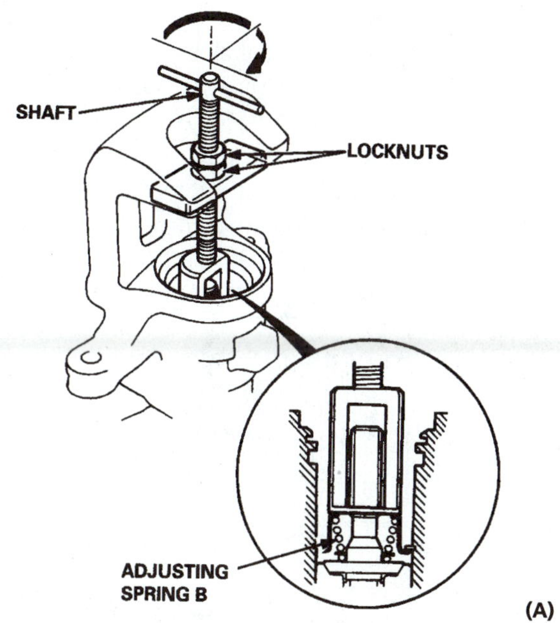

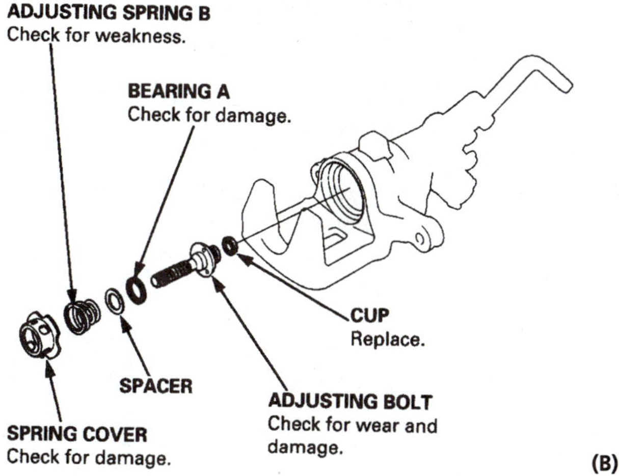

Figure 12-83 This special tool is used to compress the adjuster spring so an internal circlip/retaining ring can be removed (A). With the circlip removed, the adjusting spring and bolt can be removed for cleaning and inspection (B). (Courtesy of American Honda Motor Co., Inc.)

piston should be replaced. Lubricate a new check valve with brake fluid and install it in the piston (Figure 12-86).

b. *On Kelsey-Hayes units*, thread the thrust screw into the piston, hold the piston firmly, and pull upward on the thrust screw. It should move upward about 1/4 inches (6.3 mm), and the adjuster nut sleeve in the piston should rotate. As the adjuster screw is released, the adjuster nut should not turn. If the adjuster does not work correctly, the piston should be replaced (Figure 12-87).

If the piston and caliper are acceptable, the caliper can be assembled using the following procedure:

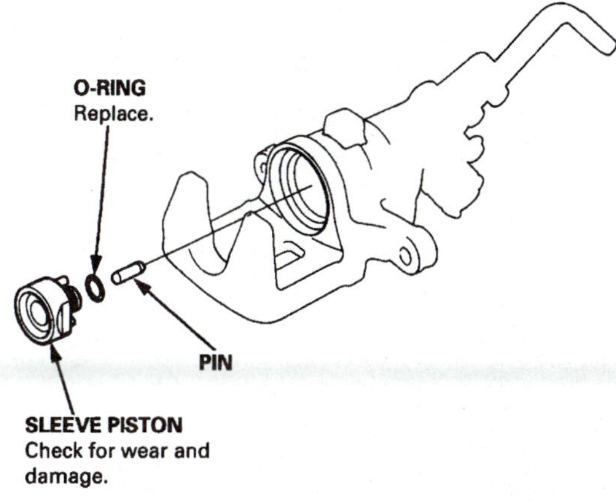

Figure 12-84 With the adjustor bolt removed, the next step is to remove the sleeve piston and pin for cleaning, inspection, and replacement of the O-ring. (Courtesy of American Honda Motor Co., Inc.)

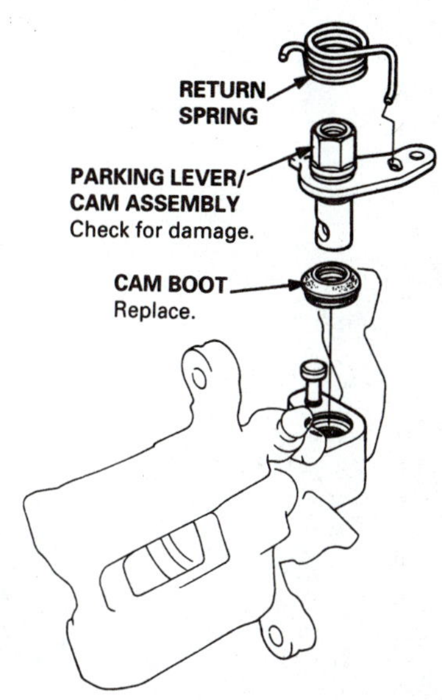

Figure 12-85 The final disassembly step on this caliper is to remove the parking lever/cam assembly. (Courtesy of American Honda Motor Co., Inc.)

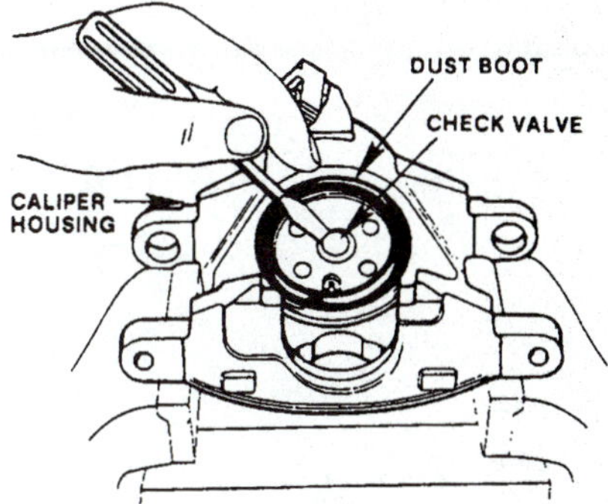

Figure 12-86 When rebuilding a Delco Moraine rear caliper, you should remove the plastic two-way check valve and inspect the inner area. If the piston is useable, a new plastic check valve should be installed. (Courtesy of Bendix Brakes, by AlliedSignal)

1. Install the O-ring in the caliper bore and lubricate the piston and cylinder with assembly fluid or brake fluid. The piston and boot should be installed in the same manner as they would be on a standard caliper with the following additions.

   a. *On Delco Moraine units*, install a new actuator shaft seal and lubricate the actuator shaft and seal with assembly fluid. Then position the actuator shaft, balance spring, and thrust bearing on the piston and start the piston in the bore (Figure 12-88). An installer tool can be used to push the piston to the bottom of the bore. As the piston enters the bore, the actuator screw should move through the back of the

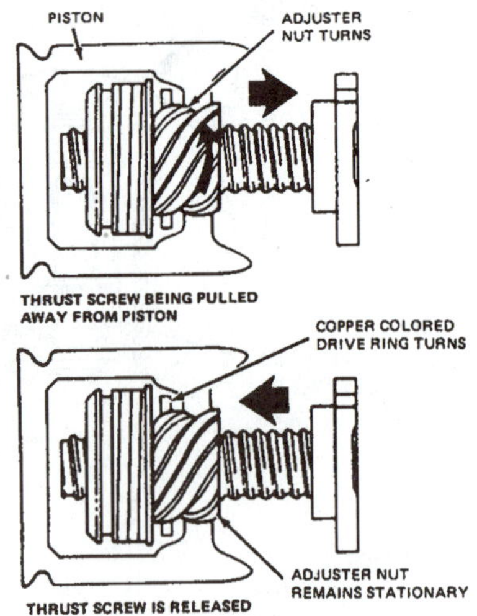

Figure 12-87 The adjuster of a Kelsey Hayes piston is checked by pulling the thrust screw. The adjuster nut should rotate as the piston is lifted, but remain stationary as the adjuster screw moves inward. (Courtesy of Ford Motor Company)

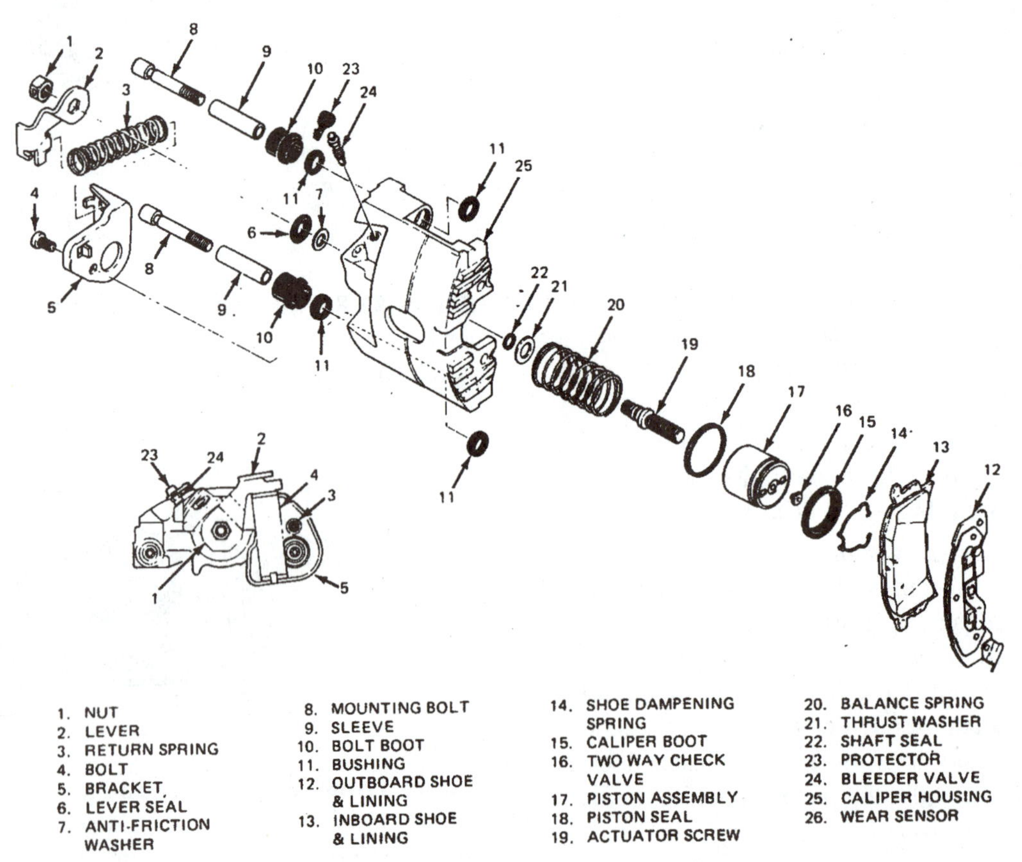

Figure 12-88 A disassembled view of a Delco Moraine rear caliper. (Courtesy of General Motors Corporation, Service Technology Group)

caliper (Figure 12-89). Seat the boot in position in the caliper bore recess using the correct boot-installing tool (Figure 12-90).

b. *On Kelsey-Hayes units*, with the boot correctly positioned, start the piston in the bore. Install a new O-ring seal on the thrust screw, and wet the seal with assembly fluid. Then thread the thrust screw into the piston until the piston is bottomed in the bore and the thrust screw is bottomed in its recess (Figure 12-91). Index the thrust screw notch and install the antirotation pin. Lubricate the three steel balls with silicone grease and place them in position. Place the operating shaft in position and lubricate the thrust bearing with silicone grease. Install a new O-ring seal on the end retainer, and tighten it to the correct torque.

It is a good practice to plug the inlet port and fill the inner cavity with brake fluid before installing the thrust screw (Figure 12-92).

2. Install the parking brake lever in the correct position and tighten the retainer nut or bolt to the correct torque.
3. Install the caliper on the car in the same manner as for a standard caliper with the additional step of connecting and adjusting the parking brake cable. This will be described in Chapter 14.
4. Bleed the brake line and caliper.

On other units, with the unit disassembled, clean and air-dry the parts and inspect them for wear, corrosion, or other damage (Figure 12-93).

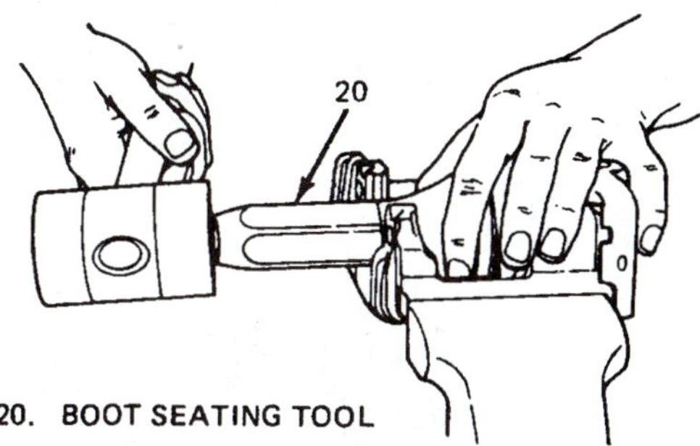

Figure 12-90 The boot of a Delco Moraine rear caliper is seated using a special tool (20), the same way as on a front caliper. (Courtesy of General Motors Corporation, Service Technology Group)

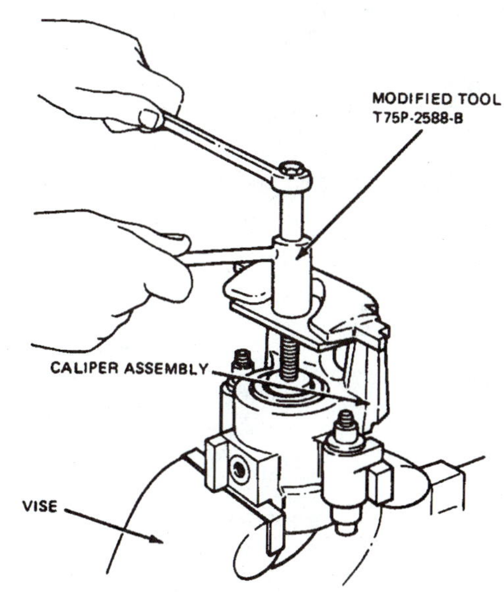

Figure 12-91 The piston of a Kelsey Hayes caliper is being pushed into the bore using this special tool. (Courtesy of Ford Motor Company)

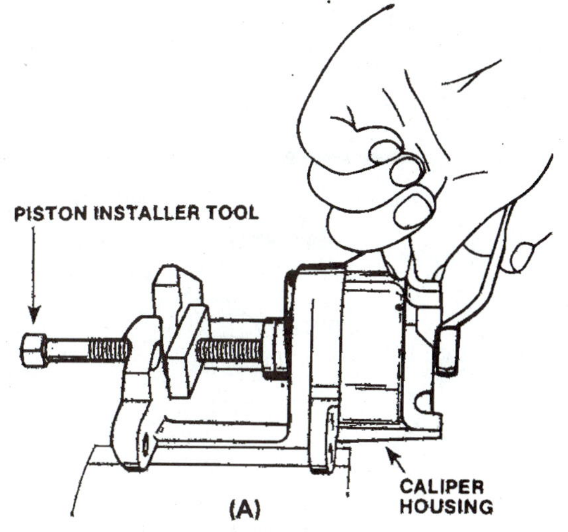

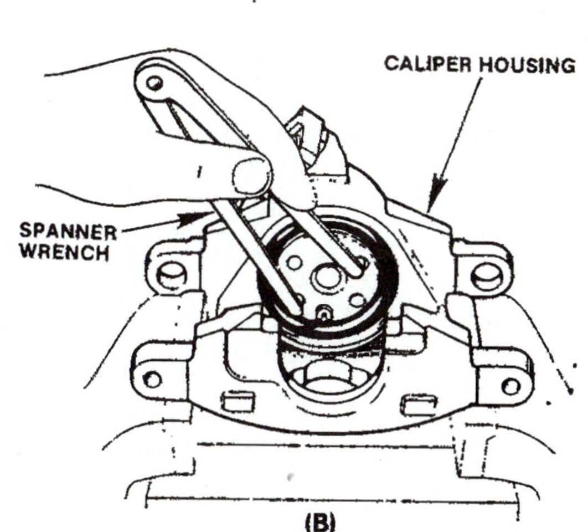

Figure 12-89 When assembling a Delco Moraine rear caliper, a piston installer can be used to push the piston through the O-ring seal and into the bore (A). After installation, the piston should be rotated so that it is correctly aligned with the inboard pad (B). (Courtesy of Bendix Brakes, by AlliedSignal)

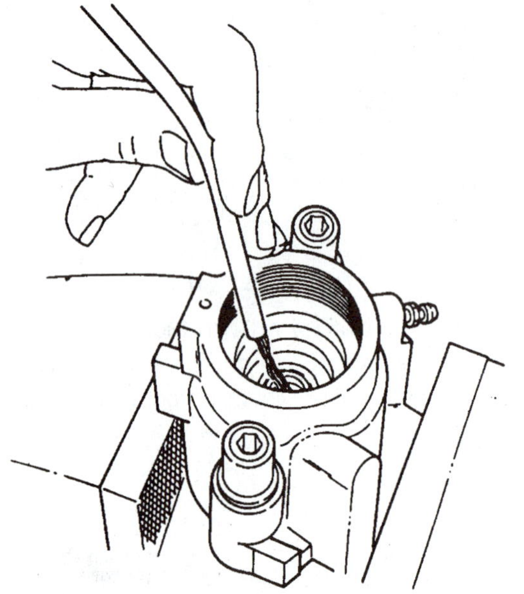

Figure 12-92 After installing the piston, it is a good practice to fill the piston chamber with brake fluid before replacing the thrust screw and parking brake mechanism. (Courtesy of Ford Motor Company)

Lubricate the various parts using silicone grease or rubber grease, and reassemble the unit following the manufacturer's recommendations (Figure 12-94).

12.7 Bleeding Brakes

The process of removing air from the hydraulic system is called **bleeding**. As mentioned earlier, air is compressible, and any air in the system will be compressed during brake pedal application and cause a spongy pedal.

Air is much lighter than brake fluid and has a natural tendency to rise above it. Anyone who has filled a bottle or glass with water has seen this occur. To facilitate bleeding to bleeder screws, actually valves, are placed in strategic locations at the top of caliper, wheel cylinder, and some master cylinder bores (Figure 12-95). Loosening these screws allows air and fluid flow through the valve. The flow will be outward as long as the internal pressure

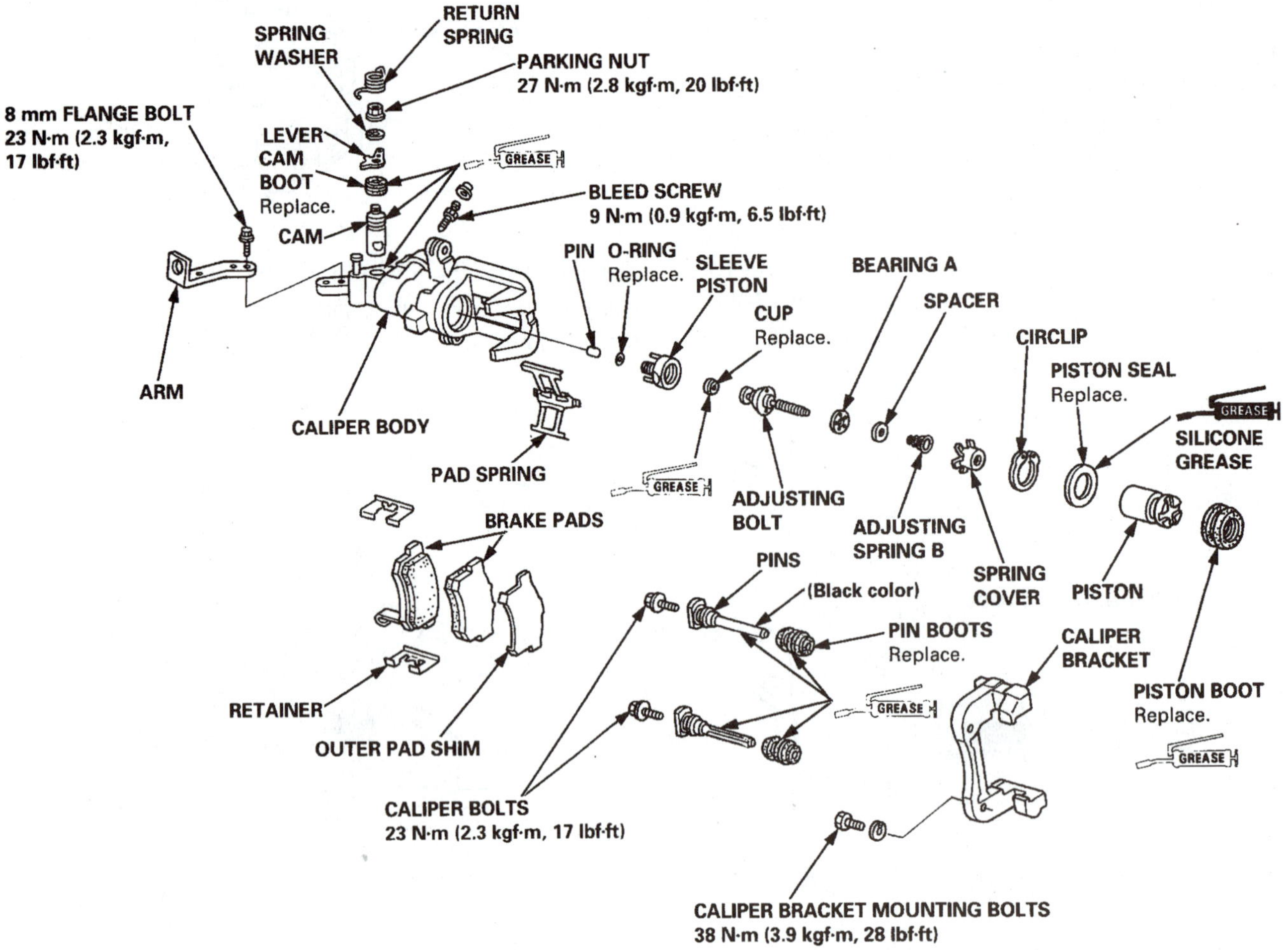

Figure 12-93 A disassembled view of a Nissan-type rear caliper; note the locations where silicone or rubber grease should be used. (Courtesy of American Honda Motor Co., Inc.)

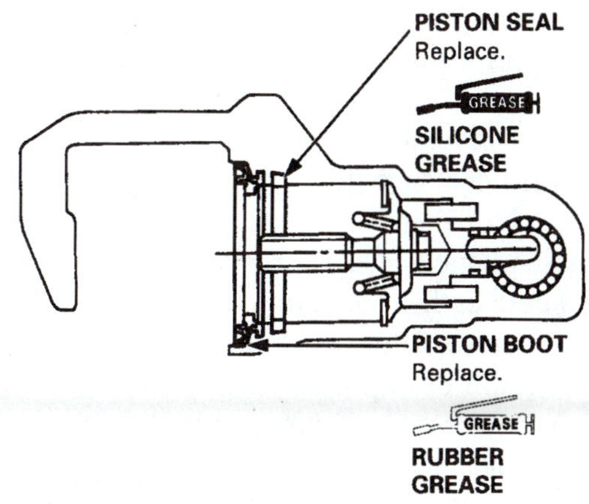

Figure 12-94 A cutaway view of an assembled Nissan-type caliper. (Courtesy of American Honda Motor Co., Inc.)

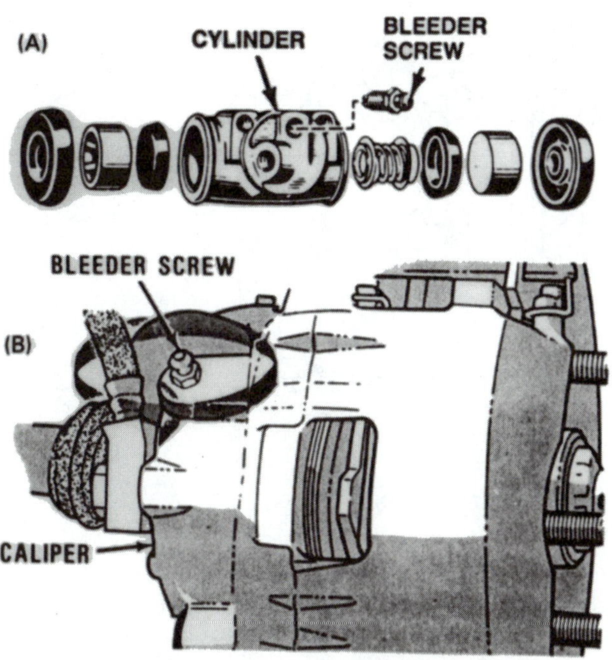

Figure 12-95 The bleeder screw is positioned in the uppermost location in a wheel cylinder (A) or caliper (B). (A is courtesy of Ford Motor Company; B is courtesy of Chrysler Corporation)

is great enough to cause fluid flow. Bleeding can also take place at any line connection by loosening the connection.

Some systems will almost bleed themselves, by gravity. Imagine a system with the master cylinder mounted in the conventional location and with a brake tube running straight in a downward direction to each caliper and wheel cylinder. As fluid is poured into the master cylinder reservoir, it will run downward through the compensating ports and through the master cylinder bore. Then it will flow out the outlet ports and down through the tubing to the caliper and wheel cylinder bores. As this happens, the air in the calipers and wheel cylinder will rise upward on a reverse path to the reservoir (Figure 12-96). Because of the small compensating port opening and the two opposite flows, this process will be slow, but reliable. Given enough time, if enough fluid is poured into the reservoir—about a pint (0.47 L)—the system will fill with fluid to the level of the line port in each caliper and wheel cylinder. If the bleeder valves are opened, the rest of the air in the cylinders will escape, and the cylinders will fill with fluid (Figure 12-97).

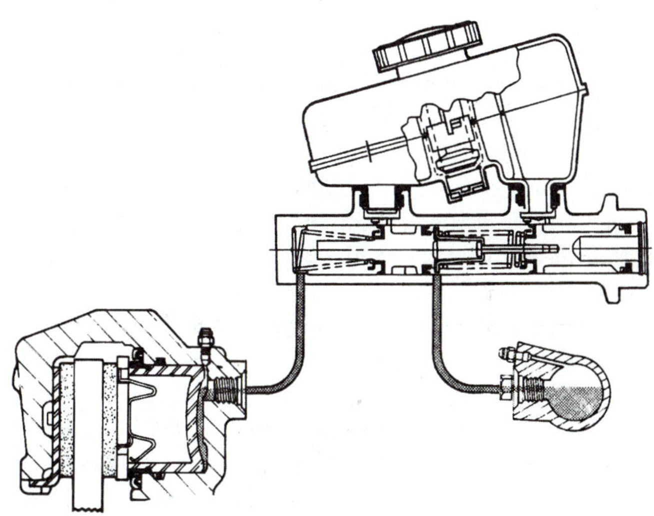

Figure 12-96 If the brake components were arranged as shown here, brake fluid could run downward through the compensating ports from the reservoir, through the brake lines, and into the wheel cylinders and calipers. The air—except for that trapped at the top of wheel cylinder or caliper—would move upward to the master cylinder reservoir.

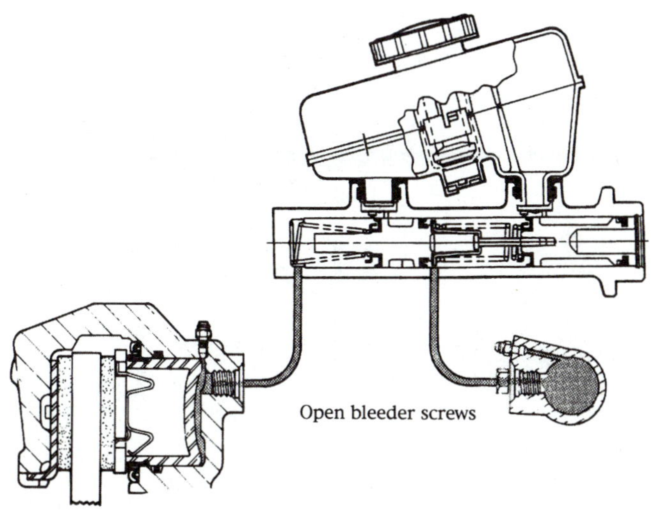

Figure 12-97 If the bleeder screws of the wheel cylinder and caliper are opened, the air can escape, along with some brake fluid.

The system just described cannot exist because the brake tubing has to be routed around many obstacles on the way to the calipers and wheel cylinders. As the tubing follows this sometimes tortuous path, it will often make several up-and-down bends. These bends trap air bubbles. These bends, plus any restrictions in control valves, can make bleeding difficult. The air bubbles must be flushed out of the bends and down to the calipers and wheel cylinders or back into the master cylinder, where they can be removed (Figure 12-98).

Bleeding can also be made difficult if fluid agitation or motion break the air bubbles into foam and still worse if molecular attraction causes small air bubbles to collect in tiny spaces such as between the caliper bore and piston. Even worse is the interior of a caliper with a parking brake mechanism inside. Each added component presents another surface for air bubbles to cling to or become trapped in. Occasionally it helps to tap the caliper firmly with a plastic hammer to dislodge these bubbles (Figure 12-99).

The metering valve can also present a bleeding problem when using pressure bleeders. Remember that this valve closes at about 10 to 15 psi (69 to 103 kPa) and reopens at about 100 to 150 psi (689 to 1,034 kPa). Most pressure bleeders operate in the pressure range that closes the valve, and the valve will shut off the flow to the front brakes. If this should occur, special tools are available for pulling on the stem of a pull-type valve or pushing on a push-type valve to hold the valve in an open position (Figure 12-100). Push-type valves are normally covered by a rubber boot whereas pull-type valves have a metal stem extending through the rubber seal. Do not use a clamp or locking pliers for this purpose because they can damage the valve.

There are several different, commonly used brake-bleeding methods including gravity, manual, pressure, and vacuum bleeding. A technician will often use one or two of these methods. Occasionally, several will be combined and used at the same time.

Bleeder screws often seize in a closed position. This can be the result of an overtightened screw seizing onto the seat, or from rust and corrosion at the threads. When you try to open a bleeder screw do not use excessive force or it will snap off; remember that the screw is small and hollow. Some technicians insert the shank of a small drill bit (the largest possible) into the bleeder screw; if the drill

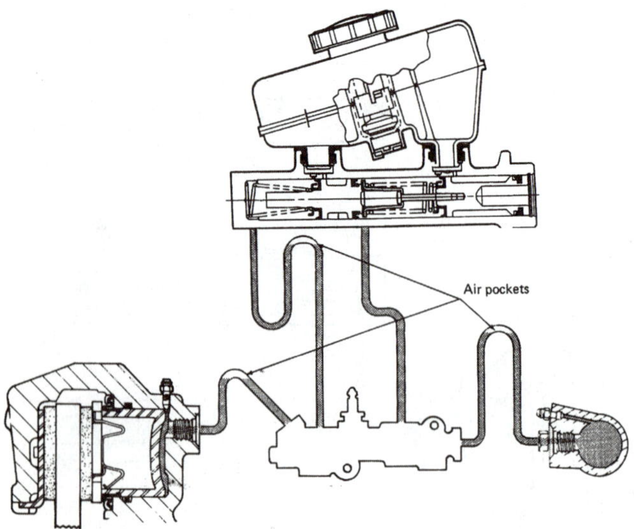

Figure 12-98 Bleeding some systems is complicated when the brake lines go upward and then downward; each upward section becomes an air trap.

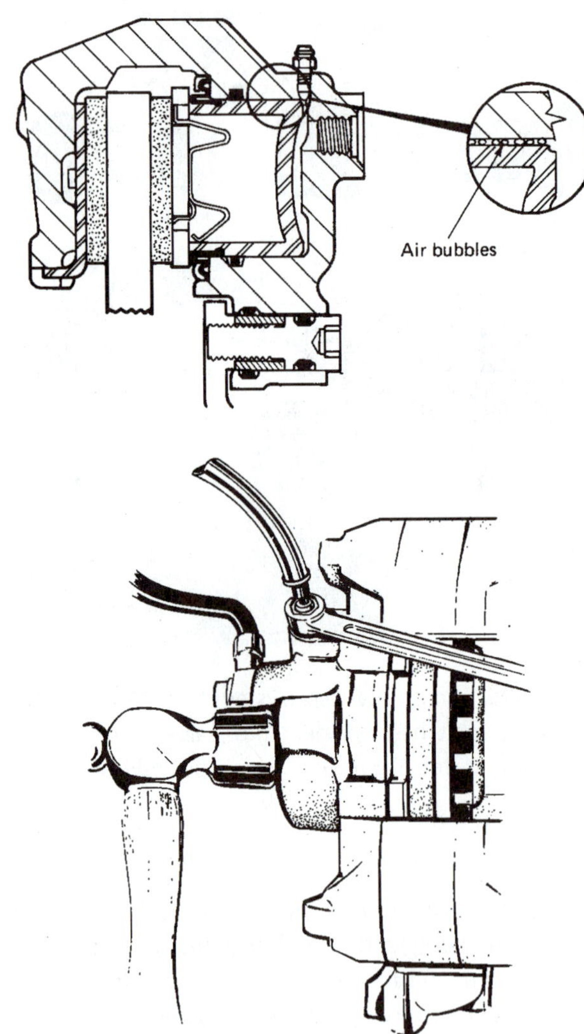

Figure 12-99 Bleeding some calipers is complicated because small air bubbles tend to cling in the tiny space between the caliper bore and piston. Tapping the caliper (a plastic hammer is preferred) will often dislodge these bubbles so they will move to the bleeder screw.

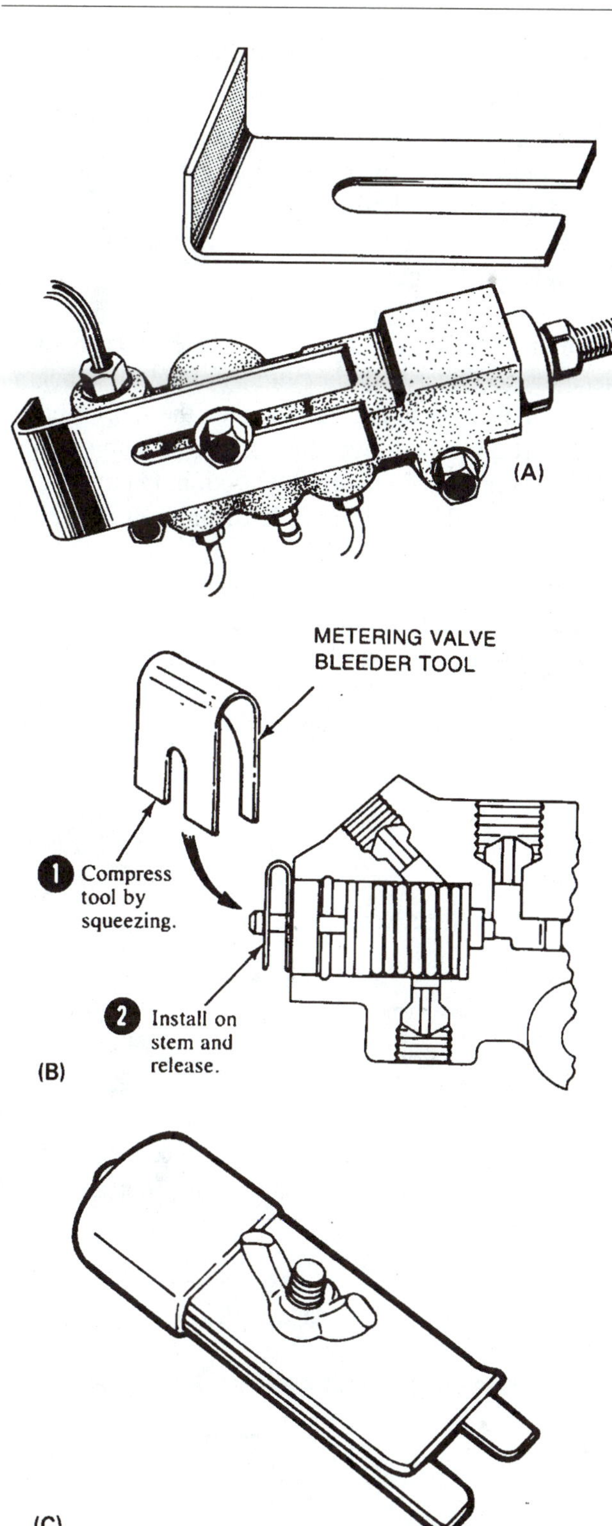

Figure 12-100 The pressure from a pressure bleeder can close the metering valve. Special tool (A) is used to compress the stem of a push-type valve; tool (B) is used to open a pull-type valve; and tool (C) depresses the valve in some RWAL systems. (A is courtesy of SPX Kent-Moore, Part # J-23709; B is courtesy of Ford Motor Company; C is courtesy of SPX/OTC)

bit locks up as you try to loosen the bleeder screw, the screw is ready to break. Some methods used to free a stuck bleeder screw are:

- Apply brake fluid to the threads and let it soak in.
- Tap the end and side of the screw using a punch and small hammer to try to break it loose.
- Heat the screw to smoking hot and touch a candle to the hot screw—the heat will expand the screw and the contraction as it cools might loosen it; the melted candle wax will be drawn into the threads and lubricate them.

When tightening a bleeder screw, remember the torque specifications and that bleeder screws are hollow. They twist apart fairly easily. Bleeder-screw torque specifications are provided by most manufacturers and are also provided in Appendix 4 of this book. It is a good practice to cap a bleeder screw so it will stay clean inside. A plugged bleeder screw can be cleaned using a small drill bit turned by hand, working the drill through both parts of the passage (Figure 12-101). A broken or damaged bleeder screw should be replaced (Figure 12-102). A 6-point box wrench should be used to loosen bleeder screws to prevent the possibility of rounding off the corners of the screws. A box wrench is also handy in that it

Figure 12-101 A small drill bit can be turned by hand to clean the passages of a plugged bleeder screw.

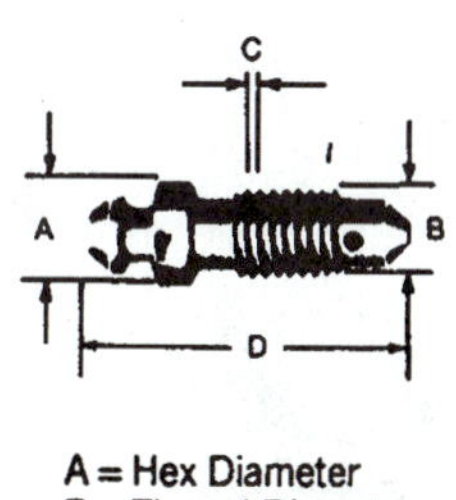

No.	A	B	C	D
3201	10mm	10mm	1.5mm	28mm
3202	3/8	3/8	24	1-33/64
3203	3/8	3/8	24	1-7/64
3204	8mm	8mm	1.25mm	24mm
3205	5/16	5/16	24	1-1/32
3206	1/4	1/4	28	15/16
3207	7/16	7/16	20	1-9/32
13209	10mm	10mm	1.25mm	33mm
13210	10mm	10mm	1mm	33mm
13211	7mm	7mm	1mm	30mm

Figure 12-102 Replacement bleeder screws are available; note the various sizes and dimensions. (Courtesy of Rolero-Omega, Division of Cooper Industries)

will hang on the bleeder screw and stay in place during the bleeding operation. Some cars require special bleeder wrenches bent in configurations necessary to reach past obstructions near the bleeder screw (Figure 12-103).

When bleeding brakes, always bleed until the fluid runs clear, making sure that you have removed all of the old, contaminated fluid, and there are no more air bubbles.

Fluid that is bled from a system must be discarded in the proper manner or recycled.

12.7.1 Bleeding Sequence

It is recommended that a set sequence be followed when bleeding a system. This sequence helps save time and ensures that the whole system is bled. The best bleeding sequence for any vehicle is the one recommended by the manufacturer, especially on vehicles equipped with ABS.

The usual sequence is to bleed the components in the following order:

1. Master cylinder, at the bleeder screw if so equipped or by loosening the lines at the outlet ports.
2. Combination valve if equipped with a bleeder screw.
3. Wheel cylinders and calipers in succession beginning with the longest brake line and ending with the shortest brake line. On most cars this sequence will be right rear, left rear, right front, left front (Figure 12-104).

With diagonal split systems, this sequence is changed to bleed the secondary master cylinder circuit first, beginning with the longer brake line and then the shorter brake line, followed by the longer line of the primary section and the shorter line of the primary section. On many cars this sequence will be right rear, left front, left rear, right front (Figure 12-105).

If a caliper has two bleeder screws, bleed the inboard section first and then the outboard section. If a drum brake has two wheel cylinders, bleed the lower one first, followed by the upper one. Most ABS vehicles require a special bleeding procedure. Some are bled by manual

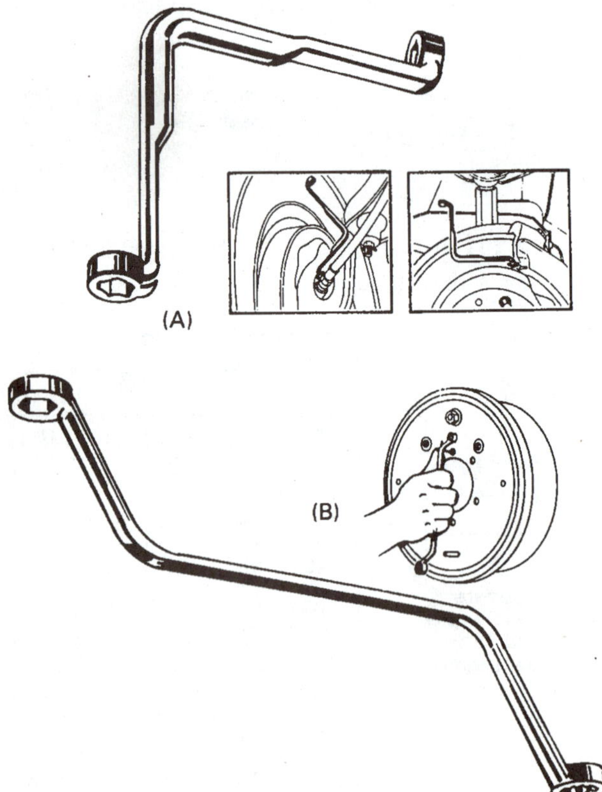

Figure 12-103 The placement of some bleeder screws requires a special wrench to get past the obstructions. (Courtesy of KD Tools)

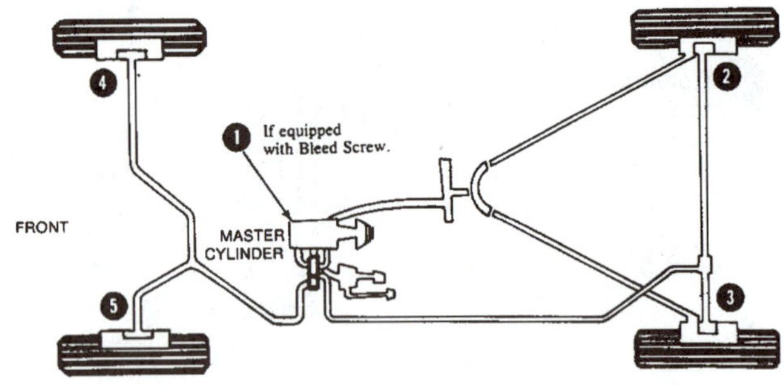

Figure 12-104 The sequence normally recommended for bleeding the brakes of a tandem split system. (Courtesy of Ford Motor Company)

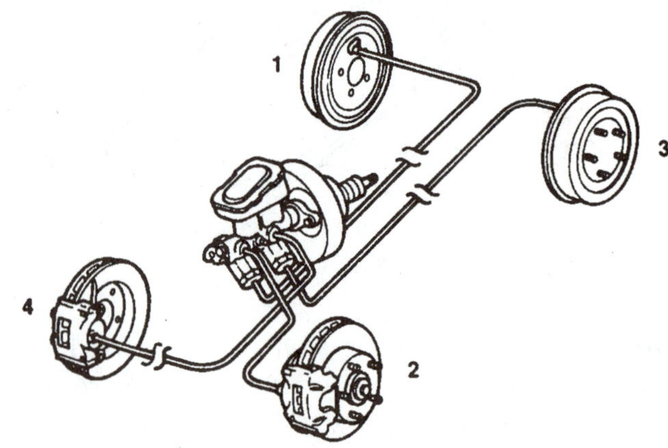

Figure 12-105 This sequence is normally recommended for bleeding a diagonal split system. (Courtesy of Bendix Brakes, by AlliedSignal)

methods, some by pressure bleeding, and some require the use of a scan tool to activate the pump or solenoids. Be sure to check the manufacturer's procedure.

12.7.2 Gravity Bleeding

The **gravity bleeding** method simply lets the fluid run downhill into the calipers and wheel cylinders and if done right, can save a substantial amount of time. Like other methods, it often becomes a step in performing a brake job. A possible drawback with gravity bleeding is that it does not always remove all the air. It sometimes must be followed with another bleeding method, but it is too simple a process to skip.

An experienced technician will arrange the sequence of a brake job so the complete hydraulic system is assembled before the brake shoes are assembled or the caliper replaced. As soon as the last brake line or hydraulic component is assembled and connected, fluid is poured into the master cylinder reservoir. Now, while the individual disc or drum brake assemblies are being put together, they can be bled. Open the bleeder screw as you begin to install the shoes, hardware, or other items on a backing plate or caliper and keep an eye on the bleeder screw. Normally, before you have completed the assembly, the bleeder screw will start to drip fluid; close the bleeder screw. This indicates that the caliper or wheel cylinder is full of fluid. Be sure to wipe up the spilled fluid.

Add fluid to the master cylinder reservoir and repeat this operation as you assemble each of the other brake units. When the last unit has been assembled and the drums installed, finish filling the master cylinder reservoir and test the brake pedal. If it is spongy, use one of the other bleeding methods to force the remaining air out of the pockets in the system.

12.7.3 Manual Bleeding

Manual bleeding uses the master cylinder and brake pedal as a pump to cause fluid flow through an open bleeder screw (Figure 12-106). This fluid flow should flush air out of any pocket or trap. Manual bleeding should be done as smoothly as possible to avoid turbulence in the fluid, which could cause foaming. Foamy fluid contains tiny air bubbles that are very hard to bleed out. Excessively fast pumping of the pedal tends to cause foaming, and if the master cylinder has not been replaced or rebuilt, the longer-than-normal piston travel can damage the rubber cups as they pass over debris or corrosion in the cylinder bore.

Manual bleeding is done with the aid of a bleeder hose and a bottle, which serve several major purposes. When the end of the hose is immersed in fluid, air bubbles are produced that are easily seen and air is prevented from flowing back into the bleeder screw. It also contains waste fluid and helps prevent a mess. A short section, about 1 foot (30 cm), of 3/16- or 1/4-inch (4.7 or 6.3 mm) rubber or plastic tubing may be used as a bleeder hose. Choose a size that will slide onto the nipple of the bleeder screw and stay in place. Clear plastic hose is preferred because it allows you to watch the air bubbles and fluid condition. Any small, half-pint or pint glass or plastic bottle can be used. Plastic is preferred because glass breaks easily (Figure 12-107).

A recent innovation is a mechanical device that can stroke the brake pedal by remote control. This unit allows the technician to be at the wheel or brake assembly and push or release the brake pedal. It can be used for diagnostic checks as well as bleeding operations.

To manually bleed a system:

1. Fill the master cylinder reservoir at the start and enough times during the operation to keep it at least half full.
2. Instruct your helper to keep a moderate, steady pressure on the pedal, to push with a slow and steady motion, to inform you when it reaches the floor, and to release it slowly when told to.
3. With pressure on the brake pedal and the bleeder hose connected to the first bleeder screw in the sequence, open the bleeder screw and observe the flow from the hose. When you are told that the pedal has reached the floor or the flow stops, close the bleeder screw and tell your helper to release the pedal. Some sources recommend a 15-second wait between the time the pedal is released and the time it is reapplied.

   Note that each time the bleeder hose is disconnected, the fluid will run out and the hose will fill with air. Bubbles will normally appear at the start

Figure 12-106 When manually bleeding a system, the pedal is pushed slowly downward while the bleeder screw is open.

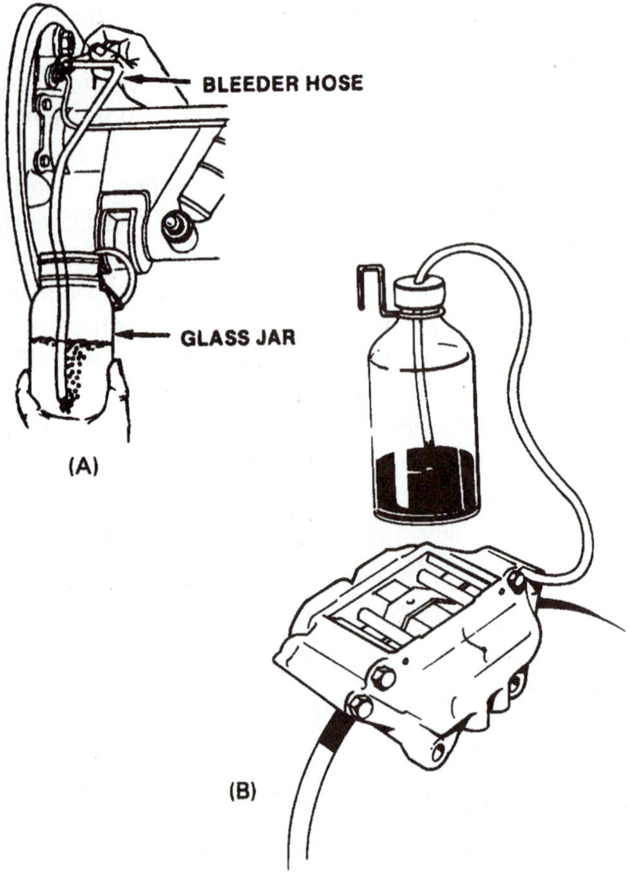

Figure 12-107 A bleeder hose is placed on the bleeder screw, with the free end in a jar partially filled with brake fluid (A). This allows the waste fluid to be collected, allows easy observation of any air bubbles, and prevents reentry of air back into the system. If a hanger is used and enough room is available, it is even better to suspend the catch container above the bleeder screw (B). (A is courtesy of Bendix Brakes, by Allied-Signal; B is courtesy of ITT Automotive)

of each bleeding step as this air is bled out of the hose.

4. Repeat step 3 until the fluid flow from the bleeder hose is clear and without bubbles. With the pedal held downward, tighten the bleeder screw to the correct torque and check the fluid level in the reservoir. Repeat this operation on the next brake in the sequence.
5. After bleeding the last brake assembly, fill the reservoir and check the pedal feel.

12.7.4 Pressure Bleeding

Pressure bleeding normally uses a pressurized tank of brake fluid to cause fluid flow through the bleeder screws (Figure 12-108). This tank should have a fluid chamber

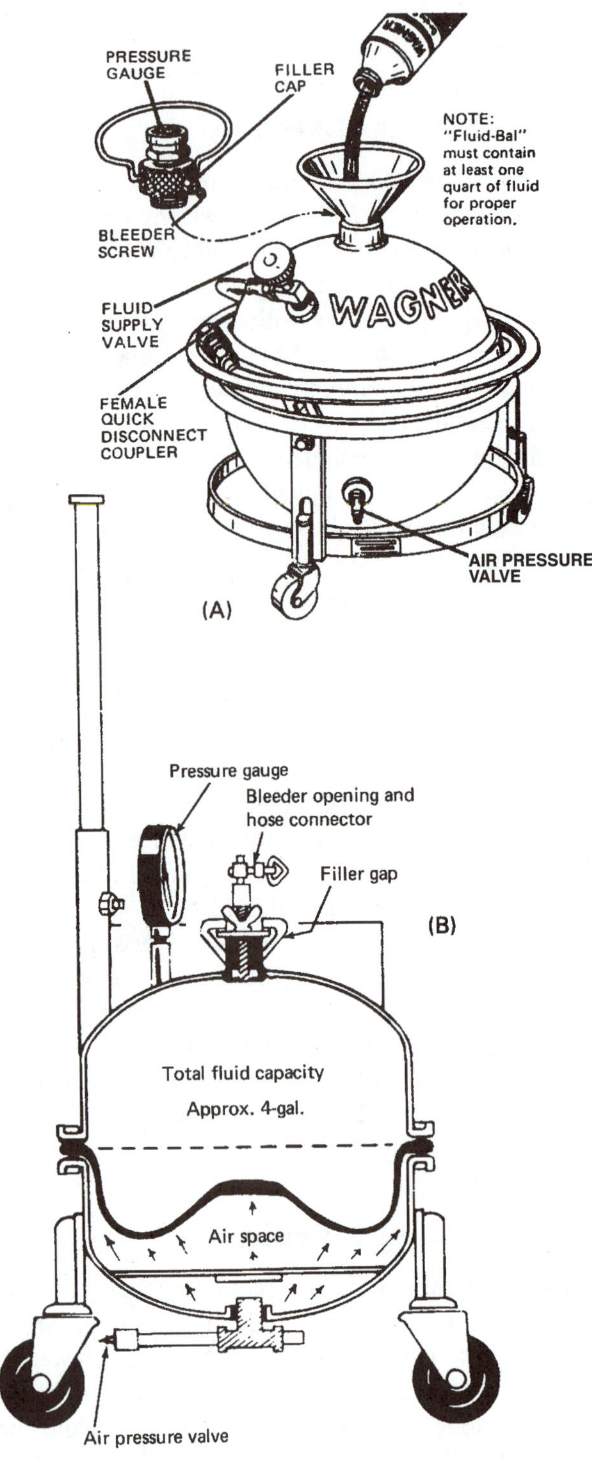

Figure 12-108 The top portion of a pressure bleeder is filled with fluid (A); note that the filler cap contains a bleeder screw so the air can be removed. Also note that the upper and lower sections are separated by a rubber diaphragm so the air pressure will not contaminate the fluid (B). (A is courtesy of Wagner Brake; B is courtesy of Branike Industries, Inc.)

that is separate from the air chamber, to prevent the fluid from being contaminated by the compressed air. The air chamber is normally pressurized to a working pressure of 10 to 15 psi (69 to 103 kPa). If using silicone fluid, the pressure should be dropped to about 5 psi (35 kPa). Higher pressures might introduce turbulence in the fluid, which could cause foaming. Pressure bleeding has the advantage that only one person is required and that the fluid level in the reservoir is continuously maintained.

Another recent innovation is a brake fluid injector. This unit is a hand-operated pump that can pump 10 to 20 milliliters (ml) (0.6 to 1.2 ci) per stroke, depending on model, at pressures up to 125 psi (862 kPa) into the brake system. Besides bleeding in the normal direction, this unit can be used to **reverse-** or **back-bleed** the system by forcing fluid into a bleeder screw and out the master cylinder. The unit can also **cross-bleed** parts of the system by forcing fluid in one bleeder screw and out another (Figure 12-109). These styles of bleeding allow the technician more flexibility in bleeding air that might become trapped in odd locations; they will be described in following sections.

A brake fluid injector can be used for pressure bleeding. Using a port adapter, this unit is used to pump fluid into the master cylinder compensating port, through the circuit, and out the bleeder valve (Figure 12-110).

Adapters are required to connect the pressure bleeder unit to the master cylinder, and a pressure-tight connection must be made to prevent fluid leakage. The adapter is normally clamped to the top of a cast-iron reservoir. Special adapters are required to fit into plastic reservoirs, because these reservoirs are not strong enough to withstand the clamping pressure (Figure 12-111).

To pressure-bleed a system:

1. Fill the master cylinder reservoir about half full.
2. Attach the correct adapter to the master cylinder (Figure 12-112).
3. Check the air pressure in the pressure bleeder unit and adjust it if necessary.
4. Connect the hose between the bleeder unit and the adapter and open the fluid supply valve.

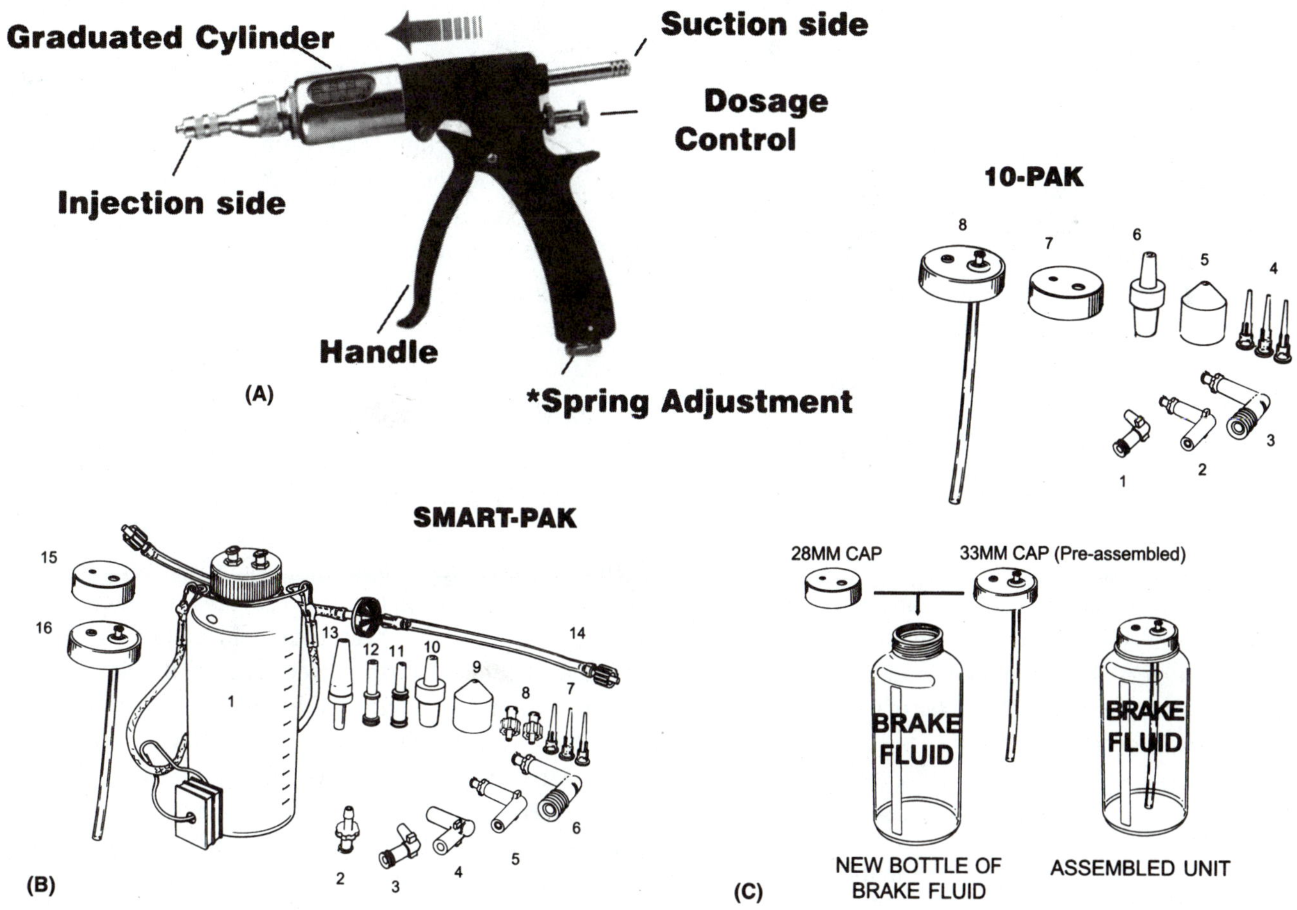

Figure 12-109 The Phoenix Injector system consists of a hand-operated injector (A), a fluid bottle, plus connector tubing and fittings (B & C). The injector can be used to create a fluid pressure or vacuum depending on how the tubing is connected. (Courtesy of Phoenix Systems, L.L.C.)

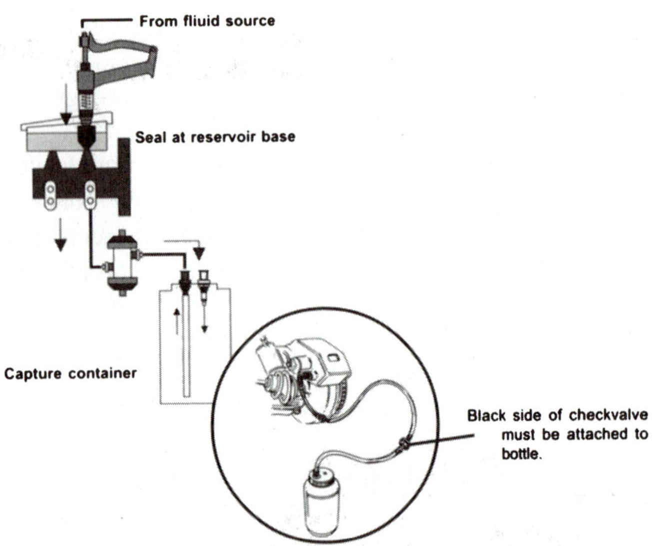

Figure 12-110 With the use of a port adapter, the Phoenix Injector can be used to pressure-bleed portions of a brake system. (Courtesy of Phoenix Systems, L.L.C.)

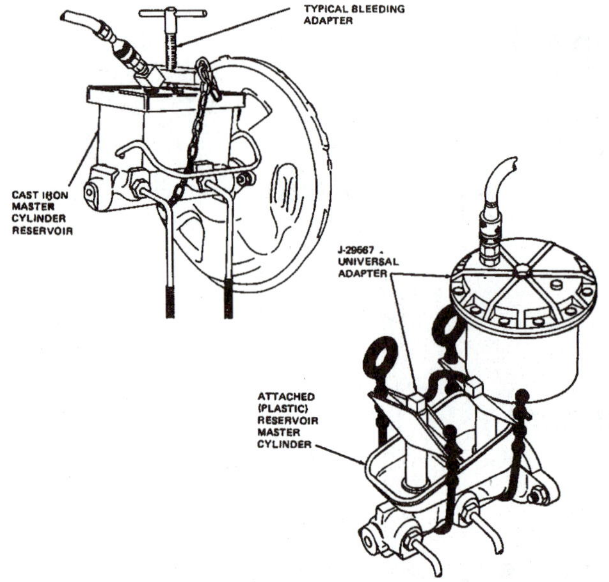

Figure 12-111 An adapter is required to attach the pressure bleeder to the master cylinder reservoir. This adapter can clamp onto the top of a metal reservoir, but it must connect to the fluid recesses of a plastic reservoir. (Courtesy of General Motors Corporation, Service Technology Group)

Check for leaks at the adapter or in the brake system.

5. Attach a bleeder hose and bottle to the first bleeder screw in the sequence. Open the bleeder screw, observe the fluid flow, and close the bleeder screw when the flow is clear and free of bubbles.

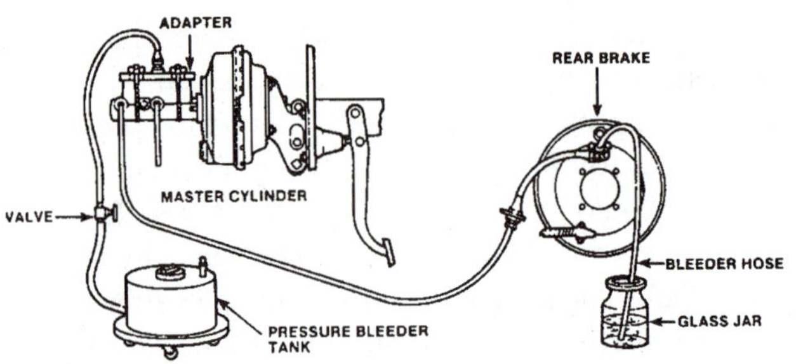

Figure 12-112 Pressure-bleeding a system. Opening the valve (at left) allows fluid under bleeder-tank pressure to flow through the system to the open bleeder screw, pushing any air and fluid into the catch container. (Courtesy of Bendix Brakes, by AlliedSignal)

6. Repeat step 5 on the last brake in the sequence.
7. After bleeding the last brake in the sequence, shut off the fluid supply valve. Disconnect the hose, remove the adapter, and wipe up any spilled fluid.
8. Fill the reservoir, if necessary, and check the pedal feel.
9. Some pressure bleeders include a system pressure gauge and allow you to shut off the pressure supply. If your unit is so equipped, turn off the pressure, note the pressure in the system, and recheck the pressure after a few minutes. A loss of pressure indicates a leak.
10. Discard the contaminated fluid in the proper manner.

12.7.5 Vacuum Bleeding

Vacuum bleeding uses a pump to pull fluid and air out of the bleeder screw (Figure 12-113). A hand- or air-powered vacuum pump can be used for this. Vacuum bleeding is a rather simple and effective operation, but several cautions should be observed. As in manual bleeding, the master cylinder reservoir will empty and let air enter the system. Also, the bleeder screw threads are not airtight. Placing a vacuum on the bleeder screw will pull air past the threads, as well as air and fluid through the screw. Air can also be pulled past the cups in a wheel cylinder. This air does no harm because it is usually removed as it enters. Bubbles will nearly always be present in the fluid flowing from the bleeder screw.

To vacuum-bleed a system:

1. Fill the master cylinder reservoir(s) at the start and enough times during the operation to keep the reservoir(s) at least one-fourth full.

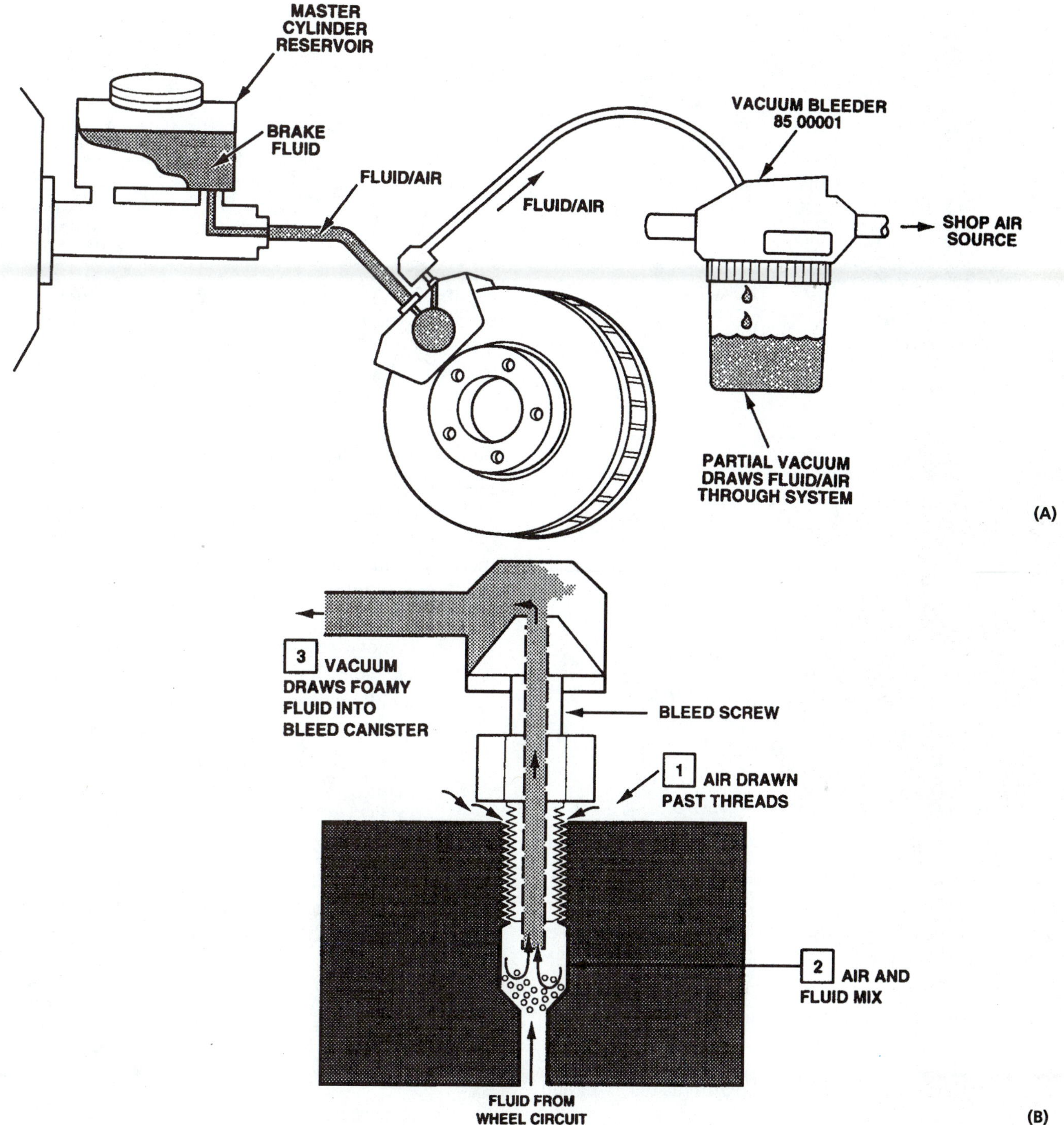

Figure 12-113 Vacuum-bleeding a system. Operating the vacuum bleeder pulls air and fluid from the open bleeder screw (A); the fluid is coming from the master cylinder reservoir. A vacuum bleeder normally removes foamy fluid because air is drawn past the threads of the bleeder screw (B). (Courtesy of General Motors Corporation, Service Technology Group)

Note that some systems have an attachment that will maintain the correct reservoir fluid level during bleeding operations (Figure 12-114).

2. Attach the bleeder unit to the first bleeder screw in the sequence. Open the bleeder screw, operate the bleeder pump, and observe the flow. After fluid flow begins and the bubble rate shows a significant drop, close the bleeder screw and stop the vacuum pump.
3. Repeat step 2 on the remaining brake units, being sure to check the reservoir fluid level after bleeding each brake.

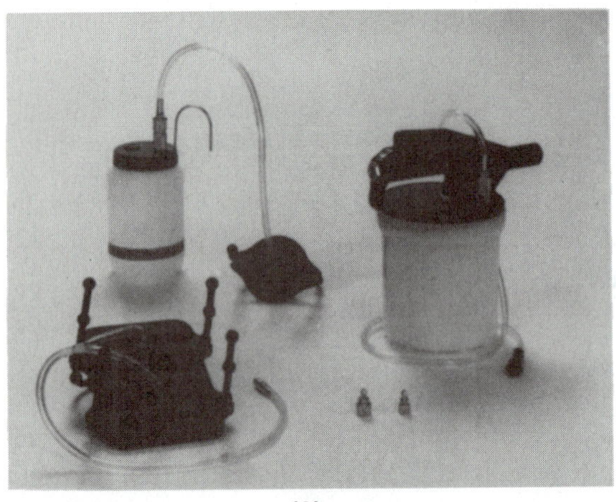

(A)

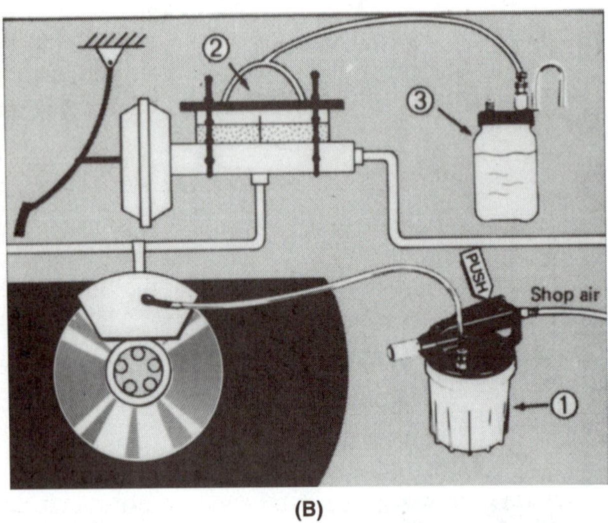

(B)

Figure 12-114 This vacuum-bleeding system includes a refiller for the reservoir (A). When using this system, the reservoir is filled, the cover (2) is installed and connected to the refiller (3). The bleeder (1) is connected to the bleeder screw, which is loosened, and the bleeder is operated to produce the bleeding operation. (A is courtesy of Vacula Automotive Products)

4. After bleeding the last brake, fill the reservoir and check the pedal feel.
5. Discard the contaminated fluid in the proper manner.

By reversing the connectors to the fluid injector, it can also be used for vacuum bleeding (Figure 12-115).

12.7.6 Reverse-Flow Bleeding

Some brake systems have the lines or brake valves arranged so there are air traps close to the master cylinder. Also, since air bubbles naturally move upward through

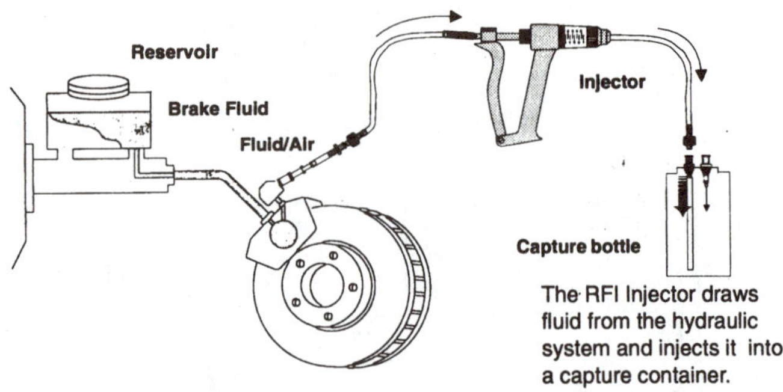

Figure 12-115 Using the proper adapters allows the fluid injector to vacuum-bleed a system. (Courtesy of Phoenix Systems, L.L.C.)

fluid, it makes more sense to bleed from the bottom up than to bleed from the top down, as in other bleeding methods. In reverse-flow or back-bleeding, fluid is forced into the bleeder valve, through the circuit, and out the master cylinder reservoir.

To reverse-bleed a system:

1. Remove enough fluid from the master cylinder reservoirs for three to ten injection strokes. Do not allow the reservoir to overfill.
2. Open the bleeder valve, and attach the injector to the bleeder valve, pumping fluid to the end of the adapter to expel all air from the hose (Figure 12-116).
3. Gently depress the handle of the injector to pump fluid into the bleeder valve in slow and steady strokes. Overly fast strokes can loosen debris and move it into the brake valves. Continue pumping for three to ten strokes.
4. Remove the injector from the bleeder valve and allow a small amount of fluid and any air to escape from the bleeder. Then tighten the bleeder valve.
5. Repeat steps 2 to 4 on the remaining brake units.
6. Discard the contaminated fluid in the proper manner.

12.7.7 Changing Brake Fluid

To ensure a maximum brake fluid boiling point and to reduce interior system corrosion, a system which uses DOT 3 or 4 brake fluid should have the fluid changed every year or every other year. As mentioned earlier, this is not commonly done, although it is a relatively simple operation. With today's cars and driving practices, chang-

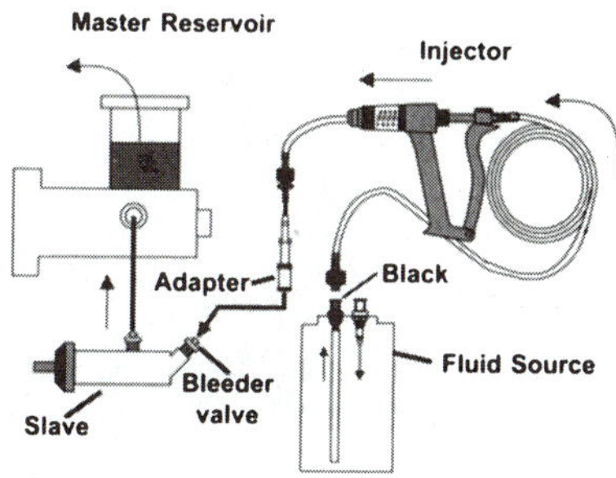

Figure 12-116 In reverse-flow bleeding, the injector is pumping fluid into the bleeder valve, and this forces fluid and air upward and out of the master cylinder. (Courtesy of Phoenix Systems, L.L.C.)

ing the brake fluid every other year can be highly recommended for two major reasons: safety and economics. Many drivers of FWD cars in heavy-traffic situations will have brake fluid that is close to the boiling point, and old, contaminated fluid has a lower-than-normal boiling point. Quick-take-up master cylinders are expensive, costing more than several hundred dollars, and ABS hydraulic modulators are very expensive with replacement costs of some units well over one thousand dollars. Old, contaminated brake fluid can easily cause improper operation and ruin these parts.

Several styles of brake fluid testers are available (see Figure 9-11). Some units test a fluid sample from the master cylinder reservoir for water contamination; others measure its boiling point. One tester uses a paper strip that is dipped into the fluid. They provide a clear indication that the fluid condition may cause a brake loss or is causing corrosion in the system.

Changing the brake fluid, also called "**flushing** a system," is the same as bleeding a brake, except that the major purpose is to remove all of the old fluid, not just the air. Most brake technicians do this when they bleed the brakes after a major brake overhaul. Most manufacturers recommend a thorough bleeding—until clean, new fluid leaves the bleeder screw—during each major brake repair to ensure that the system is filled with new, clean fluid. Any of the previously described brake-bleeding techniques can be used for this operation. Pressure bleeding is preferred by many technicians because it is quickest and maintains a supply of new fluid.

In the past, it was common to flush the dirty fluid out of a system using a flushing fluid or denatured alcohol. This is currently considered a poor practice because any leftover flushing fluid will lower the boiling point of the new brake fluid. Never flush a system with alcohol; use only clean brake fluid. Alcohol is a good cleaning agent when a unit is disassembled because it can be completely dried off the parts.

If the system is being changed over to silicone fluid, the bleeding operation should be done smoothly and slowly. The slightly higher viscosity of silicone fluid causes it to trap air bubbles, especially if it is mixed with glycol fluid. When making a fluid changeover, it is important to remove all of the old glycol fluid. Silicone fluid is lighter than glycol fluid, so the glycol fluid will stay in the bottom of the cylinders and cannot be bled out. The best time to make a fluid changeover is during a major overhaul. The calipers, wheel cylinders, and master cylinder should be rebuilt, cleaned, and emptied of old fluid.

To change brake fluid:

1. Using a syringe, remove the old fluid from the reservoir and wipe out the reservoir with a clean shop cloth to remove any contaminants and residue.
2. Connect a pressure bleeder to the reservoir and bleed the system as described in Section 12.7.4.

   Or, fill the reservoir with new fluid and bleed the system using one of the other methods which has been described.
3. Continue the bleeding operation at each bleeder screw until clean, fresh, new fluid flows from each one.

With ABS cars, changing fluid is more difficult and time consuming. An alternate method of removing most of the contaminated fluid is to cross-bleed portions of the system. The usual cross-bleeding circuits are both front brakes and rear brakes of a tandem split system and the two front-rear brake circuits of a diagonal split system. While cross bleeding, the brake pedal can be depressed to stop any flow to the master cylinder and ABS modulator valves.

To cross-bleed a system:

1. Depress the brake pedal about 1/2 to 1½ inches, and install a pedal depressor to hold it in this position.
2. Install a catch container at one end of the circuit to be bled, and open this bleeder valve.
3. Inject fluid through the bleeder valve at the other end of circuit until fluid entering the catch container runs clear (Figure 12-117).
4. Remove the old fluid from the master cylinder using the RFI injector, a syringe, or a vacuum pump (Figure 12-118).

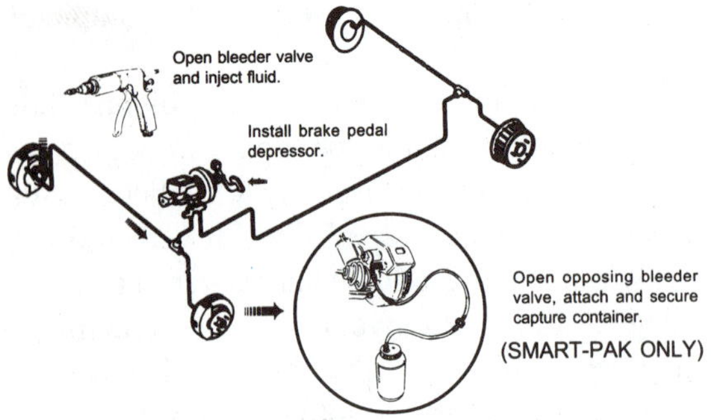

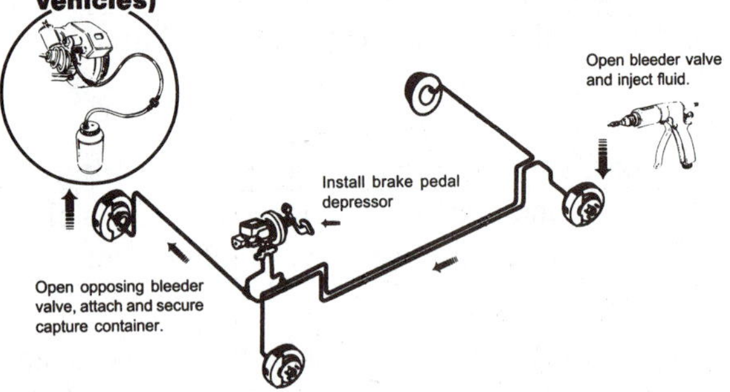

Figure 12-117 In cross bleeding, fluid is injected into the bleeder valve at one wheel while the old fluid and any air bubbles are being removed at the bleeder valve of the other wheel. (Courtesy of Phoenix Systems, L.L.C.)

5. Refill the master cylinder reservoirs with new fluid, and discard the contaminated fluid in the proper manner.

12.8 Diagnosing Hydraulic System Problems

Occasionally a problem will occur in a hydraulic brake system that cannot be corrected by the service steps previously described. Several service operations can be used to help you pinpoint the cause of the problem. The following procedures deal with these common problems.

12.8.1 Diagnosing a Spongy Brake Pedal

A spongy brake pedal is usually caused by air in the system. It can also be caused by a brake hose that is expanding under pressure or a drum that is being deflected under pressure. These two possibilities can be visually checked while an assistant applies and releases pressure on the brake pedal.

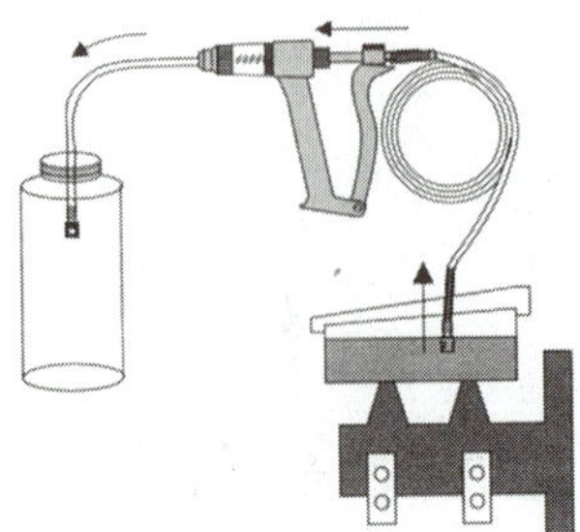

Figure 12-118 The fluid injector can be used to remove old, contaminated fluid from the master cylinder reservoir. (Courtesy of Phoenix Systems, L.L.C.)

CAUTION *Operating the brakes with the reservoir cover removed will allow fluid to spray upward, sometimes well above the master cylinder. Never have your face over or near the reservoir at this time unless protected by a face shield or fluid barrier (Figure 12-119).*

A diagnostic approach to checking for air is to have an assistant rapidly pump the brake pedal about twenty times and then keep it depressed while you remove the master cylinder reservoir cover. Release the pedal and watch the reservoir for abnormally large swirls. If there is a larger-than-normal swirl, air is probably trapped in that section of the system, and it should be re-bled (Figure 12-120). It

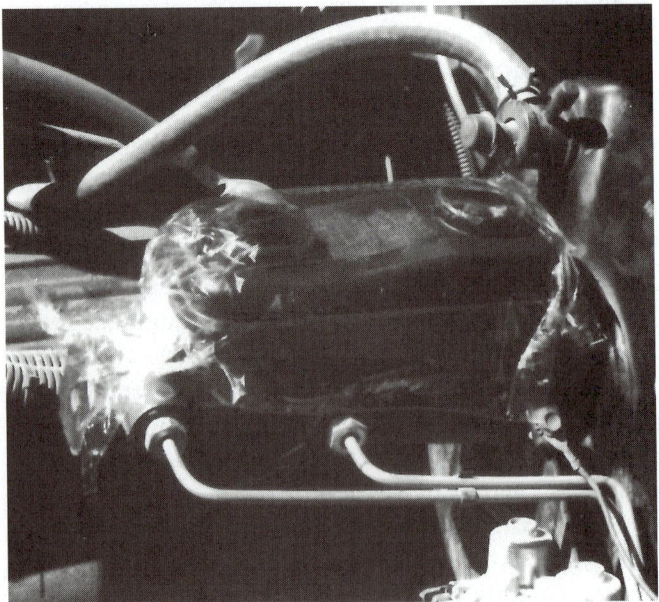

Figure 12-119 Clear plastic (food wrap) can be placed over an open master cylinder reservoir to catch any fluid sprays.

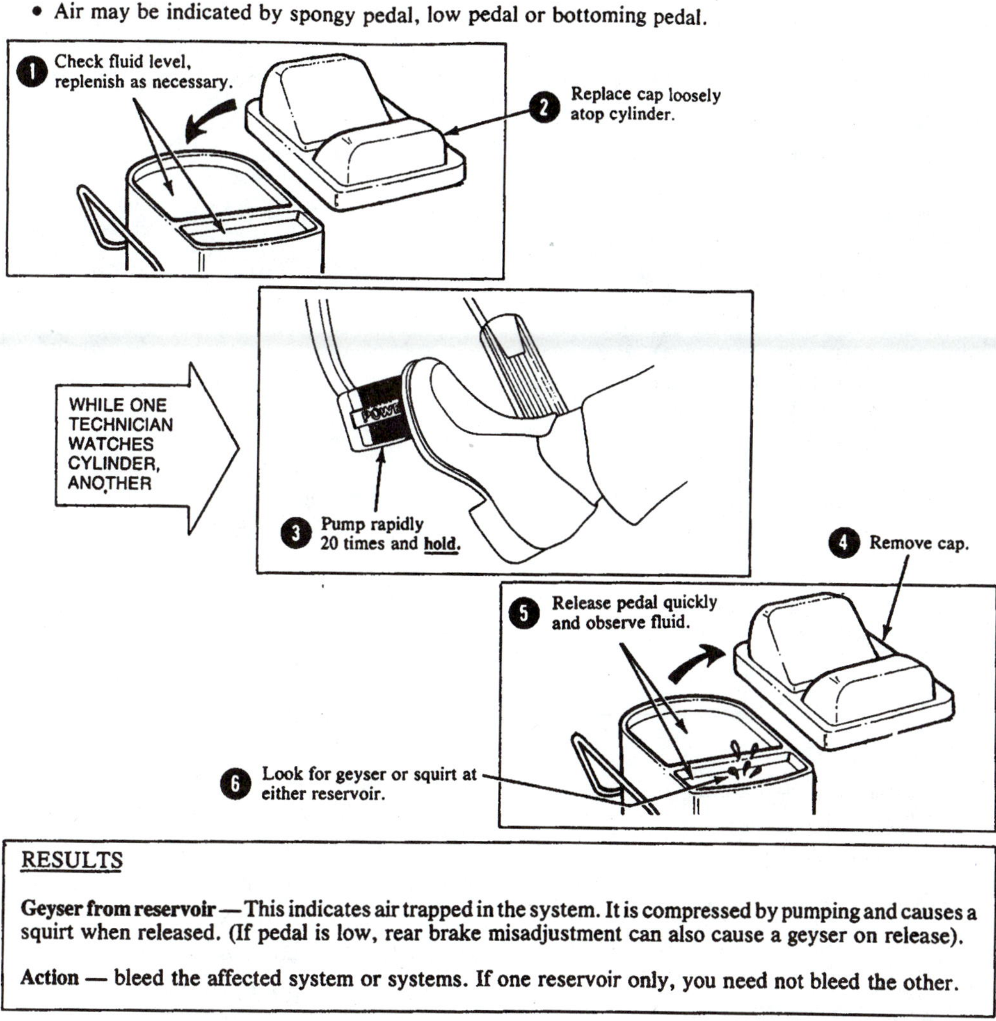

Figure 12-120 An air-entrapment test. This quick check is used to determine which portion of the brake system has air trapped in it. (Courtesy of Ford Motor Company)

is a good idea to cover the reservoir with a clear plastic film before releasing the brake pedal, to contain the fluid swirl.

If the system has been bled several times and you are sure all the air has been bled out, but the pedal is still spongy, isolate the source of the problem by closing off parts of the system. Plug up the ports of the master cylinder one at a time. If the pedal becomes firm, the problem is in the section that has just been closed off. If the pedal is still low or spongy with both ports plugged, the master cylinder is faulty. The outlet ports can be plugged using a coupler and a plug threaded onto the tube nut. Or a connection can be plugged by placing a steel ball between the tube flare and the tube seat (Figure 12-121). When using a steel ball, be careful not to damage the flare or the seat. Choose a ball large enough so that it will not become wedged in the tube (Figure 12-122). A properly sized copper rivet can be used in place of the ball. As a plug is installed, it must be bled. Loosely tighten the plug and have an assistant apply the brake pedal. Then tighten the plug as soon as the air bubbles stop and there is a clear fluid leak.

After the problem has been isolated to a particular section of the system, the exact location can be determined by moving the plug to the various connections in that section.

12.8.2 Diagnosing a Sinking/Bypassing Brake Pedal

A sinking or bypassing brake pedal is one that sinks or falls to the floor under pressure. This problem is usually caused by an internal leak in the master cylinder or an external fluid leak from the system (Figure 12-123).

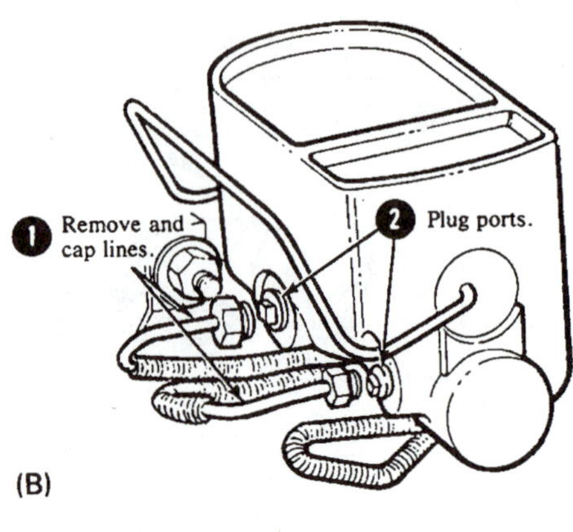

Figure 12-121 A group of inverted-flare plugs of different sizes (A). These plugs can be used to isolate sections of a hydraulic system to help locate the cause of a low or spongy pedal (B). (B is courtesy of Ford Motor Company)

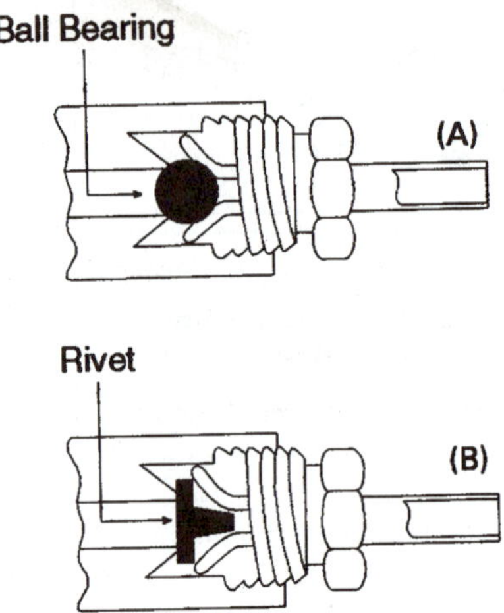

Figure 12-122 A ball bearing (A) or a rivet (B) can be used as a line plug. Be careful not to overtighten it and damage the flare seat or wedge the ball into the brake line.

Bypassing is a fluid leak from one fluid section to the other, past the secondary cup(s) on the secondary piston in the master cylinder. Although an external fluid leak is more common, bypassing is an easier check. Have an assistant watch the levels of the fluid in both reservoirs as the pedal sinks. On plain master cylinders, the level in both reservoirs should drop slightly during normal brake application. With quick-take-up master cylinders, the level in the primary reservoir will rise when the quick-take-up valve opens. Bypassing will cause a fluid transfer that lowers the level in one reservoir while raising the level in the other.

An external leak is located by looking for a leak in the system. The probable location of the leak will be a loose line fitting, a faulty hose, a faulty wheel cylinder, or a leaky caliper. As you look for the leak, follow each of the lines until fluid is located. It should be noted that the rear brake lines on some cars pass through the passenger compartment. In these cases, it is necessary to remove the rear seat to see all the lines.

If an external leak cannot be located, the probable cause of the sinking pedal is bypassing a leak past the primary cup of the primary piston or secondary cup of the secondary piston in the master cylinder. This condition will often cause a pedal to sink under light pressure but not under very heavy pressure. Sometimes it will cause fluid turbulence in the reservoir as the pedal is sinking. If you are still not sure whether a master cylinder is at fault, plug the outlets as previously described. If the pedal sinks with the ports tightly plugged, the master cylinder is clearly defective.

12.8.3 Diagnosing a Brake That Will Not Apply

If a brake unit will not apply, have an assistant exert a moderate pressure on the brake pedal while you open the bleeder screw. Be careful, because it should spurt fluid. If it does, the problem is being caused by a stuck wheel cylinder or caliper piston. If fluid does not spurt from the bleeder screw, there is a line restriction. Check for a kinked or compressed line, an internally collapsed hose, or a plugged control valve. Loosen the line connections closer to the master cylinder as an assistant maintains pedal pressure to locate the cause of the restriction.

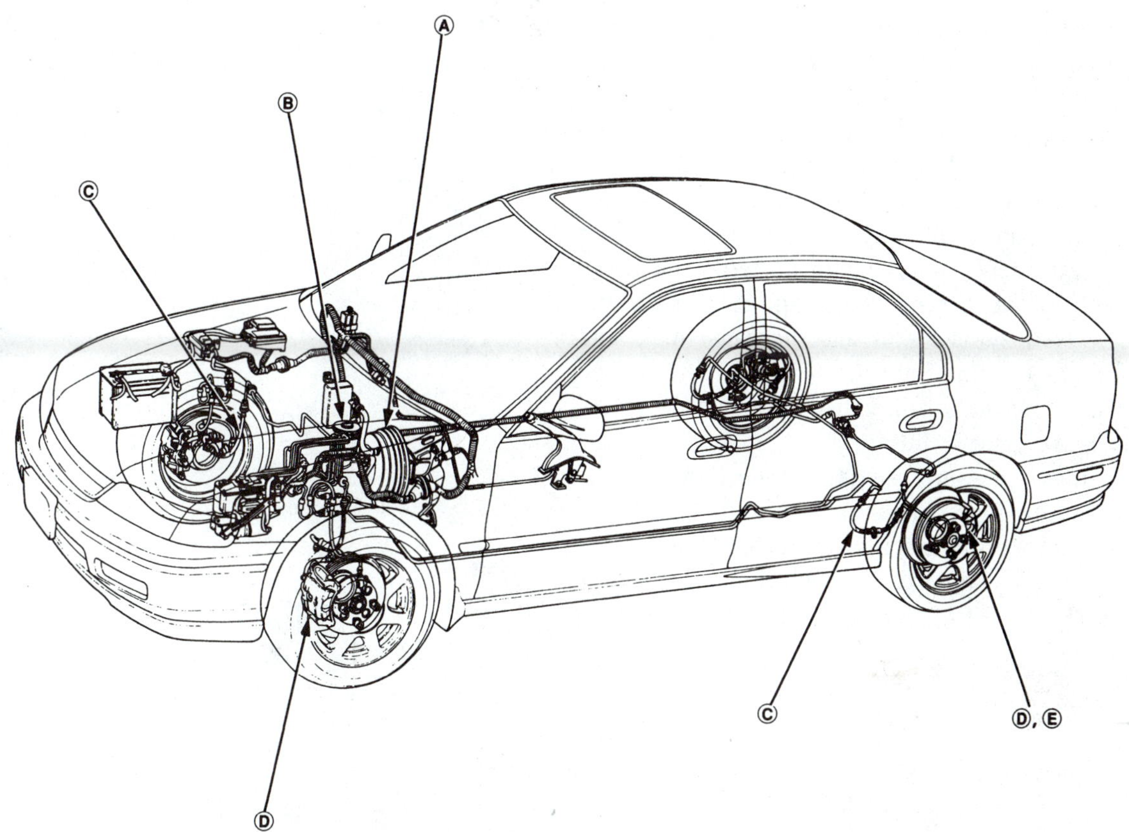

Figure 12-123 The probable locations of an external fluid leak are the master cylinder connections (B), hoses (C), calipers (D), and wheel cylinders (E). (Courtesy of American Honda Motor Co., Inc.)

12.8.4 Diagnosing a Brake That Will Not Release

If a brake drags, apply and release the brakes, and then open the bleeder screw at the dragging brake. In most modern systems, fluid should drip out. In older systems using residual valves, there should be a small spurt of fluid as the residual pressure is released. If fluid spurts out in a greater volume than normal and the brake releases, there is a restriction in the line that prevents fluid release back to the master cylinder. If there is a normal fluid release and the brake still drags, a stuck wheel cylinder or caliper piston is probably the cause.

12.8.5 Diagnosing a Rear or Front Wheel Lockup Problem

Occasionally a proportioning valve fails to split the front-to-rear pressure properly, which can result in insufficient or excessive rear brake pressure. On cars with diagonal braking systems, lockup of a single rear wheel can be caused by a malfunction in one of the proportioning valves. This problem can be diagnosed using a pair of pressure gauges (Figure 12-124).

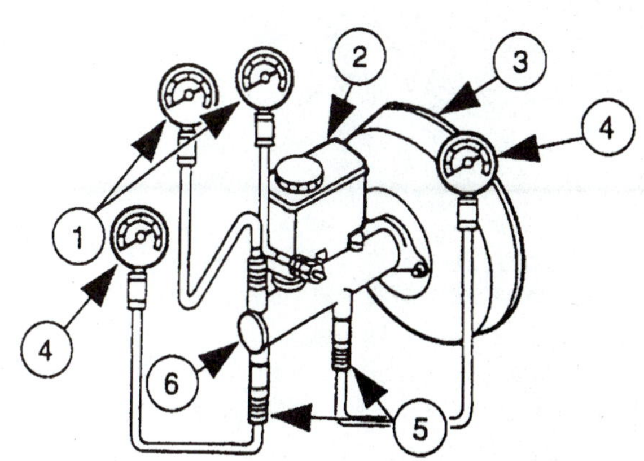

Figure 12-124 Gauges can be connected to the master cylinder (2 and 6) to determine pressure at the front brake lines (1) and rear brake lines (4). A difference between the front and rear pressures shows the operation of the differential/proportioning valves (5). (Courtesy of Ford Motor Company)

A gauge set can be assembled in the shop. Obtain two 0-to-1,000 or 0-to-2,000 psi (0-to-10,000 kPa) gauges. These gauges should have a small dial, about 2 inches in diameter, and a side-mounted port. The port will normally have a 1/8-inch National Pipe Thread (NPT) male thread. Adapters can be purchased, or shop-made from bleeder screws, that will thread into the wheel cylinder or caliper using the bleeder screw port (Figure 12-125). The gauge can also be installed in a line connection using a tee that has one 1/8-NPT female port and the same male and female ports as the line flare nut (Figure 12-126). The gauge can also be installed directly on tubing or into a fitting by using a connector fitting.

To use these pressure gauges, connect one to the bleeder port of the front caliper and one to the rear wheel cylinder or to a tee fitting installed on each side of the proportioning valve. Be sure to bleed each connection as it is being made (Figure 12-127). Next have an assistant apply

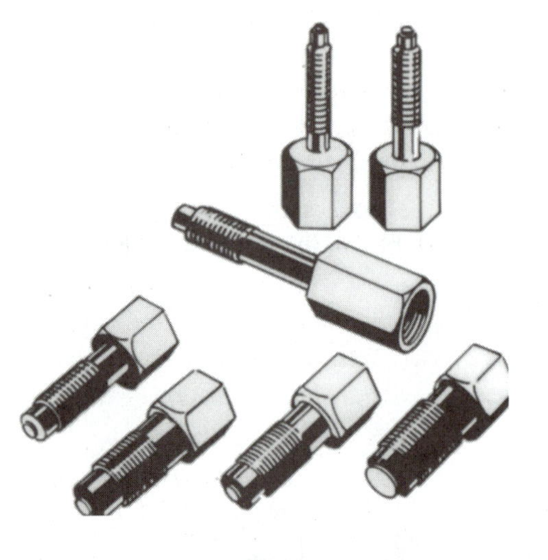

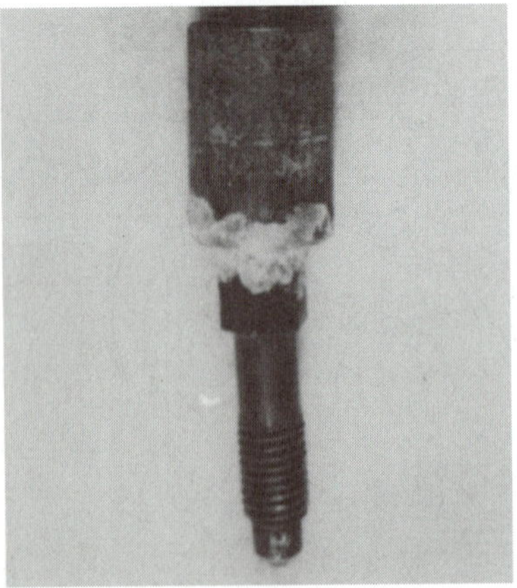

Figure 12-125 Commercial (A) or shop-made (B) adapters can be used to connect a pressure gauge to a caliper or wheel cylinder. The shop-made adapter is a bleeder screw that has a hole drilled through the end and a 1/8-inch NPT pipe coupler brazed onto it. (A is courtesy of ITT Automotive)

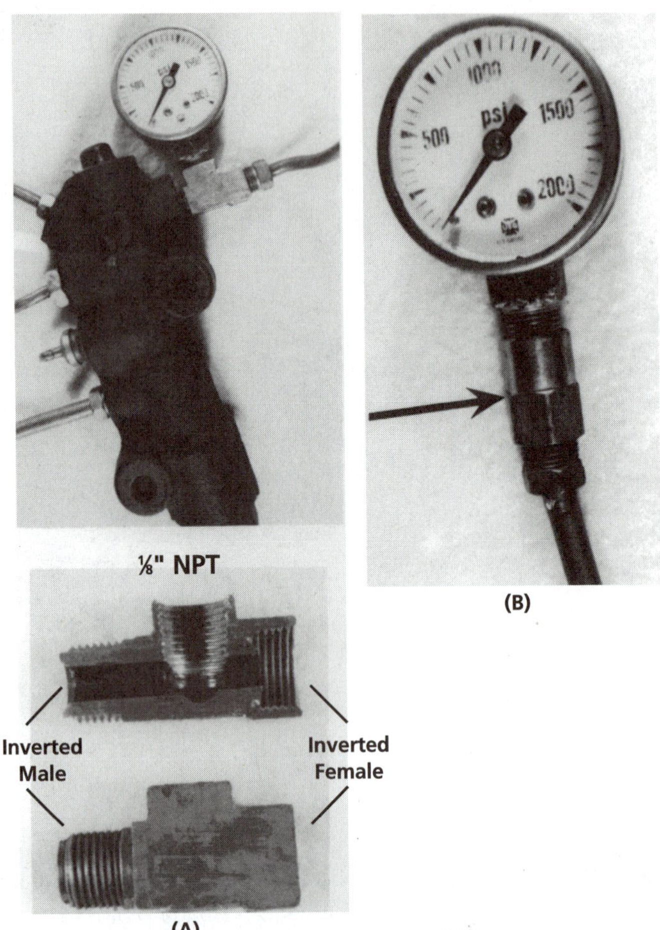

Figure 12-126 A pressure gauge can be installed in a tee placed between a component and the tubing by using a tee fitting (A), or onto the end of a brake line by using a female adapter (arrow—B).

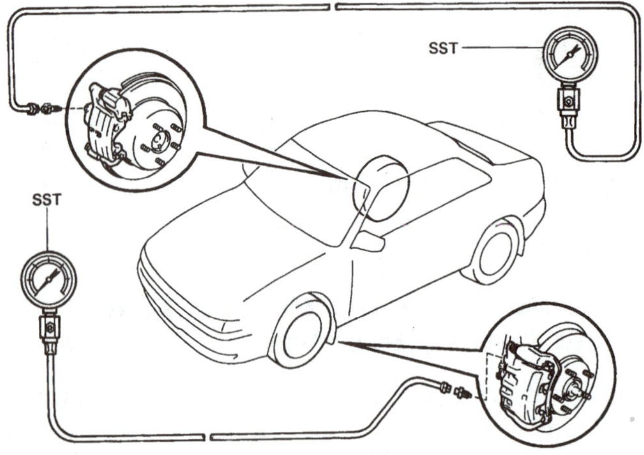

Figure 12-127 A proportioning valve can be tested by installing a gauge at the front and rear brakes. A difference in pressure with a hard brake application shows proportioning valve operation.

the brake pedal as you monitor the gauge readings. The gauge pressures should rise equally until about 300 to 700 psi (2,068 to 4,826 kPa). From that point on, the pressure in the front brakes should rise much faster than that in the rear brakes. In some systems, the metering valve will cause a lag in the front pressure increase of between 10 and 100 psi (69 to 690 kPa). Occasionally, the exact pressures are given in the manufacturer's specifications. A defective proportioning valve should be replaced. On cars with diagonal braking systems, install a gauge in each rear brake circuit and apply the brakes with a firm pedal. Each of the rear brakes should have the same pressure. A pressure difference indicates a faulty valve(s).

12.8.6 Height-Sensing Proportioning Valve Adjustment

Pickups, vans, and some cars use a height-sensing proportioning valve to control rear brake application pressure relative to vehicle height and load. This valve is also called a *pressure control valve* or a *regulated proportioning valve*.

The linkage (the mounting position) of this valve is adjustable to allow proper valve-to-axle positioning and ensure that the valve will produce the correct control pressure for a particular load. The actual adjustment will vary in different car models. The service manual should be checked to determine the correct procedure for checking and adjusting this valve (Figure 12-128).

12.9 Testing a Warning Light

When completing a brake job, it is a good practice to test the operation of the warning light system to be certain of its future operation if the brakes should fail. It is common for the switch piston to stick on-center after years of remaining stationary in this position (Figure 12-129).

To check the operation of a warning light switch:

1. First verify that the warning light operates as you crank the engine. Then test the warning light bulb circuit by turning on the ignition, disconnecting the lead from the warning light switch, and connecting a jumper wire from this lead to ground (Figure 12-130). As the jumper is being connected, the warning light should come on. If it does not, replace the bulb or repair the circuit. Reattach the lead to the switch.
2. Check the fluid level in the master cylinder reservoir to make sure it is at least half full.
3. Connect a bleeder hose and bottle to a bleeder screw.
4. With the ignition on, open the bleeder screw while an assistant applies pressure on the brake pedal. The light should come on when there is light-to-moderate pressure on the pedal. If it does not, the warning switch or valve is faulty and should be replaced. The bleeder screw should be closed before the pedal is released.

In some cases, after a system failure has been repaired, the switch piston will stick in an off-center position. This will keep the warning light on even though the system is in good condition. Also, some late 1960s and early 1970s cars have valves constructed so they cannot recenter themselves.

To center a warning light switch or valve:

1. Attach a bleeder hose and bottle to a bleeder screw in the hydraulic circuit opposite the one that has failed or was bled last.
2. Turn on the ignition, open the bleeder screw, and have an assistant push slowly on the brake pedal while watching the warning light. At the instant that the light goes out, close the bleeder screw. If the light flickers and comes back on, or does not go out at all, repeat this operation at a bleeder screw on the other hydraulic section.
3. Fill the master cylinder reservoir to the correct level.

12.10 Brake-Light Switch Adjustment

On many cars the brake-light switch position is adjustable. An improperly adjusted switch might fail to turn off the stoplights if it is too loose, or cause brake drag if it is too tight. This switch is usually closed as the pedal is depressed to apply the brakes, and opened to break the circuit as the pedal is released (Figure 12-131).

On some cars, the brake-light switch is a double switch. When the brake is applied, this switch deactivates the cruise control at the same time that the brake lights are switched on (Figure 12-132). In some cars, a separate cruise control switch is used with the brake control switch.

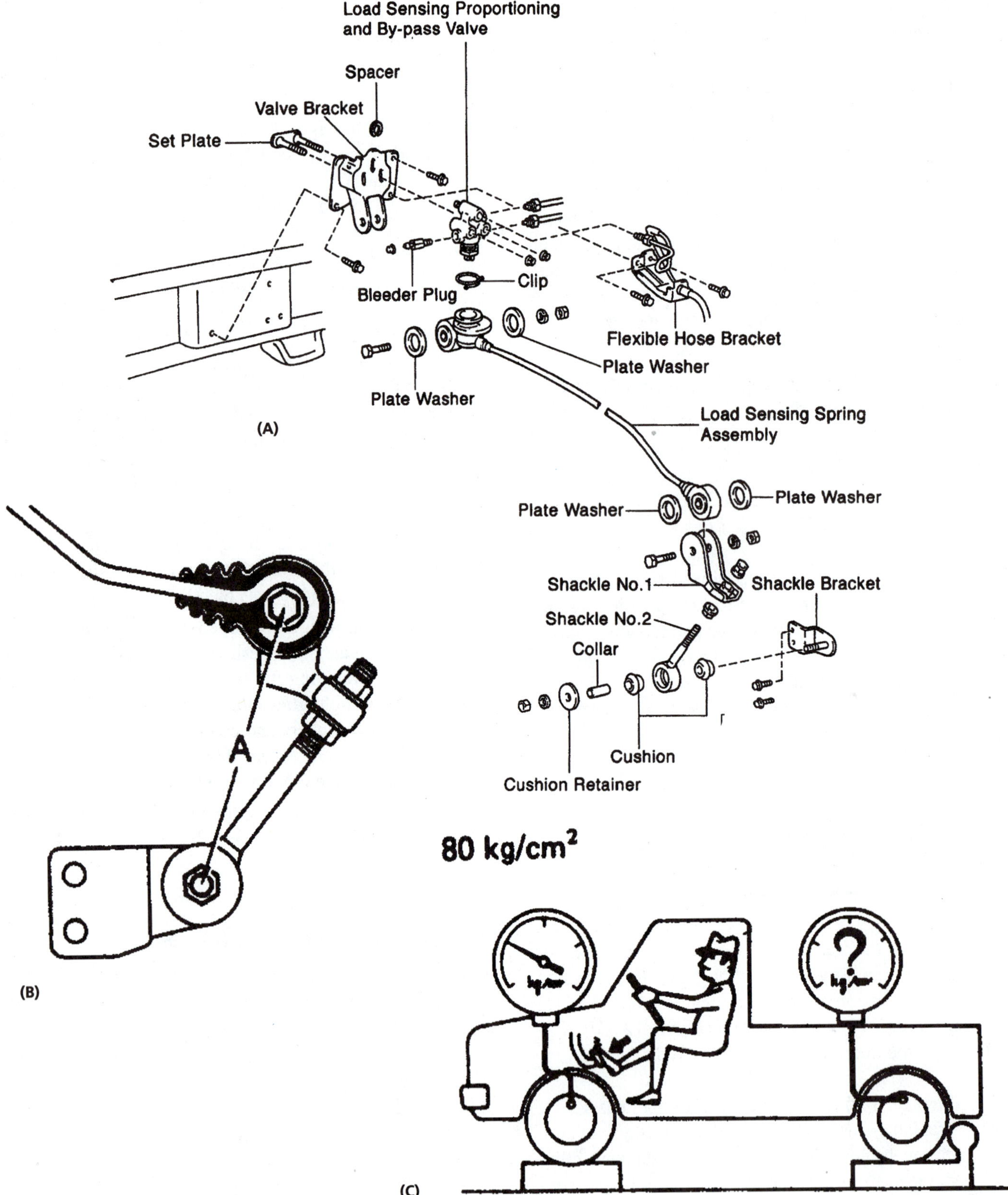

Figure 12-128 A load-sensing proportioning valve assembly (A). Its adjustment has a specification for shackle No. 2 (B), and it can be tested using two pressure gauges. Added weight to the rear of this pickup should change the valve setting and the rear pressure (C).

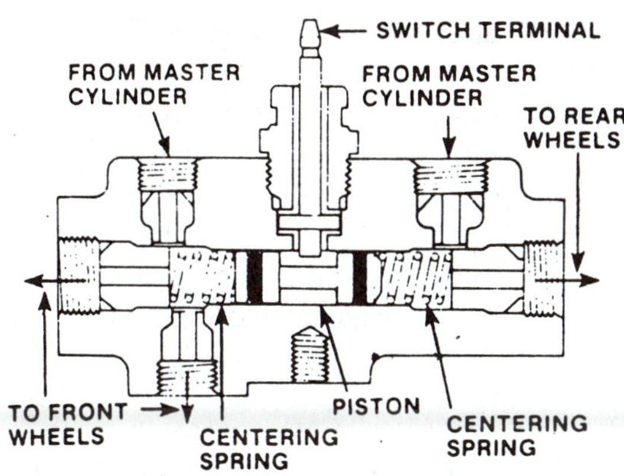

Figure 12-129 This pressure differential switch uses centering springs to keep the piston centered. A loss in pressure in the front or rear system will move the piston off-center and contact the switch. (Courtesy of Bendix Brakes, by AlliedSignal)

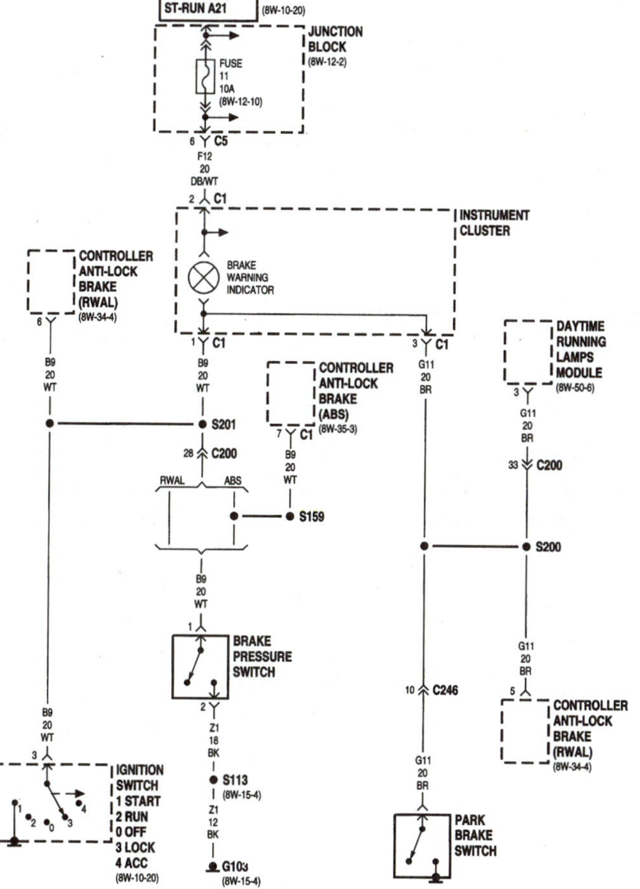

Figure 12-130 A brake warning-light circuit. The brake warning indicator should light when the ignition switch is turned to start; it can also be turned on by the brake pressure switch or the parking brake switch. It is integrated into the ABS and daytime running lights on some models. (Courtesy of Chrysler Corporation)

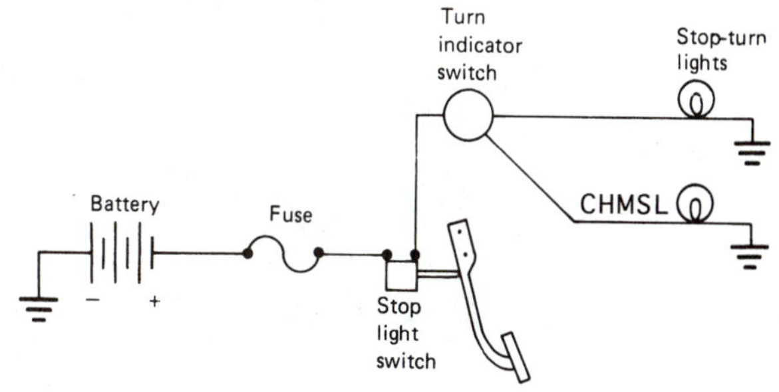

Figure 12-131 A simple brake light circuit.

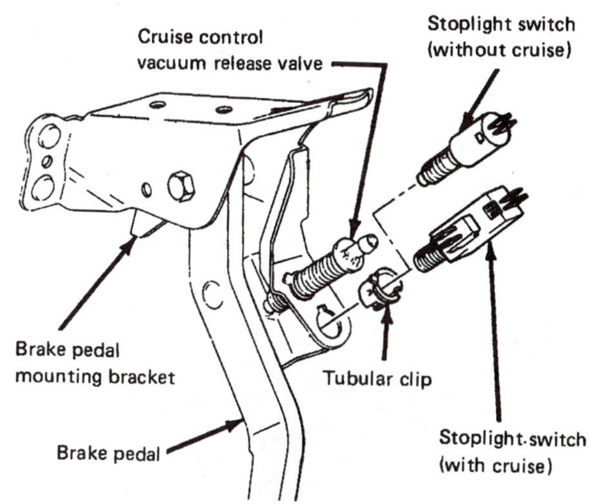

Figure 12-132 Cars equipped with cruise control use a brake switch to turn off the cruise control when the brakes are applied. Many cars use one switch for both brake lights and cruise control; if two switches are used, they should both be adjusted to operate at the right time. (Courtesy of General Motors Corporation, Service Technology Group)

On some cars with automatic transmissions, the brake-light switch is used to disconnect the torque convertor clutch during deceleration. An improperly adjusted switch will cause improper torque convertor clutch operation.

Brake-light switch adjustment will vary in different cars. Many cars have an adjustable bracket or an adjustable switch position on the bracket. Some cars have a switch position that is automatically adjusted on the bracket as the pedal is released. A service manual should be checked, especially on those cars with cruise control and/or automatic transmissions, to determine the correct adjustment method when working on a particular car (Figure 12-133). After any brake-light switch adjustment, it is important to check the amount of brake pedal free play to make certain that the adjustment has not reduced the amount of free play, which will result in brake drag.

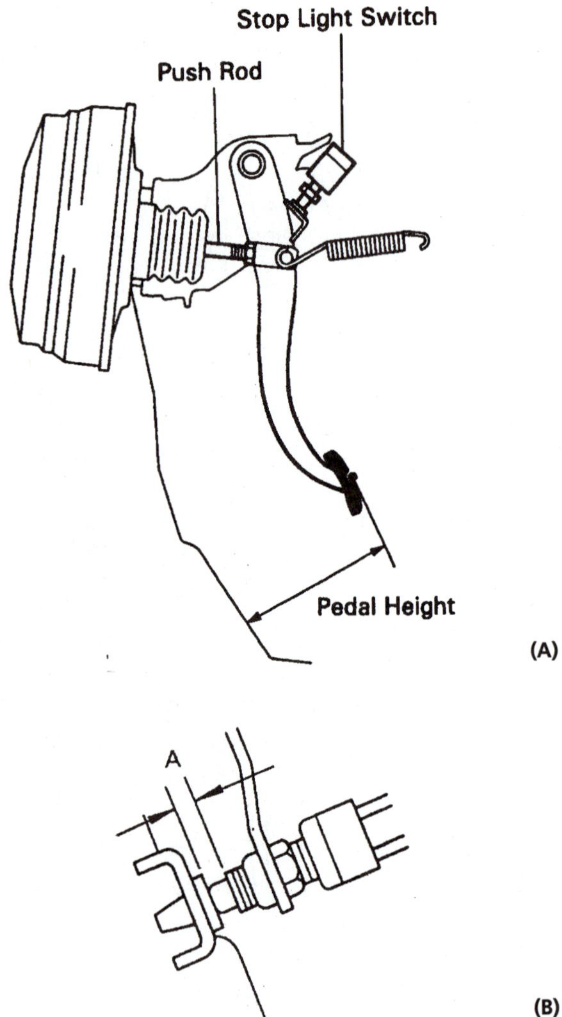

Figure 12-133 Many stoplight switches are mounted on a racket near the brake pedal (A). This switch has a specified clearance (A) between the pedal and end of the switch body (B).

12.11 Practice Diagnosis

CASE 1: The ten-year-old car was brought in with a complaint of brake drag; the right front brake does not release completely. This problem was easy to confirm because the wheel was very hot to touch. Could this be a hydraulic system problem? If so, how can you determine this? After that check, what should you do next to locate the exact cause?

CASE 2: The delivery van is driven through hilly areas, and the driver is complaining of occasional brake pedal fade. It gets bad enough that he can barely stop, and then he has to let the van sit for awhile for the pedal to come back. What is happening? What should you do to find the cause? What can you do that would probably help this vehicle?

CASE 3: The car is fifteen years old and has 140,000 miles on the odometer. The complaint is a very low brake pedal. On checking the master cylinder, you find one of the reservoirs empty and the other overflowing. What is probably wrong? What should you do to fix it? What should you do or least recommend before returning this car to the customer (besides collecting the money for the repair)?

CASE 4: The five-year-old car is brought in with a complaint of a grabby right rear brake. The customer says he doesn't know what it can be because his neighbor helped him replace the brake shoes about 1,000 miles ago. When you pull the brake drum you find fairly new lining that is wet with brake fluid from a wheel cylinder leak. What went wrong? What should you to do correct this problem?

Terms to Know

back-bleed
bench-bleeding
bleeding
cross-bleed
flushing
gravity bleeding
manual bleeding
pressure bleeding
reverse-bleed
vacuum bleeding

Review Questions

1. Two technicians are discussing brake service. Technician A says that air in the hydraulic system will cause a low but firm brake pedal. Technician B says that air is removed by bleeding the brakes. Who is right
   a. A only c. both A and B
   b. B only d. neither A nor B
2. During service, a master cylinder can be cleaned using
   a. gasoline. c. denatured alcohol.
   b. solvent. d. any of the above.
3. Technician A says that using an open-end wrench to loosen a tube nut can ruin the nut. Technician B says that during replacement, it is always a good idea to thread the nut into the connection several turns with your fingers. Who is right?
   a. A only c. both A and B
   b. B only d. neither A nor B
4. Technician A says that when you are rebuilding an aluminum wheel cylinder or master cylinder you should be very careful as you hone out the bore if it is corroded. Technician B says that the rubber cups can be damaged if a cylinder bore is too large. Who is right?
   a. A only c. both A and B
   b. B only d. neither A nor B
5. The bore size of a hydraulic brake cylinder is usually checked using
   a. a piston from the cylinder and a feeler gauge.
   b. an inside micrometer.
   c. a special plug gauge.
   d. a feeler gauge and a rubber cup.
6. If a secondary piston stop bolt is used in a master cylinder, it will enter the bore from
   a. the bottom.
   b. inside the reservoir.
   c. the side.
   d. any of the above.
7. Technician A says that removal of a stuck master cylinder piston requires the use of a special puller. Technician B says that a residual check valve can be removed from a tandem master cylinder with the aid of a self-tapping screw. Who is right?
   a. A only c. both A and B
   b. B only d. neither A nor B
8. A plastic reservoir is removed from the master cylinder body by
   a. removing the hold-down bolts.
   b. removing the retainer pin and prying it off.
   c. using air pressure to carefully blow it off.
   d. any of the above depending on the master cylinder.
9. Technician A says that the lips of all the sealing cups in a master cylinder face away from the pushrod. Technician B says that the rubber cups should be lubricated as a master cylinder is being assembled. Who is correct?
   a. A only c. both A and B
   b. B only d. neither A nor B
10. Technician A says that it is a wise practice to always bench-bleed a master cylinder before installing it. Technician B says that brake pedal free travel should always be checked after installing a master cylinder. Who is right?
    a. A only c. both A and B
    b. B only d. neither A nor B
11. A wheel cylinder can be rebuilt while it is mounted on the backing plate unless there are
    a. piston stops.
    b. deep pits in the cylinder bore.
    c. raised areas on the backing plate that prevent dust boot removal.
    d. any of the above.
12. The first step in rebuilding a wheel cylinder should be to
    a. remove it from the backing plate and hone the bore.
    b. loosen the bleeder screw.
    c. remove the boots, cups, and pistons.
    d. any of the above.
13. Technician A says that a steel caliper piston must be replaced if the chrome plating is flaking or coming loose. Technician B says that a cracked phenolic piston must never be reused. Who is right?
    a. A only c. both A and B
    b. B only d. neither A nor B
14. A caliper piston can be removed from a caliper bore with the aid of
    a. shop air pressure.
    b. the car's hydraulic brake pressure.
    c. a special puller.
    d. any of the above.

15. Technician A says that a major difference between the right and left calipers is the location of the bleeder screw. Technician B says that a caliper with a bleeder screw broken off in it must be replaced. Who is right?
    a. A only c. both A and B
    b. B only d. neither A nor B

16. Technician A says that after a brake job is completed, the brakes should be bled to remove all the air from the hydraulic fluid. Technician A says that all the old brake fluid should be bled out of the system. Who is right?
    a. A only c. both A and B
    b. B only d. neither A nor B

17. During brake-bleeding operations, fluid flow is generated by
    a. pressure from a pressure bleeder.
    b. movement of the brake pedal.
    c. a vacuum at the bleeder screw.
    d. any of the above.

18. Technician A says that the normal bleeding sequence of most cars built in the 1970s is right rear, left front, left rear, right front. Technician B says that this is the correct sequence for some cars with a diagonal split hydraulic system. Who is right?
    a. A only c. both A and B
    b. B only d. neither A nor B

19. Technician A says that the master cylinder can be damaged by improper pedal pumping during manual bleeding operations. Technician B says that a bleeder adapter should not be clamped onto the top of a master cylinder with a plastic reservoir. Who is right?
    a. A only c. both A and B
    b. B only d. neither A nor B

20. Technician A says that the only way to solve a spongy brake problem is to continue bleeding until the pedal becomes firm. Technician B says that it is normal on all modern systems for a small spurt of fluid to emit from a caliper bleeder screw as it is being loosened, after the brake pedal has been applied and released. Who is right?
    a. A only c. both A and B
    b. B only d. neither A nor B

Power Booster Service

Objectives

After completing this chapter, you should:

- ❑ Be able to test power booster operation using the brake pedal.
- ❑ Be able to check the vacuum supply using a vacuum gauge.
- ❑ Be able to check a vacuum booster and check valve for leaks and proper operation.
- ❑ Be able to check a Hydro-boost system for leaks and proper operation.
- ❑ Be able to check a Powermaster system for leaks and proper operation.
- ❑ Be able to remove and replace a power booster.
- ❑ Be able to disassemble and reassemble a power booster.
- ❑ Be able to perform the ASE Tasks for a power assist unit diagnosis and repair (See Appendix 1).

13.1 Introduction

In most cases, power booster service consists of inspection and replacement of the vacuum hose or check valve and removal and replacement of the booster. A boost—vacuum, hydraulic (Hydro-boost), and electrohydraulic (Powermaster)—can be dismantled and repaired. In the average repair shop, many technicians prefer to install a new or rebuilt replacement unit because of the special tools, skills, and parts that are required when rebuilding a unit.

Most booster service operations are undertaken because of customer complaints concerning poor or unsafe brake operation. By following an inspection procedure or a troubleshooting guide (both given in Chapter 9) or by referring to past experience, the technician will be led to the probability of faulty booster operation. In some cases booster faults are easy to see. For example, power-steering fluid leaking from a Hydro-boost unit is an obvious indication that repair or replacement is necessary. For other problems, such as a faulty control valve, other checks should be made to confirm that the booster is faulty (Figures 13-1 and 13-2).

Occasionally a faulty booster can cause brake drag. This results in rapid lining wear, especially with disc

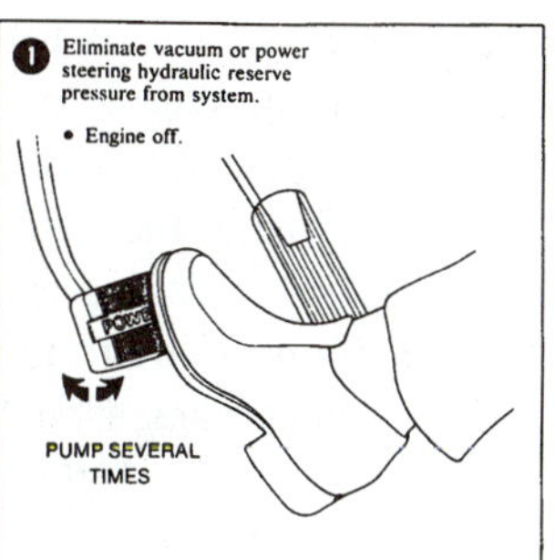

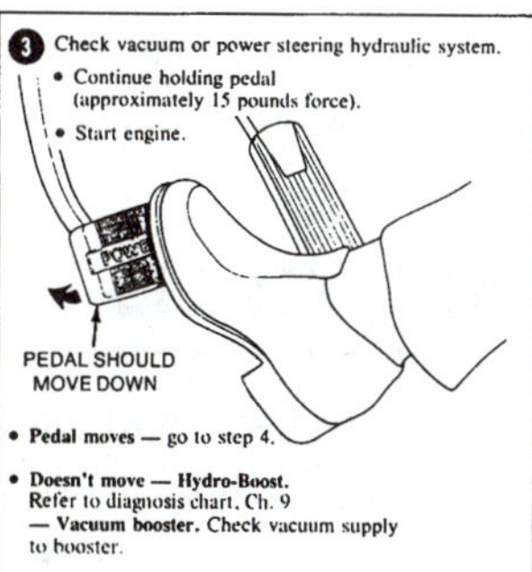

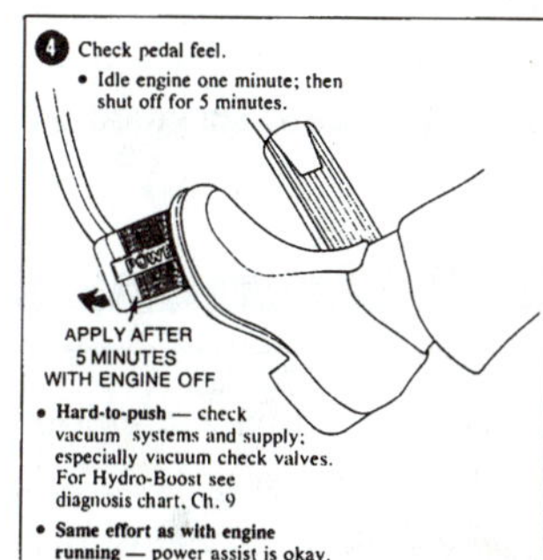

Figure 13-1 There are two basic tests for power system operation. Steps (1) and (2) make sure the booster operates properly, and Steps (3) and (4) make sure it has enough reserve to function with the engine off. (Courtesy of Ford Motor Company)

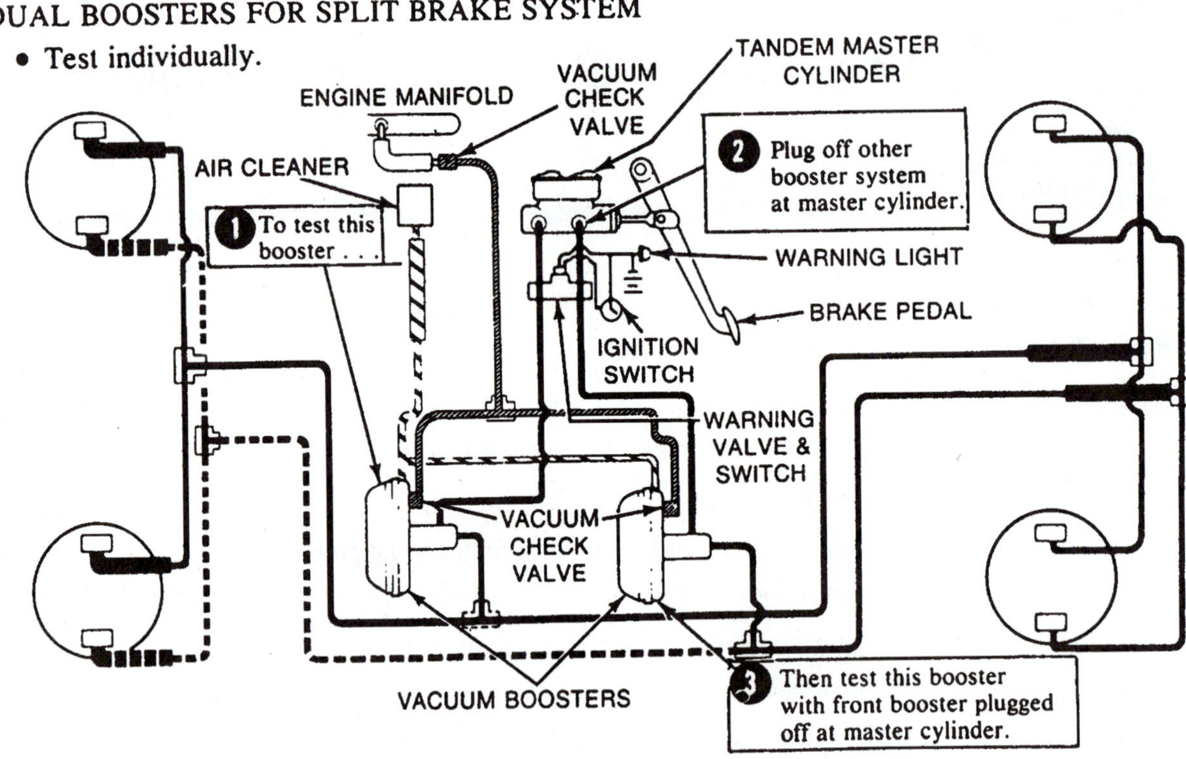

Figure 13-2 A remote booster is tested in the same manner as an integral booster. Some light truck systems use dual boosters, and these units are checked separately. (Courtesy of Ford Motor Company)

brakes. A quick check for booster-caused drag is as follows: Raise both front wheels and (with the engine off) operate the brake pedal enough times to use up the vacuum or hydraulic pressure reserve. Rotate the front wheels to check them for drag. Next start the engine and recheck for drag. An increase in the amount of drag indicates that the booster is trying to apply the brakes. Check to make sure there is free travel at the brake pedal. If there is no free travel, adjust it; if there is free travel, the booster should be replaced or rebuilt.

On Powermaster units, the booster and master cylinder sections are serviced as a single unit.

13.2 Vacuum Supply Tests

A vacuum booster cannot develop full power without a good, strong supply of vacuum. This vacuum can be checked using a vacuum gauge connected to the intake manifold as close to the booster connection as possible (Figure 13-3). With the engine running at a fast idle, there should be about 17 to 20 inches of vacuum with a minimum of 14 inches of vacuum. This reading will change as the throttle is opened and closed. With the vacuum supply clamped off, a vacuum loss greater than 1"Hg in 15 sec-

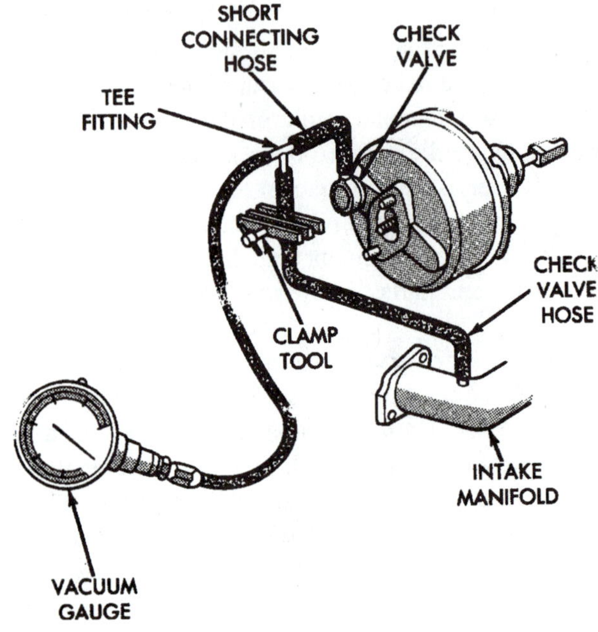

Figure 13-3 A vacuum gauge can be used to check the vacuum supply to the booster, and by adding the clamp, you can check whether it holds the vacuum. A drop of more than 1 inch in 15 seconds indicates a leak. (Courtesy of Chrysler Corporation)

onds indicates a faulty check valve and internal booster leak. Some technicians prefer to take vacuum readings while the car is being driven at cruising speeds. If the vacuum supply readings are low, check for loose or missing manifold vacuum hoses or an engine that needs a tune-up or should be rebuilt.

The booster vacuum supply hose should be inspected visually for cracks or breaks. It can be checked for restrictions by disconnecting it from the booster with the engine running. A substantial flow of air into the hose should occur, and a vacuum leak this large will cause most engines to stall. A faulty hose should be replaced. Also check the inside of the hose to see if it is wet. Signs of brake fluid indicate a faulty secondary cup (primary piston) in the master cylinder. If it is wet from engine oil, a faulty booster vacuum check valve is indicated. A faulty check valve can let crankcase fumes—passing through the positive crankcase ventilation (PCV) valve—enter the booster after engine shutoff. These fumes can ruin the rubber valve seals and diaphragm in the booster.

The vacuum check valve can be checked by prying it out of the grommet in the booster and trying to blow through it (Figure 13-4). You should be able to blow through it in the booster-toward-manifold direction but not in the opposite direction (Figure 13-5). A faulty check valve or grommet should be replaced. It is a good practice to wet the grommet and check valve with water for lubricant as they are being installed.

Internal air leaks in the booster can prevent proper booster operation. A booster with internal leaks should be replaced or rebuilt. An **internal leakage check**, also called an air-tightness test, is made by running the engine for one or two minutes and then shutting it off. Then apply the brake with normal pedal pressure and hold it for at least thirty seconds. If the pedal remains steady, the booster is probably good. If the pedal slowly rises, the booster has an internal leak.

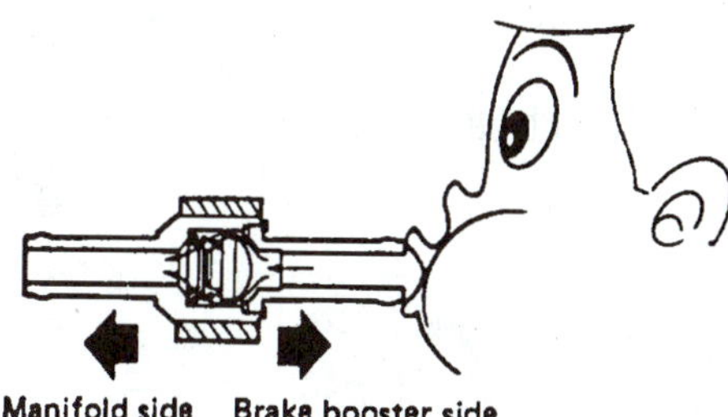

Figure 13-5 A check valve can be tested by trying to blow through it in both directions. It should allow flow only toward the manifold. (Courtesy of Nissan North America, Inc.)

Booster internal leaks can also be checked by fabricating a simple tool as shown in Figure 13-6. Choose a glass jar that has a tight-fitting lid and attach two 3/8-inch (9.5 mm) OD tubes to the lid. One tube should be long enough to reach almost to the bottom of the jar, and the other should end just inside the lid. Fill the jar about half full with water and connect the short tube to the booster supply hose from the intake manifold. Then use a short piece of hose to connect the long tube to the booster inlet. Start the engine and watch for air bubbles in the jar as you operate the brake pedal a few times. Airflow from the booster to the manifold will cause air bubbles to pass through the water in the jar. As the brake pedal is released, there should be a large flow of bubbles, but after a moment this flow should stop. With the brake pedal released or applied, there should be no flow. A constant

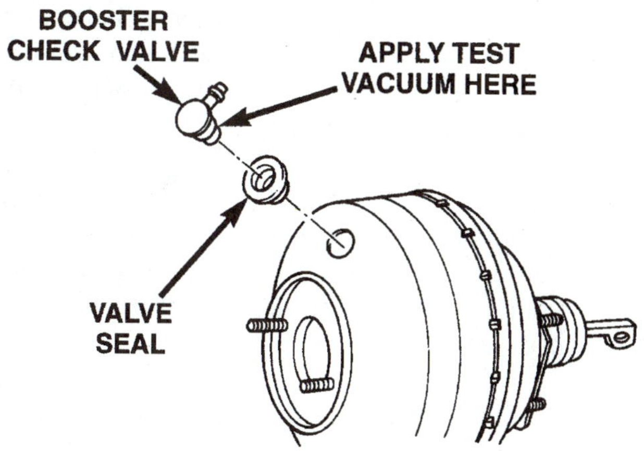

Figure 13-4 A booster check valve can be pried out for replacement of the valve or seal/grommet or to check the valve. Wet the valve with water to make removal or installation easier. (Courtesy of Chrysler Corporation)

Figure 13-6 This simple booster leak-testing tool is made by fastening two pieces of 3/8-inch tubing to the lid of a small container; note that one of the tubes extends almost to the bottom of the container.

flow of bubbles with the brake applied indicates a leaky piston diaphragm. A constant flow with the brakes released indicates a leaky control valve (Figure 13-7).

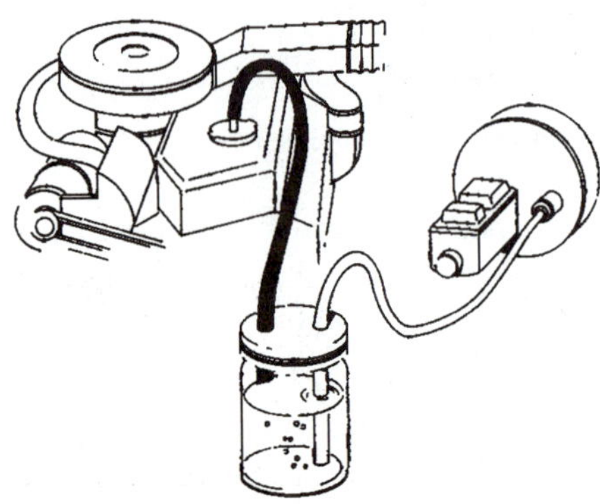

Figure 13-7 The leak tester is partially filled with water and connected between the booster and intake manifold; then the engine is started. A constant stream of air bubbles with the pedal released indicates a leaky valve in the booster, and a stream of bubbles with the pedal applied indicates a leaky booster diaphragm.

A vacuum booster can also be checked for internal leakage with a vacuum pump. Remove the vacuum hose from the check valve and connect the vacuum pump directly to the check valve or inlet fitting (Figure 13-8). With the pedal released, you should be able to draw a 17- to 20-inch vacuum, and this vacuum reading should hold steady for several minutes. A leak in the control valve or booster chamber is indicated if the drops. Next, apply the brakes with moderate pressure on the pedal. An immediate drop in the reading should occur as the pedal moves. Draw the vacuum back to 17 to 20 inches and observe the reading to make sure it does not drop more than 2 inches in the next thirty seconds. A leaky diaphragm, control valve, or vacuum chamber is indicated if it does drop.

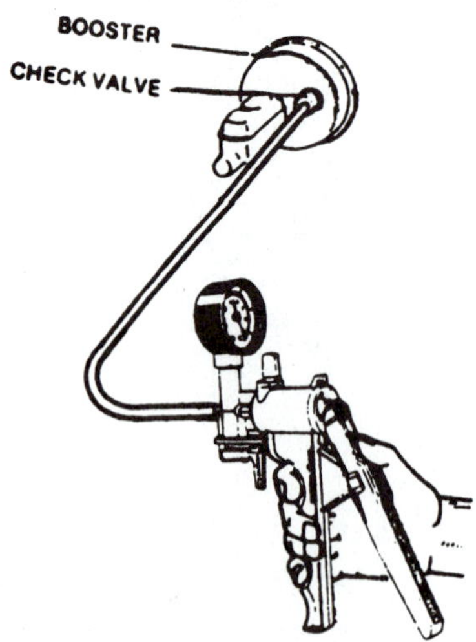

Figure 13-8 If a vacuum pump is connected to a booster, it should be possible to draw a 17- to 20-inch vacuum, and this vacuum should hold steady. Repeating this check with the pedal applied should give the same readings. (Courtesy of Mityvac/Prism)

13.3 Booster Replacement

A faulty booster can be removed for rebuilding or replacement with a new or rebuilt unit. This can be a relatively easy operation depending on the location of the bolts and nuts securing the booster to the car's bulkhead (Figure 13-9).

NOTE: On some vehicles, the brake pedal will drop and activate the brake lights when the booster is removed, and the extended operation of the center brake light can overheat the rear window or all of the lights can discharge a battery. Check the light operation after booster removal, and if the lights are lit, prop up the pedal to turn them off.

To remove a power booster:

1. On Hydro-boost and Powermaster units, first remove the pressure in the accumulator. With the engine off, apply and release the brake several times until you feel no change in the hardness of the pedal. On Powermaster units, ten to twenty pedal applications are recommended.
2. Disconnect the master cylinder from the booster. On many cars, the brake lines are long and flexible enough to allow the master cylinder to be moved away from the booster without disconnecting them. If not, remove the master cylinder as described in Chapter 12.

   NOTE: On some boosters, the master cylinder pushrod will come free. Test this by trying to pull it out, and if you can, tape the pushrod onto the front of the windshield. This will keep it from becoming lost or forgotten.

On Powermaster units, the brake lines must be disconnected because the booster and the master cylinder are combined (Figure 13-10).

3. Disconnect the booster power supply:
   a. On a vacuum booster, disconnect the vacuum line.

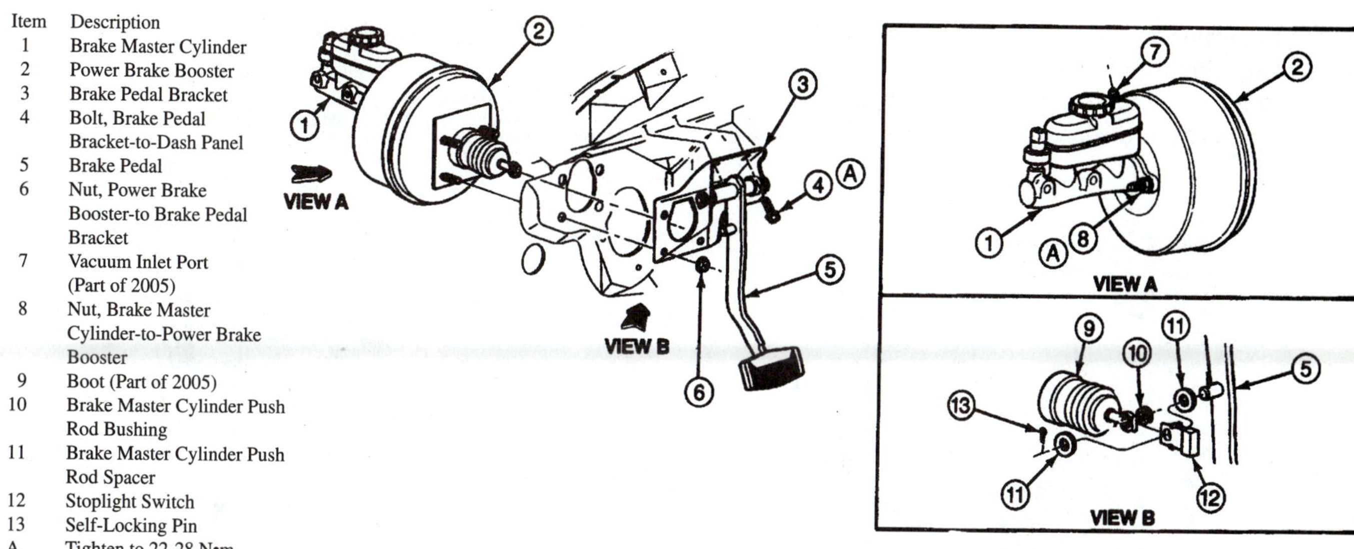

Figure 13-9 A vacuum booster can be removed by disconnecting the pushrod (view B) and the mounting nuts (6). Usually the master cylinder is disconnected (view A) and moved aside or removed. (Courtesy of Ford Motor Company)

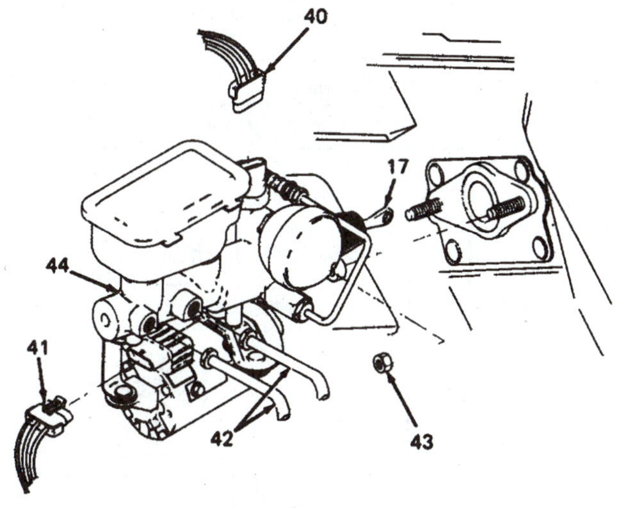

17. PUSHROD
40. ELECTRICAL CONNECTOR
41. ELECTRICAL CONNECTOR
42. BRAKE PIPE
43. NUT
44. POWERMASTER UNIT

Figure 13-10 A Powermaster unit can be removed after disconnecting the pushrod, mounting nuts, and electrical and hydraulic connections. (Courtesy of General Motors Corporation, Service Technology Group)

b. On a Hydro-boost unit, disconnect the power-steering pump and steering gear fluid lines (Figure 13-11).

c. On a Powermaster unit, disconnect the electric connectors.

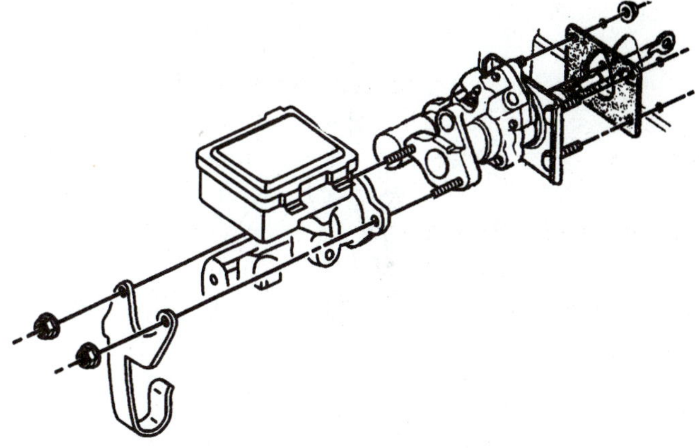

Figure 13-11 A Hydro-boost unit is removed by disconnecting the master cylinder, and pushrod, and removing the mounting nuts. (Courtesy of General Motors Corporation, Service Technology Group)

4. Disconnect the pushrod from the brake pedal.
5. Remove the nuts or bolts securing the booster to the bulkhead and remove the booster.

Booster installation is essentially the reverse of the removal procedure. During installation, check the master cylinder **pushrod adjustment**. A pushrod that is too long can position the primary piston too deep in the cylinder bore. The primary cup will then cover the compensating port and cause brake drag. A short pushrod can cause a low brake pedal. Three different procedures can be used to check this adjustment: the gauge method, the air method, and the fluid swirl method. Many boosters use an adjustable pushrod which can be lengthened or shortened as necessary (Figure 13-12).

The gauge method is the quickest. A gauge can be easily shop-made from cardboard or thin sheet metal if the dimensions are available. It is normally a two-step, "go–no go" gauge. It is placed in position with the pushrod, touching the short step of the gauge but not the long one (Figure 13-13).

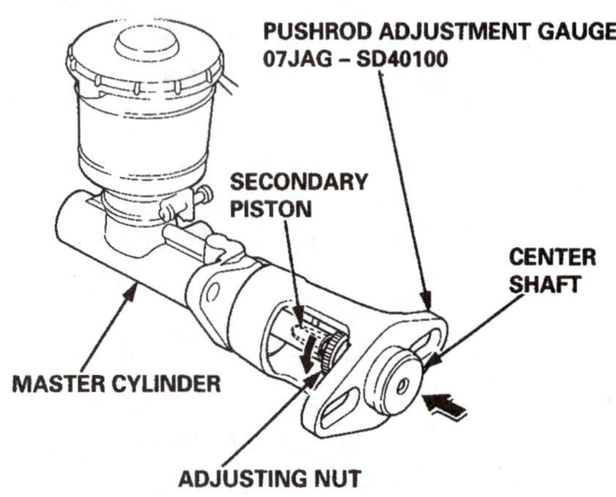

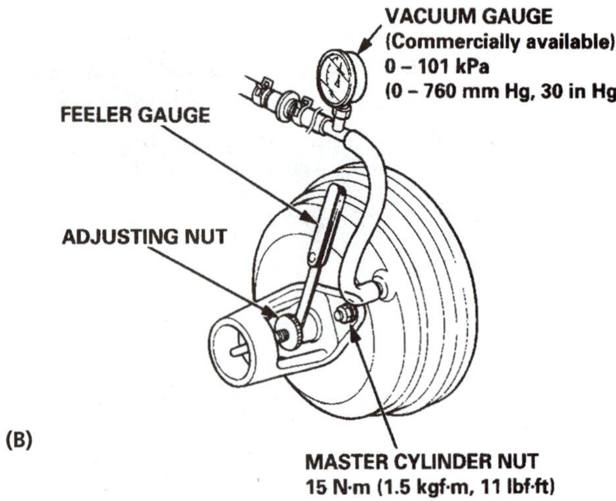

Figure 13-12 This gauge is adjusted to fit the master cylinder piston (A) and then placed onto the booster to check the pushrod adjustment (B). Note the engine is running during Step (B). (Courtesy of Ford Motor Company)

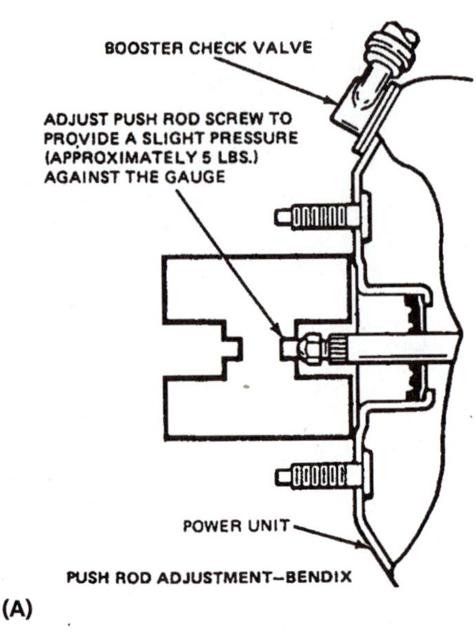

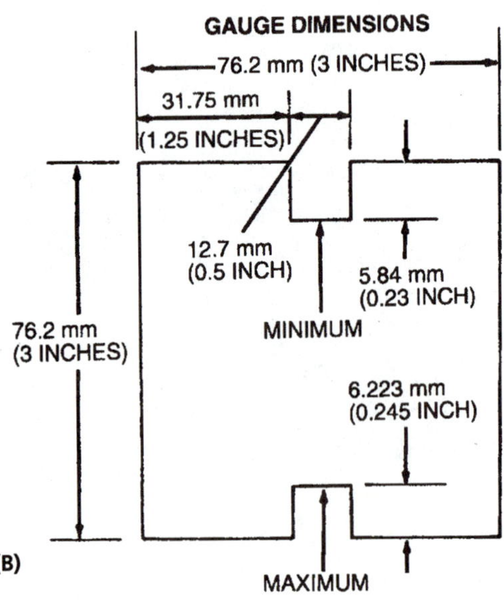

Figure 13-13 The booster pushrod can be checked using a gauge (A) to see if it is the correct length. Some manufacturers provide dimensions for making a gauge (B). (Courtesy of American Honda Motor Co., Inc.)

For the fluid swirl method, after the booster and master cylinder are mounted, have an assistant apply the brake pedal as you observe the fluid in the primary reservoir section. The reservoir should be about half full of fluid, and a clean plastic film should be placed over the reservoir to contain the fluid. A slight swirl should be seen in the fluid as the brakes are applied and also during release. No swirl on application indicates that the pushrod is too long. A larger-than-normal swirl on application indicates that the pushrod is too short. If this occurs, loosen the master cylinder mounting bolts about one-fourth of an inch and repeat this test. If a normal swirl is now present, the pushrod is too long.

With the air method, the master cylinder is mounted on the booster, and the brake line is disconnected from the primary outlet port. Then clean, low-pressure, compressed air is blown into the outlet port. If the air passes through the compensating port into the reservoir, the pushrod is not too long. Note that this method will require bleeding of the primary section of the master cylinder.

To replace a booster:

1. Place the booster in position, install the nuts or bolts securing it, and tighten them to the correct torque.

   NOTE: A Powermaster unit must be bench-bled in the master cylinder section before installation.

2. Reconnect the pushrod to the brake pedal.
3. Remount the master cylinder on the booster and tighten the nuts or bolts to the correct torque.
4. Connect the booster's power supply: *On vacuum boosters*, reconnect the vacuum hose. *On Hydro-boost units*, reconnect the lines to the power-steering pump and gear, tighten them to the correct torque, and bleed the air out of the lines. To bleed the lines:
   a. Fill the power-steering pump reservoir.
   b. Crank the engine for several seconds but do not start it.
   c. Check the power-steering reservoir and refill it as necessary.
   d. Start the engine and slowly turn the steering wheel from stop to stop twice. Do not hold the steering wheel against the stops.
   e. Stop the engine and apply and release the brake pedal several times to bleed the pressure out of the accumulator.
   f. Check the reservoir and refill it if necessary.

   On Powermaster units, reconnect the electric connector, fill all three reservoir sections with brake fluid to the correct level, and turn on the ignition. The pump should run and shut off within twenty seconds. If it is still running after twenty seconds, shut off the ignition and refer to the diagnosis chart for the solution to this problem. Powermaster pumping pressures are tested using a pressure gauge as shown in Figure 13-14.
5. Check the master cylinder reservoir and refill it if necessary. Start the engine and check the booster operation. Note that if the master cylinder has been completely removed, it will be necessary to bleed the brake hydraulic system.

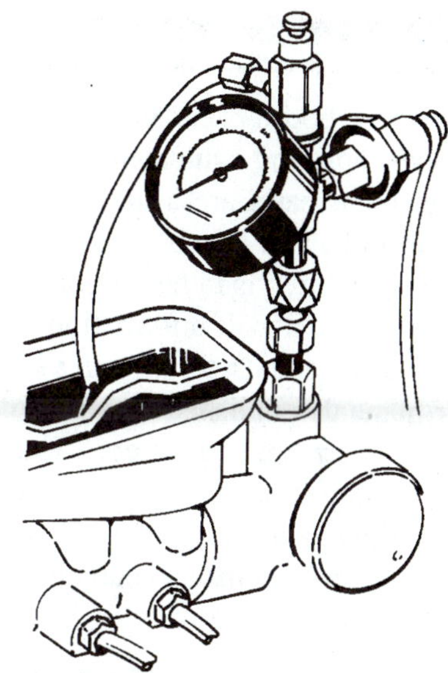

Figure 13-14 A pressure gauge has been installed on this Powermaster unit to measure pump output pressure. Abnormal pressure can indicate a faulty pump, switch, accumulator, or booster. (Courtesy of General Motors Corporation, Service Technology Group)

Note that on Powermaster units, the pump section of the reservoir will have the correct fluid level only when the accumulator is discharged. Filling the reservoir after the accumulator is charged will cause an overflow and spillage when the accumulator discharges (Figure 13-15).

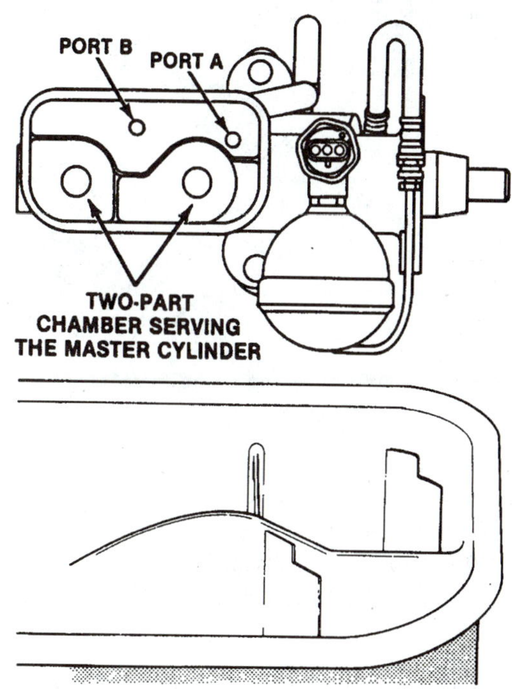

Figure 13-15 With the accumulator discharged, the Powermaster fluid level should be at the step in the reservoir baffle. (Courtesy of General Motors Corporation, Service Technology Group)

13.4 Booster Service

As mentioned earlier, boosters can be rebuilt in the repair shop if the proper tools and equipment, service information, and replacement parts are available. Most service shops prefer to install a new or rebuilt unit. In most cases, the special tooling required is rather minor, and the larger tools can be shop made. Repair information is available in some factory service manuals, technician's service manuals, and repair manuals published by some of the major aftermarket brake-part manufacturers. These manuals provide a step-by-step procedure for making the necessary repairs and adjustments. Repair parts can be obtained from some new-car dealerships and parts houses.

13.4.1 Vacuum Booster Repair

Most vacuum boosters are held together by interlocking tabs between the front and rear housings, and they can be separated by twisting one housing relative to the other (Figure 13-16). Separation is made difficult because of the drag of the rubber diaphragm that is clamped between the two sections. Another potential problem is caused by the return spring, which tries to spread the two housing sections apart. Repair of a booster will include diaphragm replacement, control valve cleaning or replacement, and air cleaner or silencer cleaning or replacement. During reassembly, the parts should be properly lubricated and the pushrod adjustment checked.

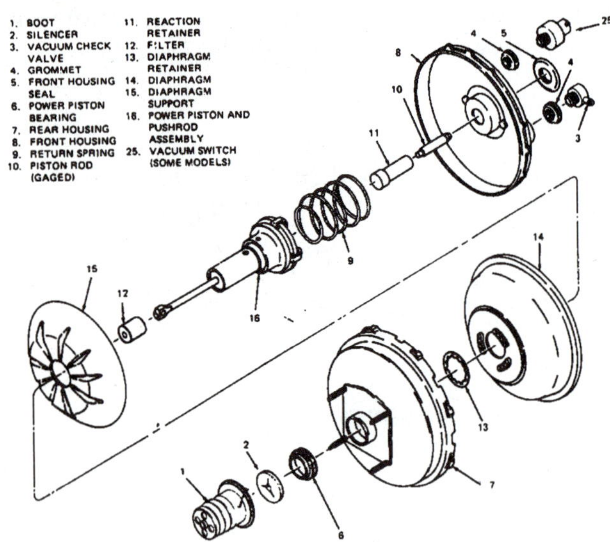

Figure 13-16 An exploded view of a vacuum booster. Note the tabs on the front housing (8) and the sockets in the rear housing (7) with which the tabs interlock. (Courtesy of General Motors Corporation, Service Technology Group)

The following method is very general. Be sure to follow the exact procedure recommended by the manufacturer. To service a vacuum booster:

1. Remove the vacuum check valve, and scribe a mark on the front and rear housing so you can properly align them during disassembly.
2. Place the booster in the disassembly fixture with the booster mounting studs entering the holes in the fixture and the long bar or wrench over the master cylinder mounting studs.

   Adjust the fixture to apply a slight downward pressure and twist the upper housing to unlock the two housings. Adjust the fixture to allow the return spring to extend and separate the parts of the booster (Figure 13-17).

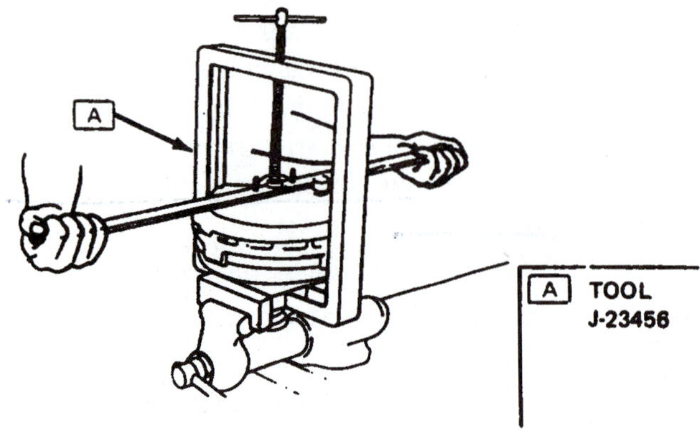

Figure 13-17 This booster is attached to a fixture while the tool is used to rotate the front housing and unlock the tabs. The clamp portion keeps the internal spring compressed during this operation. (Courtesy of General Motors Corporation, Service Technology Group)

3. Remove the power piston bearing, return spring, and piston group from the housing (Figure 13-18).

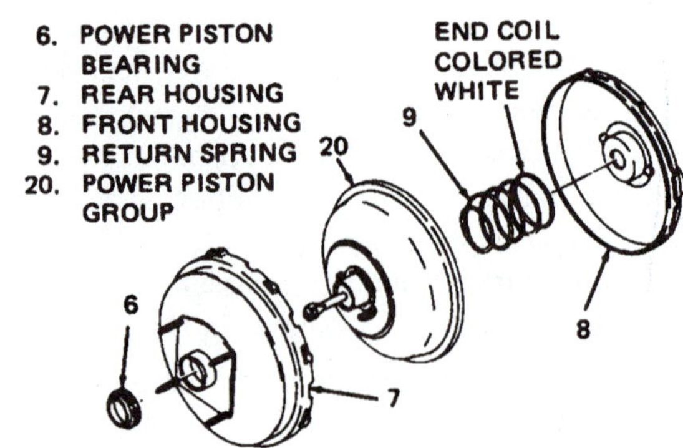

Figure 13-18 A partially disassembled vacuum booster. (Courtesy of General Motors Corporation, Service Technology Group)

4. Disassemble the piston and valve assembly to remove the diaphragm (Figure 13-19).

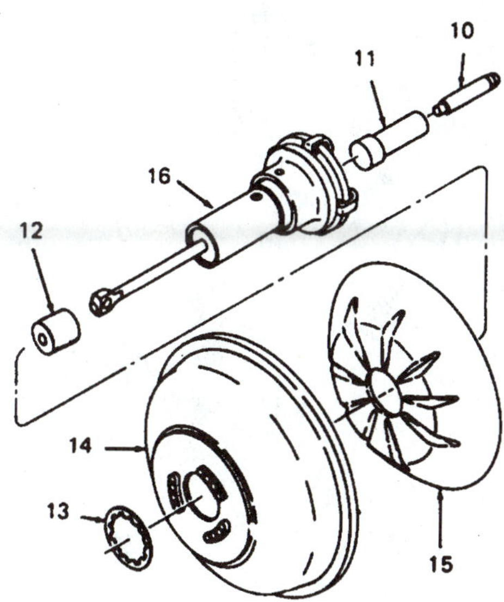

10. PISTON ROD
11. REACTION RETAINER
12. FILTER
13. DIAPHRAGM RETAINER
14. DIAPHRAGM
15. DIAPHRAGM SUPPORT
16. POWER PISTON AND PUSHROD ASSEMBLY

Figure 13-19 This power piston group has been disassembled, allowing diaphragm replacement. (Courtesy of General Motors Corporation, Service Technology Group)

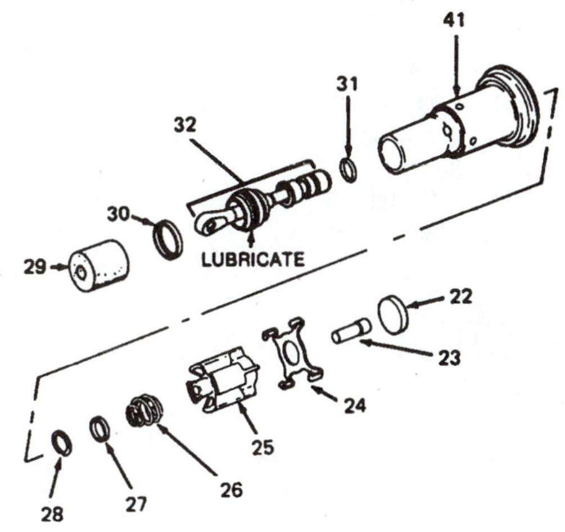

22. REACTION DISC
23. REACTION PISTON
24. REACTION BODY RETAINER
25. REACTION BODY
26. AIR VALVE SPRING
27. REACTION BUMPER
28. RETAINING RING
29. FILTER
30. RETAINER
31. O-RING
32. AIR VALVE PUSHROD ASSEMBLY
41. POWER PISTON

Figure 13-20 A disassembled vacuum booster valve assembly. (Courtesy of General Motors Corporation, Service Technology Group)

5. Disassemble the valve assembly (Figure 13-20).
6. Clean the various parts in alcohol and air-dry them. Rust deposits can be cleaned from the inside of the housings with fine sandpaper. Inspect the parts for damage. There should be no cuts or tears in any of the rubber components.
7. Lubricate the valve parts as required, and reassemble them.
8. Lubricate the components as required, and reassemble the piston and valve assembly.
9. Place the piston assembly and return spring in the housings, aligning the scribed marks. Then place them in the assembly fixture and adjust the fixture to hold the two halves together.
10. Twist the housings so the tabs interlock and stake two housing tabs to lock the assembly (Figure 13-21 here).
11. Check the pushrod adjustment.

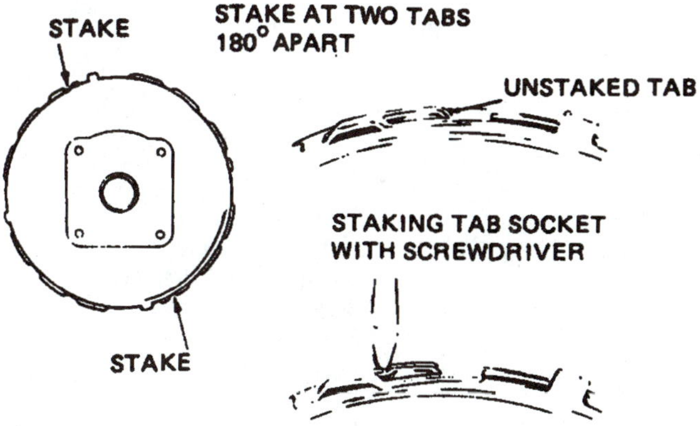

Figure 13-21 After a vacuum booster is reassembled, two tab sockets should be restaked to lock the chambers together. (Courtesy of General Motors Corporation, Service Technology Group)

13.4.2 Hydro-boost Repair

Normal Hydro-boost service includes disassembly, cleaning and inspection, and replacement of the inner seals (Figure 13-22). It is recommended that a special seal installer be used when installing the seals on the input rod. Before disassembling a Hydro-boost unit, make sure that the accumulator pressure has been released. The following method is very general. Be sure to follow the exact procedure recommended by the manufacturer (Figure 13-23).

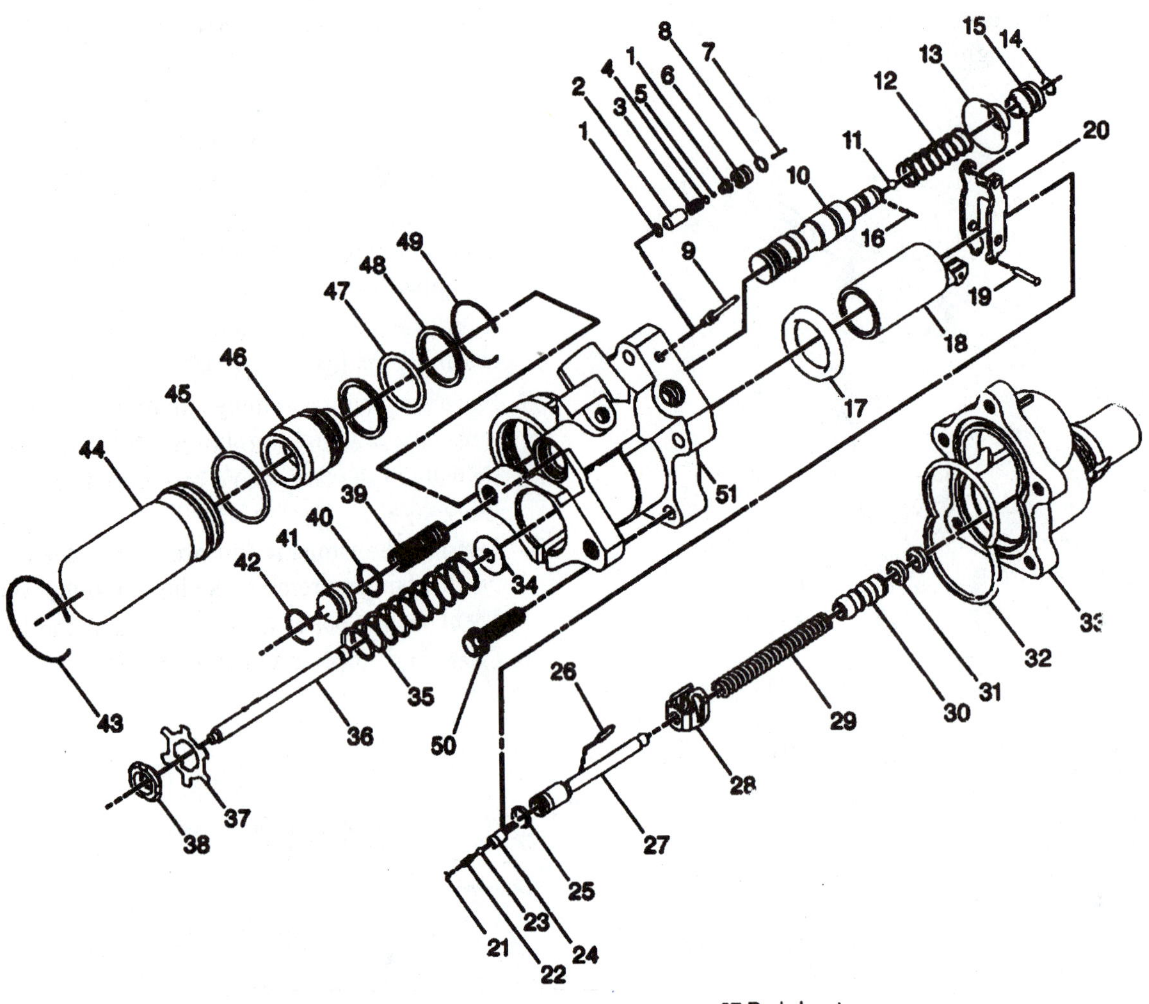

1. Seat, Check Valve
2. Body, Check Valve
3. Spring, Check Valve Relief
4. Washer, Check Valve
5. Ball, Check
6. Insert, Body
7. Plunger
8. O-ring
9. Plug, Housing
10. Valve, Spool
11. Ball, Spool Valve Check
12. Spring, Sleeve
13. Actuator
14. Ring, External Retainer
15. Sleeve, Spool
16. Pin, Spool Valve
17. Seal, Piston
18. Piston
19. Pin, Input Lever
20. Lever, Input
21. Spring, Retainer
22. Spring, Relief Valve
23. Ball, Relief Valve Check
24. Seat, Relief Valve
25. Ring, Input Rod
26. Rod and Plunger
27. Rod, Input
28. Bracket, Input Rod
29. Spring, Input Rod
30. End, Input Rod
31. Seals, Input Rod
32. Seal, Housing/Cover
33. Cover, Booster
34. Retainer, Output Push Rod
35. Spring, Piston Return
36. Push Rod, Output
37. Spring, Piston Retainer
38. Spring, Baffle Retainer
39. Spring, Spool
40. O-ring, Spool Plug
41. Plug, Spool
42. Ring, Spool Plug Retaining
43. Ring, Accumulator Retainer
44. Accumulator
45. Ring, Accumulator
46. Piston, Accumulator
47. O-ring, Accumulator Piston
48. Rings, Accumulator Piston
49. Ring, Retainer
50. Bolts, Housing/Cover
51. Housing, Booster

Figure 13-22 An exploded view of a Hydro-boost assembly. (Courtesy of General Motors Corporation, Service Technology Group)

A. INPUT SEAL LEAK	- Fluid leakage from housing cover end of booster near reaction bore. Replace input assembly kit.
B. POWER PISTON/ACCUMULATOR SEAL LEAK	- Fluid leakage from vent at front of unit near master cylinder. Replace power piston/accumulator seal kit.
C. HOUSING	- Fluid leakage between the housing and housing cover. Replace housing seal kit.
D. SPOOL VALVE SEAL	- Fluid leakage near plug area. Replace spool plug seal kit.
E. RETURN PORT FITTING SEAL	- Replace "O" Ring seal.

Figure 13-23 Areas to check for fluid leaks on a Hydro-boost unit. (Courtesy of General Motors Corporation, Service Technology Group)

To service a Hydro-boost unit:

1. Secure the unit in a vise.
2. Remove the spool valve retainer, plug, and spool valve spring. On some units, remove the spool valve sleeve (Figure 13-24).
3. Remove the power piston spring retainer if present. On Hydro-boost II units, remove the power piston spring.
4. Remove the return line fitting.
5. *On Hydro-boost* I units, remove the output rod retainer, spring, and output rod.
6. Remove the housing-to-cover bolts and separate the parts of the unit, being careful not to drop the spool valve.
7. Remove the housing-to-cover seal, detach the spool valve, and remove the input rod and seal.
   a. *On Hydro-boost II units*, remove the power piston and accumulator assembly.

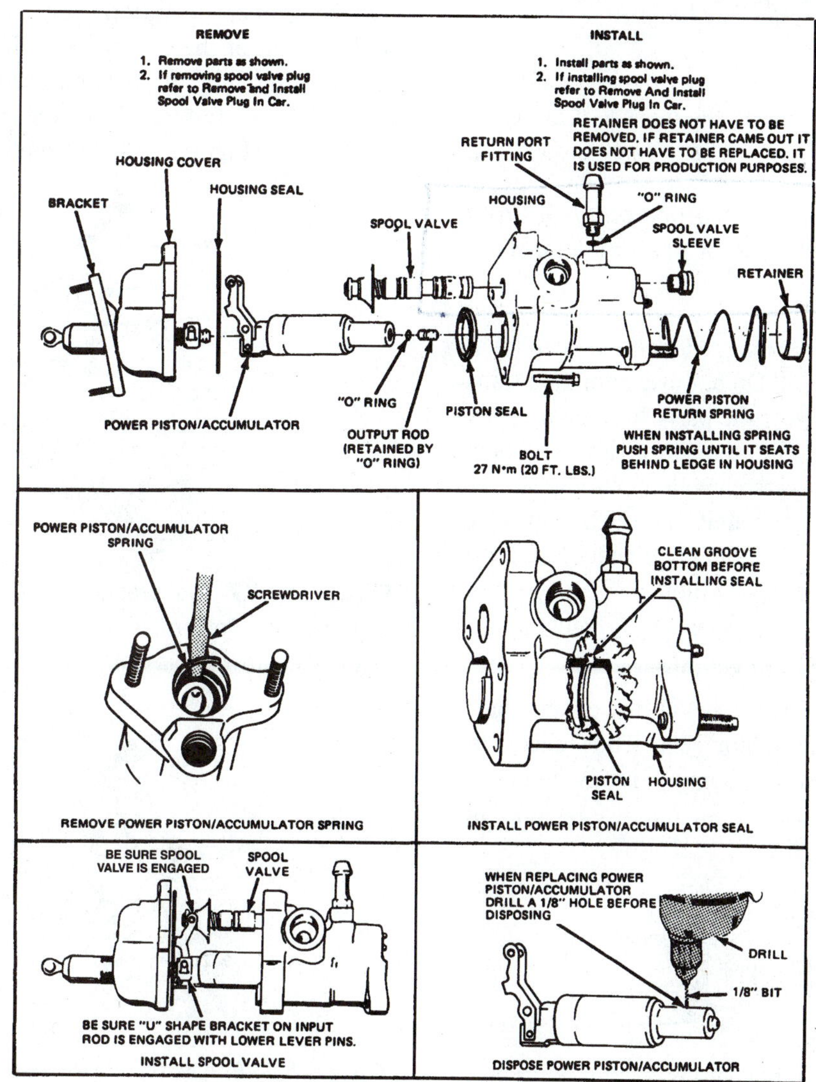

Figure 13-24 The procedure for disassembling and reassembling a Hydro-boost II unit. (Courtesy of General Motors Corporation, Service Technology Group)

b. *On Hydro-boost I units*, install a suitable clamp, remove the accumulator retainer, and loosen the clamp to remove the accumulator (Figure 13-25).

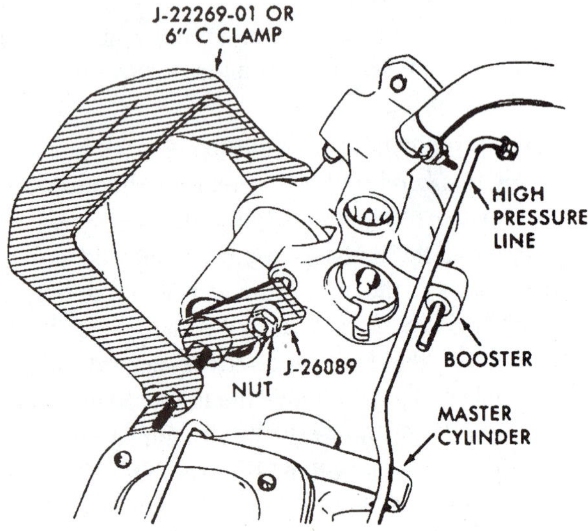

Figure 13-25 This clamp is placed over the accumulator to hold it compressed while removing the retainer. (Courtesy of General Motors Corporation, Service Technology Group)

CAUTION *Accumulators contain a strong spring or pressurized gas and can cause injury.*

8. Clean all the parts using denatured alcohol or power-steering fluid. Do not use automatic transmission fluid. Inspect the parts for damage, paying special attention to the spool valve and bore. If you locate any scratches deep enough to feel with a fingernail, the unit should be replaced. During reassembly, all the parts should be lubricated with power-steering fluid.
9. Install the return line fitting and O-ring, and tighten them to the correct torque.
10. *On Hydro-boost I units*, install the accumulator, O-ring, and accumulator seat. Then check the valve assembly and the dump valve assembly.
11. Install the input rod (with new seals) into the cover (Figure 13-26).

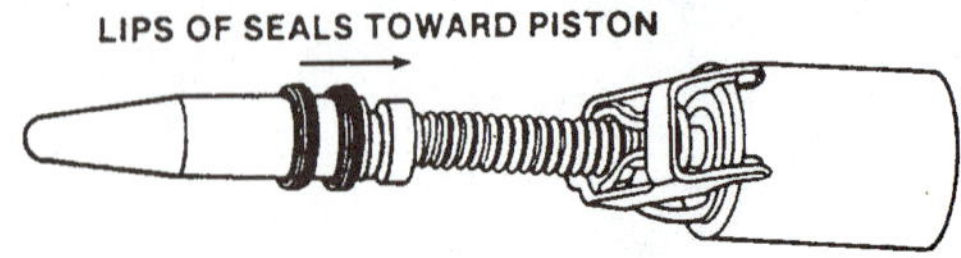

Figure 13-26 The tapered tool (left) is being used to guide the input rod seals into the proper position. (Courtesy of Bendix Brakes, by AlliedSignal)

12. Install the power piston seal, housing seal, and power piston with the spool valve attached in the housing.
13. Install the housing-to-cover bolts and tighten them to the correct torque.
14. Install the spool valve spring, spool valve, and retainer.

13.4.3 Powermaster Repair

Powermaster service includes replacement of the pressure switch, accumulator, pump assembly, or return hose, and overhaul of the unit. A unit overhaul requires removal of the assembly from the car. The other repairs can be made on the car. Before any of these operations are performed, the accumulator must be depressurized by applying and releasing the brake pedal at least ten times. Do not turn on the ignition because the pump will run and recharge the accumulator.

The on-car operations are essentially removal and replacement of the components. It is very important that new O-rings or grommets be used where necessary and that they be tightened to the correct torque during reassembly (Figures 13-27 through 13-29).

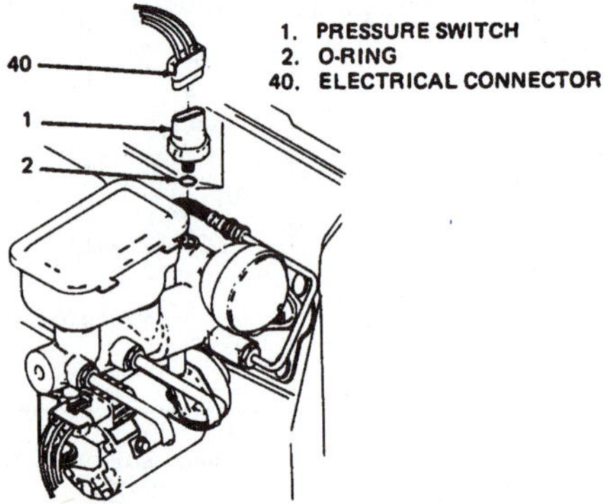

Figure 13-27 The pressure switch can be removed and replaced on a Powermaster until with it still in the car. (Courtesy of General Motors Corporation, Service Technology Group)

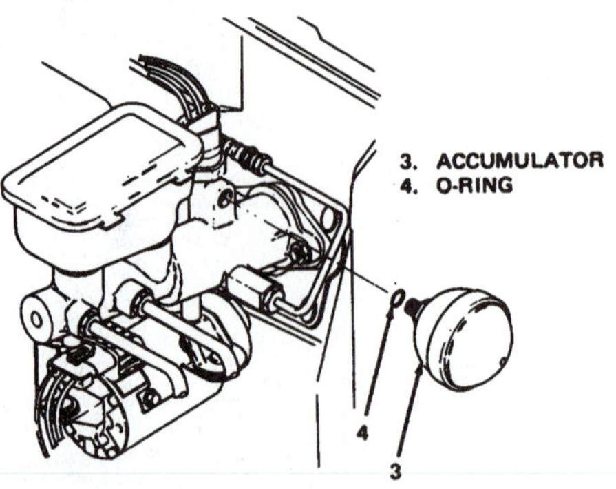

Figure 13-28 A Powermaster accumulator can be removed and replaced. (Courtesy of General Motors Corporation, Service Technology Group)

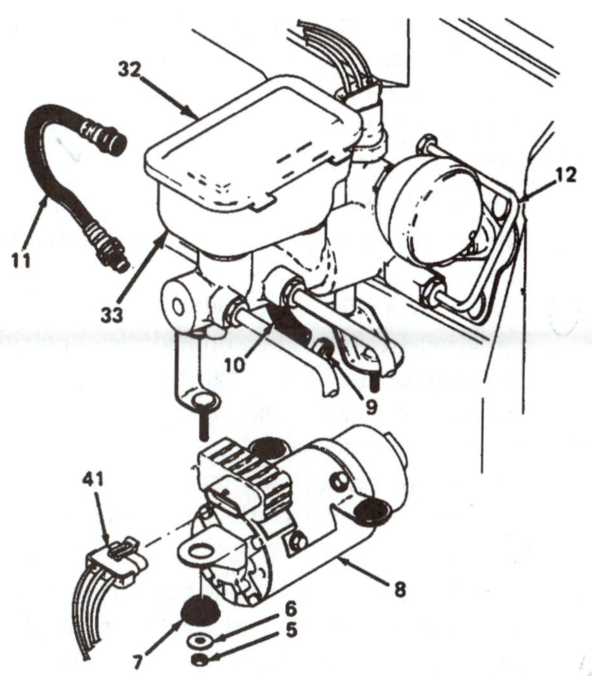

Figure 13-29 The pump and pressure hose can be removed and replaced while the unit is mounted in the car. (Courtesy of General Motors Corporation, Service Technology Group)

A unit overhaul requires disassembly, cleaning, inspection, and reassembly. The procedure is very similar to the overhauling of a master cylinder with additional components. It is very important to note the relationship and position of these additional parts (Figure 13-30). The procedure recommended by the manufacturer should be followed.

13.5 Practice Diagnosis

You are working in a brake and front-end shop and encounter these problems.

CASE 1: The customer is at a meeting in your city; he lives in another town. He is having a problem of brake drag on his 1994 Camry, and he mentions that he had to replace the vacuum brake booster the day before he left his home. You lift the car on a rack to check for drag, and you find that the wheels turn freely until the engine is started. What is probably wrong? What will you need to do to fix this problem?

CASE 2: While inspecting the brake system of a five-year-old Dodge Caravan, you find the brake pedal very hard. When you start the engine, the brake pedal drops, indicating a good booster operation, however, when you check the pedal a minute after turning the engine off, you

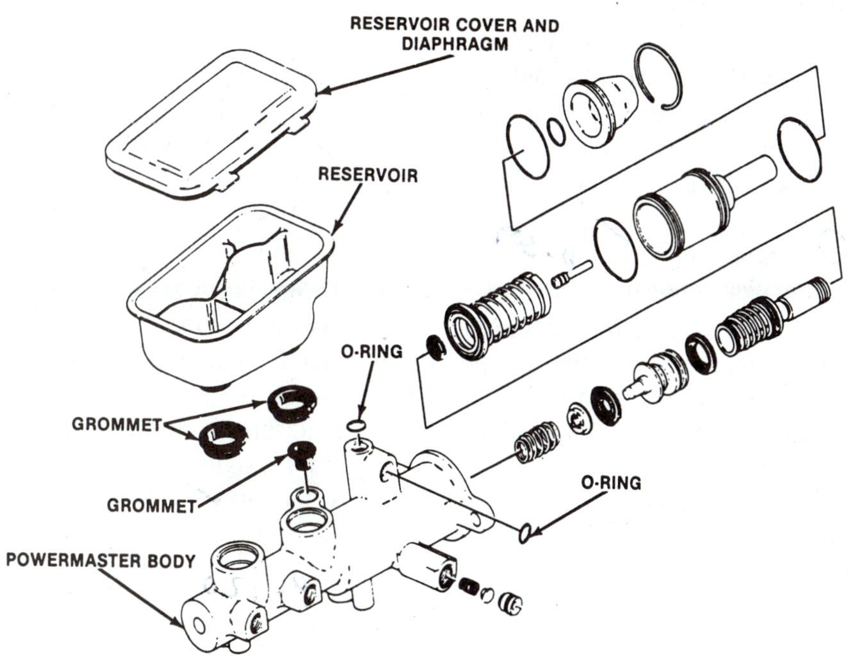

Figure 13-30 An exploded view of a Powermaster unit showing the relationship of the internal parts. (Courtesy of General Motors Corporation, Service Technology Group)

find the pedal rock hard again. What is wrong? What should you do next?

CASE 3: The customer complaints of a very hard brake pedal on his ten-year-old Chevrolet pickup. Your road test confirms the problem as you have to push very hard on the pedal of this Hydro-boost–equipped vehicle. What could be wrong? What should you do next?

Terms to Know

internal leakage check
pushrod adjustment

Review Questions

1. The standard check for power booster operation is to pump the brake pedal until the reserve is exhausted and then start the engine with the brake applied. As the engine starts, the
   a. brake pedal should rise slightly.
   b. brake pedal should fall a noticeable amount.
   c. brake warning light should come on and then off.
   d. booster should make a clicking sound.
2. Two technicians are discussing power booster checks. Technician A says that you should be able to blow through a vacuum check valve in one direction only. Technician B says that a faulty check valve can let oil fumes into the booster, which can cause damage. Who is right?
   a. A only
   b. B only
   c. both A and B
   d. neither A nor B
3. Technician A says that an airtight booster will provide an assisted brake application after the engine has been shut off. Technician B says that the engine should speed up slightly as the brakes are applied. Who is right?
   a. A only
   b. B only
   c. both A and B
   d. neither A nor B
4. If the output pushrod on a power booster is too long, the result can be
   a. an overly sensitive brake application.
   b. a very low brake pedal.
   c. brake drag, especially after a few stops.
   d. all of the above.
5. Air is bled from the hydraulic section of a Hydro-boost unit by
   a. starting the engine and slowly turning the steering wheel from stop to stop and then applying the brake pedal with the engine off.
   b. starting the engine and opening the bleeder screws on the booster.
   c. pumping the pedal slowly until it gets hard.
   d. applying the brake pedal while the engine is started.
6. Technician A says that a vacuum booster can be disassembled by turning the master cylinder end of the booster while the other housing is held stationary. Technician B says that you should be careful while doing this because the fluid can spray out. Who is right?
   a. A only
   b. B only
   c. both A and B
   d. neither A nor B
7. A faulty Hydro-boost unit will
   a. leak.
   b. not provide an assist.
   c. cause chatter on application.
   d. any of the above.
8. Technician A says that a charged accumulator can spray fluid all over as a Hydro-boost unit is being disassembled. Technician B says that the spring in the accumulator of some units can cause injury as the booster is being disassembled. Who is right?
   a. A only
   b. B only
   c. both A and B
   d. neither A nor B
9. Technician A says that the master cylinder and the booster sections of a Powermaster unit are completely separate. Technician B says that the fluid level of all three reservoir sections of a Powermaster unit should be checked with the engine running. Who is right?
   a. A only
   b. B only
   c. both A and B
   d. neither A nor B

14 Parking Brake Service

Objectives

After completing this chapter, you should:

- ❑ Be able to check a parking brake system, inspect the cables and parts for wear, rusting, or corrosion; clean or replace parts as necessary; and lubricate the assembly as needed.
- ❑ Be able to adjust the parking brake assembly and check for correct operation.
- ❑ Be able to test the parking brake indicator light, switch, and wiring and adjust the switch as necessary.
- ❑ Be able to perform the ASE Tasks relating to parking brakes (See Appendix 1).

14.1 Introduction

Normal parking brake service includes inspection and adjustment of the parking brake cable and an occasional adjustment of the warning light switch (Figure 14-1). Sometimes this service will include replacement of one or more faulty cables (the front control cable, the rear application cable(s), or both), adjustment or replacement of the lining or shoes on a separate parking brake assembly, or replacement of a faulty control assembly (Figure 14-2).

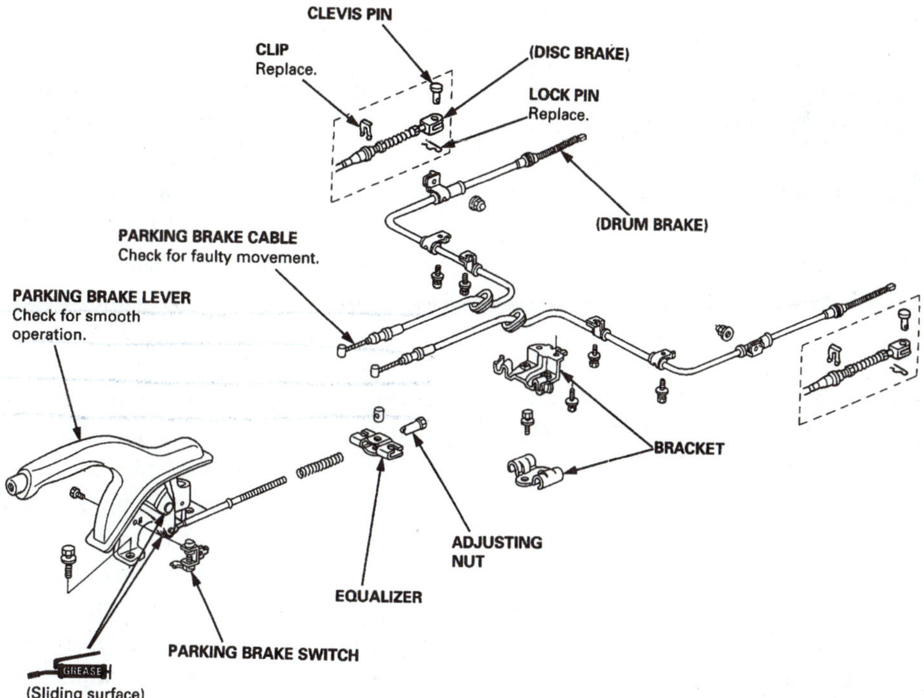

Figure 14-1 A parking brake system using a hand-operated lever, showing areas to be checked. (Courtesy of American Honda Motor Co., Inc.)

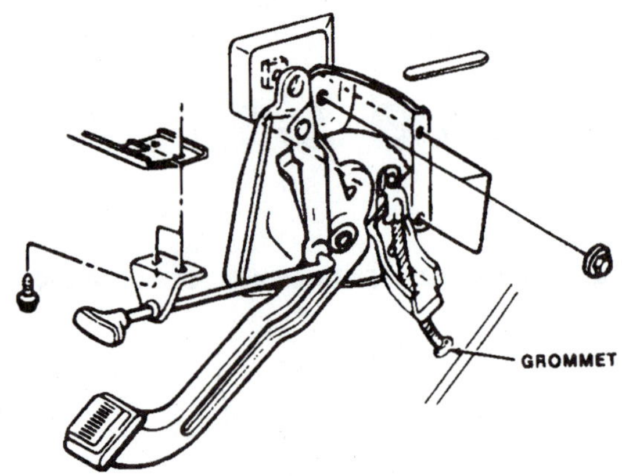

Figure 14-2 A pedal-operated parking brake control assembly. Foot pressure on the pedal pulls the cable to apply the brakes, and pulling the control handle releases the ratchet mechanism to release them. (Courtesy of Chrysler Corporation)

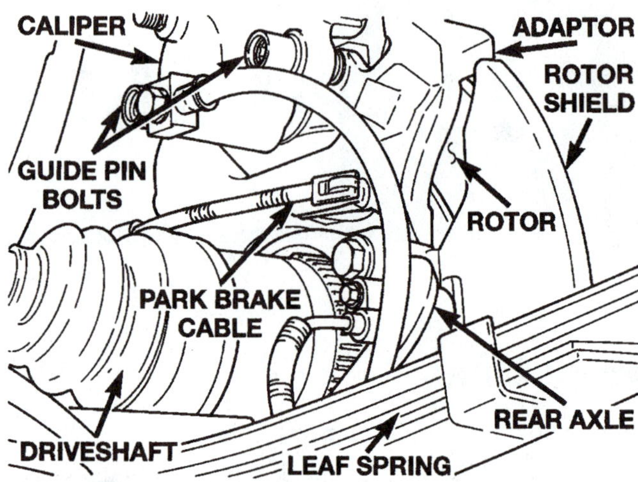

Figure 14-4 This four-wheel disc system uses a drum brake for the parking brake. Note the parking brake cable separate from the caliper. (Courtesy of Chrysler Corporation)

Most parking brakes are mounted on the rear wheels and will fall into one of three categories: drum brakes with an integral parking brake mechanism, disc brakes with an integral parking brake mechanism (Figure 14-3), or disc brakes with a separate parking brake mechanism

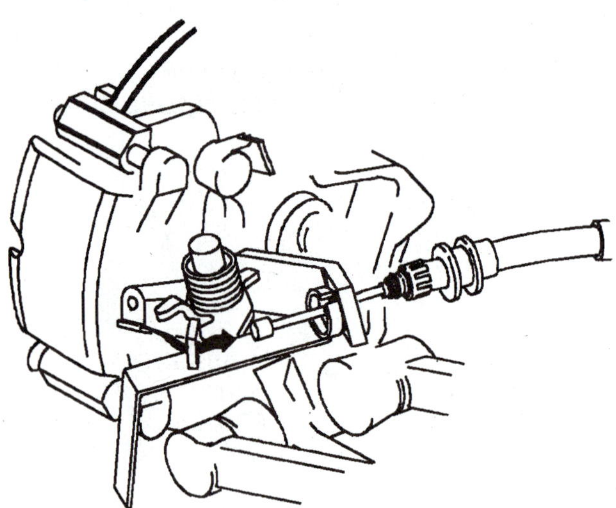

Figure 14-3 This four-wheel disc brake system uses a mechanical operation of the caliper for the parking brake. Note that the cable is being connected as the caliper is replaced. (© Saturn Corporation, used with permission)

(Figure 14-4). In the past, some cars were equipped with drum brakes with a separate parking brake mechanism mounted at the transmission output shaft. Many trucks use a separate auxiliary parking brake.

In most cases, a parking brake uses the shoes, lining, and drum or rotor of the service brake, and these portions are repaired as a brake job is done. This was described in Chapters 10 and 11. In a few cases, a parking brake has its own set of shoes, and their servicing is a separate operation. Because a parking brake normally encounters only static friction, there is very little wear. Lining replacement or even adjustment of separate parking brake assemblies is seldom done. The actual replacement of shoes on these units follows a procedure very similar to that for a drum brake assembly. This operation is described in the service manuals for each particular car (Figure 14-5).

14.2 Cable Adjustment

A parking brake is properly adjusted if it will apply completely in less than one-half the travel distance of the lever (about five clicks), and will release completely with no drag when the lever is released (Figure 14-6). Usually an adjustment is made by lengthening or shortening the cable attachment at the equalizer assembly. The major purpose of an adjustment is to compensate for cable stretch or wear of the linkage at the various contact points. Lining clearance is adjusted by the normal operation of most modern brake assemblies.

Some vehicles are equipped with automatic, self-adjusting mechanisms to maintain correct adjustment of the parking brake cable. This mechanism exerts a pull on the parking brake cable while it is released; a ratchet mechanism allows normal operation when the parking brake is applied. The auto-adjuster mechanism must be locked up with a pin when the brake control or cables are disconnected (Figure 14-7).

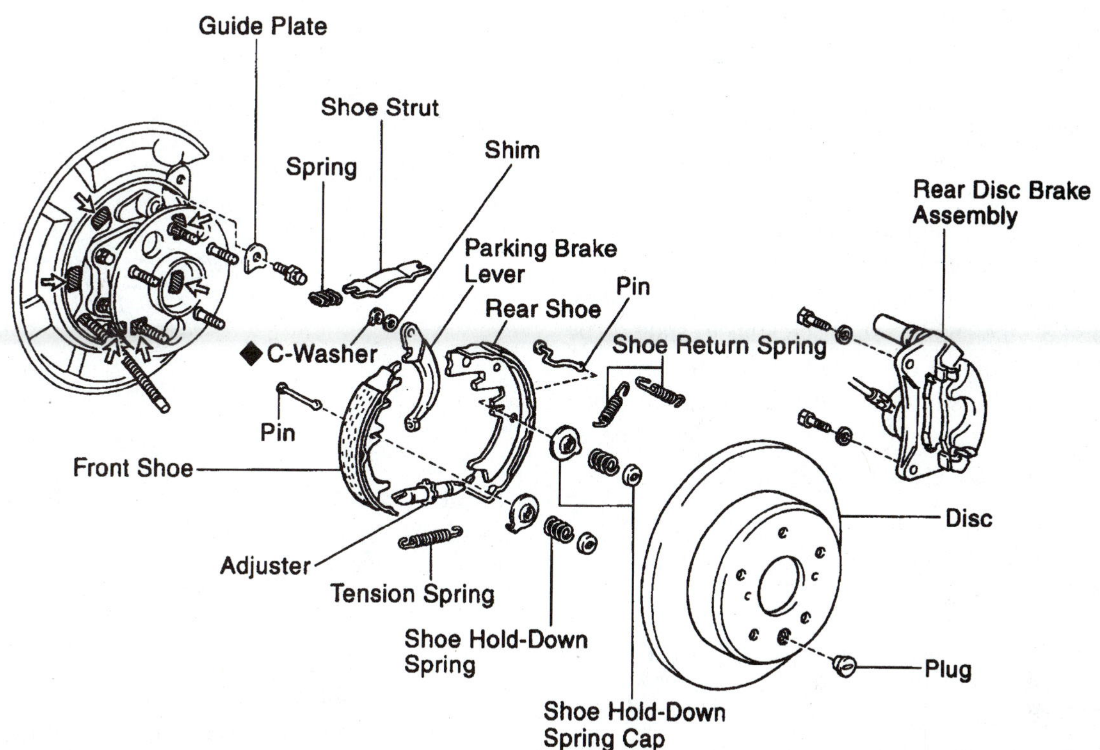

Figure 14-5 An exploded view of drum parking brake–disc service brake combination. Parking brake shoe clearance is adjusted manually.

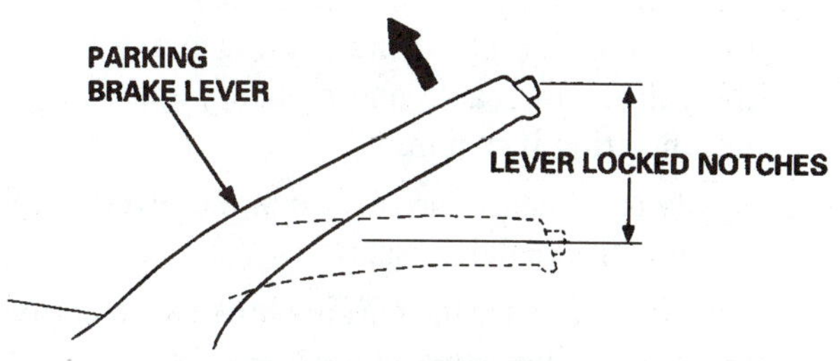

Figure 14-6 The parking brake lever should become locked when it travels four to eight notches (rear drum brakes) or seven to eleven notches (rear disc brakes). (Courtesy of American Honda Motor Co., Inc.)

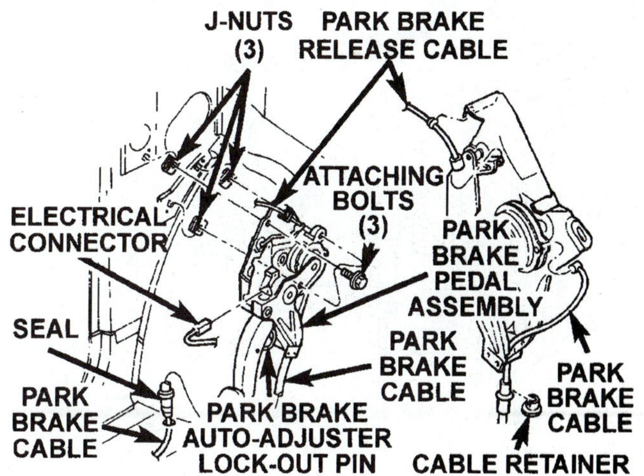

Figure 14-7 The Park Brake Auto-Adjuster Lock-Out Pin was installed in this parking brake pedal assembly to hold the cable tensioning spring from unwinding. This should be done to allow safe disconnection of a parking brake cable or removal of the pedal mechanism. (Courtesy of Chrysler Corporation)

The service brake lining should have the correct clearance when the cable adjustment is made. In most cases, this can be checked by applying the service brake. A high brake pedal indicates correct lining clearance. If the pedal is low on a car with rear-wheel drum brakes, the brake shoe adjustment should be checked as described in Chapter 10. If the car has rear-wheel disc brakes with an integral parking brake mechanism, apply and release the parking brake about five times, using a firm application pressure, and then recheck the brake pedal height. If it has increased, continue applying and releasing the parking brake until there is no further improvement. If the pedal height does not increase, have an assistant tap sharply on the caliper with a hammer while the brakes are being applied with a firm pressure on the pedal. If the pedal height still does not improve, check the self-adjuster mechanism in the rear calipers as described in Chapter 11.

If the car has rear-wheel disc brakes with a separate parking brake, the brake shoe clearance should be adjusted before changing the cable length. This operation is very similar to adjusting the brake shoe clearance on a non-self-adjusting duo-servo brake and is described in various service manuals (Figure 14-8).

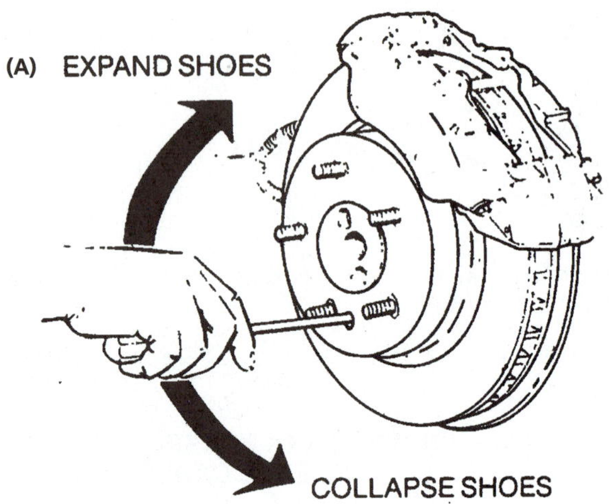

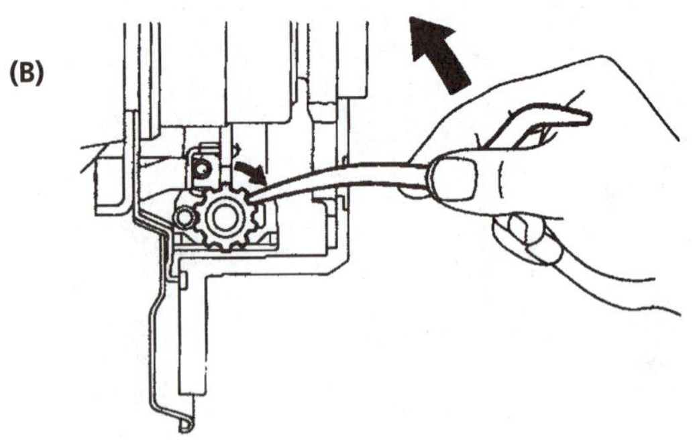

Figure 14-8 A screwdriver is being used to turn the adjuster nut to adjust the lining clearance on the older Corvette parking brake (A); a brake spoon is used to adjust the lining clearance on the Camry (B). (A is courtesy of Stainless Steel Brakes Corp.)

On some cars, there is a front cable which moves an **intermediate lever**; it is also called an **equalizer lever** or **ratio bar**. This front cable should be adjusted first if it is adjustable. A front cable adjustment is also done to compensate for cable stretch and linkage wear (Figure 14-9). It ensures complete movement of the intermediate lever without excessive play.

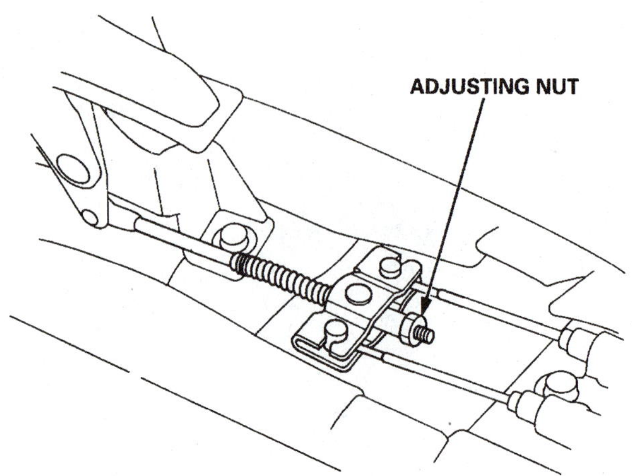

Figure 14-9 Turning this adjusting nut to tighten it will reduce lever travel. (Courtesy of American Honda Motor Co., Inc.)

To adjust parking brake linkage:

1. Inspect the cable to make sure that it can move freely in the guides and housings, has no excess wear at equalizer or other contact points, and shows no signs of broken strands (Figure 14-10). Replace any defective parts and lubricate the cable as necessary. On disc brake calipers with an integral parking brake mechanism, check for proper positioning of the parking brake lever at the caliper, the cable and housing, and the lever return spring if required.
2. Apply and release the service brake several times using a firm pedal pressure.
3. Apply and release the parking brake several times using a firm pressure on the lever.
4. Apply the parking brake lever to the first notch.
5. Adjust the cable length at the equalizer to remove all slack (Figure 14-11). On some cars, this adjustment should be made at the parking brake lever. A slight side pull on the cable should just start to cause shoe drag when the wheel is turned.
6. Release the parking brake lever and rotate the wheel. There should be no sign of drag caused by the parking brake linkage (Figure 14-12).
7. If a second jam or check nut is used on the adjuster, make sure that it is correctly tightened.

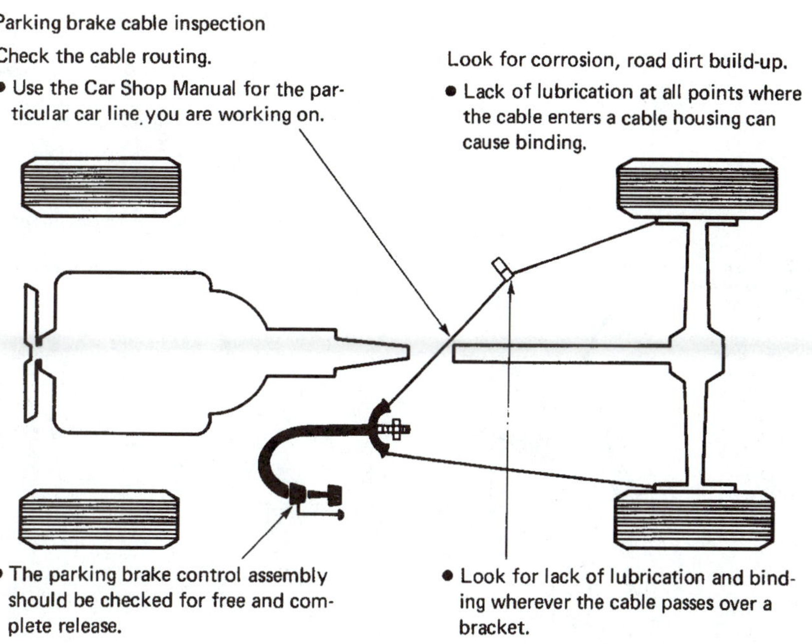

Figure 14-10 The items to check while inspecting a parking brake system. (Courtesy of Ford Motor Company)

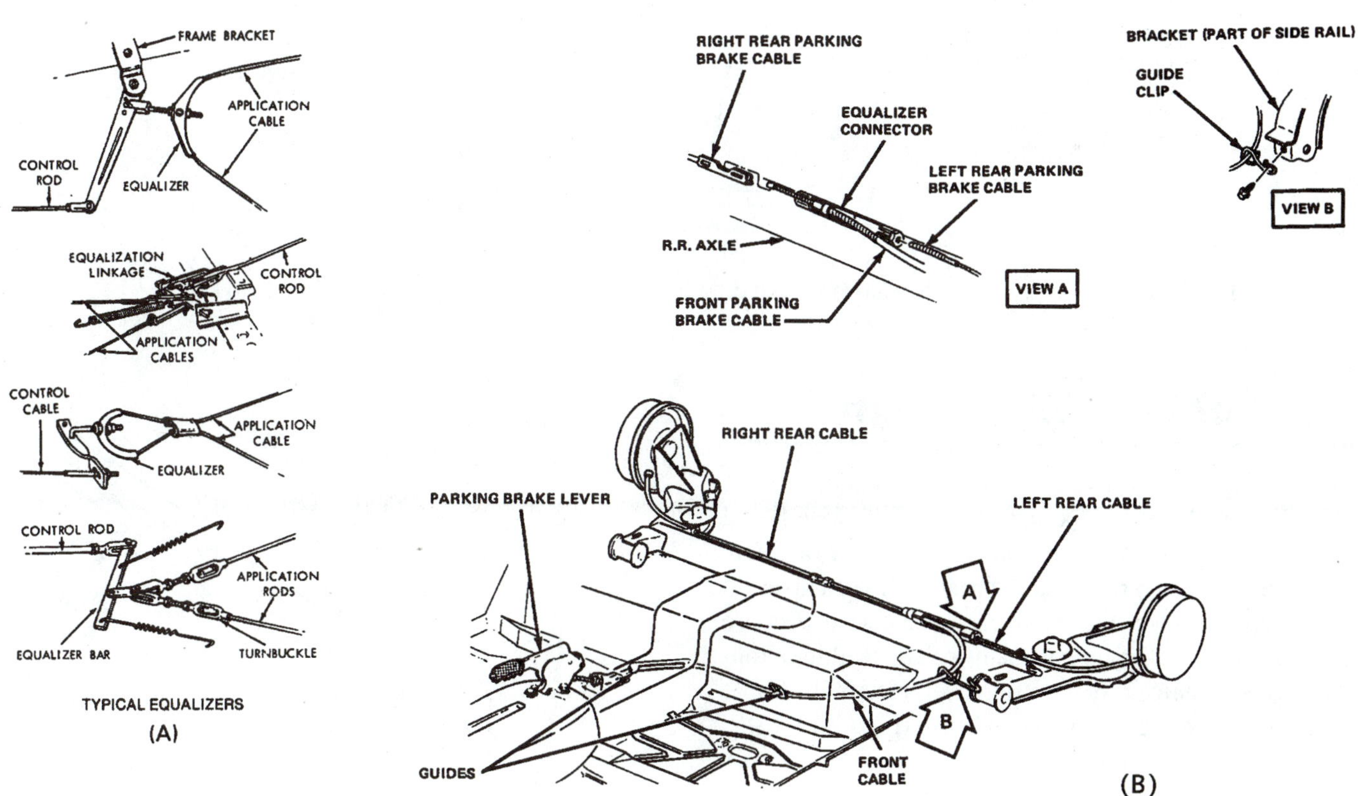

Figure 14-11 In many cars, the equalizer is at the connection between the front control cable or rod and the rear application cable (A). Some cars connect the control cable to one rear brake and the housing to the other one (B). (A is reprinted from Mitchell Anti-Lock Brake Systems, with permission of Mitchell Repair Information, LLC; B is courtesy of General Motors Corporation, Service Technology Group)

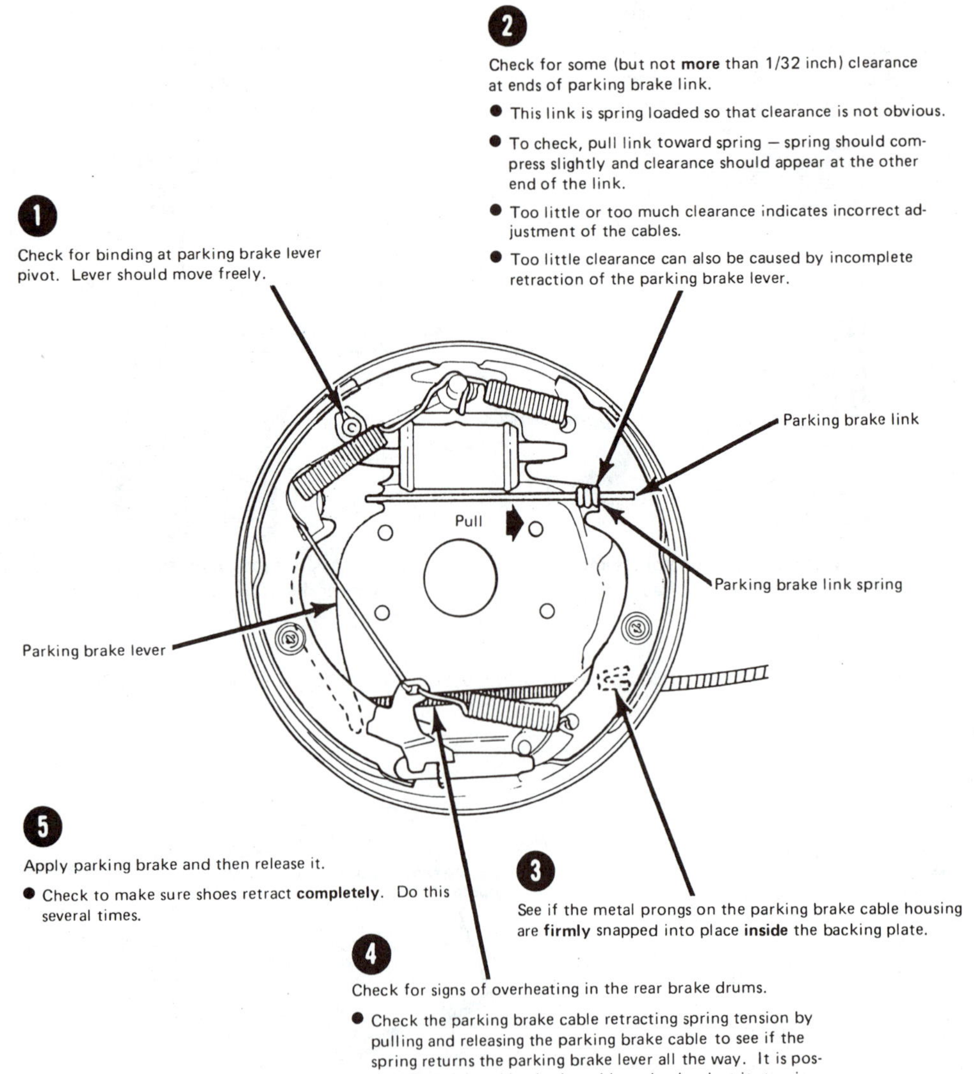

Figure 14-12 The inspection procedure for parking brake drag problems. (Courtesy of Ford Motor Company)

14.3 Cable Replacement

Occasionally it is necessary to replace a parking brake cable. Essentially this is a simple job of removing and replacing the cable and housing. On some cars, a single cable runs from one rear brake, through the equalizer, to the other rear brake. On other cars, separate cables are used that are joined by a connector or attached individually to the equalizer or control lever assembly. Cable replacement is easier in the second case (Figure 14-13).

To remove a parking brake cable:

1. Raise and support the car in a secure manner.
2. Disconnect the cable from the equalizer.
3. Remove any clips that secure the cable housing to frame or body brackets.

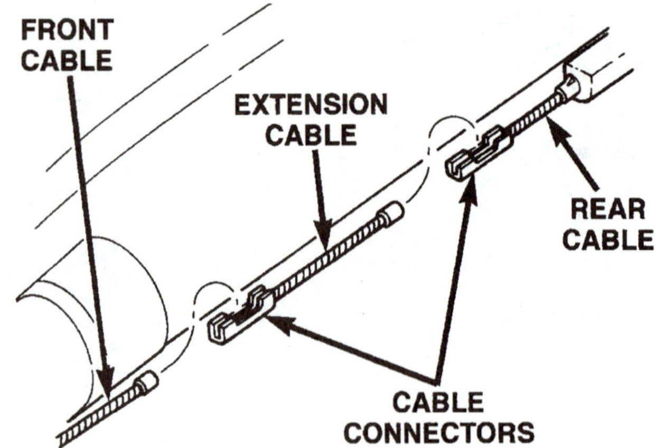

Figure 14-13 The cable connector allows cables to be connected or disconnected for replacement. (Courtesy of Chrysler Corporation)

4. On cars with disc brakes with an integral parking brake, disconnect the cable from the caliper at the lever (Figure 14-14).

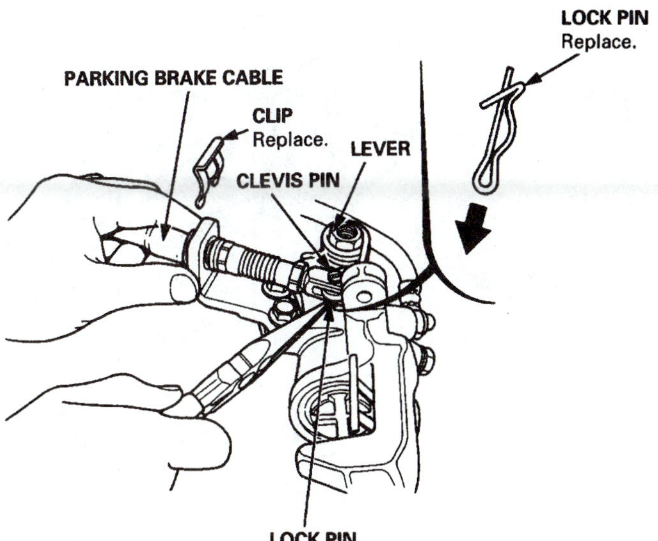

Figure 14-14 Parking brake cable housings are often attached to the cable bracket by a U-shaped clip; this cable is attached to the caliper lever by a lock pin. (Courtesy of American Honda Motor Co., Inc.)

On cars with drum brakes, follow this procedure:

a. Remove the wheel and brake drum.
b. Disengage the cable end from the parking brake lever. Side- or diagonal-cutting pliers can be used to compress the spring and grip the cable to hold the spring in a compressed position during this operation (Figure 14-15).

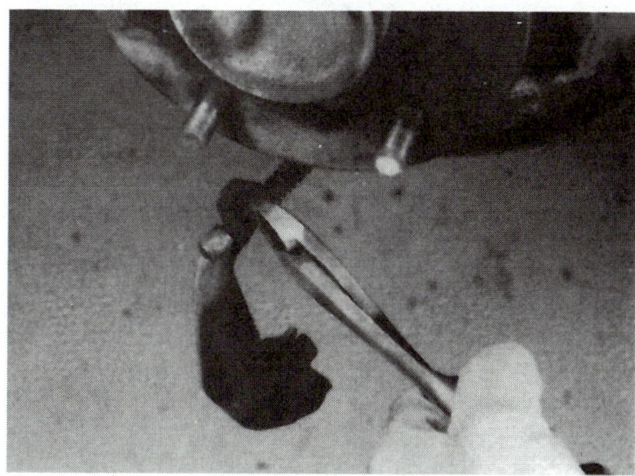

Figure 14-15 A pair of side- or diagonal-cutting pliers over the application cable. Holding the parking brake lever and moving the pliers and spring to the right compresses the spring; squeezing the pliers holds the spring compressed while the lever is disconnected.

c. Depress the cable retainer prongs using a small hose clamp and slide the cable and housing through the backing plate (Figure 14-16).

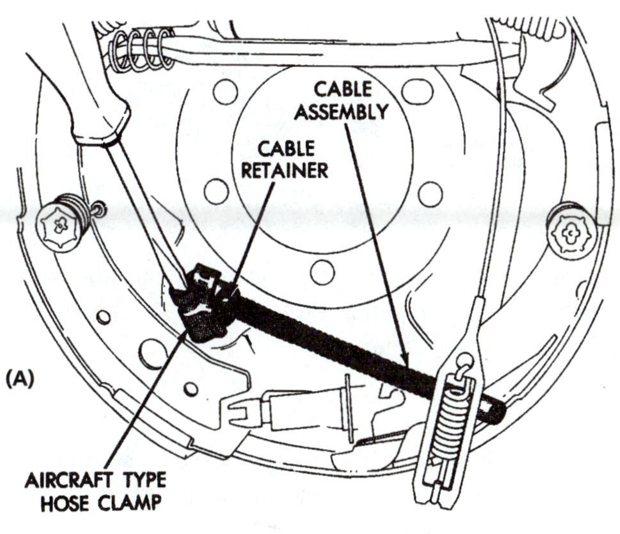

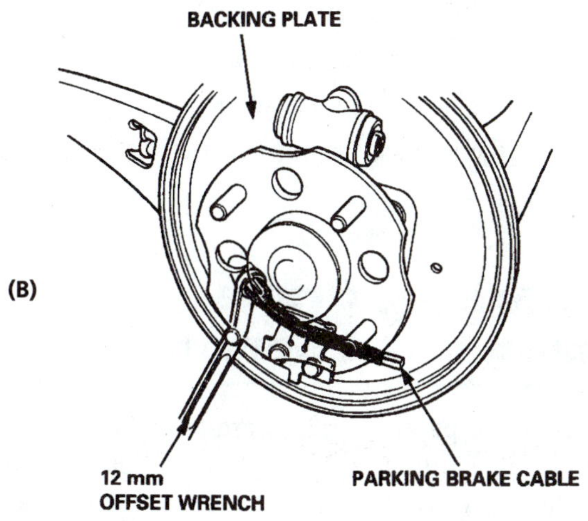

Figure 14-16 The prongs must be compressed using a hose clamp (A) or box-end wrench (B) so the parking brake cable can be removed from the backing plate. (A is courtesy of Chrysler Corporation; B is courtesy of American Honda Motor Co., Inc.)

To replace a brake cable:

1. *On cars with drum brakes:*
   a. Push the cable and housing into the backing plate, making sure that the prongs on the retainer are completely spread and locked in place (Figure 14-17).

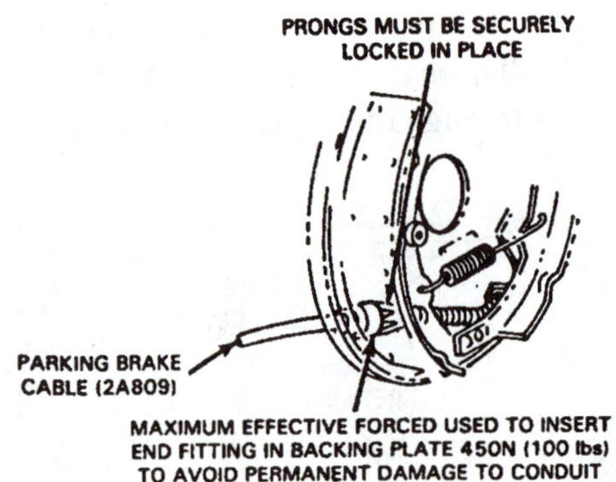

Figure 14-17 When installing a cable housing into a backing plate, make sure the prongs expand to lock it in place. (Courtesy of Ford Motor Company)

b. Slide the cable end outward or compress the spring. Then connect the cable end to the parking brake lever.

c. Install the brake drum and wheel.

On cars with disc brakes, connect the cable end to the caliper lever.

2. Route the cable and housing through the brackets or guides and connect the cable to the equalizer.
3. Replace any clips that were removed.
4. Adjust the parking brake cable.
5. Check the application and release of the parking brake.

14.4 Parking Brake Warning-Light Service

On most cars, the parking brake warning-light circuit is controlled by an additional switch and is in a parallel circuit with the brake warning-light switch on the pressure differential valve. Both these switches use the same warning-light bulb. In some cars, the parking brake warning light will use a separate bulb and switch. The switch for the parking brake warning light is mounted at the control lever. It is closed when the parking brake is applied and opened when the lever is released.

NOTE: Remember that parking brake drag can cause enough heat to boil the fluid in the wheel cylinders, and in a diagonal split system, this can cause total brake failure. Be sure that the warning light comes on when the parking brake is applied.

System faults usually result from a defective bulb, wire connection, or switch, or from an incorrect switch adjustment. In many cases, switch adjustment is controlled by the mounting bracket, and the bracket must be bent to change the adjustment (Figure 14-18).

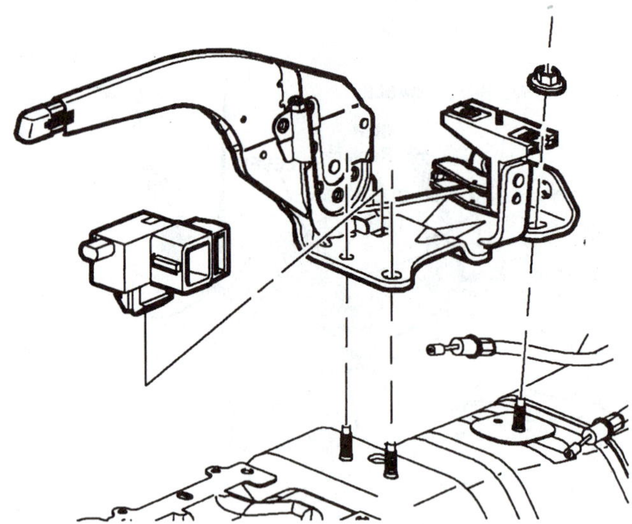

Figure 14-18 The warning-light switch is usually mounted next to the parking brake lever; this one is removed by squeezing the retaining tabs. (© Saturn Corporation, used with permission)

When troubleshooting circuit electrical problems, it is advisable to check a wiring diagram to determine how the wires are connected in the circuit (Figure 14-19). In most

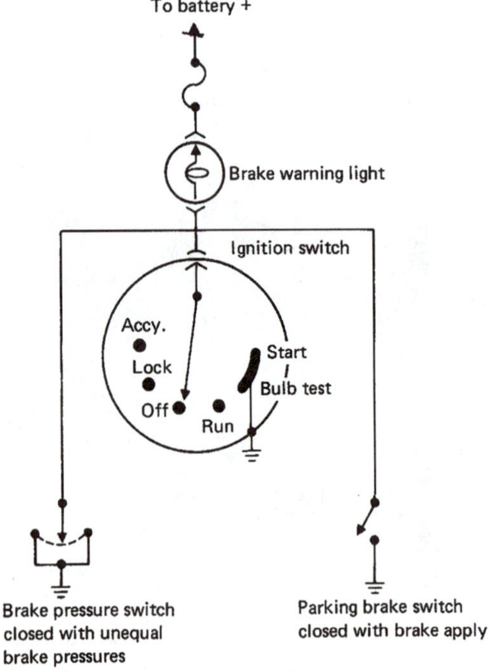

Figure 14-19 A typical parking brake warning-light circuit. Note that the switch is closed to make a path to ground and turn on the light when the brake lever is applied, and the brake pressure switch is in a parallel circuit.

cases, the closing of the switch completes the ground circuit for the warning-light bulb. In such cases, a trouble light or voltmeter can be used to make certain that there is voltage at the switch. If there is no voltage, the circuit is open. There is a break in the wiring, a bad connection, or a burned-out bulb or fuse. If there is voltage at the switch, there is probably a faulty switch or switch adjustment or there is a faulty ground connection.

14.5 Practice Diagnosis

You are working in a brake and front-end shop and encounter these problems.

CASE 1: The customer has brought in her ten-year-old 4 X 4 Nissan pickup, V-6 engine and automatic transmission, with a problem of an overheating right rear brake. The wheel is hot and smells of very hot lining. After pulling the drum, you find the lining glazed with light cracks down the center, the shoe metal slightly blued, and no clearance at the parking brake strut. What went wrong? What should you check next? What will you probably have to do to fix this problem?

CASE 2: The Corvette owner has brought in his car to have the parking brake adjusted; the brake handle goes full travel without meeting resistance. What is the probable cause of it getting this far out of adjustment? What check should you make as you adjust the cable? What other check should you recommend to the customer?

CASE 3: While making a brake inspection on a seven-year-old Olds Cutlass, you notice that the brake light does not come on as you apply the parking brake. When you crank the engine, the light does come on. What is probably wrong? What should you do to check this?

Terms to Know

equalizer lever

intermediate lever

ratio bar

Review Questions

1. Two technicians are discussing parking brake service. Technician A says that a parking brake can use the same lining and friction surfaces as the service brake. Technician B says that a parking brake must be applied by hydraulic pressure. Who is right?
   a. A only
   b. B only
   c. both A and B
   d. neither A nor B
2. The parking brake lever on a certain car travels almost its full distance when applied. Technician A says that the service brake lining might need adjustment. Technician B says that the parking brake cable needs adjustment. Who is right?
   a. A only
   b. B only
   c. both A and B
   d. neither A nor B
3. A domestic car with four-wheel disc brakes has an excessive amount of travel at the parking brake lever during application. To correct this problem, you should
   a. loosen the rear cable at the equalizer.
   b. tap the caliper body with a hammer as the brakes are applied.
   c. use "park" in the automatic transmission.
   d. none of the above will correct the problem.
4. Parking brake cable adjustments are normally made at the point where the
   a. rear cable is attached to the control cable.
   b. rear cable is attached to the parking brake lever.
   c. front cable is attached to the parking brake pedal.
   d. strut is attached to the application cable.

5. Technician A says that a parking brake must be capable of stopping a car in 45 feet from a speed of 20 mph. Technician B says that a parking brake must be able to lock up the two wheels to the limit of traction. Who is right?
   a. A only c. both A and B
   b. B only d. neither A nor B
6. Technician A says that a parking brake should apply completely in less than one-half of the available travel distance of the lever. Technician B says that a correctly adjusted parking brake cable will cause a slight amount of lining drag when released. Who is right?
   a. A only c. both A and B
   b. B only d. neither A nor B
7. On most cars, looseness of the rear cable is adjusted at the
   a. equalizer.
   b. control assembly.
   c. brake shoes.
   d. backing plate.
8. On some cars with four-wheel disc brakes, in order to completely adjust the parking brake it is sometimes necessary to adjust the
   a. slack in the rear cables.
   b. lining-to-drum clearance at the brake shoes.
   c. lining-to-rotor clearance at the brake shoes.
   d. both a and b.
9. The parking brake warning light on a certain car fails to work. Technician A says that this could be caused by a faulty switch adjustment at the application lever. Technician B says that you can sometimes determine the cause of the problem by cranking the engine. Who is right?
   a. A only c. both A and B
   b. B only d. neither A nor B
10. A parking brake cable should be replaced if it
    a. begins to fray.
    b. is frozen in the housing.
    c. has a broken end button.
    d. any of the above.

15 Antilock Brake System Service

Objectives

After completing this chapter, you should:

- ❑ Be able to check the operation of an antilock brake system.
- ❑ Be able to diagnose improper antilock brake system operation.
- ❑ Be able to check the operation of the system sensors, actuators, control module, connectors, and wiring.
- ❑ Be able to remove, replace, and adjust faulty components in an antilock brake system.
- ❑ Be able to perform the ASE Tasks (Appendix 1) relating to Antilock Brake System (ABS) diagnosis and repair.

15.1 Introduction

Normal antilock brake system service includes diagnosing, locating, and correcting problems that might occur while driving or stopping the car. Since the ABS components serve as controlling devices that prevent wheel lockup, there are essentially no wearing points requiring periodic service or maintenance. When troubleshooting brake problems, it should be remembered that an ABS is a controlling system only (Figure 15-1). Problems such as drag, pull, a low pedal, and unusual noises are probably caused by defects in the basic brake hydraulic system or wheel assemblies. A thorough inspection, like that described in Chapter 9, and proper service will usually locate and correct these faults

When diagnosing ABS problems, remember tire diameter and tread condition are important because they affect wheel-rotating speed. If the motorist has changed the tire size significantly, this can fool the electronic brake control module (EBCM) into thinking there is a fault in the system. This is an easy item to check because the original tire size is usually on a sticker in the glove compartment or on the door jamb.

ABS will use either a vacuum or an electrohydraulic power booster, and some of the electrohydraulic systems will use the hydraulic pressure from the pump to operate the rear brakes. With these systems, pump/motor failure will result in a loss of the rear brakes. The hydraulic modulator on these units is usually integral with the master cylinder. On most systems, a pair of hydraulic lines connect the master cylinder to the **hydraulic modulator**, which is also called a *hydraulic control unit, ABS actuator*, or *modulator unit*. Most modulator assemblies are not serviceable internally, even though they can be disassembled (Figure 15-2). Some manufacturers allow partial disassembly to replace subcomponents and provide repair parts; this procedure(s) is described in their service manuals. Replacement ABS components and parts are available from new-car dealerships; aftermarket suppliers also have new and rebuilt parts available.

Since ABS is essentially an electronic control system with speed sensor inputs and electric solenoids or motors as outputs, most problems will be electrical in nature. The other major parts are in the hydraulic modulator. Faulty electrical connectors are the most commonly encountered problem.

Some cautions should be observed when testing computer controls. It is important not to touch the connector pins of the EBCM. A static-electric charge can be generated in your body as you move around a car (Figure 15-3), especially over the interior carpet and seat covers. If this high-voltage charge is discharged into the EBCM, it can damage it. Some technicians will connect a ground wire between themselves and the body of the car or spray the

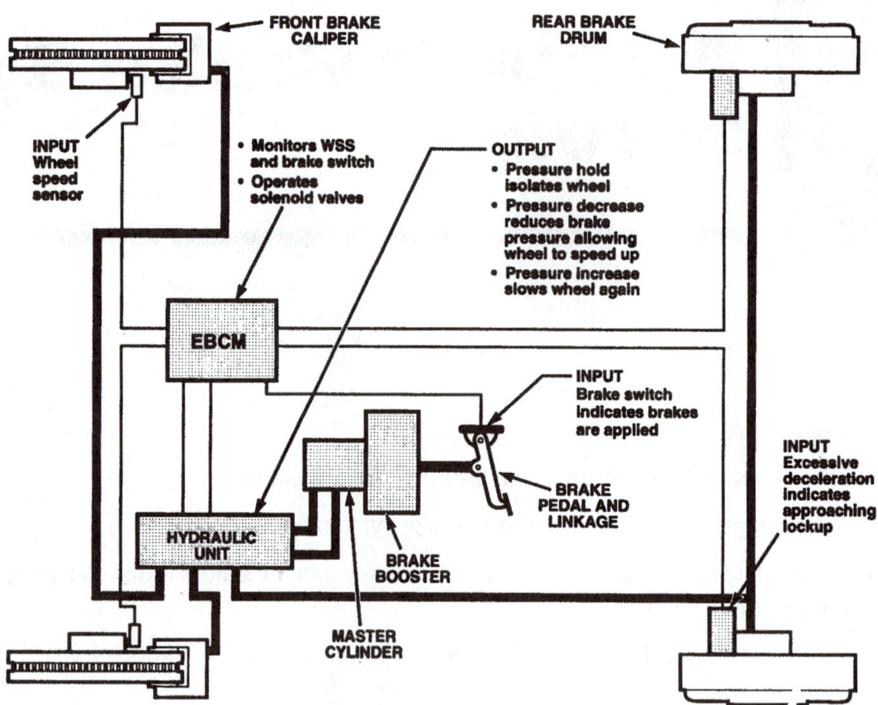

Figure 15-1 ABS adds the monitoring and control features to the base brake system. (Courtesy of General Motors Corporation, Service Technology Group)

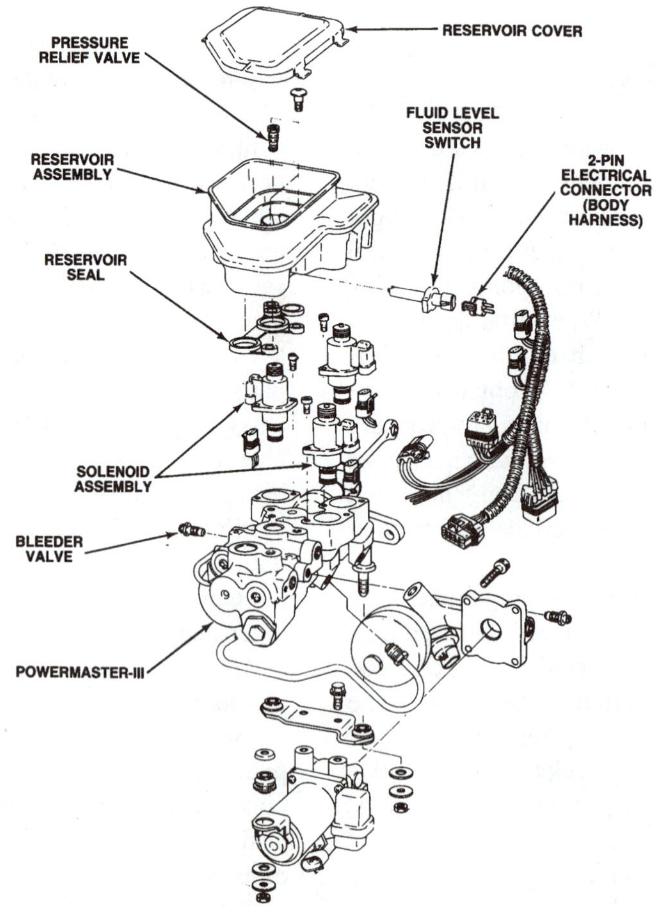

Figure 15-2 An exploded view of an ABS hydraulic unit containing the master cylinder, booster, and solenoid valves. (Courtesy of General Motors Corporation, Service Technology Group)

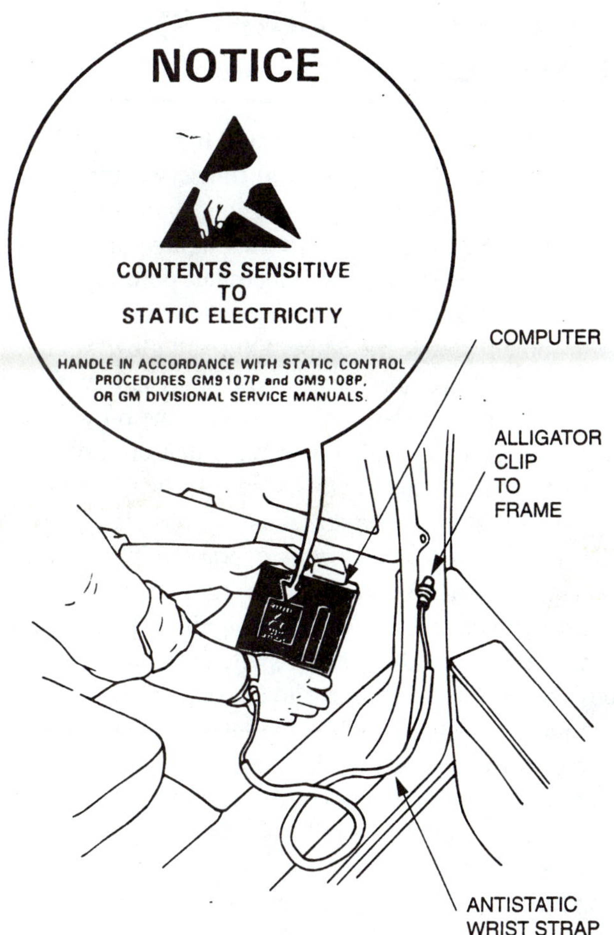

Figure 15-3 This warning symbol is used to identify components or circuits that can be damaged by electrostatic discharge (ESD). Special precautions, like not touching them unless you are grounded, should be followed when you are working with them. (Courtesy of General Motors Corporation, Service Technology Group)

carpet and seatcovers with an anti-static-cling laundry spray to help prevent this from happening. As mentioned earlier, it is also important not to use test devices, which could cause increased current flow through the computer system circuits. If electric arc welding is done on the car, the EBCM should be disconnected to prevent possible damage from a high-voltage surge. Also, if the car is to be painted and the paint cured in an oven at high temperatures, the EBCM can be damaged by the exposure to the high heat.

Extra care should be exercised when repairing ABS components to prevent the unusual or unexpected occurrence. Imagine what could happen if the wheel-speed sensor wire on a RWAL system came loose during a stop with a loaded pickup or van. The EBCM would think the rear wheels were locked up and would then reduce the rear braking pressure. This could lead to tragic consequences.

ABS units are quite complex, and there are some definite differences in the operation of the various systems. Many systems have variations, such as traction control. In a ten-year time period, one domestic manufacturer used eleven different systems, and another one had eight different systems. And some of these systems had variations between different vehicle models. Some systems are unique to one or two car models, and others, like RWAL, are very similar to several makes of pickups. An adequate service manual must be followed while diagnosing or repairing the faults in these systems. One new-car manufacturer has a eighty-page section of the service manual for one car model devoted to ABS problem diagnosis and service. It is also a good idea to familiarize yourself with the normal system operation. These units self-check each time the car is started. In some systems, if you apply the brake with a firm pedal during starts, you should feel a "bump" on the pedal and hear clicking or chunking sounds as the EBCM checks out the various components. Another design produces the brief whining noise of electric pump operation after the car reaches a certain speed, about 4 mph. On a road test, if the brake is applied with a firm-to-hard pedal, one or several chirps should be heard from the rear tires as they try to lock up. At this time, pulsations or bumps on the brake pedal are a normal sign as the ABS goes to work (Figure 15-4). While doing this, be sure that there is no traffic following closely behind.

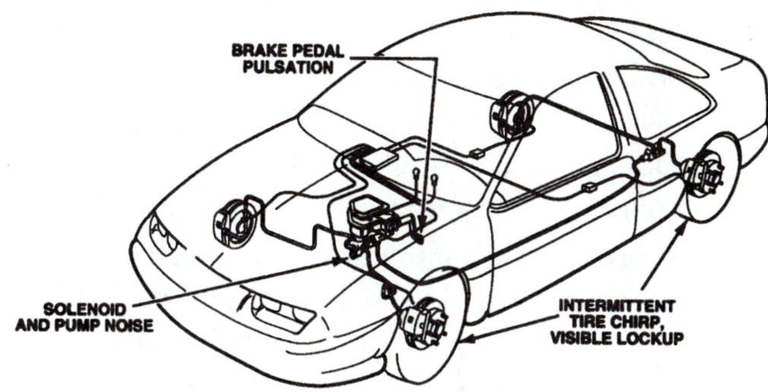

Figure 15-4 These conditions are normal in ABS-equipped vehicles, although they might appear to be problems. (Courtesy of General Motors Corporation, Service Technology Group)

The following typically encountered faults can be attributed to faulty ABS components: poor tracking during hard stops which involve antilock operation, a spongy brake pedal, and improper warning-light operation. A spongy pedal problem is corrected by bleeding the system using the correct procedure for that particular system. Some ABS systems require special bleeding procedures, and these are described in service manuals.

A problem that is becoming more common as ABS systems are getting older is leaking modulator valves.

This is being caused by debris between the valve and seat. The source of this debris has been attributed to dirty brake fluid, formation of copper crystals from chemical activity in the fluid, crud pushed up from a caliper as the piston is depressed, and material left in the system from manufacturing processes. It is highly recommended to change the brake fluid every two or three years to reduce the possibility of this problem.

When working on an ABS, a service manual for that particular car should be followed and these precautions should be observed:

- Make sure the system is depressurized before opening any hydraulic circuit. Depressurization is done by pumping the brake pedal an adequate number of times as instructed by the manufacturer of that system.
- Follow the manufacturer's directions on the use of the proper equipment when bleeding a system.
- Use only specially designed brake hoses and lines.
- Use only the recommended brake fluid; do not use silicone brake fluid in an ABS.
- Make sure the ignition is turned off before disconnecting or reconnecting any ABS electrical connectors, to prevent damage to the EBCM.
- Do not touch any of the connectors to an EBCM with your fingers or with a meter probe unless directed to by a service manual, and then only while following the directions given in the manual.
- Disconnect the EBCM and any other onboard computers if using an electric welder on the car.
- If installing a transmitting device such as a telephone or CB, make sure the electrical connections and antenna do not interfere with the ABS.
- Do not hammer or tap on speed sensors or sensor rings; they can be demagnetized, which can affect their signal accuracy.
- Use only anti-corrosion coatings on speed sensors; do not contaminate them with grease.
- Always check the sensor air gap when either a sensor or sensor ring has been replaced; this includes a rotor, axle, or CV-joint with a sensor ring on it.
- Tighten wheel lug nuts to the correct torque; overtightening can distort a rotor or drum and affect speed-sensor signals.
- If replacing the tires, do not mix the sizes; usually the diameters of all four tires must be equal and the same as the original tires.
- Do not subject the EBCM to excessive heat.

15.2 ABS Warning-Light Operation

ABS systems use two brake warning lights, a **red brake warning light** and an **amber antilock warning light**. The function of the red brake warning light is similar to that in non-ABS. The amber warning light warns of problems in the hydraulic and electrical portions of the antilock controls.

On some systems, the amber light will glow whenever the brake fluid pressure or reservoir level is low. On all systems, the amber warning light will come on when the EBCM receives or sends out signals that are not in the correct electrical range. If this light is on while the car is being driven, it indicates that the ABS has shut itself off and that the car's brakes will operate using non-ABS braking. In normal operation, this light should come on for three to six seconds when the ignition is turned on, during the time that the ignition switch is in the "crank" or "start" position, and for three to six seconds after the car starts and is running. With some integral systems, if the accumulator is completely discharged, the light may stay on longer as the pump charges the accumulator.

On some systems, the red warning light will come on whenever the hydraulic pump motor is running. If the pump motor runs longer than three minutes on some systems, this light will flash on and off. The red warning light on most systems will also come on when the pump or accumulator pressure is low, the brake fluid level is low, the parking brake is applied, and the ignition is turned to "start." The pump motor in the Teves unit, like the pump motor in the Powermaster unit, should not run longer than thirty seconds at a time. If it does, there is a problem and the excessively long operation can cause the motor to overheat and burn out. Figure 15-5 illustrates how the ABS warning light is connected into the circuits.

It is possible for inexperienced technicians to blame base brake system problems on ABS. If you are unsure of the cause of a particular problem, you can remove the ABS fuse and road-test the vehicle. With the fuse pulled, ABS will turn off and the vehicle will revert to base brakes. If the problem disappears, it was ABS related; if it is still there it is in the base brakes.

15.3 Warning-Light Sequence Test

A technician will use the improper operation of the warning lights or a diagnostic/trouble code from the EBCM to determine the correct test procedure for detecting the cause of a particular problem. The use of self-diagnosis

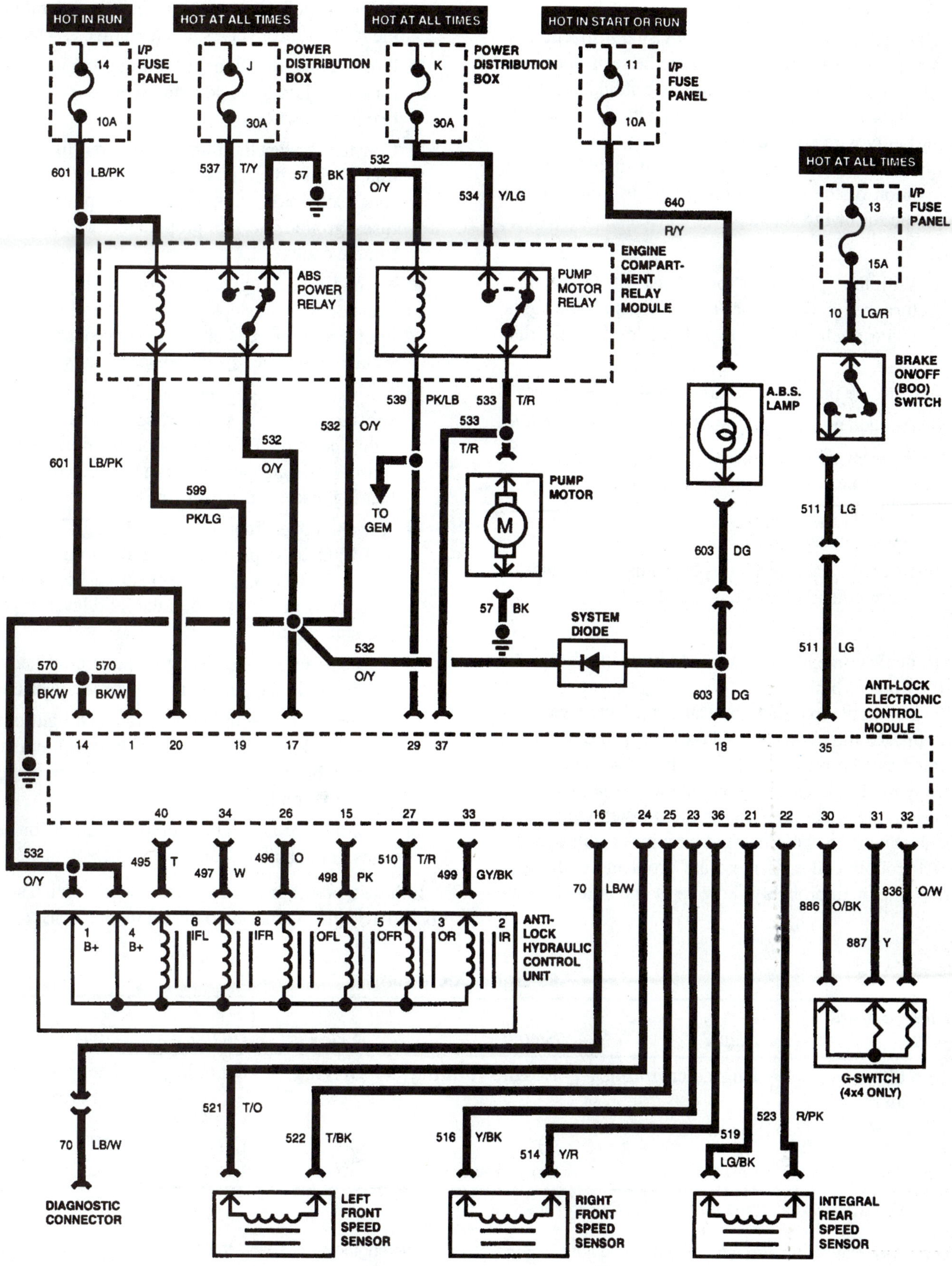

Figure 15-5 An ABS wiring diagram. Note the use of numbers to identify a wire or wire connection, and the color coding to help locate that wire. (Courtesy of Ford Motor Company)

codes will be described in the next section. A light sequence test is the second step in locating the cause of a particular problem. The first step is a visual inspection of the system to note any faults which can be seen, such as a loose wire connector, low fluid level, or a fluid leak. Thus far, the majority of ABS problems have been caused by loose or faulty wire connections.

To perform a light-sequence test, you should observe the operation of the two warning lights as you proceed through the following six steps:

1. With the ignition off for at least fifteen seconds, turn the ignition to the "run" position. If the lights come on for thirty seconds or less, repeat this step.
2. Turn the ignition to "start" and start the engine.
3. As soon as the engine starts, release the switch to the "run" position.
4. Drive the car at a minimum speed for a short distance.
5. Brake the car to a stop.
6. Place the transmission selector in "park" and let the engine idle for a few seconds.

During this time, the lights should be on or off as indicated in Figure 15-6.

Normal light operation indicates that the electrical values in both the hydraulic and the electrical portions of the brake system are normal. On integral systems using electrohydraulic boosters, this includes the hydraulic pressure in the brake lines and in the booster. On all systems, problems in the electrical circuits or components will set trouble codes. If there are problems with these systems, they will probably show up in one of the following ways:

- Normal light operation along with a stopping problem (pull, drag, etc.). Check the operation of the ABS control valves and the brake assemblies.
- Pump motor runs longer than one minute. Check for a low fluid level, low pump pressure, or defective pump switch.
- Constant amber light and normal red light operation. Check all circuits to or from the EBCM (sensors, solenoids, pump pressure, power relays, fuses, ground, etc.) for proper electrical values and trouble codes.
- Normal amber light operation in steps 1. and 2, constant light in steps 3 through 7 and normal red light operation. Check for an improper electrical value in all wheel-speed sensor circuits.
- Normal amber light operation in steps 1 through 3, constant light in steps 4 through 6, and normal red light operation. Check for a problem in the wheel sensors (wrong sensor air gap, defective wheel bearing, damaged sensor ring).
- Normal amber light in steps 1 through 3, intermittent light in steps 4 and 5, and normal red light operation. Check for loose electrical connectors in all circuits or a problem in the fluid level or pump pressure circuit.
- Constant amber and red light operation. Check for a problem in the pump or pump pressure circuit.
- Normal amber light and constant red light operation. Check for a low fluid level or a problem in the parking brake switch, fluid level switch, or pump pressure circuit.
- No amber or red light operation. Check for problems in the bulb circuits.

If these types of problems occur, the technician should proceed to self-diagnosis to measure the resistance or

ABS Light Operation

Vehicle Status ***Step 1***	***Step 2***	***Step 3***	***Step 4***	***Step 5***	***Step 6***
Engine off, ignition on	Engine cranking	Engine running	Driving	Stopping	Stopped, engine running
Light Status					
Red a	On	Off	Off	Off	Off
Amber On 3-6 sec.	On	On 3-6 sec.	Off	Off	Off

a: The light might come on for 30 seconds or less.

Figure 15-6 The amber and red brake warning lights should go on as the engine cranks, and the amber light should stay on for a brief period after the engine starts. Problems are indicated if either of them lights up at the wrong time. (Courtesy of General Motors Corporation, Service Technology Group)

voltages of the particular circuit(s) involved, to locate the faulty circuit and the problem within that circuit. Improper sensor clearance or excessive runout, end play, or damage to the sensor ring will usually cause problems that will appear when the vehicle is driven (Figure 15-7).

15.4 ABS Problem Codes and Self-Diagnosis

Most ABS systems will store a **diagnostic trouble code (DTC)**, also called a trouble or error code, in the memory of the EBCM. This code indicates the nature of the problem that occurred. When the EBCM determines a fault and turns the warning light on, it also stores the code in its memory (Figure 15-8).

The code can be a "soft" code, which is temporary, or a "hard" code, which is more permanent. A soft code will be removed from the memory if the ignition key is cycled (i.e., turned off and on). If the problem is intermittent or temporary and goes away, the code will disappear when the key is cycled. If you road-test a car, park it, and turn off the ignition, you will erase any soft codes from the road test. A hard code will stay in the memory for a certain number of ignition cycles or until cleared, depending on the system. Some systems can display the number of times a problem has occurred or the number of ignition cycles since the problem has occurred.

Symptom	*Action*
Poor tracking during antilock braking	Do test 9, the wheel valve functional tests.
Spongy brake pedal	Bleed brakes. Check mounting of hydraulic unit. Check condition of calipers and rotors.
Pump motor runs longer than 1 minute, antilock light normal, flashing brake light	Check the brake fluid level. Do test 10, the hydraulic system pressure test. Do tests 2 and 5, the pressure switches test and the pump relay test. If the brake light flashes but the pump motor turns off normally in less than 1 minute, do test 7, the timer/flasher module test.
Antilock light on solid, brake light normal	Do tests 1, 2, 3, 4, 5, 6. Start with test 1, the pin-out box test.
Antilock lights come on while moving, brake light normal	Measure the wheel speed sensor resistances and voltage in test 1, the pin-out box test. If a speed sensor voltage is missing or low, check the gap at the toothed wheel. Also check wheel bearing end play and runout.
Antilock light on solid, brake light on solid	Check brake fluid level and do test 3, the fluid level switches test. Do tests 5 and 6, the pump motor test and the pump relay test. Do test 2, the pressure switches test. If condition persists, do the hydraulic system pressure test.
Intermittent antilock light, brake light normal	Do tests 2 and 3, the pressure switches test and the fluid level switches test. Check the connectors and contacts at the wheel speed sensors.
Antilock light normal, brake light on solid	Check circuit 33 (TAN/WHT), the park brake switch, and the ignition switch for a short to ground. Disconnect the timer/flasher module. If light goes out, replace the module. Do tests 2 and 3, the pressure switches test and the fluid level test.
No antilock light during start-up, brake light normal	Check bulb. Check for battery voltage at terminal 27 of the pin-out box, test 1. Ignition on. Repair open in circuit 852 (GRY/WHT) if no voltage. Check diode, terminals 27 and 3 of breakout box.
Antilock and brake lights on while braking	Do test 10, the hydraulic system test, to check the accumulator precharge pressure. If the pump motor runs more than a few seconds after the car sits overnight, check for external leaks. If none, replace the hydraulic unit.

Figure 15-7 A symptom-and-action chart can be used to locate the cause of a problem. The tests are described in the service manual for the particular vehicle. (Courtesy of General Motors Corporation, Service Technology Group)

	None	Blink Codes	Bi-directional Scan Tool	Driver Information Center
Bosch 2U/2S	*	X	X	
Bosch III		X		X
Delco III			X	
Delco VI			X	
Kelsey-Hayes RWAL		X	X	
Kelsey-Hayes 4WAL		X	X	
Teves Mark II	*	*		*
Teves Mark IV			X	

* Depending on model year and application

Figure 15-8 Diagnostic codes are read using different methods depending on the ABS type. (Courtesy of General Motors Corporation, Service Technology Group)

The codes are displayed in various ways. On some vehicles with digital instrument panels, the code is displayed as the actual code number. On some other vehicles, the code is read by counting the number of flashes of the brake warning light. Many vehicles require that a test light, voltmeter, dedicated tester, or scanner be connected to the special diagnostic terminal. A dedicated tester is a unit developed by the vehicle manufacturer just for that particular make or model of vehicle or ABS. When using a test light or analog voltmeter, the code is read by counting the number of flashes of the test light or sweeps of the meter needle.

Handheld scan tools have become popular in diagnosing faults in engine electronic control systems. These units can comunicate with the EBCM and are very useful in reading ABS trouble codes. In some cases, they can be used on a road test to display live, real-time operation of the system, by displaying things like battery voltage, wheel-speed sensor output, and solenoid operation (Figure 15-9). Most dedicated testers are designed to plug directly into the diagnostic terminal. An aftermarket scan tool usually requires an adapter to make this connection (Figure 15-10). Some of the scan tools have the ability to indicate the proper or improper operation of the various switches and solenoids, the voltage at the control module and ignition switch, and the speed indicated at each of the wheel-speed sensors, as well as the trouble codes.

Reading a code usually requires a special operation. In some cars, this is as simple as turning the ignition on and holding the brake pedal applied for five seconds or longer. The procedure for one manufacturer is to connect a special shorting tool to the service check connector; turn on the key; and record the frequency and timing of light blinks from the ABS warning light. The light flashes indicate the codes as shown in Figure 15-11. A table given in the service manual is then consulted to interpret the codes. Most vehicles require that the diagnostic terminal

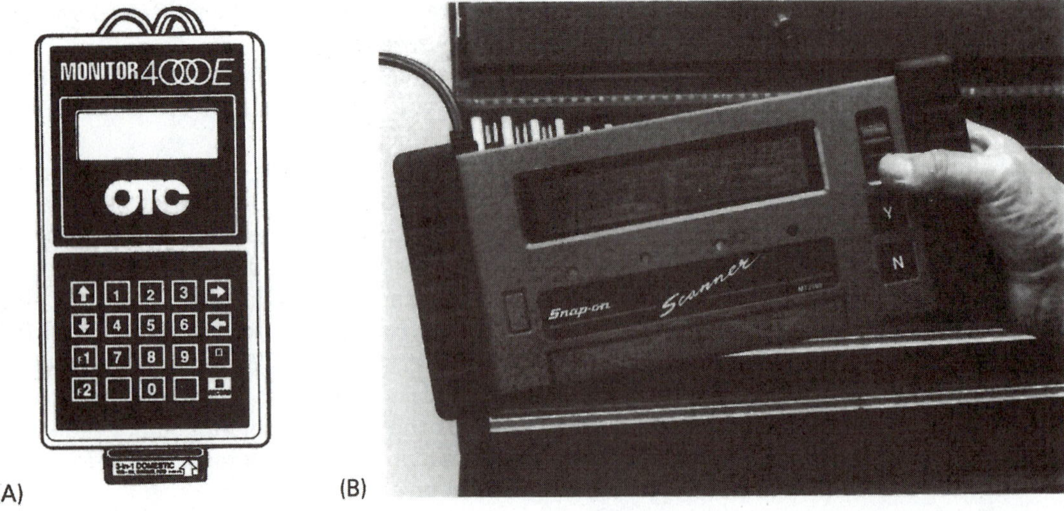

Figure 15-9 Handheld scanners can be used on some systems to read the problem codes, read ABS-component operation on some systems, and clear codes on some systems. (A is courtesy of Snap-on Tools Company, Copyright Owner; B is courtesy of SPX/OTC)

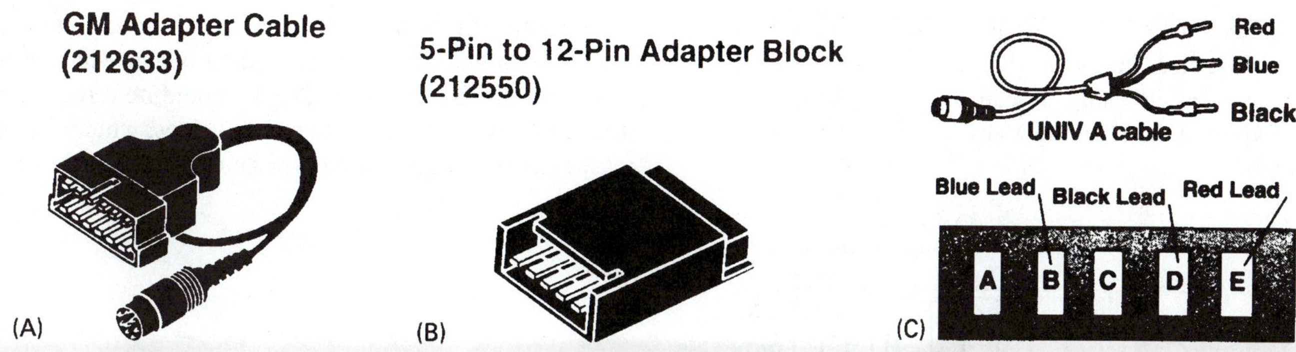

Figure 15-10 Adapter cables are used to connect an aftermarket scan tool to the vehicle's diagnostic connector. (Courtesy of SPX/OTC)

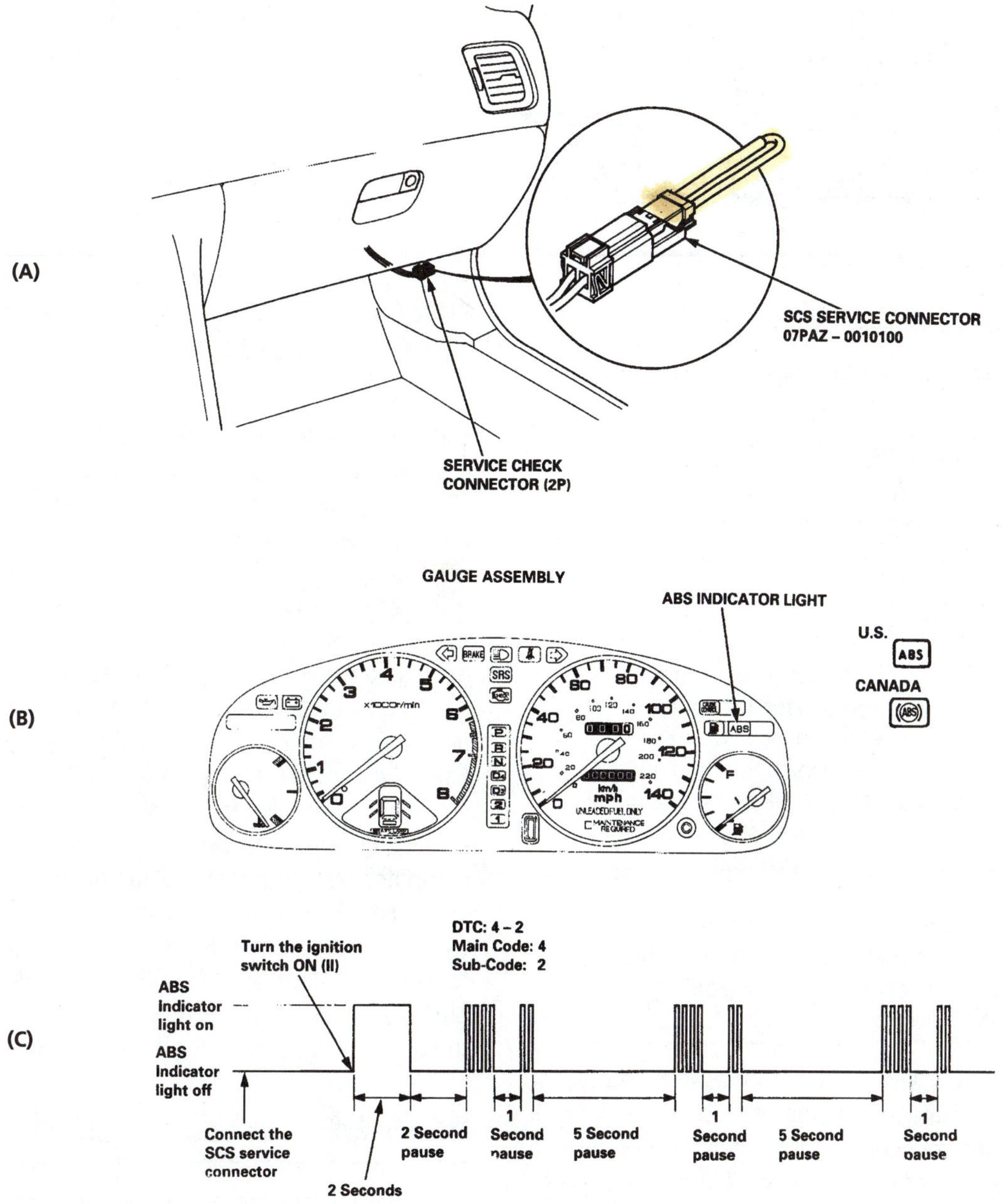

Figure 15-11 When the SCS Service Connector tool is connected to the service check connector (A) and the key is turned on, the ABS light will blink out any codes stored (B). The timing and length of the light blinks tells the code; a code 4-2 is shown here (C). (Courtesy of American Honda Motor Co., Inc.)

be connected to ground or that two terminals be connected together (Figure 15-12). Reading a code can also be accomplished using a scan tool or dedicated tester. Some systems will have as many as fifty or sixty different codes.

A chart like the one in Figure 15-13 is used to interpret the code to determine what the fault is and what further tests and repairs are necessary. It should be remembered that codes indicate faulty electrical values, which in some cases can be caused by improper mechanical operation.

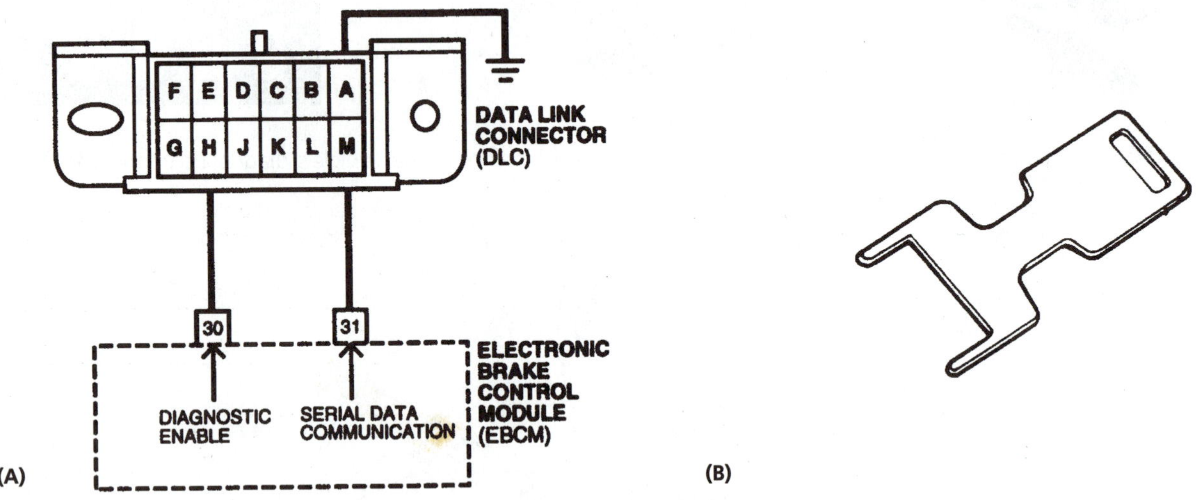

Figure 15-12 A diagnostic terminal (DLC, Data Link Connector or ALDL, Assembly Line Data Link) is manipulated to obtain diagnostic codes (A). A shorting tool (B) can be used to make the necessary connection. (A is courtesy of General Motors Corporation, Service Technology Group; B is courtesy of SPX/OTC)

ABS Diagnostic Codes

Code	*Description*
1	LF valve problem
2	RF valve problem
3	RR valve problem
4	LR valve problem
5	LF wheel speed sensor low output
6	RF wheel speed sensor low output
7	RR wheel speed sensor low output
8	LR wheel speed sensor low output
9	LF/RR diagonal wheel speed signal error
10	RF/LR diagonal wheel speed signal error
11	Replenishing valve problem
12	Valve relay error
13	Improper pressure switch signal
14	Improper travel switches sequence
15	Improper brake switch signal
16	EBCM errors

Figure 15-13 This chart identifies 16 diagnostic codes. For example, code 2 indicates that a problem exists in the electrical portion of the right front control valve. Some vehicles have up to 56 codes. (Courtesy of General Motors Corporation, Service Technology Group)

Technicians will normally make tests such as those in the next section to determine the exact cause of the problem. They will then repair the problem, clear the codes, and finally, road-test the vehicle to make sure the problem is corrected and the problem codes do not came back. Clearing a code to erase it from the memory also requires a special operation. As mentioned earlier, a soft code is cleared by simply turning off the ignition. In some cars, driving the car above a certain low speed or removing the diagnostic connector will clear all codes. In other cars, a dedicated tester or scan tool is used to clear codes.

As a last resort, codes can be cleared by removing the power fuse to the control module or disconnecting the battery cable, but this operation will erase all of the other electrical memories in the car. Some vehicles with electronic engines and transmissions will have to go through a computer relearn process if the battery or that particular control module is disconnected. Drivability problems such as rough idle, hesitation or stumble, or odd shifting patterns can be the result. It will take driving through a certain number of cycles or miles for the computer to relearn the car's operation and for things to return to normal.

15.5 Electrical Tests

To make the system electrical checks more convenient, easier, and quicker, a pin-out or breakout box is connected to the system. This unit is installed by removing the connector from the EBCM and attaching it to the pin-out box (Figure 15-14). The pin-out box provides a set of measur-

(A)

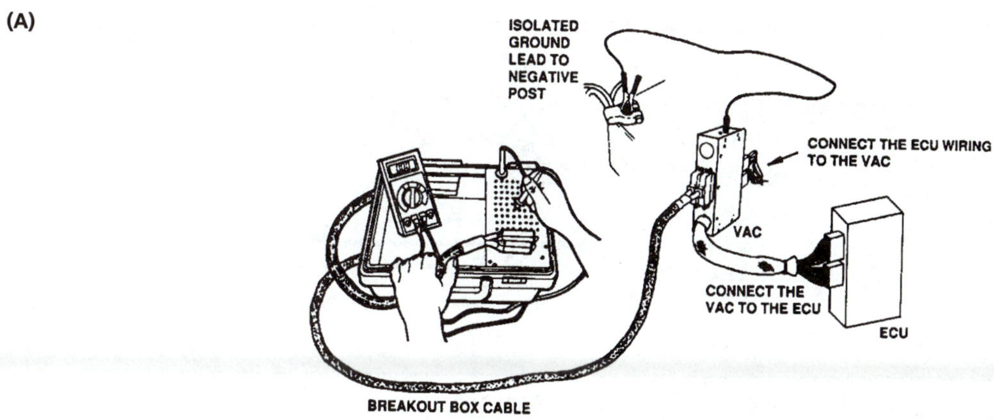

(B)

ANTI-LOCK QUICK CHECK SHEET USING ROTUNDA EEC-IV 60-PIN BREAKOUT BOX 014-00322

Item to be Tested	Ignition Mode	Measure Between Pin Numbers	Tester Scale/Range	Specification	Pinpoint Test
Battery Power	OFF or ON	1 and 13 2 and 14	dc Volts dc Volts	10 V min. 10 V min.	B B
Ignition Feed	OFF ON	15 and 13 15 and 13	dc Volts dc Volts	0 V 10 V min.	B B
LF Sensor Resistance RF Sensor Resistance	OFF OFF	5 and 6 7 and 8	kOhms kOhms	0.8-1.4 k Ohms 0.8-1.4 k Ohms	C D
LR Sensor Resistance RR Sensor Resistance Sensor Continuity To Ground LF RF LR RR	OFF OFF OFF OFF OFF OFF	9 and 10 11 and 12 5 and 13 6 and 13 7 and 13 8 and 13	kOhms kOhms Continuity Continuity Continuity Continuity	0.8-1.4 k Ohms 0.8-1.4 k Ohms No continuity No continuity No continuity No continuity	F E C D F E
Sensor voltage: Rotate wheel @ one revolution per second LF RF	 OFF OFF	 5 and 6 7 and 8	 ac mVolts ac mVolts	 100-3500 mV 100-3500 mV	 C D
LR RR	OFF OFF	9 and 10 11 and 12	ac mVolts ac mVolts	100-3500 mV 100-3500 mV	F E
Diagnostic Link	ON	28 and 13	dc Volts	10 V min.	A
ABS Warning Indicator	OFF ON	21 and 13 21 and 13	dc Volts dc Volts	0 V 10 V min.	G G
Ground Continuity	OFF	13, 14 and 22	Continuity	< 5 ohms	A

Figure 15-14 A pin-out or break-out box (A) is connected into the wiring harness at the EBCM and provides a terminal for testing each of the circuits for continuity, resistance, or voltage. A set of Quick Checks for the most common problems is shown at (B). (A is courtesy of SPX/OTC; B is courtesy of Ford Motor Company)

ing pins to which the leads of a digital volt-ohmmeter can be easily connected to measure the resistance or voltage of the various wheel-speed sensors, hydraulic modulators, and power relays in the ABS (Figure 15-15).

All electrical checks should begin at source voltage by making sure that the battery and alternator are good and can supply the proper voltage (Figure 15-16).

Next, the checks go to the circuit where a problem is indicated; let's say it is in the left front wheel sensor. For example, on a particular car, pins 5 and 23 might lead to the left wheel sensor, and this sensor should have 800 to 1,400 ohms of resistance. One ohmmeter lead is connected to pin 5, and the other to pin 23. A reading between 800 and 1,400 indicates a good sensor and a good sensor circuit. A reading less than 800 indicates a shorted sensor, and a reading over 1,400 indicates an open sensor or wire or a dirty or loose connection. If the sensor resistance is correct, the meter is then switched to AC volts and the leads are left connected to the pins 5 and 23 while the left front wheel is rotated by hand. A fluctuating reading between 0.05 and 0.7 volts indicates a good sensor, sensor ring, and air gap. No reading or a reading less than 0.05 volt indicates a fault in the sensor ring or air gap.

If the electrical value—resistance or voltage—of a component is wrong when measured at the pin-out box, the technician will then measure the value directly at the com-

Cavity	Circuit/Color	Function
1	B7 WT	RF wheel speed sensor (+)
3	B116 GY	Pump/motor relay control
4	G84 LB/BK	Traction control warning lamp
5	D1 VT/BR	C2D bus (+)
6	B6 WT/DB	RF wheel speed sensor (-)
7	A20 RD/DG	Fused battery feed
8	B28 VT/WT	Rotation sensor (-)
9	B27 RD/YL	Traction control switch sense
10	B30 RD/WT	Brake pedal travel sensor feed
11	Z1 BK	Body ground
12	Z1 BK	Body ground
13	B120 BR/WT	Switched battery feed
15	B9 RD	LF wheel speed sensor (+)
16	G19 LG/OR	ABS amber warning lamp control
17	B21 DG/WT	Low brake fluid switch #2 feed
18	G9 GY/BK	Red brake warning lamp control-CAB
19	B3 LG/DB	LR wheel speed sensor (-)
20	G83 GY/BK	Traction control function lamp
21	B29 YL/WT	Rotation sensor (+)
22	L50 WT/TN	Stop lamp switch (to lamps)
24	Z1 BK	Body ground
25	B120 BR/WT	Switched battery feed
27	D2 WT/BK	C2D bus (-)
28	B4 LG	LR wheel speed sensor (+)
29	B2 YL	RR wheel speed sensor (+)
30	B8 RD/DB	LF wheel speed sensor (-)
31	B1 YL/DB	RR wheel speed sensor (-)
32	B58 OR/BK	ABS main relay control
33	F20 WT	Ignition feed
35	B20 DB/WT	Low brake fluid level switch #2 return
36	B31 PK	Brake pedal travel sensor return
37	B120 BR/WT	Switched battery feed

Figure 15-15 The terminals of the 37-pin connector are identified; note that it is easy to get mixed up when trying to locate a particular terminal. (Courtesy of Chrysler Corporation)

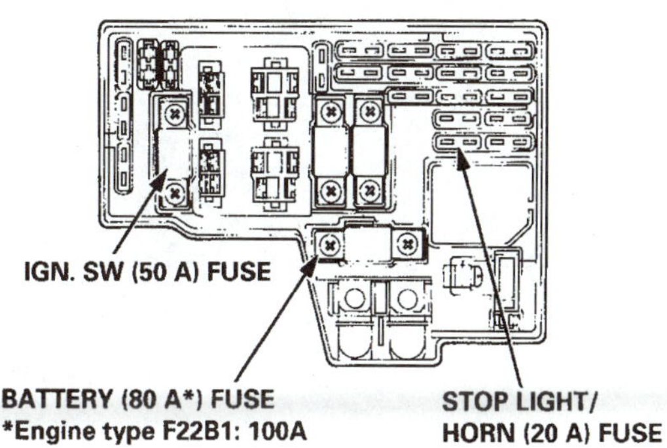

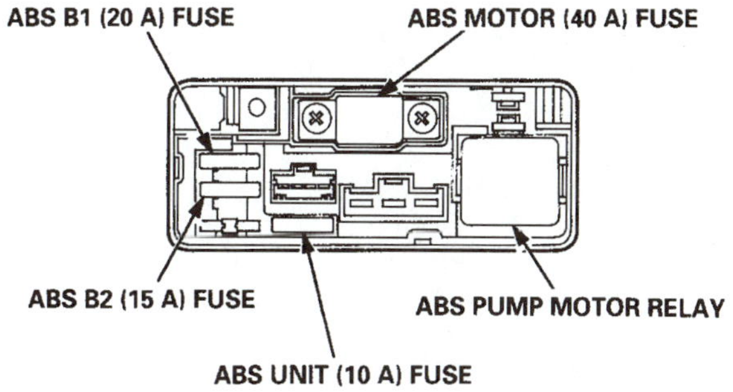

Figure 15-16 The power source for most ABS is the vehicle's power center/fuse-relay box. (Courtesy of American Honda Motor Co., Inc.)

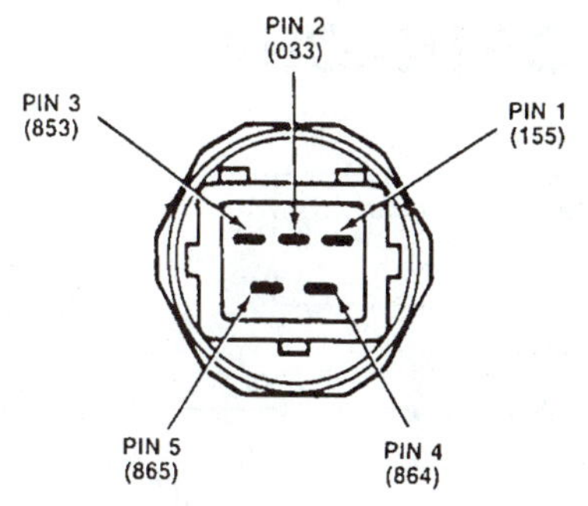

Resistance Pressure Switches–Ignition Off (System Discharged)

Pins	*Resistance 200-Ohm Range*
3 and 5	Infinite ohms
1 and 4	0 ohms
1 and 2	0 ohms

Figure 15-17 A pressure switch connector (top). An ohmmeter connected between pins (3) and (5) should read infinite resistance, and when connected between pins (1) and (2) or (4) should read 0 ohms. (Courtesy of General Motors Corporation, Service Technology Group)

ponent. If the component has the wrong value, it should be replaced. If the measurement is correct at the component but wrong at the pin-out box, faulty wiring is indicated. For example, if the resistance of the left front wheel sensor measures greater than 1,400 ohms at the pin-out box, but less than 1,400 ohms at the sensor connector, the sensor is good, but there is a bad connection somewhere between the sensor and the pin-out box (Figure 15-17).

Service manuals usually provide a detailed procedure to locate the problem cause. Figure 5-18 illustrates the circuits between the control module and the four wheel-speed sensors from the service manual of a late-model car, and this shows the wire/circuit numbers, wire color coding, and all the connectors. The diagnostic chart for

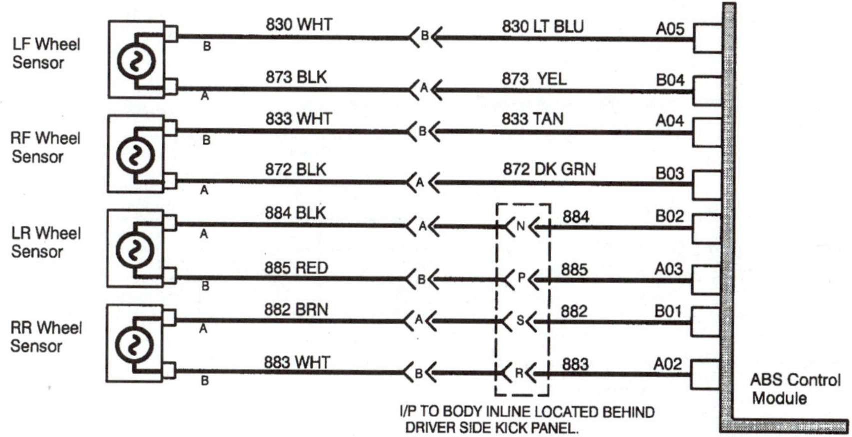

Figure 15-18 The circuit for the wheel-speed sensors identifies the wire color and number coding and the numbers for the various wire connections. It can be used along with the diagnostic chart for the same vehicle. (© Saturn Corporation, used with permission)

DIAGNOSTIC CHART

CODE 32

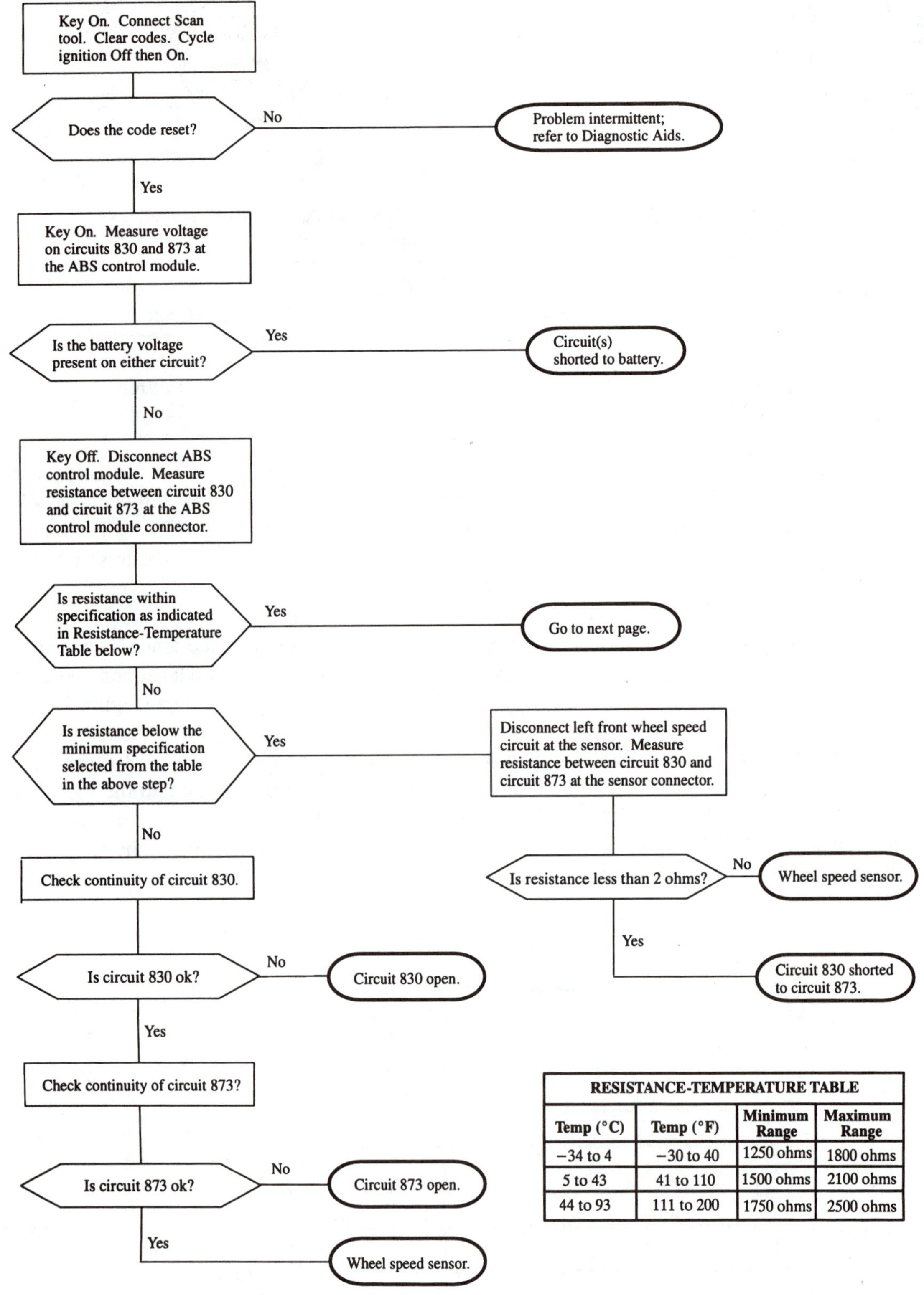

RESISTANCE-TEMPERATURE TABLE			
Temp (°C)	Temp (°F)	Minimum Range	Maximum Range
−34 to 4	−30 to 40	1250 ohms	1800 ohms
5 to 43	41 to 110	1500 ohms	2100 ohms
44 to 93	111 to 200	1750 ohms	2500 ohms

Figure 15-19 The diagnostic procedure to locate the problem causing code 32 (left wheel speed circuit fault) on this car. The circuit numbers are shown in Figure 15-18. (© Saturn Corporation, used with permission)

DIAGNOSTIC CHART

CODE 32

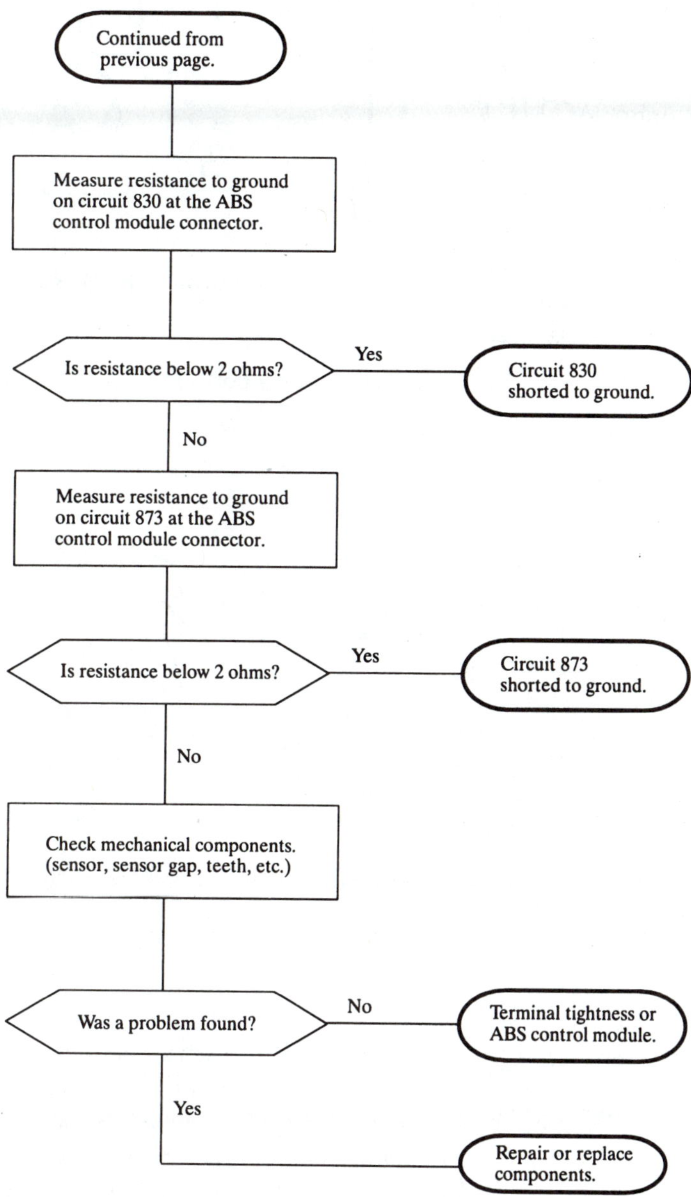

Code 32, left front wheel circuit fault, is shown in Figure 15-19. Note how this chart takes a technician through a series of diagnostic checks, using the circuit diagram, until the problem is located.

A dedicated ABS tester is available for some car models. It is attached to either a diagnostic connection or, like the pin-out box, the EBCM connector (Figure 15-20). The tester allows the technician to make electri-

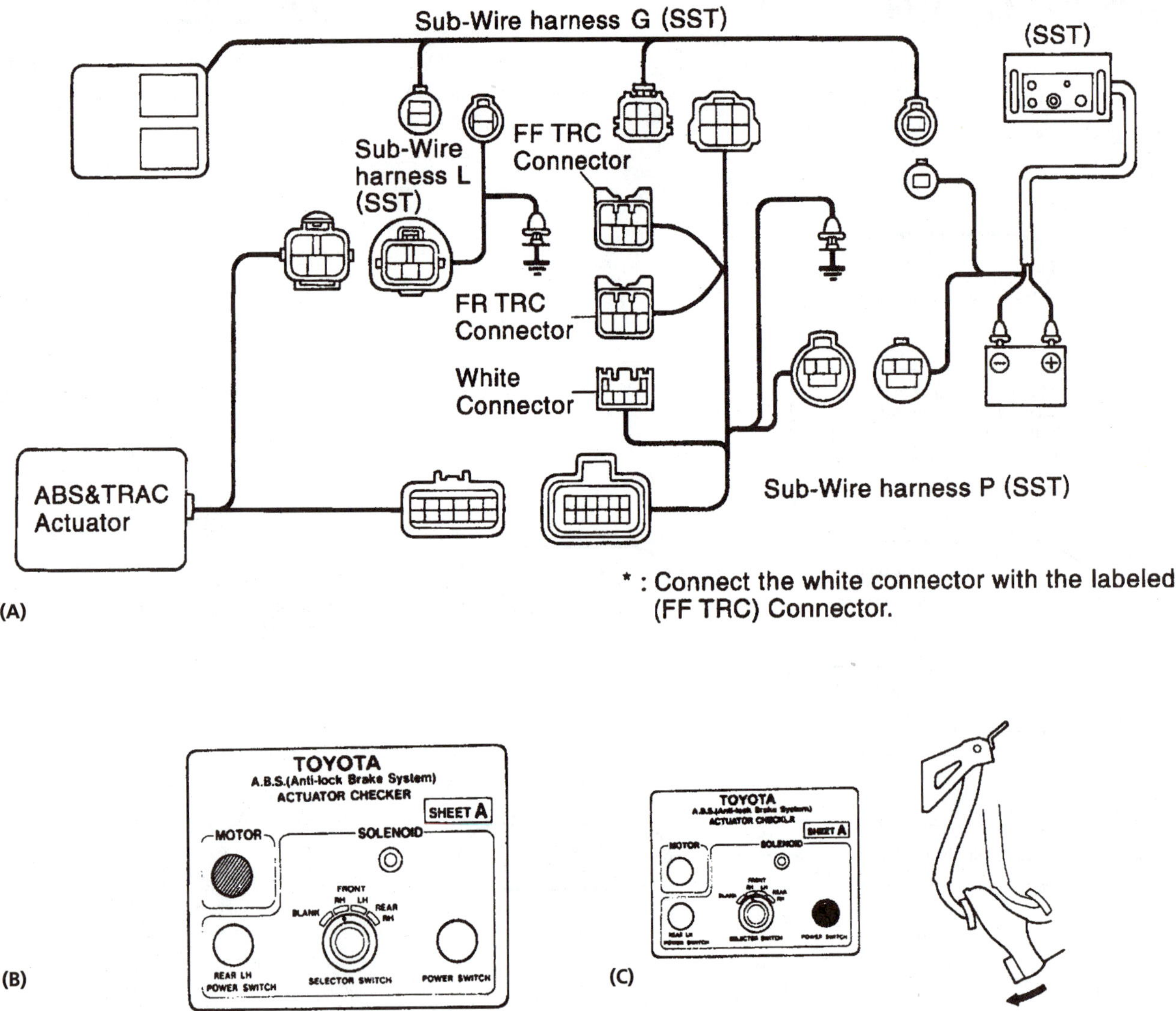

Figure 15-20 Most vehicle manufacturers have developed a tester for their ABS systems; it is connected into the diagnostic connector(s) (A) and (B) and is used to test various components (C).

cal checks for various portions of the ABS circuits, monitor pump operation, obtain output readings from the wheel-speed sensors, and operate the pressure control valves.

15.5.1 Speed Sensor Mechanical Problems

Each wheel-speed sensor generates an electrical signal from the speed of that wheel and that signal must be regular with a smooth transition as the speed and frequency increase and decrease with the speed of the car. The speeds of the different sensors must match with only a slight variation allowed between them. Something as simple as mismatched tire size can cause speed sensor problems.

Other mechanical sensor problems can be caused by a broken or chipped tooth on a sensor ring, or a ring that is either not round or is mounted so that runout occurs. Loose or faulty wheel bearings are more commonly encountered problems. Anything that can change sensor gap during rotation will cause the EBCM to shut the system down, turn on the warning light, and set a problem code. Don't forget that the center core of a sensor is a magnet, and it will pick up metal debris that can affect its operation. Sensors that are mounted into an axle or transmission can attract enough iron particles to cause an erratic signal.

When speed sensor problems that do not appear to be electrical are encountered, a careful inspection of the tire sizes, wheel bearings, sensor rings, and sensors should be conducted. Check the sensor gap using a nonmagnetic feeler gauge between the sensor pole and the toothed ring (Figure 15-21). If a steel feeler gauge is used, the magnetic pull will increase the drag and cause a false reading. Also rotate the wheel and ring, and check the gap in several locations around the ring. It should be the same in each location. When sensors are mounted down into a part so you can not measure the gap directly, measure a depth down to a tooth, and subtract the height of the sensor from this dimension (Figure 15-22).

Sensor electrical output can be checked using an oscilloscope or lab scope to get a better look at its value. This unit displays the electrical signal on a CRT/computer screen, and a wheel-speed sensor should be a smooth sine wave (Figures 15-23 and 15-24). An irregular signal indicates the type of problem.

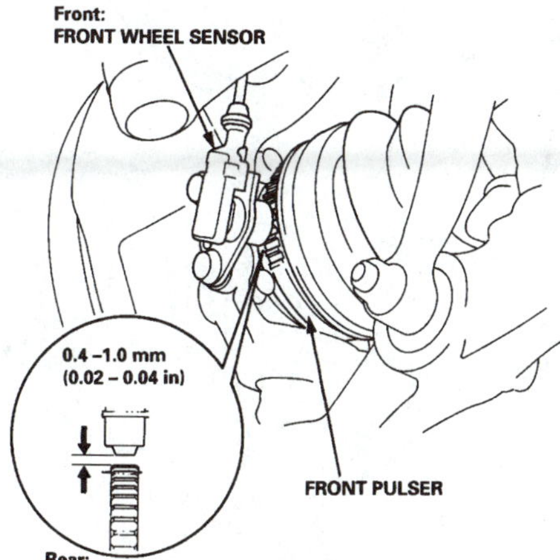

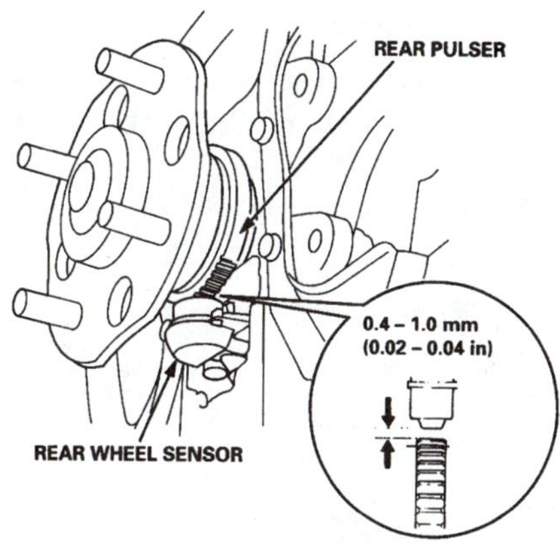

Figure 15-21 Sensor air gap is the clearance between the sensor pole and the teeth of the reluctor/sensor/pulser ring. (Courtesy of American Honda Motor Co., Inc.)

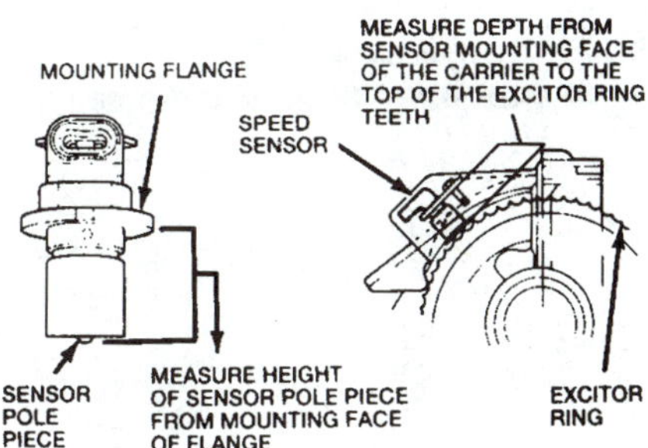

Figure 15-22 This sensor is mounted into the rear axle, so the gap cannot be measured. The amount of air gap can be determined by measuring the height of the sensor and the depth to the teeth of the sensor ring. (Courtesy of Ford Motor Company)

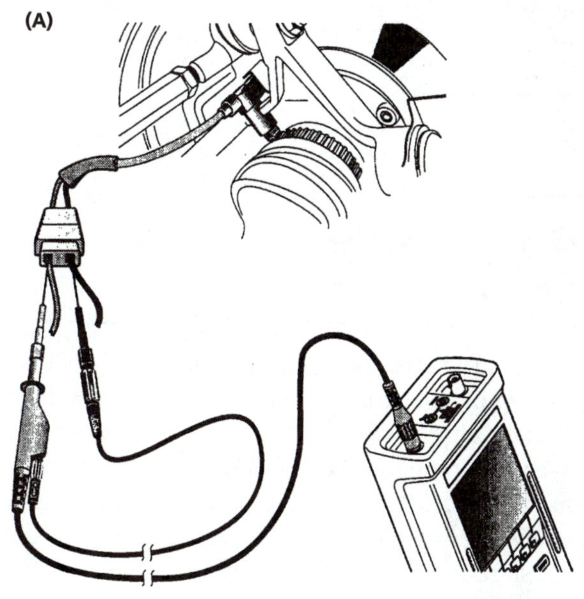

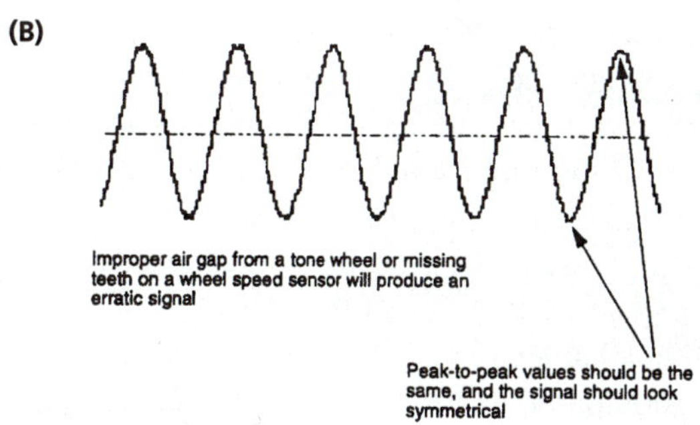

Figure 15-23 This Fluke 98 lab scope is connected into a wheel-speed sensor (A). It will display a wave form of the sensor output in which the time value is shown horizontally and the electrical voltage value is shown vertically (B). (Reproduced with permission of Fluke Corporation)

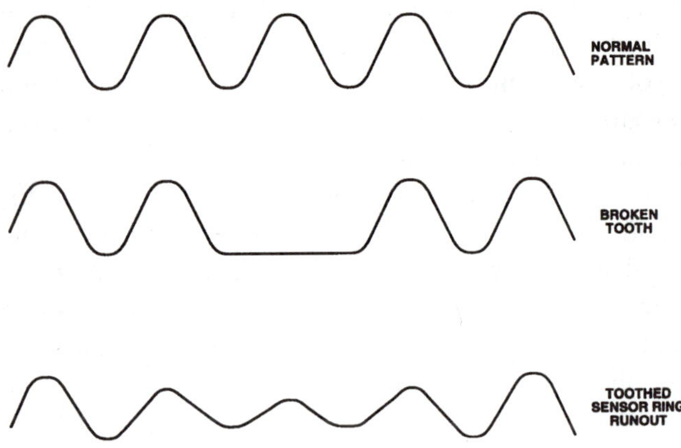

Figure 15-24 If the output of a wheel-speed sensor is displayed on an oscilloscope, it should appear as a smooth sine wave (top); a broken tooth or an uneven sine wave indicates problems. (Courtesy of General Motors Corporation, Service Technology Group)

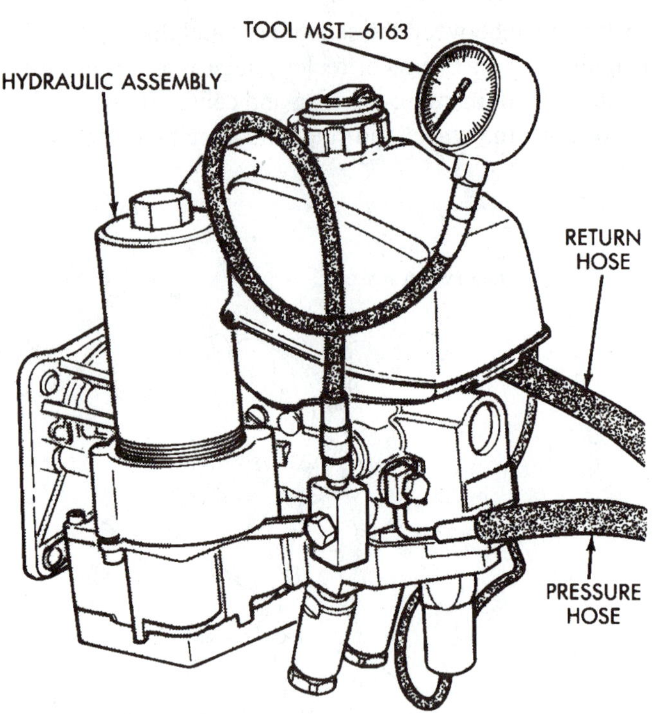

Figure 15-25 Tool MST-6163 is a pressure gauge, used to check the pump, switch, accumulator, and seals in this hydraulic assembly. (Courtesy of Chrysler Corporation)

15.6 Hydraulic Pressure Checks

Some ABS problems can be caused by poor hydraulic pump output or sticky or leaking modulator valves. Pump output can be checked by attaching a pressure gauge (Figure 15-25). A pressure gauge allows you to check the amount of pressure that the pump develops; the settings of the switch(es) to start and stop the pump; the cycle time between start and stop; and the leak rate after the pump stops. In most systems, after the pump runs, builds pressure and shuts off, the pressure should hold steady. The pump should not restart unless the brakes are applied to drop the pressure.

Some diagnostic scan tools allow the operator to cycle the modulator valves. With these units, the vehicle can be raised on a hoist, and with the brakes applied, the modulator valves can be operated while the wheels are checked to determine if they are locked up or free to rotate. A good modulator will operate correctly in each of its circuits.

A sinking brake pedal is normally diagnosed by blocking off the suspected portion of the system using plugs, as described in Chapter 12.

Most modulator valves are not serviceable and can only be repaired by replacement (Figure 15-26). If replacement becomes necessary, special bleeding procedures, as described later in this chapter, should be followed. Most experts recommend periodic brake fluid replacement to help prevent dirt and corrosion that might cause problems in these expensive units.

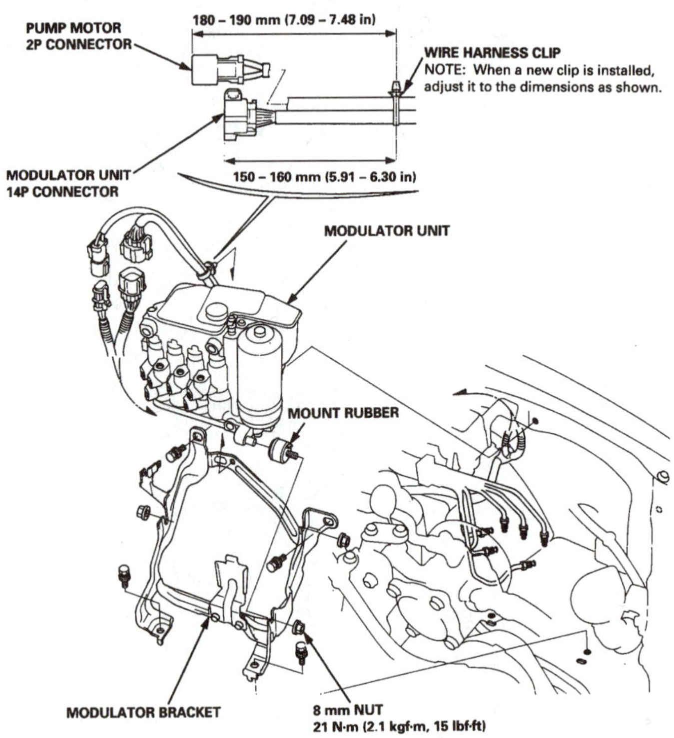

Figure 15-26 In most cases, a faulty modulator is serviced by replacing the entire unit. (Courtesy of American Honda Motor Co., Inc.)

15.7 Repair Operations

All the mechanical and electrical ABS components are serviceable to some degree. Some units allow replacement of faulty modulator components such as the motor pack, pump, or accumulator (Figure 15-27). When replacing any of these components, follow the manufacturer's procedure to prevent damage to any of the components or parts and to ensure a reliable and safe repair (Figure 15-28). Replacement is the normal service procedure for the electrical components. As mentioned earlier, the manufacturer's instructions should be followed when making replacements. Service operations will include removal and replacement of the hydraulic control unit, hydraulic accumulator, hydraulic pump motor, reservoir, EBCM, individual wheel sensors, sensor rings, or electric relays and switches. Some manufacturers require replacement of an entire wire harness rather than repair of a wire or connector fault to ensure the integrity of that electrical circuit.

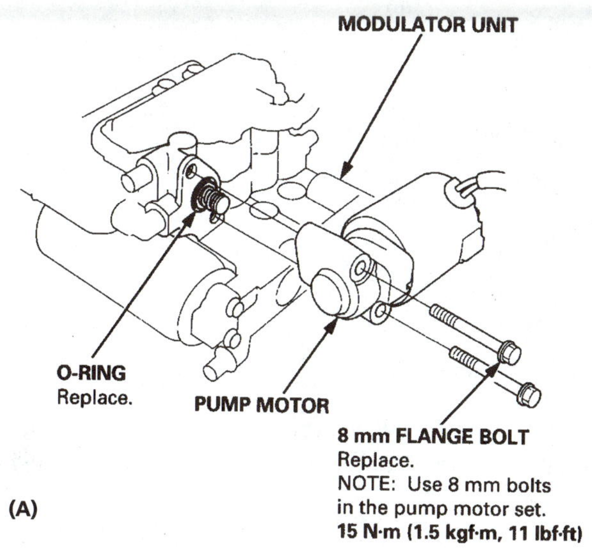

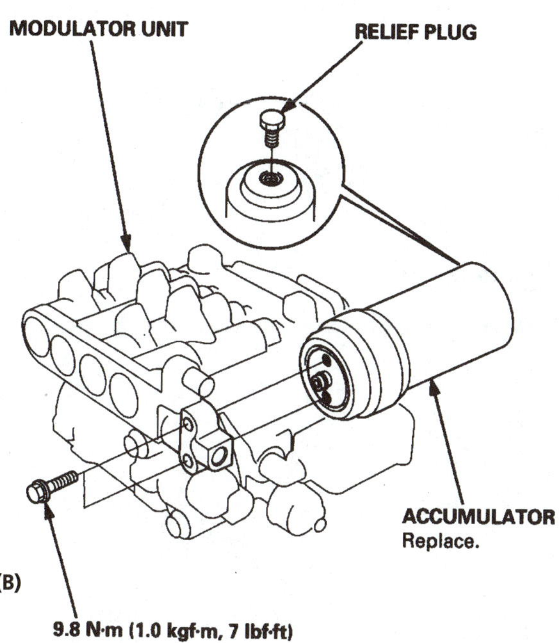

Figure 15-27 On this particular modulator assembly, a faulty pump and motor (A) or accumulator (B) can be removed and replaced. (Courtesy of American Honda Motor Co., Inc.)

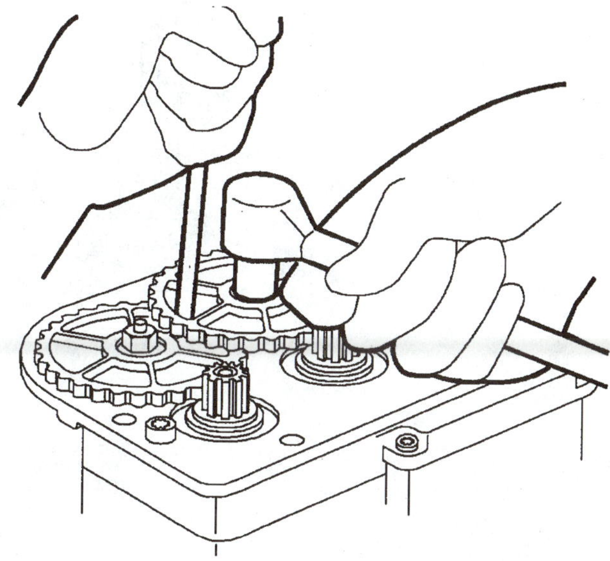

Figure 15-28 Modulator drive gears are a serviceable portion of this modulator assembly. (© Saturn Corporation, used with permission)

A faulty sensor ring can usually be replaced by pressing the old ring off of the CV-joint, rotor, or axle. The new ring should be carefully pressed into place so it is not damaged, and so it is in the correct position (Figure 15-29). Most sensors are held in place by one or two bolts and can be easily replaced; be sure to route the wires in the same manner and location as the original (Figure 15-30). When replacing sensors or sensor rings, it is necessary to adjust the air gap. On some sensors, the air gap is adjusted by installing the sensor so the paper spacer on the sensor touches a tooth of the sensor ring. A new sensor will include a paper spacer; when a used sensor is reinstalled, a new paper spacer of the correct thickness can be glued onto the sensor. The paper spacer is designed to wear off as the car is driven. When paper spacers are not used, sensor gap is checked using nonmagnetic feel gauges as previously described.

The operation of the hydraulic pump, accumulator, and control switches can be checked by measuring the pressure. Special adapters are usually required to connect a gauge to the system. A particular sequence is usually

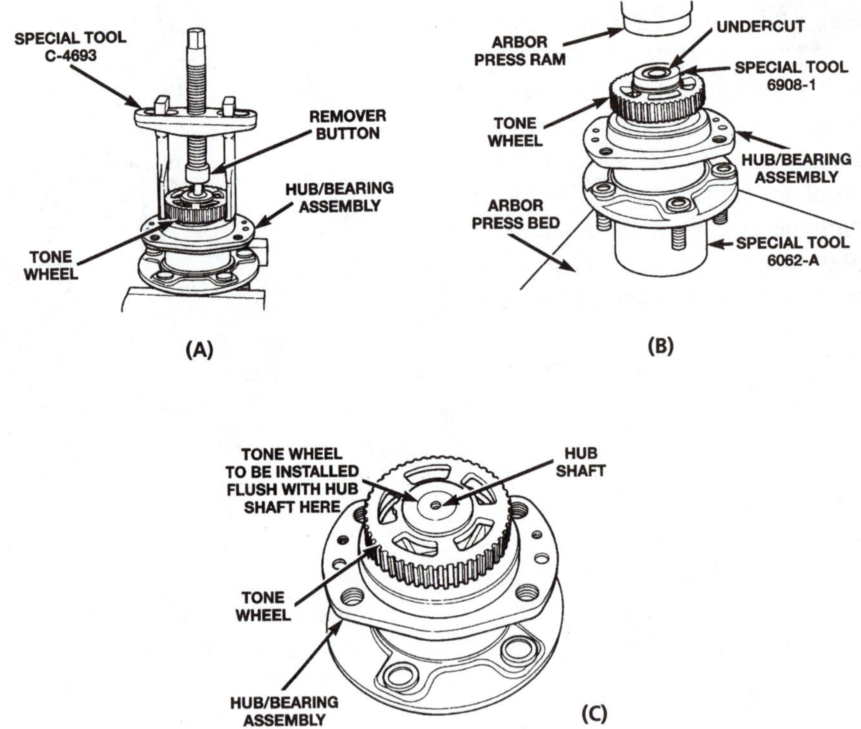

Figure 15-29 A faulty tone wheel/reluctor can be pulled or pressed off the hub (A). A new unit can be pressed into place; sometimes special press adapter tools are required (B). This new tone wheel should be flush with the end of the hub (C). (Courtesy of Chrysler Corporation)

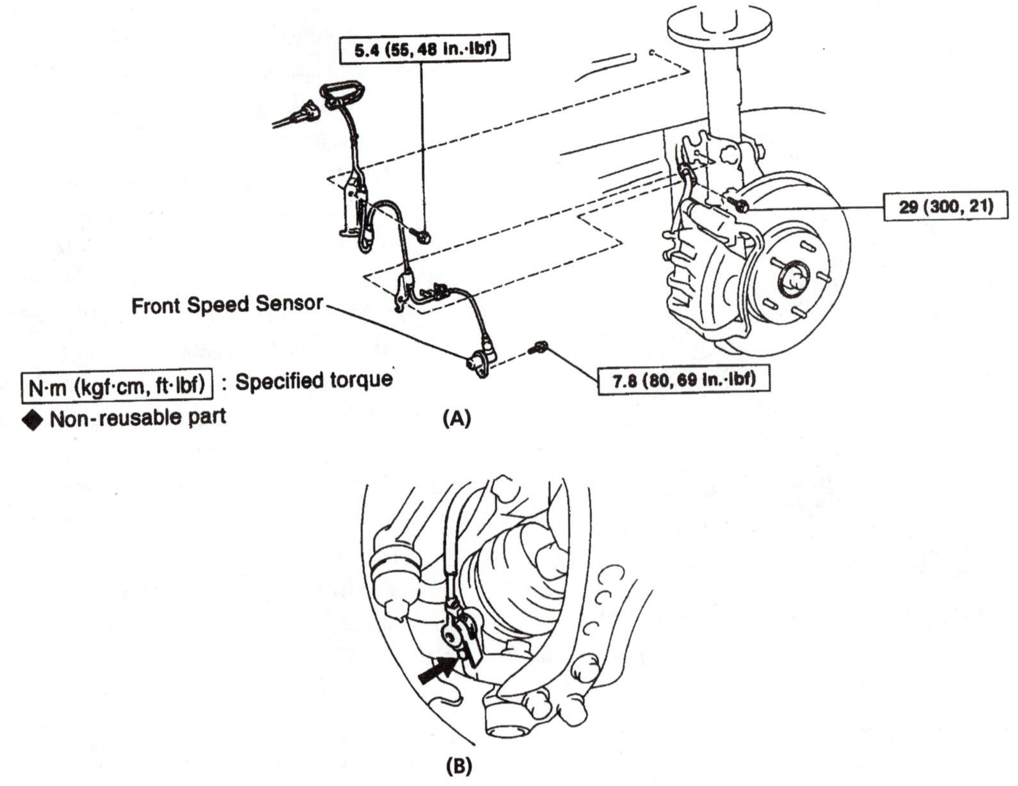

Figure 15-30 A wheel-speed sensor can be removed for testing or replacement. Note that the sensor air gap on this unit is established by a paper spacer. (© Saturn Corporation, used with permission)

required to determine what the correct pressure should be under various operating conditions. Switch problems are indicated by a failure to turn the pump on or off at the correct pressure. Pump problems are indicated by a failure to reach high-enough pressure or too fast a leak-down rate. Too fast of a leak-down rate can also be caused by a leaking valve. A faulty accumulator is indicated by a very fast pressure increase when the pump runs and a very fast pressure decrease when the brakes are operated with the pump off.

When making pressure checks or replacing the hydraulic control unit, accumulator, pump, or pressure switch, it is important to discharge the pressure from the accumulator before beginning the service operations. It is recommended that you apply and release the brake pedal twenty-five to thirty times with the key off. A definite increase in pedal resistance should be noted as the pressure is bled off.

15.7.1 Bleeding ABS

When service work on the hydraulic components is completed, a special bleeding process for that particular system is usually required. This procedure is described in the manufacturer's service manual and in various technician's service manuals from aftermarket component manufacturers.

Some systems can be bled manually or by pressure. Some of these systems include a bleeder screw at the hydraulic modulator or use a bleeding procedure like loosening a line fitting (Figure 15-31). Some systems use standard tools and procedures, but require that the key be off or on while purging air from certain locations. Some systems use flow from the electric booster pump to bleed certain portions of the system. Other systems require the use of a particular ABS tester to cycle solenoid valves for a thorough bleeding operation. These systems can usually be bled using a more lengthy procedure without the special equipment.

Remember that in an ABS that uses an electric pump and accumulator, air can become trapped in the accumulator. This air will not affect system operation until there is an ABS stop, and then the air can be forced into the system, producing a low brake pedal and a spongy operation.

Bleeding ABS is another procedure for which it is best to check the service manual for the correct procedure.

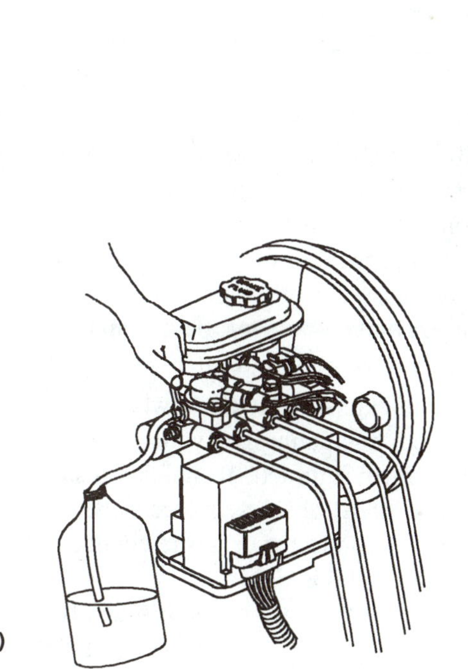

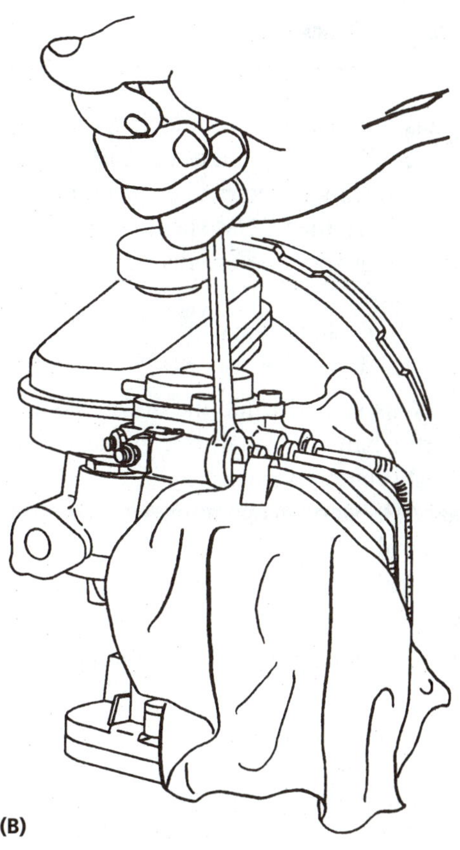

Figure 15-31 Some ABS systems require special bleeding procedures. This modulator assembly is equipped with a bleeder screw (A); line connections are bled by loosening the line fitting (B). (Courtesy of General Motors Corporation, Service Technology Group)

15.8 Completion

At the completion of ABS repair, you should check to ensure that all of the diagnostic trouble codes have been cleared. Next, take the vehicle for a test drive to make sure that all of your repairs and adjustments are working properly. After the road test, check that the red brake light and amber ABS light are working properly and that no additional trouble codes have been set.

15.9 Practice Diagnosis

You are working in a new-car dealership and encounter these problems.

CASE 1: The customer purchased a new car last month and has brought it back in with a complaint that the ABS light stays on. He mentions that he noticed it come on as he was driving away from a stop light, and it didn't go out after that. What is possibly wrong with this system? Where should you start checking for problems?

CASE 2: The customer has brought in her car with a complaint of a pulsating brake pedal. She bought her new car in the summer and had no problems until a cold, icy winter morning when she was in a hurry to get to work. On your road test, you notice normal brake and warning light operation. What should you do next? What was probably the cause of the complaint?

CASE 3: The three-year-old car had a torn CV-joint boot that caused the CV-joint to fail. You replaced the CV-joint and road-tested the car to make sure everything was right, and the ABS light came on when you got to about 10 mph. What went wrong? What checks should you make?

Terms to Know

amber antilock warning light
diagnostic trouble code (DTC)
hydraulic modulator
red brake warning light

Review Questions

1. Two technicians are discussing antilock brake system service. Technician A says that the EBCM can be ruined if you touch the electrical connector pins. Technician B says that the EBCM should be sprayed with anti-static-cling spray before disconnecting it. Who is right?
   a. A only
   b. B only
   c. both A and B
   d. neither A nor B
2. A car with ABS is started while the brake pedal is depressed, and a bump is felt in the pedal as the engine starts. This bump is caused by
   a. the normal ABS start-up operation.
   b. a faulty control module.
   c. a faulty control valve assembly.
   d. all of the above.
3. The red brake warning light in an ABS-equipped car should light
   a. during engine start-up.
   b. if the brake fluid level is low.
   c. when the parking brake is applied.
   d. all of the above.
4. In most cars, the amber warning light warns of a failure in the
   a. brake hydraulic system.
   b. EBCM or one of its circuits.
   c. mechanical portion of a control valve assembly.
   d. all of the above.
5. Technician A says that a problem in one of the wheel sensors will cause the amber warning light to come on as the car is being driven. Technician B says that an improper sensor air gap or faulty wheel bearing can cause this problem. Who is right?
   a. A only
   b. B only
   c. both A and B
   d. neither A nor B
6. Technician A says that a wheel sensor should produce a fluctuating DC voltage of about 9.5 volts as the wheel is turned. Technician B says that the resistance of a wheel sensor can be measured at the EBCM connector or at the connector closest to the sensor. Who is right?
   a. A only
   b. B only
   c. both A and B
   d. neither A nor B
7. A thorough check of an ABS electrical circuit requires a
   a. pin-out box.
   b. ohmmeter.
   c. AC voltmeter.
   d. all of the above.

16 Wheel Bearings—Theory and Service

Objectives

After completing this chapter, you should:

- ❑ Be familiar with the terms commonly used with wheel bearings.
- ❑ Be able to diagnose wheel bearing noises, wheel shimmy and vibration problems, and determine the needed repair.
- ❑ Be able to remove, clean, inspect, adjust, repack, and replace wheel bearings and races according to manufacturers' specifications.
- ❑ Be able to remove and replace axle bearings and seals.
- ❑ Be able to complete the ASE Tasks relating to wheel bearings' diagnosis and repair (See Appendix 1).

16.1 Introduction

The wheel **bearing**'s job is to allow the hub, wheel, and tire to rotate freely while holding them in alignment with the car's steering knuckle or axle. In doing this, the wheel bearing has to absorb the vertical load of the car, plus any side loads that result from cornering maneuvers, as well as other forces. In addition, on drive axles, the bearing allows the axle to rotate while transmitting driving torque to the wheel and tire.

Brake technicians must deal with wheel bearings when they service a drum or rotor mounted on a hub with serviceable wheel bearings, when loose or faulty wheel bearings cause brake problems, or when a leaky grease seal allows grease to spill on the lining or friction surfaces. There are five basic styles of wheel bearings, which are classified according to their type and function. The term **wheel bearing** is normally used for the bearings on all front wheels and nondrive-axle rear wheels. The term **axle bearing** is commonly used in relation to rear-wheel-drive axles. The various bearing types are as follows:

Nondrive-axle serviceable bearings (front wheels on RWD cars and rear wheels on FWD cars).

Nondrive-axle nonserviceable bearings (rear wheels on some FWD cars).

Drive-axle bearings on a solid axle (rear wheels on most RWD cars).

Drive-axle bearings on independent suspension (rear wheels with independent rear suspension [IRS] and front wheels on FWD cars).

Drive-axle nonserviceable bearings (front wheels on some FWD cars).

The nonserviceable styles are very similar to each other. They both support the weight of the car, and the drive-axle style is capable of transmitting torque from the axle to the wheel. As their name implies, nonserviceable bearings are not serviced. If the bearing is faulty, the bearing assembly is removed and replaced with a new one; "R and R" or R&R is the common term for this procedure.

This is also a strong similarity between nondrive-axle serviceable bearings and IRS drive-axle bearings in that they are usually a pair of tapered-roller bearings. They both require occasional cleaning, repacking with grease, and a clearance or preload adjustment. The front-wheel bearings of most four-wheel drive (4WD) utility vehicles are very similar to the nondrive-axle serviceable bearings.

16.2 Bearings and Bearing Parts

The term **bearing**, when used by most technicians and manufacturers, usually refers to ball, roller, or tapered-roller bearings. These bearing types are often described as *frictionless* bearings, because the balls or rollers, fitted between the two races, roll easily whenever either of the two races rotate. A *bushing*, commonly used for crankshaft bearings, has a sliding action (Figure 16-1). Bushing friction is usually reduced by grease or a constant flow of oil.

Frictionless bearings are usually made up of three major parts: a cone or inner race, a cup or outer race, and the balls or rollers (Figure 16-2). All of these parts are made from hardened steel alloys and are precision ground to very close size and finish tolerances. The bearing balls or rollers are usually placed in a cage or separator so they will not rub against each other. The cage also ensures that the balls or rollers stay spread out in the correct spacing for proper load distribution. Sealed bearings position a seal between the inner and outer races on one or both sides of the bearing. This seal is used to keep lubrication in the bearing and/or dirt out. A frictionless bearing must be lubricated to reduce friction and prevent wear. Lubricants also carry heat away from the bearing and protect the metal surfaces from corrosion. Dirt and other abrasive materials must be kept out to prevent bearing damage or wear.

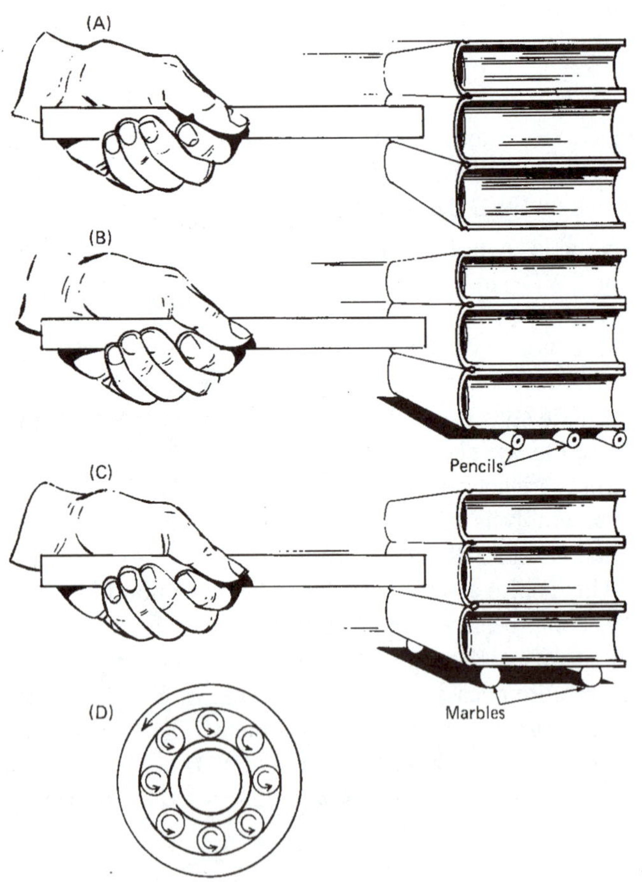

Figure 16-1 Try pushing a stack of books using a stick or ruler as shown in (A). You will notice a certain amount of drag. Place a few pencils (B) and then some marbles (C) under the books and try pushing them again. You should notice a definite reduction in the amount of drag and friction. The rolling action of the pencils and marbles is similar to the action in a bearing (D).

Figure 16-2 A disassembled tapered-roller bearing. The cup is also called the outer race and the cone is the inner race. (Courtesy of The Timken Company)

Ball bearings run in concave grooves that are ground into each of the races. A ball bearing usually has the ability to control radial movement and loads, as well as thrust movement (Figure 16-3). For example, it can support a

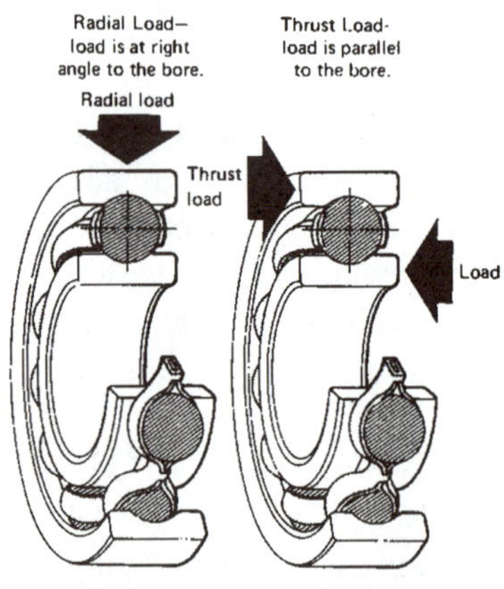

Figure 16-3 The two basic loads to which a bearing is subject are radial loads from a right angle to the bore and thrust loads parallel to the bore.

shaft and allow the shaft to rotate while keeping the shaft from moving sideways. Because the balls contact the races only in a tiny area, the amount of side or radial and/or end or thrust load that a ball bearing can support is rather limited. Note that only a few of the balls are carrying the radial load at one time (Figure 16-4). The balls on the nonloaded side of the bearing are not really doing much. Bearing failure will be discussed more thoroughly later in this chapter.

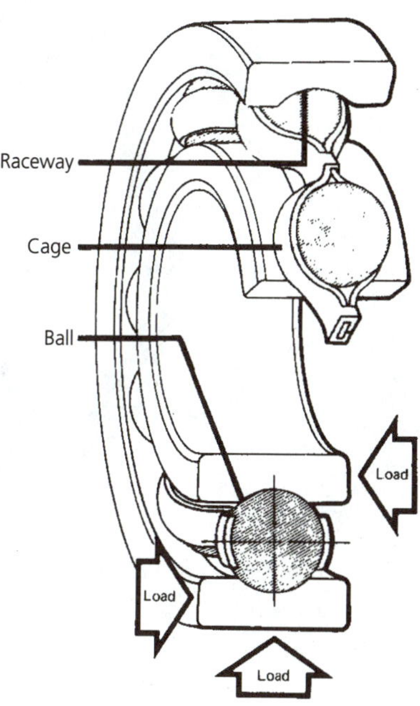

Figure 16-4 A cutaway view of a ball bearing showing the loads to which each ball is subject.

Roller bearings can carry a greater side load than ball bearings because rollers, being longer, have a much greater load-carrying surface area. A roller bearing, however, cannot control end thrust. When mounted where side thrust control is important, thrust bearings are used along with roller bearings. Very thin roller bearings are called *needle bearings*. Needle bearings can carry still more load because there are more of the thin needles and therefore a greater surface area to transfer loads.

Tapered-roller bearings are used when side thrust as well as radial loads need to be controlled. Many front-wheel bearings are a perfect example of this. The wheel bearings carry the radial load of the car's weight and the side load that is trying to take the hub off the spindle. Tapered-roller bearings are used in pairs, with the tapers of each bearing facing in opposite directions. It is these tapers that provide the sideways control (Figure 16-5).

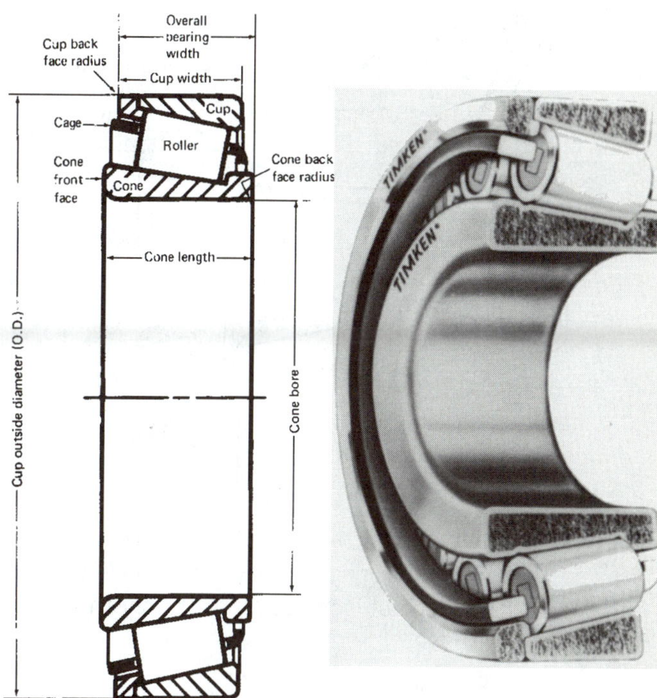

Figure 16-5 A section view of a tapered-roller bearing showing the normally used dimensions. (Courtesy of The Timken Company)

16.3 Bearing End Play—Preload

An important step when adjusting bearings is to set the correct bearing **end play**. Too much end play gives a loose, sloppy fit. Sloppy wheel bearings let the wheel wobble and change alignment angles, let the rotor or drum change position relative to the shoes and lining, or reduce the cone-to-bearing contact. As far as the bearing is concerned, it will have maximum life if it has a free-running clearance (no preload) with no appreciable end play. Free play also provides room for expansion as the bearings, shafts, and housings heat up during operation. A bearing will have end play if there is a slight clearance between the balls or rollers and the races. A bearing will have preload when there is pressure between the balls or rollers and the races (Figure 16-6).

Preload is used on bearings when the shaft must not change position. The pinion gear and shaft in a rear axle transmit a large amount of torque, and because of gear pressure, the gears try to move lengthwise as well as sideways. The pinion bearings must be preloaded to keep the gear from changing position under these heavy loads. Bearing preload increases the friction of the bearing and therefore consumes power. If possible, bearings should be adjusted to a slight end play for freer running. The exact

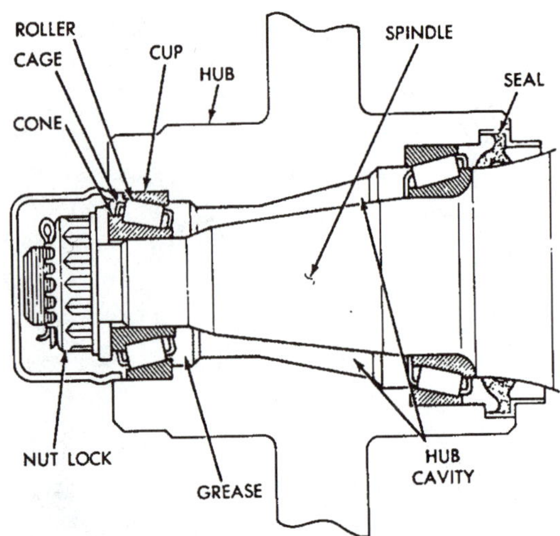

Figure 16-6 A hub, bearing, and spindle assembly as used on many nondrive wheels. Bearing end play or preload is adjusted with the nut inside the nut lock. (Courtesy of Chrysler Corporation)

amount of end play for a particular bearing set is usually specified by the vehicle manufacturer. A rule of thumb for the clearance of wheel bearings is 0.001 to 0.005 inches (0.03 to 0.13mm) of end play.

16.4 Seals

Frictionless bearings are always used with some sort of seal at each side. The seal is used to keep the lubricant in the bearing and dirt and other foreign material out. Many seals are designed to perform both functions at the same time. A lip seal is the most common type of wheel or axle bearing seal (Figure 16-7).

The sealing lip, sometimes called a *wiping lip*, is made from a flexible material, usually neoprene rubber. Different materials can be used depending on the speed of the seal, the type of lubricant, and the operating conditions where the seal is used. The flexible lip is usually molded into the seal's outer case. The seal's housing forms a static (stationary) seal to the hub or axle housing when it is pressed into position in its bore. The seal lip forms a dynamic (moving) seal by wiping oil or grease off the shaft. The open side of the seal lip always faces toward the lubricant (inside) so that any internal pressure will increase, not decrease, the wiping pressure between the seal lip and the shaft. A garter spring is often placed around the seal lip to increase the wiping pressure. Most seal failures occur when the wiping lip fails. Some seals have a second lip that faces outward, stopping dirt and grit from working their way under the inner sealing lip.

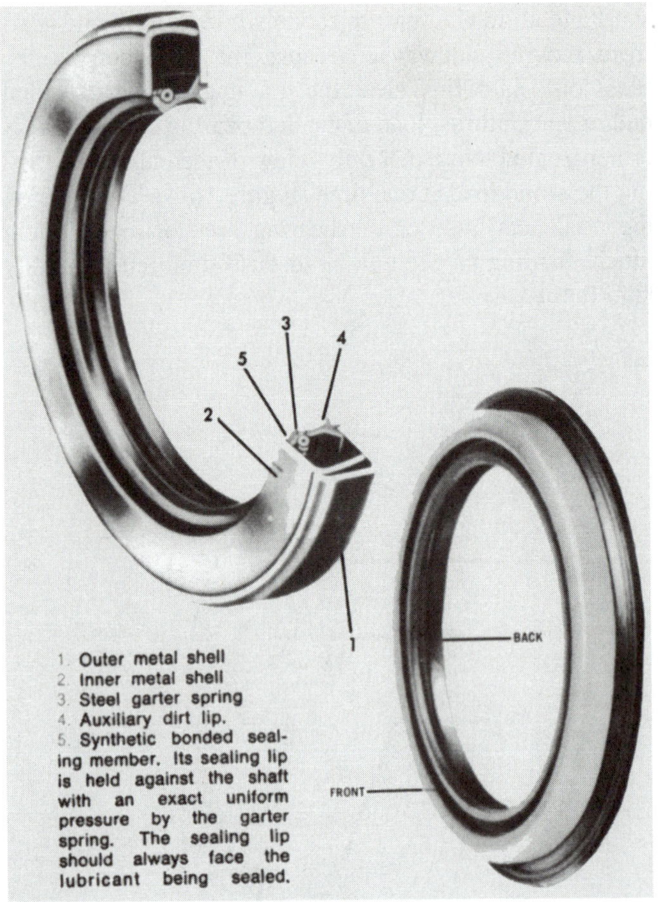

Figure 16-7 A typical seal showing its various parts. (Courtesy of Chicago Rawhide Industries)

16.5 Nondrive-Axle Serviceable Bearings

This type of wheel bearing is very common. It is used on the front wheels of most RWD cars and on the rear wheels of many FWD cars. A pair of tapered-roller bearings is placed over the spindle with the cups pressed into the hub. The bearings are positioned so the smaller diameters of the bearing tapers are toward each other, which gives the bearing a large amount of control over end play (Figure 16-8).

The larger of the two bearings normally carries most of the radial load. This bearing is positioned close to the center of the tire where most of the vehicle load is. It is also located over the larger and stronger portion of the spindle. The smaller bearing primarily serves to keep the hub from wobbling and, with the large bearing, to control end play.

The end play of most front-wheel bearings is controlled by adjusting the spindle nut. Turning the nut inward will decrease the end play, and turning it far

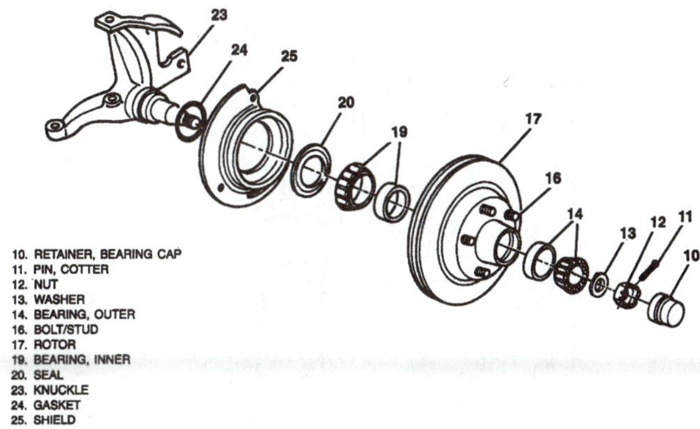

Figure 16-8 An exploded view of the hub and wheel bearing from a nondrive wheel. Note the use of a cotter pin (11) to lock the adjuster nut (12). (Courtesy of General Motors Corporation, Service Technology Group)

enough will preload the bearings. After the bearing end play is adjusted, the nut is locked in place to hold this adjustment. The usual lock is a cotter pin placed through a hole in the end of the spindle and a slot in the castellated nut or castellated-style nut lock (Figure 16-9). Other methods of locking the nut in place are:

- Using a second nut tightened against the adjusting nut; this is called a *lock nut* or *jam nut*.
- Bending or staking a portion of the adjusting nut into a groove in the spindle.
- Tightening a clamp portion of the adjusting nut onto the spindle.
- Bending a sheet-metal washer over the adjusting nut and also over the jam nut. This washer is often kept from rotating by a tang that fits a groove in the spindle.
- Tightening set screws in the lock nut against the adjusting nut. These set screws pass through slots in the washer that is tanged to the spindle (used in trucks and 4WD vehicles) (Figure 16-10).

Occasionally, the bearing clearance is adjusted by changing spacer shims or washers that are positioned between the two bearings. Thicker shims will give more end play and less preload.

Many 4WD front axles use the same style of bearing. The major differences are the hollow spindle, larger-diameter bearings, special wrenches required for the adjusting and lock nuts, and the method used to lock the bearing adjustment.

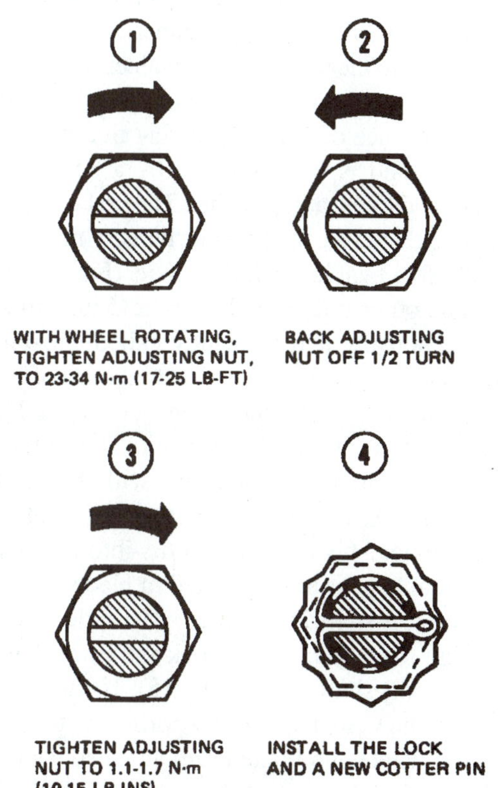

Figure 16-9 One method of adjusting the wheel bearings on a nondrive hub. (Courtesy of Ford Motor Company)

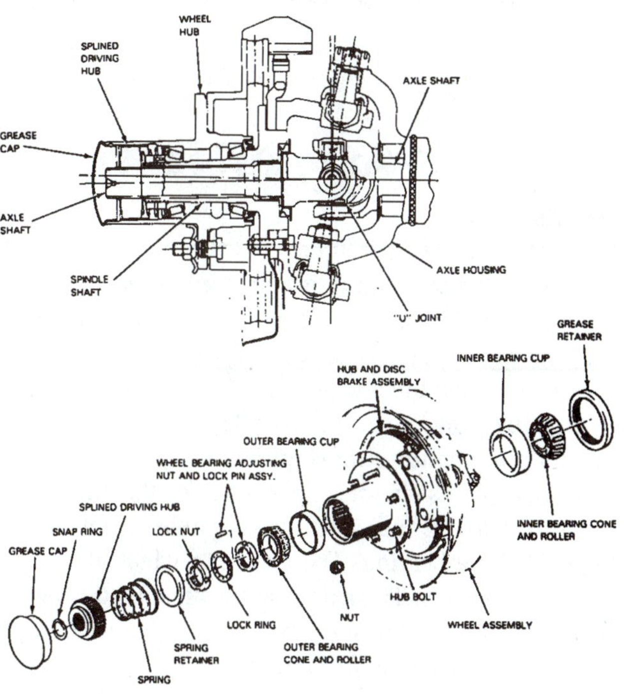

Figure 16-10 A front hub from a 4WD vehicle showing an adjusting nut lock that uses a lock pin and lock ring. (Courtesy of Ford Motor Company)

16.6 Nondrive-Axle Nonserviceable Wheel Bearings

This style of wheel bearing is used on the rear wheels of some FWD cars. This bearing assembly is permanently lubricated, adjusted, and sealed during manufacture and requires no further service. In fact, it is not possible to service it internally. This bearing assembly is bolted to the car's suspension, and the tire and wheel and the brake drum or rotor bolt to it. If this bearing becomes noisy, rough, or loose, the whole assembly must be removed and replaced with a new unit (Figure 16-11).

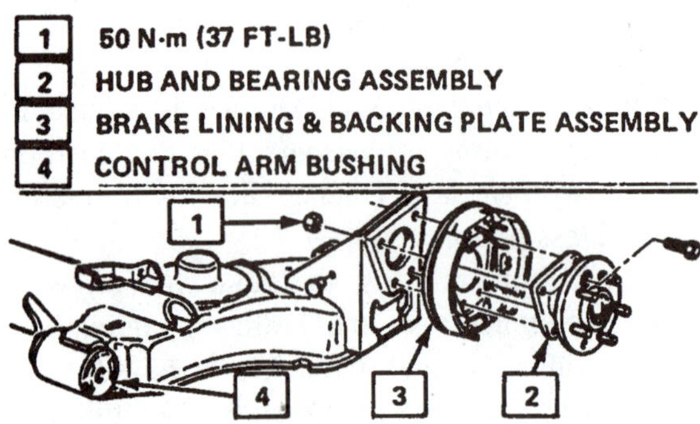

Figure 16-11 A rear suspension with brake and hub assembly from an FWD car. This hub and bearing can not be disassembled or serviced. It is replaced when a problem develops. (Courtesy of General Motors Corporation, Service Technology Group)

16.7 Solid-Axle Drive-Axle Bearings

This type of wheel bearing is commonly used on the rear end of most RWD cars. The bearing, commonly called an axle bearing, is at the outer end of the axle housing, and the axle, with the tire and wheel bolted to it, turns it.

Most passenger cars use a style of axle and bearing called a *semifloating axle* (Figure 16-12). The inner end of the axle "floats" in the axle gear inside the differential. The term "floats" indicates that the axle is not directly supported by a bearing. The gear into which the axle is splined supports it. The axle bearing supports the vertical load on the axle and in some types, the side loads. Larger pickups and trucks use *full-floating axles*; a bearing does not touch the axle at either end. The rear hubs of full-floating axles use a pair of tapered-roller bearings that are very similar to common front-wheel bearings, except that they are much larger. A full-floating axle carries no vehicle load, only torque going to the rear tires (Figure 16-13).

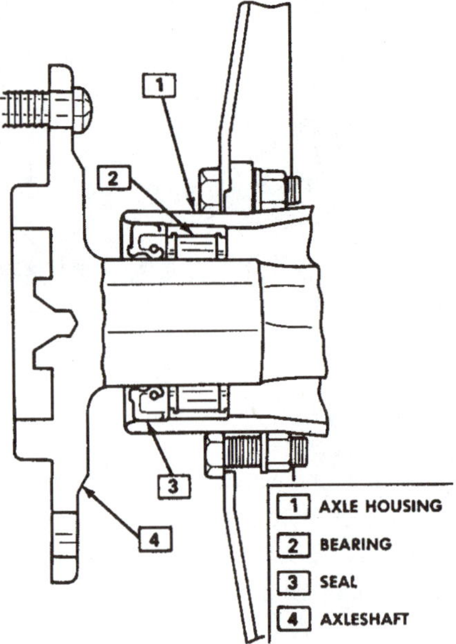

Figure 16-12 The outer end of a drive axle showing one style of axle bearing. This axle is called a semi-floating axle because its inner end is supported by a gear. The outer end supports some of the car's weight. (Courtesy of General Motors Corporation, Service Technology Group)

Two major types of bearings are used on semifloating axles, and they determine how the axle is held in the axle housing. On some axles, the bearing's inner race is pressed onto the axle, and a secondary retainer is pressed onto the shaft right next to the bearing. The retainer helps ensure that the axle does not slide out of the bearing and the axle housing. The outer race of the axle bearing fits snugly into the axle housing and is held in place by a bearing retainer, which is bolted to the axle housing. The brake backing plate is usually held in place by the same bolts. This arrangement is often called a **bearing-retained axle**. (Figure 16-14).

The other type of axle uses a C-lock to keep the axle in the housing and is usually called a **C-lock axle**. The outer end of this axle shaft, just inboard of the wheel mounting flange, is hardened and ground smooth to serve as the inner race of a roller bearing. The outer bearing race fits snugly into the axle housing with the axle passing through it. The C-lock is placed in a groove at the inner end of the axle. This C-lock also fits into a recess in the differential axle gear and is locked in place when the differential pinion shaft is installed. The axle is prevented from sliding outward by the C-lock and inward by the differential pinion shaft (Figure 16-15). If the axle breaks outboard of the C-lock, the outer portion of the axle, with the tire and wheel and brake drum attached, can slide out of the housing, at least until the tire runs into the fender.

Both types of bearings are normally lubricated with either a mist of gear oil from the axle gears or grease that was packed and sealed into the bearing during manufacture.

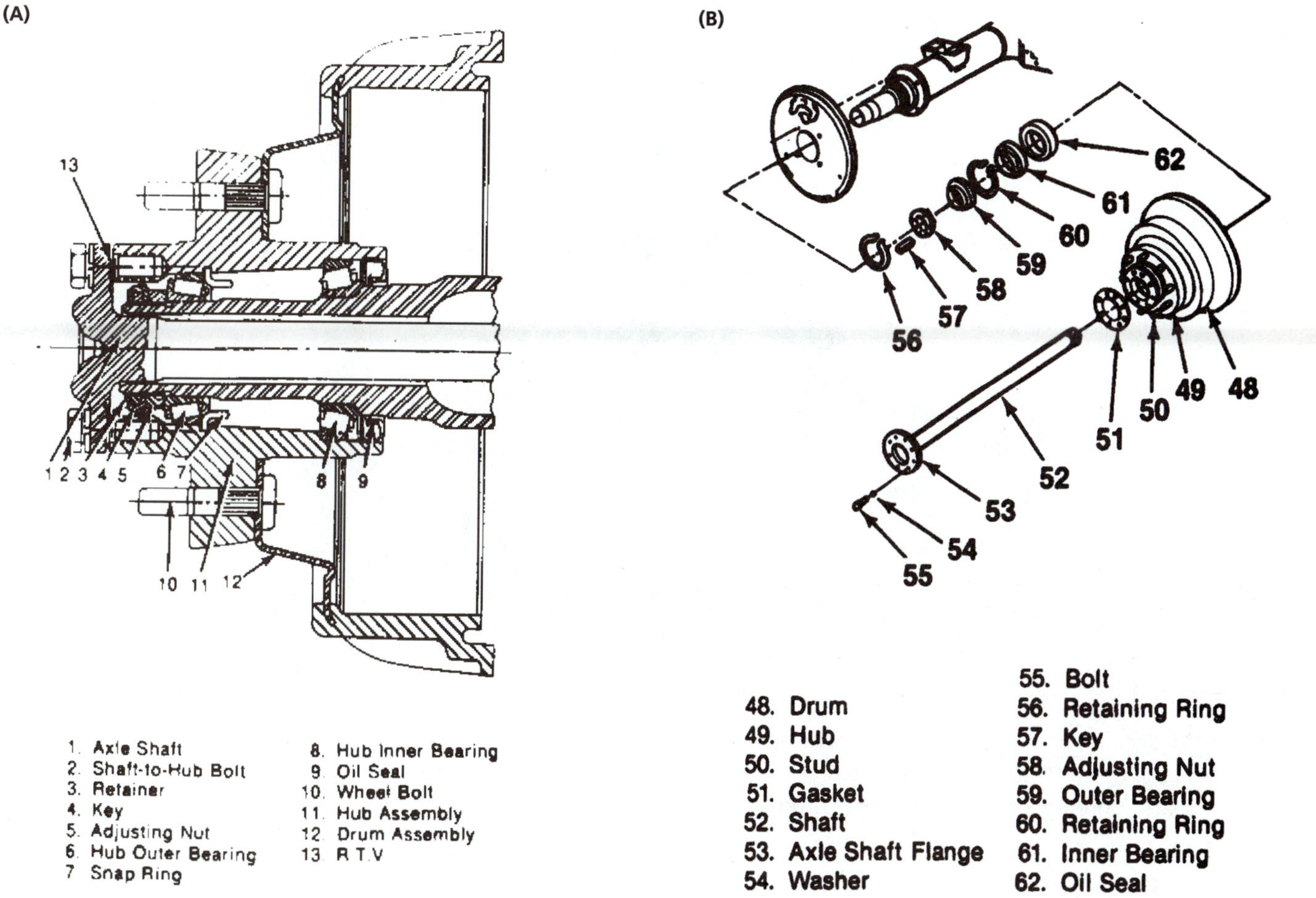

(A)
(B)
1. Axle Shaft
2. Shaft-to-Hub Bolt
3. Retainer
4. Key
5. Adjusting Nut
6. Hub Outer Bearing
7. Snap Ring
8. Hub Inner Bearing
9. Oil Seal
10. Wheel Bolt
11. Hub Assembly
12. Drum Assembly
13. R.T.V.
48. Drum
49. Hub
50. Stud
51. Gasket
52. Shaft
53. Axle Shaft Flange
54. Washer
55. Bolt
56. Retaining Ring
57. Key
58. Adjusting Nut
59. Outer Bearing
60. Retaining Ring
61. Inner Bearing
62. Oil Seal

Figure 16-13 A full floating axle (1+52) does not carry vehicle loads as shown in this cutaway (A) and exploded view (B); the hub and bearings fit over an extension of the axle housing (B). (Courtesy of General Motors Corporation, Service Technology Group)

Ring gear
Pinion gear
Left-hand axle shaft
Differential case
Right axle shaft
Drive pinion
Pinion seal
Flange
Seal
Drive pinion bearings
Bearing
Bearing retainer

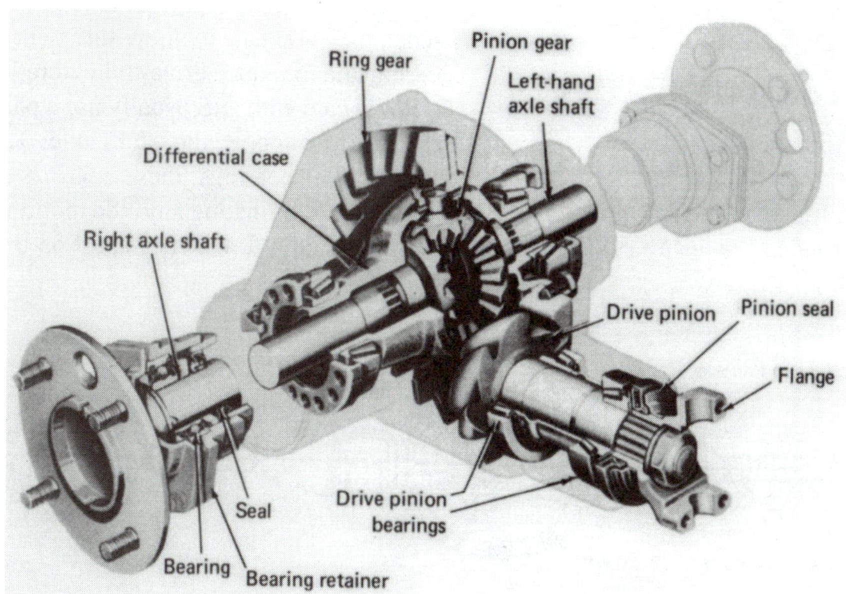

Figure 16-14 This rear axle assembly uses a bearing-retained axle. The bearing is pressed onto the axle and is secured in the housing by the outer retainer. (Courtesy of Ford Motor Company)

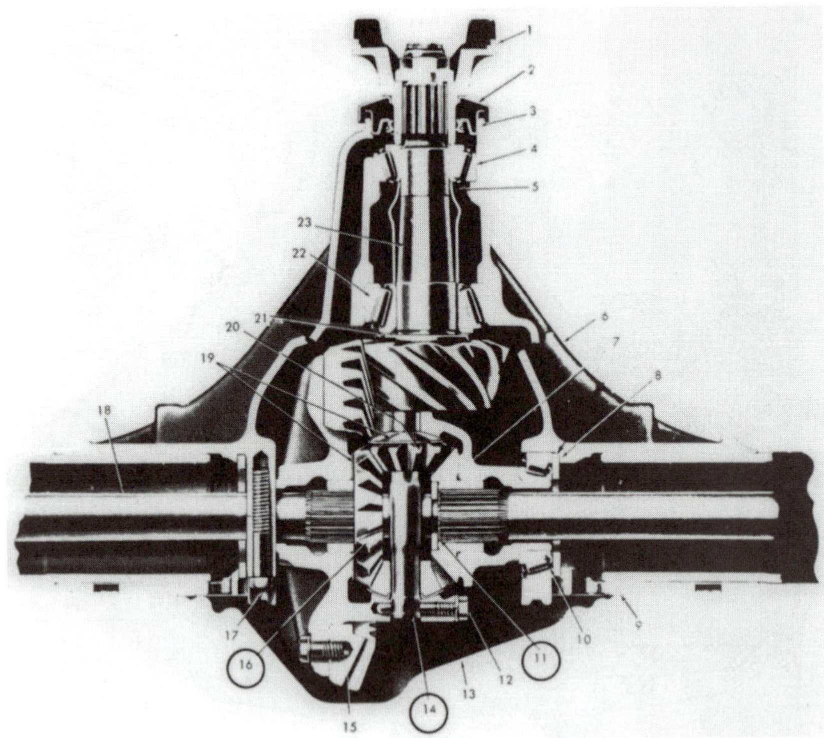

Figure 16-15 A cutaway view of a rear axle assembly showing how the C-lock (11) fits into the differential side gear (16) and is kept from coming out by the differential pinion shaft (14). The axle cannot slide outward, because of the C-lock, or inward, because of the pinion shaft. (Courtesy of General Motors Corporation, Service Technology Group)

A lip seal is often placed just inboard of the axle bearing to keep excess gear oil from passing by the bearing, out the end of the axle housing, and onto the brake shoes. Failure of this seal will allow gear oil to spill on the brake lining and ruin it.

16.8 Independent Suspension Drive-Axle Wheel Bearings

This type of wheel bearing is found on the rear drive axle of cars with IRS or on the front end of many FWD cars (Figure 16-16). Cars with independently suspended driving wheels usually transmit power to the tire through a short "stub" axle. This axle is connected to the differential by a short drive shaft or *half shaft*. (Figure 16-17). The ends of these drive shafts are connected to universal joints on most RWD cars or to constant velocity (CV) universal joints on all FWD cars and some RWD cars. **CV-joints** are needed on FWD cars to allow sufficient turning angles for steering and to reduce drive-train vibrations while turning.

RWD cars with IRS usually use a pair of tapered roller bearings to support the stub axle. Like other paired, tapered-roller bearings, these bearings allow the shaft to rotate while eliminating any side motion. The bearings are often mounted a few inches apart on the bearing support

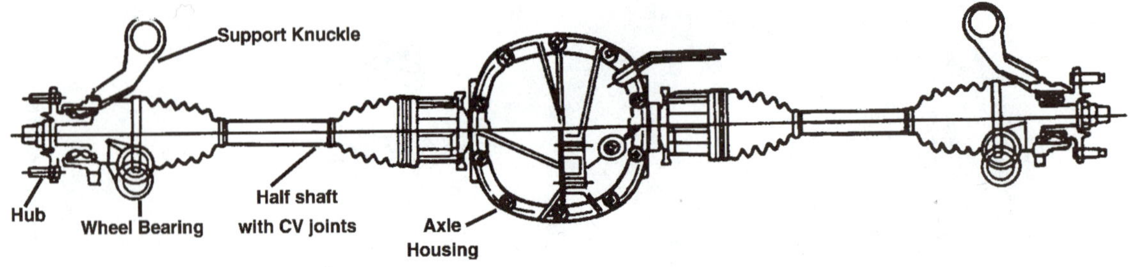

Figure 16-16 An IRS drive axle. Each half shaft/drive shaft has two CV-joints, and the wheel bearings are pressed into the support knuckle. (Courtesy of Ford Motor Company)

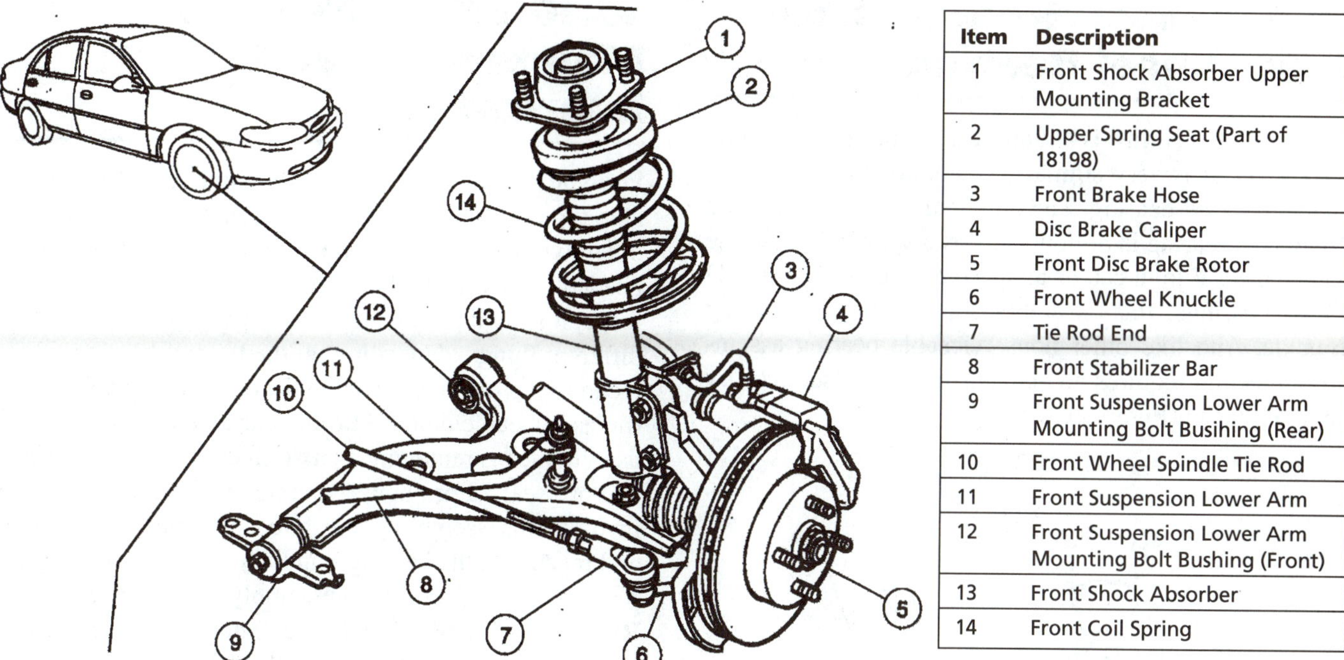

Item	Description
1	Front Shock Absorber Upper Mounting Bracket
2	Upper Spring Seat (Part of 18198)
3	Front Brake Hose
4	Disc Brake Caliper
5	Front Disc Brake Rotor
6	Front Wheel Knuckle
7	Tie Rod End
8	Front Stabilizer Bar
9	Front Suspension Lower Arm Mounting Bolt Busihing (Rear)
10	Front Wheel Spindle Tie Rod
11	Front Suspension Lower Arm
12	Front Suspension Lower Arm Mounting Bolt Bushing (Front)
13	Front Shock Absorber
14	Front Coil Spring

Figure 16-17 The front suspension for an FWD car. The wheel bearings are in the front wheel/steering knuckle (6). (Courtesy of Ford Motor Company)

to give them better leverage in controlling the shaft position. The spindle support is connected to the suspension members in such a way that it can control the alignment of the tire and wheel as the tire moves over the road surface. Lubrication of these bearings involves periodic disassembly, cleaning, and repacking with grease. Some manufacturers use permanently packed and sealed bearings. The end play or preload of this type of bearing is usually controlled by a spacer between the bearings.

Most FWD front-wheel bearings must be compact to fit in the small space provided for them. The stub axle is often the splined extension of the outer portion or housing of the CV joint. The splines of the CV-joint housing pass into the splines in the front hub, and the two splines are held together by a nut at the outer end. The hub is supported by a pair of ball or tapered roller bearings that are mounted in the steering knuckle. These bearings can be the type that is packed with lubricant and sealed during manufacture or the type that requires periodic lubrication. The vehicle manufacturer's requirements should be checked to determine the maintenance requirements for a specific car. Some manufacturers recommend that these bearings be replaced whenever the front hub is removed from the spindle (Figure 16-18). Bearing end play or preload is usually controlled by the size of the parts. As they are assembled and tightened into place, the adjustment is automatically made.

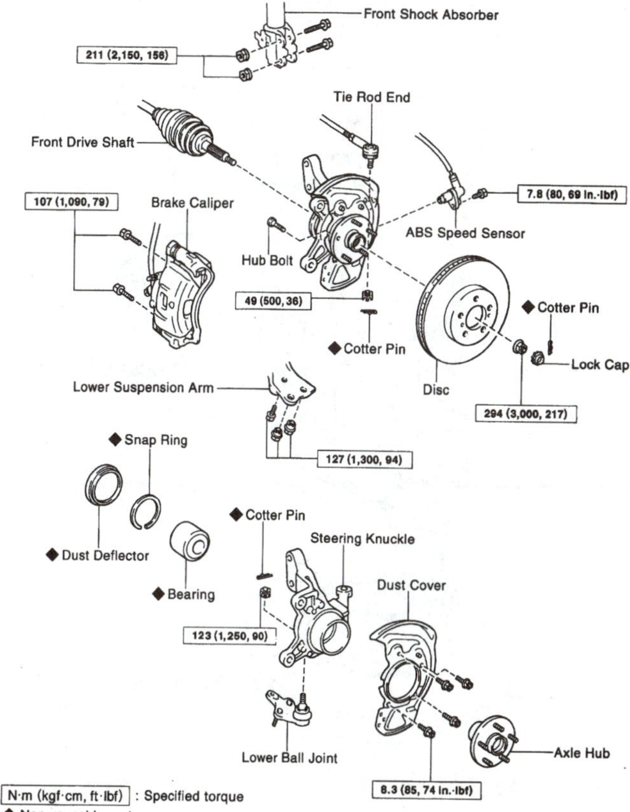

Figure 16-18 An exploded view of an FWD-car front knuckle and hub assembly. The bearing is not serviced; it is replaced when a problem develops.

16.9 Nonserviceable Drive-Axle Wheel Bearings

This type of bearing is found at the front end of some FWD cars. It is very similar to a nondrive-axle, nonserviceable wheel bearing. The only real difference is that a drive-axle bearing is hollow and has a splined hole in the hub so the CV-joint can be attached to it. Like other FWD wheel bearings, this assembly attaches to the steering knuckle. And like other nonserviceable bearing assemblies, this unit requires no maintenance and is serviced by being replaced (Figure 16-19).

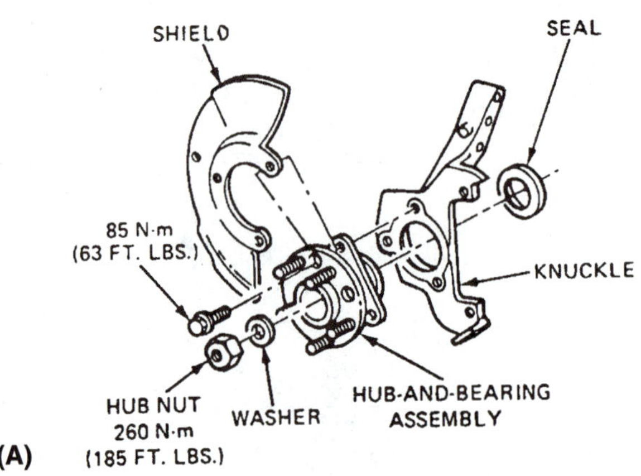

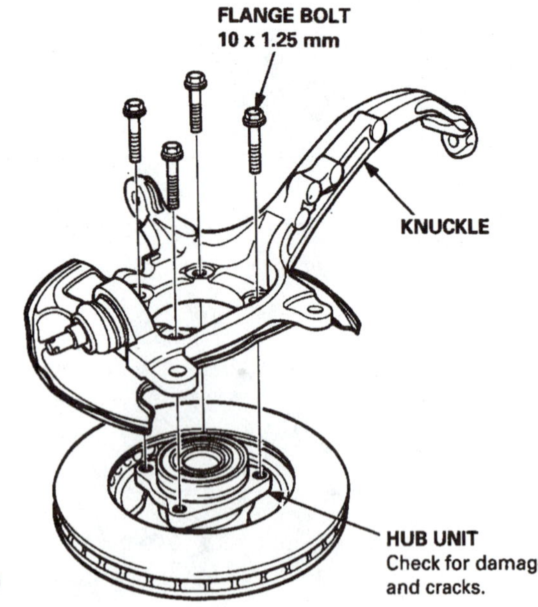

Figure 16-19 Two nonserviceable FWD front wheel bearing assemblies. The hub-and-bearing assembly (A) or the hub unit (B) is replaced if there is a problem. (A is courtesy of General Motors Corporation, Service Technology Group; B is courtesy of American Honda Motor Co., Inc.)

16.10 Wheel Bearing Maintenance and Lubrication

Periodic maintenance is normally required on the serviceable types of wheel bearings used on nondrive axles. The other types of bearings are usually serviced on an "as needed" basis. If the bearing is noisy, leaking grease, or loose, it should be serviced. The term "service" can mean different things, depending on the type of bearing. It can mean removing and replacing the entire bearing assembly (nonserviceable units). For the rear axle of a RWD car, it usually means removing an axle shaft to replace a bearing and/or seal. For nondrive-axle bearings, it usually involves disassembling the hub and bearings, cleaning and repacking the bearings, and adjusting the bearing end play or preload during reassembly. The rear axle of a RWD car with IRS usually has similar service requirements. If you are not sure of the lubrication or service requirements for a particular bearing set, check the manufacturer's or technician's service manual.

16.11 Diagnosis Procedure for Wheel Bearing Problems

Faulty or excessively loose wheel and/or axle bearings are apparent to the driver of the car as noise, road wander, wheel shake, play in the steering, cuppy tire wear, and/or a low brake pedal (disc brakes). If some or all of these problems are encountered, a systematic procedure should be followed to determine if loose or faulty wheel or axle bearings are to blame and which one(s) are faulty.

If faulty wheel bearings are suspected, raise the car on a hoist and check for loose bearings. Try pushing the tire and wheel straight up and down, pushing it straight in and out, and rocking it sideways. Depending on the bearing type, some of these motions are not permitted and some of them are. Vertical motion is not permitted in any type of wheel or axle bearing. More than barely perceptible in-and-out motion is permitted only on C-lock axle, RWD rear-axle bearings. On most of these axles, end play up to 0.010 inches (0.26 mm) is permitted, and some axles allow up to 0.030 inches (7.9 mm). Tapered-roller wheel and axle bearing sets (a pair of bearings) should have about 0.001 to 0.005 inches (0.03 to 0.13 mm of end play. Nonserviceable wheel and axle bearings are allowed up to 0.005 inches (0.13 mm) of end play (Figure 16-20).

When rocking a tire and wheel to check for bearing looseness, grip the tire at the top and bottom. Push inward with one hand while pulling outward with the other. Then

SEALED WHEEL BEARING DIAGNOSIS

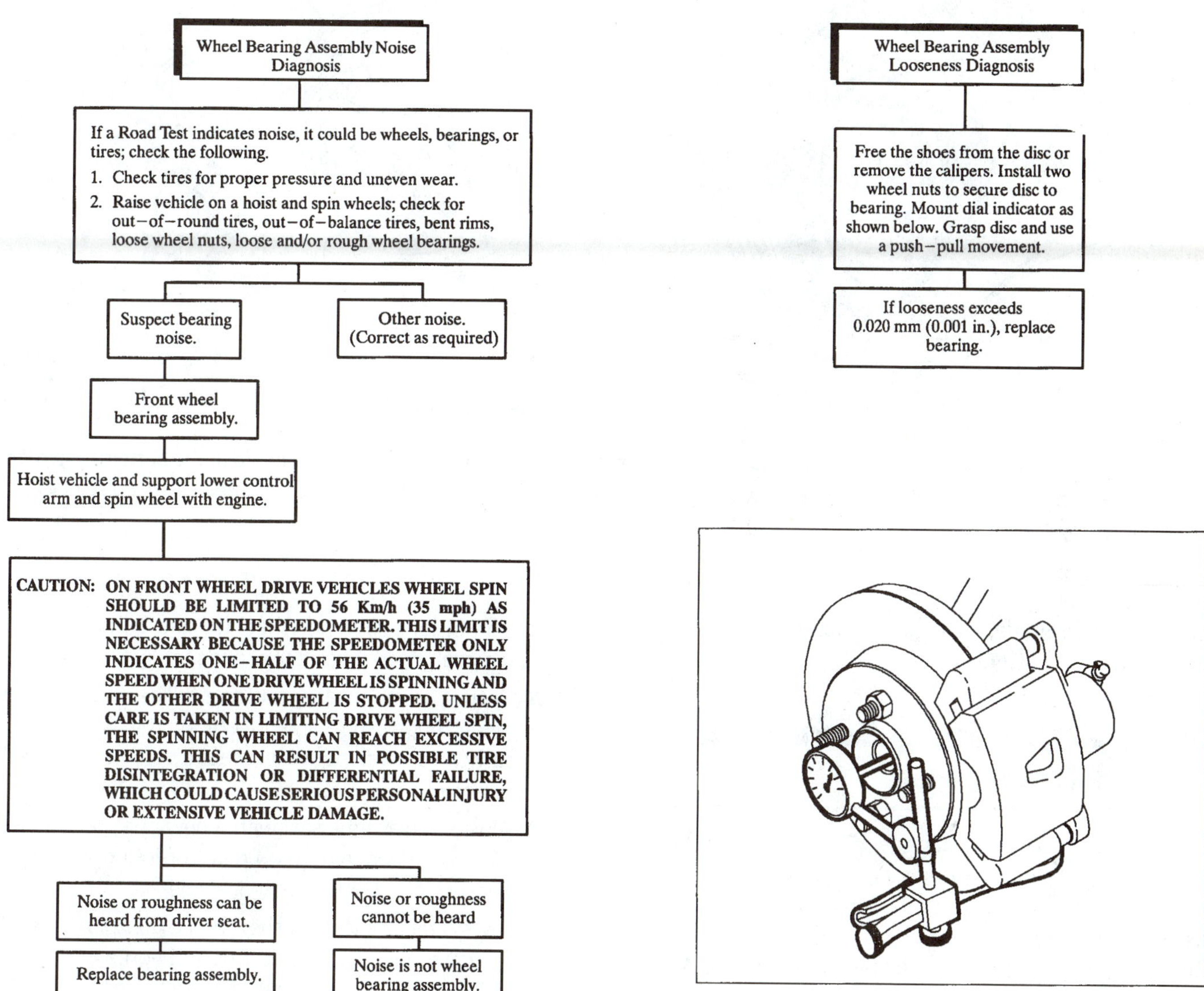

Figure 16-20 This diagnostic chart shows the procedure for checking a sealed, nonserviceable wheel bearing. Except for the repair steps, the same procedure is used for serviceable bearings. (© Saturn Corporation, used with permission)

reverse these motions (Figure 16-21). Perceptible motion is acceptable on tapered-roller wheel bearings and axle bearings on independent suspensions. More than barely perceptible rocking motion is not permitted on ball-type wheel or rear drive-axle bearings.

Next, spin a nondrive tire and wheel by hand or using a wheel balancer spinner motor. Spin a drive-axle tire and wheel with the engine. As the tire spins, listen for a harsh, grating sound. You can often confirm the sound origin and the bearing roughness by gently placing your fingertips on the steering knuckle or axle housing as close to the bearing as possible. A rough wheel or axle bearing will cause a rough, irregular feel.

SAFETY TIP: Always use caution when moving or working around a spinning tire and wheel. When spinning a tire with the engine, limit the tire speed to about 55 to 70 mph (88 to 113 kph). When one tire is stopped, the differential gears will cause the other tire to run at twice the speed shown on the speedometer; excessive speed can cause a tire explosion. Higher speeds are not necessary.

The proper repair method can be selected on after the faulty bearing has been located and the type of bearing has been determined. Some types of excessively loose bearings can be adjusted, but others will require replacement. Some types of rough bearings require replacement

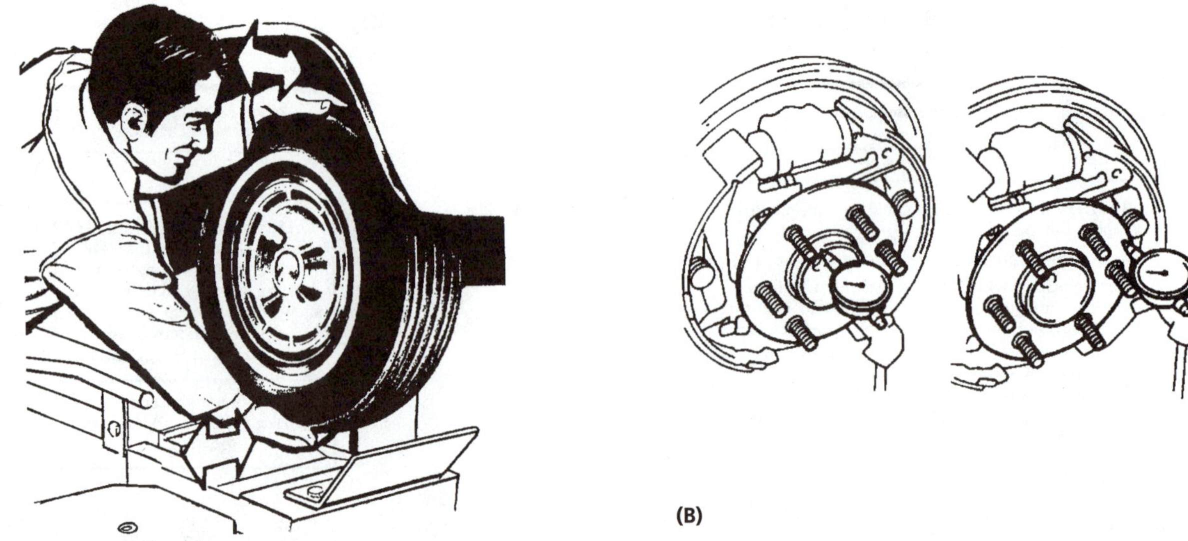

Figure 16-21 When checking for loose wheel bearings, push in at the top of the tire while pulling outward at the bottom, and then reverse these pressures while feeling for excess play (A). Hub and axle end play/bearing lash can be measured using a dial indicator (B). (A is courtesy of Hunter Engineering Company)

of the whole assembly, but on others, only the faulty bearing part needs replacement.

16.12 Repacking Serviceable Wheel Bearings

Most RWD front-wheel bearings, FWD rear-wheel bearings, and front-wheel bearings on 4WD vehicles should be repacked at regular intervals. This operation includes disassembly, cleaning, packing with grease, reassembling, and adjusting. The interval between repacking, which varies with different car manufacturers, is about 15,000 to 20,000 miles or greater. In some cases, bearing service is required only during a brake reline.

SAFETY TIP: The following good service practices should be observed when servicing wheel and/or axle bearings:

- Wear eye protection.
- Always lift the car by the correct lifting points and support it with carefully placed jack stands.
- Never "just add grease" to a bearing. Always clean, inspect, and repack it with new grease.
- Never let grease or solvent get on the braking friction surfaces. Always clean up any spilled grease, solvent, or greasy fingerprints.
- Never let brake calipers or drive shafts or axle shafts hang by their hoses or universal joints. Always support them and protect the rubber hoses and CV-joint boots.
- Do not let a tire and wheel fall and bounce during removal. Keep these heavy parts under control.
- Keep a clean, neat work area. Slide the tire and wheel under the car and out of the way during service operations that require tire and wheel removal.
- Always replace the seal with a new one. Reusing an old seal risks ruining the brakes or losing a bearing because of possible grease leakage.
- Always replace lock washers and cotter pins. Use all recommended locking devices and tighten nuts and bolts to the correct torque.
- Remember that the service operations you are performing should last for a few years and thousands of trouble-free miles.

16.12.1 Disassembling Wheel Bearings

Disassembly is the first step when repacking serviceable wheel bearings.

To disassemble wheel bearings:

1. Raise and securely support the car.
2. If disc brakes are used on the wheel, remove the wheel. If drum brakes are used, the tire and wheel with the brake drum and hub can usually be removed as a unit.
3. On disc-brake-equipped cars, remove the caliper, as described in Chapter 11, and suspend it from the steering knuckle or another convenient point, using a

bungee cord or mechanic's wire. On 4WD vehicles, remove or disassemble the front drive hub.

SAFETY TIP: Do not let the caliper hang from the hose. If the hose is bent too sharply, it can fail and cause a loss of braking power.

4. Remove the grease or dust cap from the wheel hub. A dust cap remover will do this quickly and easily, and slip-joint pliers can also be used (Figure 16-22).
5. Locate and remove the locking device for the spindle nut. The most common type of locking device is a cotter pin. Straighten the bent end(s) using a pair of dikes (diagonal- or side-cutting pliers). Grip the head of the cotter pin with the dikes and pry it out of the spindle and spindle nut (Figure 16-23). Another good tool to use for this is a cotter-pin puller. If a staked spindle nut is used, some manufacturers recommend merely unscrewing the nut. The staked portion will bend out of the way during removal. Other manufacturers suggest that you bend the staked portion of the nut upward using a small, sharp chisel. Or you can use a drill and drill bit to drill through the staked portion of the nut. Be careful not to drill into the spindle. In either case, the staked nut should be replaced.

(A)

(B)

Figure 16-22 The wheel bearing dust cover can be removed using a pair of special dust-cover pliers (A) or slip-joint (water pump/channel locking) pliers (B). (A is courtesy of Chicago Rawhide Industries; B is courtesy of Federal-Mogul Corporation)

Figure 16-23 A cotter pin can usually be easily removed by gripping its head (arrow) in the jaws of diagonal- or side-cutting pliers (dikes) and prying the pin out.

6. Remove the spindle nut. This can usually be done with your fingers. If not, use slip-joint pliers or a wrench to unscrew it.
7. Remove the washer and outer bearing and bearing cone. Rocking the tire and wheel or hub will usually work the bearing out to a point where you can grasp it (Figure 16-24).
8. Slide the hub off the spindle and pry the seal out of the back of the hub using a seal puller (Figure 16-25). Alternate methods of removing the seal are (a) tapping the inner bearing and bearing cone and seal out using a large wooden dowel and a hammer (Figure 16-26), and (b) replacing the nut on the spindle and, while sliding the hub off the spindle, catching the bearing on the spindle nut so that the bearing and seal are jerked out of the hub (Figure 16-27).

NOTE: Neither of these alternate methods is recommended as it might bend the bearing cage, especially those with plastic/nylon cages.

Figure 16-24 After the nut and washer have been removed, rock the hub to work the outer bearing out of the hub. (Courtesy of Chicago Rawhide Industries)

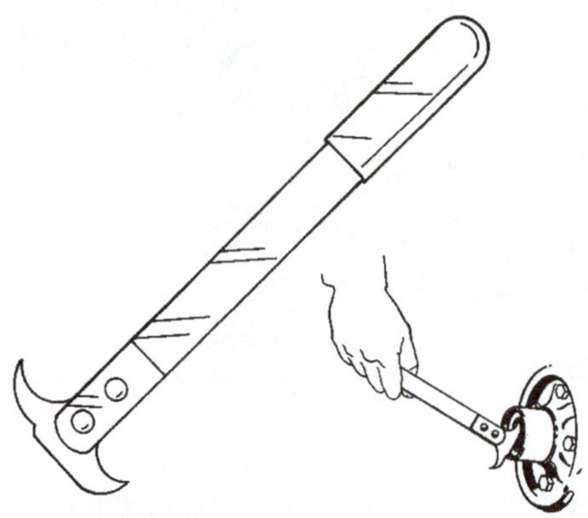

Figure 16-25 After the seal puller has been hooked into the seal, a prying action (in this case toward the left) will lift the seal out of the hub. (Courtesy of Lisle Corporation)

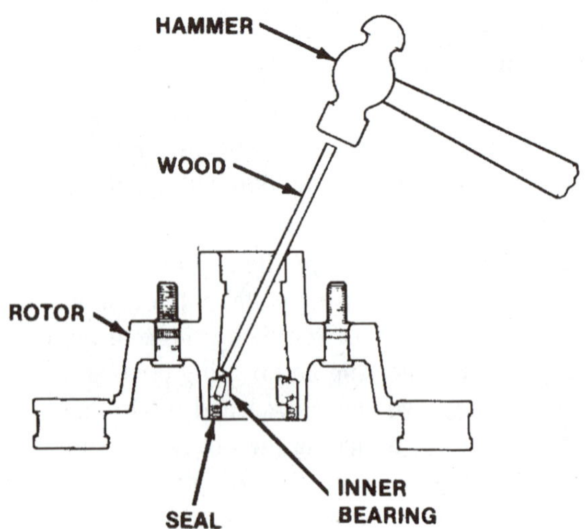

Figure 16-26 A wooden dowel or punch can be placed on the face of the inner bearing and tapped to remove the seal and inner bearing. (Courtesy of Brake Parts Inc.)

Figure 16-27 The inner bearing and seal can be removed by threading the nut on the spindle a few turns and pulling the hub outward with a quick motion so the bearing catches on the nut.

16.12.2 Cleaning and Inspecting Wheel Bearings

After disassembly, the spindle, bearings, and hub should be cleaned to get rid of all of the old grease and any dirt or metal fragments it might contain, and to allow a thorough inspection (Figure 16-28).

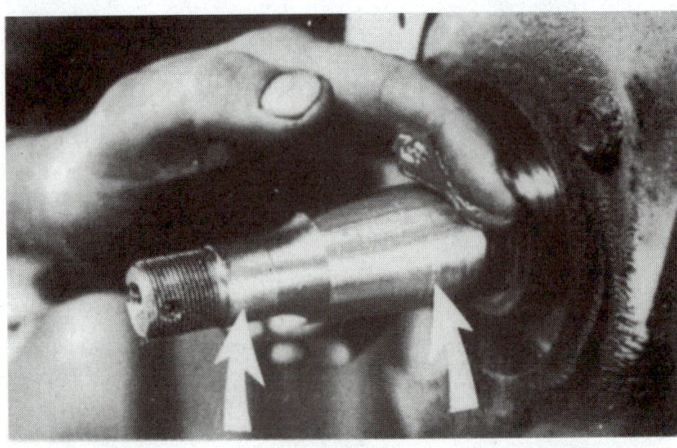

Figure 16-28 After the spindle is cleaned and inspected, coat it lightly with grease. Note the marks (arrows) where the bearing cones have been creeping on the spindle; this is a good sign. (Courtesy of Federal-Mogul Corporation)

To clean and inspect wheel bearings:

1. Wipe off the spindle and inspect it for damage. The bearing cones must have a slip fit on the spindle. This fit should leave a mark on the spindle showing that the cone has been creeping—slowly rotating at about one revolution per mile. Bearing creep keeps changing the loaded part of the cone and produces longer cone and bearing life (Figure 16-29).

Damage to the race

These dents result from the rollers "hammering" against the race. It's called brinelling.

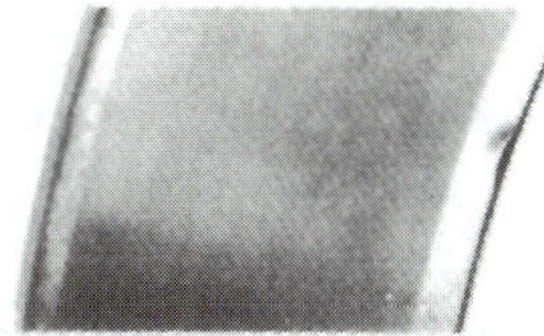

Dents like this usually come from mishandling. The bearing should be discarded.

Under load, a hairline crack like this will lead to serious problems. Discard the bearing.

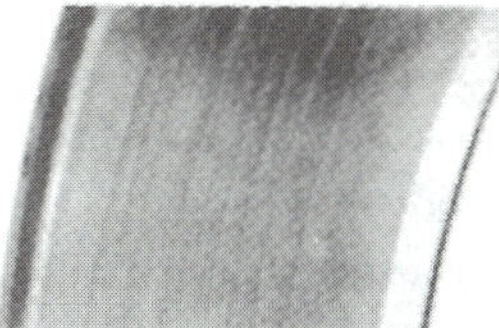

Always check for faint grooves in the race. This bearing should not be reused.

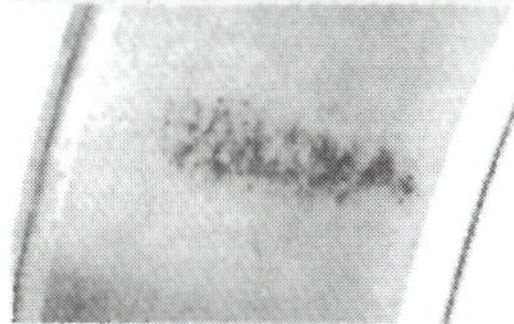

Regular patterns of etching in the race are from corrosion. This bearing should be replaced.

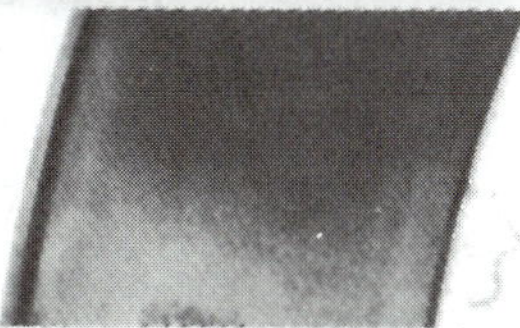

Light pitting comes from contaminants being pressed into the race. Discard the bearing.

In this more advanced case of pitting, you can see how the race has been damaged.

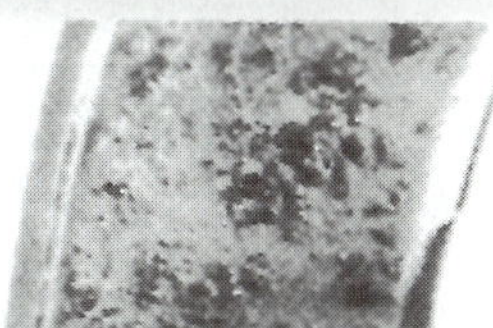

Pitting eventually leads to "spalling," a condition where the metal falls away in large chunks.

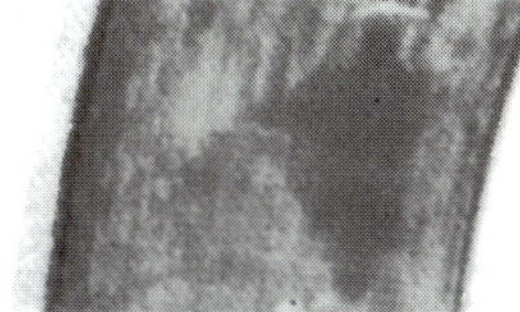

If corrosion stains haven't etched into the surface yet, try removing them with emery cloth.

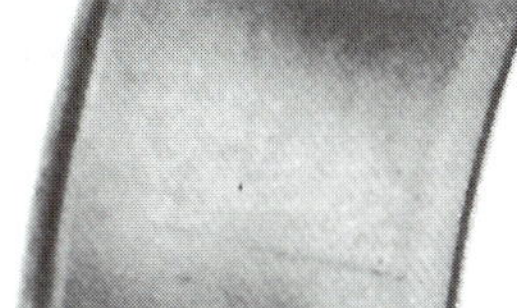

Line etching looks like cracks. If the etching can be removed, the bearing can be reused.

This condition results from an improperly grounded arc welder. Replace the bearing.

Discoloration is a result of overheating. Even a lightly burned bearing should be replaced.

Damage to the rollers

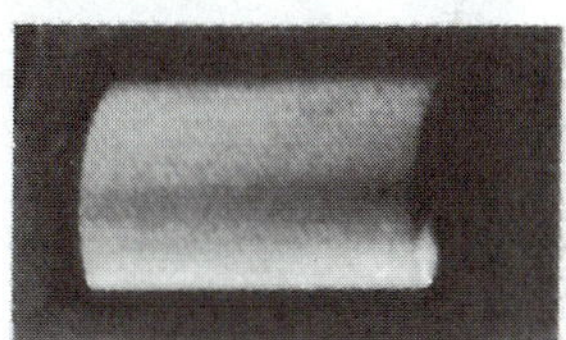

This is a normally worn bearing. If it doesn't have too much play, it can be reused.

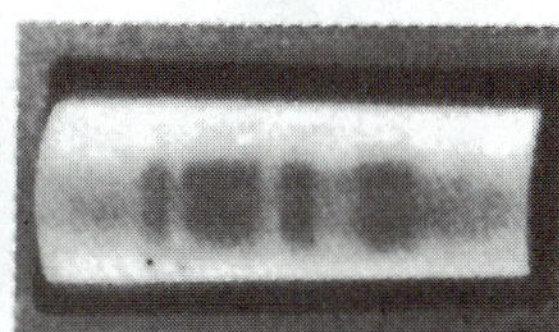

This bearing is worn unevenly. Notice the stripes. It shouldn't be reused.

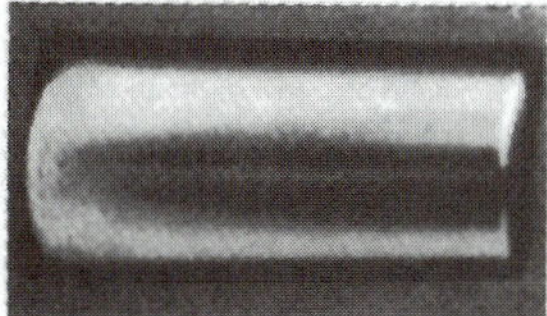

When just the end of a roller is scored, it's from excessive preload. Discard the bearing.

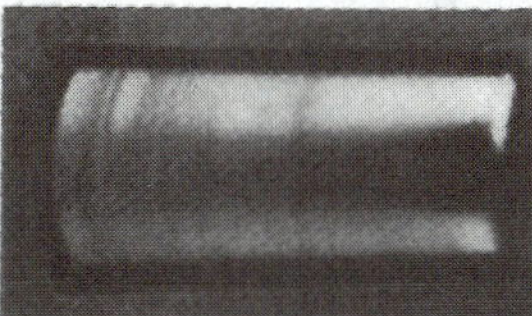

Grooves like this are often matched by grooves in the race (above). Discard the bearing.

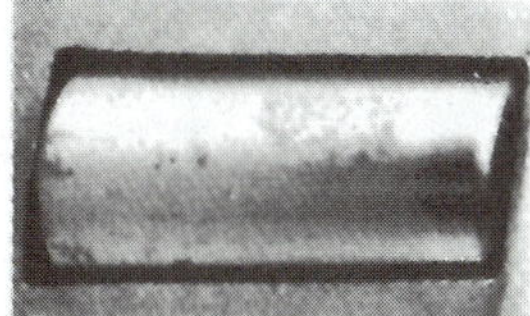

When corrosion etches into the surface of a roller or race, the bearing should be discarded.

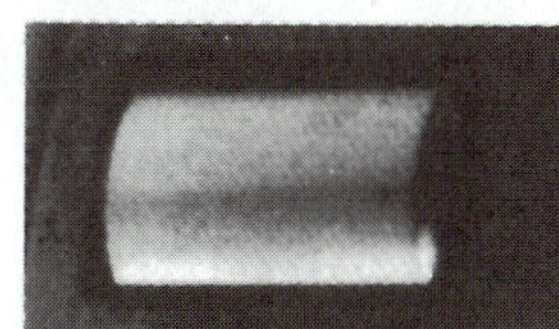

Any damage that causes low spots in the metal, renders the bearing useless.

This is a more advanced case of pitting. Under load, it will rapidly lead to "spalling."

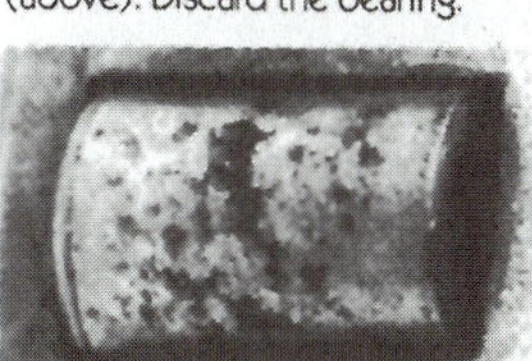

In this "spalled" roller, the metal has actually begun to flake away from the surface.

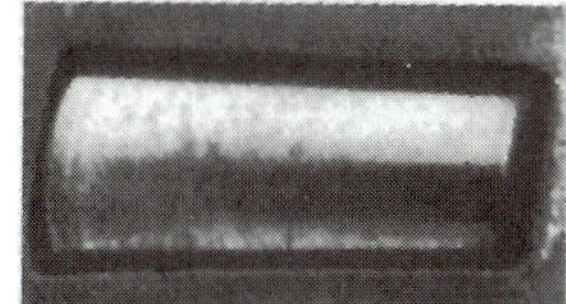

If light corrosion stains can be removed with emery cloth, the bearing can be reused.

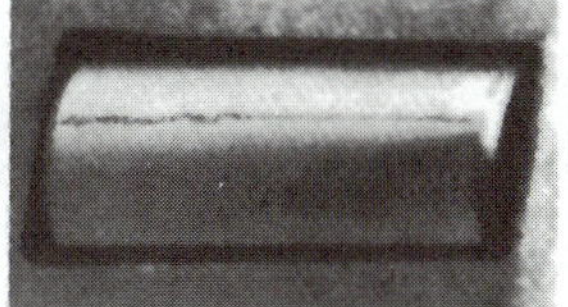

This is "line etching" from corrosion, not a crack. If you can remove it, reuse the bearing.

When "fluting" shows up on the race, it'll appear on the rollers too. Discard the bearing.

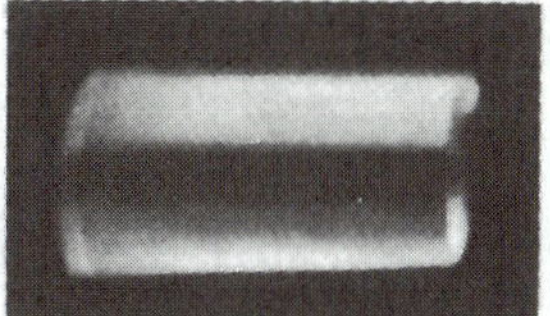

Discoloration comes from overheating. When you see it, play it safe and discard the bearing.

Figure 16-29 After the bearings and cups have been cleaned, they should be carefully inspected for possible reuse. These are the most common signs of damage. (Courtesy of Chicago Rawhide Industries)

2. Check the spindle in the area where the seal runs. It should be clean and smooth.
3. Wipe all the old grease out of the hub and wipe the bearing cups clean. Inspect the cups for pitting, spalling, or other signs of failure. If a cup is damaged, the cup and the bearing and cone should be replaced. The bearing might appear to be good, but it is possible to overlook some damage because of the cage.

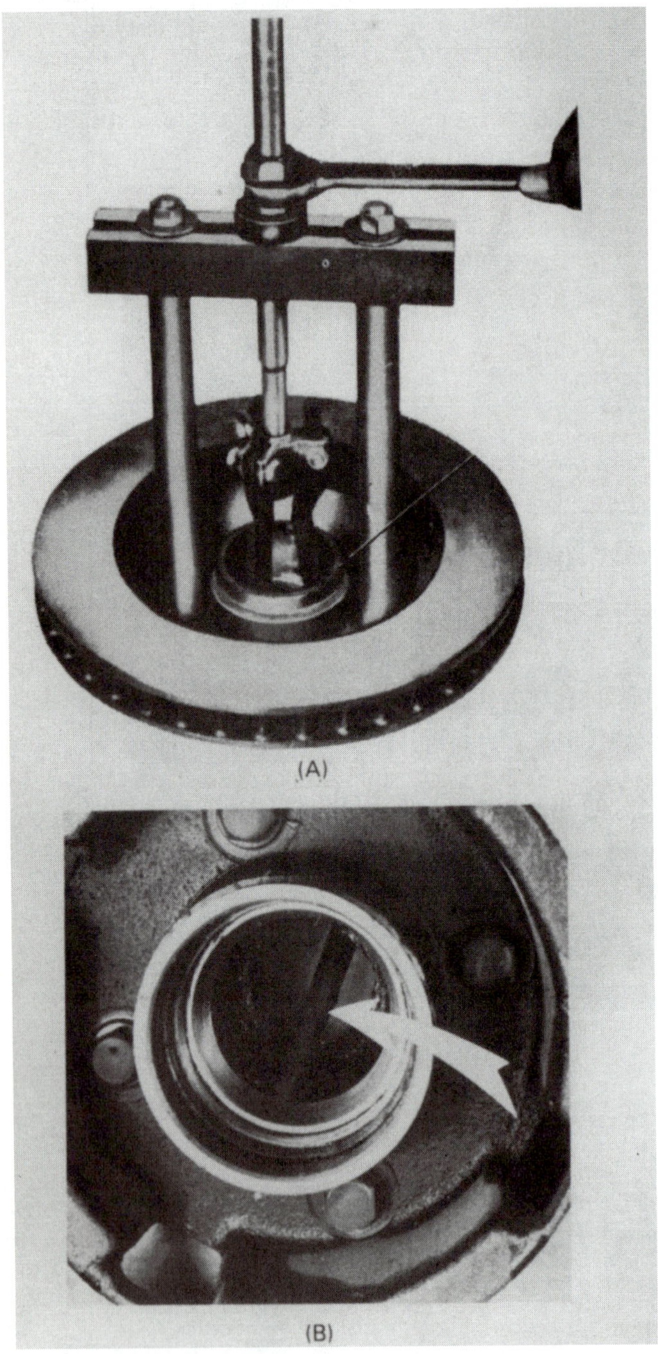

Figure 16-30 A faulty bearing cup can be removed by pulling it out using a puller (A) or driving it out using a punch and hammer (B). (A is courtesy of Ford Motor Company)

NOTE: Many technicians use the part number from the old bearing or seal to make sure they obtain the correct replacement.

4. Check the cups to be sure that they fit tightly in the hub. A loose cup usually requires replacement of the hub.
5. If a cup is damaged, it can be removed from the hub using a punch and hammer or a puller (Figure 16-30). A new cup can be installed using a cup-driving tool and a hammer. Be sure the new cup is fully seated. The hammer blows will make a different, more solid sound when the cup becomes seated (Figure 16-31).
6. Wash the bearings and cones in clean solvent and blow them dry using shop air. The direction of the air blast should always be parallel to the rollers. An air blast across the rollers does not really clean the bearing, and it tends to "spin" the bearing (Figure 16-32).

(A)

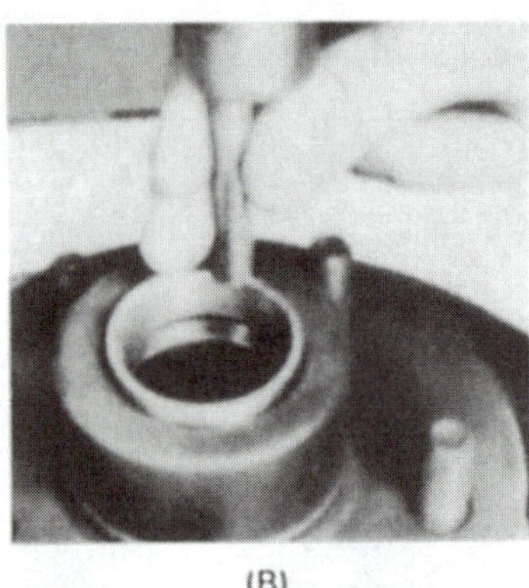

(B)

Figure 16-31 A new bearing cup is installed quickly and easily using a bearing cup driver (A). Make sure you hear the solid sound as the cup seats completely. If a driver is not available, a cup can be installed using a soft metal punch. Be very careful to drive it in straight. (A is courtesy of Chicago Rawhide Industries; B is courtesy of The Timken Company)

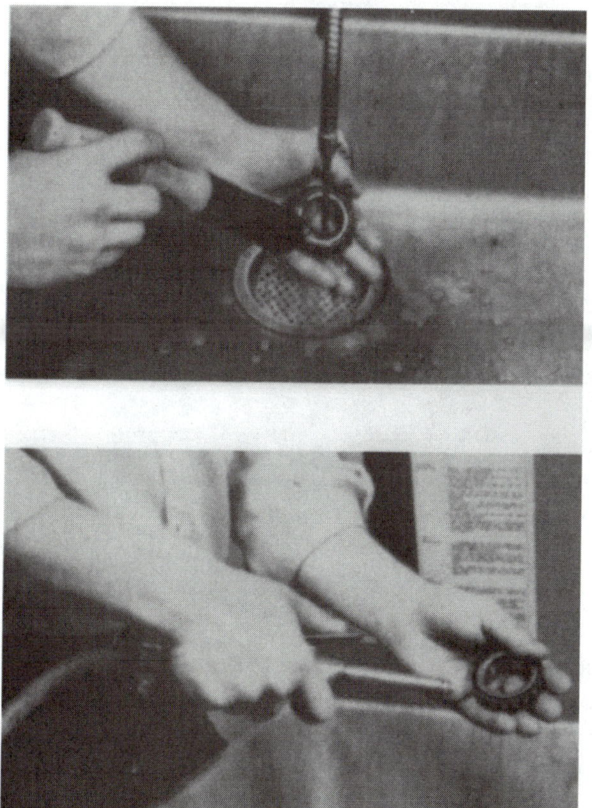

Figure 16-32 The bearing should be washed in clean solvent and then dried with clean, compressed air. Never spin the bearing with air pressure while drying it. (Courtesy of Chicago Rawhide Industries)

CAUTION *DO NOT spin a bearing. A spinning bearing might generate enough centrifugal force to explode. Also, running a bearing at high speeds without lubrication could damage it. Spinning a bearing with shop air might sound "neat," but it really doesn't do any good, only possible harm.*

7. Rewash and dry the bearings until they are thoroughly clean.
8. Inspect the bearings for roller or cage damage. The common causes of bearing failure are shown in Figure 16-29. All damaged or questionable wheel bearings should be replaced.
9. If the bearings, cones, cups, or spindle show signs of misalignment, check for a bent spindle. A spindle runout gauge can be placed over the spindle and adjusted to a sliding fit on the inner bearing (Figure 16-33). Next the dial indicator is adjusted to zero and the gauge unit rotated around the spindle. A bent spindle is indicated if the dial indicator reading changes. Another check for a bent spindle is to measure the distance from a straightedge to the location of the outer bearing cone on each side of and on top of the spindle. The measurements should be the same.

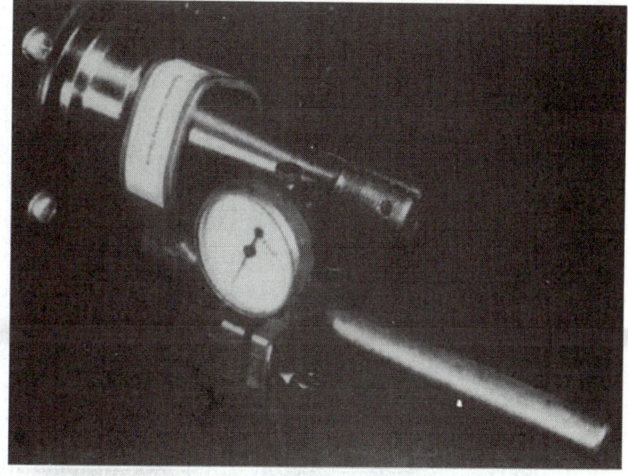

(A)

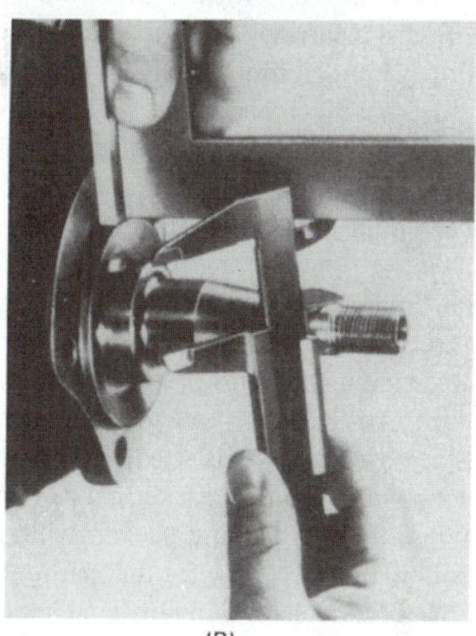

(B)

Figure 16-33 If a bent spindle is suspected, it can be checked using a spindle runout gauge (A) or a straightedge and caliper (B). The runout gauge is set and then rotated around the spindle while watching the indicator. The straightedge and caliper are used to measure the distance shown on both the top and bottom of the spindle; they should be the same. (B is courtesy of Volkswagen of America)

16.12.3 Packing and Adjusting Wheel Bearings

Wheel bearings should be packed with suitable grease and adjusted to the correct end play or preload during assembly. A good grade of chassis grease can be used to pack wheel bearings if the label states that it is suitable for this purpose. Many technicians prefer to use a grease formulated especially for wheel bearings. It is recommended that *high-temperature* or *heavy-duty grease* be used if the car is to be driven hard and/or if the brakes will get a lot of use. Some brands and types of grease are not compatible with each other. If they are mixed, "thinning" can result and the oil portion of the grease will run out of the bearing. A thorough cleanup is always important to ensure that all the old grease is removed before adding the new grease. Packing a bearing fills the area between the cage and the cone, alongside the rollers or balls, with grease.

A properly adjusted, tapered-roller wheel bearing should have 0.001 to 0.005 inches (0.03 to 0.13 mm) of end play. With this clearance, you should notice a perceptible play if you rock the tire and wheel. Ball-type wheel bearings should have zero end play. They are normally adjusted to a slight preload. Ball bearings should not have perceptible end play when the tire and wheel are rocked.

To pack and adjust wheel bearings:

1. Pack the large inner bearing first. The small outer bearing should be packed when you are ready to install it. Some technicians prefer to pack both bearings at the same time and place the small bearing on a clean shop towel until it is needed. Packing is normally done with a bearing packer but can be done by hand. When using a bearing packer, which is much faster, follow the procedure for that particular packer (Figure 16-34). When packing by hand, place a tablespoon of grease in the palm of your hand and push the bearing, open side of the cage downward, into the grease and against the palm of your hand. Repeat this step until you see grease oozing through the upper part of the cage. Work all the way around the bearing until you fill the entire cage (Figure 16-35).
2. After packing the bearing, smear a liberal coating of grease around the outside of the cage and rollers.
3. Smear a coating of grease around the inside of both bearing cups and the inside of the hub. It is also recommended that a ring of grease be placed in the hub, just inside the cups. This ring of grease will act like a dam to prevent the grease from running out of the bearing and into the hub cavity when the grease gets hot. Smear a thin film of grease on the spindle, being sure to include the area where the seal lip rubs (Figure 16-36).
4. Place the packed inner bearing in its cup.
5. Position the seal so the seal lip faces inward toward the grease and, using a suitable driver, drive the seal

Figure 16-34 The quickest way to pack a bearing is to use a bearing packer. Pushing downward on the packer forces grease into the area inside the cage (arrow). (Courtesy of SPX Kent-Moore, Part # J33067)

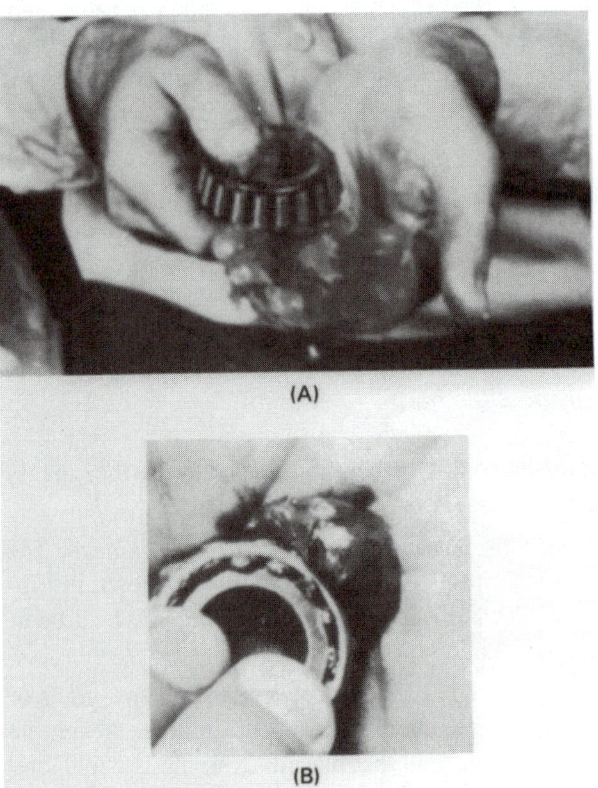

Figure 16-35 A bearing can be packed by placing grease in the palm of your hand (A) and forcing the bearing into the grease. After a few cycles, grease should come up between the rollers and cage (B). (A is courtesy of Chicago Rawhide Industries; B is courtesy of The Timken Company)

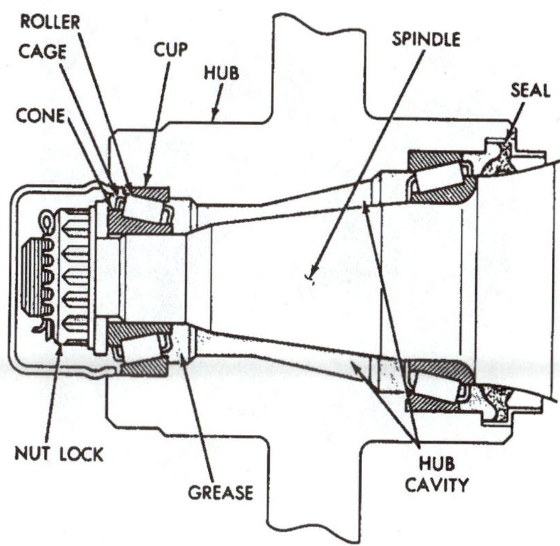

Figure 16-36 Grease should fill the shaded areas between the hub and spindle to ensure that the bearing stays lubricated. (Courtesy of Chrysler Corporation)

into the hub so it is even with the end of the hub (Figure 16-38). The back side of a bearing cup installer is often convenient to use and is of the proper size. On seals without a lip, the seal is usually positioned so the part numbers face outward.

Figure 16-37 After the cup has been greased and the inner bearing installed in the cup, the area between the cone and the cup should be filled with grease, and the top edge of the cone (arrow) should be wiped clean of grease.

6. Wipe away any grease that might be left on the outside of the seal or on the end of the bearing cone (Figure 16-37). Any grease left on the ends of the cones or on the sides of the retaining washer might cause a bearing to change adjustment. Grease that stays in these areas during adjustment can be squeezed out by vehicle side motions. It is said that each layer of grease can increase bearing clearance by 0.001 inches (0.025 mm).

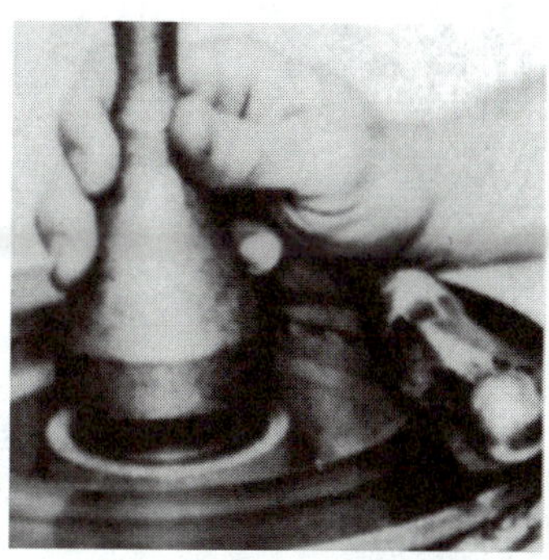

Figure 16-38 With the seal lip pointing inward, the new seal should be installed using a flat, properly sized seal driver. (Courtesy of The Timken Company)

7. Slide the hub partially onto the spindle.
8. Repeat steps 1 through 4 to pack the outer bearing, and then slide it onto the spindle and into the hub.
9. Install the washer and spindle nut.
10. Adjust the wheel bearing clearance as recommended by the car manufacturer. If the specifications are not available, a standard method of adjusting tapered-roller bearings is as follows:
    a. While rotating the hub by hand, tighten the spindle nut to about 15 to 20 foot-pounds (20 to 27 N-m) of torque. This ensures complete seating of the bearings. This is about as tight as you can tighten the nut using slip-joint pliers and normal hand pressure.
    b. Back off the spindle nut one-quarter to one-half turn.
    c. Retighten the spindle nut using your thumb and forefinger. This is equivalent to about 5 inch-pounds (0.56 N-m) of torque (Figure 16-39).
11. Replace the nut lock, if required, and position a new cotter pin through the nut or nut lock and the spindle. Using a pair of dikes, bend the long leg of the cotter pin outward, upward, and across the end of the spindle. The other leg can be cut off. Occasionally, if there is a static suppression spring in the dust cap, it is necessary to bend the two legs of the cotter pin sideways and around the spindle nut (Figure 16-40).
12. If you have doubts about the adjustment, measure the bearing or hub end play by mounting a dial indicator

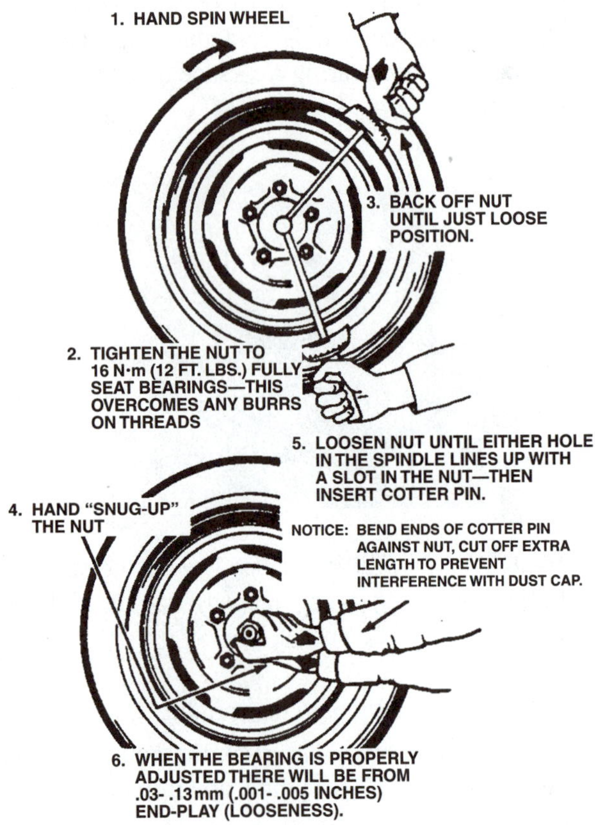

Figure 16-39 Wheel bearings are first tightened to seat the bearings and then adjusted to the correct running clearance. Always rotate the hub while seating the bearings. A commonly used bearing-adjustment procedure is shown here. (Courtesy of General Motors Corporation, Service Technology Group)

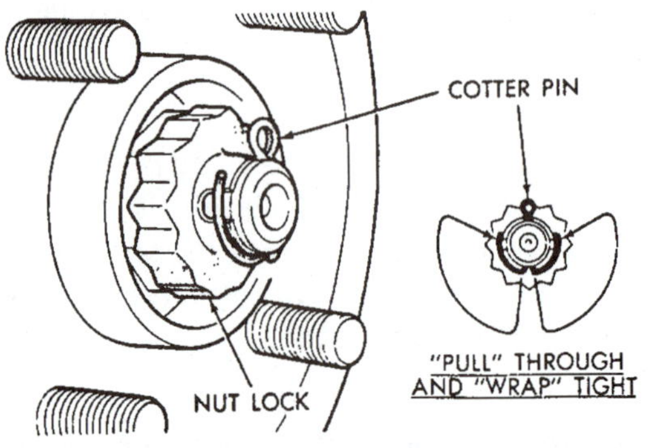

Figure 16-40 After the adjustment, the cotter pin is installed, trimmed if necessary, and bent across the end of the spindle (A) or wrapped around the spindle (B). Method (B) is required for clearance with some dust covers. (B is courtesy of Chrysler Corporation)

on the hub with the indicator stylus on the end of the spindle. Pull outward on the hub, adjust the dial indicator to zero, push inward on the hub, and read the amount of travel on the dial (Figure 16-41).

13. Replace the dust cap. Some technicians like to wipe a film of grease around the inside of the cap and/or around the lip of the cap to serve as a water seal.
14. Replace the caliper, being sure to follow the recommended procedure and tightening specifications. On 4WD vehicles, replace the front hub drive mechanism.
15. Replace the wheel, being sure to tighten the lug bolts to the correct torque using the proper pattern.

CAUTION *Before moving the car, make sure that all nuts and bolts are properly tightened and secure and that the brake pedal has a normal feel.*

Figure 16-41 Wheel bearing adjustment can be checked using a dial indicator. A correctly adjusted tapered-roller bearing set will have about 0.001 to 0.005 inches of end play. (Courtesy of The Timken Company)

16.13 Repairing Nonserviceable Wheel Bearings

As stated earlier, the only repair made on a faulty nonserviceable type of wheel bearing is removal and replacement of the hub-and-bearing assembly. This fairly simple procedure will vary slightly on different makes and models of cars. It is wise to follow the procedure described in the manufacturer's service manual.

To remove a nonserviceable hub-and-bearing assembly:

1. Raise and securely support the car.
2. Remove the wheel.
3. On drum brake cars, remove the drum. On disc brake cars, remove the brake caliper and then remove the rotor.

CAUTION *Be sure to support the caliper to prevent damage to the brake hose or tubing.*

4. Remove the bolts securing the hub-and-bearing assembly to the control arm or axle. Note that a hole is provided in the mounting flange so that a socket and extension bar can be used for removal and installation (Figure 16-42).

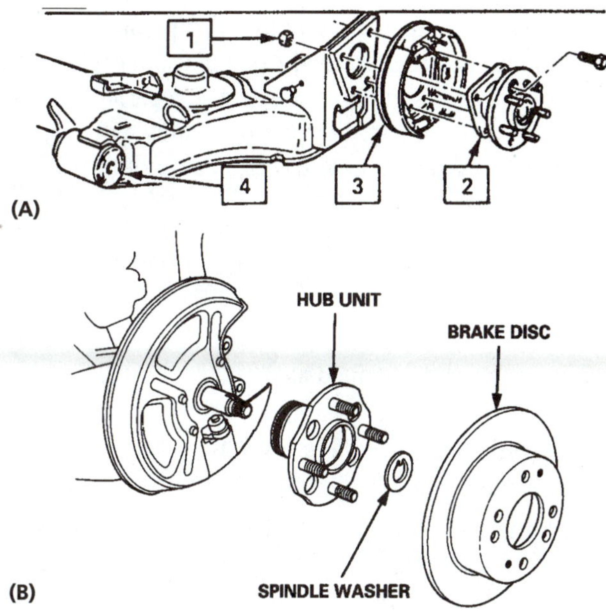

Figure 16-42 Two permanently sealed and adjusted bearing sets. (A) is bolted to the control arm; (B) slides over the control arm spindle. Either one must be replaced if it becomes faulty. (A is courtesy of General Motors Corporation, Service Technology Group; B is courtesy of American Honda Motor Co., Inc.)

To install a hub and bearing:

1. Make sure the recess in the control arm or axle, in which the hub-and-bearing assembly fits, is clean.
2. Install the hub-and-bearing assembly in position and tighten the bolts to the correct torque.
3. Install the brake drum or rotor. On disc brake cars, install the caliper and tighten the caliper mounting bolts to the correct torque.
4. Install the wheel, being sure to tighten the lug bolts correctly.

16.14 Repairing Solid-Axle Drive-Axle Bearings

Traditionally, drive-axle work has not been done by front-end technicians. Occasionally, brake technicians encounter a vehicle with oil- or grease-soaked brakes, which are the result of a faulty axle seal. Seal and sometimes bearing replacement is necessary before the brake lining is replaced. If a drive-axle bearing becomes loose, rough, or noisy, the bearing must be replaced. If a grease leak develops from the end of the axle housing, the axle seal and/or bearing (depending on the axle design) must be replaced. To replace either the bearing or the seal, the axle must be removed from the axle housing. As previously mentioned, there are two ways of securing the axle in the housing. In both designs, the inner end of the axle is supported by the axle gear in the differential unit inside the axle housing (Figure 16-43).

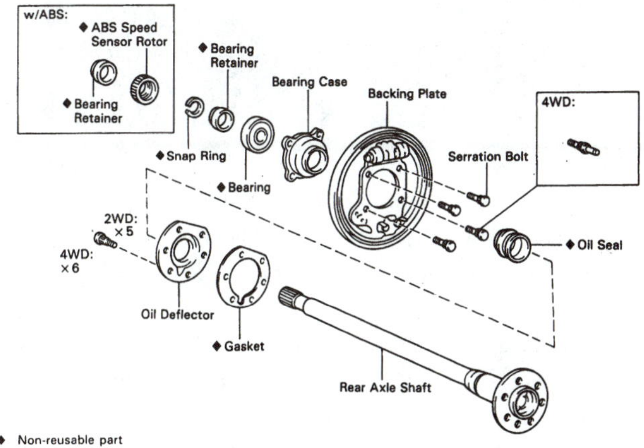

Figure 16-43 This rear axle can include a speed sensor rotor for ABS: also note the different lug bolts for 2WD and 4WD axles, and the backing plate and bearing plate relationship.

All C-lock designs use a seal that is separate from the bearing. Some bearing-retained designs use a seal incorporated in the bearing, and some use a separate seal or both. Axle bearing and seal repair procedures will vary on different cars. Always follow the procedure outlined in a technician's service manual.

16.14.1 Removing a Bearing-Retained Axle

To remove a bearing-retained axle:

1. Raise and securely support the car.
2. Remove the wheel.
3. Mark the brake drum or rotor next to the previously marked stud so that replacement is made in the same location. On disc brake cars, remove the caliper and secure it to the axle with wire.
4. Remove the brake drum or rotor.
5. Remove the nuts and bolts securing the bearing retainer and, usually, the backing plate to the end of the axle housing. Note that a hole is provided in the axle flange to allow the use of a socket and extension during removal and installation of these nuts or bolts (Figure 16-44).

Figure 16-44 When removing an axle that is retained by the bearing, the nuts holding the bearing retainer and backing plate must be removed. (Courtesy of Chicago Rawhide Industries)

6. Attach a slide hammer to the axle flange (Figure 16-45) and, using whatever force is required, remove the axle from the housing. As you slide the axle and bearing out of the housing, support the axle so it does not drag over the seal (Figure 16-46). Also make sure the bearing passes through the backing plate bore and that the backing plate stays at the end of the rear axle housing.

Figure 16-45 A special adapter connects the slide hammer to the axle, so the slide hammer can be used to pull the axle out of the housing. (Courtesy of SPX/OTC)

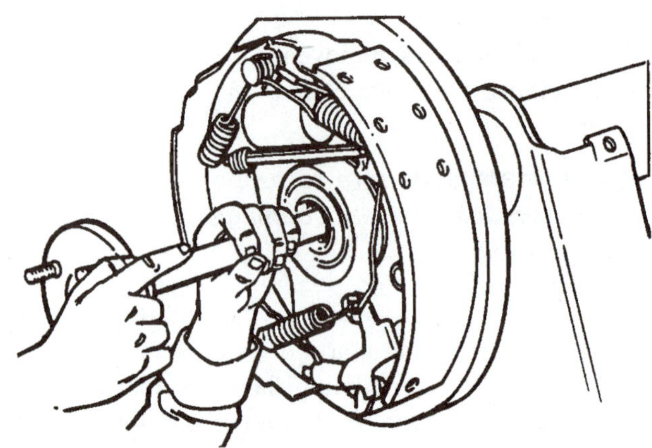

Figure 16-46 The axle should be supported as it is removed or replaced in the housing to keep it from dragging across and damaging the seal. (Courtesy of Ford Motor Company)

SAFETY TIP: Force must be applied to the axle and inner bearing race, across the balls to the outer race. This force causes an outward pressure on the outer bearing race, and this outward force can cause the race to fracture and explode violently. Newer axle-bearing-removal tools enclose the bearing in an adapter and provide some degree of protection. If the bearing is exposed during removal operations, most technicians place a shield around it during this operation to contain an explosion should one occur (Figure 16-47).

16.14.2 Removing and Replacing a Bearing on an Axle

This operation requires an axle-bearing-removal and installer set and a lot of caution. A large amount of pressure is usually required to push an axle out of a bearing, and there is only enough room to support the bearing by the outer race.

Figure 16-47 Axle bearings can explode violently as they are being pressed off the axle. A bearing separator (arrow) is positioned under the bearing and will be placed on the press bed. An old generator or starter case can be used as a scatter shield when using this style of adapter that does not enclose the bearing.

To remove an axle bearing:

1. Place the axle shaft over a large vise or anvil so it is supported under the bearing-retainer ring. Using a large chisel and hammer, slice straight into the retainer at four to six places. Usually one good blow at each location is sufficient (Figure 16-48). This action should stretch the retainer so it becomes loose on the shaft. Note that the area next to the bearing is used for a seal surface on some axles and must not be damaged. A short length of exhaust tubing or pipe can be slid over the axle to provide a shield.
2. Slide the bearing-retainer plate toward the axle flange, and attach the bearing-removal adapter to the bearing (Figure 16-49).
3. Place the axle and bearing remover in a press and press the axle out of the bearing.

CAUTION *After the axle is pushed a few inches, it will fall free. Be ready to catch it to prevent injury or any damage to the lug bolts (Figure 16-50)*

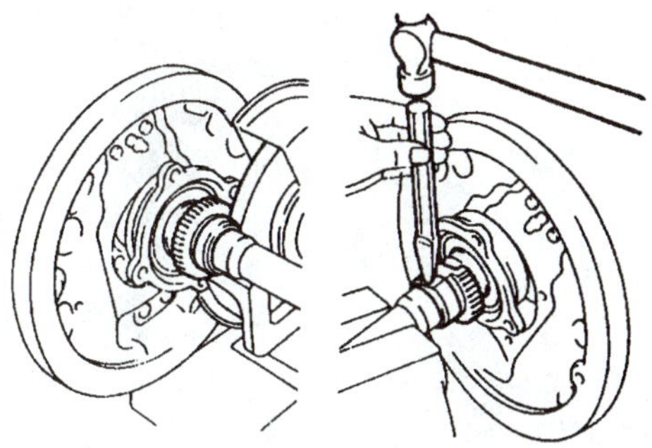

Figure 16-48 Before pressing an axle bearing off, make sure to loosen the retainer. This retainer has been ground partially through and then struck with a chisel.

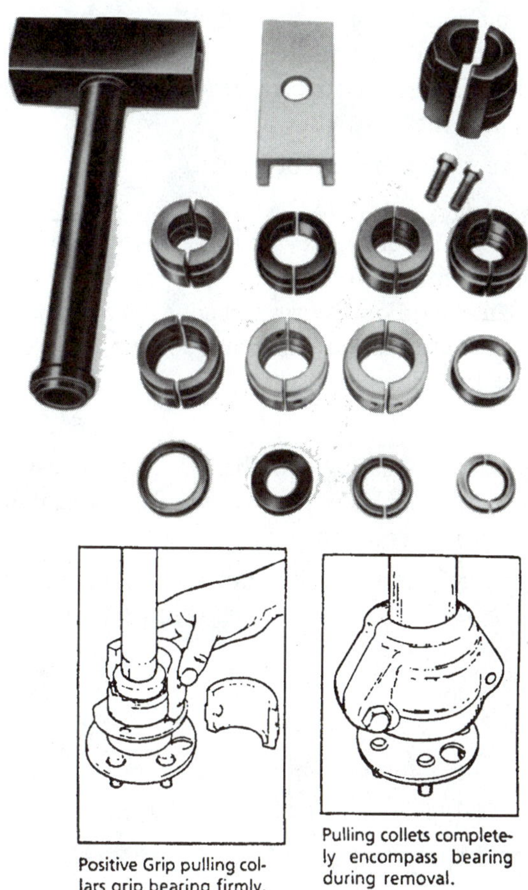

Figure 16-49 Some axle bearing puller and installer sets use a collet to enclose the bearing and attach it to the puller tube. (Courtesy of SPX/OTC)

4. Remove the bearing-retainer plate and clean the axle and retainer plate in solvent.

To install an axle bearing:

1. Place the bearing-retainer plate over the axle to the axle flange.

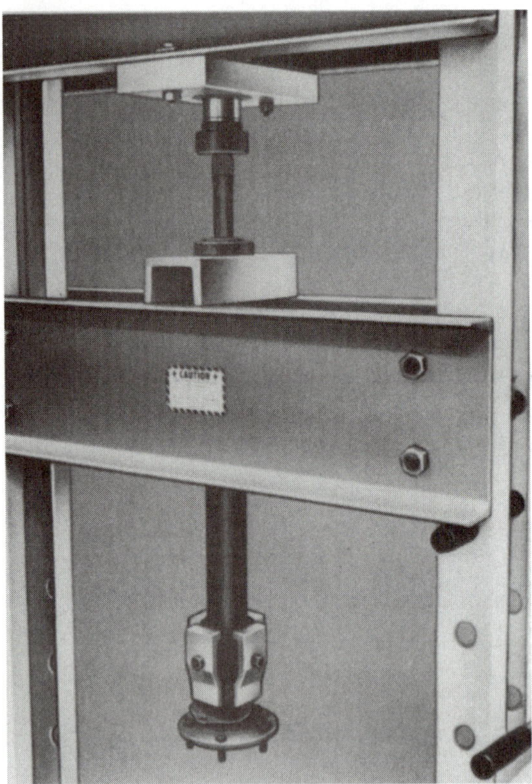

Figure 16-50 After the puller is attached to the axle, the axle and puller assembly are placed in a press to push the axle out of the bearing. (Courtesy of SPX/OTC)

2. Place the new bearing over the axle. Note that some bearings have an inner and an outer side. Be sure they are positioned correctly.
3. Support the bearing by the inner race and press the axle completely into the bearing (Figure 16-51).

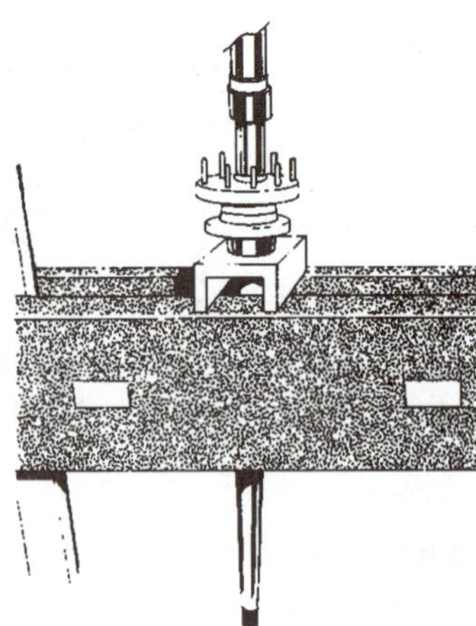

Figure 16-51 As the new bearing is pressed onto the axle, make sure that the adapters support the inner race of the bearing. A new retainer is pressed on after the bearing is installed. (Courtesy of SPX/OTC)

4. Place the new bearing-retainer ring on the axle and support the retainer ring on a press plate. Then press the axle into the ring so the ring is solidly against the bearing.

16.14.3 Installing a Bearing-Retained Axle

Axle-shaft installation is essentially the reverse of the removal process.

If seal replacement is necessary, on axles using a seal separate from the bearing, the seal should be removed and replaced while the axle is off. The procedure for replacing the seal is described later in this chapter.

To install a bearing-retained axle:

1. Slide the axle into the housing, being careful not to damage the axle seal if one is used.
2. When the axle stops a few inches from being fully installed, grip it by the flange so the inner end is lifted enough to enter the axle gear in the differential. It might be necessary to rotate the axle to align the splines. Slide the axle inward so the bearing enters the housing.
3. Install the bearing retainer flange nuts and bolts and tighten them to the correct torque.
4. Replace the brake drum or rotor and caliper.
5. Replace the wheel.
6. Check the axle lubricant level and add lubricant if needed.

16.14.4 Removing a C-Lock Axle

To remove a C-lock-retained axle:

1. Raise and securely support the car.
2. Remove the wheel.
3. Mark the brake drum or rotor next to the stud that was previously marked. On disc brake cars, remove the caliper and secure it to the axle with wire.
4. Remove the drum or rotor.
5. Remove any dirt around the rear axle cover. Position a drain pan under the cover and remove the cover-retaining bolts. Note that most of the rear axle grease drains out as you loosen the cover.
6. Remove the differential pinion-shaft lock bolt and the pinion shaft. Note that the pinion shaft won't side inward far enough for removal; it has to slide outward. If necessary, drive the axle shaft inward from the end opposite the lock pin hole. When the lock pin hole is past the differential case, insert a punch into the hole so the shaft can be pulled out (Figure 16-52).

7. Slide the axle inward until the C-lock can be removed from the axle and then remove the C-lock (Figure 16-53).
8. Slide the axle out of the housing. Be sure to support the axle to prevent damage to the seal.

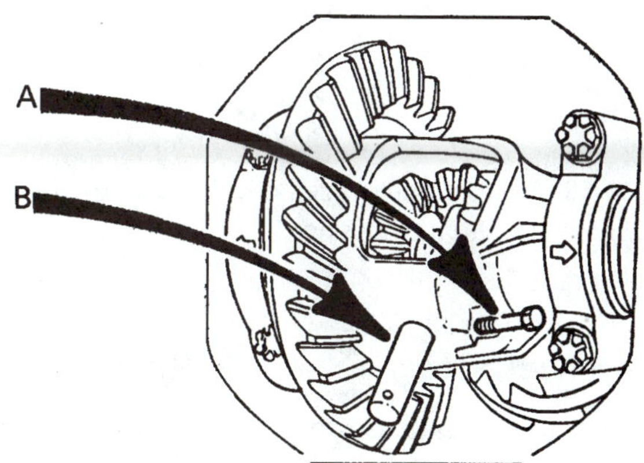

Figure 16-52 When removing an axle that is retained by a C-lock, the differential pinion-shaft locking bolt (A) is removed first, and then the differential shaft (B) is removed. (Courtesy of Ford Motor Company)

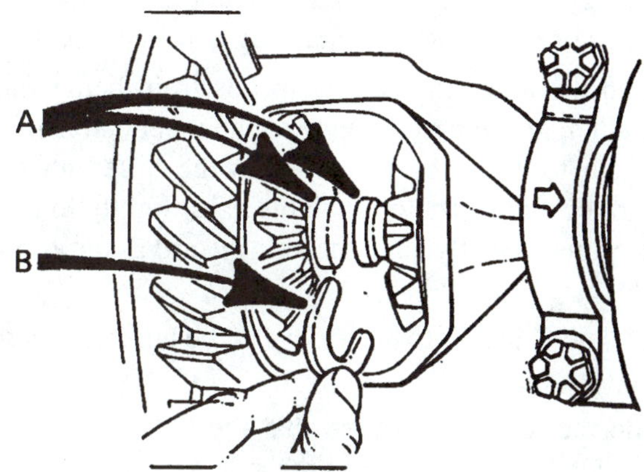

Figure 16-53 With the differential pinion shaft removed, the axles (A) can be slid inward enough to remove the C-locks (B). (Courtesy of Ford Motor Company)

16.14.5 Removing and Replacing a Bearing and Seal for a C-Lock Axle

The outer race of this bearing fits tightly into the outer end of the axle housing. The inner race is a hardened, ground portion of the axle. When an axle is removed, this portion should be inspected for damage. A rough or worn surface usually requires axle replacement. One aftermarket supplier provides a bearing that is positioned slightly differently to permit reuse of a worn axle. The axle seal is placed just outboard of the bearing with the seal lip running against the axle shaft (Figure 16-54).

To remove a bearing and/or seal:

1. Pry out the axle seal using a seal puller or pry bar (Figure 16-55).
2. Attach a bearing puller or hook to a slide hammer, position the puller inside the bearing, and pull the bearing from the housing (Figure 16-56).

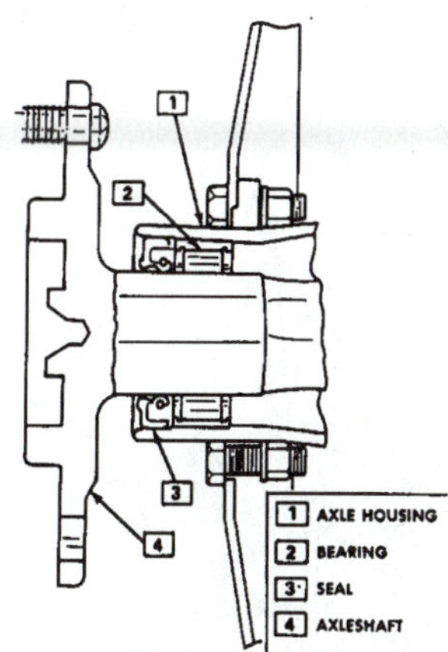

Figure 16-54 The axle serves as the inner bearing race on C-lock axles, and it will easily slide out of the bearing. If the bearing or seal surface is damaged, the axle is normally replaced. (Courtesy of General Motors Corporation, Service Technology Group)

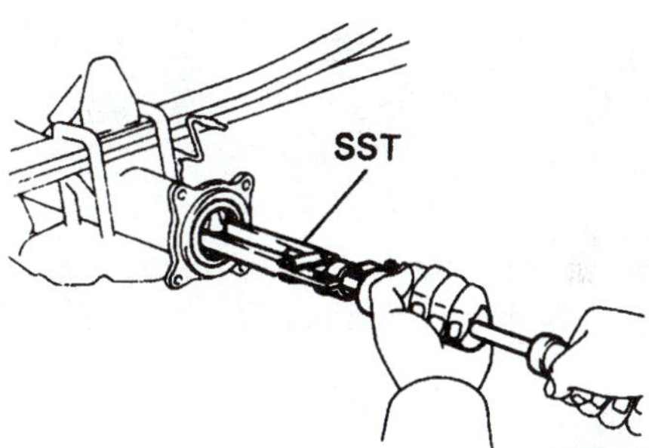

Figure 16-55 The inner housing seal can be pried out of some housings; on other axles, a puller is used to remove it.

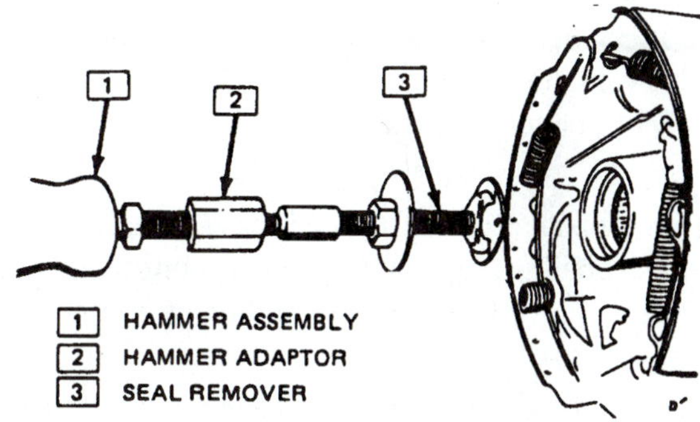

Figure 16-56 This adapter is used to connect the axle bearing to a slide hammer to pull the bearing out of the housing. (Courtesy of General Motors Corporation, Service Technology Group)

To install a new bearing:

1. Lubricate the new bearing with gear lubricant.
2. Position the bearing in the housing bore and, using a driving tool that contacts the entire side of the bearing, drive the bearing inward until it contacts the shoulder in the housing (Figure 16-57).
3. Lubricate the lip of the seal with gear lubricant (Figure 16-58).

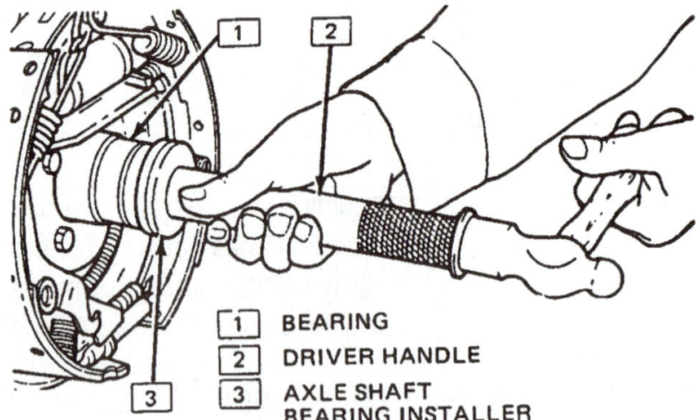

Figure 16-57 A new bearing should be driven in using a flat, properly sized driving tool. (Courtesy of General Motors Corporation, Service Technology Group)

Figure 16-58 Before a new seal is installed, it should be prelubricated with the same type of grease that it is going to seal. (Courtesy of Chicago Rawhide Industries)

4. If the outside edge of the seal case does not have a sealant coating, apply a thin film of room-temperature vulcanizing (RTV) **silicone rubber** or a nonhardening gasket sealer around the seal case.
5. Position the seal in the axle housing with the seal lip facing inward toward the lubricant. Then, using a driving tool that contacts the entire side of the seal case, drive the seal inward until it is in the proper position, usually flush with the edge of the axle housing (Figure 16-59).

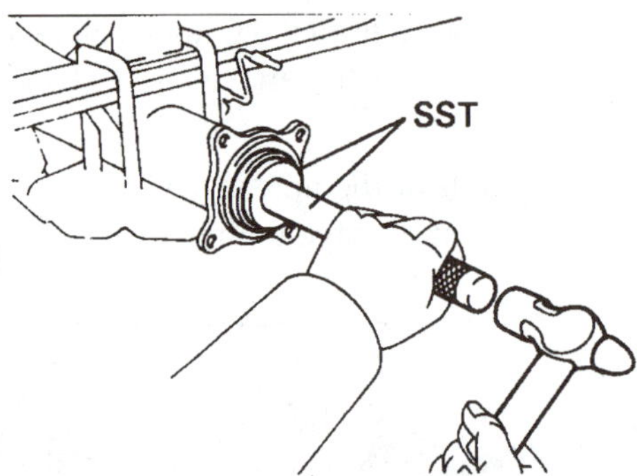

Figure 16-59 A flat, properly sized driver should be used to drive the new seal into the housing.

16.14.6 Installing a C-Lock Axle

Installing a C-lock axle uses a procedure that is essentially the reverse of the removal procedure.

To install a C-lock axle:

1. Lubricate the bearing and seal area of the axle.
2. Slide the axle into the axle housing and through the bearing and seal, being careful not to let the axle splines or rough axle surface drag across the seal.
3. When the axle stops a few inches from being fully installed, grip the axle flange so the inner end is lifted enough to enter the axle gear in the differential. It might be necessary to rotate the axle slightly to align the splines. Gently slide the axle inward as far as possible.
4. Slide the C-lock into the groove at the inner end of the axle (Figure16-60).
5. Slide the axle outward so that the C-lock is seated completely in the recess in the axle gear.

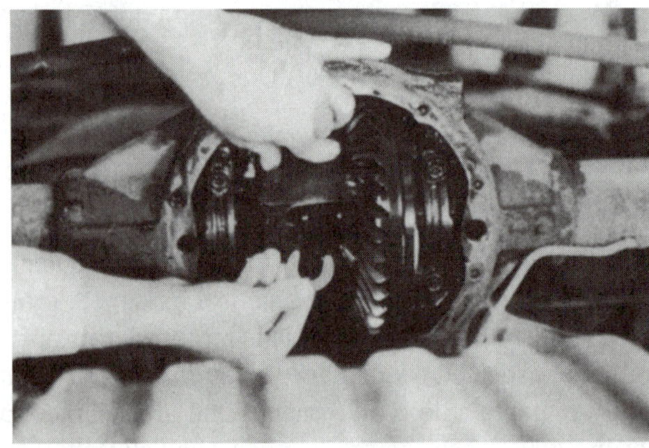

Figure 16-60 After the axle has been carefully slid into the housing, the C-locks are replaced, the differential pinion shaft is installed, and the shaft lock bolt is tightened to the correct torque. (Courtesy of Chicago Rawhide Industries)

6. Slide the differential pinion shaft into the differential case, with the shaft hole aligned with the hole for the locking bolt.
7. Install the locking bolt and tighten it to the correct torque.
8. Install a new cover gasket and the rear axle cover and tighten the retaining bolts to the correct torque. If RTV sealant is used in place of a solid paper gasket, thoroughly clean the gasket surfaces on the cover and back of the axle housing and apply a 1/16- to 1/8-inch (1.5- to 3-mm) wide bead of RTV sealant around the cover or housing gasket surface, circling each of the bolt holes (Figure 16-61).

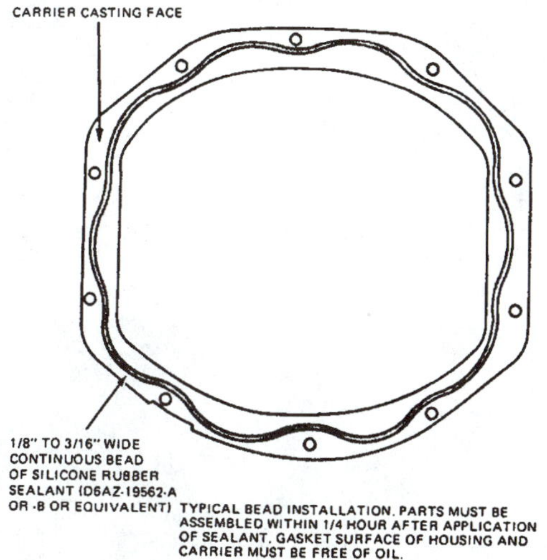

Figure 16-61 Older axle covers are sealed using gaskets. Many newer covers use a formed-in-place gasket, usually of a silicone rubber material. (Courtesy of Ford Motor Company)

9. Add lubricant through the filler hole to bring the gear oil to the correct level in the axle housing.
10. Replace the brake drum, with the marks aligned, and replace the tire and wheel.

16.15 Repairing Serviceable FWD Front-Wheel Bearings

There are several different styles of FWD front-wheel bearings—ball bearings, tapered-roller bearings, and sealed ball or roller bearings (Figure 16-62). Most of these bearings do not require periodic maintenance. The major reason for servicing is to replace a damaged hub or eliminate rough, loose, or noisy operation.

Servicing the front-wheel bearings requires removal of the outer end of the axle or CV-joint housing from the hub. In cases where a self-locking, prevailing torque hub nut is used, a new nut should be installed during replacement. In many shops, it is a common practice to remove this nut using an air-impact wrench. The hammering effect of this type of wrench can damage the front-wheel bearings if it is used to fully tighten the hub nut. This nut can usually be removed using an air-powered, impact-type wrench, but it should be installed by hand to prevent damage to the bearings. This presents a good indication of the nut's condition. Care should also be taken to prevent damage to the CV-joint or the CV-joint boot. Servicing procedures for the CV-joint and boot are described in various service manuals.

The repair procedures for FWD wheel bearings vary in different makes of cars. It is highly recommended that

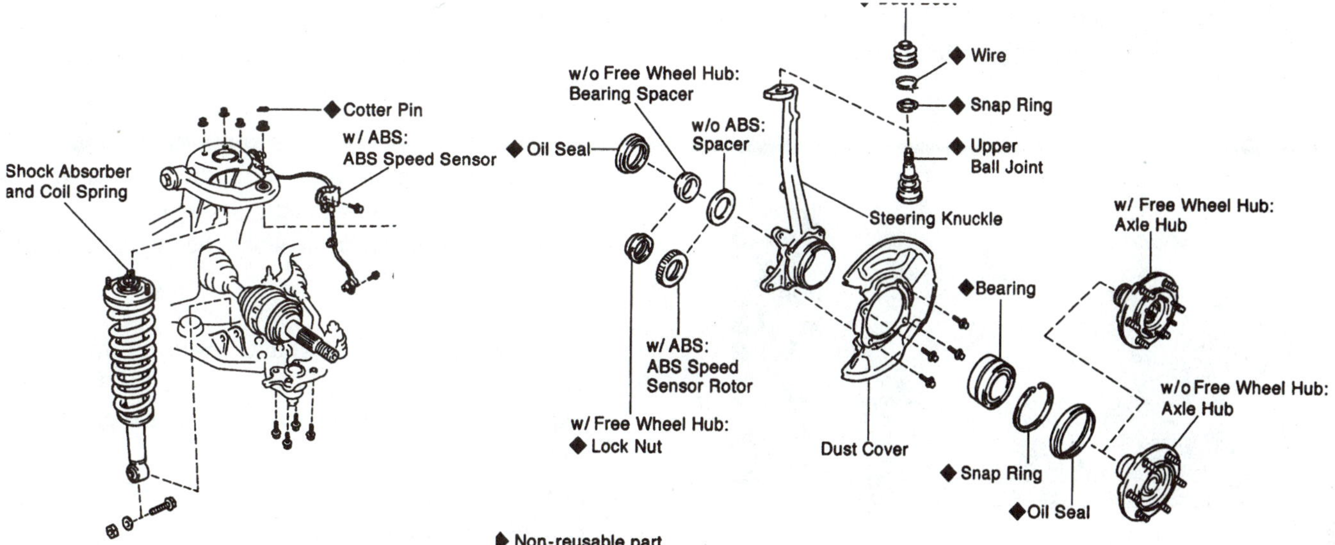

Figure 16-62 A disassembled view of the front knuckle and hub of an FWD or 4WD vehicle.

the specific repair steps for a particular car be followed when servicing the wheel bearings.

In general, to remove the front wheel bearing from a FWD car:

1. With the tires on the ground, the transmission in "park" or first gear, and the parking brake applied, remove any locking devices and loosen the hub nut (Figure 16-63).

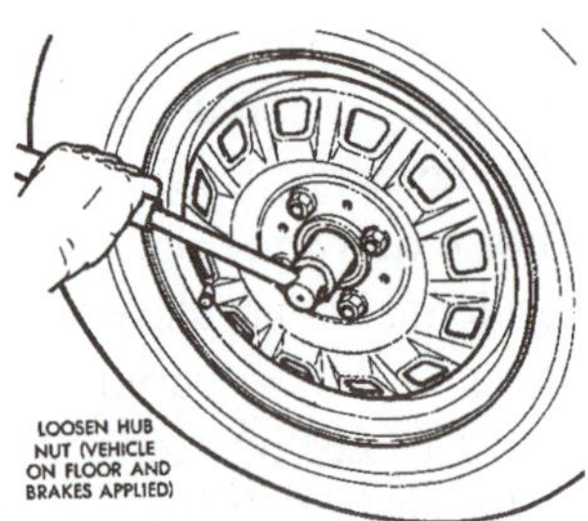

Figure 16-63 The hub nut is usually very tight; it should be loosened with the tires on the ground and brakes set. (Courtesy of Chrysler Corporation)

2. Raise and securely support the car.
3. Remove the wheel and, if available, install a boot protector over the CV-joint boots.
4. Remove the hub nut, the wheel, and the brake caliper. Using wire or a bungee cord, suspend the brake caliper from some point inside the fender.
5. Remove the hub from the steering knuckle, an operation often requiring the use of a puller. The steering knuckle must be disconnected from either the lower control arm or the strut. In some cars, the tie rod should be disconnected from the steering arm. Disconnecting these parts allows the steering knuckle freedom to move outward so the hub can be separated from the CV-joint housing (Figure 16-64).

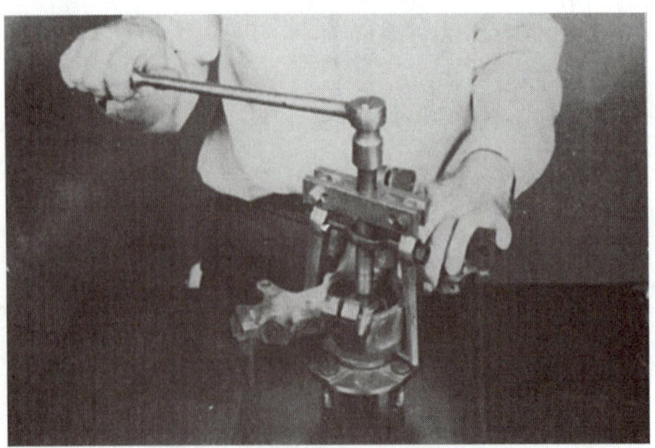

Figure 16-64 This puller is being used to pull the knuckle from the front hub. (Courtesy of Chicago Rawhide Industries)

6. Remove the bearing(s) and seal(s) from the steering knuckle. On some cars the bearing must be removed from the hub. This operation may require the use of a special puller (Figures 16-65 and 16-66).

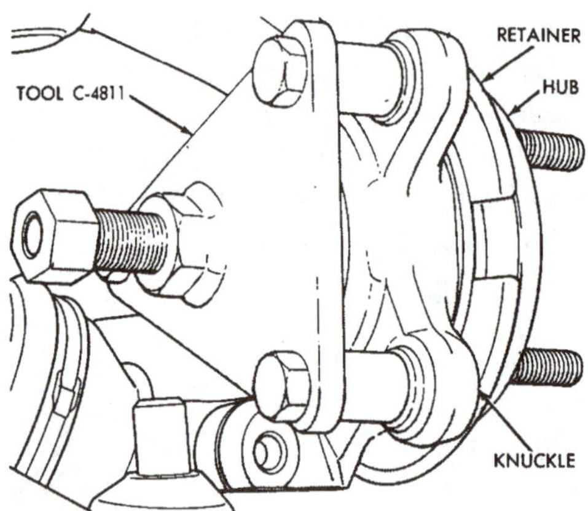

Figure 16-65 Tool C-4811 is being used to remove the hub from the knuckle. (Courtesy of Chrysler Corporation)

Figure 16-66 A socket, sized to fit against the outer bearing race, is being used to press the bearing out of the steering knuckle. (Courtesy of Chicago Rawhide Industries)

Like the disassembly procedure, the bearing installation should follow the manufacturer's instructions.

In general, to install front-wheel bearings on a FWD car:

1. Install the new bearing(s) and seal(s) in the steering knuckle. Lubricate the seal lips and the bearing, if required, with the proper amount and type of lubricant.
2. Install the hub in the steering knuckle (Figure 16-67).
3. Install the hub and steering knuckle on the axle end or CV-joint housing.

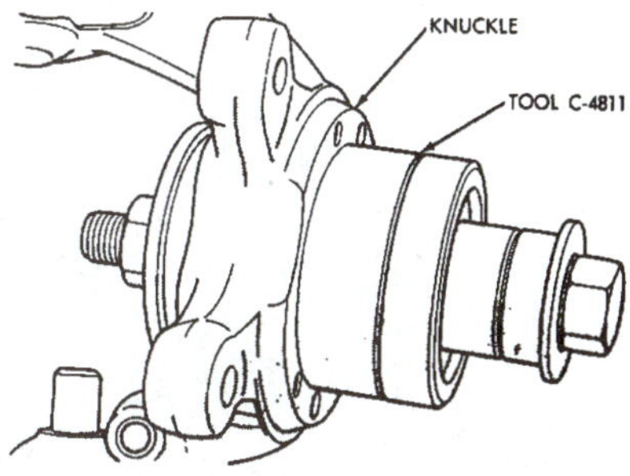

Figure 16-67 Tool C-4811 pushes against the new bearing to install it into the knuckle. (Courtesy of Chrysler Corporation)

4. Replace the hub nut and brake caliper. Tighten the hub nut and the caliper mounting bolts to the correct torque. Lock the hub nut in place as required (Figures 16-68 through 16-70).

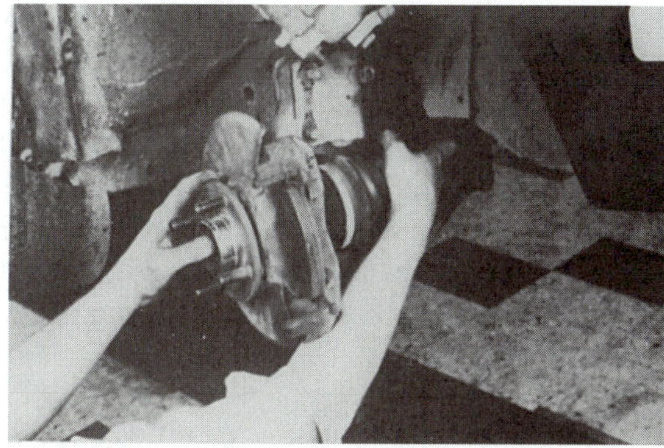

Figure 16-68 This hub is being pushed back onto the CV-joint after the bearing and hub have been installed in the knuckle. (Courtesy of Chicago Rawhide Industries)

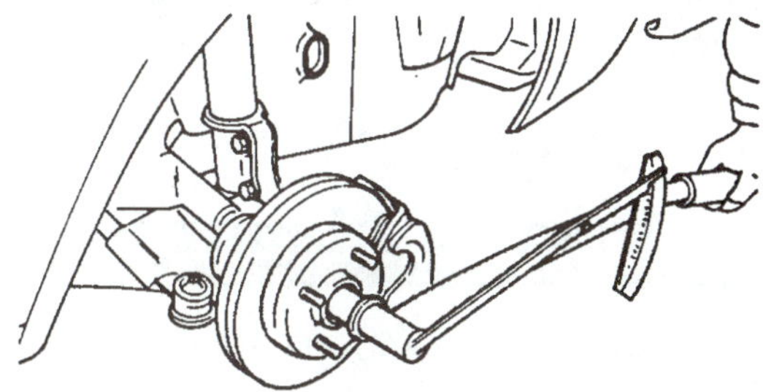

Figure 16-69 The hub nut should be tightened to the torque recommended by the vehicle manufacturer. This is usually somewhere between 100 and 250 foot-pounds. (Courtesy of Chrysler Corporation)

5. Reattach the tie rod, the lower end of the steering knuckle, and/or the steering knuckle to strut bolts if they were disconnected during disassembly. Tighten all nuts and bolts to the correct torque. Remove the CV-joint boot protector if required.

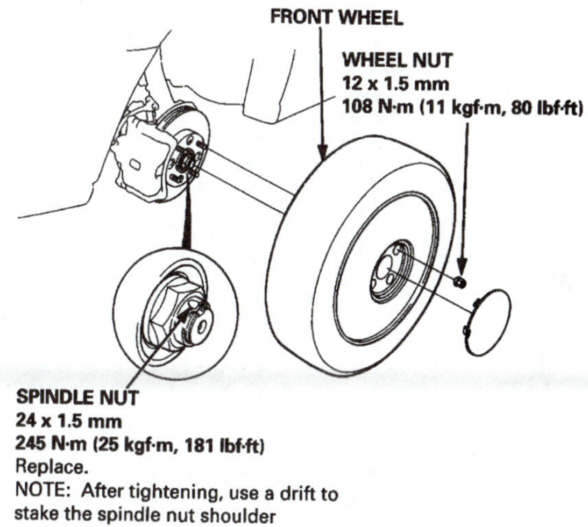

Figure 16-70 Various locking methods are used to secure the hub nut. (Courtesy of American Honda Motor Co., Inc.)

6. Replace the tire and wheel, tightening the lug nuts in the proper sequence to the specified torque.

16.16 Repairing Nonserviceable FWD Front-Wheel Bearings

The service for this type of bearing is limited to removing and replacing the hub and bearing assembly, much like the procedure for nonservicable wheel bearings. This operation requires the removal of the axle end or CV-joint housing from the hub. A boot protector or cover should be used to prevent damage to the CV-joint boot. A damaged, cracked, or torn CV-joint boot should be replaced.

To remove a nonserviceable wheel bearing:

1. With the tires on the ground, the transmission in "park" or first gear, and the parking brake applied, remove the locking device from the front axle or hub nut, if required, and loosen the hub nut.
2. Raise and securely support the car.
3. If a boot protector is available, install over the CV-joint boot.
4. Remove the hub nut and the wheel.
5. Remove the brake caliper and, using wire, suspend it from some point inside the fender. Then remove the rotor.
6. Remove the hub-and-bearing mounting bolts and the rotor splash shield (Figure 16-71).

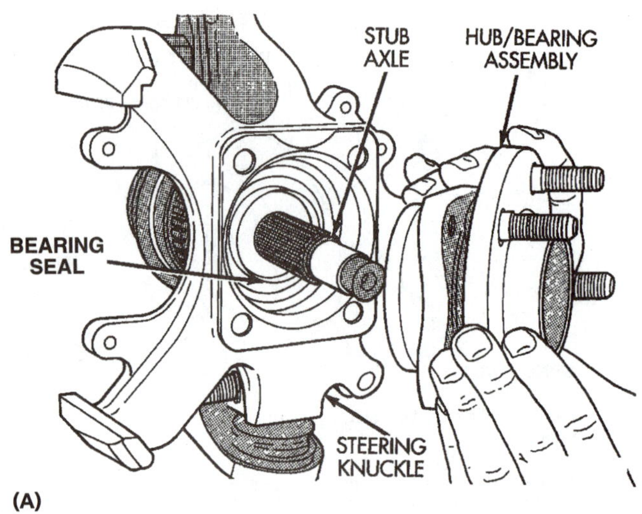

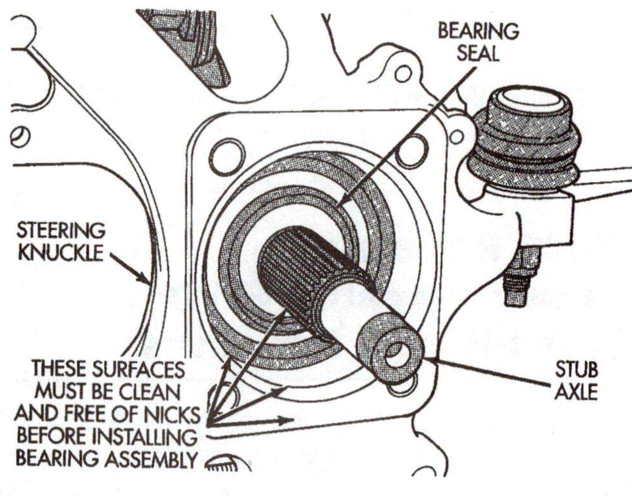

Figure 16-71 This hub/bearing assembly is permanently sealed, lubricated, and adjusted. It is replaced as an assembly if there are problems. (Courtesy of Chrysler Corporation)

7. If necessary, install a hub puller on the hub flange and pull the hub-and-bearing assembly off the end of the CV-joint housing and out of the steering knuckle (Figure 16-72).
8. Some car models also have a steering knuckle seal which should be removed at this time.

To replace a nonserviceable front-wheel bearing:

1. Clean the bore of the steering knuckle and the splined area of the CV-joint housing.

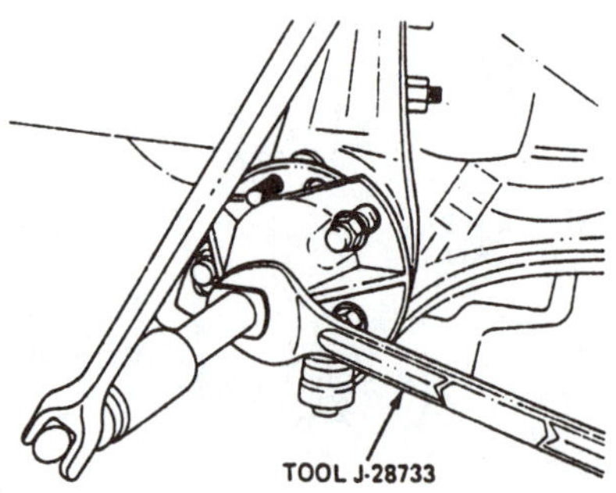

Figure 16-72 Tool J-28733 is being used to pull the hub-and-bearing assembly off the drive shaft. (Courtesy of General Motors Corporation, Service Technology Group)

2. Lubricate the lip of the new steering knuckle seal, if required, and install the seal in the steering knuckle.
3. Position the new hub-and-bearing assembly over the axle and in the steering knuckle. Install the hub nut and the hub-and-bearing-assembly mounting bolts and alternately tighten them to move the hub-and-bearing assembly into the correct location. Tighten the hub nut temporarily to about 70 foot-pounds (100 N-m) of torque and the mounting bolts to the correct torque. A long bolt can be used in place of one of the mounting bolts to keep the hub from turning while you tighten the hub nut.
4. Remove the boot protector.
5. Replace the rotor and the caliper and tighten the caliper mounting pins to the correct torque.
6. Install the wheel.
7. Lower the car to the ground and tighten the hub nut to the correct torque. Lock the hub nut in place as required.

16.17 Practice Diagnosis

You are working in a brake and front-end shop and encounter these problems:

CASE 1: The customer complains of lockup of the right rear wheel under a medium pedal pressure. The car is a mid-1980s RWD station wagon. While on a road test, you hear a slight growling sound coming from the right rear, and the brake lockup is confirmed by a skidding right rear tire when you apply the brake. In the shop, with the rear end lifted, engine running, and car in gear, you hear the same growling noise coming from inside the right rear wheel. When you remove the wheel and brake drum, you find the brake shoes are oily. Can you fix this problem? If so, what will you probably need to do?

CASE 2: The car is a mid-size, FWD General Motors "W" car, and the customer's complaint is a strange noise coming from the left rear. On a road test, you hear a growling noise that seems to be coming from the left rear, and it becomes quieter when you make left turns. What should you do, when you get the car back to the shop, to confirm this problem? What is the probable cause? What will you need to do to repair this car?

Terms to Know

axle bearing
ball bearings
bearing
bearing-retained axle
C-lock axle
CV-joints
end play
preload
roller bearings
silicone rubber
tapered-roller bearings
wheel bearing

Review Questions

1. Statement A: Most of the rear-wheel bearings, hubs, and spindles of FWD cars are similar to those on the front wheels of RWD cars.
   Statement B: Some wheel bearings cannot be adjusted. Which statement is correct?
   a. A only c. both A and B
   b. B only d. neither A nor B
2. A frictionless bearing uses as the rolling element of the bearing:
   a. balls c. needles
   b. rollers d. any of the above
3. The purpose of the cage in a ball bearing is to
   (A) keep the balls separated and properly spaced.
   (B) hold the balls between the races.
   a. A only c. both A and B
   b. B only d. neither A nor B
4. Technician A says that free movement of a hub in and out is called bearing end play. Technician B says that preload causes a drag on the bearings. Who is right?
   a. A only c. both A and B
   b. B only d. neither A nor B
5. Technician A says that the inner lip of a seal should always point toward the outside to keep dirt from entering under the seal lip. Technician B says that the sealing pressure of a lip seal is often increased by a garter spring. Who is right?
   a. A only c. both A and B
   b. B only d. neither A nor B
6. The lip seal used in a front hub of a RWD car will make a
   (A) dynamic seal with the spindle.
   (B) dynamic seal with the hub.
   a. A only c. both A and B
   b. B only d. neither A nor B
7. The front-wheel bearings of a FWD car are:
   (A) adjusted by tightening the hub nut.
   (B) repacked similarly to a RWD car's front-wheel bearings.
   a. A only c. both A and B
   b. B only d. neither A nor B
8. Technician A says that a faulty wheel bearing will usually change noise level as the car is turned in different directions. Technician B says that faulty wheel bearings will usually change noise level under different throttle conditions. Who is right?
   a. A only c. both A and B
   b. B only d. neither A nor B
9. When checking the rear axle bearings of a RWD car, in all cases, there should be
   (A) no vertical play.
   (B) no in-and-out play.
   a. A only c. both A and B
   b. B only d. neither A nor B
10. Technician A says that wheel bearing play is checked by gripping the tire at the top and bottom and trying to rock it. Technician B says the front-wheel bearing of a RWD car should have a slight amount of play when rocked. Who is right?
    a. A only c. both A and B
    b. B only d. neither A nor B

11. Technician A says that the best way to remove the inner wheel bearing from a front hub is to drive the bearing and seal out with a wooden dowel. Technician B says that the best tool for removing a cotter pin from the spindle is either a cotter-pin puller or diagonal-cutting pliers. Who is right?
    a. A only c. both A and B
    b. B only d. neither A nor B
12. When cleaning wheel bearings, spinning bearing with an air blast is
    (A) an effective way of cleaning the bearing.
    (B) a safe shop practice.
    a. A only c. both A and B
    b. B only d. neither A nor B
13. Wheel bearings are repacked by forcing clean grease into the bearing using
    (A) hand pressure.
    (B) a bearing packer.
    a. A only c. both A and B
    b. B only d. neither A nor B
14. Technician A says that front-wheel bearings of the tapered-roller type should be adjusted to have slight end play. Technician B says that after the bearings have been seated, the spindle nut should be finger tightened. Who is right?
    a. A only c. both A and B
    b. B only d. neither A nor B
15. Technician A says that you must remove the axle to repair a leaky axle seal on the rear axle of a RWD car. Technician B says that the rear axle housing cover must always be removed in order to remove the axle. Who is right?
    a. A only c. both A and B
    b. B only d. neither A nor B
16. As an axle shaft is slid into the housing, the
    a. shaft should be prevented from contacting the seal.
    b. shaft splines must be aligned with the gear in the differential.
    c. bearing should be lubricated with gear oil.
    d. all of the above.
17. Before installation, lip seals should always be
    (A) coated with sealant.
    (B) lubricated on the seal lips.
    a. A only c. both A and B
    b. B only d. neither A nor B
18. Technician A says that it is necessary to remove the CV joint from the front hub when servicing the front-wheel bearings of a FWD car. Technician B says these front-wheel bearings should be serviced every 5,000 to 10,000 miles. Who is right?
    a. A only c. both A and B
    b. B only d. neither A nor B
19. Nonserviceable wheel bearings are repaired by
    (A) removing and replacing the entire hub-and-bearing assembly.
    (B) removing, replacing, and adjusting the sealed bearing unit.
    a. A only c. both A and B
    b. B only d. neither A nor B
20. The front hub nut of a FWD car must be tightened
    (A) using an air-impact wrench.
    (B) enough to obtain the correct bearing preload.
    a. A only c. both A and B
    b. B only d. neither A nor B

17 Trailer Brake Systems

Objectives

After completing this chapter, you should:

- ❑ Be familiar with the different types of braking systems used on trailers pulled by passenger cars and light trucks.
- ❑ Be familiar with the terms commonly used with trailer brakes.
- ❑ Be able to diagnose trailer brake problems and determine the needed repair.
- ❑ Be able to test an electric brake system to determine proper operation.
- ❑ Be able to adjust shoe clearance on trailer brakes.
- ❑ Be able to service the wheel bearings used on trailers.

17.1 Introduction

Trailers of many types—recreational vehicle (RV), farm or animal, industrial, and utility—are commonly pulled by passenger cars, pickups, vans, and light trucks. The added weight of these trailers will increase the stopping distance of the tow vehicle unless the trailer is equipped with brakes. In many states, the vehicle code requires brakes on all trailers that exceed 1,500 pounds (682 kg) (Figure 17-1). Also, many states have statutes that regulate the maximum stopping distance for single vehicles and combinations of vehicles. The stopping distance requirement for a trailer–tow vehicle combination is slightly longer than that for a single vehicle.

One difficult aspect of trailer use is matching the strength of the brakes on the trailer to those on the tow vehicle. If the trailer brakes are not strong enough, the trailer will push the tow vehicle during a stop and increase the stopping distance. If the trailer brakes are too strong and lock up during a stop, the trailer will tend to skid sideways and "jackknife," especially under wet, icy, or poor road surface conditions. It is always advisable to test the brakes when driving a particular vehicle combination for the first time. Then adjust the speed downward if necessary to suit the conditions and stopping ability.

Trailers use drum brake assemblies that are operated either hydraulically or electrically. Most hydraulic brake systems use an inertia or surge type of master cylinder application. A few systems use a remote master cylinder that is operated by pressure from the hydraulic brake system of the tow vehicle.

Every trailer with brakes must also have breakaway controls. These controls will automatically apply the brakes on the trailer if it comes loose and breaks away from the tow vehicle. Breakaway controls must also be capable of holding the trailer brakes in the applied position for at least fifteen minutes.

17.2 Electric Brakes

The majority of trailer brakes are electric. Electric brakes have the advantage of easy control over the power of the trailer brakes and can easily operate two or more wheel assemblies without any lag in application. It is also possible to operate electric trailer brakes independent of the tow vehicle brakes and to install them so they will not interfere with brake operation on the tow vehicle. This is especially important on ABS-equipped vehicles. Many electric brake controllers use a hydraulic transducer, which is connected to a tee fitting installed in the tow vehicle's hydraulic system. This transducer synchronizes the trailer brakes with the tow vehicle brakes. It is quite reliable and has very little detrimental effect on the tow vehicle's brake operation. You should never tap into the hydraulic system of an ABS except for testing purposes (Figure 17-2).

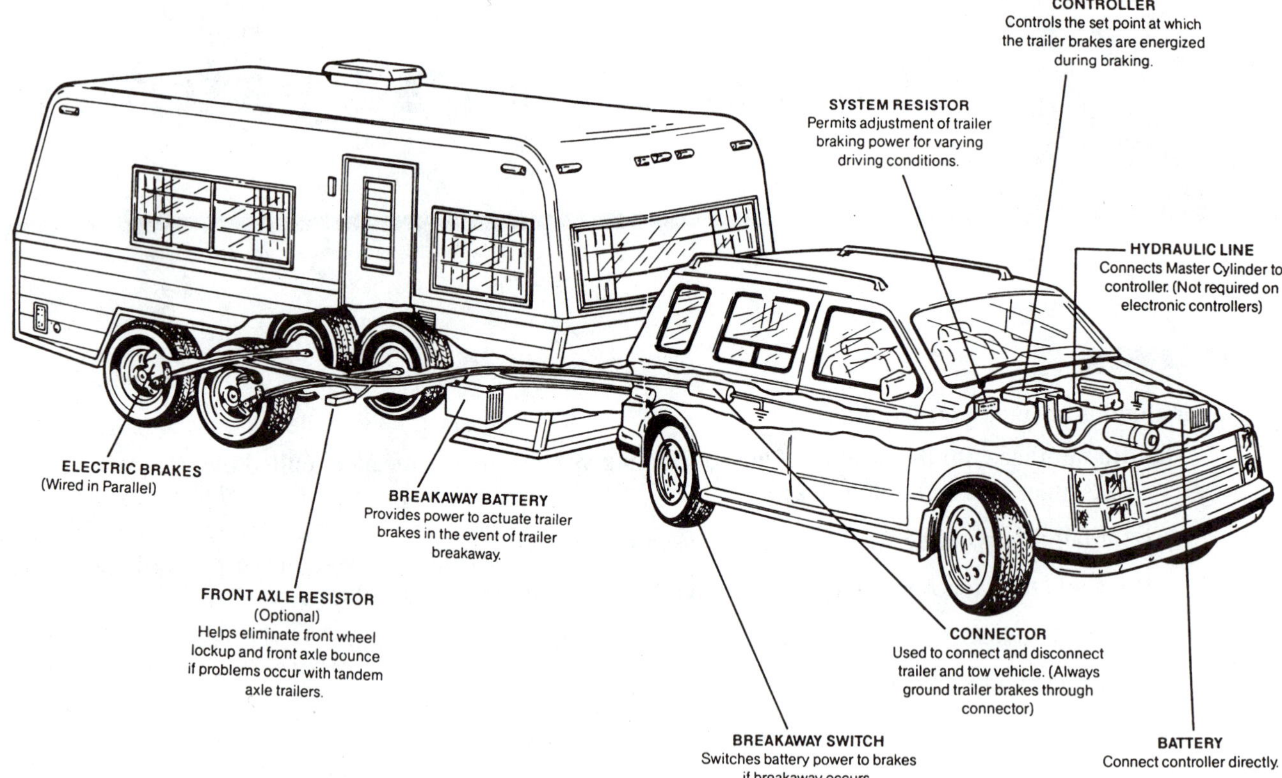

Figure 17-1 This RV trailer has electrical brakes operated by a controller in the tow vehicle.

17.2.1 Brake Assemblies

Electric brake assemblies resemble a standard duo-servo drum brake and have drums, shoes, shoe return and hold-down springs, backing plates, and anchor pins without a wheel cylinder. In addition, they have an electromagnetic coil that is mounted on a lever arm. This lever arm pivots on the backing plate on a pivot pin located below the anchor pin. Inside the brake drum is a flat, smooth armature plate (Figure 17-3).

During a stop, electric current is sent through the coil of the magnet, creating a magnetic force that attracts the coil to the rotating armature plate. The rotation of the brake drum swings the magnet and lever arm in the direction of drum rotation. This motion then applies the primary shoe during forward motion or the secondary shoe during reverse motion (Figure 17-4). From this point on servo action completes the application of the brakes. The strength of these brakes is dependent on the strength of the magnet (which is dependent on the amount of current flow) and the speed of the vehicle. A high current flow through the magnet will generate a strong magnetic pull. The coil will cling more tightly to the armature plate, which in turn will increase the shoe application pressure.

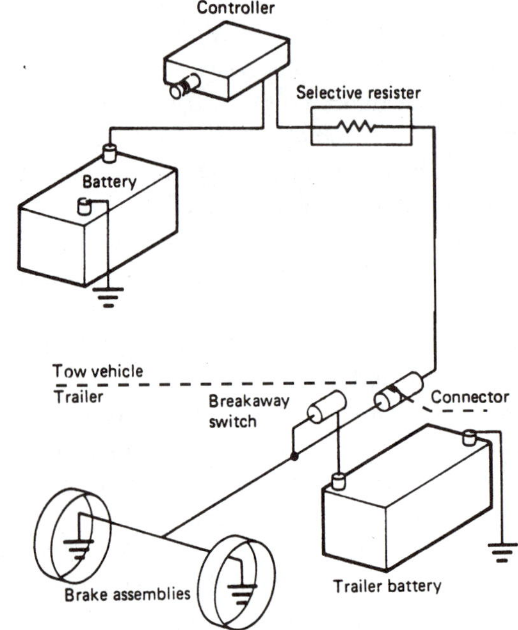

Figure 17-2 An electric brake system consists of the brake assemblies with the electromagnets that apply them, a connector to the tow vehicle, a controller, sometimes a resistor for balancing application forces, and the battery and alternator in the tow vehicle.

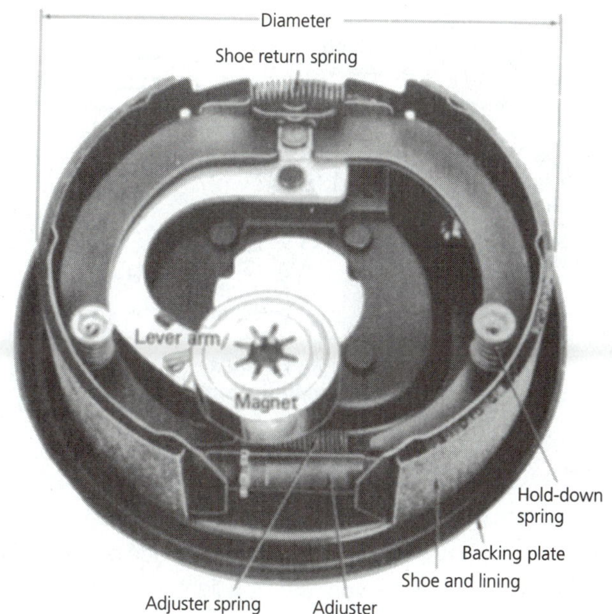

Figure 17-3 An electric brake assembly (left side) uses an actuating lever and magnet in place of a wheel cylinder.

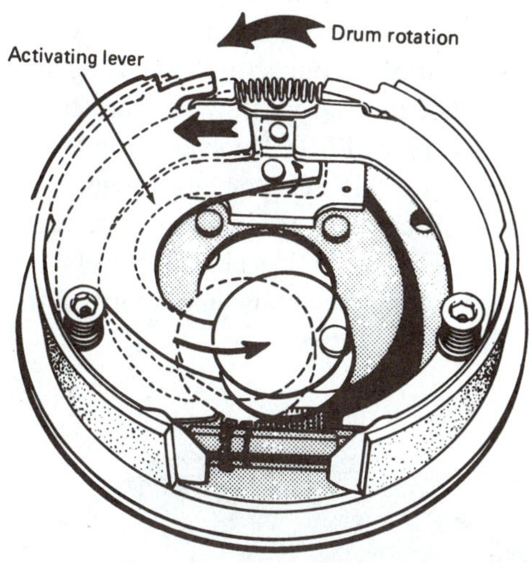

Figure 17-4 When the brake is applied, current is sent through the magnet, and this pulls the magnet against the rotating armature plate in the brake drum. The drum's rotation causes the lever to swing and apply the brake shoes.

17.2.2 Brake Controller

The tow vehicle must have a brake controller, which allows the driver to apply the trailer brakes. Trailer brake operation is automatic with many controllers. The controller consists of a switch and a variable resistor that sends current to the brake magnets and varies the current flow to adjust the rate and strength of brake application. Some controllers use either a hydraulic or a pressure transducer. Increasing the pressure on a transducer will increase the current flow. The controller is connected to the battery or close to the battery of the tow vehicle and also to the trailer connector receptacle at the rear of the tow vehicle. When the transducer is actuated, current can flow from the battery through the controller and then to the trailer connector and brake magnets. Three major types of controllers are available: foot pressure, hydraulic pressure, and inertia. Hydraulic and inertia controllers can also be applied by hand pressure. Most controllers include a device used to adjust application of the trailer brakes so that it occurs at the same time that the tow vehicle brakes are applied.

A foot-pressure controller or transducer is a pad that is clamped on top of the brake pedal pad. Foot pressure, used to apply the brakes in the tow vehicle, will compress the transducer and apply the trailer brakes. The current flow to the trailer brakes will be proportional to the driver's foot pressure.

A hydraulic controller is normally mounted under the tow vehicle's instrument panel (Figure 17-5). A hydraulic line is connected between the controller and a tee fitting installed in the tow vehicle's hydraulic system, usually at a master cylinder outlet. This line is connected to the

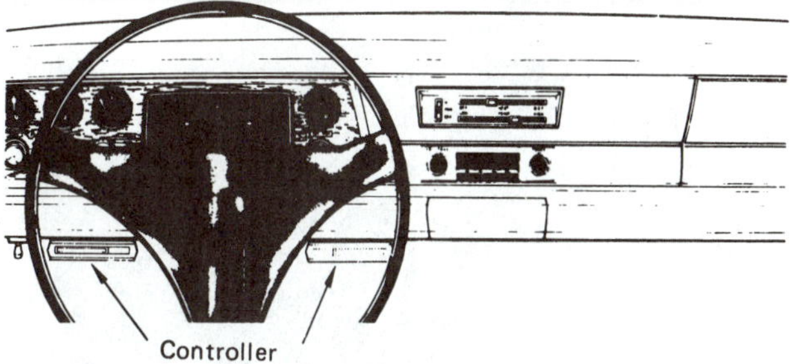

Figure 17-5 An automatic controller is attached to the instrument panel. It has a hydraulic connection to the tow vehicle brake system. (Courtesy of Hayes/Lammerz International, Inc.)

hydraulic transducer in the controller (Figure 17-6). As the tow vehicle's brakes are applied, the hydraulic transducer causes electricity to flow to the trailer brakes. The current flow will be proportional to the hydraulic pressure. A hand lever on the controller allows application of the trailer brakes independent of the hydraulic pressure.

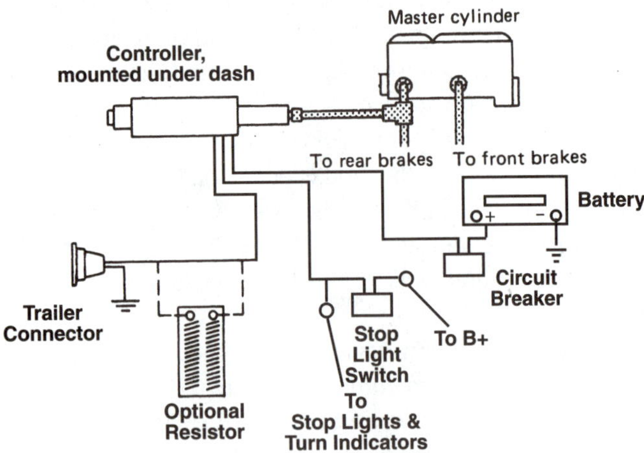

Figure 17-6 The electrical and hydraulic connections for an automatic controller; the trailer is plugged into the socket. This style of controller should not be used on tow vehicles with ABS. (Courtesy of Hayes/Lammerz International, Inc.)

An inertia controller is also mounted under the tow vehicle's dashboard (Figure 17-7). It is connected to the tow vehicle's stoplight circuit, and when the stoplights are on, it causes a current flow to the trailer brakes in direct proportion to the deceleration rate of the tow vehicle (Figure 17-8). A hand lever on this controller will allow independent hand application of the trailer brakes.

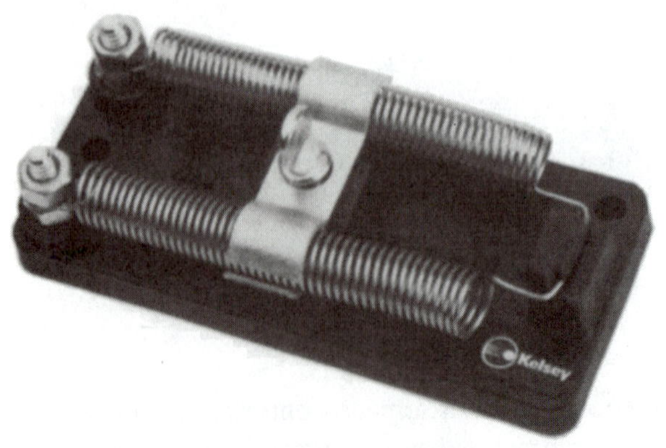

Figure 17-7 An adjustable resistor is used to obtain the correct current flow through the brake magnets, so that the trailer brakes are balanced with those on the tow vehicle.

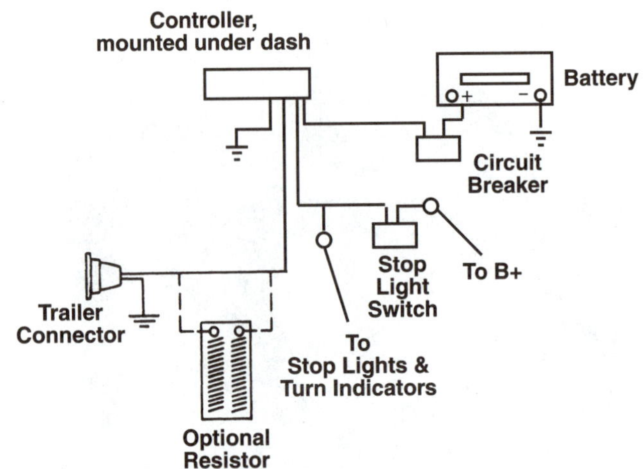

Figure 17-8 The connections to an electronic controller include the battery and circuit breaker to provide power for the trailer brakes with protection, the stop light switch so it senses when the brakes are applied, a ground, and the connections to the trailer brakes through the connector.

The power of electric brakes is adjusted, if necessary, by using a selective resistor connected between the controller and the trailer connector. Increasing the resistor value will reduce the current flow and the strength of the trailer brakes (Figure 17-9). Depending on how the wires are connected to this resistor, it provides a variable amount of resistance to balance the braking strength. Again, when the trailer brakes are completely applied, the ideal trailer braking power is just short of that which causes skidding.

Figure 17-9 This electronic controller uses a pendulum deceleration sensing device that automatically applies the trailer brakes in response to the tow vehicle slowing down. The LED indicates trailer brake application. (Courtesy of Hayes/Lammerz International, Inc.)

Power is sent to the trailer brakes through a connector cable that plugs into a receptacle on the tow vehicle. Trailers with electric brakes commonly use a five- or six-terminal connector. This connector normally has terminals for the tail lights, right-turn light and stoplight, left-turn light and stoplight, and ground. The fifth terminal will be for the brakes, and the sixth terminal, if used, provides power to recharge a battery in the trailer (Figure 17-10).

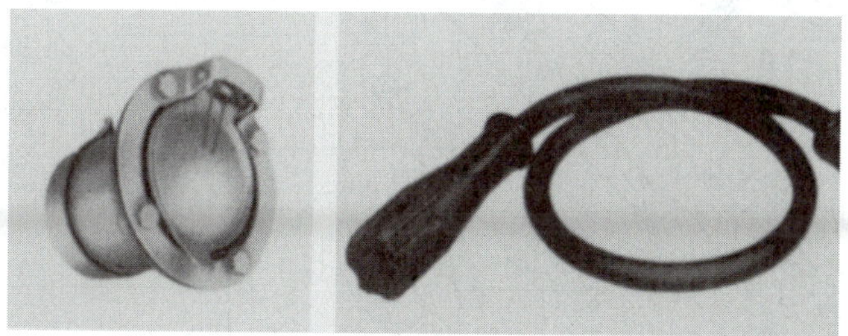

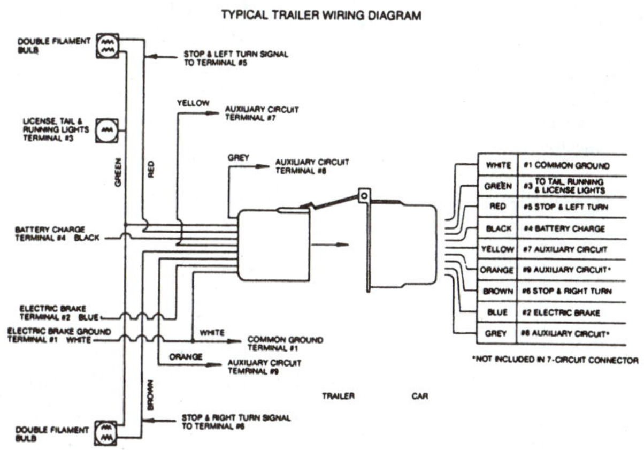

Figure 17-10 A wiring connector and socket (top) and the wiring diagram for a trailer connector. Trailers with electric brakes normally use a six- or seven-circuit (sometimes nine) connector.

17.2.3 Breakaway Control

In an electric brake system, breakaway control is provided by a switch mounted on the trailer tongue and attached to the tow vehicle by a small chain or cable. Should the trailer come loose or break away, the trailer motion will cause the chain to pull the switch to the application position. During a breakaway electric power comes from a battery mounted on the trailer. This battery can be a normal automotive battery or a 12-volt dry-cell battery. It must be able to provide enough current to stop the trailer and keep the brakes applied for a minimum of fifteen minutes.

17.3 Surge Brakes

Some trailers use a self-contained hydraulic brake system, which includes wheel drum brake assemblies, hydraulic lines, and a master cylinder. The master cylinder is applied by the inertia of the trailer as it tries to push the tow vehicle during a stop. A surge brake will always cause the trailer to push lightly on the tow vehicle in order to apply its own brakes. However, the system will provide the majority of the braking power needed to stop the trailer (Figure 17-11).

Figure 17-11 This utility trailer is equipped with surge brakes. Note the movable coupler portion of the trailer tongue that causes master cylinder application when the tow vehicle slows down. Also note the chain and lever (arrow) to apply the brakes if the trailer breaks away.

Surge brakes have a definite advantage for some trailers, such as rental units, in that the trailer can be attached to any vehicle with a trailer hitch. No brake controller is required. Possible disadvantages of a surge brake system are that there is always a slight lag in trailer brake operation, and that the trailer cannot stop itself and the tow vehicle in an emergency. Also, some systems will apply the brakes if the trailer is backed up, especially on soft ground or on slight slopes. Many systems use uni-servo brakes or are able to lock out the surge operation during these maneuvers.

17.3.1 Brake Assemblies

In most cases, the brake assemblies used with surge brakes are much the same as non-self-adjusting, passenger-car drum brakes. Many of these units use a uni-servo design with a single-piston wheel cylinder. This feature solves the problem of backing up, because a uni-servo brake does not develop very much braking power while in reverse, yet has full stopping power while going forward.

17.3.2 Coupler Assembly

The coupler at the trailer's tongue is the actuator for a surge brake system (Figure 17-12). The coupler is built with a hinged or sliding connection between the trailer tongue and the coupler assembly. A single-piston master cylinder is secured to the trailer portion of the tongue with the master cylinder pushrod attached to the coupler portion. Hydraulic dampers, which resemble shock absorbers,

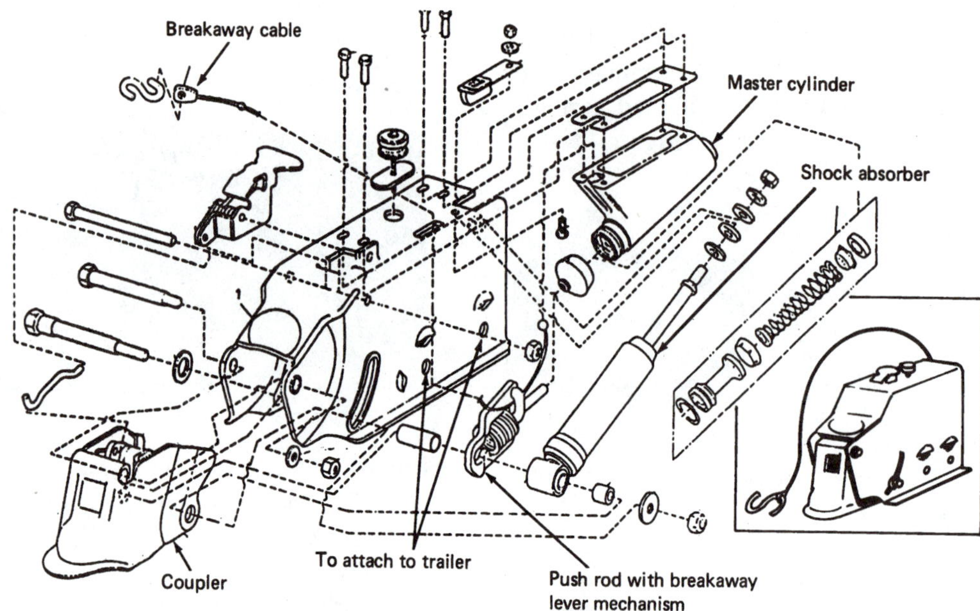

Figure 17-12 An exploded view of a surge brake coupler. Note that the coupler can move toward the right to push on the master cylinder pushrod; this occurs when the tow vehicle slows faster than the trailer.

are also mounted between the tongue and the coupler. These dampers restrict free motion and unwanted brake operation on rough roads or during jerky, rough vehicle operation.

When the tow vehicle accelerates, the coupler pulls the trailer tongue, which will lengthen the assembly to a maximum and keep the master cylinder in a released position. As the tow vehicle decelerates, inertia pushes forward on the trailer tongue, forcing the coupler and tongue together. This force generates a pressure between the pushrod and the master cylinder, and will apply the trailer brakes. The harder the deceleration force, the harder the brake application.

17.3.3 Breakaway Control

In a surge brake system, breakaway control is provided by a lever attached to the coupler assembly. This lever is hinged so that it can apply the mater cylinder. The outer end of the lever is connected to the tow vehicle by a small chain or cable. Should the trailer break away from the tow vehicle, the chain will trip the breakaway lever and apply the trailer brakes. This lever will lock in the applied position to keep the brakes on.

17.4 Slave Hydraulic Systems

A few trailers use hydraulic drum brakes, which are operated by a slave hydraulic circuit. This slave circuit uses a master cylinder, operated by a hydraulic slave cylinder and connected to the tow vehicle's hydraulic system. This system will apply the trailer brakes at the same time that the tow vehicle brakes are applied, with very little lag, and can be adjusted to generate the required amount of braking power (Figure 17-13).

This system uses a steel brake line that is connected to a tee fitting installed in the rear hydraulic brake line of the tow vehicle. This line usually is connected to a quick-disconnect fitting mounted next to the trailer hitch. The disconnect fitting is a receptacle for a coupler attached to the slave cylinder on the trailer tongue. The receptacle and coupler are designed to allow connecting and disconnecting of the trailer without introducing air or dirt into the brake lines or causing a fluid leak. Many technicians, however, do not care for this system because it adds another possibility of failure of the tow vehicle brake system. The brakes of a slave hydraulic system cannot be applied without pressure in the tow vehicle's hydraulic system.

The wheel brake assemblies used with this system are hydraulic drum brakes. Nonservo, uni-servo, and duo-servo designs are used. They are connected to a conventional single-piston master cylinder. The master cylinder pushrod is operated by a hydraulic slave cylinder. This slave cylinder uses a piston and a rubber cup, much like a brake wheel cylinder or clutch slave cylinder. As the brakes are applied in the tow vehicle, the hydraulic pressure moves the slave cylinder piston, which applies pressure on the master cylinder piston. This pressure then applies the trailer brakes.

A slave hydraulic system must never be tapped into a vehicle equipped with ABS.

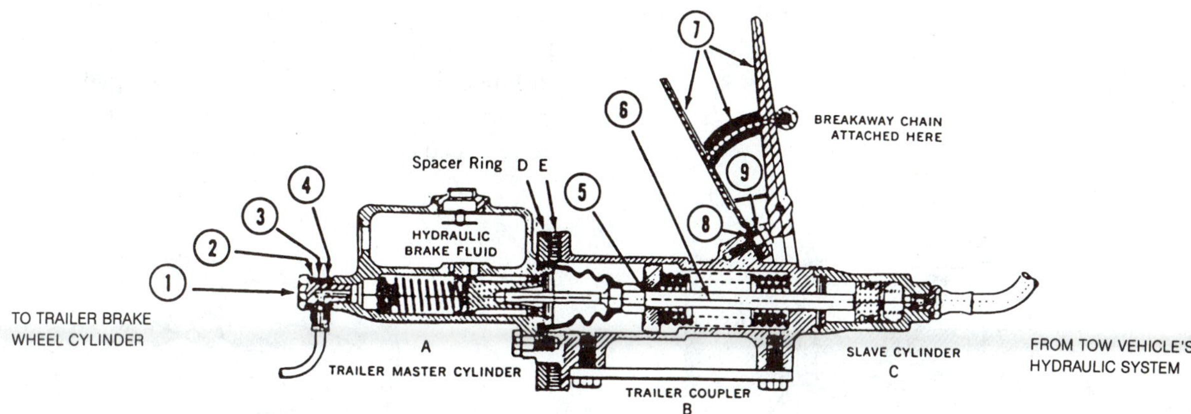

Figure 17-13 A master cylinder–slave cylinder combination is mounted on the trailer tongue, and the hose (right) is connected into the hydraulic lines of the tow vehicle. Application of the tow vehicle brakes operates the slave cylinder to apply the trailer brakes. (Courtesy of Stromberg-Hydramite Corporation)

17.4.1 Breakaway Control

Breakaway control is provided by a master cylinder lever and chain, similar to the system used with surge brakes. If the trailer breaks away, the lever will be pulled to the apply position and apply the trailer brakes. The lever will lock into the applied position to hold the trailer stationary.

17.5 Trailer Brake Service

Like other brake systems, trailer brakes require periodic service and maintenance. Normal repair operations are lining adjustments, wheel bearing service, brake lining replacement, and correction of brake problems.

On trailers that operate under very muddy or wet conditions, such as some industrial, farm equipment, or boat trailers, special attention should be given to water or dirt that might enter the wheel bearing and drum assemblies. These contaminants and the corrosion they cause lead to abnormally fast wear of the wheel bearings, brake lining, and drum friction surface. The seals and hub dust covers are very important under these conditions.

17.5.1 Brake Lining Adjustment

The procedure for adjusting the lining clearance on most trailer brake assemblies, both electric and hydraulic, is essentially the same as that used on non-self-adjusting passenger-car drum brakes. To adjust the lining clearance:

1. Raise and support the axle so the wheel will rotate freely. Note that on heavy vehicles it is a good practice to lift the axle at the ends—under the springs—to prevent it from being bent.
2. Remove the plug covering the adjustment-screw access hole.
3. Use a standard brake-adjusting tool to expand the adjuster until the wheel develops a heavy drag and is difficult to turn. The adjuster should expand when the exposed end of the adjusting tool is lifted (Figure 17-14).

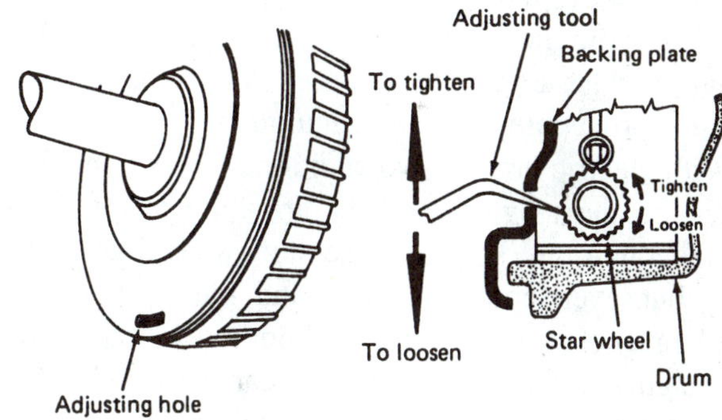

Figure 17-14 The brake shoes on most trailers are adjusted manually. Lift up on the adjusting tool handle and then back off the starwheel to obtain a slight clearance.

4. Loosen the adjuster screw until the wheel turns freely, with no noticeable drag other than the slight noise of lining-to-drum contact.
5. Replace the adjuster plug and lower the axle.

17.5.2 Wheel Bearing Service

Wheel bearing service on a trailer is identical to that for a passenger car with serviceable wheel bearings on a nondrive axle. Trailers that are subject to water or dirt should have the bearings serviced frequently enough to ensure proper, safe operation. Some wheel bearing assemblies can be lubricated with a grease gun. Pumping grease into a zerk fitting at the end of the spindle will force new grease through the bearings and flush dirty grease out (Figure 17-15).

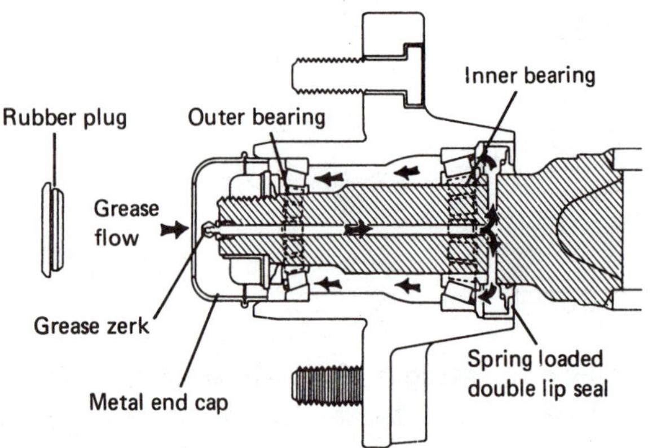

Figure 17-15 This trailer wheel bearing assembly is lubricated with a grease gun.

17.5.3 Brake Shoe or Lining Replacement

Replacing the lining on a trailer brake is much the same as the lining replacement procedure used with a non-self-adjusting, passenger-car drum brake. The drums, shoes, and springs are reconditioned or replaced using the same procedure, tools, and equipment.

With hydraulic brakes, the wheel cylinders and the master cylinder are rebuilt, replaced, and bled, also using the same procedure, tools, and equipment.

With electric brakes, two additional checks should be made—on the magnets and the armature plates. Under normal conditions, both of these items should last indefinitely, but severe contamination (sand, mud, etc.) or a distorted magnet lever can cause abnormal or rapid wear. The magnet surface will normally wear at an angle; if it is tapered and flat, it is acceptable. If the magnet is worn so that it is not flat, or is worn to the point where the brass screws that hold it together show wear, it should be replaced (Figure 17-16). If the armature plate shows excessive galling or scoring, it should be replaced. If the magnet is replaced, a new armature plate should be installed, and vice versa. In many cases, the armature plate can be removed from the drum and a new one installed. In some cases, the armature plate surface is cast as part of the drum so that the entire drum assembly must be replaced (Figure 17-17).

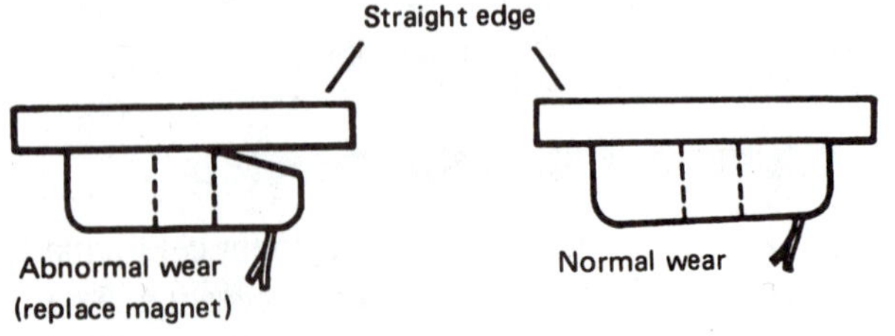

Figure 17-16 A magnet should be replaced if its rubbing surface is worn at an angle (left).

Figure 17-17 A badly scored armature plate should be replaced.

17.5.4 Trailer Brake Troubleshooting

With the exception of those involving the controls and magnets used with electric brakes, most of the problems encountered with trailer brakes are similar to those described in Chapter 9 for passenger-car drum brakes. The most common additional problems concern the relative strength of the trailer brakes, and this is often a design problem. The size of the brake assemblies, wheel cylinders, master cylinder, or slave cylinder depends on the weight of the trailer and on the number of wheels and brake assemblies.

Like other electrical devices, an electric brake system is subject to problems involving open, short, or grounded circuits. An open circuit can occur because of a broken wire or loose connection anywhere in the circuit. It can prevent current from flowing through the entire circuit or through an individual brake assembly (Figure 17-18). A short circuit can occur in the coils of the magnets, causing it to lose power with increased current flow. A grounded circuit can occur anywhere in the insulated portion of the circuit, causing greatly increased current flow, which will probably burn out a wire or the resistor in the controller.

As in passenger cars, electrical problems can be traced using a test light and a volt-ohm-meter. An ammeter—either a standard automotive type (using two leads) or a induction type—is also commonly used. An induction ammeter is a inexpensive, simple, durable piece of equipment that is placed over a wire. Either type of ammeter will measure and indicate a current flow in a wire.

The voltage measured at the connections to the magnets should begin at 0 volts (with the brakes released) and rise gradually as the brake controller is operated (Figure 17-19). When the controller is fully applied, the voltage should be about 12 volts. A lower voltage will result in a

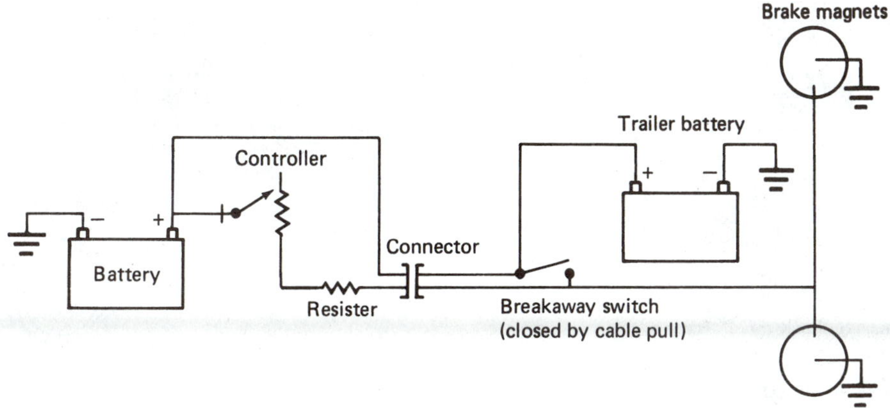

Figure 17-18 A trailer brake wiring diagram shows the insulated (ungrounded) circuit; when the brakes are applied, it should be possible to measure voltage at all points of the insulated circuit.

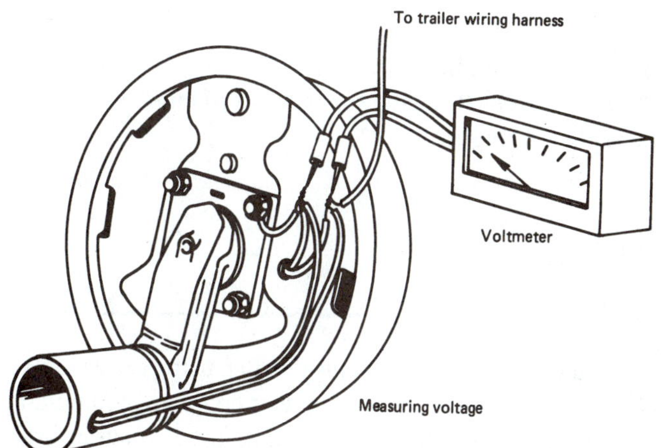

Figure 17-19 If a voltmeter is connected to the two wires that enter the backing plate, the voltage should gradually increase (dependent on brake pedal pressure) to about 12 volts.

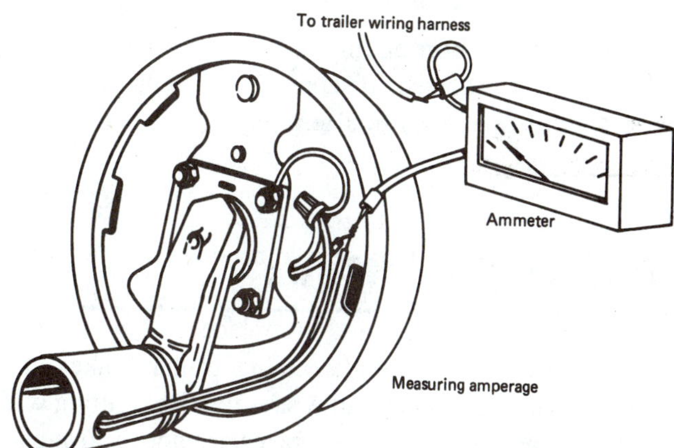

Figure 17-20 If an ammeter is connected between one of the connections at the backing plate, the current flow should generally increase to the rating of the magnets as the brakes are applied.

loss of braking power. Depending on its size, an individual magnet should draw about 2, 2.5, or 3 amps of current. The current flow for a system using two brakes should be twice this; for a four-brake system it should be four times this, and so on (Figure 17-20). This current flow is also dependent on the position of the controller and the length of the test. It is normal for the current flow to drop as the magnets heat up. In a normal two-brake system with larger magnetic coils, the brakes will begin to apply with a current of 1 to 1¾ amps. The brakes should be completely applied with a current flow of 6 to 7 amps. A two-brake system that draws over 7 amps must have a shorted magnet. If the current flow is extremely high, there must be a ground somewhere in the circuit. If the current flow in this system is less than 2 amps with the controller completely applied, there must be an open or partially open circuit.

In circuits with no current or a reduced amount of current flow, the circuit should be checked for an open circuit using a test light or voltmeter. Check for voltage at the various connectors throughout the circuit, starting from the controller's battery connection. The problem will be located between the last point showing normal voltage and the next point checked.

If the current flow for the circuit is reduced by the value of one brake magnet, check the connections to the magnets for voltage. No voltage indicates an open circuit leading to that point. Voltage with no current flow at a connection indicates an open circuit in the magnet coil or in the ground wire for that coil. A check for voltage at the magnet ground connection should show no voltage.

A magnet can be easily bench-checked to determine its condition. Connect the two leads of the magnet to a good 12-volt battery. If the magnet is good, it will become a magnet and draw the correct current (Figure 17-21). If the current flow is below specifications, check to make sure the battery is charged. If the current flow is excessive,

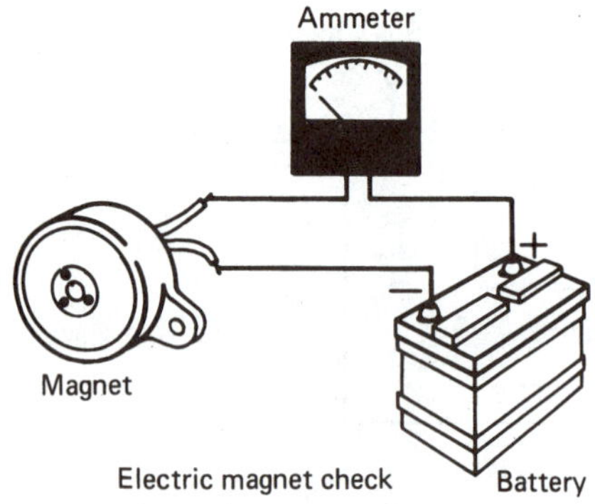

Figure 17-21 A magnet can be connected to a battery and ammeter; the magnet should become magnetized and draw 1 to 3 amps (depending on the magnet) of current.

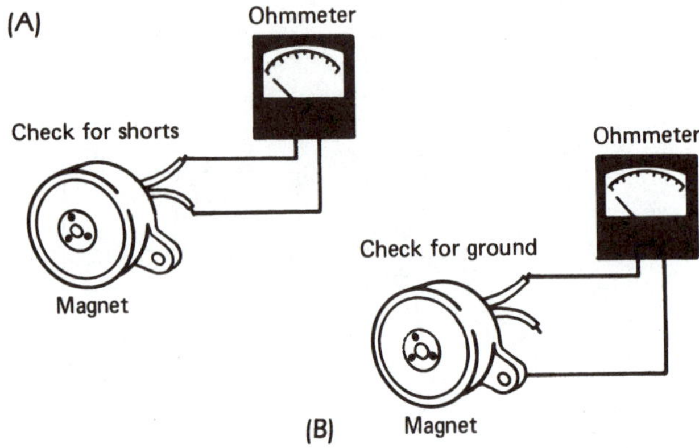

Figure 17-22 An ohmmeter can be used to check a magnet for shorts (A) or grounds (B); it should show about 5 to 10 ohms in (A) and infinite for (B).

the magnet has a short circuit and should be replaced. If the current flow is correct, disconnect one of the wires and connect a wire between that terminal of the battery and the magnet's mount. There should be no current flow. If there is current flow, the magnet is grounded and should be replaced (Figure 17-22).

Review Questions

1. Two technicians are discussing trailer brakes and their service. Technician A says that jackknifing will be the result if the trailer brakes are not as strong as those on the tow vehicle. Technician B says that the brake system on a trailer must include a system that will apply the brakes if the trailer comes loose from the tow vehicle. Who is right?
   a. A only c. both A and B
   b. B only d. neither A nor B
2. Electric brakes use
   a. an electromagnet to apply the shoes.
   b. a battery in the trailer that can power a brake application.
   c. a controller in the tow vehicle that can apply the brakes.
   d. all of the above.
3. An electric brake controller can be operated by
   a. hydraulic force. c. hand.
   b. inertia. d. any of the above.
4. A surge brake is applied by
   a. the master cylinder in the tow vehicle.
   b. inertia acting on the trailer.
   c. an electrical signal from the controller.
   d. none of the above.
5. Technician A says that a trailer using surge brakes has a special type of tongue or coupler. Technician B says that the brake assemblies used in surge brake systems are often of a uni-servo design, to provide shorter stopping distances. Who is right?
   a. A only c. both A and B
   b. B only d. neither A nor B
6. A slave hydraulic trailer brake system uses a master cylinder mounted on the trailer tongue, which is operated by
   a. inertia.
   b. an electromagnet.
   c. hydraulic force from the tow vehicle's brake system.
   d. an on-board battery.
7. Technician A says that a faulty brake magnet will be worn so that it has a flat but angled surface on its face. Technician B says that a magnet is faulty if it draws too much or too little current. Who is right?
   a. A only c. both A and B
   b. B only d. neither A nor B
8. The brake lining clearance on most trailers
   a. is adjusted automatically by self-adjusters.
   b. is adjusted automatically by electromagnetic action.
   c. is adjusted using a brake spoon.
   d. does not require adjustment.

18 Brake Systems for High-Performance Vehicles

Objectives

After completing this chapter, you should:

- ❑ Be able to test a brake system for fade and proper balance.
- ❑ Be able to reset an adjustable proportioning valve to correct improper brake balance.
- ❑ Be familiar with the differences between a passenger-car and race-car brake system.
- ❑ Be familiar with the testing procedure and the service requirements of a race-car brake system.

18.1 Introduction

Most racing vehicles have substantially better brake systems than passenger cars for several major reasons: safety and durability—to be able to stop a very fast-moving vehicle for the entire period of the race, and performance—to be able to stop quicker so the vehicle can spend more time at racing speed and less time decelerating. The major factors determining how quickly any car can stop are the amount of stopping power that can be generated by the brake assemblies and the amount of traction between the tires and the road. Race-car brake systems are affected by the same factors that affect the performance of passenger-car brakes. As mentioned earlier, the ideal brake system is powerful enough to lock up all four wheels during a stop, but the lockup of each wheel must occur at the same time and in a manner that is controllable by the driver.

Remember that stopping power is greatest just before wheel lockup. A car that is stopping on dry pavement with all four wheels locked up is stopping at a rate of about 0.7 **g**. And, because the tires are skidding, it will probably be sliding sideways or spinning. A passenger car with good tires should be able to stop at a rate of about 0.8 **g** without locking the tires, and the car will be under the control of the driver. These two stopping rates will produce stopping distances—from a speed of 55 mph (88.5 kph)—of about 144 feet (44 m) at 0.7 **g** and 126 feet (38.5 m) at 0.8 **g**. Note that the slower rate will require a stopping distance that is 18 feet (5.5 m) longer.

On most standard-production passenger cars, the braking system is adequate for normal driving, and a good-quality, complete brake job will produce a car that stops well enough for average driving conditions. There are some car enthusiasts who desire a braking performance that is better than adequate. It is the intent of this chapter to give you the ability to satisfy such drivers. This chapter will describe the braking systems used on racing and other high-performance vehicles.

You should remember that the technician is liable for any damages that result from unapproved vehicle modifications. Occasionally, a manufacturer will issue a service bulletin approving a modification such as a change in wheel cylinder size to correct a brake balance problem, and if this is the case, the manufacturer is authorizing the changes. You can make the change with no fear of lawsuits. If you recommend any changes to a brake system that are not covered by a bulletin, make sure the customer is completely informed of what you are doing and what the results should be. Also, make absolutely sure that these changes will not result in abnormal braking performance that can cause an accident.

SAFETY TIP: This chapter will describe tests and modifications that are not normally done in the automotive repair industry and, in some cases, are not approved by the vehicle manufacturer. If the road tests described here should be performed in a careful manner while observing the following precautions:

- The vehicle must be in good mechanical condition.
- The tires must be in good, sound condition.
- The braking system must be in good condition.
- The driver must wear lap and shoulder safety belts as well as a safety helmet.
- The roadway must have good, sound pavement and be clear of obstacles and other vehicles.
- Any tests conducted on public roads must comply with all the applicable traffic regulations.

If you make any of the modifications described in this chapter, remember that:

- Most of these changes are based on trial and error. They must be tested to prove their effectiveness.
- Most of these changes will interact with each other. They will usually, but not always, work better in combination.
- These changes are compromises. They will usually have a detrimental effect on some other aspect of the car's performance (eg., stopping ability with cold brakes).
- Any change made must not violate any applicable vehicle code.
- Most of these changes are based on good tire performance. Effective stopping cannot occur without good tires.

18.2 Braking System Operating Limits

Any braking system has certain operating limits. These limits can also be considered performance criteria and are as follows:

- Tire traction and brake balance.
- Brake pedal force and travel.
- Deflection of the components under pressure.
- Temperature of the components and fade.
- Wear of the linings, drums and rotors, and hydraulic seals.

Again, any brake system should have enough power to lock up all the wheels, and ideally the brakes should be balanced so that lockup occurs at the same time on all the wheels. Also, lockup should occur in a controllable manner. Tire traction should be the ultimate limit of brake performance. Modern race cars using aerodynamic devices—wings and ground effect—can increase tire loading and traction by the added down force. This provides an increase in braking power and therefore shortens stopping distances. With the aid of wings, a modern Indy or Formula 1 car can stop at a deceleration rate of 1.25 to 1.75 **g** or greater. As mentioned earlier, a car cannot achieve its highest stopping rate unless all four tires are working at the point of maximum traction, and this is at a 10- to 25-percent slip rate on the pavement. Proper brake balance will occur when all four tires are at this point at the same brake pedal pressure.

The amount of brake pedal force required to achieve wheel lockup can be easily attained by the average driver, and this should occur with a reasonable amount of pedal travel. The amount of pedal force and travel required to produce lockup should be such that the driver can modulate the pedal and accurately control the power of the brake assemblies and the point of lockup. This point must occur before the pedal contacts the floor.

Deflection of the various components increases brake pedal travel and produces a soft, spongy pedal and a lag in brake application time. Deflection can occur at the master cylinder and brake pedal mounting or as a result of a disc caliper spreading out, a brake drum elongating, air compressing in the hydraulic system, or a brake hose expanding.

If the brake lining becomes too hot the coefficient of friction will be lowered, and the lining will fade. Fading reduces the stopping power of the brakes. Also, most linings have an increased rate of wear as the temperature increases. It is extremely difficult, if not impossible, to maintain brake balance if one or more of the brake assemblies is fading.

All brake friction material wears out and the lining must be replaced. As in a passenger car, when the friction material wears out on a race car, brake performance will be lost. With passenger cars, this point usually occurs after a few years and many thousands of miles. With some race cars, such as those competing in the 24-hour race at Le Mans, this point can occur during a single race.

18.3 Passenger-Car Brake Tuning

Passenger-car brake systems are designed to meet the criteria mentioned earlier. They are strong enough to lock the wheels using a moderate amount of brake pedal effort and travel. There is some deflection that is often noticeable under hard pedal pressure in a car that is not moving, but this deflection is usually not enough to cause problems. The driving styles of most drivers is such that the brakes are not used frequently or hard enough to cause fade, and the wear of the friction material occurs gradually over a period of time. Probably the most common fault the average driver will encounter is improper balance during a very hard stop, and this is controlled in ABS-equipped vehicles.

If a very hard stop is made with the average passenger car, a lock up of either the front or rear pair of wheels will probably occur. As mentioned earlier, perfect brake balance is difficult to achieve because it is affected by weight transfer, deceleration rate, tire traction, and vehicle load. A change in any of these factors will change the front-to-rear balance of the brakes. To make this an even more difficult problem, steering and cornering loads can cause a side-to-side weight transfer, and steering loads can also reduce the amount of front tire traction available for stopping. There are too many variables under normal operating conditions to set up a car with perfect brake balance under all conditions. Antilock brake systems are the easiest way to achieve this.

The best that a brake technician can do is to set up a car for correct balance with normal straight-line stopping. The driver must be able to compensate for changes in vehicle loading, road traction characteristics, driving maneuvers, and so on, by changing his or her speed or driving habits. Then it should be unnecessary for the driver to exceed the car's stopping ability.

18.3.1 Testing

Before tuning the brakes of a passenger car, make certain that it has a good set of tires and brakes. It is a waste of time and money to try to improve a brake system that has a below-normal performance level. After determining that the lining thickness, rotor thickness, and drum diameter are all adequate or better and that the hydraulic system is sound, the car should be given a stopping test.

In most cases, no testing or measuring devices are necessary for a stopping test because the primary concern is the first of the brake performance criteria—stopping power and balance. The test is to determine if the car can lock up all four tires, lock them up equally, and lock them up in a controllable fashion.

Finding a location where this brake test can be conducted is sometimes difficult. Tests should be made from the maximum speed at which the car will be normally driven, and in many cases this will be the maximum speed limit. The test area should have good pavement, be flat and smooth, and be long enough to allow operation at the desired speeds. It must also be wide enough to allow for skids, and it must be free of traffic which would be affected by sudden, quick stops.

To perform a brake test, merely drive the car at the desired speed and then apply the brakes. The first brake application should be slow. Then gradually increase the brake pedal pressure until wheel lockup occurs or handling problems are encountered. Be prepared for vehicle darting, pulling, or skidding. If any of these occur, slow down and make a note of the condition for future action. If the stop is controllable, increase the brake pedal pressure until a two- or four-wheel lockup occurs, the pedal reaches the floor, or the vehicle comes to a stop. During the stop, you should note whether the front or rear brakes lock up early, whether the car swerves or pulls to one side, whether the deceleration rate is even throughout the stop, how much pedal pressure is being used, whether the pedal remains firm, whether there are any unusual noises, and whether there are any unusual vibrations or pedal pulsations. If the stop was controllable, bring the car back up to speed, drive for about one minute to allow a slight cooling, and repeat the stop. On this stop, try to keep the brakes at the point just before lockup—try for the shortest or best stop obtainable. During this and any additional stops, note any unusual conditions that might occur.

Depending on how severe a test you desire, repeat the stops (up to about ten stops maximum) or until you notice a change in brake performance. At the end of the test, walk around the car and carefully touch the lug bolt area of each wheel. Note any unusually high temperatures and whether the front and rear brakes are at about the same temperature. You should also check for unusual burn odors coming from the brake assemblies and to see if any grease is leaking from the wheel hubs.

18.3.2 Tuning Changes

If the stops were difficult, with darting or pulling, vibrations or pedal pulsations, unusual noises, or drag and burn odors, regular brake service is needed. Side-to-side balance of a pair of brakes is accomplished by the hydraulic system and that the two brake assemblies are equal to each other. For these problems, refer to the troubleshooting charts in Chapter 9 and make the necessary repairs as described earlier in this text. If desired, repeat the stopping test after making the repairs.

If the stops were straight and true with wheel lockup occurring at all four wheels in a controllable manner, the car has good brakes. There is not much that can be done to improve stopping ability other than obtaining a better set of tires to improve traction. However, remember that as brake performance and stopping rates increase, weight transfer will also increase. As weight transfer increases, the tendency for the rear wheels to lock up will also increase.

If the first stops were true and strong but the repeated stops required more and more pedal pressure for the same stopping rate, brake fade is indicated. The best way to correct brake fade on a passenger car is to change to a better-quality lining. Several manufacturers have high-performance linings available that have a better ability to withstand higher temperatures (Figure 18-1). If new lining is installed, it should be completely broken in before conducting a severe brake test as just described.

If the stops were straight and true but the brakes could not be made to lock up, more braking power is necessary. Increasing braking power on a passenger car is difficult. Some increase in power can be achieved by changing to a lining with a higher coefficient of friction. A greater increase in power can be achieved by installing larger brake units. Sometimes these are used on a different model of the same car (i.e., a sports version or a station wagon). Braking power can also be increased by installing a power booster if the car does not already have one, or a larger booster if it does. It should be remembered that a car with power brakes often uses a different master cylinder or brake pedal ratio. If the car does not have a booster, a stock booster from a car of the same make or a remote-type, pressure multiplier booster can be installed. A remote booster is relatively easy to install or retrofit (install after the car is made) because it does not need to be connected to the brake pedal. The only connections required are a vacuum hose and the two hydraulic line connections; this allows more flexibility in selecting a mounting location for the booster. There is a problem with a split hydraulic system, two boosters are required.

If the car had a consistent tendency to lock up both rear wheels or, on a few occasions, both front wheels early, improper brake balance is indicated. To correct this problem, the brakes must be made stronger at the end of the car that did not lock up, or weaker at the end that did. Except for adding ballast weight over an axle to increase traction, the only way to correct this problem is with a modification to the system. A brake imbalance can be expected on a vehicle which regularly carries different loads. For example, an empty pickup will probably experience rear-wheel lockup, and a fully loaded one will not. As mentioned earlier, some manufacturers install a height- and load-sensitive proportioning valve to help solve this problem. Most manufacturers use a two-wheel (rear only) ABS in their pickups.

18.3.3 Adjusting Brake Balance

Modifying a brake system can be difficult and is usually avoided by the wise brake technician. If you modify anything as important to human safety as a braking system, using methods not recommended by the manufacturer of the car, you might have to prove in a court of law that you have a better knowledge of that braking system than the designer and manufacturer of the car. When a technician makes a repair or modification that is not recommended by the manufacturer, that technician is generally liable for damages that might result from or be attributed to that change. When technicians follow a procedure recommended by the manufacturer, they are then acting as agents of the manufacturer and are not responsible if the repair methods are not correct.

Brake balance is achieved by reducing the strength of brakes that lock up early or increasing the strength of brakes that lock up late or not at all. Usually the rear wheels lock up first, so the rest of this section will deal with rear-wheel lockup problems. It is much easier to reduce the strength of a pair of brakes than to increase it. There are several ways to reduce the power of a pair of brakes, and the easiest and most reliable is to reduce the rear hydraulic pressure using an *adjustable proportioning valve*. The adjustable feature of this valve offters the advantage of being able to fine-tune brake balance, and also to easily change it in case of future load changes. On some cars, it

Brake Lining Hardness Designations

Hardness Designation	*Friction Characteristic*	*Temperature Use*
Soft	High	Low to moderate
Medium	Medium	Moderately high
Super pad	Very high (long-wearing)	Low to high
Hard	Medium	Moderate to high
Hard premium	Medium	Extremely high
Metallic	Very high	High

Figure 18-1 Different high-performance-use brake linings are available. These lining types are often classified by their hardness. (Reprinted from Mitchell Anti-Lock Brake Systems, with permission of Mitchell Repair information, LLC)

is possible to replace the rear wheel cylinders with ones of a smaller diameter. Some wheel cylinders have the same mounting provisions but different diameters, which make exchange easy. Changing wheel cylinders is a trial-and-error, nonadjustable procedure. You have to make a change and then test to see if the change produces the desired result.

Two different adjustable proportioning valves are available. One uses a threaded knob for the adjustment (Figure 18-2), and the other uses a seven-position lever (Figure 18-3). In a passenger car, the valve is normally mounted in series in the line leading to the rear brakes. The valve can be mounted in any position and anywhere in the car. In racing vehicles, it is mounted within reach of the driver. In a passenger car, it should be mounted in a clean, protected, accessible location. When mounting this valve, it is often possible to disconnect the rear brake tube from the existing proportioning or combination valve, carefully re-route this tube to the new valve, and install a replacement tubing that runs from the valve to the point where the other tube was disconnected. In a car with a diagonal split system, one valve cannot regulate the pressure of both rear brakes. Two valves must be used, and both must be adjusted to deliver the same pressure.

To adjust brake balance using an adjustable proportioning valve:

1. Determine the desired location for the valve and install it according to the instructions of the manufacturer.
2. Drive the car, make a hard stop as described earlier, and note whether rear-wheel lockup occurs.
3. If the car stops with a four-wheel lockup, you are finished. If the car stops with early front or rear lockup, readjust the valve and repeat steps 2 and 3.

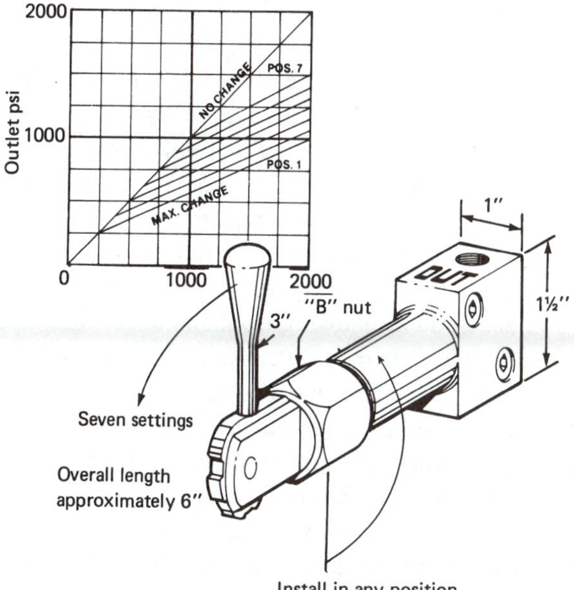

Figure 18-3 A driver-adjustable proportioning valve is placed so the driver can move the lever to make up to seven quick and distinct changes in brake balance. (Courtesy of Tilton Eng.)

18.4 Race-Car Brake Systems

Types of race cars vary greatly, and each style of racing places different demands on the brake system. In most racing vehicles, brake components represent a compromise between vehicle weight and braking power. Weight is at a premium in many race cars, and the components must be as small and light as possible. Yet in some classes,

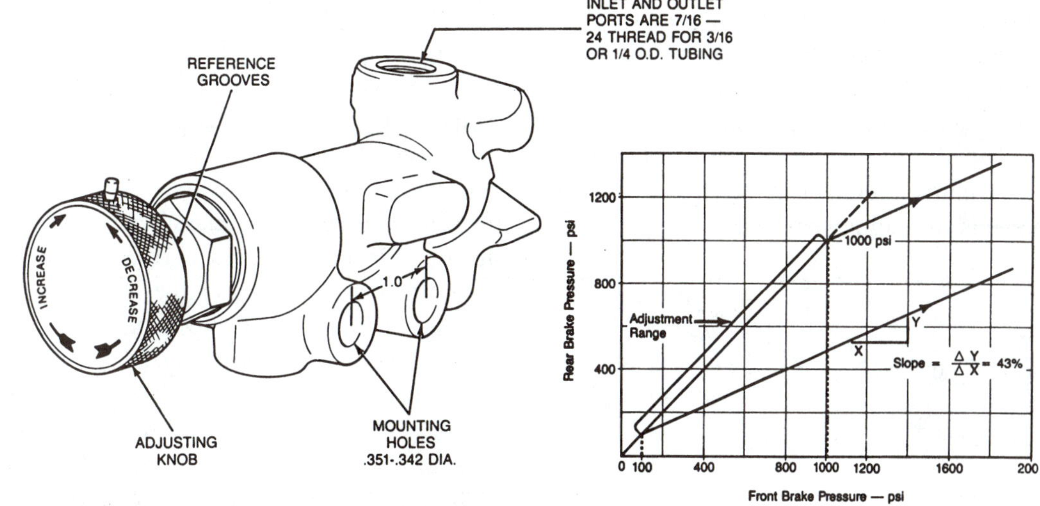

Figure 18-2 An adjustable proportioning valve can be used to change the front-to-rear brake balance. This valve has a slope of 43 percent, and the split point can be adjusted to occur between 100 and 1,000 psi by turning the adjusting knob.

the brakes must be able to absorb massive amounts of energy over extended periods of time.

Because of the high speeds involved, race-car brakes transfer much more energy than those on heavier passenger cars. These energy amounts are easily computed using the formula given in Chapter 2. A 3,500-pound (1,591 kg) passenger car coming to a complete stop from 55 mph (88.5 kph) must convert 354,097 foot-pounds (480,155 N-m) of energy into heat. A 1,500-pound (682-kg) Indy car on a road course, slowing from 180 to 20 mph (290 to 32 kph) for a turn on the racetrack, must use up 1,625,418 foot-pounds (2,176,856 N-m) of energy. And it must do this each time the car comes to this turn. A 1,950-pound (886-kg) drag-racing "funny car" must use up 3,135,451 foot-pounds (4,251,672 N-m) of energy as it makes a stop from 250 mph (402 kph). Fortunately, this car has a parachute and a cool-down time before it needs to make another stop.

In the case of drag-racing and land-speed-record cars, the brakes are used only once during every run. These systems often use a parachute for *aerodynamic braking*. A *parachute*, depending on its size and the speed of the vehicle, produces much greater stopping power than a set of conventional brakes (Figure 18-4). The limit on parachute size is determined by the amount of shock that can be delivered to the car and driver when it opens and the strength of the mounting point for the parachute. An added advantage of a parachute is that it can straighten a car out—if it is deployed early enough—and prevent it from skidding sideways during a stop. A disadvantage of a parachute is that the vehicle must be stopped and the parachute repacked before it can be used again.

Most racing cars use almost conventionally appearing brake components. They are very similar to passenger-car components. Some classes of race cars require stock brake systems. In these classes, there is little that can be done to improve brake performance that has not already been covered in this text. In most racing classes, the design of the brake components is open, and the components are selected at the discretion of the car builder. There are several manufacturers who specialize in building brake components for racing vehicles.

Figure 18-4 Very fast drag-race and land-speed-record cars usually use a parachute to provide increased stopping ability; it will also straighten out a car that is starting to spin.

18.4.1 Component Selection

Some race-car builders use stock or slightly modified passenger-car brake components. As a rule, stock parts are durable, reliable, and relatively inexpensive. They have the drawbacks of being heavy and, as in production cars, do not provide a means of adjusting brake balance.

Cars that use stock components can undergo a few modifications to improve their performance. These include upgrading the brake lining, drilling or grooving the rotor surfaces, grooving the lining, installing air ducting to the rotor, and installing an adjustable proportioning valve. These changes will be described later in this chapter. Some race-car builders will combine stock wheel-brake assemblies with twin racing-style master cylinders. This provides a possible change in the master cylinder bore diameter and, if used with a **balance bar**, a quick means of adjusting the brake balance. As in passenger cars, the master cylinder and caliper bore sizes are matched to produce the hydraulic pressures and fluid volume flows necessary for sufficient application force at the shoes with a normal amount of pedal travel.

All-out race cars normally use specially designed, racing brake components in the entire system. Racing components offer the major advantage of being lightweight. They can also be sized correctly to do the job of supplying enough braking power over extended periods of time. Most modern race-car brake units are disc brake designs because of the massive amounts of heat which are generated. As a rule of thumb, most road racing cars use the largest rotor and caliper combinations that will fit inside the wheel. This will provide the largest heat sink and cooling area possible (Figure 18-5).

18.4.2 Calipers

Racing calipers resemble production-car calipers. They are often made of aluminum to reduce their weight. Fixed calipers (two- or four-piston) and floating designs are both used. Most racing calipers have a rather massive appearance compared to the other lightweight components around them. The massiveness helps prevent caliper flex or deflection, which can occur if the unit is too light. Like other deflections, caliper flex can cause a lower, spongy brake pedal. Calipers are available in different sizes, which offer various-sized pistons, use different-sized pads, and provide different amounts of stopping power (Figure 18-6). This allows selection of the correct size to deliver sufficient stopping power. One racing caliper design uses a rather unique kidney-shaped piston to ensure an even piston-to-pad pressure along the length of the pad.

Figure 18-5 This brake assembly is used on the left rear on an Indy car. Note the large rotor and caliper, the drilled rotor, and the air duct to bring air into the rotor.

Another possible source of deflection is the caliper mount. This part is often constructed by a race-car builder. It must hold the caliper in alignment, centered over the rotor, and be able to withstand braking torque that tries to pull the caliper along with the rotor. Deflection or bending of weak caliper mounts will cause misalignment of the pads with the rotor, which will tend to increase caliper flexing (Figure 18-7). Test this by applying and observing the brake with the car stationary. The caliper should not move, only the shoes and pistons.

Front calipers are usually mounted in the conventional outboard position on the steering knuckle or spindle. In a few older racing cars, they were mounted inboard, inside the body shell. Inboard mounting requires a shaft to couple the rotor to the wheel hub. Front inboard brakes are much more complicated, more expensive, and heavier than outboard-mounted brakes. However, they offer the advantages of moving braking loads away from the front-end suspension components and also reduce unsprung weight. Because of their increased complexity and weight, inboard brakes are not commonly used. On most race cars, the rotors and calipers are mounted in the conventional outboard position inside the wheel.

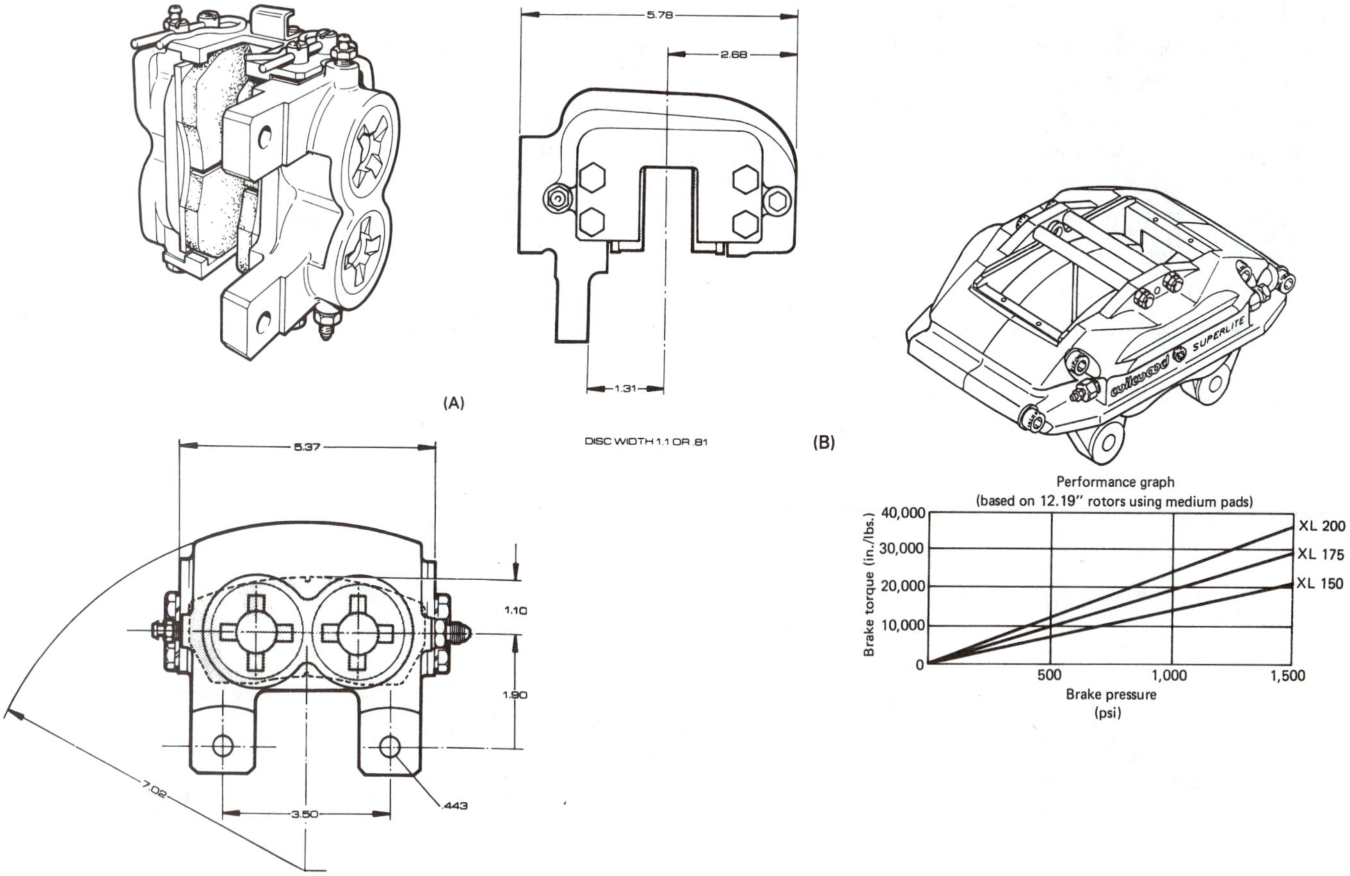

Figure 18-6 Calipers developed especially for race cars are available in different sizes and styles to meet the needs of a particular car. (A is courtesy of Tilton Eng.; B is courtesy of Wildwood Engineering)

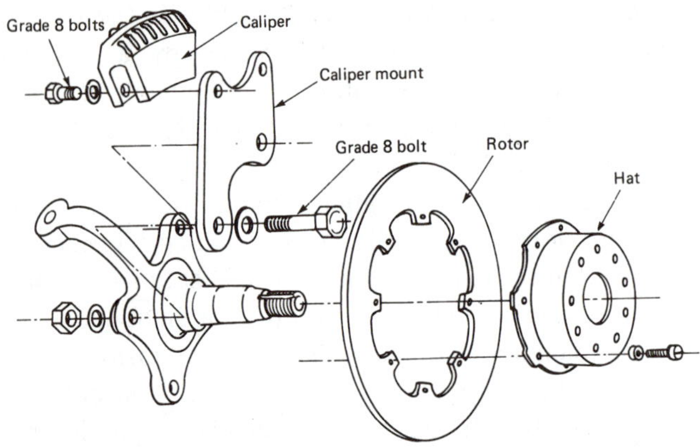

Figure 18-7 The race-car builder often constructs the caliper mounting bracket. It must be strong enough to ensure deflection-free braking during the hardest stops and use Grade 8 bolts.

Figure 18-8 This formula Ford race car uses rear inboard brakes; note the rigid caliper mounting points at the transaxle case.

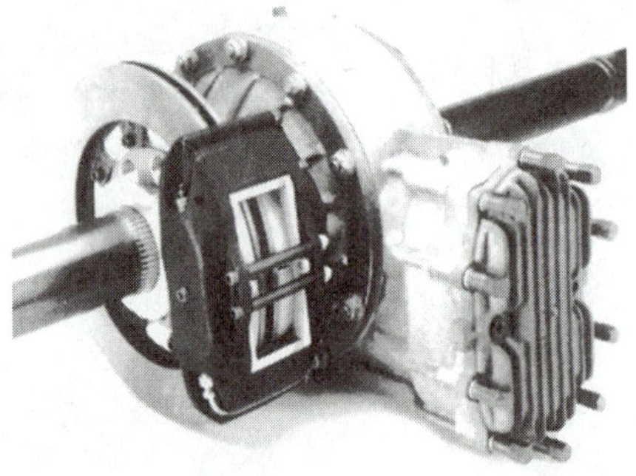

Figure 18-9 A sprint-car rear-end center section with a single rotor and caliper. Because the rear end is "locked," it will stop both rear tires. (Courtesy of Wildwood Engineering)

Rear calipers are often mounted inboard on cars using independent rear suspension. The drive shafts run from the transaxle to the wheel so the rotor can be easily installed anywhere along it. Mounting the caliper directly on the transaxle housing provides a convenient, rigid, strong mounting point (Figure 18-8). Outboard mounting requires that the caliper be mounted inside the wheel, and this limits the diameter of the brake assembly and the airflow to the rotor and caliper. Dirt-track and drag-race cars normally use a "locked" rear end; there is no differential. Since both rear tires must turn together and there is no large brake requirement for the rear tires on these particular cars, a single brake assembly is often used at the rear. The rotor is often mounted inboard, with the caliper mounted on the gear housing (Figure 18-9). If more braking power is required, two calipers can be mounted on a single rotor (Figure 18-10).

Recent developments to reduce caliper temperature are liquid cooling, heat shielding, and fluid transfer. One new caliper design has internal passages through which a

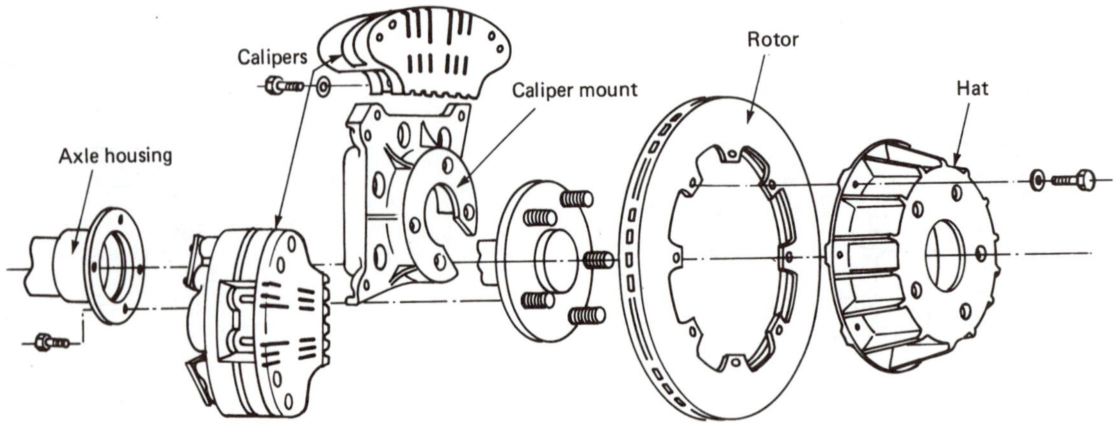

Figure 18-10 Two calipers can be mounted on a single rotor to produce more stopping power; this will also increase heat and the possibility of wheel lockup.

cooling fluid can be circulated. A pump transfers this fluid through a heat exchanger, which is like a small radiator, and then back to the calipers. Heat shielding in the form of insulating paint or other coatings are being put on the inner portion of the caliper to reduce the radiant heat flowing from the pads and rotor. A *dynamic bleeding system* has been developed to circulate fluid through the caliper (Figure 18-11). In a normal system, the fluid stays in the calipers and gets hotter and hotter. In a dynamic bleed system, the hot fluid returns to the master cylinder while the brakes are released and new, cooler fluid is used for the next application. Valves are used so that fluid to apply the brakes travels from the master cylinder piston to the caliper, but the fluid flows from the caliper back to the

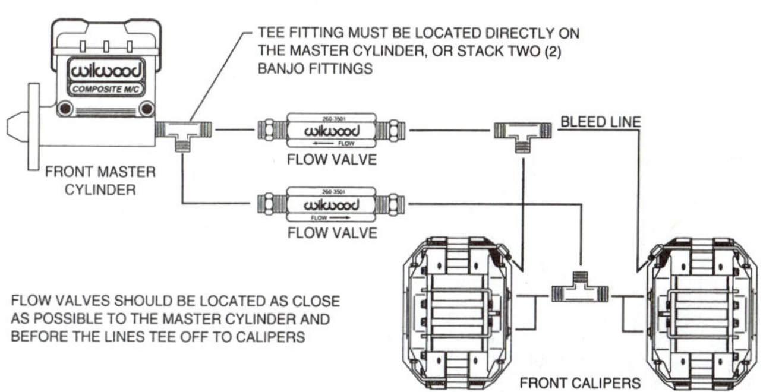

Figure 18-11 This Dynamic Bleed System uses two flow/check valves. Fluid flows through the lower valve into the normal caliper inlets to apply the brakes, and it flows out the bleeder line through the other valve during release. This circulates a small amount of fluid through the caliper on each brake application, cooling the caliper and removing any gases. (Courtesy of Wildwood Engineering)

master cylinder during release from the bleeder screw. This flow also bleeds gasses formed by localized boiling from the caliper.

18.4.3 Rotor or Disc

Race-car rotors are often two-piece units. The rotor can be bolted directly to the hub or to a *"hat"* connected to the hub (Figure 18-12). The hat is used to offset the rotor from the hub by the amount needed to align the rotor with the caliper. It also creates a longer heat path, which helps reduce the amount of heat flowing into the hub and wheel bearings. Too much brake heat can cook wheel bearings and cause early failure. The better rotor-and-hat combinations are designed so that heat expansion of the rotor does not cause distortion of the rotor or hat. They use a splined connection that removes most of the brake torque from

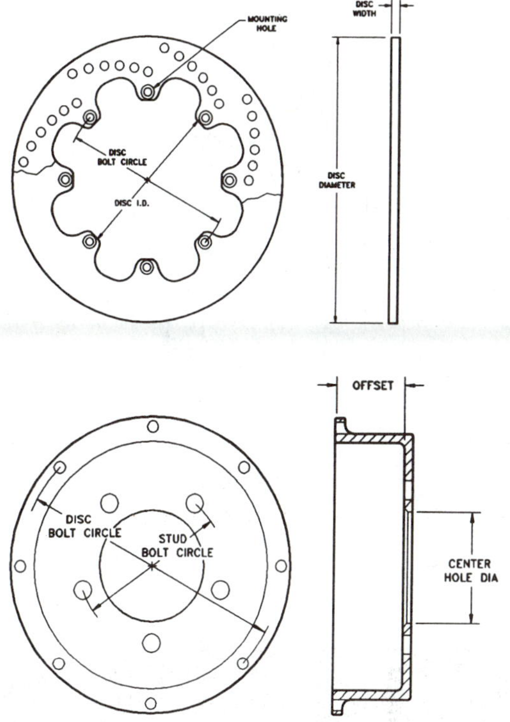

Figure 18-12 A variety of rotors (A) are available to meet the stopping requirements of race cars, and a variety of hats (B) are available to connect the rotor to the hub. (Courtesy of Wildwood Engineering)

the connecting bolts and also provides a means of allowing rotor expansion without distorting the hat.

Rotors are available in various diameters and thicknesses and in solid or vented configurations (Figure 18-13). Most race-car rotors are made from high-strength gray iron. Some of the rotors used with Formula 1 cars are made from a space-age *carbon-graphite composite* to reduce weight and allow operation at extremely high temperatures; however, these rotors are very expensive. It is common for a race-car builder to select the largest-diameter rotor that, with the caliper, will fit inside the wheel. Larger rotor diameters provide more swept area, which increases braking power and lining life. Rotor width is selected to provide a sufficient heat sink for braking loads. Thicker rotors provide a larger heat sink, which will reduce the rate of temperature increase during a stop. A light car that is frequently braked can use thin, nonvented rotors to reduce weight. A car that makes many stops from very high speeds must use heavier, vented rotors to remove the heat.

Many rotors are drilled or have grooves cut across them. These holes or grooves allow gas and debris to escape from between the pad and the rotor. Gas is produced as the binders holding the lining materials together burn up from brake heat, and the particles are the bits of

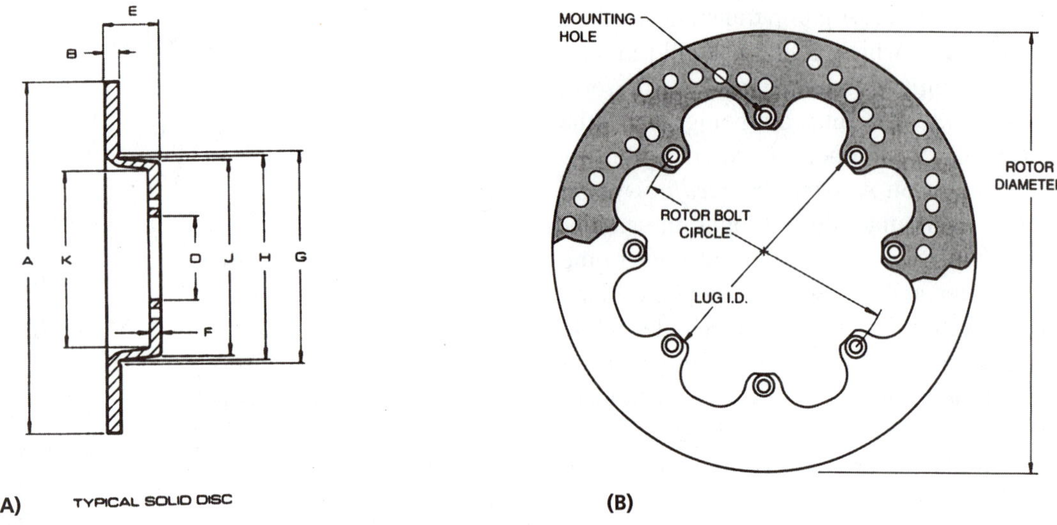

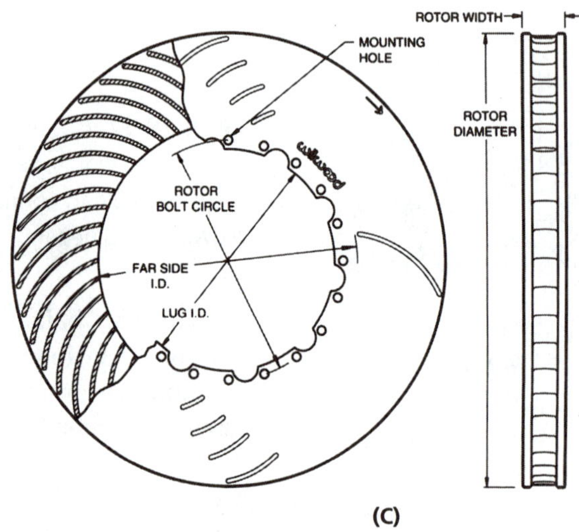

Figure 18-13 Rotors are available in solid (A) and vented (B) styles; the "sphericone" design (C) cools better than a vented disc. Note that rotor (A) connects to a stock hub, (B) uses a castellated junction with the hat, and (C) uses a bolted junction with the hat. (A is courtesy of Tilton Eng.; B and C are courtesy of Wildwood Engineering)

lining material that wear away. They reduce the frictional grip of the lining and contribute to fade. The holes help reduce fade, but they can also lead to cracking of the rotor from thermal stress. When drilling a rotor, a pattern should be selected that will space the holes fairly equally across the friction surface, around the rotor to retain the balance, and not so close to each other that the rotor is weakened. The outer, friction-surface edge of each hole should be beveled slightly to reduce the sharp corner that tends to increase pad wear.

18.4.4 Cooling Ducts

Most road race cars use ductwork to force air over the rotors to help cool them. The air duct used with open-wheel cars is often a short, rigid air scoop that extends into the air stream and is often mounted directly on the brake assembly (Figure 18-14). Full-bodied cars must use inlet openings in the front of the body work and large-diameter, flexible hose to connect these openings to the brake assemblies (Figure 18-15). The opening for the ducts can be closed or the duct hose removed on cold days if brake temperatures are too low.

On vented rotors, the air duct should direct the air to the eye of the rotor so it will be carried through the rotor

Figure 18-14 This Indy-car front brake uses a scoop to collect air and direct it into the eye of the rotor. Note that the rotor is drilled and vented.

Figure 18-16 This formula car uses a solid rotor, so the air duct brings cooling air to an enclosure to distribute the air to both sides of the rotor.

Figure 18-15 This full-bodied car has a front air duct and flexible tubing to bring air to the rotor eye; note the grooved and vented rotor.

Figure 18-17 This Corvette wheel has a set of directional fins that are designed to pump air through the wheel and past the brakes. The wheels for the other side of the car have vanes curved in the opposite direction.

by the internal fins. Directing the air to one side can cause warping or distortion if one side of the rotor runs hotter than the other. On nonvented rotors the air duct should end at the outer edge of the rotor and be split so that equal amounts of air will flow over both sides of the rotor (Figure 18-16).

Evaporative cooling kits are available if still more cooling is needed. These units consist of a small reservoir and pump, usually operated as the brakes are applied, that sprays water into the brake ducts. The water evaporates in the air stream and cools the air. It is a poor practice to spray water directly into the eye or onto the braking surfaces of a hot rotor. The severe thermal stresses and shock caused by the water drops can lead to cracking and fractures.

Airflow past the brake assemblies can also be increased by changing the design of the wheel or wheel cover. Some wheels and wheel covers are designed to pump air from the inside of the wheel, past the brakes, to the outside. Brake dust on the outer wheel rim is evidence that this is occurring. A good example of a wheel designed for brake-cooling airflow is the production wheel used on many Corvettes (Figure 18-17).

18.4.5 Brake Lining

Brake lining is available in several grades to suit the various operating-temperature conditions race cars encounter. The temperature characteristics of the lining are closely related to its hardness (Figure 18-18). A soft lining will work well at relatively low temperatures—up to 750°F (400°C)—and does not require warming up. It will usually have a higher coefficient of friction and be more powerful with a light pedal pressure. It will also *"bed in"* or break

Brake Lining Characteristics

Pad Compound	*Friction Characteristics*	*Optimum Operating Temperature*	*Application Conditions*	*Fade at High Temperature*
Soft	High	Low	Less severe	Greater
↕	↕	↕	↕	↕
Hard	Low	High	Most severe	Lesser

Figure 18-18 This chart compares the five major operating characteristics of brake lining. Race-car brake lining should operate close to its optimum operating temperature. (Courtesy of The Brake Man Inc., 2455 Blanchard Rd., Camarillo, CA 93012, 805-491-2185. Makers of High Performance Brake Components)

in easily and quickly. The disadvantage of a soft lining is that it will fade at relatively low temperatures and wear rapidly as the temperature increases.

A very hard lining shows less tendency to fade and has the best wear characteristics at high temperatures—up to about 1,100°F (600°C). But it will require more pedal pressure, because of a lower coefficient of friction, and will not work very well when cold. Hard linings must often be warmed up by repeated pedal applications before they reach full power. Temperatures this high should be avoided if possible because of the effect on the fluid and the rubber materials used in the caliper.

There are one or two lining grades between soft and very hard to suit different conditions. The type of lining is selected by observing the operating temperatures of the brakes of a particular car on a particular racetrack. Low-temperature brakes can use a soft lining, but higher-temperature conditions require a harder lining.

Harder linings must be bedded in (worn in) before they are used under racing conditions. To bed in a lining, the car is driven at moderate speeds, and the brakes are applied with light but frequent pedal applications until they reach operating temperatures or start to fade slightly. At that point, they should be cooled down, preferably with the car in motion. They are then ready for use.

A lining can be cooled slightly if grooves or slots are cut across its surface. These grooves or slots also serve to let gas or debris escape, much like holes drilled in a rotor. The grooves should be about 1/8-inch (0.3 cm) wide and about one-half the depth of the lining. When grooving a lining, be careful because of the asbestos dust, and do not cut it too deeply or cut too many grooves. This can severely weaken the lining, causing cracks or lining breakup. Many racing teams will only groove lining as a last resort (Figure 18-19).

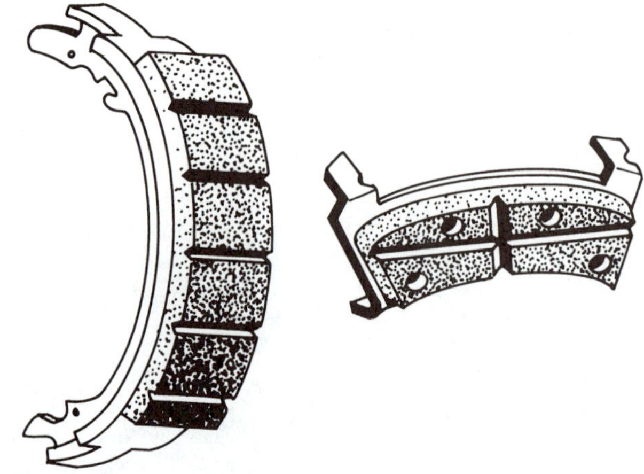

Figure 18-19 Grooves can be cut across the lining to help remove dust and gas. Many racers prefer not to do this because it weakens the lining.

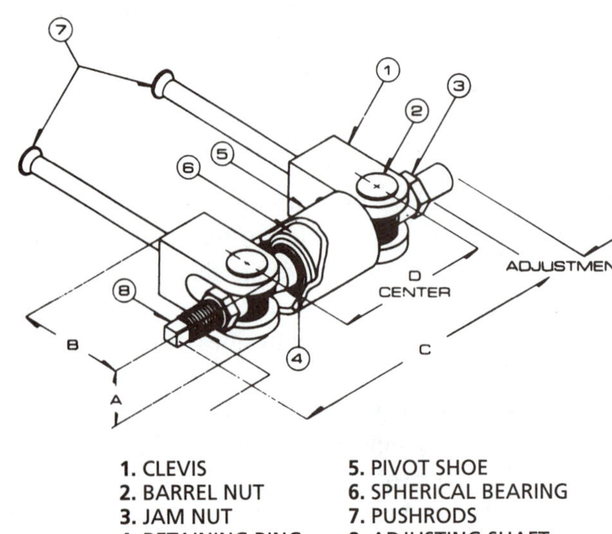

Figure 18-20 A balance bar has two pushrods (7) to operate two master cylinders; the brake pedal pushrod pushes against the pivot sleeve (5) at the center. Turning the adjusting shaft (8) moves the spherical bearing (6) inside the pivot sleeve to change the relative amount of application force on each master cylinder pushrod. (Courtesy of Tilton Eng.)

18.4.6 Master Cylinders and Balance Bars

Many racing cars use two single-piston master cylinders that are connected to the brake pedal through a balance bar (Figure 18-20). One master cylinder is connected to the brakes at the front of the car, and the other to the brakes at the rear. The balance bar can be adjusted sideways so the force from the brake pedal can be split equally between the two master cylinders, or unequally so that one receives more application force than the other. This allows a quick and easy means of changing brake balance. Many cars have a driver control allowing brake balance to be changed during a race. If the decreasing fuel load or increased track and tire temperatures cause rear-wheel lockup, the driver can dial in or adjust for more front master cylinder pressure and braking power and less rear master cylinder pressure and braking power (Figure 18-21).

Race cars that use a stock, tandem two-piston master cylinder usually use an adjustable proportioning valve for adjusting brake balance as previously described.

Most racing master cylinders are the same internally as single-piston passenger-car units (Figure 18-22). They are available in different bore sizes, to allow selection of a master cylinder of the correct diameter for the system and a combination of master cylinders to provide close to the correct balance. The master cylinder is sized relative to the caliper or wheel cylinder pistons to provide enough pressure and fluid displacement to correctly operate the brakes. Some master cylinders have remote or removable reservoirs of various sizes. The reservoir must contain enough fluid to supply the system's needs. Remote mounting places the reservoir in a more convenient location. In some formula cars, remote mounting can position the reservoir above the brake units, which is a definite benefit when bleeding.

Figure 18-21 This cable control can be attached to the balance bar so the driver can easily change brake balance. (Courtesy of Tilton Eng.)

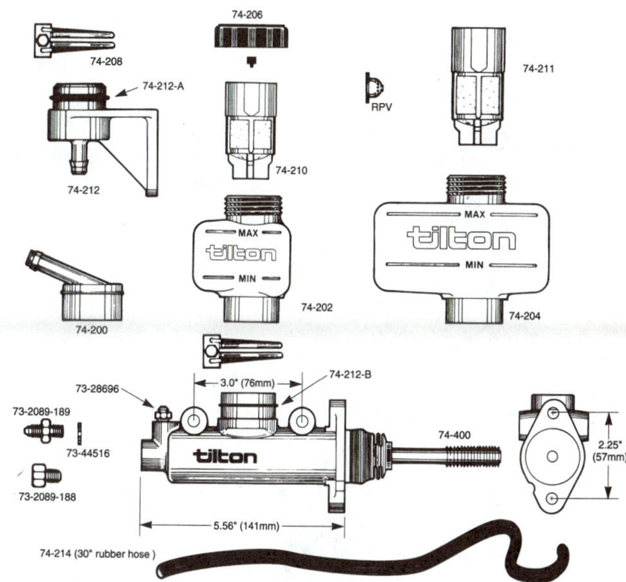

Figure 18-22 This race-car master cylinder is available in five bore sizes and two different reservoir sizes for integral or remote mounting. (Courtesy of Tilton Eng.)

When a race car is constructed, the master cylinder and pedal should be mounted securely, usually on the front bulkhead and sometimes on a roll bar or roll-bar brace. In a panic situation, a driver can exert over 300 pounds (136 kg) of force on the brake pedal. The master cylinder and pedal mounting, as well as the brake pedal, must be able to withstand this pressure without flexing or bending. Several companies market pedal assemblies that combine master cylinder mounts and pedal pivots into one solid, lightweight unit (Figure 18-23).

When constructing a race car, the master cylinder should be mounted above the calipers if possible. It is much easier to bleed the brakes and keep air from working its way into the system if the master cylinder, or at least the reservoir, is higher.

18.4.7 Brake Lines

All brake lines and hoses in a race car are 3/16-inch (often referred to as -3, "dash three") or smaller. Larger lines tend to balloon under high pressures. Rigid lines use conventional, better-quality steel tubing, available for passenger-car use, or annealed stainless-steel tubing. Flexible lines usually have stainless steel braided on the outside to keep the hose from expanding. The steel braiding also gives added protection from possible injury. Flexible lines should be kept to a minimum and used only where necessary because of their tendency to expand under pressure (Figure 18-24).

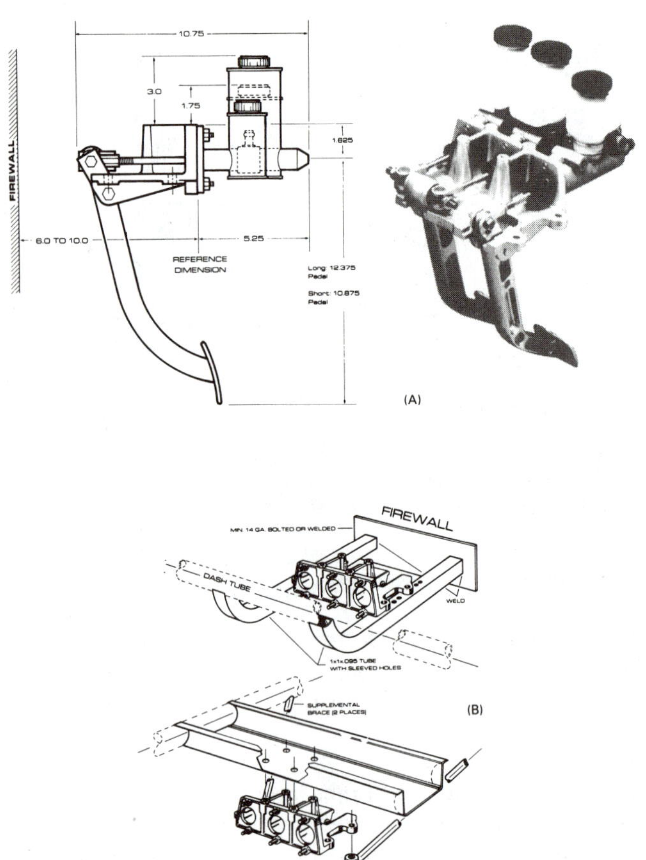

Figure 18-23 The pedal and master cylinder assembly (A) includes a clutch pedal and master cylinder, and a brake pedal, balance bar, and pair of brake master cylinders. The assembly must be securely mounted to a strong portion of the car (B). (Courtesy of Tilton Eng.)

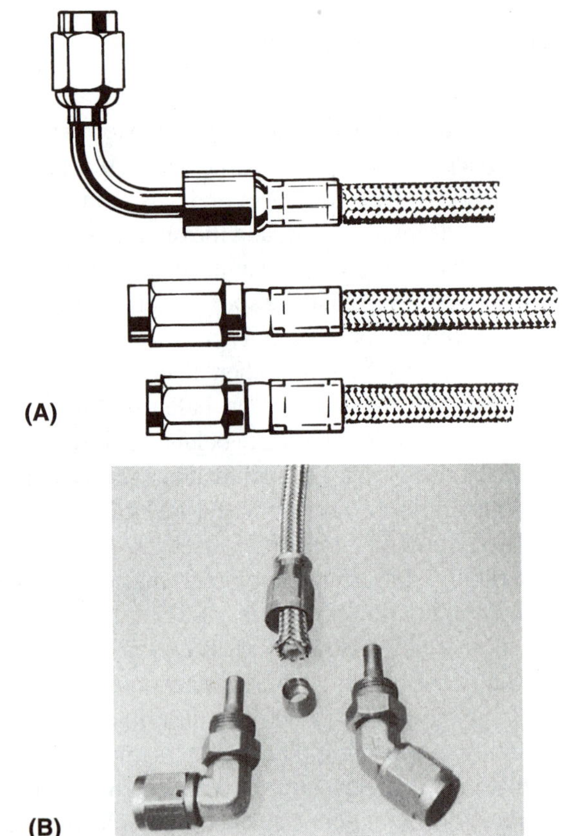

Figure 18-24 Most race-car flexible hoses use a braided stainless-steel outer lining for protection. The end fittings can be swaged or crimped in place or be shop-made using a kit (B).

Hose and tubing ends have a standard SAE double flare or an *Army-Navy (AN)* or *military standard (MS)* flare (Figure 18-25). These two flares are not interchangeable, but adapter fittings are available for joining one type to the other (Figure 18-26).

Brake lines should be secured to the car's frame at regular intervals and at each end of a flexible line. Vehicle vibration can cause continuous line flex, and this repeated flexing can lead to cracks in and failure of the line. When they are mounted, flexible lines should be checked to ensure that they will not interfere with the tire or other parts during steering and suspension travel (Figure 18-27).

18.4.8 Brake Fluid

Racing grades of DOT 3 brake fluid are available, which have a dry boiling point of 550°F (288°C). This fluid can resist boiling under most racing conditions. There is little need for a higher boiling point because temperatures this high can damage the rubber components in the hydraulic

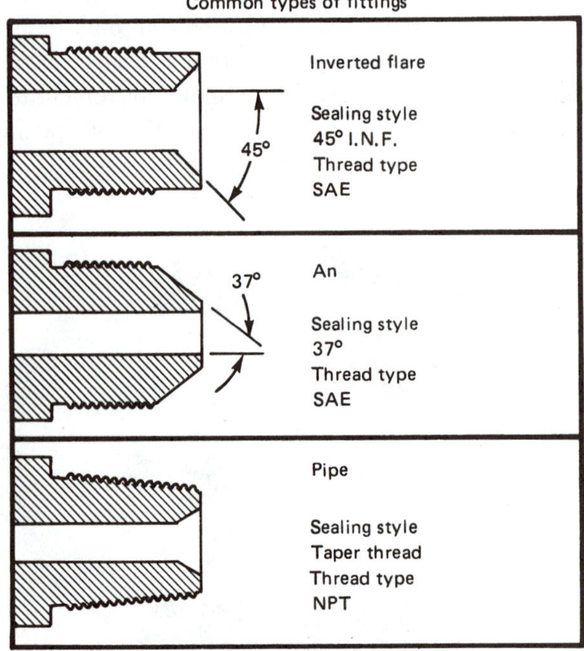

Figure 18-25 An SAE inverted flare, an AN flare, and a pipe thread. Inverted and AN flare fittings seal on a tapered seat; pipe threads seal at the tapered threads.

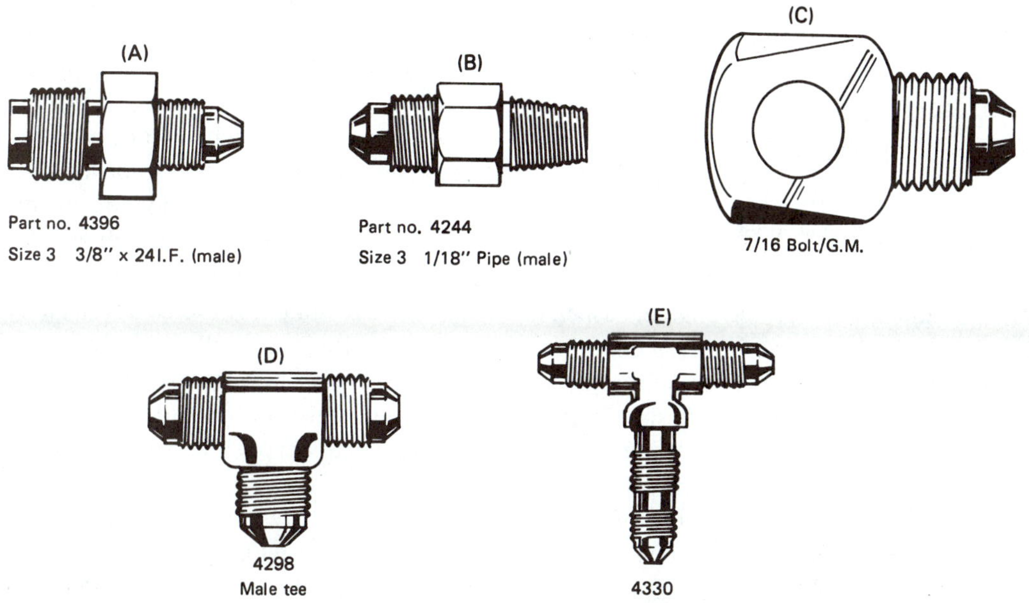

Figure 18-26 Adapter fittings are available to connect AN flares to SAE, pipe, or banjo fittings, as well as to other AN fittings.

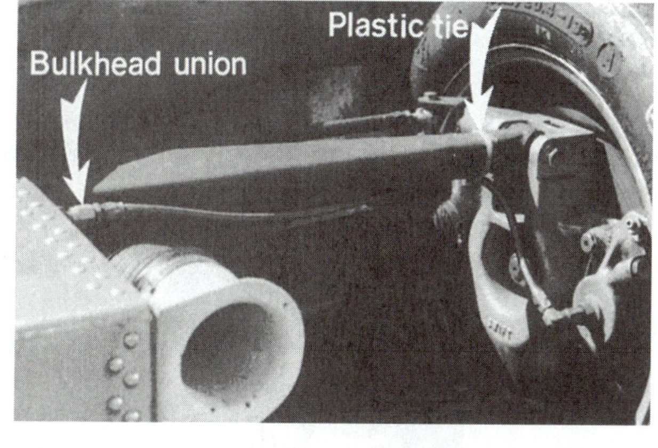

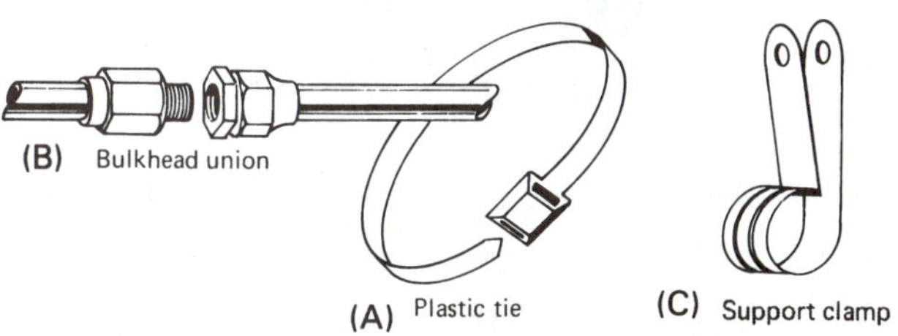

Figure 18-27 The flexible line is connected to a bulkhead union at the body and to the caliper; note the plastic tie holding the line to the control arm.

system. The wet boiling point of these fluids is much lower, within the range required of a DOT 3 fluid. The lower, wet boiling point is not too important, because the fluid in a racing brake system should be changed frequently, often for each race.

Silicone fluid is not normally used in all-out race cars because it tends to be compressible when hot. This will cause the pedal to fall noticeably, sometimes completely, after the brakes heat up. Silicone fluid also expands considerably when it gets hot, and this can cause the reservoir to overfill and develop a hydraulic lock.

18.4.9 Drum Brakes

Drum brakes are used on some race cars, and there are a few modifications that can be made to improve their performance in a racing situation. These changes are very similar to those made on disc brakes. They include the use of a race-quality brake lining of a temperature or hardness that suits race conditions, and the practice of ducting air into the brake assembly to help cool the components. The backing plate is often opened up for air circulation and a scoop can be added to force air into it. In some cases, the drum can be drilled in the friction surface area and the lining cross-grooved or diagonally grooved to provide a means of removing gas and debris from between the lining and the drum. Increasing the airflow through the backing plate also increases the amount of dirt and debris that can enter the lining area, and therefore the rate of lining and drum wear. Drilling a drum or grooving the lining will also increase the possibility of thermal cracks and breakage.

If necessary, it is often possible to replace the wheel cylinder with one of a larger or smaller bore diameter. Many cars have a wheel cylinder that uses the same mounting as a larger or smaller cylinder, so an exchange is fairly easy. Changing the wheel cylinder diameter is a quick way to change the strength of a drum brake unit. Don't forget that it also changes the amount of fluid that is required for a brake application.

18.4.10 Testing and Adjusting

When a race car is assembled and prepared for a race, it undergoes a series of adjustments on the engine, suspension, and brake system. Brake testing involves checking and adjusting the brake balance, checking brake assembly operating temperatures, and checking pad or shoe lining wear. Most brake testing takes place on a racetrack, preferably the same track on which the next race will take place. Most major teams do a certain amount of brake testing and adjusting during the practice sessions for each race. If problems are encountered, they are diagnosed and cured using a process just like that used with passenger cars (Table 18-1).

Cars that use a balance bar or adjustable proportioning valve are relatively easy to adjust for brake balance. The driver can often tell when front or rear lockup is occurring, and merely turns the knob or swings the lever to move more of the braking load to the end of the car not undergoing lockup. In some high-speed situations, brake lockup can occur very quickly, and a tire can be flat-spotted and ruined in the blink of an eye. Flat-spotting wears the tire in a small area when the tire locks up. In such cases, observers on each side of the track and often video cameras and recorders are used to determine exactly what is happening so the correct adjustments can be made.

Temperature testing shows how high the braking temperatures are, providing a good indication of how well the brakes and the brake cooling systems are working under

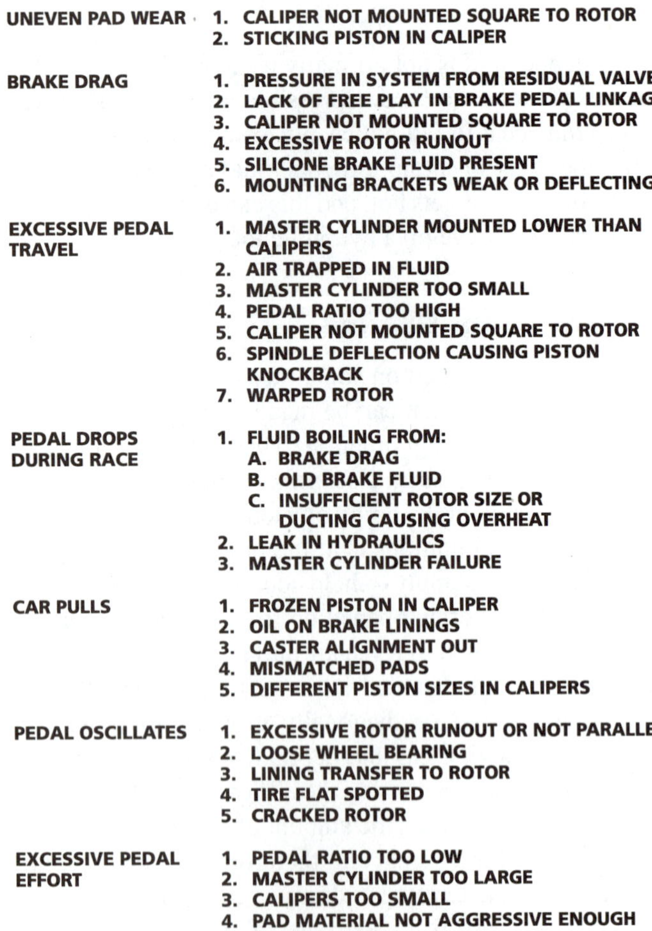

SYMPTOM	CHECK THE FOLLOWING:
UNEVEN PAD WEAR	1. CALIPER NOT MOUNTED SQUARE TO ROTOR 2. STICKING PISTON IN CALIPER
BRAKE DRAG	1. PRESSURE IN SYSTEM FROM RESIDUAL VALVE 2. LACK OF FREE PLAY IN BRAKE PEDAL LINKAGE 3. CALIPER NOT MOUNTED SQUARE TO ROTOR 4. EXCESSIVE ROTOR RUNOUT 5. SILICONE BRAKE FLUID PRESENT 6. MOUNTING BRACKETS WEAK OR DEFLECTING
EXCESSIVE PEDAL TRAVEL	1. MASTER CYLINDER MOUNTED LOWER THAN CALIPERS 2. AIR TRAPPED IN FLUID 3. MASTER CYLINDER TOO SMALL 4. PEDAL RATIO TOO HIGH 5. CALIPER NOT MOUNTED SQUARE TO ROTOR 6. SPINDLE DEFLECTION CAUSING PISTON KNOCKBACK 7. WARPED ROTOR
PEDAL DROPS DURING RACE	1. FLUID BOILING FROM: A. BRAKE DRAG B. OLD BRAKE FLUID C. INSUFFICIENT ROTOR SIZE OR DUCTING CAUSING OVERHEAT 2. LEAK IN HYDRAULICS 3. MASTER CYLINDER FAILURE
CAR PULLS	1. FROZEN PISTON IN CALIPER 2. OIL ON BRAKE LININGS 3. CASTER ALIGNMENT OUT 4. MISMATCHED PADS 5. DIFFERENT PISTON SIZES IN CALIPERS
PEDAL OSCILLATES	1. EXCESSIVE ROTOR RUNOUT OR NOT PARALLEL 2. LOOSE WHEEL BEARING 3. LINING TRANSFER TO ROTOR 4. TIRE FLAT SPOTTED 5. CRACKED ROTOR
EXCESSIVE PEDAL EFFORT	1. PEDAL RATIO TOO LOW 2. MASTER CYLINDER TOO LARGE 3. CALIPERS TOO SMALL 4. PAD MATERIAL NOT AGGRESSIVE ENOUGH 5. FROZEN PISTON IN CALIPER 6. BRAKE PAD FADE 7. ROTOR DIAMETER TOO SMALL

Table 18-1 Troubleshooting chart. (Courtesy of The Brake Man Inc., 2455 Blanchard Rd., Camarillo, CA 93012, 805-491-2185. Makers of High Performance Brake Components)

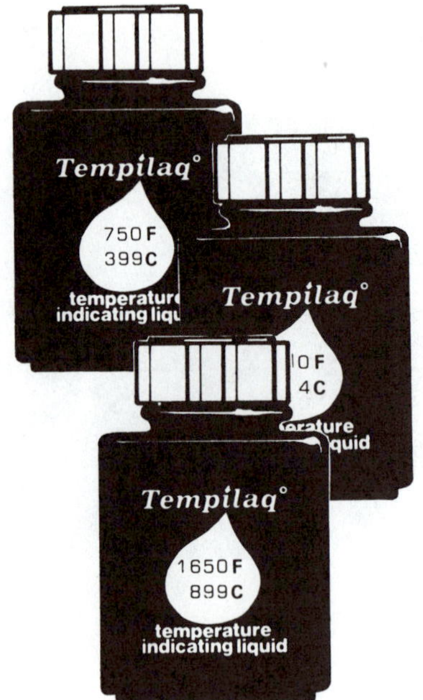

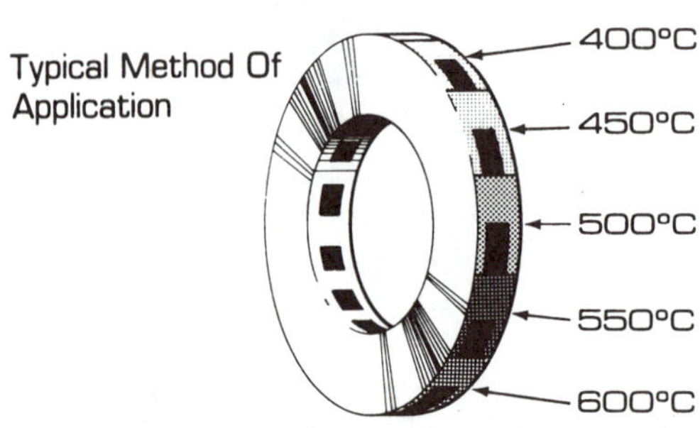

Figure 18-28 Temperature-indicating fluids are calibrated to different temperatures. The fluid is painted on the surface, and it will melt and change appearance when its specific temperature is reached. (Courtesy of Tilton Eng.)

actual operating conditions. Temperature indicator fluids make this a fairly easy operation (Figure 18-28). Samples of these fluids are painted on the edge of the rotor, on the caliper, or on any rubber, plastic, or metal surface that is to be checked. When applied, these paints have a dull appearance. Each fluid is compounded to melt at a different specific temperature between 250°F (121°C) and 1850°F (1010°C). When they melt, they take on a glossy, sometimes black appearance. After a test, the technician simply checks the painted patches and determines which ones have melted. The temperature reached was equal to or above that of the highest temperature mark that melted, but less than the marks that did not melt. Welders' crayons are very similar. Stick-on labels are also available, which can be placed on a caliper or other device to indicate the highest temperature reached.

Lining thickness is measured to determine the rate and evenness of lining wear. The pads or shoes are removed and measured at six points—the inner and outer edges at the center and at both ends of the lining (Figure 18-29). Checking the thickness at several intervals will give an indication of the rate of wear for the amount of braking done in that interval. With this knowledge, it is fairly easy to project the life of the lining. Braking problems are indicated if the lining wear is not even across the pad or shoe. Except for a normal increase in wear from the leading edge to the trailing edge of the lining, uneven wear indicates a pad that is not aligned with the rotor or a shoe that is not aligned with the drum. The trailing end of the lining is usually hotter, so it will normally have a faster wear rate.

18.5 Conclusion

It is the intent of this author to introduce you to the aspects, possibilities, and maybe some of the excitement of brake

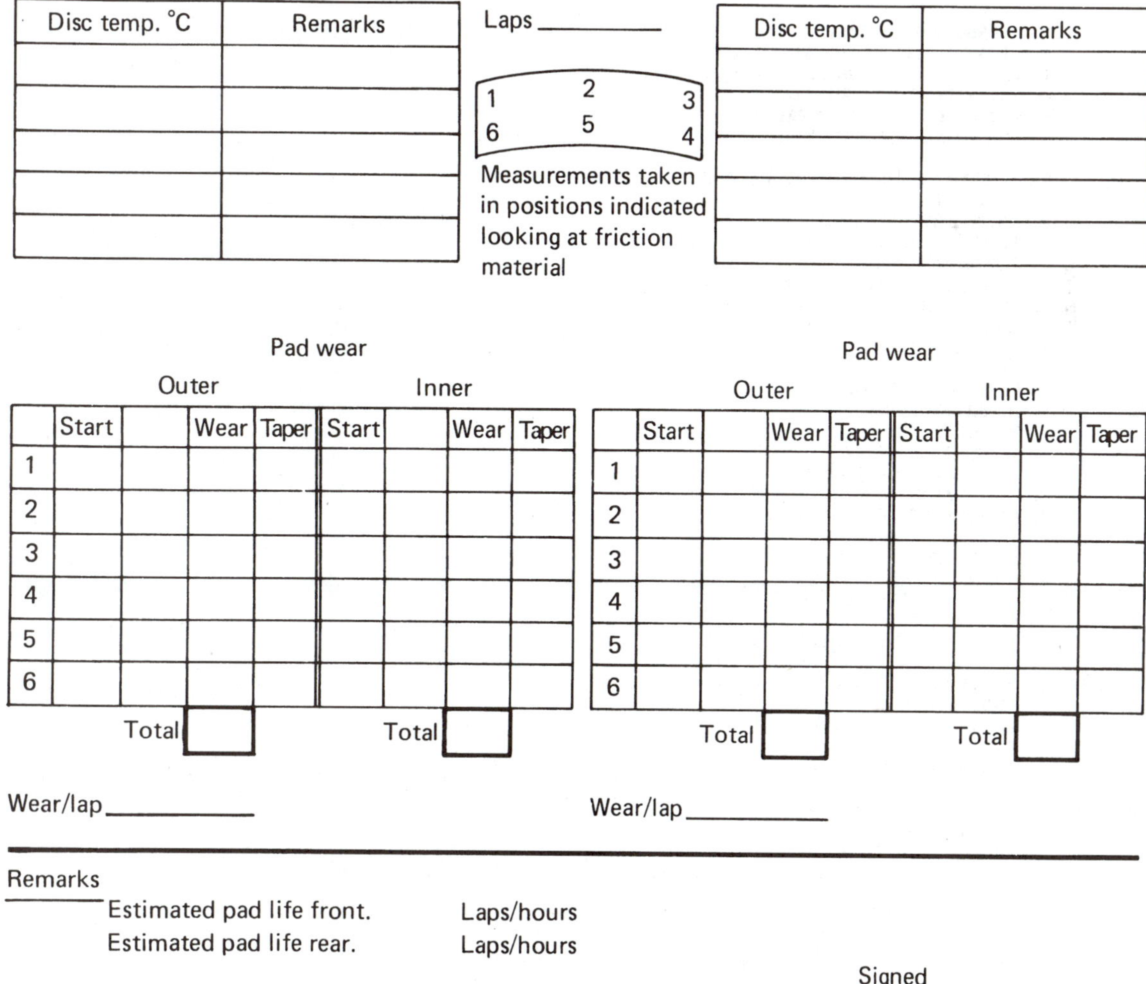

Disc temp. °C	Remarks

Laps ________

1 2 3
6 5 4

Measurements taken in positions indicated looking at friction material

Disc temp. °C	Remarks

Pad wear

	Outer				Inner			
	Start		Wear	Taper	Start		Wear	Taper
1								
2								
3								
4								
5								
6								

Total [ ] Total [ ]

Wear/lap ________

Pad wear

	Outer				Inner			
	Start		Wear	Taper	Start		Wear	Taper
1								
2								
3								
4								
5								
6								

Total [ ] Total [ ]

Wear/lap ________

Remarks

Estimated pad life front. Laps/hours

Estimated pad life rear. Laps/hours

Signed

Figure 18-29 This brake record chart is used to record the operating temperatures and lining wear as a car is set up for a particular track; it is used to choose the correct lining and estimate the probable lining life. (Courtesy of Tilton Eng.)

tuning. It is impossible to cover all the information needed to develop a brake system for all of the various types of cars (racing and production) in a book of this type. Excellent specialized books are available that cover brake tuning and race-car construction; these books describe brake development and modification or tuning in much greater detail.

Brake tuning or modification is not normally done in the average brake shop because of the time consumed, the skill required, and the liability involved. This type of work is usually left to highly specialized shops, builders of race cars, or highly knowledgeable hobbyists.

Terms to Know

balance bar

Glossary

ABS: See Antilock Brake system.

Accumulator: A chamber used to store pressurized hydraulic fluid.

Adapter: A bracket that is used to attach the caliper to the spindle, steering knuckle, or axle.

Adjustable Anchor: An anchor pin that is mounted on the backing plate so that it can be adjusted up or down.

Adjusting Cam: A bolt that passes through the backing plate and is used to adjust brake lining clearance.

Air Brakes: A brake system, commonly used on large trucks and busses, that uses air pressure to activate the brakes.

Air Chamber: The component in an air brake system that converts air pressure into brake shoe application force.

Amber Antilock Warning Light: An amber colored light mounted in the instrument panel to warn the driver of a problem in the ABS.

Ampere: The unit used to measure electrical current flow rate.

Anchor: That part of a drum brake assembly that prevents the shoe from rotating with the drum.

Anchor Pin: The steel pin, attached to a backing plate, that is used to prevent the brake shoes from turning with the drum.

Anodize: To use an electrochemical process to harden the surface of aluminum and increase its ability to resist corrosion.

Anti-Dive Suspension: A suspension design to reduce the amount of dive during braking.

Antilock Brake System (ABS): A system that prevents lockup of the tires under hard brake applications.

Antirattle Springs, Clips, or Washers: Parts that prevent brake pads, shoes, or shoe parts from rattling while they are released.

Aramid: A synthetic fabric used for brake linings. It is sold under the brand name of Kevlar or Flexten.

Arcing Shoes: A procedure of regrinding the lining on brake shoes to the correct arc to match the arc of the drum.

Asbestos: A noncombustible, fibrous substance used for brake linings. Caution should be used when working around asbestos as it is a health hazard.

Atmospheric Pressure: The pressure on the surface of the Earth created by the weight of the air in the atmosphere. This pressure is 14.7 psi at sea level.

Automatic Adjuster: A mechanism that will automatically adjust lining clearance when necessary.

Automotive Service Excellence (ASE): A group that promotes excellence in the automotive service industry through the voluntary testing and certification of competent technicians.

Axle Bearing: The bearing used to allow drive wheels to turn freely.

Backing Plate: The mounting plate upon which the components of a drum brake are assembled.

Ball Bearing: A type of bearing that uses a series of steel balls to reduce friction.

Banjo: A type of hydraulic fitting in which the end of the hose has a ring-like opening.

Barrel-Shaped Drum: See Concave Drum.

Bearing: A device that allows rotation or linear motion with a minimum of friction.

Bearing Retained Axle: A drive axle that is held in the housing by a C-shaped lock at the inner end.

Bell-Mouthed Drum: A brake drum that has been deformed so that the open end has a larger diameter than the closed end.

Bench-Bleeding: A procedure used to remove all of the air from the hydraulic portion of a component before mounting that component on the car. Master cylinders are normally bench-bled before installation.

Bleeder Screw: A small, hollow screw that is loosened to "open" it, so air and fluid can flow out of a hydraulic component.

Bleeding: A procedure that removes all of the air from a hydraulic system.

Bonded Lining: A brake lining that is attached to a shoe using an oven-cured adhesive.

BOO: Brake on/off switch. See Brake Light Switch.

Boot: See Dust Boot.

Bore: The diameter and walls of a cylinder.

Brake: A mechanism that converts energy, in order to slow or stop a vehicle.

Brake Balance: The relative amount of stopping force generated by the different axles of a vehicle.

Brake Band: A flexible metal band that is lined with friction material and wraps around the outside of a brake drum.

Brake Block: A short lining segment that is curved and drilled so it can be bolted or riveted onto a shoe.

Brake Booster: A device that uses engine vacuum, power-steering pump pressure, or electricity to reduce the amount of brake pedal pressure necessary to stop a car.

Brake Disc: See Brake Rotor.

Brake Drum: The rotating portion of a drum brake assembly against which the shoes push.

Brake Dynamometer: A devise used to measure braking power.

Brake Fluid: A special fluid that is used to transmit application pressure from the master cylinder to the pistons in the wheel cylinders and calipers.

Brake Hose: A reinforced, flexible hose that is used to transfer hydraulic pressure, from the rigidly mounted steel lines to a caliper or axle that moves or rotates.

Brake Light Switch: A mechanically or hydraulically operated switch used to turn the stoplights on when the brakes are applied.

Brake Line: A rigid steel tube that conducts hydraulic pressure to the various parts of the brake system.

Brake Lining: A special friction material that is attached to the brake shoes and pads. It withstands high temperatures and pressures.

Brake Pad: The lining and backing assembly used in disc brake units that is pressed against the rotor to cause braking action.

Brake Pedal: The foot-operated lever against which the driver pushes to cause brake application.

Brake Rotor: The rotating portion of a disc brake assembly against which the shoes press.

Brake Shoe: The curved metal part of the brake to which the lining is attached and which is pushed against the inside of a drum to cause braking action.

Brake Spoon: A tool, somewhat resembling a screwdriver, that is used to adjust brake shoe clearance.

Brake Torque: The stopping power of a brake assembly.

Brake Warning Light: A light, mounted on the dash, that lights up when there is a pressure loss in one of the hydraulic circuits.

Bridge Bolts: The bolts that are used to secure the two halves of a fixed caliper together.

Build Cycle: In ABS, the action of increasing braking pressure.

Burnishing: The process in which the lining and the friction surface of the drum or rotor wear to conform to each other.

Bypass Port: The port that allows flow from the master cylinder reservoir to the system during brake release.

Caliper: The C-shaped housing of a disc brake assembly, that fits over the disc and holds the pads and piston(s).

Caliper Hardware: See Hardware.

Caliper Mount: The mounting point for a caliper.

Center of Gravity (CG): The balance point of a car.

Changeover: See Split Point.

Check Valve: A valve that allows fluid or gas flow in only one direction.

Circuit: A path for electrical flow.

C-Lock Axle: A drive axle that is held in the housing by the axle bearing.

Coefficient of Friction: The ratio between the amount of force required to slide an object over another and the amount of force holding the objects together. The coefficient of friction of typical brake lining is 0.35 to 0.45.

Combination Valve: A brake warning-light valve in combination with a proportioning and/or metering valve.

Compensating Port: One or more passages in the master cylinder that are open between the cylinder bore and the reservoir, when the master cylinder is in the released position.

Composite: The result of a manufacturing method for brake drums and rotors, that combines a cast-iron friction surface with a stamped-steel center section.

Computerized Plate Tester: A device using moveable plates under each wheel to measure brake power.

Concave Drum: A brake drum that is deformed so that the diameter is larger in the center than at the ends.

Control Cable: A cable used to transmit application force for a parking brake.

Controller: A device used to control the current flow that applies electric brakes.

Convex Drum: A brake drum that is deformed so that the diameter is smaller in the center than at the ends.

Corrosion: The deterioration of metal by chemical or electrochemical action.

Cup: The common hydraulic piston seal used in wheel cylinders and master cylinders. Also, the outer bearing race.

Cup Expanders: A washer-like device or spring shape that places an outward pressure on the lip of a cup.

Decay Cycle: In ABS, the action of reducing braking pressure.

Decelerate: To reduce speed or slow down.

Deceleration Rate: The rate of the reduction in speed.

Decelerometer: A device used to measure deceleration rate.

Deenergized: In a drum brake, the state of brake shoes when shoe application pressure is reduced by the rotation of the drum. Trailing shoes are deenergized.

Denatured Alcohol: Ethyl alcohol that contains methyl alcohol and is used to clean a hydraulic brake system (not fit for human consumption).

Department of Transportation (DOT): A government agency that sets standards for brakes as well as other automotive units.

Diagnostic Trouble Code (DTC): A series of numbers or letters, generated by an on-board diagnostic system, that indicate a fault condition.

Diagonal Split System: A hydraulic system that is split so one front wheel and the opposite rear wheel are in each part.

Dial Indicator: A precision instrument used to measure linear movement.

Diaphragm: A flat, flexible rubber membrane used to separate two areas in a chamber.

Digital: An electrical signal that has only two possible values, on or off.

Directional Finish: A machined pattern in which the scratches form a parallel pattern.

Disc: See Brake Rotor.

Disc Brake: A brake design that generates stopping power from friction caused by two pads pressing against the sides of a rotating disc.

Dive: The lowering of the front of the car during braking. Dive is the result of weight transfer.

Double Flare: A type of flare, used at the ends of a brake line, in which the end of the tubing is doubled over.

Drum: See Brake Drum.

Drum Brake: A brake design that generates stopping power from friction caused by shoes pushing against the cylindrical portion of the drum.

Dual Master Cylinder: A master cylinder, with two pistons and pressure chambers, that is used with split brake systems.

Duo-servo: A drum brake design that generates servo action in both forward and reverse directions.

Dust Boot: A flexible rubber part that keeps dirt and foreign material from entering the bore of a wheel cylinder or caliper.

Dynamic Seal: A seal that allows movement between it and the object that it seals.

Eccentric: Off-center; two circles having different centers.

Edge Brand: See Edge Code.

Edge Code: A series of letters and numbers on the edge of the brake lining that identifies the manufacturer, specific lining, and coefficient of friction.

Electronic Brake Control Module (EBCM): A computer-like unit that operates the hydraulic brake module when needed to prevent wheel lockup.

Electronic Controller: See EBCM.

Electronic Control Module (ECM): See EBCM.

Emergency Brake: See Parking Brake.

End Play: The distance a shaft or hub can move sideways.

Energized: In a drum brake, the state of brake shoes when shoe application pressure is increased by the rotation of the drum. Primary and leading shoes are energized during forward rotation.

Equalizer Lever: A device used with parking brake cables to ensure equal pull on each cable.

Fade: A loss in braking power usually caused by excessive heat.

Finish Cut: The final cut made while turning a drum or rotor; usually at a shallow depth and slow feed rate in order to obtain a smooth surface finish.

Fixed Anchor: An anchor pin that is riveted or welded to the backing plate; it is not adjustable.

Fixed Caliper: A caliper design mounted securely to the steering knuckle, spindle, or axle, that uses one or two pairs of pistons to apply the pads on each side of the disc.

Floating Caliper: A caliper design that uses a single piston to apply both brake pads; the caliper floats sideways on pins or bolts to apply the outboard pad.

Floating Valve: A valve in which the outer bore and inner spool move.

Flushing: The operation used to replace the brake fluid in a system.

FMVSS: Federal Motor Vehicle Safety Standard.

Forward Shoe: See Leading shoe.

Friction: The resistance as one surface slides over another.

Friction Material: See Lining.

Front Wheel Drive (FWD): A vehicle that drives from front wheels.

Gravity Bleeding: A method of removing air from a wheel cylinder or caliper by using the force of gravity to cause a fluid flow.

Ground Circuit: An electrical path that uses the vehicle body or frame to return the current to the source.

Grounded: An electrical connection to ground.

Guide Pins or Bolts: The pins that control the sideways motion of a floating caliper.

Hard Spots: Areas on the surface of a brake drum or disc that are harder than the metal around them.

Hardware: Clips, springs, guides, and other parts used to ensure proper action of the brake shoes or calipers.

Heat Sink: A mass with the ability to absorb heat.

Heel: The end of the brake shoe where it pivots, opposite the end that is applied.

Heel Drag: A condition in which the edges of a cup get caught by a piston.

Height Sensing Proportioning Valve: A proportioning valve that changes rear brake pressure relative to vehicle height and load.

Hold Cycle: In ABS, the action of maintaining a steady braking pressure.

Hold-Down Spring: A spring used to hold a brake shoe against the backing plate.

Hone: A set of fine abrasive stones used to true and refinish a cylinder.

Hub: A rotating unit, mounted on bearings, that supports the wheel and brake drum or disc.

Hydraulic Brake System: A braking system that uses hydraulics to transfer the motion and pressure from the driver's foot to the brake shoes and pads.

Hydraulic Modulator: See Modulator Valves.

Hydraulic Pressure: The pressure within a hydraulic system measured in the amount of force on a unit of area, commonly in pounds per square inch (psi).

Hydraulics: A system that uses fluid under pressure to transfer motion and pressure.

Hydro-boost: The brand name for a brake booster that uses the hydraulic pressure from the power steering pump as a source of power.

Hygroscopic: The ability to readily absorb moisture.

Inches of Mercury, "Hg": A system for measuring vacuum.

Index: To align a part, such as a rotor or wheel, to marks applied before removal, or to realign parts to the best possible position.

Inertia: The physical tendency of a body at rest to remain at rest, and a body in motion to remain in motion and travel in a straight line.

Integral: Made together in one unit.

Integral Parking Brake: A caliper with mechanism that allows mechanical parking brake application.

Internal Leakage Check: A test for an air leak through a vacuum booster.

International Standards Organization (ISO) Flare: The shape of the end of the brake tubing designed by the ISO.

Inverted Flare: The type of flare fitting commonly used on brake hydraulic tubing.

Jackknifing: An undesirable result of lockup of the brakes of a trailer, or of the rear axle of a tow vehicle, which causes the trailer to skid forward into the rear corner of the tow vehicle.

Jacobs Brake: A device, used with diesel engines, that can open the exhaust valve during part of the compression stroke to increase the amount of braking force created by the engine during deceleration.

Kinetic Energy: The energy contained in an object that is in motion.

Knock-Back: An undesirable result of excessive rotor runout, which increases pad clearance and produces a lower brake pedal.

Lateral Runout: A distortion of a brake disc that will cause it to wobble from side to side as it rotates.

Leading Shoe: A shoe that is ahead of the axle, or a shoe that is mounted so the drum rotation will cause it to push harder against the drum.

Leading-Trailing Shoe: See Nonservo.

Lining: The actual friction material that is attached to the brake shoe or pad.

Lining Brake-In: The process used to help the lining mate with the drums and rotors.

Loaded Caliper: A reconditioned caliper that comes complete with new parts, shoes, and hardware.

Lockup: The point where braking power overcomes traction and the wheel stops turning; skidding will normally occur.

Low-Drag Caliper: A caliper design that retracts the piston farther, so it will have more clearance between the disc and the lining and less friction drag in the released position.

Machining: In servicing brakes, the procedure of mounting a drum or rotor on a lathe and recutting the friction surfaces to remove imperfections.

Major Brake Adjustment: An adjustment of the anchor to position the curvature of the shoe to the drum.

Manual Bleeding: A bleeding procedure that requires two people—one to pump the master cylinder while the other performs the bleeding operation.

Master Cylinder: The source of hydraulic pressure in a brake system. Operated by the brake pedal, it supplies fluid pressure to the calipers and wheel cylinders.

Maximum Diameter: The largest size allowable for a brake drum or a cylinder bore.

Metering Valve: A hydraulic valve used in some systems to cause a slight delay in operation of the front disc brakes.

Minimum Thickness: The thinnest allowable width for a brake rotor.

Modulation: In ABS, the process of turning a valve on and off at a rapid rate to maintain proper braking action.

Modulator: See Modulator Valves.

Modulator Valves: The ABS valves that control the power of the individual brakes during an ABS stop.

Motorist Assurance Program (MAP): An organization that has developed guidelines for automotive service that include brake systems.

Mounting Hardware: Parts used to connect a caliper properly to its mount.

National Highway Traffic Safety Administration (NHTSA): A government body that establishes regulations for motor vehicles.

National Institute of Automotive Service Excellence (NIASE): See Automotive Service Excellence (ASE).

NIOSH: National Institute of Occupational Safety and Health.

Nonasbestos: A brake lining compound that does not include asbestos.

Nonasbestos Organic (NAO): Use old Nonasbestos.

Nondirectional Finish: A machined surface in which the scratches do not form any noticeable pattern.

Nonservo: A brake shoe design that applies the two shoes independently of each other.

Occupational Safety and Health Administration (OSHA): A government body that regulates working conditions as they relate to personal safety.

Ohm: The unit used to measure electrical resistance.

On-Board Diagnostic (OBD): A system that monitors all computer input and output signals to determine any faults that might occur.

Open: An electrical circuit that is broken.

Original Equipment (OE): The parts that were originally used on the vehicle.

Out-of-round: A brake drum that has two different diameters; it is not round.

Pad: The brake shoe for a disc brake. It consists of the lining and the backing to support it.

Pad Wear Indicator: A device that warns the driver when the lining is almost worn out.

Parallelism: A variation in the thickness of the friction surfaces of a rotor.

Parking Brake: A mechanically applied brake system designed to hold a car stationary.

Pascal's Law: Physical law governing action of fluids

Phenolic: A plastic material that is used to make caliper pistons.

Pitting: Surface irregularities that result from rust or corrosion.

Pounds Per Square Inch (PSI): A measurement of fluid pressure.

Power Assist: See Power Brakes.

Power Booster: See Brake Booster.

Power Brakes: A brake system that uses a power booster.

Preload: The amount of load added to a bearing, less than zero lash.

Pressure: The amount of force acting on a specific area.

Pressure Bleeder: A container of brake fluid that is pressurized. It will supply brake fluid under pressure to the master cylinder for bleeding purposes.

Pressure Bleeding: A procedure of bleeding the brakes using a pressure bleeder to force fluid through the system.

Pressure Differential Switch: A switch, operated by a pressure difference in the two hydraulic circuits, that turns on the brake warning light to indicate a faulty hydraulic circuit to the driver.

Pressure Differential Valve: A valve that senses pressure in each part of a split system and operates the Pressure differential switch if one side looses pressure.

Primary Cup: In a master cylinder, the cup that pumps fluid to the system.

Primary Piston: In a tandem master cylinder, the piston that is operated by the pushrod.

Primary Shoe: The forward shoe; in a duo-servo brake, the primary shoe is energized by the rotation of the drum and applies pressure on the secondary shoe.

Proportioning Valve: A hydraulic valve that is used to improve brake balance by reducing the pressure applied to the rear brakes.

Pull: The tendency of a car to self-steer to one side during a stop.

Pushrod Adjustment: An adjustment to ensure the proper length of the pushrod between the vacuum booster and the master cylinder.

Quick-Take-Up Master Cylinder: A dual master cylinder design that uses a stepped primary piston to displace more fluid and take up the clearance of low-drag calipers.

RABS: Rear antilock brake system.

Reaction Time: The amount of time that it takes a driver to recognize a problem and apply the brakes.

Red Brake Warning Light: See Brake Warning Light.

Regenerative Braking: Braking that uses an electrical or hydraulic system to remove and store the kinetic energy of a vehicle to reduce its speed; the energy removed is later used to propel the vehicle.

Reservoir: The portion of a master cylinder where fluid is stored.

Residual Pressure Check Valve: A hydraulic check valve used in some master cylinders to maintain a slight pressure in the lines while the brakes are released.

Return Spring: The spring used to return a brake shoe to a released position. Also called pull-back spring or retracting spring.

Reverse Shoe: See Trailing Shoe.

Riveted Lining: A lining that is attached to the steel backing with rivets.

Roller Bearing: A type of bearing that uses a series of steel rollers to reduce friction.

Roller Burnishing: The process to produce a very smooth cylinder bore.

Room Temperature-Vulcanizing (RTV): A material used for formed-in-place gaskets.

Rotor: The flat, rotating disc against which the brake pads press in a disc brake assembly.

Roughing Cut: The first cut made while turning a drum or rotor; usually made fairly deep and at a fast feed rate.

Runout: See Lateral Runout.

Scoring: Irregular grooves that are worn into the friction surface of a rotor or drum.

Seal: A device used to prevent foreign material or water from entering an area; a grease seal also retains grease or oil inside an area.

Secondary Cup: On a master cylinder piston, the cup that is used to prevent fluid from running out the end of the bore or from running between the primary and secondary sections.

Secondary Piston: In a tandem master cylinder, the piston that is applied by hydraulic pressure from the primary piston.

Secondary Shoe: The rearward shoe, in a duo-servo brake; the secondary shoe is applied by the energized primary shoe.

Self-Adjusting: A drum brake mechanism that automatically adjusts the lining clearance when necessary.

Semimetallic: A nonasbestos lining compound that uses powdered metal or strands of steel wool.

Sensor: A device that senses a value and converts it into an electrical signal.

Service Brake: The primary brake system used to stop a vehicle under normal driving conditions.

Servo Brake: A drum brake design that uses the rotational force generated by one shoe to apply the other one.

Shoe: On drum brakes, the lining with its curved, steel backing.

Shoe Anchor: See Anchor.

Shoe Return Spring: See Return Spring.

Short Circuit: An undesired connection between a circuit and any other point.

Shorted: See Short Circuit.

Skid: The act of a tire sliding without rotating; this usually results from a failure of the tire to grip the roadway with enough force to overcome the stopping or cornering loads placed on it. A skid often results in a loss of vehicle control.

Skid Control: See Antilock Brake System.

Sleeves: Tubular items on which a caliper slides sideways.

Sliding Caliper: A floating caliper that moves sideways on machined ways or keys.

Slope: The rate of difference between front and rear brake pressures above the split point.

Society of Automotive Engineering (SAE): A group that sets standards used in the automotive industry.

Solid Rotor: A nonventilated rotor.

Solid State: A term that describes an electronic device with no moving parts that performs a variety of operations.

Split Point: The pressure at which the proportioning valve begins to reduce the pressure increase to the rear brakes.

Spongy Pedal: A condition where the pedal will feel springy under pressure instead of having the normal solid feel.

Starwheel: A coglike attachment of the adjuster screw that allows it to be rotated by a tool or lever.

Static Seal: See Stationary Seal.

Stationary Seal: A seal that does not move, such as the O-ring seal in many calipers; the piston moves within the seal.

Stoplight Switch: See Brake Light Switch.

Stroking Seal: A hydraulic seal that moves with the piston.

Swaging: A method of deforming the metal to permanently lock an item in place.

Swept Area: The total amount of area on the drums and rotors on which the brake lining rubs.

Swirl Finish: See Nondirectional Finish.

Tandem Diaphragm: A power booster that uses two diaphragms.

Tandem Hydraulic System: A hydraulic brake system that is split front and rear.

Tandem Master Cylinder: A master cylinder that uses two pistons, one mounted behind the other.

Tapered Roller Bearing: A type of bearing that uses a series of tapered steel rollers to reduce friction. This bearing type is used in pairs to control end play.

Tee: A hydraulic fitting that is used to connect three lines.

Thermodynamics: The physical principles that govern mechanical actions and heat energy.

Thickness Variation: See Parallelism.

Toe: The end of a brake shoe that contacts the drum first.

Traction: The frictional grip between the tires and the road surface.

Trailing Shoe: A shoe that is behind the axle; in a nonservo brake, it is the deenergized shoe.

Turning: The process of using a special lathe to machine the friction surface of a drum or rotor so it is like-new.

Two-Leading Shoe: A drum brake design using two leading shoes.

Union: A hydraulic fitting that is used to connect two lines.

Uni-servo: A drum brake design that generates servo action in one direction only.

Vacuum Bleeding: A bleeding method that pulls fluid and air from the bleeder valve using vacuum.

Vacuum Booster: A brake booster that uses engine vacuum as a power source.

Vehicle Speed Sensor (VSS): A sensor that provides vehicle speed information.

Vented Rotor: A rotor that has internal air passages between the friction surfaces.

Viscosity: The thickness or body of a fluid as it relates to the ability to flow.

Volt: The unit used to measure electrical force or pressure.

Warning Light: A light that lights up to warn the driver of a braking problem.

Watts: Electrical power determined by multiplying volts times amps.

Wear Sensor: See Pad Wear Indicator.

Weight Transfer: The amount of weight that moves laterally across the car because of cornering forces, or lengthwise because of braking or acceleration forces.

Wheel Bearing: The bearing used to allow the non-drive wheels to turn freely.

Wheel Cylinder: A component in drum brakes that converts hydraulic pressure into a mechanical force to push the shoes against the drum.

Wheel Slip: Tire rotational speed relative to vehicle speed; 100-percent slip indicates wheel lockup.

Wheel-Speed Sensor: An ABS sensor that measures the speed of a wheel.

APPENDIX 1

ASE Certification

Many automotive technicians have taken tests to become certified by ASE, the National Institute for Automotive Service Excellence. These tests are voluntary in that technicians decide on their own tests. Certification has become a status symbol that indicates better technicians.

One of the eight automotive tests is A5, Brakes; the medium/heavy truck category also has eight tests, with one of them being T4, Brakes; the school-bus category has six tests, with one of them being S4, Brakes. The A5 test has 55 questions that are divided into these content areas:

A. Hydraulic System Diagnosis and Repair		14
1. Master Cylinders	(3)	
2. Fluids, lines, and Hoses	(3)	
3. Valves and Switches	(4)	
4. Bleeding, Flushing, and Leak Testing	(4)	
B. Drum Brake Diagnosis and Repair		6
C. Disc Brake Diagnosis and Repair		13
D. Power Assist Units Diagnosis and Repair		4
E. Miscellaneous Diagnosis and Repair		7
F. Antilock Brake System Diagnosis and Repair		11
		55

These content areas are further divided into a group of tasks to aid the technician in preparing for the certification test. As you look over the task list, you can compare it with the table of contents for this text and locate where the information for each task. If you are preparing to take a certification test, you can get an up-to-date task list and information concerning the test by calling ASE at 703-713-3800.

BRAKES TASK LIST

A. Hydraulic System Diagnosis and Repair (14 questions)

1. Master Cylinders (Non-ABS) (3 questions)

Task 1 - Diagnose poor stopping or dragging caused by problems in the master cylinder; determine needed repairs.

Task 2 - Diagnose poor stopping, dragging, high or low pedal, or hard pedal caused by problems in the step bore master cylinder and internal valves (e.g. volume control devices, quick-take-up valve, fast-fill valve, pressure regulating valve); determine needed repairs.

Task 3 - Measure and adjust pedal pushrod length.

Task 4 - Check master cylinder for defects by depressing brake pedal; determine needed repairs.

Task 5 - Diagnose cause of master cylinder external fluid leakage.

Task 6 - Remove master cylinder from vehicle; install master cylinder; test operation of hydraulic system.

Task 7 - Bench bleed (check for function and remove air) all non-ABS master cylinders.

2. Fluids, Lines, and Hoses (3 questions)

Task 1 - Diagnose poor stopping, pulling, or dragging caused by problems in the brake fluid, lines, and hoses; determine needed repairs.

Task 2 - Inspect brake linings and fittings for leaks, dents, kinks, rust, cracks, or wear; tighten loose fittings and supports.

Task 3 - Inspect flexible brake hoses for leaks, kinks, cracks, bulging, or wear; tighten loose fittings and supports.

Task 4 - Fabricate and/or replace brake lines (double flare and ISO types), hoses, fittings, and supports.

Task 5 - Select, handle, store, and install brake fluids (includes silicone fluids).

Task 6 - Inspect brake lines and hoses for proper routing.

3. Valves and Switches (Non-ABS) (4 questions)

Task 1 - Diagnose poor stopping, pulling, or dragging caused by problems in the hydraulic system valve(s); determine needed repairs.

Task 2 - Inspect, test, and replace metering (hold-off), proportioning, pressure differential, and combination valves.

Task 3 - Inspect, test, replace, and adjust load- or height-sensing-type proportioning valve(s).

Task 4 - Inspect, test, and replace brake warning-light system switch and wiring.

4. Bleeding, Flushing, and Leak Testing (Non-ABS Systems) (4 questions)

Task 1 - Bleed (manual, pressure, vacuum, or surge) and/or flush hydraulic system.
Task 2 - Pressure-test brake hydraulic system.

B. Drum Brake Diagnosis and Repair (6 questions)
Task 1 - Diagnose poor stopping, pulling, or dragging caused by drum brake hydraulic problems; determine needed repairs.
Task 2 - Diagnose poor stopping, noise, pulling, grabbing, dragging, or pedal pulsation caused by drum brake mechanical problems; determine needed repairs.
Task 3 - Remove, clean, inspect, and measure brake drums; follow manufacturers' recommendations in determining need to machine or replace.
Task 4 - Machine brake drum according to manufacturers' procedures and specifications.
Task 5 - Using proper safety procedures, remove, clean, and inspect brake shoes/linings, springs, pins, self-adjusters, levers, clips, brake backing (support) plates, and other related brake hardware; determine needed repairs.
Task 6 - Lubricate brake shoe support pads on backing (support) plate, adjuster/self-adjuster mechanisms, and other brake hardware.
Task 7 - Install brake shoes and related hardware.
Task 8 - Pre-adjust brake shoes and parking brake before reinstalling brake drums or drum/hub assemblies and wheel bearings.
Task 9 - Reinstall wheel, torque lug nuts, and make final check and adjustments.

C. Disc Brake Diagnosis and Repair (13 questions)
Task 1 - Diagnose poor stopping, pulling, or dragging caused by disc brake hydraulic problems; determine needed repairs.
Task 2 - Diagnose poor stopping, noise, pulling, grabbing, dragging, or pedal pulsation caused by disc brake mechanical problems; determine needed repairs.
Task 3 - Retract integral parking brake piston(s) according to manufacturers' recommendations.
Task 4 - Remove caliper assembly from mountings; clean and inspect for leaks and damage to caliper housing.
Task 5 - Clean and inspect caliper mountings and slides for wear and damage.
Task 6 - Remove, clean, and inspect pads and retaining hardware; determine needed repairs, adjustments, and replacements.
Task 7 - Disassemble and clean caliper assembly; inspect parts for wear, rust, scoring, and damage; replace all seals, boots, and any damaged or worn parts.
Task 8 - Reassemble caliper.
Task 9 - Clean and inspect rotor; measure rotor with a dial indicator and micrometer; follow manufacturers' recommendations in determining need to machine or replace.
Task 10 - Remove and replace rotor.
Task 11 - Machine rotor, using on-car or off-car method, according to manufacturers' procedures and specifications.
Task 12 - Install pads, calipers, and related attaching hardware; bleed system.
Task 13 - Adjust calipers with integrated parking brakes according to manufacturers' recommendations.
Task 14 - Fill master cylinder to proper level with recommended fluid; inspect caliper for leaks.
Task 15 - Reinstall wheel and torque lug nuts, and make final check and adjustments.

D. Power Assist Units Diagnosis and Repair (4 questions)
Task 1 - Test pedal free travel with and without engine running to check power booster operation.
Task 2 - Check vacuum supply (manifold or auxiliary pump) to vacuum-type power booster.
Task 3 - Inspect the vacuum-type power booster unit for vacuum leaks and proper operation; inspect the check valve for proper operation; repair, adjust, or replace parts as necessary.
Task 4 - Inspect and test Hydro-boost system and accumulator for leaks and proper operation; repair, adjust, or replace parts as necessary.

E. Miscellaneous (Wheel Bearings, Parking Brakes, Electrical, etc.) Diagnosis and Repair (7 questions)
Task 1 - Diagnose wheel bearing noises, wheel shimmy and vibration problems; determine needed repairs.
Task 2 - Remove, clean, inspect, repack wheel bearings or replace wheel bearings and races; replace seals; adjust wheel bearings according to manufacturers' specifications.
Task 3 - Check parking brake system; inspect cables and parts for wear, rusting, and corrosion; clean or replace parts as necessary; lubricate assembly.
Task 4 - Adjust parking brake assembly; check operation.
Task 5 - Test service and parking brake indicator and warning light(s), switch(s), and wiring.
Task 6 - Test, adjust, repair or replace brake stoplight switch, lamps, and related circuits.

F. Antilock Brake System (ABS) Diagnosis and Repair (11 questions)
Task 1 - Follow accepted service and safety precautions during inspection, testing, and servicing of ABS hydraulic, electrical, and mechanical components.
Task 2 - Diagnose poor stopping, wheel lockup, pedal feel and travel, pedal pulsation, and noise problems caused by the ABS; determine needed repairs.

Task 3 - Observe ABS warning light(s) at startup and during road test; determine if further diagnosis is needed.
Task 4 - Diagnose ABS electronic control(s) and components using self-diagnosis and/or recommended test equipment; determine needed repairs.
Task 5 - Depressurize integral (high pressure) components of the ABS following manufacturers' recommended safety procedures.
Task 6 - Fill the ABS master cylinder with recommended fluid to proper level following manufacturers' procedures; inspect system for leaks.
Task 7 - Bleed the ABS hydraulic circuits following manufacturers' procedures.
Task 8 - Perform a fluid pressure (hydraulic boost) diagnosis on the integral (high pressure) ABS; determine needed repairs.
Task 9 - Remove and install ABS components following manufacturers' procedures and specifications; observe proper placement of components and routing of wiring harness.
Task 10 - Diagnose, service, test, and adjust ABS speed sensors and circuits following manufacturers' recommended procedures (includes voltage output, resistance, shorts to voltage/grounds, and frequency data).
Task 11 - Diagnose ABS braking problems caused by vehicle modifications (tire size, curb height, final drive ratio, etc.) and other vehicle mechanical and electrical/electronic modifications (communication, security, and radio, etc.)
Task 12 - Repair wiring harness and connectors following manufacturers' procedures.

(Courtesy of the National Institute for Automotive Service Excellence)

APPENDIX 2

English-Metric Conversion

The following conversion factors can help you convert a dimension from one measuring system to another. Simply multiply the dimension you have by the factor to get the dimension you want.

Unit	Multiply	By	To Get
LENGTH	inch	25.4	millimeter (mm)
	foot	0.305	meter (m)
	yard	0.914	meter
	mile	1.609	kilometer (km)
	millimeter	0.04	inch
	centimeter	0.4	inch
	meter	3.28	feet
	kilometer	0.62	mile
AREA	$inch^2$	645.2	$millimeter^2$ (mm^2)
	$foot^2$	0.093	$meter^2$ (m^2)
	$millimeter^2$	0.0016	$inch^2$
	$centimeter^2$	0.16	$inch^2$
	$meter^2$	10.76	$foot^2$
VOLUME	$inch^3$	16,387	$millimeter^3$ (mm^3)
	quart	0.164	liter (l)
	gallon	3.785	liter
	$millimeter^3$	0.000061	$inch^3$
	liter	1.06	quart
	liter	0.26	gallon
WEIGHT	ounce	28.4	gram (g)
	pound	0.45	kilogram (kg)
	ton	907.18	kilogram
	gram	0.035	ounce
	kilogram	2.2	pound
FORCE	kilogram	9.807	newton (N)
	ounce	0.278	newton
	pound	4.448	newton

(continued)

Unit	Multiply	By	To Get
PRESSURE	in. of water (H_2O)	0.2488	kilopascals (kPa)
	pounds/inch^2	6.895	kilopascals
	kilopascals	0.145	pounds/in.^{2}
	kilopascals	0.296	in. of mercury (Hg)
POWER	horsepower	0.746	kilowatt (kw)
	kilowatts	1.34	horsepower
TORQUE	inch-pound	0.113	newton-meter (N-m)
	foot-pound	1.356	newton-meter
	newton-meter	8.857	inch-pound
	newton-meter	0.737	foot-pound
SPEED	miles per hour	1.609	kilometer per hour (km/h)
	km per hour	0.621	miles/hour
ACCELERATION/ DECELERATION	feet/sec.^{2}	0.345	meter/sec.^{2}
	inch/sec.^{2}	0.025	meter/sec.^{2}
	meter/sec.^{2}	3.28	feet/sec.^{2}
FUEL ECONOMY	miles/gallon	0.425	km/liter (km/l)
	km/liter	2.35	miles per gal.
TEMPERATURE	Fahrenheit, degree	0.556 (°F −32)	Celsius, deg. (°C)
	Celsius, degree	1.8 (°C+32)	Fahrenheit, deg. (°F)

APPENDIX 3

Bolt Torque-Tightening Chart

Torque-tightening values for a bolt will generally vary depending on the diameter of the bolt, the grade of the bolt material, the pitch of the bolt thread, whether the threads are lubricated and the type of lubricant used, and the material into which the bolt is threaded. Tightening a bolt too tightly can stretch the bolt to the yield point where the bolt might break, or it might cause stripping of the threads of the bolt or nut. Tightening a bolt to too low a torque value might allow the bolt to come loose prematurely.

If the tightening torque for a particular bolt cannot be located, the values in the following table can be used as a guide.

Grade:								
	S.A.E.	*1 & 2*		*5*		*8*		
	Metric		*5*		*8*		*10*	*12*
Size/Diameter								
U.S.	Metric							
	6		5		9		11	13
1/4		5		7		10		
5/16		9		14		22		
	8		12		21		26	32
3/8		15		25		37		
	10		23		40		50	60
7/16		24		40		60		
	12		40		70		87	105
1/2		37		60		90		
	14		65		110		135	160

Note: All torque values are given in foot-pounds and for clean, lubricated bolts. The values given are for steel-to-steel threads using motor oil for a lubricant.

To convert these values to inch-pounds, multiply them by 12. To convert them to newton-meters, multiply them by 1.356

APPENDIX 4

Torque-Tightening Chart for Line Connections and Bleeder Screws

Torque-tightening values for a tube nut, banjo bolt, or bleeder screw will generally vary depending on the diameter of the bolt, whether the threads are lubricated, whether a sealing washer is used, and the material into which the tube nut or bleeder screw is threaded. Tightening a connection too much can stretch the parts to the yield point where the bleeder screw or banjo bolt might break, or it might cause stripping of the threads of the bolt, nut, or component to which it is being connected. Tightening a bolt to too low a torque value might allow a leak or cause the bolt to come loose prematurely. Dry threads should be lubricated with brake fluid.

It is recommended that the torque values provided by the vehicle manufacturer always be used. If the tightening torque for a particular screw or bolt cannot be located, the values in the following table can be used as a guide.

Bleeder Screws

Screw Size	Torque Inch-Pounds	Newton-Meters
1/4 in.	65	7.3
7 mm	70	7.9
5/16 in.	80	9
8 mm	80	9
3/8 in.	80	9
10 mm	80	9

Banjo Bolts

Bolt Size	Torque Foot-Pounds	Newton-Meters
10 mm.	25	34
7/16 in.	30	40

Tube Nuts

Nut Size	Torque Foot-Pounds	Newton-Meters
6 mm	2	3
7 mm	3	4
8 mm	6	8
3/8 in.	15	20
10 mm	11	15
7/16 in.	15	20
1/2 in.	15	20
9/16 in.	15	20

Note: All torque values are given for clean, steel-to-steel threads lubricated with brake fluid.

To convert a value from foot-pounds to inch-pounds, multiply it by 12.

To convert a value from foot-pounds to newton-meters, multiply it by 1.356.

APPENDIX 5

Shoe Size Chart

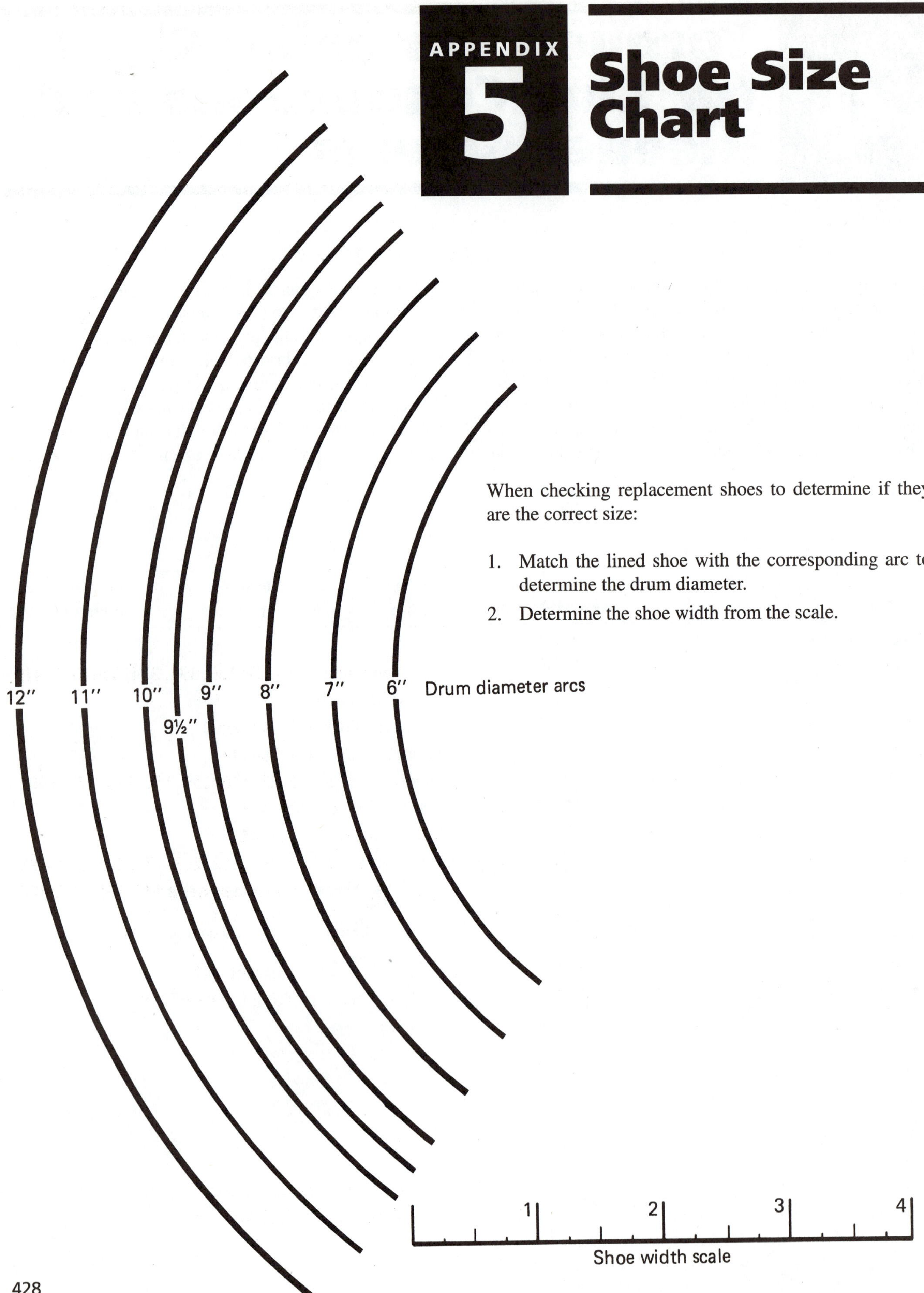

When checking replacement shoes to determine if they are the correct size:

1. Match the lined shoe with the corresponding arc to determine the drum diameter.
2. Determine the shoe width from the scale.

Index

A

ABS
(*See* Antilock Brake Systems)
Acceleration Slip Regulation (ASR), 142
Accumulator, 116–117, 119–121, 126, 127, 313, 318, 334, 349, 351, 413
Adjusters, brake shoe, 42
Ampere, 413
Anti-dive suspensions, 31
Antilock brake systems
air brakes, 413
Bosch, 133, 135, 136, 137
Delco ABS-VI, 133, 136, 141
electrical tests, 340–347
faults in, 337–338, 340
hydraulic control unit, 132, 331, 349, 351
problem codes, 337–338, 340
repair operations, 349–351
servicing, 331–352
Teves, 133, 135, 137, 138, 139, 334
warning light, 126, 127, 130, 131, 334, 336–337, 419
Anti-skid braking
(*See* Antilock brake systems)
Apportioning valve, 98–99
Asbestos, 13, 20–21, 23, 146, 151, 152, 178, 179, 190, 191, 204, 406, 413
ASE certification, 421–423
Assembly fluid, 199, 261, 262, 263, 270, 275, 282–283
ATE/Teves antilock brake system, 133, 135, 137, 138, 139, 334
Audible sensor, 57
Automatic traction control, 142
Auxiliary drum parking brake, 67–68
Axle bearings
(*See* Wheel bearings)

B

Balance bars, high-performance vehicles, 407
Ball bearings, 354–355, 370, 379

Bearing-retained axle, 358, 359, 374, 376
Bedding-in
(*See* Burnishing)
Beehive spring, 195
Bench bleeding, 263–265
Bendix brake spring tool, 192
Bendix Hydrovac, 112
Bleeder screw, 86–87, 88, 91–92, 97, 214, 245, 272, 285, 286–288, 413, 427
Bleeding brakes
ABS, 288–289, 295–296, 351
cross, 295–296
gravity, 289, 416
manual, 289–290, 417
pressure, 290–292, 417
reverse flow, 294
sequence for, 288–289
vacuum, 292–294, 419
Boot, caliper, 90–91
Boot, cylinder bore, 86
Brake balance, 99, 398–399, 407, 410, 414
Brake drag, 81, 82, 83, 87, 95, 105, 154, 307–308, 311, 326, 328, 410
Brake drum
arbor, 184
backing plate, 3, 34, 35, 39, 173, 176, 198, 199, 200, 409, 413
cleaning and inspection, 198–199
composite, 46, 414
frozen, 175, 410
hard spots, 179, 180, 416
heat sink, 18, 46–47, 416
high-performance vehicles, 409–410
inspecting, 179–184
measuring diameter of, 180–183
micrometer (mike), 180
post-servicing checks, 207
removal, 174–179
scored, 179, 392
servicing, 173–207
turning, 179, 184–189, 419
Brake fade, 21, 398
Brake fluid
changing, 294–296
contamination of, 104
high-performance vehicles, 408–409
inspecting, 148
level switch, 101
standards, 101–103
Brake hoses
failure of, 158
high-performance vehicles, 407–408
Brake light switch
adjusting, 301, 300
Brake lines and hoses, high-performance vehicles, 407–408
Brake lining
attachment of, 38–39
block, 38, 414
bonded, 38, 57, 414
break-in, 163–165
disc brakes, 56–57
edge code, 22–23, 37, 415
friction, 19–23
hardness designations, 23, 398
high-performance vehicles, 405–406
ingredients of, 20–21
inspecting, 152–153
integral-molded, 38
mold-bonded, 38, 57
riveted, 38, 57, 418
semimetallic, 20–22, 57, 418
types of, 20–22
wear, 38, 153
Brake lockup, 1, 8, 28, 98, 410
Brake pads
removing and replacing, 240–244
wear, 57
wear indicator, 57, 417
Brake pedal
failure to apply, 163, 298
failure to release, 163, 299
feedback, 31
sinking, 163, 297–298, 348
spongy, 47, 74, 90, 105, 147, 158, 163, 181, 182, 284, 296–297, 333, 351, 396, 419
travel sensor, 131, 132
Brake power
measuring, 26–27
Brake record chart, 411

Brake service, checklist for, 148
Brake shoe
adjusters, 42–46
anchors, 39–40, 190, 200
cleaning, 196, 198
clearance check, 203
drum brake, 37–38
energizing/deenergizing, 3, 35
heel, 37
installation, 204–207
leading (forward), 3, 35–36, 416
preinstallation checks, 201–203
primary, 3–4, 36, 418
regrinding, 204
removal, 196–197
return spring, 41–42, 418
rim, 37
secondary, 3–4, 36, 418
toe, 37, 419
trailing, 3, 35–36, 419
two-leading, 3, 36, 419
two-trailing, 3, 36
web, 37, 38
Brake spoon, 42, 414
Brake spring
inspecting, 155
removing/replacing, 190–196
Brake torque, 24–27, 39, 52, 414
Braking efficiency, 15, 16
Breakaway control, on trailer brakes, 389
Burnishing, 77, 163–165, 414
Bypass port, 75, 77, 414

C

Caliper
abutment clearance, 217, 238, 239
boot, 90–91
Delco Moraine, 68–69, 91, 276, 278–283
fixed, 5, 87–88, 214–215, 217, 238, 241, 400, 415
floating, 5, 52–56, 87–88, 215–218, 238, 241, 400, 415
high-performance vehicles, 400–403
installing, 245–247
Kelsey-Hayes, 68, 69, 91, 278–279, 281, 283
loaded, 213, 272, 416
with mechanical parking brake, 278–284
mount, 52–55, 91, 414
mounting hardware, 57–59, 417
pistons and sealing rings, 88–90
reconditioning, 272–284
removing, 213–218
retracting piston, 214–215
seals, 88–90
servicing, 272–284
sliding, 5, 55, 86–87, 216–218, 238, 241, 418
Center of gravity, 28, 414
Changeover point
(*See* Proportioning valve, split point)
Check valve, 78–80, 113, 115, 117, 133–134, 258, 307, 309, 414
C-lock axle
installing, 378–379
removing, 376–377
removing/replacing bearing and seal, 377–378
Coefficient of friction, tire-to-road, 24
Combination valve, 6, 7, 95, 414
Compensating port, 75, 76, 81, 414
Constant Velocity (CV) universal joint, 360
Contamination, brake fluid, 104
Controller, 129–132, 386, 387–389, 415
Control valve, 98, 115, 120, 126, 301, 307, 310, 314
Cooling ducts, high-performance vehicles, 404–405
Corvette, 61, 68, 128, 136, 243, 324, 405
Cup, primary/secondary, 86
Cutter bit, 180, 184, 226
CV-joint
(*See* Constant velocity (CV) universal joint)

D

Dash lamp switch, 96
Deceleration, measuring rates of, 16, 17
Decelerometer, 16, 415
Delco Moraine, 23, 68–69, 86, 91, 135–136, 215, 276, 278–283
Department of Transportation, 101, 415
Diagnostic trouble codes (DTCs), 337–340

Diagonal split system, 6, 64, 100, 288, 295, 328, 399, 415
Directional/nondirectional finish, 164
Disc balancing valve, 97
Disc brakes
advantages and disadvantages of, 50–51
caliper (*See* Caliper)
mounting hardware, 57, 59, 417
pads (*See* Brake pads)
post-servicing checks, 247
rear wheel, 5, 61–62
rotor (*See* Rotor)
sensors, 57, 241
service of, 211–247
troubleshooting, 159
turning disc, 219, 228–236
Dive, 28, 30, 31, 158, 163, 415
Duo-servo brake, 3, 36, 155, 205, 324, 415
Duplex brake, 36
Dynamometer, 26–27, 414

E

Edge code
(*See* Brake Lining, edge code)
EIS Sur-Bleed, 264
Electrical circuits, problems in, 166
Electrical system, diagnosis/repair of, 165–170
Electricity
measuring, 167–169
principles of, 165
Electronic brake control module (EBCM), 129, 331, 415
Electronic controller, 124, 129–131, 388, 415
Electronic control module (ECM), 129, 415
Electronic modulator, 132–134
Emergency brake, 8, 64, 415
Energized shoes, 3, 35
Energy of motion, 17–18
Equilibrium Reflux Boiling Point (ERBP), 102
External contracting brake
(*See* Parking brake)

F

Failure warning light, 7, 66, 81, 95, 97, 120
Fasteners
(*See* Torque specifications)
Federal Motor Vehicle Safety Standard (FMVSS), 13
Flare, ISO/SAE
(*See* Tubing)
Flex-Hone, 256
Floating drum, 174–175, 177
Floating valve, 114, 116, 415
Friction
defined, 72
and heat energy, 18, 419
tire, 24
Friction materials, 2, 19, 23
Friction Materials Standards Institute (FMSI), 37
Frozen drum, 175–176
Full-floating axle, 358

G

Gravity, 14, 28
Grease
wheel bearing, 362, 364, 369–370
Grounded circuit, 166, 392
Guide bolts, 58, 217, 416
Guide pins, 54, 55, 58, 217, 238, 416

H

Heat energy, 18
Height-sensing proportioning valve, adjusting, 99, 301
High Efficiency Particulate Air (HEPA) vacuum, 23, 178
High-performance brake systems
balance bars, 407
brake fluid, 408–409
brake lines and hoses, 407–408
brake lining, 405–406
calipers, 400–403
component selection, 400

cooling ducts, 404–405
drum brakes, 409–410
master cylinder, 407
operating limits, 396
rotor, 403–404
testing and adjusting, 410–411
troubleshooting, 397, 410
tuning, 396–399, 412
Hold-down spring, 43, 193–196, 416
Hold-off valve, 97
Hoses, 13, 92–95, 158, 407–408
Hydraulic brake system
control valves and switches, 95–101
diagnosing, 296–301
servicing, 250–303
split, 82–83
tandem, 80–82, 419
Hydraulics, principles of, 71–74
Hydraulic System Mineral Oil (HSMO), 102
Hydro-boost power booster
repairing, 315, 317–318
Hygroscopic, 102–103, 251, 416

I

Inertia, 15, 27–28, 30, 385, 387, 388, 416
Inspection, of braking system, 147–159
Intake (replenishing) port, 75
Internal expanding brake, 34
International Standard Organization (ISO), 92
ISO or bubble flare
(*See* Tubing, ISO flare)

J

Jackknifing, 28, 395, 416

K

Kelsey-Hayes Co., 68, 69, 91, 135, 278–279, 281, 283
Kinetic energy, 17–18, 416
Knock-back, 416

L

Lateral runout, 60, 221–223, 416
Light Emitting Diode (LED), 167
Linkage booster, 112, 123
Loaded calipers
(*See* Caliper, loaded)
Load-sensing proportioning valve, 302
Locating pins, 217
Lockheed brake spring tool, 193

M

Mandrel, 184, 226
Master cylinder
basics and components, 75
bench bleeding, 263–265
construction of, 76–78
dual, 79–80, 84–85, 415
high-performance vehicles, 407
inspecting, 148
operation, 75–76
quick-take-up, 83–84, 150, 214, 259, 236, 264, 295, 298, 418
rebuilding, 258–263
removal of, 257
replacing, 265–266
reservoir, 76, 77, 418
residual pressure, 78–79, 418
servicing, 256–266
step bore, 83, 85, 263
tandem, 80–82, 419
Metering valve, 7, 95, 97–98, 286, 301, 417
Modulator valve, 125, 126, 127, 132–134, 138
Motor Vehicle Safety Standards
brake fluid, 6, 13
brakes—stopping, 13
parking brakes, 13

N

National Highway Traffic Safety Administration (NHTSA), 13, 101, 417

National Pipe Thread (NPT), 300
Noise, troubleshooting, 162
Nonasbestos organic lining (NAO), 21, 417
Nondirectional finish, of rotor, 223
Nonservo brake, 35

O

Occupational Safety and Health Administration (OSHA), 23, 32, 417
Ohmmeter, 167, 169, 341, 343
Open circuit, 166, 169, 170, 392, 393
O-ring, 54, 83, 88–90, 254, 272, 275, 318

P

Parking brake
auxiliary, 67–68
cable
adjustment, 322–325
inspecting, 324
replacement, 326–328
disc, 68–69
equalizer, 64–65
inspecting, 325
integral, 67, 416
intermediate lever, 324, 329
lever, 66
ratio bar, 324, 329
reconditioning caliper, 278–283
servicing, 321–329
warning light, servicing, 328–329
Pascal's law, 72, 417
Peening, 175
Pistons, in tandem master cylinder, 80–81
Power brakes
electrohydraulic booster, 118–121
hydraulic booster, 116–118
Hydro-boost type
repairing, 305, 317–318
replacing, 310–313
troubleshooting, 161
Powermaster type
repairing, 318–319
replacing, 310–313
vacuum booster
repairing, 314–315
replacing, 310–313
testing vacuum supply, 308–310
troubleshooting, 161
Pressure bleeder, 98, 287, 290–292, 417
Pressure control valve, 98, 301
Pressure differential valve, 7, 82, 95, 417
Pressure holding, 125
Pressure multiplier, 112, 398
Pressure-ratio valve, 98
Pressure-reducing valve, 98
Pressure-regulating valve, 98
Proportioning valve
height-sensing, 99–100, 301
load-sensing, 302
slope, 98–100, 399, 418
split point, 98–100, 399, 419
Pull-back spring, 418

Q

Quick-take-up master cylinders, 83–84, 150, 214, 259, 263, 264, 295, 298, 418

R

Race cars
(*See* High-performance brake systems)
Reaction group, 120
Reaction time, 30, 124, 418
Rear-wheel lockup, 28, 36, 37, 95, 398–399, 407
Repair recommendations, 159, 163
Reservoir, master cylinder, 76, 77
Residual pressure, 78–79, 299, 418
Resistance, electrical, 165–166
Retracting spring, 41, 418
Riveting, 38, 175
Roller bearings, tapered, 355, 419
Roller burnishing, 77, 418
Root Mean Square (RMS), 164

Rotor
composite, 59, 219, 227, 403, 414
directional, 60, 405
finish cut, 189, 229, 231–232, 415
high-performance vehicles, 403–404
inspecting, 219–233
machining off-car, 228–233
machining on-car, 233–236
measurements, 61, 220–223
micrometers, 182
parallelism, 60, 219, 221, 222, 417
refinishing, 223–228
replacement, 236–238
resurfacing, 236
roughing cut, 189, 229, 232, 418
runout, 60, 221–223
swirl finish, 223, 419
thickness variation of, 221
truing, 179, 219
Runout
lateral, 60, 221–223, 416
power booster, 114
Rear Wheel Antilock (RWAL), 125

S

Safety tip, 146, 151, 152, 157, 159, 174, 177, 185, 190, 204, 213, 228, 233, 236, 363, 365, 374, 395
Scoring, 38, 163, 219, 222, 223, 392, 418
Scratch cut, 188, 230
Scrub radius, 6, 7, 82
Seals
Caliper, 88–90
dynamic, 88, 89, 251, 356, 415
labyrinth, 35
piston, 88–90
static, 88, 89, 272, 356, 419
stationary, 88, 89, 356, 419
Secondary shoe, 3, 4, 36, 41, 43, 67, 201, 386, 418
Self-adjusting brakes, 42
Semifloating axle, 358
Service, of braking systems, 145–170
Servo action, 3, 4, 35–37, 41, 43, 50, 51, 386
Servo brake, 4, 418
Short circuit, 166, 167, 392, 394, 418
Short-to-ground, 166
Silicone brake fluid, 102, 334
Simplex brake, 36, 48
Skid, 28–30, 418
Society of Automotive Engineers (SAE), 13, 22
Speed sensor, 126, 129, 331, 346–347, 373
Spindle, 184, 221, 229
Split point
(*See* Proportioning valve, split point)
Spool valve, 117
Step bore, 83, 85, 86, 263
Stop Control System (SCS), 126
Stoplight
switch, 101
switch adjustment, 301, 303
Stopping distances, 14, 124, 395
Stopping sequence, 30
Surface finish, 163–165, 184, 186, 219, 223, 225, 227
Swaging, 175, 419
Swept area, 26, 403, 419
System effectiveness switch, 96

T

Tactile sensor, 57
Tire friction, 24
Torque specifications, 157, 287
Traction, defined, 24, 419
Trailer brakes
electric, 385–389
servicing, 391–394
slave systems, 390–391
surge type, 389–390
troubleshooting, 392–394
Troubleshooting charts, 159–162, 397
Tubing
double flare, 92, 253–254, 415
failure of, 93–94
ISO flare, 92, 254, 416
replacing, 253–254

U

Uni-servo brake, 36, 85, 389
Universal joint, 360
Upsetting, 175

V

Vacuum, defined, 110
Vacuum boosters
(*See* Power brakes)
Visual sensor, 57
Voltmeter, 167, 169, 329, 338, 393
Volt-ohm-meter, 167, 168, 392

W

Warning light
antilock brake system, 334–337
low brake fluid, 95, 101
parking brake, 66, 328–329
testing, 301, 334–337
Washer, expander, 87
Wear indicator, 57, 417, 419
Weight transfer, 27–28, 29, 36, 82, 98, 397, 398, 419
Wheel bearing end play, 355–356
Wheel bearing preload, 355–356
Wheel bearings
ball type, 354–355
C-lock axle, 358, 362, 346–378
drive-axle, nonserviceable, 353, 362
frictionless, 354, 356
front-wheel-drive
nonserviceable, 381–382
serviceable, 379–381
independent suspension, drive-axle, 360–361
needle type, 355
nondrive-axle, nonserviceable, 353, 358, 362
nondrive-axle, serviceable, 353, 356–357
roller type, 355
solid-axle drive-axle, 358, 359, 373
tapered-roller type, 355
trailer, 391–392
Wheel bearings, servicing
cleaning and inspecting, 366–369
diagnosing, 362–364
disassembling, 364–365
inspecting, 366–369
maintenance and lubrication, 362
packing and adjusting, 369–372
repacking, 364–372
repairing
nonserviceable, 373
nonserviceable front-wheel-drive, 381–382
serviceable front-wheel-drive, 379–381
solid-axle drive-axle, 373–379
trailer, 391–392
Wheel bearing seals, 356
Wheel cylinder
bleeder screw, 87
cup, 86
faulty, 266
inspecting, 148
pistons, 86
reconditioning, 269–271
removing, 267–268
replacing, 271–272
servicing, 266–272
step bore, 86
straight bore, 85–86
Wheel lockup, diagnosing, 160, 162, 299–301
Wheel removal, 364
Wiring diagrams, 335